Handbook of
Experimental Pharmacology

Volume 118

Springer

Berlin
Heidelberg
New York
Barcelona
Budapest
Hong Kong
London
Milan
Paris
Santa Clara
Singapore
Tokyo

Pharmacological Aspects of Drug Dependence

Toward an Integrated Neurobehavioral Approach

Contributors

R.L. Balster, P.B.S. Clarke, D.R. Compton, L.A. Dykstra
G.I. Elmer, M.W. Fischman, R.A. Glennon, D.J. Greenblatt
R.R. Griffiths, L.S. Harris, K. Hellevuo, J.E. Henningfield
P.L. Hoffman, J.R. Hughes, C.-E. Johanson, R.M. Keenan
H.D. Kleber, T.R. Kosten, M.J. Kreek, M.J. Kuhar
M.C. LaBuda, F.R. Levin, A.H. Lichtman, B.R. Martin
L.G. Miller, G.K. Mumford, R.W. Pickens, C.R. Schuster
B. Tabakoff, G.R. Uhl, F.J. Vocci, E.A. Wallace, J. Willetts

Editors

Charles R. Schuster and Michael J. Kuhar

 Springer

Charles R. Schuster, Ph.D.
Wayne State University School of Medicine
University Psychiatric Center
2751 E. Jefferson Avenue
Detroit, MI 48207
USA

Michael J. Kuhar, Ph.D.
Director, The National Institute on Drug Abuse
Addiction Research Center
P.O. Box 5180
Baltimore, MD 21224
USA

With 29 Figures and 15 Tables

ISBN 3-540-58989-9 Springer-Verlag Berlin Heidelberg New York

Library of Congress Cataloging-in-Publication Data. Pharmacological aspects of drug dependence: toward an integrated neurobehavioral approach/contributors, R.L. Balster . . . [et al.]; editors, Charles R. Schuster and Michael J. Kuhar. p. cm.—(Handbook of experimental pharmacology; v. 118) Includes bibliographical references and index. ISBN 3-540-58989-9 (hardcover) 1. Drugs of abuse. 2. Drug abuse. 3. Neuropsychopharmacology. I. Balster, Robert L. II. Kuhar, Michael J. III. Schuster, Charles R. IV. Series. [DNLM: 1. Substance Dependence — genetics. 2. Brain — physiology. 3. Nerve Tissue — chemistry. 4. Pharmacokinetics. 5. Behavior, Addictive. 6. Drug Therapy. W1 HA51L v. 118 1996/ WM 270 P5361 1996] QP905.H5 vol. 118 [RM316] 615'.1 s — dc20 [615'.78] DNLM/DLC for Library of Congress 96-36826

© Springer-Verlag Berlin Heidelberg 1996
Printed in Germany

The use of general descriptive names, registered names, trademarks, etc. in this publication does not imply, even in the absence of a specific statement, that such names are exempt from the relevant protective laws and regulations and therefore free for general use.

Product liability: The publisher cannot guarantee the accuracy of any information about dosage and application contained in this book. In every individual case the user must check such information by consulting the relevant literature.

Cover design: Springer-Verlag, Design & Production

Typesetting: Best-set Typesetter Ltd., Hong Kong

SPIN: 10099792   27/3136/SPS – 5 4 3 2 1 0 – Printed on acid-free paper

Preface

In spite of a "war on drugs" that spans years, and in spite of increases in law enforcement efforts and to a lesser extent treatment, substance abuse and dependence continue. While the number of people who experiment with drugs has decreased in recent years, those who use drugs repeatedly, perhaps several times a week, a measure of "hard core" drug abusers, has not changed, and emergency room visits associated with substances abuse continues to rise. Considering both licit and illicit drug abuse, the number of drug abusers in, for example, the United States is very large and, accordingly, the cost to society is great. A recent Institute of Medicine study suggests that the overall cost is about $66 billion per year (1990 costs). Certainly when assessing the total cost of substance abuse, we must include of the costs of disease and behavioral disorders that are intimately associated with substance abuse. For example, sharing of needles among illicit drug abusers has been and is a critical vector in the spread of drug resistant TB, AIDS, and hepatitis. Further, the use of drugs and alcohol leads to unsafe sexual practices with the attendant risk of AIDS and other sexually transmitted diseases. Tragically, infants are born with drugs in their system and with infections and disease transmitted from the mothers who are drug abusers or their sex partners who are drug abusers. Crime has been shown to be intimately connected with substance abuse; a recent Institute of Justice survey found that individuals recently arrested for crime had cocaine metabolites in their urine at a very high frequency, greater than 50% in many cities. Thus, the chronic, relapsing disease of substance abuse costs society in many ways. It is, therefore, imperative that we conduct research that will contribute to the control of this problem. For these and other reasons, we have decided to produce this volume. We will state the rationale for the topics covered in subsequent chapters, and we will discuss some additional relevant topics as well.

While we often think of drug abuse as a problem with illegal drugs, it is also clearly a problem with the licit substances of alcohol and tobacco. The use of these substances has a tremendous impact on public health in our society. The number of premature deaths caused by smoking tobacco, which is initiated by peer pressure and continued primarily because of nicotine dependence, is estimated at 3 million per year worldwide – the largest number due to any abused substance. In addition, the morbidity and illnesses associated with chronic tobacco use is a tragedy for the individual and family and is a major cost for our health care system. The cost of alcohol abuse in terms of health care and loss of productivity is also staggering. It is estimated that a large percentage of hospitalized patients are there for a disorder caused by or exacerbated by alcohol. Alcohol abuse is also a major contributor to accidents in the home or workplace and on the highways as well as to violence. Fortunately, there continue to be major advances in our understanding of the abuse of these substances as well as new pharmacotherapies for treatment. Therefore, we have included chapters on ethanol as well as nicotine in this volume.

Although all illicit drugs are associated with morbidity and mortality, those that currently have the biggest impact on public health are cocaine and heroin. As mentioned above, AIDS, hepatitis, drug-resistant tuberculosis, violence, and crime are associated with groups abusing these substances. Thus, we include chapters on cocaine and opiates. Because the abuse of other drugs is associated with morbidity, mortality, and accidents, we have included chapters on other drug classes as well.

Although caffeine has long been recognized to have psychoactive properites, the recognition of caffeine as a substance with potential for dependence is recent. It is clear from controlled laboratory studies with humans that it produces reinforcing effects, discriminative effects, tolerance, and dependence, and also has effects on performance and functioning. Similarly, recent work has more clearly delineated the physiological effects of nicotine and its withdrawal. It is clear that we should regard tobacco as a nicotine delivery device and the inability to stop smoking as nicotine dependence. Important developments in treating smoking/nicotine dependence are the nicotine patch and chewing gum in conjunction with behavioral interventions which reduce relapse to smoking. Nicotinic cholinergic receptors are one of the best understood of the chemically gated ion channel receptors, and this understanding could lead to new pharmacologic treatments for tobacco dependence. Work on hallucinogens has progressed steadily; it is thought that stimulation of $5HT_{2A}$ receptors are a major mechanism of their pharmacological action. The sedatives benzodiazepines and barbituates are well known to have abuse liability. Mechanisms underlying their actions have been increasingly clarified in the last two decades. Inhalants continue to be a problem in society with little known about their mechanisms of action. PCP, a unique drug of abuse, is known to affect glutamate receptors. In the case of marijuana, the "gateway drug," receptors for its psychoactive constituent, THC, have been cloned and endogenous ligands have been identified. Chapters on all of these substances have been included.

A significant achievement is that receptors for all drugs of abuse have been cloned, allowing a variety of molecular mechanisms to be explored as well as providing tools for molecular genetic studies. As already mentioned, one of the most recent receptors to be cloned is that for the cannabinoid group of drugs. Receptors for the three major subtypes (and subtypes of subtypes) of opiate drugs have also been cloned. This has important implications for developing novel analgesics as well as for understanding addiction mechanisms. Transporters, targets for psychostimulants, have also been cloned, with implications for antidepressant development as well as for understanding mechanisms of psychostimulant action.

In part, the study of drug abuse is a changing field. The introduction from time to time of "designer" drugs presents us with new problems and new toxicities and new issues. Well-known examples of drugs with significant toxicities include MPTP, which causes a Parkinson-like syndrome, and MDMA or "ecstasy," which is known to cause a degeneration of serotonin-containing nerve terminals in animals, and whose toxicity in humans is suspected but not yet proven.

An exciting area of technical advances includes brain imaging. These techniques allow us to identify changes in regional activity or receptors, all in the living brain in situ, under conditions relevent to behavior and problems in substance abuse. For these reasons, findings from imaging techniques have been integrated into chapters wherever possible.

In the behavioral arena, new models and new findings continue to emerge. The interaction of environment with organism can be studied; for example, the finding that food deprivation enhances drug self-administration is an interesting one that presumably applies to human environments. Control over drug taking is important; it has been shown that passive administration of doses of cocaine can be lethal, while self-administration of the same doses is not and is in fact reinforcing. Similarly, the neurochemical effects of self-administered drugs are different from those of passively administered drugs.

One of the most interesting findings has been the issue of polydrug abuse. It is felt that it is becoming more prevalent and it certainly complicates research and treatment. Drug interactions can create toxicities. Also, the importance of comorbidity in the etiology of substance abuse has been established in the last decade. Although it is not clear which comes first, the psychiatric problem or the substance abuse problem, treatment of both conditions is needed. In any case, while polydrug abuse and comorbidity makes research into substance abuse and understanding substance abuse more difficult, they both have significant implications for producing effective treatments. It is clear that treatment needs to be more encompassing and multileveled to deal with these complexities.

Vulnerability to drug abuse is a very important issue. If we could clearly identify vulnerable populations, we would be able to focus prevention efforts and funds at these smaller groups and presumably be more cost effective. While the specific influences of genetic factors in substance abuse are debated, there is little doubt that they contribute directly or indirectly to the problems. Interesting findings in the last several years show that certain behaviors such as increased locomotor activity in animals are associated with a predisposition to drug abuse. These issues are currently being explored in human populations. It has been found, for example, that the coincidence or cooccurrence of both ADHD and conduct disorders strongly predisposes individuals to substance abuse. This is clearly one of the most important areas for research and justifies chapters in this volume.

While it is not the only nor the most important aspect of treatment, pharmacotherapy is proven to have an important role in treatment. Pharmacotherapy of addiction really only began in mid-sixties with the use of methadone in opiate addicts. Recently, there have been new medications as well as new techniques for their delivery. For example, LAAM and naltrexone now join the armamentarium along with methadone as treatments for opiate abusers. Nicotine replacement in conjunction with behavioral therapies has been shown to be a useful treatment for nicotine abusers. A recent surprise has been the finding that naltrexone is useful for treatment of alcoholism. Of course, more and better pharmacotherapies are needed. Medications for different stages of treatment are needed: these include medications for acute withdrawal, for chronic treatment, and for comorbid illnesses. For example, we have no accepted and proven pharmacotherapy for cocaine abusers at this time although many studies are under way. The abuse potential of medications themselves must be considered and weighed, particularly when given to a population of patients with substance abuse problems. In addition to medications, new methods of delivery have proven effective. For example, the nicotine patch is effective and there are depot injections of naltrexone for long-lasting effects. These recent advances are included in this volume.

The goal of research in pharmacotherapy of substance abuse and in other aspects of substance abuse is not to compile papers, handbooks, or interesting findings. The goal is to further understand this chronic relapsing disease and to produce new medications and treatments. For this reason, treatment implications as well as mechanisms underlying the pharmacodynamic effects of drug abuse are included in the various chapters. Treatment standards must be developed so that caregivers can effectively apply the therapies. Treatment development and improvement is extremely important given its relative effectiveness in reducing drug use. A recent study showed that treatment is 23-fold less costly than controlling the source of drugs in foreign countries, and much less expensive than domestic crime enforcement. This is particularly striking since the increase in spending for law enforcement has outpaced the increase in spending for treatment several fold, at least up until 1993. Fundamental to new strategies for treatment include novel ways of measuring and treating craving to determine its functional significance in the maintenance of drug abuse. Indeed, fundamental studies in the neurochemistry of addiction will hopefully lead to many new strategies for new medications. Hopefully, with advances and improvements, treatment will become more understood and accepted by policy and law-makers as well.

Detroit, MI, USA   CHARLES R. SCHUSTER
Baltimore, MD, USA   MICHAEL J. KUHAR
August 1995

List of Contributors

BALSTER, R.L., Department of Pharmacology and Toxicology and Center for Drug and Alcohol Studies, Virginia Commonwealth University, P.O. Box 980310, Richmond, VA 23298-0310, USA

CLARKE, P.B.S., Pharmacology and Therapeutics, McGill University, 3655 Drummond Street, Rm. 1325, Montreal, Quebec, Canada H3G 1Y6

COMPTON, D.R., Department of Pharmacology and Toxicology, Medical College of Virginia, Virginia Commonwealth University, Box 980027, Richmond, VA 23298-0027, USA

DYKSTRA, L.A., Department of Psychology and Pharmacology, University of North Carolina, Campus Box 3270, Davie Hall, Chapel Hill, NC 27599-3270, USA

ELMER, G.I., Intramural Research Program, National Institute on Drug Abuse, National Institutes of Health, P.O. Box 5180, 4940 Eastern Avenue, Baltimore, MD 21224, USA

FISCHMAN, M.W., College of Physicians and Surgeons of Columbia University, Department of Psychiatry, Unit 66, 722 W. 168th Street, New York, NY 10032, USA

GLENNON, R.A., Department of Medicinal Chemistry, Medical College of Virginia, Virginia Commonwealth University, Box 980540, Richmond, VA 23298-0540, USA

GREENBLATT, D.J., Department of Pharmacology and Experimental Therapeutics, Tufts University School of Medicine, 136 Harrison Avenue, Boston, MA 02111, USA

GRIFFITHS, R.R., Department of Psychiatry and Behavioral Sciences, Behavioral Biology Research Center, Johns Hopkins University School of Medicine, 5510 Nathan Shock Drive, Suite 3000, Baltimore, MD 21224-6823, USA

HARRIS, L.S., Department of Pharmacology and Toxicology, Medical College of Virginia, Virginia Commonwealth University, Box 980027, Richmond, VA 23298-0027, USA

HELLEVUO, K., Department of Pharmacology, University of Colorado School of Medicine, 4200 E. Ninth Avenue, Box C236, Denver, CO 80262, USA

HENNINGFIELD, J.E., National Institute on Drug Abuse, Addiction Research Center, P.O. Box 5180, 4940 Eastern Avenue, Baltimore, MD 21224, USA

HOFFMAN, P.L., Department of Pharmacology, Office of the Chairman, University of Colorado School of Medicine, 4200 E.Ninth Avenue, Box C236, Denver, CO 80262, USA

HUGHES, J.R., Department of Psychiatry, University of Vermont, 38 Fletcher Place, Burlington, VT 05401-1419, USA

JOHANSON, C.-E., Department of Psychiatry and Behavioral Neurosciences, Wayne State University, 2751 E. Jefferson, Detroit, MI 48207, USA

KEENAN, R.M., National Institute on Drug Abuse, Addiction Research Center, P.O. Box 5180, 4940 Eastern Avenue, Baltimore, MD 21224, USA

KLEBER, H.D., Columbia University, Department of Psychiatry, 722 West 168th Street, New York, NY 10032, USA

KOSTEN, T.R., Division of Substance Abuse, Department of Psychiatry, Yale University School of Medicine, 34 Park Street-Room S-205, New Haven, CT 06519, USA

KREEK, M.J., Department of Biology and Addictive Diseases, The Rockefeller University, 1230 York Avenue, New York, NY 10021, USA

KUHAR, M.J., The National Institute on Drug Abuse, Addiction Research Center, P.O. Box 5180, Baltimore, MD 21224, USA

LaBUDA, M.C., Division of Child Psychiatry, Department of Psychiatry and Behavioral Science, Johns Hopkins University School of Medicine, Baltimore, MD 21287, USA

LEVIN, F.R., Columbia University, Department of Psychiatry, 722 West 168th Street, New York, NY 10032, USA

LICHTMAN, A.H., Department of Pharmacology and Toxicology, Medical College of Virginia, Virginia Commonwealth University, Box 980027, Richmond, VA 23298-0027, USA

MARTIN, B.R., Department of Pharmacology and Toxicology, Medical College of Virginia, Virginia Commonwealth University, Box 980027, Richmond, VA 23298-0027, USA

MILLER, L.G., Department of Pharmacology and Experimental Therapeutics, Tufts University School of Medicine, 136 Harrison Avenue, Boston, MA 02111, USA

MUMFORD, G.K., Department of Psychiatry and Behavioral Sciences, Behavioral Biology Research Center, Johns Hopkins University School of Medicine, 5510 Nathan Shock Drive, Suite 3000, Baltimore, MD 21224-6823, USA

PICKENS, R.W., Intramural Research Program, National Institute on Drug Abuse, National Institutes of Health, P.O. Box 5180, 4940 Eastern Avenue, Baltimore, MD 21224, USA

SCHUSTER, C.R., Department of Psychiatry and Behavioral Neurosciences Wayne State University School of Medicine, 2751 E. Jefferson Avenue, Detroit, MI 48207, USA

TABAKOFF, B., Department of Pharmacology, University of Colorado School of Medicine, 4200 E.Ninth Avenue, Box C236, Denver, CO 80262, USA

UHL, G.R., Intramural Research Program, National Institute on Drug Abuse, National Institutes of Health, P.O. Box 5180, 4940 Eastern Avenue, Baltimore, MD 21224, USA

VOCCI, F.J., Medications Development Division, National Institute on Drug Abuse, National Institutes of Health, 5600 Fishers Lane, Rm. 11A-55, Rockville, MD 20857, USA

WALLACE, E.A., Department of Psychiatry, Emory University School of Medicine, 1365 Clifton-Room 5314, Atlanta, GA 30322, USA

WILLETTS, J., Nelson Institute of Environmental Medicine, New York University Medical Center, Tuxedo, NY 10987, USA

Contents

CHAPTER 2

Integrative Neurobehavioral Pharmacology: Focus on Cocaine
M.J. KUHAR and C.R. SCHUSTER. With 8 Figures   53

**II. Molecular, Behavioral, and Human Pharmacology
     of Dependence and Consequences**

CHAPTER 3

Marihuana
D.R. COMPTON, L.S. HARRIS, A.H. LICHTMAN, and B.R. MARTIN
With 2 Figures   83

CHAPTER 4

Cocaine

CHAPTER 5

Opioid Analgesics

CHAPTER 6

CHAPTER 7

CHAPTER 8

Nicotine
J.E. HENNINGFIELD, R.M. KEENAN, and P.B.S. CLARKE 271

CHAPTER 9

**Caffeine Reinforcement, Discrimination, Tolerance
and Physical Dependence in Laboratory Animals and Humans**
R.R. GRIFFITHS and G.K. MUMFORD. With 2 Figures   315

III. Advances in the Pharmacotherapy of Addiction

CHAPTER 12

Pharmacotherapy of Addiction: Introduction and Principles

CHAPTER 13

Development of Medications for Addictive Disorders
F.I. Vocci. With 3 Figures ....

CHAPTER 14a

**Long-Term Pharmacotherapy for Opiate (Primarily Heroin)
Addiction: Opioid Agonists**
M.J. KREEK 487

CHAPTER 14b

Long-Term Pharmacotherapy for Opiate (Primarily Heroin) Addiction: Opioid Antagonists and Partial Agonists

CHAPTER 15

Pharmacotherapy of Nicotine Dependence

CHAPTER 16

Pharmacotherapies for Cocaine Dependence
E.A. WALLACE and T.R. KOSTEN 627

Contents

I. Research in the Study of Drug Action and Addiction

CHAPTER 1
Genetic Vulnerability to Substance Abuse

R.W. Pickens, G.I. Elmer, M.C. LaBuda, and G.R. Uhl

A. Heterogeneity of Drug Abuse

Considerable heterogeneity is present in drug abuse. People differ in tendency to use drugs, age of onset and duration of use, pharmacokinetic and pharmacodynamic response to drugs, liability to and expression of dependence, and consequences of use. In the general population, drug use falls on a continuum ranging from abstinence to chronic daily use. While most people never take illicit drugs nor engage in nonmedical use of psychotherapeutics, an estimated 37% of the population reports having done so at least once in their lifetime (Substance Abuse and Mental Health Services Administration 1993). Population prevalence of illicit drug use varies with drug availability as well as a number of demographic variables (e.g., sex, occupation, education, marital status, employment, age, race, and geographical region) (Flewelling et al. 1992).

Some individuals initiate drug use early in life while others begin much later. Across individuals, probability of initiating drug use increases during the teen years and then decreases as the person enters adulthood (Kandel and Logan 1984; Kandel and Raveis 1989). Onset of drug use before age 15 is associated with an increased risk of severe drug disorders (Robins and Przybeck 1985). After initial use, some people elect never to use drugs again. For others, drug use may be limited to discrete periods in their lives (e.g., college). For still others, drug use may continue over the greater part of adulthood and become an integral part of their lifestyle. The probability of continued drug use is the lowest for psychedelics (8%) and highest for alcohol (93%) and cigarettes (60%) (Raveis and Kandel 1987).

Among drug users, considerable interindividual differences are evident in frequency and pattern of drug use (Heath et al. 1991). For example, the majority of people who use marijuana and cocaine do so less often than once a month, but a significant number of individuals take these drugs one or more times a week (National Institute on Drug Abuse 1992). For infrequent users, drug use typically occurs episodically and is largely limited to social occasions. Among more frequent users, drug use may occur independent of discrete environmental events and be regulated by the individual to achieve a constant pharmacological state. While increased medical and legal risks are associated with higher rates of drug use (Brown et al. 1993), certain people are believed to be more susceptible to adverse drug consequences as a result of population variations in biological functioning (Wolf 1991; Burgio 1993). For example, low plasma cholinesterase activity may make some people more susceptible to adverse reactions to cocaine (Om et al. 1993; Hoffman et al. 1990).

Pharmacokinetic and pharmacodynamic profiles of drug action also differ across individuals (Alvan 1983). For example, significant interindividual differences are evident in metabolic half-life for cocaine (Isenschmid et al. 1992) and in heart rate changes following chronic cocaine administration (Hatsukami et al. 1994). Interindividual differences in drug response have also been reported for other abused drugs (Busto and Sellers 1986). In terms of subjective effects, some people report "liking" the same dose of a drug that others report not liking and, when given a choice of self-administering the drug, do so on the basis of the reported subjective effect (de Wit et al. 1986; Chutuape and de Wit 1994). In addition, interindividual differences are evident in withdrawal symptoms that occur following abrupt discontinuation of drug use (Bixler et al. 1985).

Drug users also differ in liability of becoming drug dependent. While all abused drugs have the potential for producing dependence, not every person who uses these drugs becomes dependent. Among individuals reporting that they took a drug at least once a month during the past year, variability is seen in incidence of dependence. For example, only 58% of regular cocaine users report any symptoms of cocaine dependence (National Institute on Drug Abuse 1991). In a recent national survey, about 15% of illicit drug and alcohol users were found to be dependent, while the dependence/use ratio for tobacco was somewhat higher (32%) (Anthony et al. 1994).

Individuals also differ in the expression of dependence. This is reflected in the nosological development of different subtypes of dependence (e.g., Cloninger et al. 1981) and problem severity indicators (McLellan et al. 1985). In addition, drug-dependent individuals differ in the nature and severity of symptoms presented at treatment admission (Horn and Wanberg 1969), signs and symptoms of substance use withdrawal syndromes (Greenblatt et al. 1990; McMicken 1990), general life styles (Flaherty et al. 1984), degree of criminal involvement (Nurco et al. 1981), and ability to quit drug use, either on their own (Hajek 1991) or following formal drug abuse treatment (Hser et al. 1993; Vaillant 1973).

Finally, people differ in conditions that are associated with drug use/ dependence. Many but not all drug abusers have coexisting mental illness. Individuals with an alcohol disorder are 2.3 times more likely to have a mental disorder than individuals without an alcohol disorder, and individuals with other drug disorders are 4.5 times more likely to have a mental disorder than individuals without another drug disorder (Regier et al. 1990). While mental illness in some cases is secondary to the drug use, in other cases it occurs prior to onset of drug use and is thought to contribute to the onset of drug use or dependence (SCHUCKIT and MONTEIRO 1988).

I. Etiological Influences

A number of factors contribute to this heterogeneity in drug use. Broadly, these can be divided into genetic and environmental influences. Genetic influences involve the expression of genes that are passed down from parent to child. To date, while no genes have been specifically identified that contribute to drug use (or dependence), evidence of genetic involvement is indicated by the results of both adoption and twin studies with humans and from animal research. Environmental influences represent the effects of past experience and current environmental conditions. At present, more is known about environmental than genetic influences in drug use.

1. Risk Factors

At the human level, most research on etiology of drug use has focused on identifying risk factors. For the most part, risk factors are higher level concepts that represent the interaction of basic genetic and environmental influences (PLOMIN et al. 1990). A large number of risk factors have been associated with drug use. They include poverty, racial discrimination, parental abuse, family instability, inadequate parenting, having an alcoholic parent, learning disabilities, inability to cope with stress, early aggressiveness in combination with shyness, low self-esteem, personality characteristics, school failure, rebelliousness, negative peer pressure, etc. Altogether, over 70 risk factors for substance use have been identified (OFFICE OF SUBSTANCE ABUSE PREVENTION 1991; HAWKINS et al. 1992).

People are thought to differ in number and types of risk factors they possess for drug use. Some risk factors are believed to play a greater role than others in determining drug use, and the probability of drug use is thought to increase with number of risk factors. At least some evidence suggests simple linear summation of risk factors is related to initiation of general substance use by adolescents (BRY et al. 1982; NEWCOMB et al. 1986; NEWCOMB and FELIX-ORTIZ 1992). Risk factors that contribute to the initiation of drug use may be somewhat different from those that contribute to heavy problematic drug use (SCHEIER and NEWCOMB 1991) or to the development of drug dependence (GLANTZ and PICKENS 1992).

Finding that likelihood of drug use increases with number of risk factors is not unlike the etiology of cardiovascular disease and mental illness, in which the likelihood of these disorders also increases with the number of disease-specific risk factors (NEWCOMB et al. 1986). In behavior genetics, this concept is termed the "liability-threshold model" (FALCONER 1989), which postulates than many genetic and environmental factors underlie the liability to drug use/dependence and that a certain level of liability must be exceeded for the condition to develop (Fig. 1).

In contrast to risk factors, protective factors contribute to drug abstinence and guard against the development of regular use patterns and/or drug dependence (BROOKS et al. 1989a,b). Examples of protective factors include having a middle to upper socioeconomic family background, high quality health care, a structured and nurturing family, high intelligence, social adeptness, an internal locus of control, and high self-esteem (OFFICE OF SUBSTANCE ABUSE PREVENTION 1991). Absence of a risk factor is not necessarily the same as a protective factor for drug use, as risk factors and protective factors may exist along different dimensions (NEWCOMB and FELIX-ORTIZ 1992).

2. Genetic Influences

While a large number of risk factors are associated with drug use, in many cases the etiological significance of this association is not clear (PICKENS and SVIKIS, in press). Rather than being causally related to drug use, some risk factors may be a consequence of drug use, or merely a state that is correlated with drug use. Others may represent multiple expressions of a common underlying mechanism. For example, a number of risk factors may be reduced to their impact on drug availability. Drug availability varies across individuals, with 51% of the population reporting access to marijuana com-

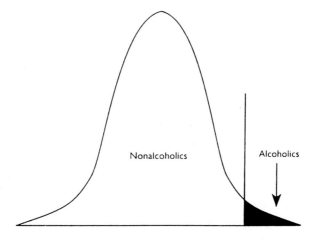

Alcoholism Liability

Fig. 1. The liability threshold model as applied to alcoholism. Underlying the categorical distinction between alcoholics and nonalcoholics is an hypothesized quantitative liability. Alcoholism occurs whenever an individual's combined liability exceeds a fixed threshold value. The inheritance of the categorical designation is accounted for entirely by the inheritance of liability. (From McGUE 1994)

pared to only 5% reporting access to heroin. Of those reporting access, the likelihood of drug use also differs accross individuals, with 65% of those having access to marijuana reporting use compared to only 16% of those having access to heroin (NATIONAL INSTITUTE ON DRUG ABUSE 1991).

Similarly, genetic influences may underlie a number of risk factors for drug use and dependence. For example, the risk factors of rebelliousness, delinquency, and aggressiveness may result from a basic personality disorder that is genetically influenced, while the risk factors of poor school performance, unemployment, and low self-esteem may result from a neuropsychological deficit that is genetically influenced. Genes may also influence the expression of specific risk factors. For example, genes may be expressed to enhance sensation seeking or risk taking personality characteristics, which would make the individual more likely to engage in drug use, or they may be expressed to produce chronic dysphoric mood states (or performance deficits that indirectly lead to such states), which are relieved by drug administration.

In addition, genes may influence vulnerability more directly – by altering an individual's response to drugs. They may enhance the reinforcing effects of drugs themselves (through pharmacokinetic or pharmacodynamic mechanisms), or the reinforcing effects of other stimuli (e.g., peer pressure) that contribute to drug use (MATTOX et al., in press). Alternatively, they may decrease aversiveness of drug withdrawal symptoms (or decrease sensitivity to social influences not to use drugs) which tend to reduce drug consumption. Finally, they may be expressed to facilitate the acquisition process, resulting in more rapid learning of drug taking behavior. Taken together, genetic influences have the potential for explaining many of the interindividual differences that are evident in drug abuse.

This review will examine vulnerability to substance abuse from a behavioral genetic and molecular genetic perspective, employing findings that come from both animal and human studies. Most research with humans has been conducted with alcoholism and relatively little has involved other types of drug abuse. Twin and adoption studies have determined that both genetic and environmental influences are involved in substance abuse and have indicated the extent of the involvement. Genes for many of the drug receptors have been cloned, and linkage and association studies have attempted to identify the specific genes involved. Animal studies have shown that a variety of species can be bred for drug accepting preferences and have attempted to identify likely candidate genes for these and other drug abuse symptoms. Together, animal and human studies provide useful complementary approaches for the study of genetic and environmental factors in substance abuse (PICKENS and SVIKIS 1991).

Although the review will focus primarily on genetic influences, it is clear that environmental factors are also involved in substance abuse. For example, the concordance rate for drug abuse among monozygotic twins is less than unity. Also, under experimental conditions, genetically undifferentiated animals from various species will come to self-administer most drugs that humans abuse (JOHANSON et al. 1987). Several studies have shown that changes in laws regulating alcohol availability affect alcohol consumption (LEVY and SHEFLIN 1985; DECKER et al. 1988). An often unrecognized environmental influence for dependence is the behavior of drug taking. Unless a person engages in the behavior of drug taking, drug dependence is not possible. Thus, drug taking is a necessary but not sufficient condition for a person to become drug-dependent.

Another environment influence is the manner in which a drug is taken. For dependence to develop, a drug must be taken in a manner that will allow operant conditioning to occur (drug reinforcement of drug taking behavior). This involves establishing a temporal contingency between the drug taking behavior and the drug's effects. Another environment influence is type of drug. Drugs differ considerably in their reinforcing effects, with their abuse liability being reflected by their tendency to establish and maintain regular patterns of self-administration. Other environmental factors, such as cost (DEGRANDPRE et al. 1993) and absence of other reinforcers (CARROLL et al. 1989), have also been experimentally shown to be related to drug taking.

However, there is evidence that, at least in humans, experience with drugs alone is not a sufficient factor for the development of chronic dependence. Although many patients are prescribed drugs with abuse potential in general medical practice, relatively few patients receiving those drugs later show evidence of dependence (PORTER and JICK 1980). Some people smoke tobacco for decades without developing dependence (SHIFFMAN 1991). In the Vietnam War, a large number of servicemen became dependent on opiates, ostensibly because of their availability, cheapness, and purity (environmental factors). Yet, when those servicemen were detoxified as a condition for returning home, 1 year later only 5% were found to be opiate-dependent. Three years after returning home, only 12% of those addicted in Vietnam had been addicted at any time since returning to the U.S. (ROBINS 1993). These results suggest nonpharmacological environmental factors were responsible for the drug dependence that occurred.

B. Clinical Studies

In this review we will focus on genetic influences on the clinical problem of drug dependence (including alcoholism), which falls at the upper end of the drug use continuum. A number of studies have investigated genetic influences on lifetime prevalence or quantity and frequency of drug use in general. With rare exceptions (e.g., PEDERSEN 1981), this work has focused on alcohol and tobacco use (KAPRIO et al. 1982; HUGHES 1986; CARMELLI et al. 1990; SWAN et al. 1990; HEATH and MARTIN 1993). With tobacco smoking, in addition to finding genetic influences on ever smoking, genetic influences were also found in quitting smoking (CARMELLI et al. 1992; HEATH and MARTIN 1993).

As in other research areas, variability in results is sometimes found across genetic studies of drug dependence. Since the definition of drug dependence has changed over the years, at least some of this variability may be due to this nosological variable. In general, most definitions of drug dependence include evidence of legal, health, job, or medical problems associated with excessive (pathological) drug use (AMERICAN PSYCHIATRIC ASSOCIATION 1980, 1994). Additional evidence of tolerance and physiological dependence is sometimes required. Recently, the dependence syndrome first described by EDWARDS and GROSS (1976) has been incorporated into many diagnostic approaches, which also includes evidence of loss of control over drug use.

I. Family Genetic Studies

Studies of relatives can provide evidence of the proportionate contributions of genes and environment to a behavioral disorder such as drug dependence. Each methodology is based on a set of assumptions and results must therefore be interpreted given the limitations of each method (Table 1). The most convincing evidence for a genetic influence in drug dependence comes from the convergence of information from methodologically different studies.

1. Alcoholism

Most genetic studies of substance abuse in families have focused on alcoholism. Several well-controlled family, twin, and adoption studies have identified increased risk for alcoholism in the biological relatives of alcoholics (MERIKANGAS 1990). In one of the most complete family studies, BLEULER (1955) systematically assessed alcoholism in first- and second-degree relatives of 50 alcoholics ascertained as such through a treatment facility and found strong familial aggregation for alcoholism in the first-degree relatives when compared to controls. Furthermore, the relative risk for alcoholism to second-degree relatives was approximately half that of first-degree relatives.

A two-fold increase in risk for first-degree relatives of alcoholics compared to second-degree relatives is not inconsistent with a purely genetic, Mendelian pattern of inheritance determined by a single gene locus; however, results from other data suggest a more complicated etiology for alcoholism. These data come mainly from two sources: studies of half-siblings and risk to offspring based upon spousal concordance for alcoholism. The risk for alcoholism to half siblings was equal to or greater than that of full siblings in the work of BLEULER (1995) and SCHUCKIT and colleagues (1972). These data are inconsistent with purely genetic contributions to alcoholism susceptibility and demonstrate the importance of environmental factors in the etiology of alcoholism. Data from other family studies suggest

Table 1. Methodologies for studying the etiology of alcohol and drug abuse

Method	Purpose	Problems and limitations[a]
Family studies	To determine if family members of substance abusers are at increased risk for substance abuse To provide evidence consistent with but not conclusive of a genetic basis for a disorder	Generational/secular trends in drug use. Problems with non-paternity. No separation of genetic and family environmental effects.
Twin studies	To determine if within-pair similarity for substance abuse is greater in genetically identical twins than in less genetically similar fraternal twins	Requires accurate zygosity detemination. Assumes equal "trait-relevant" environments in identical and fraternal twins. Rarity of subjects.
Adoption studies	To separate the effects of genes and environment by studying the similarity between biological parents and their adopted-away chidren and adopted-away children and their adoptive parents	Assumes no selective placement. Places prenatal environmental effects in the "genetic" component. Rarity of subjects.
Segregation analysis	To use diagnostic status of family members to determine the relative likelihoods of alternative modes of transmission for a familial disorder (i.e., polygenic, single major locus, mixed)	Affected by statistical non-normality. No direct test of single, major locus model vs polygenic model of transmission.
Linkage analysis	To use family data to statistically link a trait to a genetic marker on a particular chromosome	Multiple statistical comparisons. Assumes particular mode of transmission. Rare multigenerational drug abuse pedigrees. Generational and secular trends.
Association studies	To compare genetic variation at a specific site to behavioral variation within affected and unaffected individuals	Appropriate control group.

[a] Problems common to all methods include: diagnostic imprecision, variable age of onset, etiological heterogeneity, assortative mating, and ascertainment bias.

that both risk for alcoholism and the clinical course of the disorder in offspring show a significant linear trend dependent upon the number of alcoholic parents (MCKENNA and PICKENS 1981; MERIKANGAS 1990). Whether this is due to increased genetic loading, a deleterious combination of environmental factors, or both cannot be determined from family study data. Nevertheless, these data indicate the importance of this familial influence on the development of alcoholism.

Family studies analyze transmission of substance abuse disorders without distinguishing between genetic and environmental contributions. Twin and adoption studies provide data that can allow separation of these influences (Table 2). In five of seven studies assessing concordance for alcoholism in male twins, monozygotic (MZ) twins were significantly more concordant than dizygotic (DZ) twins. In these studies with males, MZ/DZ ratio for

Table 2. Clinical studies of genetic influences in alcoholism

Twin studies
  Males
    Positive
        KAIJ (1960)
        HRUBIC and OMENN (1981)
        PICKENS et al. (1991)
        CALDWELL and GOTTESMAN (1991)
        McGUE et al. (1992)
    Negative
        GURLING et al. (1984)
        ALLGULANDER et al. (1991)
  Females
    Positive
        KENDLER et al. (1992)
    Negative
        GURLING et al. (1984)
        PICKENS et al. (1991)
        CALDWELL and GOTTESMAN (1991)
        ALLGULANDER et al. (1991)
        McGUE et al. (1992)
Adoption studies
  Males
    Positive
        GOODWIN et al. (1973)
        CLONINGER et al. (1981)
        CADORET et al. (1985)
    Negative
        ROE (1944)
  Females
    Positive
        BOHMAN et al. (1981)
        CADORET et al. (1986)
    Negative
        ROE (1944)
        GOODWIN et al. (1977)

alcohol dependence ranged from 1.4 to 3.1, with heritability estimates fairly consistently ranging from 0.50–0.60 (McGue 1994). Furthermore, in all but the oldest and smallest study of male adoptees (Roe 1944), adoptee risk for alcoholism was significantly greater for males with a positive biological history of alcoholism than for males without a positive biological history of alcoholism, with the risk ratio associated with having a biological alcoholic parent ranging from 1.6 to 3.6 times that of not having a biological alcoholic parent (McGue 1994). Based on results from an adoption study conducted in Sweden (Bohman et al. 1981), Cloninger and colleagues (1981) described two subtypes of alcoholism: type I alcoholism is comprised of equal numbers of males and females and is hypothesized to be mildly to moderately heritable whereas the type II subtype is predominantly male and is highly heritable.

Considerably fewer females than males have been involved in twin and adoption studies of alcoholism. There is great variability in the estimate from twin studies of the influence of genetic factors in alcoholism in females. Of the six twin studies involving females, only two found significantly higher MZ than DZ concordance rates, with heritability estimates ranging from 0% to 56% (McGue 1994). Similarly, there is notable variability in relative risk for alcoholism in female adoptees with a positive biological history of alcoholism compared to female adoptees without a positive biological history of alcoholism. Only two of four studies found greater risk for alcoholism in the daughters of alcoholics than in the daughters of nonalcoholics. In general, it seems likely that genetic factors may be less important in the development of alcoholism in females than in males. However, this conclusion must be viewed with caution, as many fewer women than men have been employed in twin and adoption studies (Svikis et al. 1994). Significant cross-sex transmission of alcoholism in opposite sex twins indicates that totally separate etiological mechanisms for alcoholism in males and females are not plausible (McGue and Slutske 1993).

2. Other Drug Abuse

There are fewer family, twin, or adoption studies of substance abuse other than alcoholism. In the past several years, only two well-conducted family studies of drug abuse have been published. Mirin and colleagues (1991) examined the rates of drug abuse in first-degree family members of 350 adult, in-patient, drug abusers. The probands were classified in one of three groups according to the drug primarily used: (1) cocaine, (2) opioids, and (3) sedative-hypnotics. Probands and their family members were assessed by a structured (but nonstandard) clinical interview and diagnoses were determined based on DSM-III criteria. Overall, family members showed high rates of drug abuse other than alcohol (9.3%). In addition, differences in familial risk were found between the proband groups. For example, male relatives of cocaine and opioid abusers were at significantly increased risk for drug abuse compared to male relatives of sedative-hypnotic abusers (relative risks equal to 7.7 and 4.9 for the relatives of cocaine and opioid abusers, respectively, compared to the sedative-hyponotic group). Male relatives of cocaine abusers were 1.5 times as likely to be diagnosed with drug abuse as male relatives of opioid abusers, but this difference failed to reach statistical significance.

The MIRIN study (MIRIN et al. 1991) is remarkable both in terms of the number of subjects involved (350 probands and 1478 relatives) and the high rate of first-degree family members directly interviewed (45%). Results are limited, however, due to lack of a control/comparison group assessed by similar methods and the fact that assessments were not conducted blindly.

A second, large-scale family study of drug abusers was published by ROUNSAVILLE and colleagues (1991). In this study, 201 opioid addicts from in-patient and out-patient treatment programs were assessed using a modification of the Schedule for Lifetime Affective Disorders and Schizophrenia (SADS-L) and Research Diagnostic Criteria (RDC). In addition, 877 first-degree relatives received diagnoses. Of these diagnoses, 27% were based on direct interview; the others were based on proband report only (59%) or a combination of proband report and other relative informant(s) (41%). In the full sample, 12% of parents and 37% of siblings received diagnoses of drug abuse. These rates are significantly elevated over the rate of drug abuse in the siblings and children of controls from an independent, family genetic study of affective illness (3.3%). The familial risk estimates are also increased when compared to community estimates from the Epidemiological Catchment Area study (5.7%) (HELZER and PRYZBECK 1988).

Studies of vertical transmission of substance abuse are significantly limited by differences in availability of drugs across generations. Both geographical differences and secular trends can influence parent-offspring resemblance for drug use and dependence. For example, a recent report found fathers and mothers of cocaine addicts to be at 5.8% and 4.3% risk for drug abuse, respectively, whereas risk for drug misuse in brothers and sisters of cocaine addicts equaled 40.3% and 22.3%, respectively (LUTHAR and ROUNSAVILLE 1993). Similar generational differences in risk were found for parents and siblings of opioid abusers (LUTHAR et al. 1992a,b). These results suggest familial aggregation of drug abuse is best examined by restricting the analyses to persons of the same cohort (i.e., siblings or twins).

As with family studies, there are few twin and adoption studies of drug abuse. Two twin studies assessed concordance for a variety of illicit substance use and/or dependence. In males, evidence was found of genetic influence in use of sedatives, stimulants, analgesics, hallucinogens, and cannabis, with heritability estimates ranging from 0.39 to 0.62 (GOLDBERG et al. 1993). Evidence was also found in males of genetic influence in DSM-III clinical diagnosis of any substance abuse and/or dependence (excluding alcohol and tobacco), with a heritability estimate of 0.31 (PICKENS et al.

1991; LaBuda et al. 1993). Data from females, available only from one twin study (PICKENS et al. 1991), showed no significant genetic influence on clinical diagnosis of any substance abuse and/or dependence (excluding alcohol and tobacco) (LaBuda et al. 1993). Since probands in the PICKENS et al. (1991) study were alcoholic as well as drug abusers/dependent, the extent to which these findings generalize to pure drug abusers is unknown.

At least for males, there was evidence that environmental factors shared within a twin pair (e.g., neighborhood characteristics, religious affiliation, common friends) were a significant cause of resemblance for drug usage (PICKENS and SVIKIS 1991). In contrast, for females, it appeared that environmental factors unique to the individual were more important predictors of liability to drug abuse. In a more direct examination of environmental influences, MZ twins were found to be more similar than DZ twins in degree of contact and other measures of closeness. This within-pair environmental measure, although not significantly related to alcohol dependence, was related to drug abuse and/or dependence. Nevertheless, the MZ/DZ difference in concordance for drug abuse and/or dependence remained significant when the effects of within-pair closeness were partialled out (LaBuda and PICKENS 1994).

CADORET and colleagues (1986) studied 40 adopted, drug-abusing individuals and found evidence of two alternative genetic pathways to drug abuse: (1) through an inherited vulnerability to antisocial personality which is then correlated with drug abuse in the adoptee, and (2) in individuals who do not exhibit antisocial personality, through an inherited vulnerability to alcohol abuse correlated with drug abuse in the adoptee. Because substance abuse other than alcoholism was not assessed in biological parents, it is not possible to test for the specific transmission of drug abuse separate from alcohol abuse in this sample. In summary, although the evidence for genetic components to drug abuse vulnerability is not as dense as that indicating a genetic basis for alcoholism, the convergence of familial, adoption, and twin data provides support for the hypothesis that drug abuse has significant genetic components.

II. Levels of Genetic Analysis

As described earlier, there are several points at which genes may influence vulnerability to substance abuse. They may increase vulnerability by influencing other risk factors, by altering drug response (pharmacodynamics) or drug metabolism (pharmacokinetics), via personality or mood variables (e.g., sensation seeking), or by eliminating factors that protect individuals from substance abuse, among others. The level of analysis of vulnerability to drug dependence provides answers to different questions and therefore has different implications for treatment and prevention. Whereas most human genetic substance abuse research focuses on phenotypic resemblance for abuse or dependence, there is important human genetic work being conducted (mostly in the area of alcoholism) with regard to factors affecting use (as opposed to abuse or dependence, metabolism, and subjective and objective effects of substances). Each of these areas is discussed below.

1. Factors Affecting Use

Extensive research has been conducted on personal, familial, and societal factors which affect exposure to alcohol and drugs and lead to initial experimentation with these substances, which is a necessary but not sufficient condition for the development of drug dependence. As previously discussed, many studies have focused on environmental factors, including laws affecting the minimum drinking age, drug availability, and other neighborhood characteristics. Although vulnerability to alcohol and drug abuse is most plausibly due to a combination of genetic and environmental factors, studies of interpersonal and family characteristics affecting substance use often do not take into account genetic vulnerability and, similarly, genetic studies of vulnerability most often ignore specific environmental influences.

The impact of both genetic and environmental effects on teenage alcohol use were simultaneously estimated in a study of Australian twins (HEATH and MARTIN 1988). While this study did not assess specific genetic or specific environmental influences, it did find both genetic and environmental factors shared within twin pairs important in determining teenage abstinence for both males and females (although to differing extents). Additionally, genetic factors were found to be a major contributor to the observed variance in age of onset of alcohol use in females whereas shared environmental factors were much more influential in males. Contrastingly, genetic factors were significant contributors to average weekly consumption in both sexes. Just as environmental factors important in exposure to alcohol and drugs may differ from environmental factors important in first use or escalating use to abuse, the genetic factors important in the degree of alcohol consumption are important only once alcohol use has started but have little to do with the determinants of age of onset. Because this study involved nonalcoholic individuals, there is no evidence to determine whether the genetic factors affecting alcohol use are the same or different from those genetic factors which are involved in alcoholism.

2. Metabolism

The pathway by which most alcohol metabolism occurs involves two enzymes: alcohol dehydrogenase (ADH) and aldehyde dehydrogenase (ALDH). Genetic variants of each of these enzymes have been characterized (see SMITH 1986, for summary) and certain variants have been shown to occur more frequently in alcoholics than controls (THOMASSON et al. 1993). The presence of the mutant ALDH2 allele slows alcohol metabolism and is responsible for the characteristic "flushing" response noted in Asians (THOMASSON et al. 1993).

Another genetically influenced enzyme is cytochrome P450 2D6 (EICHELBAUM and GROSS 1990), which is responsible for metabolism of a large number of drugs. Due to gene deletion or mutations, the enzyme is absent in approximately 7% of Caucasians (GONZALEZ and MEYER 1991), making such individuals poor metabolizers of drugs metabolized by the enzyme. When subjective effects of an opiate drug were compared in poor metabolizers and extensive metabolizers, more "good opiate effects" and less "bad opiate effects" were reported by the extensive metabolizers, suggesting genetic variation in the enzyme may influence the reinforcing effects of certain drugs (OTTON et al. 1993). Similar effects may also occur with abused drugs metabolized by other enzyme systems, such as the metabolism of cocaine by cholinesterase (HOFFMAN et al. 1990; OM et al. 1993).

Twin studies have also provided evidence for genetic influences on alcohol metabolism. Studies of adult, nonalcoholic twins have demonstrated heritable differences in absorption and elimination of a standard dose of alcohol (KOPUN and PROPPING 1977; MARTIN et al. 1985a,b; VESELL 1972; WILSON and PLOMIN 1985). Conceivably, similar genetic contributions to drug metabolism could help to explain components of the genetic contribution to substance abuse vulnerability.

3. Subjective and Objective Effects

Studies in this area investigate whether genetic vulnerability (as assessed by positive family history) is related either to subjective or objective effects of substances. These aspects have been approached in alcoholism, in which results are somewhat contradictory. Some studies have found individuals with a positive family history report less intense feelings of intoxication after alcohol administration (Moss et al. 1989; O'MALLEY and MAISTO 1985; POLLOCK et al. 1986; SCHUCKIT 1984a,b). Offspring of alcoholics have also exhibited less sensitivity than controls on other alcohol-related measures such as body sway (SCHUCKIT 1985) and cortisol and prolactin response (SCHUCKIT 1984b; SCHUCKIT et al. 1987a,b; Moss et al. 1989). In contrast, other studies found family history to be unrelated to measures of sensitivity (McCAUL et al. 1990; NAGOSHI and WILSON 1987). NEWLIN and THOMSON (1990) attempted to account for these discrepancies by relating findings to the time after drinking when the measures were obtained.

4. Drug Specificity

Alcoholism and drug abuse show substantial comorbidity both in clinical samples and in the general population. Abuse of cocaine, sedatives, opiates, hallucinogens, and amphetamines, for example, was found to be ten times higher in alcoholics than nonalcoholics in the Epidemiological Catchment Area survey (HELZER and PRYZBECK 1988). Many family pedigrees contain both alcohol and drug abusers and it is difficult to determine whether these behaviors are independently or jointly transmitted.

In an early study based upon the family history method with unspecified diagnostic criteria, HILL and colleagues (1977) examined alcohol and drug problems in relatives of probands with alcoholism only, opiate addiction without alcoholism, or both alcohol and opiate abuse. Evidence was found for specific clustering of disorders (i.e., there was more alcoholism in relatives of alcoholic probands than in relatives of opiate addicts and there was more drug abuse in relatives of opiate addicts than in relatives of alcoholics). Two other family studies of drug abusers (ROUNSAVILLE et al. 1991; MIRIN et al. 1991) used a similar analytical approach and found more alcoholism in relatives of probands with alcohol and drug abuse than in relatives of probands with drug abuse only. Nevertheless, in all three family studies, it is clear that the risk for alcoholism in relatives of probands with drug abuse only (or conversely, the risk for drug abuse in relatives of alcoholic probands) is non-zero. It is not known if this risk is significantly different from general population estimates using similar diagnostic assessment.

MELLER et al. (1988) used a multivariate approach to study the independence of transmission of alcohol and drug problems by fitting a logistic regression model to data hypothesized to be predictive of proband drug abuse. Parental drug abuse was found to be the strongest predictor of proband drug abuse, even stronger than sibling drug abuse, which was also significant. In contrast, sibling alcohol abuse was a much weaker predictor and parental alcohol abuse was not significant once other variables were entered into the prediction model. Unfortunately there is not a model presented either incorporating or predicting proband alcohol abuse. Nevertheless, the analysis offers an intriguing way to approach the issue of specificity of transmission of alcohol and drug abuse.

5. Alternative Modes of Transmission

There have been two attempts to characterize the transmission of the underlying genetic vulnerability to alcoholism (GILLIGAN et al. 1987; ASTON and HILL 1990). Both studies ascertained families through alcoholic probands. The results of the studies were consistent in finding evidence to support a major effect in males, although the specific nature of this major effect differed in the two studies. Whereas the study by GILLIGAN and coworkers (1987) found evidence to support a recessive single major locus in families of male alcoholics, ASTON and HILL (1990) concluded that the major effect transmitted within the families of male alcoholics in their study was either non-Mendelian or a major environmental effect with a polygenic background. No evidence for a major effect of either genes or environment was found in families of female alcoholics (GILLIGAN et al. 1987). Multifactorial transmission (i.e., the combination of many small genetic and environmental effects) was determined to be the most likely mode of transmission in these families.

6. Phenocopies

Both genetic and environmental influences contribute to drug dependence. While the relative contribution of each has been estimated at the population level, at present it is not possible to determine the relative contribution of genetic and environmental factors to the etiology of a given person's drug dependence. Discovering highly genetically influenced phenotypes will significantly enhance our ability to identify candidate genes for drug abuse, as it will allow linkage and association studies to focus on individuals who are most likely to have those genes. Similarly, discovering highly environmentally influenced phenotypes will enhance our ability to identify individuals who are at risk for drug dependence for environmental reasons. At present, no genes specific for drug abuse or alcoholism have been confirmed, possibly because highly genetically influenced subtypes of the disorders have not been identified and employed in studies.

C. Molecular Studies

Genetic markers at a single gene locus can be identified as *linked* with a disease if, in different generations of families displaying a genetic familial disorder, one particular gene marker form is present in family members with the disease but not in those displaying normal phenotypes. Genetic markers at a gene locus can also be *associated* with a disease in population if they are present more often in unrelated individuals displaying the disease than in unaffected individuals.

The ability of genetic markers to adequately indicate either genetic linkage or allelic association is based on considerations of genetic variation and chromosomal recombination. Chromosomes are the linear arrangements of genes. There are two copies of each chromosome in a normal cell but allelic variants exist (such that there can be many forms of several genes) such that the copies of the chromosomes are not identical. Cell division, which produces the chromosomes for each human generation, results in "crossing over" or recombinant events which exchange sequences from one member of the chromosomal pair with those of the other member of the chromosomal pair. As a result, new chromosomal arrangements are formed. The average rate of this process allows 1% of chromosomal loci separated by 1 million base pairs of DNA to recombine in each generation. However, this average rate of recombination varies dramatically across different chromosomal loci. In general, the closer two genetic loci are to each other, the less often crossing-over or recombination occurs.

For a genetic marker to indicate linkage, chromosomal recombination between that marker and a postulated gene leading to familial substance abuse must occur infrequently within the several generations assessed in the family. Markers located within serveral million base pairs of the functional gene defect causing substance abuse would thus be likely to reveal genetic linkage if substance abuse were due to a gene of major effect. However, for allelic association to be detected, chromosomal recombination between a polymorphic marker and a postulated gene contributing to substance abuse in the population must occur infrequently within the many generations separating the inheritance of the population sampled. A useful polymorphic marker should thus lie very close to the functional gene defect contributing to substance abuse vulnerability. Alternatively, such a marker could be further away but in linkage disequilibrium with the functional gene defect. Linkage disequilibrium is a term used to define a process with poorly understood mechanisms that results in significantly lower than expected rates of recombination observed between some chromosomal loci. Linkage disequilibrium can thus allow a genetic marker used in an allelic association study to provide information about not only whether closely adjacent DNA contains a functional gene defect, but also about the possibility that DNA thousands of bases removed from the polymorphic genetic marker, but in linkage disequilibrium with it, could also contain a functional gene defect.

I. Selecting Genetic Markers

The process of drug dependence begins by a drug occupying a brain receptor. Individual differences in brain receptor genes could contribute to individual differences in drug abuse vulnerability. Significant data now support the idea that virtually every abused drug can induce behavioral reinforcing properties, at least in part, by altering function in brain dopamine circuits arising from the ventral midbrain. Genes important in the mesolimbic/ mesocortical dopaminergic pathways are thus candidate genes for possible contributions to individual differences in substance abuse vulnerability. The dopamine D_2 receptor (DRD2) gene represents one such candidate gene. The linkage disequilibrium that has now been documented at the DRD2 gene locus also provides a plausible rationale for the study of allelic association between this locus and substance abuse vulnerability.

Recent molecular cloning studies have identified the genes encoding many other drug receptors, including the dopamine transporter that is the pharmacologically defined cocaine receptor (SHIMADA et al. 1991; KILTY et al. 1991), G-linked opiate receptors that are the heroin/morphine receptors (KIEFFER et al. 1992; EVANS et al. 1992; WANG et al. 1993), the G-protein-linked canabanoid receptor that mediates marijuana action (MATSUDA et al. 1990), the NMDA-glutamate receptor ligand-gated ion channel receptors that mediate phencyclidine actions (MORIYOSHI et al. 1991; DURAND et al. 1992), the nicotinic acetylcholine receptor ligand-gated ion channel that is the site of action of nicotine (MISHINA et al. 1984; BOULTER et al. 1987), and the GABA receptors ligand-gated ion channels that mediate actions of barbiturates and benzodiazepines (OLSEN and TOBIN 1990; CUTTING et al. 1991). GABA and NMDA receptors are also strong candidate loci for acute ethanol effects.

Although several drug receptor genes have been characterized, processes directly modulating receptor function have not been able to account for significant fractions of the biochemical bases of addiction (NESTLER 1992). Direct genetic approaches which are not dependent on biochemical hypotheses may be more likely to identify genes that could contribute to interindividual differences in substance abuse vulnerability than candidate gene approaches. A large number of richly polymorphic genetic markers are now available for use. These markers include variable number simple sequence polymorphisms such as CA repeats and restriction fragment length polymorphisms (RFLPs). The increasing number of such markers, distributed on each segment of each human chromosome, provides increasing potential power for genetic studies.

II. Linkage Analysis

There have been two genetic linkage studies of alcoholism reported (HILL et al. 1988; TANNA et al. 1988). In the study of HILL and colleagues (1988), families were ascertained in which only alcoholism was segregating (i.e., there were no cases of any other psychiatric disorders in probands or their first- or second-degree relatives). A lod score of 2.02 was obtained for linkage of alcoholism to the MNS blood group. Linkage to the MNS blood group was excluded in the linkage study by TANNA et al. (1988) which included families of probands who were either alcoholic or depressed. A lod score of 1.64 was obtained between alcoholism and esterase-D in this study; however most of this is accounted for by a single family. In subsequent work (WESNER et al. 1991), linkage of esterase-D and alcoholism was not confirmed. No linkage studies of substance abuse other than alcoholism have been reported. The results from the linkage studies of alcoholism should be interpreted with caution given that neither the linkage with MNS nor esterase-D reached the generally accepted minimal statistical criteria for genetic linkage (i.e., a lod score equal to or exceeding 3.0) and that neither result has been replicated.

1. Dopamine D₂ Receptor Locus

BLUM et al. (1990) provided the first evidence that the DRD2 gene might display population variants associated with alcoholism. Although this gene is interesting because it codes for a protein expressed abundantly in dopaminergic circuits important for behavioral reward, the BLUM et al. report was viewed with several cautions (GELERNTER et al. 1991; UHL et al. 1992). The probability, a priori, of identifying vulnerability enhancing alleles of one of the many thousands of genes expressed in the human brain was low. Subsequent studies of families in which alcohol abuse appeared to pass from generation to generation revealed no linkage in 14 informative families (BOLOS et al. 1990; PARSIAN et al. 1991). Another caution in acceptance of this association was based on evidence for variable representation of the DRD2 alleles in different human populations (O'HARA et al. 1993). The frequency of the allele associated with alcoholism in the BLUM et al. report is significantly different in white individuals of primarily European descent, for example, than in black, Asian and American Indian populations (GELERNTER et al. 1993; O'HARA et al. 1993; UHL et al. 1992). Sample stratification, disproportionate sampling of abusers or controls from population subgroups displaying atypical allelic frequencies, could thus yield spurious results not indicative of true association between DRD2 allelic status and substance abuse (LANDER and SCHORK 1994).

The initial findings of a significant allelic association spurred interest in a number of laboratories in the roles that DRD2 gene might play in alcoholism as well as polysubstance abuse. Several polymorphic sites within and near the DRD2 gene have been found (UHL et al. 1993). The "A" site is located in the 3' flanking region of the DRD2 gene and is a *Taq* I A restriction fragment length polymorphism (RFLP). This is the allelic variant studied by BLUM et al. (1990). A *Taq* I B RFLP lies more 5' and a *Taq* I C RFLP (termed "D" in the initial report) provides a polymorphic marker for a site lying between *Taq* I A and B. Several studies have examined both substance abusers and controls obtained from similar populations. Other studies have examined either substance abusers or control populations only (reviewed in UHL et al. 1993). These analyses appear to support several conclusions concerning the association between DRD2 alleles marked by *Taq* I RFLPs and substance abuse, as detailed below.

A general conclusion from the DRD2 studies is that the *Taq* I A1 and B1 DRD2 RFLPs are interesting reporters for events at the DRD2 gene locus in Caucasians. As noted above, the *Taq* I A and B RFLP sites are separated by chromosomal distances sufficiently large to render the two sites randomly associated with each other, or with the structural or regulatory regions of the dopamine receptor gene. Maintaining association would require linkage disequilibrium to preserve the chromosomal connection between *Taq* I A and B RFLPs and coding or regulatory sequence DRD2 gene variants. Strong linkage disequilibrium between these 3' and 5' gene markers does exist. In Caucasians, more than 95% of possible linkage disequilibrium between the 3' and 5' markers is present (O'HARA et al. 1993). These findings point to much less recombination between these two gene locus markers than expected at most chromosomal loci separated by thousands of base pairs of DNA. This striking linkage disequilibrium provides support for the idea that the *Taq* I A and B genotypes *could* reliably mark a structural or functional gene variant that could be directly involved in altering behavior.

With regard to the association between the DRD2 variants and drug and alcohol abuse, combined analysis of data from several studies indicates that the A1 and B1 markers appear to be more frequent in polysubstance abusers and possibly even in alcoholics than in controls (COMINGS et al.

1994; Gelernter et al. 1993; Noble et al. 1993; O'Hara et al. 1993; Smith et al. 1992). Furthermore, the most severe drug abusers may manifest higher A1 and B1 DRD2 gene marker frequencies (Uhl et al. 1993), while "control" comparison groups, screened carefully to eliminate individuals with significant use of any addictive substance, appear to display lower A1 and B1 frequencies than unscreened control populations.

Meta-analytic approaches to combining data from different studies have limitations but can yield the statistical power required to reliably identify modest effect sizes when large environmental influences are expected a priori. No data available to date derive from true population based sampling techniques. Although the possibility of false-positive error based on sample stratification thus remains (Lander and Schork 1994), combining available data from nonpopulation-based studies of Caucasians sampled in several North American and European centers would serve to minimize effects due to unplanned stratification. Meta-analysis of this combined data reveals statistically significant differences between A1 and B1 frequencies in substance abusing and control populations (Tables 3, 4). On average, the relative risk for *severe* substance abuse in a Caucasian individual with the A1 RFLP marker is estimated to be 2.2-fold greater than controls (based upon an estimated 10% of the population abusing at least one addictive substance at this level) (Uhl et al. 1993).

Table 3. D2 dopamine receptor gene *Taq*I Al and B1: data

Studies that collected data for:	Controls	Drug users
Amadeo et al. (1992)	7/43 = 16.3%	21/49 = 42.9%
Blum et al. (1990)	4/24 = 16.7%	14/22 = 63.6%
Blum et al. (1991)	6/31 = 19.4%	42/89 = 47.2%
Bolos et al. (1990)	21/62 = 33.9%	15/40 = 37.5%
Comings et al. (1991)	24/108 = 22.2%	63/156 = 40.4%
Gelertner et al. (1991)	24/69 = 35.3%	19/44 = 43.2%
Parsian et al. (1991)	3/25 = 12%	13/32 = 40.6%
Schwab et al. (1991)	22/69 = 31.9%	11/45 = 24.4%
Smith et al. (1992)	14/56 = 25%	96/232 = 41.4%
Studies that collected data for:		Drug users only
Noble et al. (1993)		27/53 = 50.9%
George et al. (1993)		34/89 = 38.2%
Turner et al. (1992)		9/47 = 19.1%
Studies that collected data for:	Controls only	
Grandy et al. (1989)	16/43 = 37.2%	
Nothen et al. (1992)	31/69 = 44.9%	
O'Hara et al. (1993)	31/104 = 29.8%	
Parsian et al. (1991)	15/46 = 32.6%	
Totals	218/748 = 29.1%	364/898 = 40.5%

Table 4. Meta-analysis of *Taq*I A1 and B1 studies examining abusers and controls

Comparison	Number of studies	Odds ratio	Significance
A1			
Any user vs any controls	9	2.09	$p < 0.0001$
Any user vs assessed controls	6	3.48	$p < 0.0001$
More severe vs any controls	6	2.91	$p < 0.0001$
More severe vs assessed controls	4	3.64	$p < 0.0001$
B1			
Any user vs assessed controls	2	3.02	$p < 0.002$
More severe vs assessed controls	2	3.49	$p < 0.0001$

2. Other Genes

Even if the association between DRD2 markers and substance abuse is found to be valid, other genetic and environmental influences are likely to determine the majority of variance in individual vulnerability to substance abuse. Calculations based on twin data indicate that genes may influence up to 60% of the liability variance to severe substance abuse. Thus, at most ¼ of the risk for *severe* substance abuse could potentially be attributed to DRD2 variants marked by the A1 genotype (UHL et al. 1993). If these influences on substance abuse phenotype were largely additive, an assumption of several current working models of substance abuse genetics, then comparisons between estimates of the magnitudes of these effects can be made. Such comparisons indicate that even if the DRD2 variants represent the prominent single gene determinants of susceptibility to severe substance abuse, other genes and the environment still play the largest roles.

Association studies for polymorphic markers at loci for other dopaminergic genes, including the dopamine transporter and tyrosine hydroxylase, have not yielded positive associations with substance abuse (BLUM et al. 1990; NOBLE et al. 1991; PERSICO et al. 1993a). However, the limitations of these negative data need to be kept clearly in mind, since the linkage disequilibrium that may make the *Taq I* A and B DRD2 gene markers accurately reflect DRD2 gene variants at regulatory or coding sequences is not documented at these other dopaminergic gene loci.

D. Animal Studies

Virtually all drugs abused by humans will serve as positive reinforcers across a number of animal species. In addition, drug-related phenotypes such as dependence, tolerance, sensitization and conditioned drug effects can also be experimentally analyzed in animal models to a degree unavailable in clinical settings. Since the addiction process is likely to involve a constellation of environmental and genetic variables, animal models can help to selectively isolate or explore dependent variables or interactions that are important in the development of drug abuse and dependence.

Several points should be considered when reviewing findings from animal studies. First, each paradigm must demonstrate pharmacologically relevant drug intake during discrete time intervals or use pharmacologically relevant doses of the abused drug to accurately assess the results in the context of vulnerability. Second, a two-strain comparison provides no conclusive evidence for underlying mechanisms or genetic covariance unless the data are used to rule out a hypothesis. A correlation between two points has zero degrees of freedom. If two strains differ widely on a biochemical parameter yet do not differ in the phenotype, a tentative conclusion that the biochemical trait does not govern the behavior can be made. Third, the absolute quantity of drug intake or the dose that maintains maximum response rate does not necessarily indicate the relative *efficacy* or magnitude of the drug to serve as a reinforce (KATZ 1990). Parameteric methods such as systematic manipulation of response requirement, use of progressive ratio, or application of behavioral economic analysis may be required to address the efficacy or persistence of drug seeking behavior under intermittent schedules of reinforcement. Fourth, evidence for a biochemical/genetic marker relationship to reinforcement depends on the behavioral genetic and behavioral pharmacology technique used to determine the association. The generality of the association will be determined by assessing the degree of genetic correlation across numerous behavioral pharmacology tests (e.g., McCLEARN 1968). Fifth, the phenotypic outcome of molecular modification of genotype (transgenic, anti-sense, homologous recombination) will depend upon the host strain or background genotype (e.g., COLEMAN and HUMMEL 1975). The target manipulation must still act in the neurobiological system of the host strain, thus significantly affecting transduction of the initial stimulus event (pharmacological or environmental) into a behavioral outcome.

I. Animal Models of Drug Taking Behavior

1. Behavioral Pharmacology Approach

There are many methods that can be used to study drug self-administration and drug-reinforced behavior. Two-bottle choice procedures, conditioned place preference (CPP), intracranial self-stimulation (ICSS), and operant self-administration are four of the most commonly used methods (BOZARTH 1987). As with clinical methods, each animal method is based on a set of assumptions, and the results must be interpreted given the limitations of each method (Table 5).

The method used to analyze drug self-administration and drug-reinforced behavior will, in part, determine which aspect of the addiction process is

Table 5. Methodologies for studying the behavioral pharmacology of alcohol and abused drugs

Technique	Method	Advantages	Disadvantages
Two-bottle choice	Determine relative oral intake of drug vs vehicle solution	Rapid assessment Large scale screening	Taste confounds Low plasma concentration Adulterated solutions
Conditioned place preference	Classically condition previously neutral stimuli with drug administration, then measure approach behavior	Rapid assessment Large scale screening Test not confounded by direct drug effects Measures conditioned reinforcing properties	Unequal baseline preferences Differential stimulus intensity Inconsistent dose-response data
Intracranial self-stimulation	Electrically stimulate brain region following operant, then measure stimulation threshold changes following drug administration	May assess innate thresholds Consistent baselines Less subject to direct drug effects	Labor intensive
Operant self-administration	Administer drug (oral, i.v., i.g., i.c.v, i.m.) following emission of specific operant	Drug serves as a primary reinforcer Direct comparisons to clinical literature Direct comparisons to other reinforced behavior Controlled manipulation of environmental context	Labor intensive Possible learning confounds Subject to large environmental influence

being modeled. There are potentially distinct components that contribute to drug addiction that are modeled in each paradigm. For example, a loose distinction could be drawn among the four approaches so that the two-bottle choice, CPP, ICSS and operant self-administration methodologies depict drug self-administration, classically conditioned reinforcement, reinforcement threshold and operant reinforced behavior, respectively. Each dependent variable may mirror important but distinct aspects of the addiction process and reflect a subset of variables that contribute to vulnerability. There are genetic and environmental convariates to each phase of addiction that, when taken as a whole, have more predictive value than each individual phenotype (MCCLEARN 1993).

2. Behavioral Genetics Approach

There are as many behavioral genetic methods for determining genetic contributions to drug-related effects as there are behavioral pharmacology methods for determining drug-reinforced behavior (Table 6). Inbred strain

Table 6. Methodologies for studying the behavioral genetics of alcohol and abused drugs

Technique	Method	Purpose
Inbred strain correlation or within subjects correlation in a heterogenous population	Determine correlation between two phenotypes across a panel of inbred strains or in a heterogeneous population	Determine degree of genetic covariance between the two phenotypes
Classical analysis	Produce F1 and F2 generations from parental strains, then backcross F1 generation to parental strains	Determine the mode of inheritance for drug related behaviors
Selective breeding	Select high and low phenotypic responders to breed for each drug effect into high and low lines while randomizing irrelevant genes	Segregate the gene(s) responsible for the observed phenotype
Quantitative trait loci	Inbreed F2 generation to form recombinant inbred strains, then determine correlation coefficient between gene marker(s) and phenotype	Map the location of gene(s) responsible for the observed drug effect

correlations, classic genetic analysis, selective breeding, and recombinant inbred strain analysis (single gene, quantitative trait loci) are four of the most commonly use methods (McClearn 1991; Crabbe et al. 1994).

Despite the large array of tools available to behavioral geneticists, most studies investigating vulnerability have focused either on individual variability in drug seeking behavior within a randomly inbred population or variability in drug seeking behavior across a range of defined genotypes. Studies using the former approach have typically segregated a randomly inbred population into two groups based on phenotypic response (e.g., Sudakov et al. 1991; Piazza et al. 1989). These groups are then tested for the ability of the primary phenotypic score to predict subsequent drug seeking behavior. Most studies investigating vulnerability, however, have utilized genotypically defined subjects. This is because the contribution of genotype in acute response to drugs of abuse, the mechanistic utility of specific inbred or selectively bred genotypes, and the ability to determine the contribution of genetic versus environmental variance in the observed phenotype are known. In particular, since the neurobiological factors important in initial drug use are not necessarily the same as those involved in the maintenance of the drug as a reinforcer or could change following drug exposure, investigation of vulnerability must be determined, in part, in experimentally naive individuals. Inbred strains provide an opportunity to essentially conduct a "within-subject" investigation of the central nervous system (CNS) mechanisms responsible for vulnerability to the acquisition of drug-reinforced behavior. Since inbred strains are >99% homogeneous, neurological function can be considered constant in animals used for invasive measures of CNS function and those used for investigation of the acquisition of drug taking behavior.

II. Genetic Variation in Drug Self-Administration and Drug-Reinforced Behavior

1. Two-Bottle Choice

This procedure produced seminal observations demonstrating clear genetic influences in animal drug taking behavior. McClearn and Rodgers (1959) demonstrated large differences in ethanol preference as a function of genotype in several commonly available inbred mouse strains. This finding has been replicated a number of times in various laboratories and stands tribute to the consistency of the results obtained using inbred strains. Since the initial demonstration, a number of studies employing two-bottle choice and home-cage drinking methods have demonstrated a significant effect of genotype on ethanol (for review see Phillips and Crabbe 1991), opioid (Berrettini et al. 1994; Belknap et al. 1993; for review see Belknap and Crabbe 1992) and cocaine intake (George and Goldberg 1988; Seale 1991; Alexander et al. 1993). In general, large quantitative and qualitative differences in drug taking behavior exist as a function of genotype. For example, C57BL/6J mice consume large amounts of morphine, ethanol and cocaine solutions in preference to vehicle, while DBA/2J mice consistently avoid these solutions.

Preference studies are most often used in behavioral genetic methods that require a large number of subjects or strains, such as selective breeding or quantitative trait loci (QTL) analysis. A number of selective breeding programs have produced lines of rats that differ by as much as tenfold in ethanol preference ratios (P/NP, HAD/LAD, Li et al. 1981, 1991; AA/ANA, Eriksson 1968; HChA/UChB, Mardones et al. 1983; sP/sNP, Fadda et al. 1989). These selected lines have been used in a large number of studies to determine heritability and genetic covariance of ethanol-related traits. Selective breeding for many ethanol-related phenotypes, such as hypothermia, withdrawal, low-dose stimulation and narcosis (e.g., Phillips et al. 1989; Deitrich 1993), complement these selection programs and provide ethanol consumption models with one of the most comprehensive set of behavior genetic tools.

Ongoing QTL analysis of ethanol and morphine preference reveals several important aspects of vulnerability research (Plomin and McClearn 1993). First, the behavioral pharmacology measure used to investigate vulnerability may significantly influence the results of behavioral genetic analysis. Conversely, the behavioral genetic analysis can be used to discriminate differences between behavioral pharmacological methods. For example, two seemingly related phenotypes, preference and acceptance (forced choice), are in fact only marginally genetically related. The genetic correlation between ethanol preference and acceptance is only 0.38. Only two of the four chromosome regions implicated in acceptance (*Acrg*, chromosome 1 at 40 cM and *Pmv-42*, on chromosome 15 at 62 cM) were similarly implicated in preference (Plomin and McClearn 1993). These data highlight the fact that different genetic processes may be responsible for seemingly closely related phenotypes. In general, methods such as QTL can be used to discriminate the molecular basis for the proposed covariance of two or more behavioral measures. For example, previous selective breeding data suggested a relationship between ethanol and opioid preference (Nichols and Hsiao 1967). To date, QTL analysis does not indicate chromosomal loci common to both phenotypes. QTL analysis of morphine + saccharin vs quinine + saccharin in 20 BXD RI lines and 15 inbred strains demonstrates a significant negative association with the *Es-1* region of chromosome 8 (Belknap and Crabbe 1992). Morphine preference in the F_2 generation of the B6 × D2 intercross implicates three QTLs on chromosome 1, 6, and 10 (Berrettini et al. 1994). Ethanol preference or acceptance measures do not implicate these regions. The required addition of saccharin and/or quinine to engender drug intake and dissimilarities in drug exposure may have a significant influence on the observed relationship.

2. Conditioned Place Preference

Conditioned place preference measures the presumed motivational effects of a drug via classic conditioning methods. Several single-dose studies have found qualitative differences in CPP as a function of genotype (DYMSHITZ and LIEBLICH 1987; SEALE and CARNEY 1991). Inbred rat strains and selectively bred rat lines have shown large differences in CPP. F344 and Lewis rats differ in cocaine (15 mg/kg) and morphine (4 mg/kg) induced place preference (GUITART et al. 1992). Lewis show more than twofold greater preference for both drugs and may, like C57BL/6J mice, be a strain that consistently shows drug preference regardless of pharmacological class or behavioral paradigm. In rats selectively bred for high and low ICSS rates, only the rats selectively bred for high rates of self-stimulation behavior (LC2-Hi) demonstrated a significant change in place preference upon opioid conditioning (DYMSHITZ and LEIBLICH 1987). Finally, rats selectively bred from a heterogeneous N/Nih stock for differential place preference in a biased procedure differed by sixfold in the place preference produced by 2.5 mg/kg cocaine (SCHECHTER 1992). These lines do not differ in the discriminative stimulus properties when trained to a much higher dose of cocaine (10 mg/kg). Thus, while there is a significant difference in the ability of cocaine to induce place preference, the discriminative cue produced by a higher dose of cocaine does not differ between these lines.

The particular advantage of CPP is the capacity to measure the *dose-dependent* ability of a drug to engender place preference in the absence of direct rate-altering drug effects. As a result, differences in the potency and efficacy of the drug as a function of genotype can be determined without confounding drug effects at the time of testing. Clear dose-dependent CPP has not been demonstrated in a number of studies, however, it is an essential step in verifying qualitative differences in the motivational effects of a drug (CARNEY et al. 1991; CUNNINGHAM et al. 1992). One study, in particular, demonstrates the utility of dose-effect curves and the potential confound of phenocopies in determining mechanisms underlying behavior. The potency of morphine to induce CPP does not differ in the CXBK mice compared to ddY mice (SUZUKI et al. 1993) despite a significant insensitivity to most morphine-related phenotypes (analgesia, locomotor stimulation) and significantly less μ-receptors in the CXBK mice (BARAN et al. 1975; MOSKOWITZ and GOODMAN 1985). Despite similar dose-effect curves, different mechanisms may underly the observed phenotype; D2 receptor antagonists block opioid-induced place preference in the ddY mice but not in the CXBK mice (SUZUKI et al. 1993).

3. Intracranial Self-Stimulation

The stimulation parameters required to elicit ICSS behavior and the response rate once ICSS behavior is established differ across inbred mouse strains in a region-specific manner (CAZALA et al. 1974; ZACHARKO et al. 1987, 1990). For example, stimulation of the mesocortex maintains nearly eight times the amount of behavior in BALB/cByJ and C57BL/6J mice than in DBA/2J mice, whereas stimulation of the nucleus accumbens maintains similar rates in the DBA/2J and C57BL/6J mice and approximately 1.5 times greater rates in the BALB/cByJ mice (ZACHARKO et al. 1987, 1990). CAZALA et al. (1974; CAZALA 1976; CAZALA and GUENET 1980) examined the response rate and threshold for lowest intensity of stimulation eliciting lateral hypothalamic ICSS behavior in BALB/cByJ, C57BL/6J and DBA/2J mice. These three strains differed in the threshold, response rate and temporal distribution of ICSS behavior. In general, the lower the minimal intensity necessary to elicit responding, the higher the response rate. Analysis of backcross and recombinant inbred strains developed from BALB/c × DBA/2 mice suggested the existence of a major gene effect on the performace of ICSS behavior (GAZALA and GUENET 1980). In the only study to determine the effect of an abused drug on ICSS behavior in inbred mice, the rank order for sensitivity to the rate increasing effect (and presumably lowered threshold) of amphetamine was BALB/cByJ > C57BL/6J > DBA/2J mice (CAZALA 1976).

The technique of selective breeding has generated separate lines of rats having high and low hypothalamic self-stimulation rates (LEIBLICH and OLDS 1971; LEIBLICH et al. 1978). Initially, two populations of rats were used to generate the selected rat lines. One population contained a mixture of HUC, Holtzman and Lewis rats (LC1-P) and the other contained only HUC and Holtzman (LC2-P). Interestingly, the two control and selected lines from the separate parental populations differed significantly. The stimulation rates for control and selected LC1-Hi and LC1-Lo lines were significantly lower than for the control and LC2-Hi and LC2-Lo lines. The inclusion of the Lewis rats in the LC1-P population is proposed to be responsible for the difference, as Lewis rats have very low rates of self-stimulation behavior (LIEBLICH et al. 1978). These data are of interest in light of the consistent demonstration of drug-reinforced behavior in Lewis rats vs other rat strains (SUZUKI et al. 1988, 1992; GEORGE and GOLDBERG 1989; AMBROSIO et al., in press). In addition, Δ^9-THC lowered the threshold for medial forebrain bundle self-stimulation only in Lewis rats and not in Sprague-Dawley, Long-Evans, or F344 rats (GARDNER and LOWINSON 1991). Unfortunately, baseline ICSS rates or stimulation thresholds were not given in these reports.

4. Operant Self-Administration

Operant drug-reinforced behavior in defined genotypes provides an ideal paradigm to pursue questions related to genetic and/or environmental factors important in drug use and dependence. Operant paradigms place defined genotypes in a rich theoretical framework that is uniquely suited to address historical and contextual factors in the acquisition, extinction, and relapse of drug seeking behavior. Under these conditions, genetic factors can be held constant while environmental conditions are manipulated as the independent variable in order to investigate environmental factors. Conversely, the environmental conditions can be held constant while genotype is manipulated as the independent variable in order to investigate genetic factors.

Unique to the preclinical operant self-administration paradigm is the ability to demonstrate heterogeneity in drug use and dependence that exists on a similar continuum as evidenced clinically. First, qualitative differences in the likelihood of acquiring drug-maintained behavior is significantly influenced by genetic and environmental conditions (ELMER et al. 1990; HYYTIA and SINCLAIR 1990). For example, Lewis rats self-administer opioids across a wide range of environmental conditions (training, access schedule, route) whereas F344 rats do not readily acquire opioid self-administration behavior when the drug is given orally (SUZUKI et al. 1993) but will when the drug is given intravenously (AMBROSIO et al., 1995). Second, once drug-reinforced behavior is established, differences in the amount of drug intake exists as a function of genotype. LS/Ibg mice consume significantly greater amounts of ethanol and achieve greater blood ethanol concentrations across a broad range of environmental conditions than even C57BL/6J mice (ELMER et al. 1990). Etonitazene serves as a reinforcer in CXBK/ByJ, CXBH/ByJ, BALB/ByJ, and C57BL/6J mice but drug intake differs significantly across strain (ELMER et al. 1995b). Third, the persistence of drug seeking behavior under increasing work load differs across genotype. When drug is available under simple fixed ratio 1 schedules of reinforcement, NP and HAD rats exhibit similar levels of drug-maintained behavior. However, under increasing fixed ratio requirements, NP rats persist in drug seeking behavior at a significantly higher rate than HAD rats (RITZ et al. 1994). Fourth, in addition to differences in acquisition and maintenance of drug-reinforced behavior, there are genetic differences in extinction patterns following vehicle substitution (AMBROSIO et al., 1995; ELMER et al. 1995b). These differences in extinction may be due to genetic differences in the rate-depressant or classically conditioned effects of the drug and offer an avenue for exploring different components of the addiction process. Fifth, toxic side effects and withdrawal signs differ significantly as a function of genotype even when drug intake approaches equivalent levels (AMBROSIO et al., 1995; ELMER et al. 1995b). Finally, heterogeneity of drug use across drug class can be delineated in animal models without the confound of comorbid psychiatric conditions often found clinically. To this end, several rodent strains stand out in their drug taking behavior across a wide range of pharmacological classes (see GEORGE 1991). Lewis rats and C57BL/6J mice readily self-administer ethanol, etonitazene and cocaine. In addition, multiple measures of drug taking and conditioned reinforcing procedures have demonstrated relatively consistent avoidance in some strains (DBA/2J) while some strains vary in drug intake as a function of drug class (i.e.,

BALB/cJ). Clear demonstration of the degree of commonality across multiple drug class awaits further systematic investigation within and across behavior pharmacology paradigms.

The ability of preference, CPP, or ICSS methods to predict genetic differences in operant drug self-administration can be assessed by investigating genetic rank-order relationships across the phenotypes or correlated measures in selected lines. Selection for differential ethanol preference in two-bottle choice procedures has generally produced lines that differ in operant ethanol self-administration (RITZ et al. 1986, 1989a,b; MURPHY et al. 1989; WALLER et al. 1984), although environmental history (i.e., training procedure) can influence the outcome (CARROLL et al. 1986; SAMSON et al. 1989). In general, ethanol drinking in the preference paradigm is not highly predictive of ethanol-reinforced behavior in the operant paradigm (GEORGE and RITZ 1993). For example, lines of rats selected for low ethanol preference (NP) maintain operant self-adminstration behavior at levels similar to rats selected for high ethanol preference (HAD) (RITZ et al. 1994). The differences in ethanol preference and operant ethanol self-administration are consistent in C57BL/6J and BALB/cByJ mice (McCLEARN and RODGERS 1959; ELMER et al. 1987, 1988) but not in LS/Ibg and SS/Ibg mice (ELMER et al. 1990). Preference for other drugs of abuse and the ability to predict self-administration behavior has not been examined. A few single dose CPP studies have demonstrated significant differences in cocaine and opioid place preference (GUITART et al. 1992) that are consistent with results found in operant self-administration studies (SUZUKI et al. 1992), however more work is needed before conclusions can be drawn.

Although the self-administration technique is labor intensive and time consuming, this procedure can address important issues in vulnerability research. For example, there are several components of drug seeking behavior that, when broken down, may independently influence the addiction process. As an example, RITZ et al. (1994) demonstrated significant differences in the persistence of drug seeking behavior under intermittent schedules of reinforcement that may begin to differentiate an important distinction between the primary reinforcing effects and motivational effects of abused drugs. Several other elements in operant self-administration may lead to further distinctions (i.e., rate of acquisition may suggest initial vulnerability, extinction rates may indicate strong conditioned drug seeking behavior, and the rate of reacquisition or the ability of stimuli to reestablish drug seeking behavior may indicate vulnerability to relapse). Decomposing the addiction process into its constituent parts may help us better understand addiction and vulnerability.

III. Drug-Naive Behaviors as Predictors of Vulnerability

Any number of inherited drug-naive behaviors could influence an individual's vulnerability to drug addiction. A number of ethologically relevant behaviors are available for examination; the question posited to preclinical research is which behavior(s) best predicts vulnerability. Proposals for behavioral phenotypes that predict vulnerability can be suggested from clinical literature or extrapolated from candidate neuroanatomical sites or transmitter systems.

Significant genotypic influence in drug-naive behaviors have been described for a number of phenotypes, including locomotor activity (DeFries et al. 1978), response to stress (Glowa et al. 1992), aggression (Jones and Brain 1985), and learning (Upchurch and Wehner 1988, 1989). These replicable differences in innate behaviors can be used to investigate predictive factors of an individual's vulnerability to drug addiction. For example, there are significant genetic differences in the level of drug-naive locomotor behavior and rates of habituation (within and across days) following exposure to novel environmental stimuli (Crusio and Schwegler 1987; DeFries et al. 1978; Lipp et al. 1987; van Abeelen 1989). These data demonstrate an inherited neurobiological difference responsible for the observed behavioral effects. More importantly, they offer an opportunity to explore the neurobiological and behavioral relationship between a proposed phenotype and vulnerability via independent manipulation of the genotype. For example, recent studies have suggested that rats at risk for psychomotor stimulant self-administration may be identified from their innate locomotor reaction to a novel environment (Piazza et al. 1989). Animals segregated from a population of randomly inbred rats that have a high locomotor response to a novel environment acquire amphetamine self-administration behavior more rapidly than those with a low locomotor response (Piazza et al. 1989).

A behavioral genetics approach to the same hypothesis can establish the degree of variability in acquisition due to genetic and environmental components and utilize a stable genetic source to pursue the neurobiological substrates underlying the observed genetic differences. For example, rats selectively bred for high open-field emotional reactivity, MR/Har/Lu, were the only line to demonstrate morphine preference ratios greater than 0.5 (vs MNR/Har/Lu:low reactivity and RCA/Lu:control line) in a two-bottle choice procedure (Satinder 1977). Conversely, rats selectively bred for high ethanol preference (P and HAD) are more active in response to a novel environment than low preference rats (Li et al. 1991; Krimmer and Schecter 1992). Selective breeding for animals with either high activity or preference results in high preference or high activity, respectively, thus suggesting linkage, covariance or pleiotropy. However, this relationship is not universal and is dependent on the vulnerability phenotype. Using the CPP method, Cunningham et al. (1992) simultaneously monitored locomotor activity and place preference in C57BL/6J and DBA/2J mice. While C57BL/6J mice exhibited greater locomotor activity (saline rates), DBA/2J mice showed greater morphine- and ethanol-induced place preference.

Recent studies using operant self-administration techniques support a relationship between locomotor activity and positive reinforcement (Ambrosio et al., 1995; Elmer et al. 1995b). In mice, etonitazene maintained significantly greater rates of self-administration behavior than vehicle on a limited access schedule of reinforcement for CXBH/ByJ, C57BL/6J, BALB/ByJ, and CXBK/ByJ mice. There was a positive relationship between baseline locomotor activity and drug intake for these four strains. Lewis, F344, ACI, and NBR rats were used in an unlimited access protocol to determine the degree of variability in morphine self-administration behavior that can be accounted for by environmental and genetic factors (Ambrosio et al., 1995). The correlation within an inbred strain between locomotor activity and rate of acquisition varied between 0.02 and 0.50. The genotypic correlation across the four inbred strains was 0.98. In two inbred rat strains, F344 and WAG/GSto, there was no significant difference in locomotor activity (other than rearing behavior) despite large differences in morphine-reinforced behavior (Sudakov et al. 1993). Importantly, by using inbred strains, the neurobiology underlying the observed genetic differences can be investigated in drug-naive rats, since the degree of variability in CNS neurochemistry within genotype is small and highly replicable.

IV. Neurobiological Markers as Predictors of Vulnerability

Potential neurobiological markers of vulnerability can be posited from molecular genetic studies (Uhl et al. 1993; Gora-Maslak et al. 1991), pharmacological correlational analysis (Woods et al. 1981; Ritz et al. 1987) and behavioral pharmacology techniques (Koob and Bloom 1988; Bertalmio et al. 1993; Shippenberg 1993). To complement these hypotheses, genetic difference in B_{max}, K_d, or distribution of most of the major neurotransmitters has been demonstrated (see Harris and Crabbe 1991). Thus, the role of any one of these systems in vulnerability can be investigated by choosing the appropriate genotype of the subject.

As an example, μ-opiate receptors have been implicated via classic biochemical and behavior pharmacology studies as a primary component of opioid-reinforced behavior (Bertalmio and Woods 1989; Young et al. 1981). Logically, it might be presumed that variation in μ-receptors would account for individual differences in acquisition of opioid self-administration behavior. Fortunately, the wide array of mouse genotypes that are available provides tools to explore this hypothesis. In particular, the CXBK/ByJ mice have been used effectively in a number of studies to illuminate the role of μ-receptors in various phenotypes due to their significant decrease in μ-receptor concentration (Baran et al. 1975; Moskowitz and Goodman 1985; Vaught et al. 1988; Pick et al. 1993). The CXBK/ByJ mice have been compared to several other inbred mouse strains in opioid-reinforced behavior and CPP. The potent opioid etonitazene served as a positive reinforce (Elmer et al. 1995b) and engendered significant morphine-induced CPP (Suzuki et al. 1993) qualitatively similar to other inbred mouse strains. Thus, μ-receptor density may not account for individual variability in the *acquisition* of opioid-reinforced behavior or CPP under the conditions described in these experiments. These results suggest that the neurotransmitter systems essential for the maintenance of drug-reinforced behavior or CPP may not be the neurotransmitter systems that account for individual differences in vulnerability to acquisition of drug taking behavior. Genetic differences in cocaine self-administration suggest a similar finding. Lewis and F344 differ significantly in cocaine self-administration behavior (GEORGE and GOLDBERG 1989) yet do not differ in B_{max} or K_d of the primary cocaine receptor (dopamine transporter) thought to account for cocaine's reinforcement properties (RITZ et al. 1987; RITZ 1991).

Another dopaminergic marker of potential interest is alleles of the DRD2 gene (discussed earlier). This gene has been located on mouse chromosome 9 between 25 and 28 cM (SMITH et al. 1992) and found to be polymorphic in the BXD progenitors. Given the obvious role of dopaminergic function in drug-reinforced behavior (KOOB and BLOOM 1988; KUHAR et al. 1991), the potential role of DRD2 heterogeneity as a risk factor (UHL et al. 1993) and the syntenic relationship between the mouse and human chromosomes holding the DRD2 gene, the relationship between this region and a number of drug-related phenotypes will prove interesting. Several studies thus far have confirmed this region as a "hot spot" using QTL analysis (see CRABBE et al. 1994); five ethanol-related phenotypes, including preference and CPP, map to this region (9:27–31 cM), as does methamphetamine consumption (9:20 cM). Another study to address genetic polymorphisms in DRD2 used radioligand binding, solution hybridization/ nuclease protection and Southern blot hybridization techniques in three randomly outbred and five inbred rat strains (LUEDTKE et al. 1992). Two polymorphisms associated with *Xba*I and *Msp*I restriction endonucleases differentiated the Sprague-Dawley and Brown-Norway rats from Wistar, Long-Evans, Buffalo, DA, Lewis, and F344 rats. No significant binding or affinity differences in striatal DRD2 was associated with these polymorphisms.

Alternative measures of the dopaminergic system have been implicated in psychostimulant abuse. Using a within-subjects approach in randomly outbred rats, individuals at risk for psychomotor stimulant self-administration have been identified by a specific neurochemical pattern in the CNS: high dihydroxyphenylacetic acid/dopamine (DOPAC/DA) ratios in the nucleus accumbens and striatum and a low DOPAC/DA ratio in the prefrontal cortex (PIAZZA et al. 1991). Stress, in the form of social isolation, increases locomotor response to novelty, increases propensity to self-administer amphetamine, and induces the type of DOPAC/DA ratios in the nucleus accumbens, striatum and prefrontal cortex seen in rats predisposed to self-administer amphetamine (PIAZZA et al. 1991). Thus, it is proposed that decreased dopaminergic activity in the prefrontal cortex and increased dopaminergic activity in the ventral and dorsal striatum may mediate predisposition to amphetamine self-administration behavior.

The hypothesis suggests that genetic differences in dopaminergic balance should predict the acquisition rate of self-administration behavior. For example, it is possible that the observed differences between DBA and C57 mice in measures of drug intake and reinforcement depend upon baseline differences in dopamine metabolism. The dopaminergic balance between nucleus accumbens (DOPAC/DA) and frontal cortex (DOPAC/DA) is higher in the C57 mice (0.553) than in the DBA mice (0.317) (BADIANI et al. 1992). These results suggest that differential basal dopaminergic balance may account for the some of the observed behavioral differences in these two inbred strains. However, a two-point genetic correlation has zero degrees of freedom and should be pursued in numerous inbred strains. In addition, dynamic measures of drug-induced release of dopamine may be more revealing than absolute content. Three separate examples support a significant role of drug-induced DA release in vulnerability: (1) Δ-9-THC increases dopamine release in the only rat strain (Lewis) to demonstrate a drug-induced lowering of ICSS threshold (GARDNER and LOWINSON 1991), (2) there is a greater amphetamine-stimulated release of dopamine in rats that rapidly acquire self-administration behavior (ROUGÉ-PONT et al. 1993), and (3) there is a significant corresponding difference in the potency of morphine to stimulate dopamine release and engender CPP in two randomly outbred rat populations (SHOAIB et al. 1995). However, the potency of morphine-induced dopamine release from the nucleus accumbens is significantly greater in the F344 than in the Lewis rat (SCHUTZ et al. 1994), contrary to most vulnerability measurements (see above).

In general, a broad range of biochemical values will be needed to ascertain the relative importance of each marker in vulnerability. In addition to dopaminergic mechanisms, behavioral genetic studies have postulated serotonin as a mediator of the motivational effects underlying drug seeking behavior (GEORGE and RITZ 1993) and neurofilament structure in the nucleus accumbens as an important factor in observed differences between Lewis vs F344 and P vs NP (GUITART et al. 1992, 1993). In a system as complex as the CNS and a behavior as complicated as drug addiction, no one transmitter is likely to account for a single behavior. Each marker is potentially modulated by hundreds of other transmitter systems prior to transducing the initial biological stimulus into a behavioral effect (see KENAKIN 1993). As such, the phenotypic expression of the proposed markers will be influenced by the background genotype (COLEMAN and HUMMEL 1975; MCCLEARN 1993). In all likelihood, a set of neurochemical markers will best predict the phenotypic outcome.

V. Response to Abused Drugs as Predictors of Vulnerability

There are quantitative and in some cases qualitative genetic differences in the behavioral effects of drugs of abuse. Vulnerability research has primarily focused on determining which drug-induced phenotypes are predictive of drug-reinforced behavior in hopes of determining genetic covariance and as a means to understand the underlying mechanisms of drug-reinforced behavior.

An example in alcohol research has been the investigation of the relationship between acute neurosensitivity to ethanol and ethanol-reinforced behavior. Genetically determined acute neurosensitivity to ethanol has been suggested to be a factor in alcoholism, as determined in both clinical (SCHUCKIT 1985) and preclinical studies (SPUHLER and DEITRICH 1984). Inbred strains of mice that have increased ethanol preference ratios are generally less sensitive to the acute narcotic effects of ethanol. Preclinical use of specific genotypes provides a specific test of this hypothesis. Mice selectively bred for increased (LS/Ibg) or decreased (SS/Ibg) neurosensitivity to the acute narcotic effects of ethanol were tested for acquisition and maintenance of operant ethanol-reinforced behavior (ELMER et al. 1990). Despite greater than 20-fold differences in neurosensitivity, the opposite results were found than those initially proposed. The LS mice self-administered greater amounts of ethanol over a greater range of environmental conditions than SS mice.

Recent within-subjects studies using randomly inbred rats have demonstrated that the acute locomotor stimulant effects of amphetamine are predictive of the rate of acquisition of amphetamine self-administration behavior (PIAZZA et al. 1989). Amphetamine-induced locomotor behavior is significantly higher in animals that rapidly acquire drug taking behavior. The behavioral genetics approach to the same hypothesis does not consistently yield the same results. Qualitative and quantitative genetic differences in response to the acute locomotor stimulant effects of cocaine, etonitazene, and morphine have been determined in our laboratory and by others (BELKNAP and CRABBE 1992; ELMER et al. 1995a, 1995b; unpublished observations). Up to eightfold differences in potency (ED_{50}) and sixfold differences in the efficacy of cocaine- and opioid-induced locomotor activity exist. If the acute locomotor stimulant effects of abused drugs are predictive of the rate of acquisition of drug self-administration behavior, there should be a positive correlation between the potency of drug-induced locomotor behavior and the rate of acquisition of drug taking behavior. The psychomotor stimulant theory of reward (WISE and BOZARTH 1987) also states that the strength of the psychomotor stimulant properties of the drug should predict the strength of the reinforcing action. Thus, the relative efficacy of a drug as a stimulant may predict the efficacy of the drug as a reinforcer in each of the genotypes. Thus far, the potency or efficacy of ethanol-, cocaine-, or opioid-induced stimulation of locomotor activity does not consistently predict drug-reinforced behavior (i.e., GEORGE and GOLDBERG 1989; GEORGE et al. 1991b; SANCHEZ et al. 1993; ELMER et al. 1995b).

An additional aspect to the pharmacological profile of a drug's effects are the acute versus chronic effects of the drug. Tolerance to and repeated withdrawal from drugs of abuse are important aspects to addiction. The chronic effects of a drug may best represent drug-related behaviors which are present in the maintenance and extinction phase of drug use and are argued to be the most relevant for preclinical examination (O'BRIEN 1993).

Several lines of selectively bred mice are available to test specific hypotheses (see CRABBE et al. 1994). As an example, initial studies using selectively bred lines of mice suggest the genes that determine ethanol withdrawal severity are involved in the CPP effects of ethanol (CRABBE et al. 1992). However, no differences in operant ethanol self-administration were found in these same selected lines (GEORGE and RITZ 1993).

Using a within-subjects design, the degree of sensitization following repeated administration of amphetamine has been shown to be predictive of the rate of acquisition of amphetamine self-administration behavior (DEROCHE et al. 1992; PIAZZA et al. 1989, 1990). Repeated administration of 1.5mg/kg amphetamine significantly increases locomotor response to a challenge dose of amphetamine (sensitization) in low activity rats (LR) that do not acquire self-administration behavior but not in high activity rats (HR) that readily acquire self-administration behavior (PIAZZA et al. 1989). The sensitization process enhances the rate of acquisition of amphetamine self-administration behavior in LR rats to the level demonstrated in HR rats (PIAZZA et al. 1989, 1990). Thus, the degree of locomotor response at the time of initial access to amphetamine determines the rate of acquisition of drug taking behavior.

Individual differences in sensitization to repeated administration of psychomotor stimulants are due, in part, to the genotype of the subject. For example, repeated administration of the same dose of cocaine to seven lines of the recombinant C57BL/ByJ × BALB/ByJ inbred strains results in various degrees of sensitization depending upon the genotype (SHUSTER et al. 1977). However, since previous studies have demonstrated that the degree of sensitization is proportional to the training dose of cocaine (HOOKS et al. 1991; SHUSTER et al. 1977), genetic differences in sensitization may be a result of a single dose of cocaine having different locomotor stimulant effects in each strain. When inbred mice are given repeated injections of an equiactive dose of cocaine, significant genetic differences in context-specific sensitization to cocaine remain (ELMER et al. 1995a). The strain that most readily shows a context-specific sensitization response to cocaine (C57BL/6J) and strong conditioned effects of opioids (ELMER et al. 1993) is the strain for which cocaine, ethanol, and opioids maintain significant amounts of drug-reinforced behavior. Since behavioral sensitization in laboratory animals may predict vulnerability to drug self-administration, it is of considerable importance to study these genotype-dependent differences in the sensitization processes and their relationship to the acquisition of drug-reinforced behavior.

E. Summary

A large number of risk factors (demographic, social, environmental, and physiological) have been reported for drug use. However, it is not clear in a number of case whether these risk factors represent etiologial influences or sequelae of drug use. In addition, it is not clear if a number of the risk factors are functionally independent or can be reduced to more basic factors. In particular, genetic influences are thought to underlie a number of risk factors for substance abuse and represent a separate risk factor in its own right.

Evidence suggests genetic influences are involved in human alcohol and drug abuse. However, considerable genetic heterogeneity is present, with genes estimated to account for as much as 60% of liability variance for certain subtypes of the disorder in certain groups and for as little as 0% of liability variance for other subtypes in other groups. Genes can enhance vulnerability at many levels, including altered drug response and personality characteristics that bring people into contact with reinforcing drugs. Linkage and association studies offer a means for identifying specific genes associated with substance abuse. While genes encoding many drug receptors have been cloned and characterized, to date no genes increasing susceptibility to substance abuse have been identified. Research on the DRD2 gene, which has been putatively associated with alcoholism and other drug abuse, may provide a model for future studies in the area.

Animal research holds great promise for understanding genetic influences in substance abuse. Animal models can isolate specific phenotypes for parametric investigations and allow drug dependence to be broken into component parts for genetic analysis. To this end, phenotypes that are predictive of future drug abuse behavior may represent etiological risk factors whereas those that do not may represent sequelae of drug use. Dismantling the drug abuse phenotype into component parts will enable specific testing of putative neurobiological and environmental risk factors that will help to pinpoint the nature of genetic contribution to drug abuse.

Knowledge of genetic influences in substance abuse will contribute significantly to our understanding of substance abuse as well as to the development of new and improved prevention strategies.

Acknowledgements. The preparation of this chapter was supported by intramural research funding from the National Institute on Drug Abuse. The technical assistance of Kim Steinberg and Carol Sneeringer is gratefully acknowledged.

References

Alexander RC, Duda J, Garth D, Vogel W, Berrettini WH (1993) Morphine and cocaine preference in inbred mice. Psychiatr Genet 3:33–37

Allgulander C, Nowak J, Rice JP (1991) Psychopathology and treatment of 30344 twins in Sweden. II. Heritability estimates of psychiatric diagnosis and treatment in 12884 twin pairs. Acta Psychiatr Scand 83:12–14

Alvan G (1983) Individual differences in the disposition of drugs metabolized in the body. In: Gibaldi M, Prescott L (eds) Handbook of clinical pharmacokinetics. AIDS Health Science Press, New York, pp 133–155

Amadeo S, Abbar M, Gorwood PH, Chignon JM, Fourcade M, Castelnau D (1992) Association between a D2 dopamine receptor gene polymorphism and alcoholism. Clin Neuropharmacol 15 [Suppl 1, b]:300B

Ambrosio EA, Goldberg SR, Elmer GI (in press) Behavior genetic investigation of the relationship between locomotor activity and the acquisition of morphine self-administration behavior. Behav Pharmacol

American Psychiatric Association (1980) Diagnostic and statistical manual of mental disorders, 3rd edn. American Psychiatric Association, Washington DC

American Psychiatric Association (1994) Diagnostic and statistical manual of mental disorders, 4th edn. American Psychiatric Association, Washington DC

Anthony JC, Warner LA, Kessler RC (1994) Comparative epidemiology of dependence on tobacco, alcohol, controlled substances, and inhalants: basic findings from the National Comorbidity Survey. Exp Clin Psychopharmacol 2:244–268

Aston CE, Hill SY (1990) Segretation analysis of alcoholism in families ascertained through a pair of male alcoholics. Am J Hum Genet 46:879–887

Badiani A, Cabib S, Puglisi AS (1992) Chronic stress induces strain-dependent sensitization to the behavioral effects of amphetamine in the mouse. Pharmacol Biochem Behav 43:53–60

Baran A, Shuster L, Eleftheriou BE, Bailey DW (1975) Opiate receptors in mice: genetic differences. Life Sci 17:633–640

Belknap JK, Crabbe JC (1992) Chromosome mapping of gene loci affectign morphine and amphetamine responses in the BXD recombinant inbred mice. In: Kalivas PW, Samson HH (eds) The neurobiology of drug and alcohol addiction. New York Academy of Sciences, New York, pp 311–323

Belknap JK, Crabbe JC, Riggan J, O'Toole LA (1993) Voluntary consumption of morphine in 15 inbred mouse strains.Psychopharmacology 112:352–358

Berrettini WH, Ferraro TN, Alexander RC, Buchberg AM, Vogel WH (1994) Quantitative trait loci mapping of three loci controlling morphine preference using inbred mouse strains. Nature Genet 7:54–58

Bertalmio A, Woods JH (1989) Reinforcing effect of alfentanil is mediated by mu opioid receptors: apparent pA_2 analysis. J Pharmacol Exp Ther 251:455–460

Bertalmio AJ, France CP, Woods JH (1993) Establishing correlations between the pharmacodynamic characteristics of opioid agonist and their behavioral effects. In: Herz A (ed) Opioids II. Springer, Berlin Heidelberg New York, pp 449–472

Bixler EO, Kales JD, Kales A, Jacoby JA, Soldatos CR (1985) Rebound insomia and elimination half-life: assessment of individual subject responses. J Clin Pharmacol 25:115–124

Bleuler M (1955) Familial and personal background of chronic alcoholics. In: Diethelm O (ed) Etiology of chronic alcoholism. Thomas, Springfield, pp 110–116

Blum K, Noble EP, Sheridan PJ, Montgomery A, Ritchie T, Jagadeeswaran P, Nogami H, Briggs AH, Cohn JB (1990) Allelic association of human dopamine D2 receptor gene in alcoholism. JAMA 263:2055–2060

Blum K, Noble EP, Sheridan PF, Finley O, Montgomery A, Ritchie T, Ozkaragoz T, Fitch RJ, Sadlack F, Sheffield D, Dahlmann T, Haldardier S, Nogami H (1991) Association of the A1 allele of the D2 dopamine receptor gene with severe alcoholism. Alcohol 8:409–416

Bohman M, Sigvardsson S, Cloninger CR (1981) Maternal inheritance of alcohol abuse. Arch Gen Psychiatry 38:965–969

Bolos AM, Dean M, Lucas-Derse S, Ramsburg M, Brown GL, Goldman D (1990) Population and pedigree studies reveal a lack of association between the dopamine D2 receptor gene and alcoholism. JAMA 264:3156–3160

Boulter J, Connolly J, Deneris E, Goldman D, Heinemann, S Patrick J (1987) Functional expression of two neuronal nicotine acetylcholine receptors from cDNA clones identifies a gene family. Proc Natl Acad Sci USA 84:7763–7767

Bozarth MA (1987) Methods of assessing the reinforcing properties of abused drugs. Springer, Berlin Heidelberg New York

Brooks JS, Nomura C, Cohen P (1989a) A network of influences on adolescent drug involvement: neighborhood, school, peer, and family. Genet Soc Gen Psychol Monogr 115:125–145

Brooks JS, Nomura C, Cohen P (1989b) Prenatal, perinatal, and early childhood risk factors and drug involvement in adolescence. Genet Soc Gen Psychol Monogr 115:223–241

Brown LS, Hickson MJ, Ajuluchukwu DC, Bailey J (1993) Medical disorders in a cohort of New York city drug abusers: much more than HIV disease. J Addict Dis 12:11–27

Bry BH, McKeon P, Pandina RJ (1982) Extent of drug use as a function of number of risk factors. J Abnorm Psychol 91:273–279

Burgio GR (1993) Biological individuality and disease. Acta Biotheor (Leiden) 41:219–230

Busto U, Sellers EM (1986) Pharmacokinetic determinants of drug abuse and dependence. A conceptual perspective. Clin Pharmacokinet 11:144–153

Cadoret RJ, O'Gorman T, Troughton E, Heywood E (1985) Alcoholism and anti-social personality: interrelationships, genetic and environmental factors. Arch Gen Psychiatry 42:161–167

Caldwell CB, Gottesman II (1991) Sex differences in the risk for alcoholism: a twin study. Paper present at Behavior Genetic Association meeting, St Louis, MO

Carmelli D, Swan GE, Robinette D, Fabsitz RR (1990) Heritability of substance use in the NAS-NRC twin registry. Acta Genet Med Gemenol 39:91–98

Carmelli D, Swan GE, Robinette D, Fabsitz R (1992) Genetic influences in smoking – a study of male twins. N Engl J Med 327:829–833

Carney JM, Cheng M-S, Cao W, Seale TW (1991) Issues surrounding the assessment of the genetic determinants of drugs as reinforcing stimuli. J Addict Dis 10: 163–177

Carroll ME, Pederson MC, Harrison RG (1986) Food deprivation reveals strain differences in opiate intake of Sprague-Dawley and Wistar Rats. Pharmacol Biochem Behav 24:1095–1099

Carroll ME, Lac ST, Nygaard SL (1989) A concurrently avialable nondrug reinforcer prevents the acquisition or decreases the maintenance of cocaine-reinforced behavior. Psychopharmacology 97:23–29

Cazala P (1976) Effects of d- and l-amphetamine on dorsal and ventral hypothalamic self-stimulation in three inbred strains of mice. Pharmacol Biochem Behav 5:505–510

Cazala P, Guenet J-L (1980) The recombinant inbred strains: a tool for the genetic analysis of differences observed in the self-stimulatin behavior of the mouse. Physiol Behav 24:1057–1060

Cazala P, Cazals Y, Cardo B (1974) Hypothalamic self-stimulation in three inbred strains of mice. Brain Res 81:159–167

Chutuape MA, de Wit H (1994) Relationship between subjective effects and drug preferences: ethanol and diazepam. Drug Alcohol Depend 34:243–251

Cloninger CR, Bohman M, Sigvardsson S (1981) Inheritance of alcohol abuse: cross fostering analysis of adopted men. Arch Gen Psychiatry 38:861–868

Coleman DL, Hummel KP (1975) Influence of genetic background on the expression of mutations at the diabetes locus in the mouse. Isr J Med Sci 11:708–713

Comings DE, Comings BG, Muhleman D, Dietz G, Shahbahrami B, Tast D, Knell E, Kocsis P, Baumgarten R, Kovacs BE, Levy DL, Smith M, Borison RL, Evans DD, Klein DN, MacMurray J, Tosk JM, Sverd J, Gysin R, Flanagan SD (1991a) The dopamine D2 receptor locus as a modifying gene in neuropsychiatric disorders. JAMA 266:1793–1800

Comings DE, Mulhleman D, Ahn C, Gysin R, Flanagan SD (1994) The dopamine D2 receptor gene: a genetic risk factor in substance abuse. Drug Alcohol Depend 34:175–180

Crabbe JC, Phillips TJ, Cunningham CL, Belknap JK (1992) Determinants of ethanol reinforcement. In: Kalivas PW, Samson HH (eds) The neurobiology af drug and alcohol addiction. New York Academy of Sciences, New York, pp 302–310

Crabbe JC, Belknap JK, Buck KJ (1994) Genetic animal models of alcohol and drug abuse. Science 264:1715–1723

Crusio WE, Schwegler H (1987) Hippocampal Mossy fiber distribution covaries with open-field habituation in the mouse. Behav Brain Res 26:153–158

Cunningham CL, Niehus DR, Malott DH, Prather LK (1992) Genetic differences in the rewarding and activating effects of morphine and ethanol. Psychopharmacology 107:385–393

Cutting GR, Lu L, O'Hara BF, Kasch LM, Montrose-Rafizadeh C, Donovan DM, Shimada S, Antonarakis SE, Guggino WB, Uhl GR, Kazazian HH (1991) Cloning of the gamma-aminobutyric acid (GABA)ρl cDNA: a GABA receptor subunit highly expressed in the retina. Proc Natl Acad Sci USA 88:2673–2677

Decker MD, Graitcer PL, Schaffner W (1988) Reduction in motor vehicle fatalities associated with an increase in the minimum drinking age. JAMA 260:3604–3610

DeFries JC, Gervais MC, Thomas EA (1978) Response to 30 generations of selection for open-field activity in laboratory mice. Behav Genet 8:3–13

DeGrandpre RJ, Bickel WK, Hughes JR, Layng MP, Badger G (1993) Unit price as a useful metric in analyzing effects of reinforcer magnitude. J Exp Anal Behav 60:641–666

Deitrich RA (1993) Selective breeding for initial sensitivity to ethanol, Behav Genet 23:153–162

Deroche V, Piazza PV, Casolini P, Maccari S, Le Moal M, Simon H (1992) Stress-induced sensitization to amphetamine and morphine psychomotor effects depend on stress-induced corticosterone secretion. Brain Res 598:343–348

de Wit H, Uhlenhuth EH, Johanson CE (1986) Individual differences in the reinforcing and subjective effects of amphetamine and diazepam. Drug Alcohol Depend 16:341–360

Durand GM, Gregor P, Zheng X, Bennett MVL, Uhl GR, Zukin SR (1992) Cloning of an apparent splice variant of the rat N-methyl-D-asparate receptor NMDAR1 with altered sensitivity to polyamines and activators of protein kinase C. Proc Natl Acad Sci U S A 89:9359–9363

Dymshitz J, Lieblich I (1987) Opiate reinforcement and naloxone aversion, as revealed by place preference paradigm, in two strains of rats. Psychopharmacology 92:473–477

Edwards G, Gross MM (1976) Alcohol depiendence: provisional description of a clinical syndrome. Br Med J 1:1058–1061

Eichelbaum M, Gross AS (1990) The genetic polymorphism of debrisoquine/sparteine metabolism – clinical aspects. Pharmacol Ther 46:377–394

Elmer GI, Meisch RA, George FR (1987) Mouse strain differences in operant self-administration of ethanol. Behav Genet 17:439–451

Elmer GI, Meisch RA, Goldberg SR, George FR (1988) A fixed ratio analysis of oral ethanol reinforced behavior in inbred mouse strains. Psychopharmacology 96:431–436

Elmer GI, Meisch RA, Goldberg SR, George FR (1990) Ethanol self-administration in Long Sleep and Short Sleep mice: evidence for genetic independence of neurosensitivity and reinforcement. J Pharmacol Exp Ther 254:1054–1062

Elmer GI, Mathura CB, Goldberg SR (1993b) Genetic factors in conditioned tolerance to the analgesic effects of etonitazene. Pharmacol Biochem Behav 45:251–254

Elmer GI, Goldberg SR, Gorelick DA, Rothman RB (1995a) Acute sensitivity versus context-specific sensitization to cocaine as a function of genotype. Pharmacol Biochem Behav (in press)

Elmer GI, Pieper JO, Goldberg SR, George FR (1995b) Opioid operant self-administration, analgesia, stimulation and respiratory depression in μ-deficient mice. Psychopharmacology 117:23–31

Eriksson K (1968) Genetic selection for voluntary alcohol consumption in the albino rat. Science 159:739–741

Evans CJ, Keith DEJ, Morrison H, Magendzo K, Edwards RH (1992) Cloning of a delta opioid receptor by functional expression. Science 258:1952–1955

Fadda F, Moscal E, Colombo G, Gessa GL (1989) Effect of spontaneous ingestion of ethanol on brain dopamine metabolism. Life Sci 44:281–287

Falconer DS (1989) Introduction to quantitative genetics, 3rd edn. Longman Scientific and Technical. Wiley, New York

Flaherty EW, Kotranski L, Fox E (1984) Frequency of heroin use and drug users' life-style. Am J Drug Alcohol Abuse 10:285–314

Flewelling RL, Rachal JV, Marsden ME (1992) Socioeconomic and demographic correlates of drug and alcohol use. US Government Printing Office, Washington DC

Gardner EL, Lowinson JH (1991) Marijuana's interaction with brain reward systems: update 1991. Pharmacol Biochem Behav 40:571–580

Gelernter J, O'Malley S, Risch N, Kranzler HR, Krystal J, Merikangas K, Kennedy JL, Kidd KK (1991) No association between an allele at the D2 dopamine receptor gene (DRD2) and alcoholism. JAMA 266:1801–1807

Gelernter J, Goldman D, Risch N (1993) The A1 allele at the D2 dopamine receptor gene and alcoholism: a reappraisal. JAMA 269:1673–1677

George FR (1991) Is there a common biological basis for reinforcement from alcohol and other drugs? J Addict Dis 10:127–139

George FR, Goldberg SR (1988) Genetic differences in response to cocaine. NIDA Res Monogr 88:239–249

George FR, Goldberg SR (1989) Genetic appproaches to the analysis of addiction. Trends Pharmacol Sci 10:73–83

George FR, Ritz MC (1993) A psychopharmacology of motivation and reward related to substance abuse treatment. Exp Clin Psychopharmacol 1:7–26

George FR, Ritz MR, Elmer GI (1991a) The role of genetics in drug dependence. IN: Pratt J (ed) The biological basis of drug tolerance and dependence. Academic, New York, pp 265–290

George FR, Porrino LJ, Ritz MC, SRG (1991b) Inbred rat strain comparisons indicate different sites of action for cocaine and amphetamine locomotor stimulant effects. Psychopharmacology 104:457–462

George S, Israel Y, Nguyen T, Cheng R, O'Dowd BF (1993) Polymorphism of the D4 dopamine receptor alleles in chronic alcoholism. Biochem Biophys Res Commun 196:107–144

Gilligan SB, Reich R, Cloninger CR (1987) Etiologic heterogeneity in alcoholism. Genet Epidemiol 4:395–414

Glantz M, Pickens R (1992) Vulnerability to drug abuse. American Psychological Association, Washington DC

Glowa JR, Geyer MA, Gold PW, Sternberg EM (1992) Differential startle amplitude and corticosterone response in rats. Neuroendocrinology 56:719–723

Goldberg J, Lyons MJ, Eisen SA, True WR, Tsuang M (1993) Genetic influence on drug use: a preliminary analysis of 2674 Vietnam era veteran twins. Behav Genet 23:552 (abstract)

Gonzalez FJ, Meyer UA (1991) Molecular genetics of the debrisoquin-sparteine polymorphism. Clin Pharmacol Ther 50:233–238

Goodwin DW, Schulsinger F, Hermansen L, Guze SB, Winokur G (1973) Alcohol problems in adoptees raised apart from alcoholic biological parents. Arch Gen Psychiatry 29:238–243

Goodwin DW, Schulsinger F, Knop J, Mednick S, Guze SB (1977) Alcoholism and depression in adopted-out daughters of alcoholics. Arch Gen Psychiatry 34:751–755

Gora-Maslak G, McClearn GE, Crabbe JC, Phillips TJ, Belknap JK, Plomin R (1991) Use of recombinant inbred strains to identify quantitative trait loci in psychopharmacology. Psychopharmacology 104:413–424

Grandy DK, Litt M, Allen L, Bunzow JR, Marchionni M, Makam H, Reed L, Magenis Re, Civelli O (1989) The human dopamine D2 receptor gene is located on chromosome 11 at q22–q23 and identifies a TaqI RELP. Am J Hum Genet 45:778–785

Greenblatt DJ, Miller LG, Shader RI (1990) Benzodiazepine discontinuation syndromes. J Psychiatr Res 24 [Suppl 2]:73–79

Guitart X, Beitner JD, Marby DW, Kosten TA, Nestler EJ (1992) Fischer and Lewis rat strains differ in basal levels of neurofilament proteins and their regulation by chronic morphine in the mesolimbic dopamine system. Synapse 12:242–253

Guitart X, Lumeng L, Li T-K, Nestler EJ (1993) Alcohol-preferring and nonpreferring rats display differenet levels of neurofilament proteins in the ventral tegmental area. Alcoholism 17:586–591

Gurling HMD, Oppenheim BE, Murray RM (1984) Depression, criminality and psychopathology associated with alcoholism: evidence from a twin study. Acta Genet Med Gemenol 33:333–339

Hajek P (1991) Individual differences in difficulty quitting smoking. Br J Addict 86:555–558

Harris RA, Crabbe JC (1991) The genetic basis of alcohol and drug action. Plenum, New York

Hatsukami DK, Pentel PR, Glass J, Nelson R, Brauer LH, Crosby R, Hanson K (1994) Methodological issues in the administration of multiple doses of smoked cocaine-base in humans. Pharmacol Biochem Behav 47:531–540

Hawkins JD, Catalano RF, Miller JY (1992) Risk and protective factors for alcohol and other drug problems in adolescence and early adulthood: 'implications for substance abuse prevention. Psychol Bull 112:64–105

Heath AC, Martin MG (1988) Teenage alcohol use in the australian twin register: genetic and environmental determinants of starting to drink. Compr Psychiatry 12:735–741

Heath AC, Martin NG (1993) Genetic models for the natural history of smoking: evidence for a genetic influence on smoking persistence. Addict Behav 28:19–34

Heath AC, Meyer J, Eaves LJ, Martin NG (1991) The inheritance of alcohol consumption patterns in a general population twins sample: I. Multidimensional scaling of quantity/frequency data. J Stud Alcohol 52:345–351

Helzer JE, Pryzbeck TR (1988) The co-occurrence of alcoholism with other psychiatric disorders in the general population and its impact on treatment. J Stud Alcohol 49:219–224

Hill SY, Cloninger CR, Ayre FR (1977) Independent transmission of alcoholism and opiate abuse. Alcohol Clin Exp Res 1:335–342

Hill SY, Aston C, Rabin B (1988) Suggestive evidence of genetic linkage between alcoholism and the MNS blood group. Alcohol Clin Exp Res 12:811–814

Hoffman RS, Henry GL, Weisman RS, Howland MA, Goldfrank LR (1990) Association between plasma cholinesterase activity and cocaine toxicity. Ann Emerg Med 19:467

Hooks MS, Jones GH, Neill DB, Justice JB (1991) Individual differences in amphetamine sensitization: dose-dependent effects. Pharmacol Biochem Behav 41:203–210

Horn JL, Wanberg KW (1969) Symptom patterns related to excessive use of alcohol. Q J Stud Alcohol 30:35–58

Hrubec Z, Omenn GS (1981) Evidence of genetic predisposition to alcoholic cirrhosis and psychosis: twin concordances for alcoholism and its biological endpoints by zygosity among male veterans. Alcohol Clin Exp Res 5:207–212

Hser YI, Anglin D, Powers K (1993) A 24-year follow-up of California narcotics addicts. Arch Gen Psychiatry 50:577–584

Hughes JR (1986) Genetics of smoking: a brief review. Behav Ther 17:335–345

Hyytia P, Sinclair JD (1990) Differential reinforcement and diurnal rhythms of lever pressing for ethanol in AA and Wistar Rats. Alcohol Clin Exp Res 14:375–379

Isenschmid DS, Fischman MW, Foltin RW, Caplan YH (1992) Concentration of cocaine and metabolites in plasma of humans following intravenous administration and smoking of cocaine. J Anal Toxicol 17:318–319

Johanson CE, Woolverton WL, Schuster CR (1987) Evaluating laboratory models of drug dependence. In: Meltzer HY (ed) Psychopharmacology, the third generation of progress. Raven, New York, pp 1617–1625

Jones SE, Brain PF (1985) An illustration of simple sequence analysis with reference to the agonistic behavior of four strains of laboratory mouse. Behav Proc 11:365–388

Kaij L (1960) Alcoholism in twins. Almqvist and Wiksell, Stockholm

Kandel DB, Logan JA (1984) Patterns of drug use from adolescence to young adulthood: I. Periods of risk for initiation, continued use, and discontinuation. J Publ Health 74:660–666

Kandel DB, Raveis VH (1989) Cessation of illicit drug use in young adulthood. Arch Gen Psychiatry 46:109–116

Kaprio J, Hammar N, Koskenvuo M, Floderus-Myrhed B, Langinvainio H, Sarna S (1982) Cigarette smoking and alcohol use in Finland and Sweden: a cross-national twin study. Int J Epidemiol 11:378–386

Katz J (1990) Models of relative reinforcing efficacy of drugs and their predictive utility. Behav Pharmacol 1:288–301

Kenakin TP (1993) Pharmacologic analysis of drug-receptor interaction, 2nd edn. Raven, New York

Kendler KS, Heath AC, Neale MC, Kessler RC, Eaves LJ (1992) A population based twin study of alcoholism in women. JAMA 268:1877–1882

Kieffer BL, Befort K, Gaveriaux-Ruff C, Hirth CG (1992) The α-opioid receptor: Isolation of a cDNA by expression cloning and pharmacological characterization. Proc Natl Acad Sci USA 89:12048–12052

Kilty J, Lorang D, Amara SG (1991) Cloning and expression of a cocaine-sensitive rat dopamine transporter. Science 254:78–79

Koob GF, Bloom FE (1988) Cellular and molecular mechanisms of drug dependence. Science 242:715–723

Kopun M, Propping P (1977) The kinetics of ethanol absorption and elimination in twins and supplementary repetitive experiments in singleton subjects. Eur J Clin Pharmacol 11:337–344

Krimmer EC, Schecter MD (1992) HAD and LAD rats respond differently to stimulating effects but not discriminative effects of ethanol. Alcohol 9:71–75

Kuhar MJ, Ritz MC, Boja JW (1991) The dopamine hypothesis of the reinforcing properties of cocaine. Trends Neurosci 14:299–302

LaBuda MC, Pickens RW (in press) Sex differences in twin-pair closeness and concordance for alcoholism. In Harris L (ed) NIDA research monograph. US Government Printing Office, Washington DC

LaBuda MC, Pickens RW, Svikis DS (1993) Concordance for drug abuse in alcoholic twins. In: Harris L (ed) NIDA research monograph. US Government Printing Office, Washington DC, p 191

Lander ES, Schork NJ (1994) Genetic dissection of complex traits. Science 265: 2037–2048

Levy D, Sheflin N (1985) The demand for alcoholic bevergaes: an aggregate time-series analysis. J Publi Policy Marketing 4:47–54

Li T-K, Lumeng L, McBride WJ, Waller MB (1981) Indiana selection studies on alcohol-related behaviors. In: McClearn GE, Deitrich RA, Erwin VG (eds) Devleopment of animal models as pharmacogenetic tools. National Institute on Alcohol Abuse and Alcoholism, Rockville MD, pp 171–191 (2 DHEW publication no (ADM) 78–847)

Li T-K, Lumeng L, Doolittle DP, Carr LG (1991) Molecular associations of alcohol-seeking behavior in rat lines selectively bred for high and low voluntary ethanol drinking. Alcohol Alcohol [Suppl] 1:121–124

Lieblich I, Olds J (1971) Selection for the readiness to respond to electrical brain stimulation of the hypothalamus as a reinforcing agent. Brain Res 27:153–161

Lieblich I, Cohen E, Beiles A (1978) Selection for high and for low rates of self-stimulation in rats. Physiol Behavi 21:843–849

Lipp H-P, Schwegler H, Hausheer-Zarmakupi Z (1987) Strain-specific correlations between hippocampal structural traits and habituation in a spatial novelty situation. Behav Brain Res 24:111–123

Luthar SS, Rounsaville BJ (1993) Substance misuse and comorbid psychopathology in a high-risk groups: a study of siblings of cocaine misusers. Int J Addict 28:415–434

Luthar SS, Anton SF, Merikangas KR, Rounsaville BJ (1992a) Vulnerability to drug abuse among opioid addict's siblings: individual, familial, and peer influences. Compr Psychiatry 33:190–196

Luthar SS, Anton SF, Merikangas KR, Rounsaville BJ (1992b) Vulnerability to substance abuse and psychopathology among siblings of opioid abusers. J Nerv Ment Dis 180:153–161

Luedtke RR, Artymyshyn RP, Jonks BR, Molinoff PB (1992) Comparison of the expression, transcription and genomic organization of D2 dopamine receptors in outbred and inbred strains of rats. Brain Res 584:45–54

Mardones J, Segovia-Riquelme N (1983) Thirty-two years of selection of rats for ethanol preference: UChA and UChB strains. Neurobehav Toxicol Teratol 5:171–178

Martin NG, Oakeshott JG, Gibson JB, Starmer GA, Perl J, Wilks AV (1985a) A twin study of psychomotor and physiological responses to an acute dose of alcohol. Behav Genet 15:305–347

Martin NG, Perl J, Oakeshott JG, Gibson JB, Starmer GA, Wilks AV (1985b) A twin study of ethanol metabolism. Behav Genet 15:93–109

Matsuda LA, Lolait SJ, Brownstein MJ, Young AC, Bonner TI (1990) Structure of a cannabinoid receptor and functional expression of the cloned cDNA. Nature 346:561–564

Mattox AJ, Hohanson CE, Schuster CR (in press) Effects of conditioned reinforcement on mood states. In: Harris LS (eds) NIDA research monograph. US Government Printing Office, Washington DC

McCaul ME, Turkkan JS, Svikis DS, Bigelow GE (1990) Alcohol and secobarbital effects as a function of familial alcoholism: acute psychophysiological effects. Alcohol Clin Exp Res 14:704–712

McClearn GE (1968) The use of strain rank orders in assessing equivalence of technique. Behav Res Methods Instrument 1:49–51

McClearn GE (1991) The tools of pharmacogenetics. In: Crabbe JC, Harris RA (eds) The genetic basis of alcohol and drug actions. Plenum, New York, pp 1–24

McClearn GE (1993) Genetics, systems, and alcohol. Behav Genet 23:223–230

McClearn GE, Rodgers DA (1959) Differences in alcohol preference among inbred strains of mice. Q J Stud Alcohol 20:691–695

McGue M (1994) Genes, environment and the etiology of alchoholism. NIAAA, Res Monogr 26:1–40

McGue M, Slutske W (1993) The inheritance of alcoholism in women. Paper presented at National Institute on Alcohol Abuse and Alcoholism Working Group for Prevention Research on Women and Alcohol, Bethesda

McGue M, Pickens RW, Svikis DS (1992) Sex and age effects on the inheritance of alcohol problems: a twin study. J Abnorm Psychol 101:3–17

McKenna T, Pickens R (1981) Alcoholic children of alcoholics. J Stud Alcohol 42:1021–1029

McLellan AT, Luborsky L, Cacciola J, Griffith J, Evans F, Barr HL, O'Brien CP (1985) New data from the addiction severity index: reliability and validity in three centers. J Nerv Ment Dis 173:412–423

McMicken DB (1990) Alcohol withdrawal syndromes. Emerg Med Clin North Am 8:805–819

Meller WH, Rinehart R, Cadoret RJ, Troughton E (1988) Specific familial transmission in substance abuse. Int J Addict 23:1029–1039

Merikangas KR (1990) The genetic epidemiology of alcoholism. Psychol Med 20:11–22

Mirin SM, Weiss RD, Griffin ML, Michael JL (1991) Psychopathology in drug abusers and their families. Compr Psychiatry 32:36–51

Mishina M, Kurosaki T, Tobimatsu T, Morimoto Y, Noda M, Yamamoto T, Terao M, Lindstrom J, Takahashi T, Kuno M, Numa S (1984) Expression of functional acetylcholine receptor from cloned cDNAs. Nature 307:604–608

Moriyoshi K, Masu M, Ishii T, Shigemoto R, Mizuno N, Nakanishi S (1991) Molecular cloning and characterization of the rat NMDA receptor. Nature 354:31–37

Moskowitz AS, Goodman RR (1985) Autoradiographic analysis of mu1, mu2, and delta opioid binding in the central nervous system of C57BL/By and CXBK (opioid receptor-deficient) mice. Brain Res 360:108–116

Moss HB, Yao JK, Maddock JM (1989) Responses by sons of alcoholic fathers to alcoholic and placebo drinks: perceived mood, intoxication, and plasma prolactin. Alcohol Clin Exp Res 13:252–257

Murphy JM, Gatto GJ, McBride WJ, Lumeng L, Li T-K (1989) Operant responding for oral ethanol in the alcohol-preferring P and alcohol non-preferring NP lines of rats. Alcohol 6:127–131

Nagoshi CT, Wilson JR (1987) Influence of family alcoholism history on alcohol metobolism, sensitivity, and tolerance. Alcohol Clin Exp Res 11:392–398

National Institute on Drug Abuse (1991) National Household Survey on Drug Abuse: main findings 1990. US Department of Health and Human Services, Washington DC (DHHS publication no (ADM) 91-1788)

National Institute on Drug Abuse (1992) National Household Survey on Drug Abuse: population estimates 1991, revised 20 Nov. 1992. US Department of Health and Human Services, Washington

Nestler ER (1992) Molecular mechanisms of drug addiction. J Neurosci 12:2439–2450

Newcomb MD, Felix-Ortiz M (1992) Multiple protective and risk factors for drug use and abuse: cross-sectional and prospective findings. J Person Soc Psychol 63:280–296

Newcomb MD, Maddahian E, Pentler PM (1986) Risk factors for drug use among adolescents: concurrent and longitudinal analyses. J Publ Health 76:525–531

Newlin DB, Thomson JB (1990) Alcohol challenge with sons of alcoholics: a critical review and analysis. Psychol Bull 108:383–402

Nichols JR, Hsiao S (1967) Addiction liability of albino rats: breeding for quantitative differences in morphine drinking. Science 157:561–563

Noble EP, Blum K, Ritchie T, Montgomery A, Sheridan PF (1991) Allelic association of the D2 dopamine receptor gene with receptor-binding characteristics in alcoholism. Arch Gen Psychiatry 48:648–654

Noble EP, Blum K, Khalsa ME, Ritchie T, Montgomery, A Wood RC, Fitch RJ, Ozkaragoz T, Sheridan PJ, Anglin MD, Paredes A, Treiman LJ, Sparkes RS (1993) Allelic association of the D2 dopamine receptor gene with cocaine dependence. Drug Alcohol Depend 33:271–285

Nothen MM, Erdman J, Korner J, Lanczik M, Fritze J, Fimmers R, Grandy DK, O'Dowd B, Propping P (1992) Lack of association between dopamine D1 and D2 receptor gene and bipolar affective disorder. Am J Psychiatry 149:199–201

Nurco ND, Cisin IH, Balter MB (1981) Addict careers. II. The first ten years. Int J Addict 16:1327–1356

O'Brien CP (1993) Opioid addiction. In: Hertz A (ed) Opioids II. Springer, Berlin Heidelberg New York, pp 803–824 (Handbook of experimental pharmacology, vol 104/2)

Office of Substance Abuse Prevention (1991) Breaking new ground for youth at risk: program summaries. OSAP technical report 1. DHHS publication no (ADM) 91-1658, Washington DC

O'Hara BF, Smith SS, Bird G, Persico A, Suarez B, Cutting GR, Uhl GR (1993) Dopamine D2 receptor RFLPs, haplotypes and their association with substance use in black and Caucasian research volunteers. Hum Hered 43:209–218

Olsen RW, Tobin AJ (1990) Molecular biology of GABA-A receptors. FASEB 4:1469–1480

Om A, Ellahham S, Ornato JP, Picone C, Theogaraj J, Corretjer GP, Vetrovec GW (1993) Medical complications of cocaine: possible relationships to low plasma cholinesterase enzyme. Am Heart J 125:1114–1117

O'Malley SS, Maisto SA (1985) Effects of family drinking history and expectancies on responses to alcohol in men. J Stud Alcohol 46:389–297

Otton SV, Schadel M, Cheung SW, Kaplan HL, Busto UE, Sellers EM (1993) CYP2D6 phenotype determines the metabolic conversion of hydrocodone to hydromorphone. Clin Pharmacol Ther 54:463–472

Parsian A, Todd RD, Devor EJ, O'Malley KL, Suarez BK, Reich T, Cloninger CR (1991) Alcoholism and alleles of the human D2 dopamine receptor locus. Studies of association and linkage. Arch Gen Psychiatry 48:655–663

Pedersen N (1981) Twin similarity for usage of common drugs. In: Gedda L, Parisi P, Nance WE (eds) Twin research 3: part C, epidemiological and clinical studies. Liss, New York, pp 53–59

Persico AM, Vandenbergh DJ, Smith SS, Uhl GR (1993a) Dopamine transporter gene polymorphisms are not associated with polysubstance abuse. Biol Psychiatry 43:265–267

Persico AM, O'Hara BF, Gysin R, Farmer S, Flanagan SD, Uhl GR (1993b) Dopamine D2 receptor gene TaqI "A" locus map including A4 variant: relevance for alcoholism and drug abuse. Drug Alcohol Depend 31:229–234

Phillips TJ, Crabbe JC (1991) Behavioral studies of genetic differences in alcohol action. In: Crabbe JC, Harris RA (eds) The genetic basis of alcohol and drug actions. Plenum, New York, pp 25–104

Phillips TJ, Feller DJ, Crabbe JC (1989) Selected mouse lines, alcohol and behavior. Experientia 45:805–827

Piazza PV, Deminiere J-M, Le Moal M, Simon H (1989) Factors that predict individual vulnerability to amphetamine self-administration. Science 245: 1511–1513

Piazza PV, Deminiere JM, Le Moal M, Simon H (1990) Stress- and pharmacologically-induced behavioral sensitization increases vulnerability to acuisition of amphetamine self-administration. Brain Res 514:22–26

Piazza PV, Rouge'-Pont F, Deminiere JM, Kharoubi M, Le Moal M, Simon H (1991) Dopaminergic activity is reduced in the prefrontal cortex and increased in the nucleus accumbens of rats predisposed to develop amphetamine self-administration. Brain Res 567:169–174

Pick CG, Nejat RJ, Pasternak GW (1993) Independent expression of two pharmacologically distinct supraspinal Mu anagesic systems in genetically different mouse strains. J Pharmacol Exp Ther 265:166–171

Pickens R, Svikis D (1991) Genetic influences in human drug abuse. J Addict Dis 10:205–213

Pickens R, Svikis DS (in press) Vulnerability to drug use. In: Jaffe JH (eds) Encyclopedia of drug and alcohol abuse. Macmillan, New York

Pickens RW, Svikis DS, McGue M, Lykken DT, Heston LL, Clayton PJ (1991) Heterogeneity in the inheritance of alcoholism: a study of male and female twins. Arch Gen Psychiatry 48:19–28

Plomin R, McClearn GE (1993) Quantitative trait loci (QTL) analysis and alcohol-related behaviors. Behav Genet 23:197–212

Plomin R, DeFries JC, McClearn GE (1990) Behavioral genetics: a primer, 2nd edn. Freeman, New York

Pollock VE, Teasdale TW, Gabrielli WF, Knop J (1986) Subjective and objective measures of response to alcohol among young men at risk for alcoholism. J Stud Alcohol 47:297–304

Porter J, Jick H (1980) Addiction rare in patients treated with narcotics. N Engl J Med 302:123 (letter)

Raveis VH, Kandel DB (1987) Changes in drug behavior from the middle to the late twenties: initiation, persistence, and cessation of use. J Publ Health 77:607–611

Regier DA, Farmer ME, Rae DS, Locke BZ, Keith SJ, Judd LL, Goodwin FK (1990) Comorbidity of mental disorders with alcohol and other drug abuse. JAMA 264:2511–2518

Ritz MC (1991) Biochemical genetic differences in vulnerability to drug effects: is statistically significant always physiologically important and vice versa? J Addict Dis 10:189–204

Ritz MC, George FR, deFiebre CM, Medisch RA (1986) Genetic differences in the establishment of ethanol as a reinforcer. Pharmacol Biochem Behav 24:1089–1094

Ritz MC, Lamb RJ, Goldberg SR, Kuhar MJ (1987) Cocaine receptors on dopamine transporters are related to the self-administration of cocaine. Science 237:1219–1223

Ritz MC, George FR, Meisch RA (1989a) Ethanol self-administration in Alko rats: I. Effects of selection and concentration. Alcohol 6:227–233

Ritz MC, George FR, Meisch RA (1989b) Ethanol self-administration in Alko rats: II. Effects of selection and fixed-ratio size. Alcohol 6:235–239

Ritz MR, Garcia JM, Protz D, Rael A-M, George FR (1994) Ethanol-reinforced behavior in P, NP, HAD and LAD rats: differential genetic regulation of reinforcement and motivation. Behav Pharmacol 5:521–531

Robins LN (1993) Vietnam veterans' rapid recovery from heroin addiction: a fluke or normal expectation? Addiction 88:1041–1054

Robins LN, Przybeck TR (1985) Age of onset of drug use as a factor in drug and other disorders. In: Jones CL, Battjes RJ (eds) Etiology of drug abuse: implications for prevention. DHHS publ no (ADM) 87-1335. US Government Printing Office, Washington DC, pp 178–192 (NIDA research monograph 56)

Roe A (1944) The adult adjustment of children of alcoholic parents raised in foster homes. Q J Stud Alcohol 5:378–393

Rougé-Pont F, Piazza PV, Kharouby M, Le Moal M, Simon H (1993) Higher and longer stress-induced increase in dopamine concentrations in the nucleus accumbens of animals predisposed to amphetamine self-administrati on. A microdialysis study. Brain Res 602:169–174

Rounsaville BJ, Kosten TR, Weissman MM, Prusoff B, Pauls DL, Anton SF, Merikangas K (1991) Psychiatric disorders in relatives of probands with opiate addiction. Arch Gen Psychiatry 48:33–42

Samson HH, Tolliver GA, Lumeng L, Li T-K (1989) Ethanol reinforcement in the alcohol nonpreferring rat: initiation using behavioral techniques without food restriction. Alcohol Clin Exp Res 13:378–385

Sanchez FP, Page SL, Dickinson L, George FR (1993) Relationship between ethanol-reinforced behavior and locomotor stimulation in FAST and SLOW selectively bred mice. Proc Western Pharmacol Soc 36:367

Satinder KP (1977) Oral intake of morphine in selectively bred rats. Pharmacol Biochem Behav 7:43–49

Schechter MD (1992) Rats bred for differences in preference to cocaine: other behavioral measurements. Pharmacol Biochem Behav 43:1015–1021

Scheier LM, Newcomb MD (1991) Differentiation of early adolescent predictors of drug use versus abuse: a developmental risk-factor model. J Subst Abuse 3:277–299

Schuckit MA (1984a) Subjective responses to alcohol in sons of alcoholics and control subjects. Arch Gen Psychiatry 41:879–884

Schuckit MA (1984b) Differences in plasma crotisol after ingestion of ethanol in relatives of alcoholics and controls: preliminary results. J Clin Psychiatry 45:374–376

Schuckit MA (1985) Ethanol-induced changes in body sway in men at high alcoholism risk. Arch Gen Psychiatry 42:375–379

Schuckit MA, Monteiro MG (1988) Alcoholism, anxiety, and depression. Br J Addict 83:1373–1380

Schuckit MA, Goodwin DW, Winokur G (1972) A study of alcoholism in half siblings. Am J Psychiatry 128:1132–1136

Schuckit MA, Gold E, Risch C (1987a) Serum prolactin levels in sons of alcoholics and control subjects. Am J Psychiatry 144:854–859

Schuckit MA, Gold E, Risch C (1987b) Plasma cortisol levels following ethanol in sons of alcoholics and controls. Arch Gen Psychiatry 44:942–945

Schutz CG, Ambrosio E, Shippenberg TS, Elmer GI, Heidreder C (1994) Morphine-induced locomotor activity overflow in the nucleus accumbens in Lewis and Fischer 344 inbred rat strains: a comparative study. In: Louilot A, Durkin T, Spampinato U, Cador M (eds) Monitoring molecules in neuroscience. 6th international conference on in vivo methods, France, pp 163–165

Schwab S, Soyka M, Niederecker M, Ackenhei M, Scherer J, Wildernauner DB (1991) Allelic association of human dopamine D2 receptor DNA polymorphism ruled out in 45 alcoholics. Proceedings of the 8th international congress on human genetics.

Seale TW (1991) Genetic differences in response to cocaine and stimulant drugs. In: Crabbe JC, Harris RA (eds) The genetic basis of alcohol and drug actions. Plenum, New York, pp 279–321

Seale TW, Carney JM (1991) Genetic determinants of susceptibility to the rewarding and other behavioral actions of cocaine. J Addict Dis 10:141–161

Shimada S, Kitayama S, Lin CL, Nanthakumar E, Gregor P, Patel A, Kuhar MJ, Uhl GR (1991) Cloning and expression of a cocaine-sensitive dopamine transporter cDNA. Science 254:576–578

Shippenberg TS (1993) Motivational effects of opioids. In: Herz A (ed) Opioids II. Springer, Berlin Heidelberg New York, pp 633–652 (Handbook of experimental pharmacology, vol 104/2)

Shiffman S (1991) Refining models of dependence: variations across persons and situations. Br J Addict 86:611–615

Shoaib M, Spanagel R, Stohr T, Shippenberg TS (1995) Strain differences in the rewarding and dopamine-releasing effects of morphine in rats. Psychopharmacology 117:240–247

Shuster L, Yu G, Bates A (1977) Sensitization to cocaine stimulation in mice. Psychopharmacology 52:185–190

Smith M (1986) Genetics of human alcohol and aldehyde dehydrogenases. Adv Hum Genet 15:249–290

Smith SS, O'Hara BF, Persico AM, Gorelick DA, Newlin DB, Vlahov D, Solomon L, Pickens R, Uhl GR (1992) Genetic vulnerability to drug abuse: the dopamine D2 receptor TaqI B1 RFLP is more frequent in polysubstance abusers. Arch Gen Psychiatry 49:723–727

Spuhler K, Deitrich RA (1984) Correlative analysis of ethanol-related phenotypes in rat inbred strains. Alcohol Clin Exp Res 8:480–484

Substance Abuse and Mental Health Services Administration (1993) National House-hold Survey on Drug Abuse: preliminary estimates from 1992, Advance report number 3, June

Sudakov SK, Konstantinopolsky MA, Surkova LA, Tyurina IV, Borisova EV (1991) Evaluation of Wistar rats' individual sensitivity to the development of physical dependence on morphine. Drug Alcohol Depend 29:69–75

Sudakov SK, Goldberg SR, Borisova EV, Surkova LA, Turina IV, Rusakov DJ, Elmer GI (1993) Differences in morphine reinforcement property in two inbred rat strains: associations with cortical receptors, behavioral activity, analgesia and the catalepatic effects of morphine. Psychopharmacology 112: 183–188

Suzuki T, George FR, Meisch RA (1988) Differential establishment and maintenance of oral ethanol reinforced behavior in Lewis and Fischer 344 inbred rat strains. J Pharmacol Exp Ther 245:164–170

Suzuki T, George FR, Meisch RA (1992) Etonitazene delivered orally serves as a reinforcer for Lewis but not Fischer 344 rats. Pharmacol Biochem Behav 42:579–586

Suzuki T, Funada M, Narita M, Misawa M, Nagase H (1993) Morphine-induced place preference in the CXBK mouse: characteristics of mu opioid receptor subtypes. Brain Res 602:45–52

Svikis DS, Valez M, Pickens RW (1994) Genetic aspects of alcohol use and dependence in women. Alcohol Health Res World 18:192–196

Swan GE, Carmelli D, Rosenman RH, Fabsitz RR, Christain JC (1990) Smoking and alcohol consumption in adult male twins: genetic heritability and shared environmental influence. J Subst Abuse 2:39–50

Tanna VL, Wilson AF, Winokur G, Elston RC (1988) Possible linkage between alcoholism and esterase-D. J Stud Alcohol 49:472–476

Thomasson HR, Crabb DW, Edenberg HJ, Li TK (1993) Alcohol and aldehyde dehydrogenase polymorphisms and alcoholism. Behav Genet 23:131–136

Turner E, Ewing J, Shilling P, Smith TL, Irwin M, Schuckit M, Kelsoe JR (1992) Lack of association between an RFLP near the D2 dopamine receptor gene and severe alcoholism. Biol Psychiatry 31:285–290

Uhl GR, Persico AM, Smith SS (1992) Current excitement with D2 dopamine receptor gene alleles in substance abuse. Arch Gen Psychiatry 49:157–160

Uhl GR, Blum K, Smith SS (1993) Substance abuse vulnerability and D2 receptor genes. Trends Neurosci 16:83–88

Upchurch M, Wehner JM (1988) Difference between inbred strains of mice in Morris water maze performance. Behav Genet 18:55–68

Upchurch M, Wehner JM (1989) Inheritance of spatial learning ability in inbred mice: A classical genetic analysis. Behav Neurosci 103:1251–1258

Vaillant GE (1973) A 20-year follow-up of New York narcotic addicts. Arch Gen Psychiatry 29:237–241

van Abeelen JHF (1989) Genetic control of hippocampal cholinergic and dynorphinergic mechanisms regulating novelty-induced exploration. Experientia 45:839–845

Vaught JL, Mathiasen JR, Raffa RB (1988) Examination of the involvement of supraspinal and spinal mu and delta opioid receptors in analgesia using the mu receptor deficient CXBK mouse. J Pharmacol Exp Ther 245:13–16

Vesell ES (1972) Ethanol metabolism: regulation by genetic factors in normal volunteers under a controlled environment and the effect of chronic ethanol administration. Ann N Y Acad Sci 197:79–88

Waller MB, McBride WJ, Gatto GJ, Lumeng L, Li T-K (1984) Intragastric self-infusion of ethanol by ethanol-preferring and -nonpreferring lines of rats. Science 225:78–80

Wang JB, Imai Y, Eppler CM, Gregor P, Spivak CE, Uhl GR (1993) μ opiate receptor: cDNA cloning and expression. Proc Nat Acad Sci USA 90: 10230–10234

Wesner RB, Tanna VL, Palmer PJ, Thompson RJ, Crowe RR, Winokur G (1991) Close linkage of esterase-D to unipolar depression and alcoholism is ruled out in eight pedigrees. J Stud Alcohol 52:609–612

Wilson JR, Plomin R (1985) Individual differences in sensitivity and tolerence to alcohol. Soc Biol 32:162–184

Wise RA, Bozarth MA (1987) A psychomotor stimulant theory of addiction. Psychol Rev 94:469–492

52   R.W. Pickens et al.: Genetic Vulnerability to Substance Abuse

Wolf CR (1991) Individuality in cytochrome P450 expression and its association with the nephrotoxic and carcinogenic effects of chemicals. IARC Sci Publ 115: 281–287

Woods JH, Katz JL, Young AM, Medzihradsky F, Smith CB (1981) Correlations among certain behavioral, physiological, and biochemical effects of narcotic agonists. In: Harris LS (ed) Problems of Drug Dependence. DHHS publication, Washington DC, pp 43–57

Young AM, Swain HH, Woods JH (1981) Comparison of opioid agonists in maintaining responding and in suppressing morphine withdrawal in rhesus monkeys. Psychopharmacology 74:329–335

Zacharko RM, Lalonde G, Kasian M, Anisman H (1987) Strain-specific effects of inescapable shock on intracranial self-stimulation from the nucleus accumbens. Brain Res 426:164–168

Zacharko RM, Gilmore W, MacNeil G, Kasian M, Anisman H (1990) Stressor induced variations of intracranial self-stimulation from the mesocortex in several strains of mice. Brain Res 533:353–357

CHAPTER 2
Integrative Neurobehavioral Pharmacology: Focus on Cocaine

M.J. KUHAR and C.R. SCHUSTER

A. Introduction

In the past 10–15 years, science has produced remarkable new technologies and discoveries. From the discovery of "dark matter," a new kind of matter in the universe, to the use of specific genes as "medications" to cure genetic defects, there have been advances with staggering implications for our future. In the neurosciences, brain and behavior studies show such promise in unraveling the mysteries of mood, cognition and psychopathology that the 1990s have been named the "Decade of the Brain" in the hope of focusing energies and resources to realize the potential of the area. Much of the progress involves advances in understanding neuronal communication and interaction. Twenty years ago, there were only a few substances that satisfied strict criteria for being neurotransmitters; today there are dozens. In recent years, peptides have emerged as important neurochemical mediators. The appreciation of the colocalization of peptides with more classical neuro-transmitters in single neurons has expanded our appreciation of complexities in neuronal interactions. The identification of receptors for drugs and neurotransmitters and cloning of cDNAs for these and other proteins such as ion channels have revealed much about how these proteins function at the molecular level. Positron emission tomography (PET) and single photon emission computed tomography (SPECT) imaging, marvelous technologies, allow us to measure, in living humans, drug receptors and to identify brain regions functionally associated with specific physiologies. Concurrently, laboratory studies of behavior have made significant advances in under-standing how environmental variables interact with biological substrates to control behavior. This understanding of the behavioral determinants of both normal and abnormal behaviors has led to the development of therapeutic behavioral interventions for the treatment of a variety of alcohol, drug abuse and mental disorders. These are only a few of the many advances that could be mentioned. The goal of this chapter is to summarize some of the more recent developments of technologies and concepts which have advanced the field of addiction. Interdisciplinary studies from molecular genetics to the experimental analysis of human behavior will be utilized to document advances in understanding drug action. Cocaine is viewed by the public as the major drug of abuse responsible for many of society's ills such as violence, homelessness and the deterioration of families and communities. In part, at least because of these societal concerns, there has been a great deal of research advances in recent years in our understanding of the behavioral and biological mechanisms underlying cocaine abuse and dependence. For this reason, there will be a partial focus in this chapter on cocaine as a drug, and on techniques in molecular biology, neuropharmacology, brain imaging and behavior analysis as approaches to understanding the biological and behavioral mechanisms mediating the dependence producing properties of cocaine.

B. Approaches to Drug Abuse and Dependence

Drug abuse is a highly complex problem involving sociological, behavioral, and physiological factors and a broad array of tragic consequences to the individual, the family and the community. The relative importance of these factors in the etiology of drug abuse/dependence varies with the individual. For example, some may use cocaine for the rush and euphoria it produces and others for the social reinforcers of group participation in drug taking rituals. Others may use cocaine as a means of attenuating aversive subjective states associated with psychopathology. Because of the need for an understanding of the effects of drugs of abuse at all levels of integration, from the subcellular to the behavioral, the focus in this chapter and book will be on the relationship between behavior and physiology and an integration of knowledge from a variety of disciplines from the most molecular to the macro. An understanding of these relationships is crucial in advancing the field not only for developing a new understanding of drug abuse and dependence, but also for developing useful medications as treatments for these disorders.

The behavioral aspects of substance abuse include: seeking and taking the drug even though it is having adverse consequences to the health of the individual and to his/her functioning in professional, societal or family relationships; failed efforts to stop or curtail its use; and, with some drugs, the development of tolerance and withdrawal signs which appear when drug intake is stopped or reduced. In the case of some drugs of abuse, such as opiates, the withdrawal symptoms are aversive and generate increased drug-seeking behaviors. It is known at a neuropharmacological level that administration of drugs of abuse results in many neurochemical changes in brain including stimulation of various receptor sites, production of second messengers, inhibition or stimulation of various physiological processes and alteration of gene expression. Since drugs of abuse produce a variety of effects in the central nervous system, it must be determined which of the many changes that are observed are critically related to dependence. For example, the central nervous system has a large number of and a wide distribution of opiate receptors and a question is which of these receptors in which area is associated with the abuse of the opiate substances. Similarly, cocaine also has many actions in the brain, as it inhibits the uptake of the neurotransmitters dopamine, norepinephrine and serotonin and at higher concentrations can block sodium channels (the sodium channel blocking ability of cocaine is believed to be the basis of its local anesthetic properties). Which of these effects is associated with the reinforcing properties of cocaine? What are the strategies by which we can attempt to answer these questions?

A useful first step for testing relationships between neurochemical mechanisms and behavior is a correlational analysis. These are many examples of this type of analysis in pharmacology. For example, we can search for a correlation between activity at various cocaine binding sites and the reinforcing effects of cocaine. In our group, we utilized a variety of drugs that share with cocaine the ability to inhibit the reuptake of various neurotransmitters. Neurotransmitter reuptake is the major process which terminates the synaptic transmission for some neurotransmitters by removing the transmitter from the synapse. The reuptake occurs via a transporter (or "reuptake carrier") molecule which moves the neurotransmitter back inside the terminal of the nerve that released it. We compared the potencies of these cocaine analogs in binding to the various transporters to their potency as reinforcers in laboratory animal drug self-administration studies. The site at which binding affinities correlated best with the reinforcing potency of these drugs is presumably the site where cocaine interacts to produce its reinforcing effects (or at least the initial steps of reinforcement). Because there are many different sites at which cocaine can interact, a correlational approach is made more complex and it is appropriate to analyze the influence of multiple sites at the same time. An example of this work is given in more detail below. Further, the correlational approach is suggestive but not conclusive, since it is well recognized that a correlation between two factors does not establish causality. Therefore, it is not possible to say that cocaine's action at a given site causes the reinforcing properties of the drug even if they correlate, but it certainly supports a hypothesis that this is the case. Certainly, if two measures are causally related, then a correlation between the two must exist. Thus, while this approach has its limitations, it has proven to be valuable and useful for identifying cellular and molecular events that may then be more directly investigated for their role in various physiological and behavioral events.

C. A Cocaine Receptor

It is known that cocaine has many binding sites in brain and it is widely acknowledged that the dopamine tramsporter, one of these binding sites, is somehow related to its reinforcing effects in animals. The following is a description of some of the evidence for this hypothesis and its derivation.

I. Behavioral Models for Drug Abuse

1. Drug Self-Administration

The study of the pharmacological basis of drug abuse and dependence was markedly advanced by the finding that certain psychoactive drugs could serve as reinforcers for the behavior of laboratory animals (Thompson and Schuster 1964; Yanagita et al. 1965; Weeks 1962; Davis 1966). The finding that laboratory animals would readily learn to self-administer certain psychoactive drugs led to the development of a sophisticated animal model of human drug abuse which has been extremely useful for a variety of research purposes (see other chapters in this book). Most importantly, this model has allowed the study of the pharmacological and environmental variables that are important in controlling the initiation, persistence and pattern of drug-seeking and drug-taking behavior (Johanson and Schuster 1981). This behavior has been productively analyzed within the context of the principles of behavior analysis, the fundamental tenet of which is that behavior is controlled by its consequences. What has emerged from the study of psychoactive drugs as controlling consequences of behavior (i.e., reinforcers) is that drug-mainatined behaviors are controlled by the same classes of environmental variables as other reinforcers such as food, water or sex. Further, recent advances have been made in determining the neuro-pharmacological mediators of the reinforcing effects of certain drugs. It is increasingly recognized that these two lines of investigation must ultimately coalesce. A complete understanding of the neurobiological basis of drug reinforcement must account not only for the actions of these drugs on the relevant neural cells and circuitry, but as well, for the neurochemical mediators responsible for the modulation of the drugs reinforcing effects by biological and environmental variables.

Although the reinforcing effects of drugs have been studied using a variety of routes of drug administration, the most common is the intravenous route. This is because many drugs of abuse (e.g., heroin) are self-administered intravenously and also because this route of administration assures the rapid onset of the drug's reinforcing effects to enhance the animal's learning the drug self-administration behavior. Figure 1 shows the typical laboratory setting in which such studies are conducted. In this case, a rhesus monkey has been surgically prepared with a chronic indwelling venous catheter which exits from the skin in the back where it is attached to tubing protected by the swivel arm which restrains the animal. The tubing is connected to an automatic pump which can be energized by the monkey pressing the lever switch shown on the wall of the experimental chamber. Programming equipment is used to control the schedule of reinforcement (i.e., the relationship between responding and the delivery of the drug reinforcer) and the stimulus lights which are illuminated when the drug reinforcer is available. The salient feature of this system is that a controlled

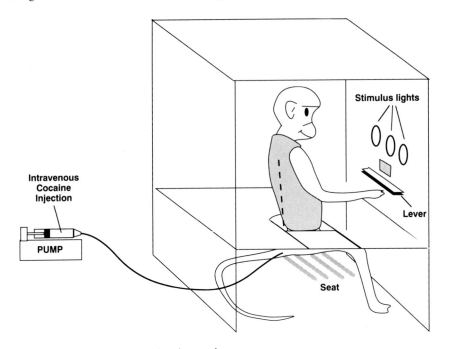

Fig. 1. Drug self-administration in monkey

amount of a drug can be administered with minimal delay contingent upon the operant response of lever pressing. It should be pointed out that other routes of administration can be used for the study of drugs as reinforcers. FALK, for example (1993), has studied a variety of drugs of abuse using an oral route of administration. There have been a series of studies showing that substances such as nitrous oxide, toluene, chloroform and ether, which readily volatilize, will serve as positive reinforcers in several species (WOOD et al. 1977; WOOD 1979; WOOD and WEISS 1978). This latter route of administration has achieved increasing prominence as the smoking of "crack" (cocaine base) and "ice" (methamphetamine) has become a prevalent part of the drug abuse scene.

The reinforcing effects of drugs have been increasingly studied in humans (JOHANSON and DE WIT 1989), usually in conjunction with the more traditional measurement of drug-induced mood changes such as "euphoria," which are felt by some to be the basis of why drugs are self-administered. However, there are several advantages to the use of animals for such research, the foremost of which are the types of manipulations and the range of variables which can ethically be studied. Clearly, it would be impossible to conduct many of the studies involving direct measurement of neurochemical effects of drugs of abuse in humans. The use of animal models of drug abuse to gain an understanding of the neurobiological basis of drug reinforcement in humans is based upon at least two assumptions: that animals and humans are self-administering drugs for the same reasons and that there is homology in the neurobiological substrates of drug action across the different species. Thus, the "validation" of the relevance of animal models of drug abuse to the problems of human drug abuse is of utmost importance. The validity of the drug self-administration model is supported by the fact that the same drugs that serve as reinforcers are commonly abused by humans (Johanson and Balster 1978). Drugs which are aversive in humans (phenothiazines, for example) will maintain avoidance or escape behavior in animals. Further, at least some of the same variables that effect the reinforcing effects of drugs in humans have similar effects in animals (Johanson and Schuster 1981). These data have led to the acceptance of the drug self-administration animal model of human drug abuse for a wide variety of purposes (Johanson and Schuster 1981). In addition to the drug self-administration model, there are several other indirect ways to study the dependence producing properties of drugs of abuse, each of which has its own unique advantages.

2. Place Preference

Another commonly used procedure for investigating the reinforcing actions of pharmacological agents is the place preference and aversion procedure (Stolerman 1992). In this procedure, administration of a drug, such as cocaine, is paired with a distinct environment, and vehicle administration is associated with a different environment. Following several pairings of the drug and vehicle with the two different environments, the amount of time the animal chooses to spend in each environment is measured. If the animal spends more time in the environment associated with cocaine than the one associated with the vehicle, then cocaine is said to produce a place preference and by inference is serving as a positive reinforcer. If, however, the animal spends more time in the environment associated with the vehicle, then the drug is said to produce place aversion, and the drug is presumed to be producing aversive effects.

One of the advantages of the place preference procedure is that neuro-chemicals being investigated for their ability to block the positive (or negative) reinforcing effects of a drug, such as cocaine, can be given during the acquisition phase and are not present during the test phase. Thus, this procedure prevents the confounding of nonspecific effects (e.g., sedation, motor incapacitation) of the neurochemical manipulation on the behavior used to measure any alteration of the reinforcing efficacy of cocaine.

3. Alteration of Thresholds for Electrical Brain Stimulation Reinforcement

In 1954, Olds and Milner reported that electrical stimulation of certain regions of the brain would serve as a reinforcer for lever pressing behavior of rats (Olds and Milner 1954). Over the next several decades, research in this area defined the regions of the brain and the neurochemical systems which mediate reinforcing intracranial self-stimulation (ICSS). It is currently believed that ICSS activates dopaminergic brain systems that normally mediate the reinforcing actions of food, water, sex, etc. Thus, this area of the brain has been termed a "reward pathway." Part of the evidence that brain stimulation reinforcement was mediated by brain dopamine systems came from the observation (WANQUIER and NIEMEGEERS 1974; ESPOSITO et al. 1978; KORNETSKY et al. 1979; KORNETSKY and DURAUCHELLE 1994) that psychomotor stimulant drugs decreased the threshold for electrical stimulation of the brain and dopamine antagonists raised the threshold. These observations have led to a series of experiments by KORNETSKY and colleagues which have shown that drugs of abuse decrease the threshold for electrical brain stimulation (KORNETSKY 1994). More recently, PHILLIPS et al. (1992) have shown that in rats intracranial self-stimulation produces dopamine efflux in the nucleus accumbens as measured by both voltametry and dialysis techniques. It therefore would appear that the threshold lowering effects of drugs of abuse on ICSS is caused by their producing a higher basal level of synaptic dopamine thus decreasing the amount of electrical stimulation necessay to release sufficient dopamine to serve as a reinforcing event. It would also seem logical that drugs which lower the threshold for ICSS would presumably sensitize the animal to the reinforcing effects of normal biological reinforcers.

Cocaine and other psychomotor stimulant users often self-administer the drug in "binges" ranging in duration from a few hours to several days. Following such binges, a "crash" occurs characterized by depression, anhedonia, and irritability (GAWIN et al. 1985). Animal research has shown that following the chronic administration of amphetamines the threshold for ICSS is elevated (LEITH and BARRETT 1976) for a period of several hours to several days depending on the dose and duration of the chronic drug regimen. An elegant study by MARKOU and KOOB (1991), showed that the threshold for ICSS was elevated following periods in which rats were allowed to self-administer cocaine. The magnitude and the duration of the threshold increase in ICSS were found to be related to the length of time that animals were allowed to self-administer cocaine prior to the withdrawal period. These authors suggest that the increase in ICSS is a useful animal model for the anhedonia associated with cocaine withdrawal. If it can be validated, such a model might be extremely useful for determining the neurochemical mechanisms associated with the period of cocaine withdrawal induced anhedonia.

4. Drugs of Abuse and Endogenous Reward Systems

The brain reinforcement stystems obviously do not exist for activation by drugs of abuse, but presumably to reinforce activities that have survival value. Thus, it is not surprising that the dopaminergic mesolimbic system has been linked to eating, drinking and sexual activity, activities obviously essential for survival. For example, Fibiger and colleagues (PFAUS et al. 1990; DAMSMA et al. 1992), using microdialysis procedures in rats, examined dopamine levels in the nucleus accumbens and striatum during various events associated with and including copulation. They found that dopamine efflux in nucleus accumbens increased significantly in male rats when they were presented with a sexually receptive female. Efflux increased further during copulation. By contrast, dopamine efflux in striatum did not change during the presentation of a sexually receptive female and increased to a lesser degree during copulation. Additional control experiments showed that locomotor activity alone, exposure to a novel chamber or exposure to sex odors alone did not increase dopamine efflux in either brain region. The investigators noted that their results are in agreement with the idea that dopaminergic transmission in nucleus accumbens is associated with or mediates activity with reinforcement value. Recently, this notion has been restated in terms of "salience;" ROBINSON and BERRIDGE (1993) suggest that the mesolimbic dopaminergic system activates incentive salience which transform stimuli, making them significant and attractive.

II. Receptor Binding

The method for finding physiologically important receptors for a drug is, in principle, easy to describe. Radiolabeled drug is mixed with membranes so that drug can bind to receptor molecules in the membranes. Drug-receptor-membrane complex is separated from excess drug by filtration in which excess drug passes through the filter but the membranes with drug receptor complex do not. A problem is that radiolabeled drugs will bind to many, many sites, only some of which are physiologically relevant receptors. For example, some drugs will bind to the filter paper itself or the glass containers. Fortunately, the criteria for identifying physiologically relevant sites are well understood. These criteria include high affinity, saturability and pharmacological similarity (YAMAMURA et al. 1990). High affinity indicates that the ligand will bind to the receptor tightly enough and long enough so that the drug-receptor complex can be measured practically. Saturability is a property confirming that the receptor site is present in measurable, finite concentrations (non-specific or non-receptor binding sites are present in very large, almost "infinite," concentrations). Pharmacological similarity implies that the pharmacology of the receptor site is similar to the pharmacology of some related physiological processes. For example, it is known that the analgesic properties of opiate drugs show stereospecificity, e.g., the minus isomer levorphanol produces analgesia while the plus isomer dextrorphan does not. Therefore, the receptor mediating the analgesic properties of opiates must show similar stereospecificity, i.e., show a high affinity for levorphanol but not for dextrorphan, which it does. Using this approach hundreds of relevant receptors have been identified for various drugs. For example, it can be inferred that ^{125}I labeled RTI-55 binds to dopamine transporters in rat striatal membranes because various drugs inhibit the binding of ^{125}I-labeled RTI-55 with the same relative potency that they inhibit dopamine uptake (Fig. 2). This inference is currently regarded as fact because of much new data. For example, placement of a transporter gene inside a cell originally lacking both binding activity and uptake activity confers the expression of both factors in the cell (SHIMADA et al. 1991; BOJA et al. 1992c). It is also clear that receptor sites are directly behaviorally relevant to addiction; injections of a μ opiate antagonist (which blocks μ receptors) into opiate dependent subjects results in the immediate precipitation of a withdrawal syndrome. Also, naltrexone blocks the reinforcing effects of opiates (JAFFE 1990).

Many cocaine binding sites have been found and include sites in the liver and placenta, sodium channels in neuronal membranes, and monoamine transporters in monoaminergic nerve terminals (see RITZ et al. 1987; MADRAS et al. 1989; CARROLL et al. 1992 for reviews). Given all of these binding sites, it can be inferred that cocaine has multiple actions at multiple sites. How do we decide which of these cocaine binding sites is related to its various behavioral effects? Can one of these sites be shown to correlate with the reinforcing potency of cocaine and its analogs? This is where we turn to

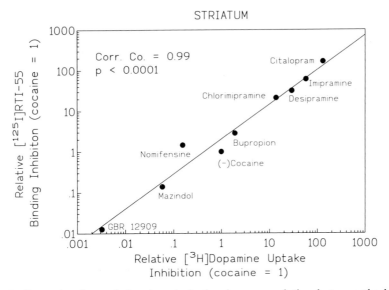

Fig. 2. Example of correlational analysis showing a correlation between the log of relative potencies (cocaine = 1) of various drugs in striatal ^{125}I-labeled RTI-55 binding assay with potencies in [³H]dopamine uptake assay. Values for inhibition of dopamine uptake are from HYTTEL et al. (1982) and ANDERSEN (1987). Because such an excellent correlation exists, the authors concluded that RTI-55 binding is to the dopamine transporter. (From BOJA et al. 1992b)

correlational analysis and attempt to relate the properties of a given binding site for cocaine to the properties exhibited by an appropriate animal model for the reinforcing properties of cocaine and related compounds.

The drug self-administration animal model described above has been used extensively by many laboratories. Many groups have utilized cocaine and related compounds in assessing the doses required in animals to maintain drug self-administration. As expected, these drugs vary in their potencies in maintaining self-administration behavior. Some are weaker than cocaine and some are more potent than cocaine. We can now take these same compounds and measure their potency at the various binding sites for cocaine. Measuring potencies at these sites is straightforward and involves competition experiments with radioactive cocaine or with some other radioactive cocaine-like molecule. These studies have been carried out and it is possible to show that, with a selected series of compounds, there is a correlation between drug self-administration behavior and binding to the dopamine transporter but not to the other binding sites (Fig. 3). This correlation suggests that the initial site responsible for the reinforcing action of these compounds is the dopamine transporter and that inhibition of dopamine uptake, rather than some other function, is the initial event that somehow ultimately produces the reinforcing effects of the drug. Since the binding site appears to be the initial site of action of the drugs for producing reinforcement, the dopamine transporter is referred to as the "cocaine receptor," and the idea is referred to as the dopamine hypothesis (Fig. 4). In this instance, we refer to the dopamine hypothesis as being related specifically to cocaine, but it has been shown that mesolimbic dopamine appears to be involved in the action of other abused drugs as well (DiChiara and Imperato 1988; Wise and Bozarth 1987; Koob and Bloom 1988; Koob 1992).

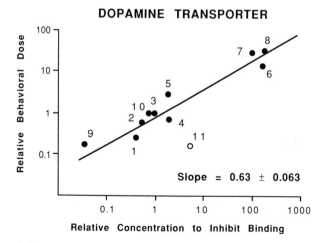

Fig. 3. Correlation between drug binding at dopamine transporter and potency in drug self-administration. (From Ritz et al. 1987)

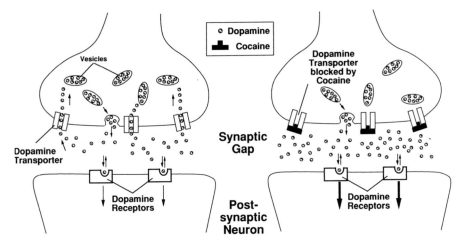

Fig. 4. The dopamine hypothesis of cocaine reinforcement. Cocaine binds to the dopamine transporter and blocks the reuptake of dopamine in the mesolimbic pathway. This potentiates dopaminergic neurotransmission and initiates the sequence of events that ultimately cause the rewarding effects of the drug. (From Kuhar et al. 1991)

III. Interdisciplinary Support for a Dopamine Hypothesis

Again, a correlation does not indicate causality, although it is suggestive. This association of reinforcing behavior with the mesolimbic dopamine system has been examined in many other experiments and cumulative data are substantial and support the notion that there is a causal relationship between transporter inhibition and reinforcing behavior. Some of these data can be summarized as follows.

Using the technique of microdialysis, it can be clearly shown that inhibition of dopamine uptake by cocaine occurs in vivo. A microdialysis probe which is small in size and permeable at its tip (Fig. 5) can be inserted directly into brain tissue. Dopamine from the tissue can diffuse into a buffer solution inside the probe and can then be measured in aliquots of the buffer. An acute injection of cocaine causes an increase in dopamine efflux from the tissue into the probe (DiChiara and Imperata 1988; Pettit and Justice 1989; Kalivas and Duffy 1990; Hurd et al. 1989; Carboni et al. 1989; Bradberry and Roth 1989), indicating that cocaine does indeed prevent reuptake of dopamine into nerve terminals. Careful in vivo and in vitro studies show that cocaine, unlike amphetamine, does not release dopamine but only inhibits its uptake. However, chronic cocaine results in an attenuation of this effect indicating some adaptive change in the neuron with repeated treatment with cocaine (Hurd et al. 1989).

A number of laboratories are examining dopaminergic parameters after chronic cocaine administration. For example, after withdrawal from repeated

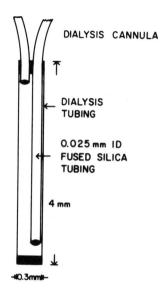

Fig. 5. Microdialysis probe. The probe can be implanted chronically. Various biochemicals can be sampled in the effusate. (From Wages et al. 1986)

cocaine: (1) dopamine transporters tend to increase in number (at about 3 days) and then decrease below basal levels (at about 10 days) for a long period (Pilotte et al. 1994). (2) The sensitivity of autoreceptors for dopamine decreases and then slowly returns to normal (Henry et al. 1989; Ackerman and White 1990). (3) The number of spontaneously firing cells first increase and then decrease (Ackerman and White 1990). (4) The basal release of dopamine is altered over this period as well (Imperato et al. 1992; Rossetti et al. 1992). These results show that dramatic biochemical changes occur during the withdrawal period and an intriguing consideration is that these may be related to craving and relapse to drug use.

A number of brain lesion studies support the view that mesolimbic dopaminergic neurons support psychostimulant self-administration since destruction of these neurons, for example with the neurotoxin 6-hydroxydopamine, results in a disruption of drug self-administration (Goeders et al. 1986; Roberts et al. 1980; Roberts and Koob 1982; Martin-Iverson et al. 1986; Lyness et al. 1979). Also, direct dopamine receptor agonists substitute for cocaine in intravenous self-administration studies (Yokel and Wise 1978; Woolverton et al. 1984), and blockade of dopamine receptors attentuates the reinforcing properties of both cocaine and amphetamine in animals (DeWitt and Wise 1977; Ettenberg et al. 1982; De Lagarza and Johanson 1982; Wilson and Schuster 1972; Goeders et al. 1986; Yokel and Wise 1978; Reisner and Jones 1976; Davis and Smith 1975). Thus, these pharmacological experiments, which involve augmenting and attenuating dopaminergic signals in brain and which result in a corresponding effect on drug self-administration behavior, support the findings from receptor binding studies.

In an effort to extend the pharmacological studies with the animal model of the self-administering monkey, dopamine receptor blocking drugs were administered to human subjects before administration of psychostimulant drugs. Both cocaine and *d*-amphetamine were used in these studies as they are quite similar in their actions and effects. Since dopamine receptor blockers disrupt psychostimulant self-administration in subhuman primates, such blockers might affect human cocaine abuse. In an open label study, flupenthixol decanoate caused a decreased craving for cocaine and appeared to increase the retention time that patients remained in treatment (GAWIN et al. 1989a). However, this study purported to show that it was the low doses of flupenthixol which were effective, and the authors suggested that low doses would preferentially block presynaptic sutoreceptors, thus increasing dopamine levels in the synapse. If this interpretation is accurate, this would not be a good illustration of the dopamine receptor blocking effects. It would be more like replacement therapy. Another report (GUNNE et al. 1972) indicated that euphoria induced by amphetamine was reduced or blocked or blocked by the dopamine receptor blocking drugs chlorpromazine and pimozide. But more recent studies utilizing haloperidol indicated that the blocker had no effects on the "rush" following intravenous administration of cocaine and had only a limited effect on some of the subjective effects on drug liking (SHERER et al. 1989). In another study, while there was a decrease in psychotic symptoms of patients, chlorpromazine did not block euphoria or increase abstinence from cocaine (GAWIN 1986). Thus, the dopamine receptor blocking drugs do not obviously and clearly disrupt cocaine use in humans. However, these drugs differ markedly in their receptor specificity and additional studies are needed. Direct or indirect acting dopamine agonists have also been examined. Methylphenidate (GAWIN et al. 1985), amantadine (GAWIN et al. 1989b; WEDDINGTON ·et al. 1991; KOSTEN et al. 1992), desipramine (WEDDINGTON et al. 1991; KOSTEN et al. 1992; ARNDT et al. 1992), bromocriptine (PRESTON et al. 1992) and mazindol (PRESTON et al. 1993) have been used but they do not appear to have marked benefit in reducing cocaine use, at least at the doses utilized. *a*-Methyl-P-tyrosine, a blocker of dopamine synthesis, blocked amphetamine-induced euphoria at higher doses (GUNNE et al. 1972). These and other studies indicate the conflicting and preliminary nature of these findings in human subjects. While many human studies are compatible with the dopamine hypothesis, additional work is needed to assess its relevance to human drug abusers.

An interesting challenge to the dopamine hypothesis is whether or not useful treatment medications can be developed from this hypothesis. For example, if the reinforcing action of cocaine is indeed through the dopamine mesolimbic system, could dopamine agonists be used as "substitute" medications for treating cocaine addicts? Also, could dopamine receptor blockers be used to block the effects of reinforcing compounds like cocaine? These substitutes or blockers would be analogous to methadone and naltrexone which are used for the treatment of heroin dependence. Because of the evidence suggesting that many drugs of abuse utilize dopaminergic systems as a final common pathway (DiChiara and Imperata 1988), the hypothesis may have wide ranging implications for treating drug abuse in general. However, as the previous paragraph indicates, we have not yet found how to use blockers or agonists as treatment for cocaine abuse, nor do we know if this is possible.

D. Brain Imaging

Brain imaging is an area in which dramatic advances have been made in recent years. Imaging techniques allow us to measure brain structures as well as key biochemical entities and processes in various brain regions, sometimes in living humans. In a field such as drug addiction, which is in part a "brain-behavior" problem, the ability to measure processes in the brains of humans is essential.

One of the earliest techniques that was used with much success is autoradiography, which shows the distribution of radioactive compounds in brain regions although not in living tissues. This is usually done at the light microscopic level, although it can be extended to the electron microscopic level. The approach has been used to measure glucose utilization, blood flow, and drug receptor densities in various brain regions. In some studies, radioactive compounds such as 2-deoxyglucose are injected into animals under conditions in which brain regional activities of glucose metabolism can be related to various treatments or functional tasks. The accumulation of 2-deoxyglucose-6-phosphate in various brain regions is measured by sectioning the brain and apposing the sections to an autoradiographic film. An example of this approach is the work of Dworkin et al. (1992), who found increased glucose utilization in limbic areas in animals self-administering cocaine. Blood flow studies can be carried out in the same general manner. Blood flow is increased in brain areas where there is an increase in activity; when the blood carries a radiolabeled marker that can move into tissues, increased flow is reflected in increased tissue radioactivity. Changes in autoradiographic grain density in specific brain areas reflect neuronal activity and we can therefore relate regions and functions. Because of the need to section the brain and appose the slices to film, these approaches are limited to animal studies.

In order to work with human tissues, in vitro labeling receptor autoradiography was developed specifically for localization of drug receptors. After removing the brain at autopsy, tissue sections can be subjected to procedures designed to bind ligands to receptors and to minimize or eliminate diffusion of the ligand. Regional localizations can be measured by autoradiography. In vitro labeling autoradiography has many advantages compared to injection approaches because it allows the utilization of ligands that do not cross the blood-brain barrier or that are degraded in the blood, and it allows the use of post mortem human tissues. This procedure, developed mainly in the early 1980s, utilized binding techniques that had been developed for biochemical homogenate binding studies. Slide mounted tissue sections were shown to bind ligands in the same manner and with largely the same properties as fresh tissue homogenates (ROTTER et al. 1979; YOUNG and KUHAR 1979; HERKENHAM and PERT 1980). This realization and achievement, especially with reversible binding ligands (YOUNG and KUHAR 1979), coupled with the apposition of the labeled sections to dry emulsion coated surfaces (ROTH et al. 1974; EHN and LARSSON 1979), allowed essentially the unlimited mapping of the post mortem human brain for drug receptors at the light microscopic level.

Since drug receptors are often neurotransmitter receptors and since these receptors are often localized to discrete neuronal populations, receptor imaging can be used to study neuronal degenerative diseases. For example, SHAYA et al. (1992) showed that a loss of dopamine containing neurons in response to an injection of neurotoxins can be detected readily by SPECT imaging. Further, since drug receptors are biochemically regulated in response to drug agonists and antagonists, the loss or increase of such receptors can be readily determined at the light microscopic level and used to measure aspects of neuronal plasticity (KUHAR and UNNERSTALL 1990).

Magnetic resonance imaging (MRI) is a very powerful technique that allows imaging of brain structures (and of course other structures in the body) very rapidly and with a high degree of spatial resolution. The essence of the technique is the utilization of a magnetic field along with modulated radio frequency waves to study the distribution of various atomic particles in the brain. Because of the influences of the varying compositions of different brain regions, it is possible to generate an image showing "texture" differences that correspond to the structure of various brain regions and nuclei. It is quite feasible to detect brain degeneration or brain tumors by this approach and it is in fact a major diagnostic tool in neurological medicine. Gadolinium salts are used in conjunction with the MRI apparatus as contrast generating reagents. By using higher magnetic fields to amplify various small signals, it is possible to generate functional as well as structural data regarding the brain. For example, changes in contrast in visual cortex have been generated in response to flashing lights (BELLIVEAU et al. 1991). This procedure, as mentioned above, is fairly rapid, has high resolution and holds great promise for future studies of brain structure and function.

There are several techniques, which include PET and SPECT, that measure radioactivity emitted at various positions around the head. After data collection, planar images showing the location and quantity of radio-isotope within the head are reconstructed by computer and presented as an image in the form of a brain slice. These images of radioactivity distribution determined by PET or SPECT can be superimposed on a brain slice structural image such as that from an MRI. Then, regions of interest can be selected and regional radioactivity can be quantified. With such imaging, "functional" studies which measure more global brain functions such as glucose metabolism or blood flow can be carried out; as mentioned above, regional changes in metabolism or blood flow reflect relative changes in activity in these brain regions. Also, in vivo binding to specific molecules in brain can be used to image the distribution and quantity of these molecules such as receptors, transporters, enzymes or other physiologically useful and important binding sites (Kuhar et al. 1986; Kuhar and DeSouza 1989).

In PET, pairs of gamma rays produced by the annihilations of positrons and electrons are measured. The positrons are emitted from unstable nuclei of carbon (carbon-11), for example. Since drugs and various other biochemicals can be prepared with carbon-11 substituted for carbon-12, it is possible to localize the distribution of these various biochemicals in the brain or other organs. For example, the utilization of receptor binding ligands that are labeled with carbon-11 can be used to image various drug and neurotransmitter receptors in brain. Also, the utilization of fluorine-18 labeled fluoro-deoxyglucose can be used to identify metabolically active regions in brain. A major advantage of PET is that one can image carbon-11, oxygen-15 or fluorine-18. The advantage of the first two atoms is that they are normally found in all organic compounds in nature, and therefore a wide variety of naturally occurring biochemicals can be prepared with positron emitting atoms. Therefore, these key biochemicals can be studied in living human brains by external imaging techniques. A disadvantage of PET is the relatively limited resolution of the technique. the theoretical maximal limit of resolution depends on the atom which is the source of the positron, but in general, the lower limit of resolution of the approach is about 2–3 mm. This corresponds to the distance that the positron travels before it meets an electron and annihilates. This limit in resolution creates problems in that it is difficult to identify changes in small brain regions where high spatial resolution is important.

Brain imaging studies utilizing PET have shown that there is a similarity between the time course of cocaine's occupancy of the binding site at the dopamine transporter and the time course of the drug's psychological effects (Fowler et al. 1989). Also, some studies indicate that postsynaptic D_2 dopaminergic receptors in basal ganglia are decreased in cocaine abusers (Fig. 6) (Volkow et al. 1993). In the frontal lobes, regions receiving a dopaminergic innervation, decreases in glucose metabolism are associated with chronic cocaine use and these effects may persist even after detoxification for 3–4 months (Volkow et al. 1991, 1992a,b, 1993; also see below). These findings are clearly compatible with the involvement of dopaminergic systems in human cocaine use. Brain imaging has proven invaluable in extending hypotheses from animal studies to human subjects and will become

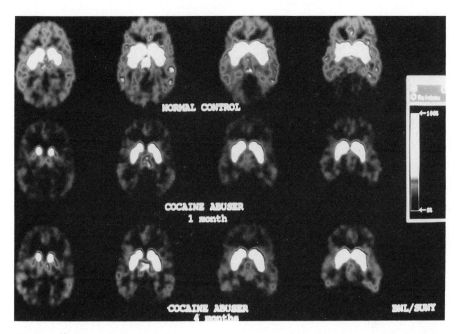

Fig. 6. [^{18}F]N-methylspiroperidol images in a normal control and in a cocaine abuser tested 1 month and 4 months after last cocaine use. The images correspond to the four sequential planes where the basal ganglia are located. The intensity scale has been normalized to the injected dose. Notice the lower uptake of the tracer in the cocaine abuser when compared with the normal control. Notice the persistence of the decreased uptake even after 4 months of cocaine discontinuation. (From VOLKOW et al. 1993)

of increasing importance as more ligands for additional functionally important binding sites become available.

In other studies, PET has been used to examine the effects of cocaine on cerebral blood flow and metabolism. For example, London and colleagues (LONDON et al. 1990a) administered 40 mg of cocaine hydrochloride, a dose that produces euphoria, to human drug abusers. It reduced brain glucose utilization globally by about 14%. Most, but not all, brain regions showed a decrement as well (Fig. 7). In similar studies with morphine, similar results were found (LONDON et al. 1990b). There was a decrease in whole brain glucose utilization and in several cortical areas as well. Other abused drugs such as ethanol also decrease glucose metabolism (LONDON and MORGAN 1993). Thus a reduction in glucose metabolism is associated with drug-induced euphoria, but the significance of this is unknown. At the least, it can be said that drugs of abuse produce measurable changes in glucose metabolism in many brain regions and that they therefore have a significant impact on brain chemistry.

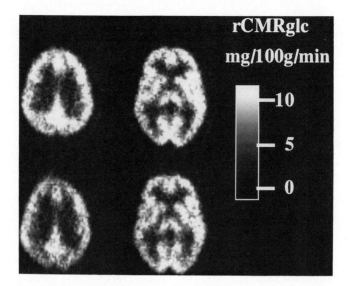

Fig. 7. Representative pseudogray-coded Position emission tomography (PET) scans of a human volunteer showing the effects of 40 mg cocaine (*bottom slices*) vs placebo (*top slices*) on cerebral metabolic rates for glucose. *Lighter shades of gray* indicate higher rates of cerebral glucose metabolism. Intravenous cocaine administration decreases cerebral glucose metabolism globally. (Figure courtesy of Drs. E. London and A. Kimes)

Imaging has also been used to examine brain changes during cocaine withdrawal. For example, in the first week of withdrawal, higher metabolic activity was found in basal ganglia and orbitofrontal cortex (Volkow and Fowler 1992b). This hypermetabolism correlated positively with intensity of craving and negatively with the length of time since cocaine was last used. Studies of brain metabolism at later times of withdrawal produced different results. Significant reductions in frontal metabolic activity were found compared with normal controls. These changes persisted at 3 and 4 months after detoxification. Also, the relative changes were correlated with magnitude of depression found in the patients (Volkow and Fowler 1992b). Thus, these results document changes in brain chemistry during cocaine withdrawal in humans. These changes paralleled other biochemical changes noted in animal studies (see Pilotte et al. 1994, for discussion).

An older and simpler approach compared to PET is SPECT. A SPECT camera measures single photons and, by using mathematical techniques, the distribution of emitting atoms inside the brain can be constructed. Unfortunately, while SPECT is easier to use and is very widespread and available, the atoms used to generate measurable photons include technetium-99, indium, and iodine-123. Thus, we are limited to utilizing probes whose behavior and properties are not significantly altered by the utilization of these atoms. Significant successes have been made, for example, with radioactive iodine-containing cocaine analogues which are known to bind to dopamine transporters (BOJA et al. 1994). The loss of these transporters in degenerative diseases such as Parkinson's disease is quite dramatically shown by PET and SPECT and can be utilized to diagnose Parkinson's disease (SHAYA et al. 1992; TEDROFF et al. 1988; INNIS et al. 1993; FARDE et al. 1994; BRUCKE et al. 1993; WULLNER et al. 1994; FROST et al. 1993; MADRAS 1994).

E. Molecular Actions of Cocaine

I. Cloning the Dopamine Transporter/Cocaine Receptor

Any given cell is defined by the proteins that constitute it and therefore by the genes that are being expressed. Over the past decade or so, techniques have been developed so that the messenger RNAs for a given cell can be converted to a complementary DNA, which can be cloned and sequenced so that the coding sequence of the messenger RNA can be determined. These cloning approaches have been very powerful tools in elucidating the molecular structure and properties of a number of drug receptors. Within the last 3 or 4 years, the sodium- and chloride-dependent neurotransmitter transporters, which are the reuptake sites for many neurotransmitters, have been cloned (AMARA and KUHAR 1993; UHL and HARTIG 1992; GIROS and CARON 1993).

In most fields, there are often critical periods when significant breakthroughs occur. In transporter research, this has certainly been true. In 1990, GUASTELLA et al. (1990) reported the cloning of a cDNA for a GABA transporter. These investigators noted that, while the protein had the same general properties as those of other transporters, the exact amino acid sequence was not similar to any other known transporter, whether from mammalian cells or bacteria. This led the investigators to speculate that they had found a new family of transporter proteins. Some months later, PACHOLCZYK et al. (1991) reported the cloning of a cDNA for the human norepinephrine transporter. A most significant finding was that there was a high degree of homology between the norepinephrine transporter and the GABA transporter. Taken together, these two papers defined a new family of transporter proteins and began an exciting era of neurotransmitter transporter research. Very soon thereafter, a large number of transporters from this family were cloned and reported. The dopamine transporter (DAT), sensitive to cocaine, was cloned by sveral laboratories for several species (SHIMADA et al. 1991; CIROS et al. 1991; USDIN et al. 1991; KILTY et al. 1991).

The dopamine transporter from the rat (Fig. 8) has, like other transporters, 12 postulated transmembrane regions, a large extracellular loop between transmembrane regions three and four containing cosensus glycosylation sites, and NH_2-terminal and COOH-terminal portions within the cell.

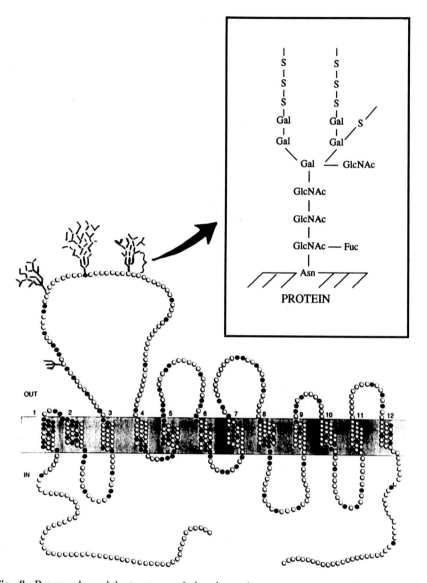

Fig. 8. Proposed model structure of the dopamine transporter protein and other transporters in this family showing 12 transmembrane regions of the aminoacid chain and the N-linked carbohydrate containing sialic acid(s), galactose residues (*gal*) and β-N-acetylglucosamine (*GlcNac*). The glycosylation was predicted by GRIGORIADIS et al. (1989) and LEW et al. (1991) in biochemical studies. (From BOJA et al. 1994 and adapted from AMARA and KUHAR 1993 and SMITH and WOOD 1991)

This hallmark achievement of DAT cloning very quickly led to a new level of understanding about cocaine's molecular actions. For example, it had never been clear whether or not the dopamine transporter is composed of multiple or distinct polypeptide subunits. The cloning results show that a single cDNA for a single protein has the power to confer dopamine uptake on cells. Thus, this single polypeptide has transporter activity.

Also, it had never been clear whether or not cocaine actually bound to the dopamine transporter or some accessory protein. The cloning results show that transfecting a cell with a cDNA for a single protein confers upon that cell both dopamine transport and cocaine binding (SHIMADA et al. 1991). These results indicate that the dopamine transporter and the cocaine binding protein are one and the same.

Several workers have shown that cocaine binding to brain tissues has multiple components. In other words, the binding curves could be explained by assuming that there were two binding sites for cocaine, one with a high affinity and low capacity and a second with a low affinity and high capacity (BOJA et al. 1994). There were many speculations as to whether this meant there were two different binding proteins for cocaine or two different states of the same binding protein. It has been shown that transfection of cells with a cDNA for the single DAT protein confers upon that cell the two binding sites for cocaine. Thus, the two binding sites must reflect something fundamental about the way the dopamine transporter protein is produced and processed. While there has been some speculation that the high affinity site represents the outward facing confirmation of the transporter and the low affinity site represents the inward facing of the transporter, no one really understands the full significance of these two binding sites.

Figure 8 shows a two-dimensional depiction of the dopamine transporter, although it is clear that the transporter protein must be folded into a three-dimensional conformation which we cannot simply deduce from the two-dimensional structure. By a series of clever site-directed mutations, UHL and colleagues (KITAYAMA et al. 1992) showed that dopamine probably binds to the transporter by forming bonds within an aspartate residue in transmembrane region one and serine residues in transmembrane region seven. The amino group of dopamine presumably interacts with the negative carboxyl group of aspartate and the hydroxyl groups on the catechol presumably interact with the serine residues by hydrogen bonding. These data imply that transmembrane regions one and seven must conform to be within an optimal space of one another, so that dopamine can properly insert. It has been further speculated that the negatively charged caboxyl group from the aspartate is attacted to the nitrogen of cocaine (KITAYAMA et al. 1992).

Thus, substantial progress has been made in understanding the molecular machinery that cocaine utilizes. While we have come to some understanding of the structure, much remains to be done. Site-directed mutagenesis studies have their limitations and structural data by X-ray crystalography would be very helpful. Transporter research is making great progress and our understanding of the binding of cocaine to the transporter will undoubtedly progress hand in hand with other advances.

An interesting issue is the regulation of the dopamine transporter. Members of this family show highly conserved consensus phosphorylation sites. The data suggest that phosphorylation is an important regulator of the protein although definitive evidence has not yet appeared. Also, it has been shown that chronic administration of cocaine causes changes in the messenger RNA for the dopamine transporter (Xia et al. 1992; Cerruti et al. 1994). However, these studies are just beginning and much new information is needed to determine the functional significance of these changes.

II. Effect of Chronic Cocaine Administration and Withdrawal

In human subjects, cocaine has long been known for its lack of a dramatic physical withdrawal syndrome. Nevertheless, the current literature suggests that phenomena such as craving and behavioral conditioning underlie relapse. The occurrence of biochemical changes in the brains of humans have only been inferred from brain imaging studies (see above). However, studies in animals have provided an intriguing series of observations indicating that chronic administration of cocaine can have profound effects on the neurochemistry of dopaminergic neurons as well as in postsynaptic neurons. This has been discussed above.

If cocaine is administered for a prolonged period of time, several changes can be observed in the brain in rat studies. For example, neurofilament proteins are decreased in dopaminergic neurons and tyrosine hydroxylase is increased in cell body regions but decreased in nerve terminal regions. The change in tyrosine hydroxylase could be due to the changes in neurofilament proteins as they may be involved in the axonal transport of tyrosine hydroxylase to the nerve terminal. In postsynaptic regions such as the nucleus accumbens, changes in adenylate cyclase, protein kinase A and AP-1 binding proteins have been found. Also, changes in postsynaptic receptors in these regions have been noted. As these studies have recently been summarized (Nestler 1992), we will not explore these topics further here. Since these biochemical changes are found during a withdrawal period, they are quite interesting as they may underlie craving and other phenomena associated with relapse to drug abuse which have been described above. However, much remains to be done before this association can be made.

F. Conclusions

Due to the availability of a variety of marvelous and sophisticated techniques and animal models, the opportunity for interdisciplinary work in addiction is greater than ever. Interesting hypotheses, some involving mesolimbic dopamine and reward pathways, are available to guide directions.

Questions for the future are obvious but difficult. What brain regions are involved in cocaine (and other drugs) abuse and dependence? What are the neuronal circuits in these regions? What are the biochemical changes that occur, and can we develop treatments for drug abuse and dependence based on this knowledge? Integrated techniques from different disciplines will be essential in the pursuit of answers to these questions.

References

Ackerman JM, White FJ (1990) A10 somatodendritic dopamine autoreceptor sensitivity following withdrawal from repeated cocaine treatment. Neurosci Lett 117:181–187

Amara S, Kuhar MJ (1993) Neurotransmitter transporters: recent progress. Annu Rev Neurosci 16:73–93

Anderson PH (1987) Biochemical and pharmacological characterization of [^{3}H]GBR 12935 binding in vitro to rat striatal membranes: labeling of the dopamine uptake complex. J Neurochem 48:1887–1896

Arndt IO, Dorozynsky L, Woody GE, McLellan AT, O'Brien CP (1992) Desipramine treatment of cocaine dependence in methadone-maintained patients. Arch Gen Psychiatry 49:888–893

Belliveau JW, Kennedy DN, McKinstry RC, Buchbinder BR, Weisskoff, RM, Cohen MS, Vevea JM, Brady TJ, Rosen BR (1991) Functional mapping of the human visual cortex by magnetic resonance imaging. Science 254:716–719

Boja JW, McNeill RM, Lewin AH, Abraham P, Carroll FI, Kuhar MJ (1992a) Selective dopamine transporter inhibition by cocaine analogs. Neuroreport 3:984–986

Boja JW, Mitchell WM, Patel A, Kopajtic TA, Carroll FI, Lewin AH, Abraham P, Kuhar MJ (1992b) High affinity binding of [^{125}I]RTI-55 to dopamine and serotonin transporters in the rat brain. Synapse 12:27–36

Boja JW, Markham L, Patel A, Uhl G, Kuhar MJ (1992c) Expression of a single-dopamine transporter cDNA can confer two cocaine binding sites. Neuroreport 3:247–248

Boja, JW, Vaughan R, Patel A, Shaya EK, Kuhar MJ (1994) The dopamine transporter. In: Niznik H (ed) Dopamine receptors and transporters. Dekker, New York, pp 611–644

Bradberry C, Roth R (1989) Cocaine increases extracellular dopamine in rat nucleus accumbens and ventral tegmental area as shown by in vivo dialysis. Neurosci Lett 103:97–102

Brucke T, Kornhuber J, Angelberger P, Asenbaum S, Frassine H, Podreka I (1993) SPECT imaging of dopamine and serotonin transporters with [^{123}I]β-CIT. Binding kinetics in the human brain. J Nural Transm 94:137–146

Carboni E, Imperato A, Perezzani L, DiChiara G (1989) Amphetamine, cocaine, phencyclidine and nomifensine increase extracellular dopamine concentrations preferentially in the nucleus accumbens of freely moving rats. Neuroscience 28:653–661

Carroll FI, Lewin AH, Boja JW, Kuhar MJ (1992) Cocaine receptor: biochemical characterization and structure-activity relationships of cocaine analogues at the dopamine transporter. J Med Chem 35:969–981

Cerruti C, Pilotte NS, Uhl G, Kuhar MJ (1994) Reduction in dopamine tranpsorter mRNA after cessation of repeated cocaine administration. Mol Brain Res, 22:132–138

Damsma G, Wenkstern D, Pfaus JG, Phillips A, Fibiger HC (1992) Sexual behavior increases dopamine transmission in the nucleus accumbens and striatum of male rats: comparison with novelty and locomotion. Behav Neurosci 106(1):181–191

Davis JD (1966) A method for chronic intravenous infusion in freely moving rats. Exp Anal Behav 9:385–387

Davis WM, Smith SG (1975) Effect of haloperidol on (+)-amphetamine self-administration. J Pharm Pharmacol 27:540–542

De la Garza R, Johanson CE (1982) Effects of haloperidol and physostigmine on self-administration of local anesthetics. Pharmacol Biochem Behav 17:1295–1299

de Wit H, Wise RA (1977) Blockade of cocaine reinforcement in rats with the dopamine receptor blocker pimozide, but not with the noradrenergic blockers phentolamine or phenoxybenzamine. Can J Psychol 31:195–203

Di Chiara G, Imperato A (1988) Drugs abused by humans preferentially increase synaptic dopamine concentrations in the mesolimbic system of freely moving rats. Proc Natl Acad Sci USA 85:5274–5278

Dworkin SI, Porrino LJ, Smith JE (1992) Importance of behavioral controls in the analysis of ongoing events. In: Frascella J, Brown RM (eds) Neurobiological approaches to brain-behavior interaction. NIDA Res Monogr 124:173–188

Ehn E, Larsson B (1979) Properties of an antiscratch-layer-free X-ray film for the autoradiographic registration of tritium. Science Tools 26:24–29

Esposito RU, Motola AHD, Kornetsky C (1978) Cocaine: acute effects on reinforcement thresholds for self-stimulation behavior to the medical forebrain bundle. Pharmacol Biochem Behav 8:437–439

Ettenberg A, Pettit HO, Bloom FE, Koob GF (1982) Heroin and cocaine intravenous self-administration in rats: mediation by separate neural systems. Psychopharmacol Bull 78:204–209

Falk JL (1993) Schedule-induced drug self-administration. In: Van Haarew F (ed) Methods in behavioral pharmacology. Elsevier, Amsterdam

Farde L, Halldin C, Muller L, Suhara T, Karlsson P, Hall H (1994) PET study of [^{11}C]β-CIT binding to monoamine transporters in the monkey and human brain. Synapse 16:93–103

Fowler JS, Volkow ND, Wolf AP, Dewey SL, Schyler DJ, MacGregor RR, Hitzemann R, Logan J, Bendreim B, Gatley SJ, Christman D (1989) Mapping cocaine binding sites in human and baboon brain in vivo. Synapse 4:371–377

Frost JJ, Rosier AJ, Reich SG, Smith JS, Ehlers MD, Snyder SH, Ravert HT, Dannals RF (1993) Positron emission tomograpic imaging of the dopamine transporter with ^{11}C-WIN 35,428 reveals marked declines in mild Parkinson's Disease. Ann Neurol 34:423–431

Gawin FH (1986) Neuroleptic reduction of cocaine-induced paranoia but not euphoria? Psychopharmacology 90:142–143

Gawin FH, Riordan C, Kleber HD (1985) Methylphenidate treatment of cocaine abusers without attention deficit disorder: a negative report. Am J Drug Alcohol Abuse 11:193–197

Gawin FH, Allen D, Humblestone B (1989a) Outpatient treatment of "crack" cocaine smoking with flupenthixol decanoate. Arch Gen Psychiatry 46:322–325

Gawin FH, Morgan C, Kosten TR, Kleber HD (1989b) Double-blind evaluation of the effect of acute amantadine on cocaine craving. Psychopharmacology 97:402–403

Giros B, Caron MG (1993) Molecular characterization of the dopamine transporter. TiPS 14:43–49

Giros B, Mestikawy SE, Bertrand L, Caron MG (1991) Cloning and functional characterization of a cocaine-sensitive dopamine transporter. FEBS Lett 295 [1,2,3]:149–154

Goeders NE, Dworkin SI, Smith JE (1986) Neuropharmacological assessment of cocaine self-administration into the medial prefrontal cortex. Pharmacol Biochem Behav 24:1429–1440

Grigoriadis DE, Wilson AA, Lew R, Sharkey JS, Kuhar MJ (1989) Dopamine transport sites selectively labeled by a novel photoaffinity probe: ^{125}I-DEEP. J Neurosci 9:2664–2670

Guastella J, Nelson N, Nelson H, Czyzyk L, Keynan S, Miedel MC, Davidson N, Lester HA, Kanner BI (1990) Cloning and expression of a rat brain GABA transporter. Science 249:1303–1306

Gunne LM, Anggard E, Jonsson LE (1972) Clinical trials with amphetamine blocking drugs. Psychiatry Neurol Neurochir (Amst) 75:225–226

Henry DJ, Greene MA, White FJ (1989) Electrophysiological effects of cocaine in the mesoaccumbens dopamine system: repeated administration. JPET 251:833–839

Herkenham M, Pert CB (1980) In vitro autoradiography of opiate receptors in rat brain suggests loci of "opiatergic" pathways. Proc Natl Acad Sci 77:5532–5536

Hurd YL, Weiss F, Koob GF, And N-E, Ungerstedt U (1989) Cocaine reinforcement and extracellular dopamaine overflow in rat nucleus accumbens: an in vivo microdialysis. Brain Res 498:199–203

Hyttel J (1982) Citalopram – pharmacological profile of a specific serotonin uptake inhibitor with antidepressant activity. Prog Neuropsychopharmacol Biol Psychiatry 6:277–295

Imperato A, Mele A, Scrocco MG, Puglisi-Alegra S (1992) Chronic cocaine alters limbic extracellular dopamine. Neurochemical basis for addiction. Eur J Pharmacol 212:299–300

Innis RB, Seibyl JP, Scanley BE, Laruelle M, Abi-Dargham A, Wallace E, Baldwin RM, Zea-Ponce Y, Zoghbi S, Wang S, Gao Y, Neumeyer JL, Charney DS, Hoffer PB, Marek KL (1993) Single-photon emission-computed tomography imaging demonstrates loss striatal dopamine transporters in Parkinson disease. Proc Natl Acad Sci USA 90:11965–11969

Jaffe JH (1990) Drug addiction and drug abuse. In: Gilman AG, Rall TW, Nies AS, Taylor P (eds) The pharmacological basis of therapeutics. Permagon, Oxford, pp 522–573

Johanson CE, Balster RL (1978) A summary of the results of a drug self-administration study using substitution procedures in rhesus monkeys. Bull Narc 30:43–54

Johanson CE, de Wit H (1989) The use of choice procedures for assessing the reinforcing properties drugs in human. In: Fishman MW, Mello NK (eds) Assessing the abuse liabilaty of drugs in humans. National Institute on Drug Abuse Monography Series, US Govt. Printing Office, Washington, vol 92, pp 171–209

Johanson CE, Schuster CR (1981) Animal models of drug self-administration In: Mello NK (ed) Advances in substance abuse: behavioral and biological research, vol II. JAI Press, Greenwich, pp 219–297

Kalivas PW, Duffy P (1990) Effect of acute and daily cocaine treatment on extracellular dopamine in the nucleus accumbens. Synapse 5:48–58

Kilty J, Lorang D, Amara SG (1991) Cloning and expression of a cocaine-sensitive rat dopamine transporter. Science 254:578–579

Kitayama S, Shimada S, Xu H, Markham L, Donovan DM, Uhl GR (1992) Dopamine transporter site-directed mutations differentially alter substrate transport and cocaine binding. Proc Natl Acad Sci USA 89:7782–7785

Koob GF (1992) Drugs of abuse: anatomy, pharmacology, and function of reward pathways. TIPS 13:177–184

Koob GF, Bloom FE (1988) Cellular and molecular mechanisms of drug dependence. Science 242:715–723

Kornetsky C, Duvauchelle C (1994) Dopamine, a common substrate for the rewarding effects of brain-stimulation reward (BSR), cocaine, and morphine. In: Neurobiological models for evaluating mechanisms underlying cocaine addiction and potential pharmacotherapies for treating cocaine abuse. NIDA Research Monographs (to be published)

Kornetsky C, Esposito RU, McLeau S, Jacobson O (1979) Intracranial self-stimulation thresholds. A model for the hedonic effects of drugs of abuse. Arch Gen Psychiatry 36:289–292

Kosten TR, Morgan CM, Falcione J, Schottenfeld RS (1992) Pharmacoltherapy for cocaine-abusing methadone-maintained patients using amantadine or desipramine. Arch Gen Psychiatry 49:894–898

Kuhar MJ, DeSouza EB (1989) Autoradiographic imaging: localization of binding sites other than neurotransmitter receptors. In: Ottoson D, Rosten W (eds) Visualization of brain functions. Stockton, pp 57–66

Kuhar MJ, Unnerstall JR (1990) Receptor autoradiography. In: Yamamura HI et al. (eds) Methods in neurotransmitter receptor analysis. Raven, New York, pp 177–218

Kuhar MJ, DeSouza EB, Unnerstall JR (1986) Neurotransmitter receptor mapping by autoradiography and other methods. Annu Rev Neurosci 9:27–59

Kuhar MJ, Ritz MC, Boja JW (1991) The dopamine hypothesis of the reinforcing properties of cocaine. Trends Neurosci 14:299–302

Leith NJ, Barrett RJ (1976) Amphetamine and the reward system: evidence for tolerance and post-drug depression. Psychopharmacology 46:19–25

Lew R, Grigoriadis DE, Wilson A, Boja JW, Simantov R, Kuhar MJ (1991) Dopamine transporter: deglycosylation with exo- and endoglycosidases. Brain Res 539:239–246

London ED, Morgan MJ (1993) Positron emission tomographic studies on the acute effects of psychoactive drugs on brain metabolism and mood. In: &: London ED (ed) Imaging Drug Action in the brain. CRC, Boca Raton, pp 265–280

London ED, Cascella NG, Wong DF, Phillips RL, Dannals RF, Links JM, Herning R, Grayson R, Jaffe JH, Wagner HN (1990a) Cocaine-induced reduction of glucose utilization in human brain. Arch Gen Psychiatry 47:567–574

London ED, Broussolle EPM, Links JM, Wong DF, Cascella NG, Dannals RF, Sano M, Herning R, Snyder FR, Rippetoe LR, Toung TJK, Jaffe JH, Wagner HN (1990b) Morphine-induced metabolic changes in human brain. Arch Gen Psychiatry 47:73–81

Lyness WH, Friedle NM, Moore KE (1979) Destruction of dopaminergic nerve terminals in nucleus accumbens: effect on d-amphetamine self-administration. Pharmacol Biochem Behav 11:553–556

Madras BK (1994) ^{11}C-WIN 35428 for detecting dopamine depletion in mild Parkinson's disease. Ann Neurol 35(3):376–379

Madras BK, Fahey MA, Bergman J, Canfield DR, Spealman RD (1989) Effects of cocaine and related drugs in nonhuman primates. I. [^{3}H] Cocaine binding sites in caudate-putamen. J PET 251:131–141

Markou A, Koob GF (1991) Postcocaine anhedonia: an animal model of cocaine withdrawal. Neuropsychopharmacology 4:17–26

Martin-Iverson MT, Szostak C, Fibiger HC (1986) 6-Hydroxydopamine lesions of the medial prefrontal cortex fail to influence intravenous self-administration of cocaine. Psychopharmacology 88:310–314

Mestler EJ (1992) Molecular mechanisms of drug addiction. J Neurosci 12:2439–2450

Olds J, Milner P (1954) Positive reinforcement produced by electrical stimulation of septal area and other regions of rat brain. J Comp Physiol Psychol 47:419–427

Pacholczyk T, Blakely RD, Amara SG (1991) Expression cloning of a cocaine- and antidepressant-sensitive human noradrenaline transporter. Nature 350:350–354

Pettit H, Justice J Jr (1989) Dopamine in the nucleus accumbens during cocaine self-administration as studied by in vivo microdialysis. Pharmacol Biochem Behav 34:899–904

Pfaus JG, Damsma G, Nomikos GG, Wenkstern DG, Blaha CD, Phillips AG, Fibiger HC (1990) Sexual behavior enhances central dopamine transmission in the male rat. Brain Res 530:345–348

Phillips AG, Goury A, Fiorino D, LePiane FG, Brown E, Fibiger HC (1992) Self-stimulation of the ventral tegmental area enhances dopamine release in the nucleus accumbens: a microdialysis study. In: Kalivas PW, Samson HH (eds)

The neurobiology of drug and alcohol addiction. New York Academy of Sciences, New York (Ann NY Acad Sci, vol 256)

Pilotte NS, Sharpe LG, Kuhar MJ (1994) Withdrawal of repeated intravenous infusions of cocaine persistently reduces binding to dopamine transporters in the nucleus accumbens of lewis rats. J PET 269(3):963–969

Preston KL, Sullivan JT, Strain EC, Bigelow GE (1992) Effects of cocaine alone and in combination with bromocriptine in human cocaine abusers. JPET 262:279–291

Preston KL, Sullivan JT, Berger P, Bigelow GE (1993) Effects of cocaine alone and in combination with mazindol in human cocaine abusers. JPET 267:296–307

Peisner ME, Jones BE (1976) Role of noradrenergic dopaminergic processes in amphetamine self-administration. Pharmacol Biochem Behav 5:477–482

Ritz MC, Lamb RJ, Goldberg SR, Kuhar MJ (1987) Cocaine receptors on dopamine transporters are related to self-administration of cocaine. Science 237:1219–1223

Roberts DCS, Koob GF (1982) Disruption of cocaine self-administration following 6-hydroxydopamine lesions of the ventral tegmental area in rats. Pharmacol Biochem Behav 17:901–904

Roberts DCS, Koob GF, Klonoff P, Fibiger HC (1980) Extinction and recovery of cocaine self-administration following 6-hydroxydopamine lesions of the nucleus accumbens. Pharmacol Biochem Behav 12:781–787

Robinson T, Berridge KC (1993) The neural basis of drug craving: and incentivesensitization theory of addiction. Brain Res Rev 18:247–291

Rossetti ZL, Hmaidan Y, Gessa GL (1992) Marked inhibition of mesolimbic dopamine release: a common feature of ethanol, morphine, cocaine and amphetamine abstinence in rats. Eur J Pharmacol 221:227–234

Roth LJ, Diab IM, Watanabe M, Dinerstein RJ (1974) A correlative radioautographic, flourescent, and histochemical technique for cytopharmacology. Mol Pharmacol 10:986–998

Rotter A, Birdsall NJM, Burgen ASV, Field PM, Hulme EC, Raisman G (1979) Muscarinic receptors in the central nervous sustem of the rat. I. Techniques for autoradiographic localization of the binding of [^{3}H]propylbenzibylcholine mustatd and its distribution in the forebrain. Brain Res Rev 1:141–166

Shaya EK, Scheffel U, Dannals RF, Ricaurte GA, Carroll FI, Wagner HN, Kuhar MJ, Wong DF (1992) In vivo imaging of dopamine reuptake sites in the primate brain using single photon emission computed tomography (SPECT) and iodine-123 labeled RTI-55. Synapse 10:169–172

Sherer MA, Kumor KM, Jaffe JH (1989) Effects of intravenous cocaine are partially attenuated by haloperidol. Psychiatry Res 27:117–125

Shimada S, Kitayama S, Lin C-L, Patel A, Nathankumar E, Gregor P, Kuhar M, Uhl G (1991) Cloning and expression of a cocaine-sensitive dopamine transporter complementary DNA. Science 254:576–578

Smith CA, Wood EJ (1991) Biological molecules. Chapman and Hall, New York

Stolerman I (1992) Drugs of abuse: behavioral principals, methods and terms. TIPS 13:170–176

Tedroff J, Aquilonius SM, Hartvig P, Lundquist H, Gee AG, Uhlin J, Langstrom B (1988) Monoamine re-uptake sites in the human brain evaluated in vivo by means of ^{11}C-nomifensine and positron emission tomography: the effects of age and Parkinson's disease. Acta Neurol Scand 77:192–201

Thompson T, Schuster CR (1964) Morphine self-administration, food-reinforced and avoidance behaviors in rhesus monkeys. Psychopharmacology 5:87–94

Uhl GR, Hartig PR (1992) Transporter explosion: update on uptake. TiPS 13(12):421–425

Usdin TB, Mezey E, Chen C, Brownstein MJ, Hoffman BJ (1991) Cloning of the cocainesensitive bovine dopamine transporter. Proc Natl Acad Sci 88:11168–11171

Volkow ND, Fowler JS, Wolf AP, Hitzemann R, Dewey S, Bendreim B, Alpert R, Hoff A (1991) Changes in brain glucose metabolism in cocaine dependence and withdrawal. Am J Psychiatry 148:621–626

Volkow ND, Hitzemann R, Wang G-J, Fowler JS, Wolf AP, Dewey S, Handlesman L (1992a) Long-term frontal brain metabolic changes in cocaine abusers. Synapse 11:184–190

Volkow ND, Fowler JS (1992b) Neuropsychiatric disorders: investigation of schizophrenia and substance abuse. Semin Nucl Med 22(4):254–267

Volkow ND, Fowler JS, Wang G-J, Hitzemann R, Logan J, Schyler DJ, Dewey SL, Wolf AP (1993) Decreased dopamine D2 receptor availability is associated with reduced frontal metabolism in cocaine abusers. Synapse 14:169–177

Wages SA, Church WH, Justics JB Jr (1986) Sampling considerations for on-line microbore liquid chromatography of brain dialysate. Anal Chem 58:1649–1656

Wangquier A, Niemegeers CJE (1974) Intracranial self-stimulation in rats as a function of various stimulus parameters. Psychopharmacologia 38:201–210

Weddington WW Jr, Brown BS, Hartsen CA, Hess JM, Mahaffey JR, Kolar AF, Jaffe JH (1991) Comparison of amantadine and desipramine combined with psychotherapy for treatment of cocaine dependence. Am J Drug Alcohol Abuse 17:137–152

Weeks JR (1962) Experimental morphine addiction: method for automatic intravenous injections in unrestrained rats. Science 138:143–144

Wilson MC, Schuster CR (1972) The effects of chlorpromazine on psychomotor stimulant self-administration in the rhesus monkey. Psychopharmacology 26:115–126

Wise RA, Bozarth MA (1987) A psychomotor stimulant theory of addiction. Psychol Rev 94:469–492

Wood RW (1979) Reinforcing properties of inhaled substances. Test methods for definition of effects of toxic substances on behavior and neuromotor function. neurobehav toxicol 1:67–72

Wood RW, Weiss B (1978) Volatile anesthetic self-administration by the squirrel monkey. In: Sharp and M Carroll (eds) Voluntary inhalation of industrial solvents. DHEW Publication no (ADM)79-779. US Government Printing Office, Washington, pp 355–362

Wood RW, Grubman J, Weiss B (1977) Nitrous oxide self-administration by the squirrel monkey. JPET 202:2491–499

Woolverton WL, Goldberg LI, Ginos JZ (1984) Intravenous self-administration of dopamine receptor agonists by rhesus monkeys. JPET 230:678–683

Wullner U, Pakzaban P, Brownel AL, Hantraye P, Burns L, Shoup T, Elmaleh D, Petto AJ, Spealman RD, Brownell GL, Isacson O (1994) Dopamine terminal loss and onset of motor symptoms in MPTP-treated monkeys: a positron emission tomography study with ^{11}C-CFT. Exp Neurol 126:305–309

Xia Y, Goebel DJ, Kapatos G, Bannon MJ (1992) Quantitation of rat dopamine transporter mRNA. Effects of cocaine treatment and withdrawal. J Neurochem 59:1179–1182

Yamamura HI, Enna SJ, Kuhar MJ (eds) (1990) Methods in neurotransmitter receptor analysis. Raven, New York

Yanagita T, Deneau GA, Seevers MH (1965) Evaluation of pharmacologic agents in the monkey by long-term intravenous self- or programmed administration. Excerpta Med Int Congr Ser 87:453–457

Yokel RA, Wise RA (1978) Amphetamine-type reinforcement by dopaminergic agonists in the rat. Psychopharmacology 58:289–296

Young III WS, Kuhar MJ (1979) A new method for receptor autoradiography: [3]H opioid receptor labelling in mounted tissue sections. Brain Res 179:255–270

II. Molecular, Behavioral, and Human Pharmacology of Dependence and Consequences

CHAPTER 3
Marihuana

D.R. COMPTON, L.S. HARRIS, A.H. LICHTMAN, and B.R. MARTIN

A. Introduction

Among the drugs of abuse which are regulated under the United States Controlled Substances Act and the International Conventions, none has created more intense public debate and controversy than marihuana. Marihuana is one of many names given to the leaves and flowering tops of the plant *Cannabis sativa*. The plant grows in all temperate regions of this planet and has been used commercially as a source of fiber and oil. Wherever the plant grows people have learned to ingest the material for its introxicating effects. The usual routes of administration are by mouth or smoking. It has been estimated that, worldwide, more than one hundred million individuals are regular users of the plant material. However, accurate data on this situation are not available. In the United States, use data have been collected on a regular basis using two large surveys, the National Household Survey on Drug Abuse (SUBSTANCE ABUSE AND MENTAL HEALTH SERVICES ADMINISTRATION 1994) and the Monitoring the Future Survey (JOHNSTON et al. 1994) which covers eighth, tenth, and twelfth grade students in public and private schools. Figure 1 presents the data over time for lifetime, annual, 30 day, and daily use of marihuana among high school seniors in the United States (JOHNSTON et al. 1994). As can be seen, use peaked from 1978 to 1980 and had been declining slowly up to 1992. From 1992 to 1993 there was a significant increase in all use categories. Even more disturbing was the fact that very similar trend data were reported for eighth and tenth graders. This was matched in 1992 and 1993 by a decrease in the reported "perceived risk" from use of the drug.

Data from the Household Survey revealed similar findings among those 12–17 years old. It is estimated that in 1993, 600000 individuals in this age group used marihuana weekly. There was also an increase in reporting that "obtaining marihuana is fairly or very easy." These trends should give us early warning of increased future public health problems. This is especially true when one looks at the increasing concentrations of Δ^9-tetrahydrocannabinol (Δ^9-THC), the psychoactive principal found in confiscated samples of the plant material. This is illustrated in Fig. 2. During the years of peak use in the United States (1976–1980) the average Δ^9-THC content of confiscated cannabis was about 1.5%. There was a steady increase

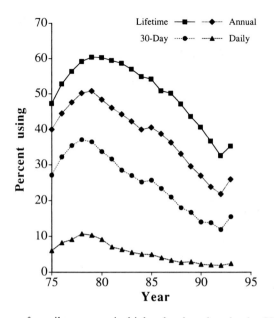

Fig. 1. Prevalence of marihuana use in high school seniors in the U.S. Incidence is defined as having used marihuana at least once during their lifetime, the last year, or the last 30 days or by daily use in the last 30 days

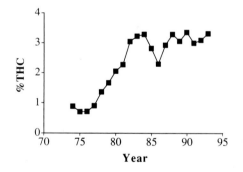

Fig. 2. Δ^9-Tetrahydrocannabinol (Δ^9-THC) content in confiscated marihuana in the U.S. The data are expressed as an average of the percentage in individual samples

to the mid-1980s, when it stabilized at 3.0%–3.5%, double that of earlier years (ELSOHLY and ROSS 1994). It should be noted, however, that seized samples of buds and sinsemilla (flowering tops of the female *Cannabis* plant) have considerably higher concentrations of Δ^9-THC. Indeed, concentrations of Δ^9-THC as high as 20%–30% have been reported in individual samples. Thus, marihuana with a very high concentration of psychoactive Δ^9-THC is regularly appearing on the street in the United States. We have every reason to believe that a similar situation exists internationally.

The general structure of the active principle of cannabis was elucidated in the 1940s. Indeed, very potent psychoactive cannabinoids were synthesized by ADAMS and colleagues. However, it remained for MECHOULAM and colleagues, in the 1960s, to first isolate in pure form and identify $(-)$-*trans*-Δ^9-tetrahydrocannabinol (Δ^9-THC) as the molecule primarily responsible for the psychoactive properties of the plant material. For decades the pharmacological activity of the cannabinoids was attributed to some nonspecific mechanism usually associated with its lipid solubility and disruption of cell membranes. A large body of structure-activity data and the demonstration of stereoselectivity gradually led to the postulation of more specific mechanisms. In recent years there has been an explosive advance in our knowledge of cannabis action. Specific binding sites have been demonstrated, receptors cloned and sequenced, purported endogenous ligands isolated and identified and, most recently, a competitive antagonist has been reported. This review will provide a relatively extensive and critical examination of these recent findings and will attempt to put them in context with our previous knowledge of the fascinating properties of this natural product.

B. Cellular and Molecular Effects

I. Neurochemistry

1. Effects on Neurotransmitters

a) Traditional Monoamine and Cholinergic Systems

Cannabinoids affect a wide variety of neurotransmitter pathways in the central nervous system and several of these share common second messenger systems, thus providing potential common sites for biochemical interactions mediated by the cannabinoid receptor. That cannabinoids can potentiate the actions of norepinephrine or acetylcholine by altering their receptors or second messenger systems has been the subject of numerous reviews (DEWEY 1986; PERTWEE 1990, 1992). However, neither adrenergic, dopaminergic, serotonergic or cholinergic agents (agonist or antagonist) bind directly to the CB_1 receptor (HOWLETT et al. 1992). The CB_1 receptor is the cannabinoid receptor primarily located in brain neuronal tissue, discussed in detail below.

Δ^9-THC decreases the release of acetylcholine from frog nerve via a presynaptic action, which was proposed to occur due to a decrease in the influx of calcium into nerve terminals (KUMBARACI and NASTUK 1980). Cannabinoids either suppress (NIEMI 1979), enhance (TURKANIS and KARLER 1986), or produce biphasic effects on neuronal transmission (TRAMPOSCH et al. 1981). Cannabinoids reportedly interact synergistically with cholinergic agonists in the production of catalepsy, tremor, circling, salivation, lacrimation, hypothermia and drinking (PERTWEE 1990).

b) Dopamine and the Reward System

Cannabinoids have been shown to enhance the formation of dopamine (DA) and reported to stimulate the release of DA from rat corpus striatum, nucleus accumbens and medial prefrontal cortex. Enkephalinergic neurons synapse upon DA neurons in the nucleus accumbens, the site proposed to modulate the reward system for all addicting drugs. Thus, drugs that alter opioid activity in this region alter the release of DA, which may in turn underlie the rewarding properties of the drugs. Theoretically, naloxone should attenuate the rewarding properties of all addicting drugs. The interaction of cannabinoids with opioids in reward mechanisms in the brain has recently been reviewed (GARDNER 1992; GARDNER and LOWINSON 1991). Data indicate that cannabinoids interact with opioids allosterically, either presynaptically on the enkephalinergic neuron or on the opioid receptor of the DA neuron, to produce the reinforcing effects. This research endeavor has been particularly provocative in that traditionally this reward system has been closely linked to agents with strong reinforcing properties, such as morphine and cocaine. The reinforcing properties of cannabinoids have been more difficult to characterize and quantitate. Hence, these findings provide an additional avenue of pursuing the etiology of marihuana self-administration and placing it in the context of agents with high abuse potential. Thus, opioid/cannabinoid interactions may play an important role in the subjective effects of the cannabinoids, in addition to other pharmacological effects, as also discussed below.

Dopaminergic regulation of cannabinoid receptor mRNA levels in rat caudate putamem has been observed (MAILLEUX and VANDERHAEGHEN 1993a). Furthermore, recent desciptions of the interactions between the DA system and the endogenous cannabinoid, anandamide (discussed below), provide further credence for the existence of an interrelationship between the dopaminergic and cannabinoid systems (CHEN et al. 1993; GARDNER and LOWINSON 1991; NAVARRO et al. 1993; RODRIGUEZ DE FONSECA et al. 1992b).

c) Amino Acid Transmitters

A variety of cannabinoid-mediated effects have been attributed to modulation of amino neurotransmitter systems (PERTWEE 1990, 1992). Cannabinoids have been reported to enhance the turnover of γ-aminobutyric acid (GABA). Interpretation of the actions of cannabinoids on amino acid neurotransmitter synthesis has not been straightforward, because there is evidence that they inhibit as well as stimulate neurotransmitter reuptake. Evidence also suggests that cannabinoids can potentiate the actions of GABA by altering receptors or second messenger systems. Cannabinoids reportedly interact with GABA agonists in the production of catalepsy, excitement, hypothermia and antinociception (PERTWEE 1990). Recently, it has been reported that anandamide acts in a fashion similar to that of other cannabinoids to enhance GABAergic transmission (WICKENS and PERTWEE 1993).

d) Opioid Interactions

The cannabinoids produce effects which have much in common with the opioids, such as antinociception, hypothermia, catalepsy (in rats), cross-tolerance to morphine, and attenuation of naloxone-precipitated withdrawal from opiates. With regard to withdrawal symptoms, the interaction of the cannabinoids with the opiates is very ambiguous. As early as 1942, opioid withdrawal in humans was reported to be attenuated by marihuana administration (ADAMS 1942). Conversely, the irreversible μ antagonist chlornaltrexamine has been shown to decrease tolerance to Δ^9-THC (TULUNAY et al. 1981). Blockade by cannabinoids of naloxone-precipitated withdrawal (jumping in opioid-tolerant animals) has been summarized (MARTIN 1986; PERTWEE 1992). However, in morphine-tolerant mice, a series of both psychoactive and nonpsychoactive cannabinoids prevented withdrawal jumping (BHARGAVA 1976). Similarly, in morphine-tolerant rats, Δ^9-THC and cannabinol (CBN), but not cannabidiol (CBD, a marihuana constituent lacking pscyhoactive properties), attenuated withdrawal scores of a variety of behaviors (CHESHER and JACKSON 1985). While these findings should not be ignored, they raise the question whether this is a specific action of cannabinoids on opioid withdrawal and strongly suggest that the cannabinoid receptor is not involved. In a morphine-tolerant guinea-pig ileum model, Δ^9-THC attenuated withdrawal both in vivo and in vitro (using an ileum preparation). The in vitro effects of Δ^9-THC are presumably due to a decrease in acetylcholine release. Blockade of naloxone-precipitated withdrawal jumping has been shown to occur when cannabinoids are administered up to 24 h prior to the precipitated event, and specificity suggested by the fact that cannabinoids failed to alter the withdrawal (jumping behavior) following chronic amphetamine treatment (BHARGAVA 1978). However, it is unclear why Δ^9-THC was effective 24 h prior, while other cannabinoids were only effective for up to a 2 h pretreatment period. In morphine-tolerant rats, Δ^9-THC attenuated naloxone-precipitated "wet dog" shakes and defecation, but failed to attenuate any other behaviors associated with opioid withdrawal (HINE et al. 1975). Attenuation of morphine withdrawal signs by Δ^9-THC, nantradol, and nabilone have also been observed in the dog (GILBERT 1981). The mechanism by which cannabinoids attenuate opioid withdrawal in humans or in other animal species is unclear. However, cannabinoid-induced blockade of the release of various mediators of opioid withdrawal, such as acetylcholine and norepinephrine, has been proposed. Another possibility includes interactions of the cannabinoids and opioids with a common, possibly non-CB_1 related, second messenger system.

As in withdrawal studies, investigations of the cross-tolerance between the cannabinoids and opioids have also resulted in ambiguity. Although symmetrical cross tolerance between the opioids and cannabinoids have been shown in some studies (HINE 1985; KAYMAKCALAN and DENEAU 1972), it was not observed in other studies measuring analgesia and/or heart rate. Cross-tolerance was not observed in pigeons (McMILLAN and DEWEY 1972; McMILLAN et al. 1971), or in rats in nonanalgesic evaluations (NEWMAN et al. 1974), or in shock avoidance measures (NEWMAN et al. 1974). Asymmetrical cross-tolerance was observed whereby, in Δ^9-THC-tolerant mice, tolerance to the hypothermic effect of morphine was observed, though this was not so for antinociception (BLOOM and DEWEY 1978). In morphine-tolerant mice, cross-tolerance to the antinociceptive effects of Δ^9-THC has been demonstrated, but there was no cross-tolerance to the hypothermic effect of Δ^9-THC. Similar asymmetric cross-tolerance was observed using measures of motor activity in rats (TULUNAY et al. 1982).

Similarities between the opioids and cannabinoids include their antinociceptive properties. Although there were early reports of cannabinoid-induced antinociceptive properties (BLOOM and DEWEY 1978; SOFIA et al. 1973), it became clear that route of administration plays a critical role in the expression of this pharmacological property (MARTIN 1985a). Early results on the antinociceptive effects of the cannabinoids following injection into spinal sites (GILBERT 1981; YAKSH 1981), which has recently been investigated more extensively (WELCH 1993; WELCH and STEVENS 1992), point to the participation of spinal sites in this cannabinoid action. The antinociceptive effect of intrathecal (i.t.) CP-55,940, a potent cannabinoid (JOHNSON and MELVIN 1986), was attenuated in spinal-transected rats, indicating that cannabinoid-induced antinociception was mediated at both spinal and supraspinal sites (LICHTMAN and MARTIN 1991). However, Δ^9-THC (i.t.) produced the same degree of antinociception in mice that were spinal-transected as those that had the spinal cord intact (SMITH and MARTIN 1992), suggesting that the effects of i.t. administered Δ^9-THC in mice are spinally mediated. The possibility remains that central components may also play a role following administration by other routes.

In vivo, the opioid antagonist naloxone has been shown to attenuate the antinociceptive effects of 11-OH-Δ^9-THC (WILSON and MAY 1975). However, many investigators have shown that naloxone fails to block the effects of various parenterally administered cannabinoids (CHESHER et al. 1973; CHESHER and JACKSON 1985; MARTIN 1985a; SANDERS et al. 1979). Naloxone, administered i.t., subcutaneously (s.c.), or intracerebroventricular (i.c.v.) also failed to block the antinociception induced by a variety of i.t., i.c.v. or spinally administered cannabinoids (GILBERT 1981; WELCH 1993; WELCH and STEVENS 1992; YAKSH 1981). However, the irreversible μ antagonist chlornaltrexamine was reported to attenuate the antinociceptive and hypothermic effects of Δ^9-THC (TULUNARY et al. 1981).

It has been shown that the antinociceptive effect of Δ^9-THC and morphine are additive following intravenous (i.v.) administration, thus implying distinct mechanisms of action (GENNINGS et al. 1993). In vitro the effects of Δ^9-THC on adenylyl cyclase have been shown to be insensitive to naloxone blockade and additive with the decrease in adenylyl cyclase observed with morphine (BIDAUT-RUSSELL and HOWLETT 1988). In rat striatum, a potent cannabinoid agonist was not found to be additive with either morphine or the dopamine agonist LY 171555 in decreasing cAMP levels (BIDAUT-RUSSELL and HOWLETT 1991). Results of another study using opioids and cannabinoids, alone and in combination, indicate that the cannabinoids and opioids may alter cAMP levels via similar mechanisms. The common final pathway of cAMP modulation by cannabinoids and opioids may be the phosphorylation of similar proteins, such as synapsins I and II, which are involved in the release of neurotransmitters (CHILDERS et al. 1992). The binding of μ and δ opioids has been shown to be displaced by the cannabinoids in brain, but only at high concentrations (VAYSSE et al. 1987), whereas δ opioid binding is not displaced by cannabinoids (DEVANE et al. 1986). The cannabinoid receptor was shown to be dense in the striatum (HERKENHAM et al. 1990), which is an area also associated with a dense population of opioid receptors (YAKSH et al. 1988). It is intriguing that, despite the data suggesting independent mechanisms of action, the effects of morphine have been found to be enhanced by orally administered Δ^6- and Δ^9-THC (MECHOULAM et al. 1984).

Intrathecal administration of several cannabinoids leads to synergism with i.t. and i.c.v. administered morphine in the production of antinociception in mice (SMITH and MARTIN 1992; WELCH and STEVENS 1992). Although pretreatment with morphine ehnanced the effects of Δ^9-THC, pretreatment of the mice with naloxone (s.c. or i.t.) failed to block the antinociceptive effects of the cannabinoids indicating that the cannabinoid-induced antinociception does not occur via interactions with the μ opioid receptor. Pretreatment of mice with Δ^9-THC significantly enhanced the potency of i.t. administerred morphine. Parallel shifts in morphine dose-response curves were produced not only with Δ^9-THC, but also with 11-OH-Δ^9-THC, Δ^8-THC and levonantradol, but interestingly not by CP-55,940. Thus, the antinociceptive effects of i.t. administered morphine are enhanced by the pretreatment with some, but not all, cannabinoids active at the CB_1 receptor (WELCH and STEVENS 1992).

Recently, the blockade of cannabinoid antinociception by the κ opioid antagonist, nor-binaltorphimine (nor-BNI) has been reported (WELCH 1993). Antinociception produced by Δ^9-THC and Δ^8-THC (i.v., ED_{80} doses) was blocked by the κ antagonist nor-BNI, and the dose-effect curve for Δ^9-THC was shifted to the right in a parallel fashion. Specificity was suggested by the fact that the δ antagonist ICI 174,864 (i.t.) was without effect. Though Δ^9-THC activity was additive to that of a κ agonist (U50,488H), it was unexpected that Δ^9-THC would produce a parallel 37-fold shift to the left in the dose-effect curve of a δ agonist (DPDPE). The AD50 values for nor-BNI vs i.t. administered Δ^9-THC, Δ^8-THC, levonantradol, and CP 55,940 ranged from 1.1 to 4.5 μg/mouse. Interestingly, the effects of i.v. CP-55,940 were blocked by i.v. nor-BNI, but not i.t. or i.c.v. nor-BNI, suggesting a locus of action for nor-BNI vs i.v. cannabinoids outside the central nervous system. Selectivity is indicated by the fact that nor-BNI blocks the antinociceptive effects of i.t. Δ^9-THC, without altering responses of catalepsy, hypothermia, or hypoactivity (SMITH et al. 1993). However, the inability of naloxone to block cannabinoid-induced antinociception raises questions as to the involvement of opioid receptors in the nor-BNI effects, since all κ opioid antinociceptive effects described can be blocked by naloxone (despite the need for high doses of antagonist). In addition κ opioid binding has been shown to remain unaltered by cannabinoids (VAYSSE et al. 1987), and neither nor-BNI nor the κ antagonist U50,448H bind to the CB_1 receptor (WELCH 1993). Thus, the nature of the nor-BNI effect is unclear, though it may be related to its ability to block the antinociceptive effects of ketoralac, which acts via a reduction in the synthesis of prostaglandins. (UPHOUSE et al. 1993). The interaction of nor-BNI with prostanoid formation has not been evaluated but might provide an alternative mechanism by which nor-BNI blocks the cannabinoid antinociception.

To summarize, the interactions between opioids and cannabinoids most likely involve a combination of indirect interactions mediated through numerous neurotransmitter systems as well as direct interactions between endogenous cannabinoid and opioid systems. At present, the relationship between these two systems under normal physiological circumstances is unclear. However, recruitment of either system through either pathological pain or opioid withdrawal can be manipulated by the other system. Past efforts to meld these two drug classes into a single entity represents an oversimplification and a misunderstanding of their biochemical and cellular actions.

2. Receptors

a) Pharmacological Characteristics

The structure-activity relationship (SAR) for cannabinoids has been reviewed extensively elsewhere (RAZDAN 1986), so only a brief statement of some aspects of the structural requirements for cannabinoid activity are presented. These and other data suggested the existence of specific cannabinoid receptors before identification by ligand binding and confirmation via cloning techniques.

Enantioselectivity is an important criterion for drug-receptor interactions because enantiomers share the same physicochemical characteristics. Initial studies with Δ^9-THC failed to demonstrate complete enantioselectivity. However, almost complete enantioselectivity (MECHOULAM et al. 1988) can be achieved when highly pure enantiomers are obtained, as demonstrated pharmacologically with 11-OH-Δ^8-THC-dimethylheptyl (LITTLE et al. 1989). It had generally been assumed that an intact three-ring structure, based upon Δ^9-THC, was essential for activity since CBD (a bicyclic structure) lacks psychoactivity (RAZDAN 1986). However, a bicyclic derivative of 9-*nor*-9β-hydroxyhexahydrocannabinol, which also had a dimethylheptyl side chain (rather than the traditional pentyl group) at the C3 position (MELVIN et al. 1984), proved to have a pharmacological profile similar to that of Δ^9-THC, though much more potent. This synthetic strategy led to the development of CP 55,940 which proved to be 4–25 times more potent than Δ^9-THC depending upon the pharmacological measure (LITTLE et al. 1988). CP-55,940 and related novel bi- and tricyclic analogs have subsequently been referred to as nonclassical cannabinoids. CP-55,940 was radiolabeled in an attempt to discover a cannabinoid binding site (DEVANE et al. 1988).

The systematic approach taken in the development of cannabinoid antinociceptive agents (JOHNSON and MELVIN 1986) helped define many of the structural determinants of cannabinoid action and produced extremely potent agonists. Some of these nonclassical analogs are as much as 700 times more potent than Δ^9-THC (LITTLE et al. 1988). Other investigators prepared 11-OH-Δ^8-THC-DMH (MECHOULAM et al. 1988), which also proved to be several hundred times more potent than Δ^8-THC in several behavioral evaluations (LITTLE et al. 1989), as well as 11-OH-Δ^9-THC-DMH, which exhibited similarly high potency (MARTIN et al. 1991). In addition, a hexa-hydro-analog of the 11-OH-THC-DMH has proven to be potent and useful in ligand binding assays (DEVANE et al. 1992a).

Attempts to develop nonulcerogenic nonsteroidal anti-inflammatory drugs led unexpectedly to the discovery of yet another class of cannabinoid compounds, the aminoalkylindole (AAI) drugs, which are structurally distinct from both the traditional and nonclassical cannabinoids, yet bind to the cannabinoid receptor (KUSTER et al. 1993) and exhibit cannabinoid behavioral effects (COMPTON et al. 1992a). WIN-55,212-2 (the prototypic AAI cannabinoid) was one of a series of analogs whose antinociceptive properties could not be explained by inhibition of either cyclooxygenase or by opioid mechanisms. Results in both radiolabeled CP-55,940 and WIN 55,212 ligand binding assays indicate similar rank potencies and suggest identical binding sites (PACHECO et al. 1991). That the AAI analogs share a common pharmacological profile with Δ^9-THC is indicated by the fact that the (+)-enantiomer and several related analogs exhibited ED_{50} values in the range of those of Δ^9-THC for producing hypoactivity, antinociception, hypothermia and ring immobility in mice. Additionally, they generalized from the Δ^9-THC cue in the rat drug discrimination paradigm despite considerable response rate suppression (COMPTON et al. 1992a). The (−)-isomer was inactive up to the highest dose tested.

b) Ligand Binding and Biochemical Characteristics

Convincing evidence for a receptor binding site for the cannabinoids did not emerge until the late 1980s. The highly lipophilic nature of Δ^8- and Δ^9-THC produced a large degree of nonspecific and nonsaturable binding which, coupled with their relatively low receptor affinities, provides the most likely explanation for failure of earlier investigators to characterize a cannabinoid binding site in brain (HARRIS et al. 1978; ROTH and WILLIAMS 1979), though they were able to demonstrate high-affinity, saturable binding of $[^3H]\Delta^8$-THC to hepatoma cells in culture. Attempts to circumvent lipophilicity problems with the hydrophilic cannabinoid $[^3H]5'$-trimethylammonium-Δ^8-THC were also unsuccessful in that this ligand labeled a site which interacted both with pharmacologically active and inactive cannabinoids and which was later identified as a myelin basic protein (NYE et al. 1984, 1985). However, radiolabeling the potent bicyclic analog CP-55,940 proved to be a successful strategy for characterizing a cannabinoid binding site in brain homogenates (DEVANE et al. 1988).

Studies with CP-55,940 were the first to provide convincing evidence that a cannabinoid receptor existed. In rat brain cortical membranes, reported K_D values for CP 55,940 range from 0.13 to 5 nM and B_{max} values on the order of 0.9–3.3 pmol/mg protein (COMPTON et al. 1993; DEVANE et al. 1988; WESTLAKE et al. 1991). A limited series of analogs exhibited an excellent correlation between antinociceptive potency and affinity for this binding site (DEVANE et al. 1988). Subsequently, this correlation was extended to include a large number of cannabinoids and several behavioral measures (COMPTON et al. 1993). A high degree of correlation was found between the K_I values and in vivo potency in the mouse for depression of spontaneous locomotor activity, and for production of antinociception, hypothermia, and catalepsy. Similarly high correlations were demonstrated between binding affinity and in vivo potency in both the rat drug discrimination model and for psychotomimetic activity in humans. Thus, these studies suggest that the requirements for activation of the cannabinoid receptor are similar across different species and that this receptor is sufficient to mediate many of the known pharmacological effects of cannabinoids. This binding site has also been characterized with $[^3H]$1-OH-hexahydro-cannabinol-DMH (DEVANE et al. 1992a), $[^3H]$11-OH-Δ^9-THC-DMH (THOMAS et al. 1992), and $[^3H]$WIN-55,212-2, and the findings are consistent with those reported for $[^3H]$CP-55,940.

Autoradiographic studies of the cannabinoid receptor have shown a heterogeneous distribution in brain that is conserved throughout a variety of mammalian species, including humans, with most of the sites in the basal ganglia, hippocampus and cerebellum (HERKENHAM et al. 1990, 1991b). Binding sites are also abundant in the cerebral cortex and striatum. It is interesting to speculate that these sites correlate with some of the pharmacological effects of marihuana, for example, cognitive impairment (hippocampus and cortex), ataxia (cerebellum), catalepsy (basal ganglia), hypothermic and endocrine effects (hypothalamus), and even relatively low toxicity (paucity of receptors in the brainstem). Similar results have been obtained in studies conducted with [^{3}H]WIN 55,212 (JANSEN et al. 1992) and [^{3}H]11-OH-Δ^9-THC-DMH (THOMAS et al. 1992).

With regard to the existence of cannabinoid receptors in peripheral tissues, an examination of [^{3}H]CP 55,940 binding in all major organs of the rat resulted in detectable binding only in the immune system (LYNN and HERKENHAM 1994). Binding was detected in B lymphocyte-enriched areas (marginal zone of the spleen, cortex of the lymph nodes and nodular corona of Peyer's patches) but not in T lymphocyte-enriched areas (thymus and periarteriolar lymphatic sheaths of the spleen) and macrophage-enriched areas (lung and liver). Cannabinoid receptor binding in mouse spleen was consistent with Δ^9-THC inhibition of forskolin-stimulated cAMP accumulation in this tissue (KAMINSKI et al. 1992). Enantioselective immune modulation was observed with CP-55,940 and 11-OH-Δ^8-THC-DMH. In both cases, the ($-$) enantiomer demonstrated greater immunoinhibitory potency than the ($+$) isomer. Scatchard analysis of [^{3}H]CP 55,940 binding suggested a single binding site with a K_D of 910 pM and a B_{max} of approximately 1000 receptors/spleen cell. It is unclear why other sites were not found since cannabinoids apparently directly inhibit neuronal activity in peripheral sites (KUMBARACI and NASTUK 1980) as well as directly affect various smooth muscle preparations (e.g., vas deferens, ileum). A likely explanation is the lack of highly selective ligands for receptor subtypes.

Further validation of a receptor is often derived from manipulation of the endogenous system. One reaction of neuronal systems to the continued presence of agonist is receptor down-regulation. Temporally, in most systems, this process follows desensitization and is characterized by a loss of ligand binding at cell membrane receptors. Chronic exposure to Δ^9-THC results in the development of tolerance to the behavioral effects of Δ^9-THC (DEWEY 1986). In mice, tolerance has been shown to occur for most Δ^9-THC-induced behaviors (COMPTON et al. 1990a). Long-term exposure to Δ^9-THC (90 days) apparently does not irreversibly alter the cannabinoid receptor, since 60 days after cessation of the treatment the receptor affinity and number were unaltered (WESTLAKE et al. 1991). Downregulation of receptor density has been observed in discrete brain regions of animals tolerant to Δ^9-THC (OVIEDO et al. 1993; RODRIGUEZ DE FONSECA et al. 1994) but not in whole brain homogenates (ABOOD et al. 1993).

c) Receptor Cloning

Although SAR and receptor binding provide a compelling argument for existence of a cannabinoid receptor, the cloning of the protein provided definitive evidence. Homology screening with an oligonucleotide probe based on the structure of a G-protein coupled receptor (substance K) led to the isolation of a unique clone from a rat brain library (Matsuda et al. 1990). Subsequently, data from autoradiographic studies indicated that the distribution of the mRNA for the clone closely corresponded to that of the cannabinoid receptor. Thus, a ligand for this "orphan receptor" was eventually identified following the screening of many candidate ligands (opioids, neurotensin, angiotensin, substance P, neuropeptide Y, and others) when cannabinoids were found to act via this molecule. CP-55,940 and Δ^9-THC and other psychoactive cannabinoids (but not CBD and CBN) were found to inhibit adenylyl cyclase in cells transfected with the clone.

After the discovery of the rat cannabinoid receptor sequence, a human cannabinoid receptor cDNA was identified (Gérard et al. 1990). The nucleic acid sequences of these two clones were 90% identical, while the respective receptor proteins were 98% identical at the amino acid level. The human clone was expressed in COS cells and specific binding was demonstrated with [^{3}H]CP-55,940 (Gérard et al. 1990). The message for this receptor was also detected in the brains of the dog, rat, and guinea pig, but not found in dog stomach, spleen, kidney, liver, heart, or lung. Interestingly, the message was also identified in human testis, with a trace amount present in dog testis. The discrepancies between these results and those from receptor binding (Lynn and Herkenham 1994) include the failure to detect mRNA in the spleen (an organ which exhibits receptor binding) and the failure to detect binding in testis (an organ which contains message).

A peripheral receptor has been identified that is structurally distinct from the brain receptor (Munro et al. 1993). This receptor is expressed in macrophages in the marginal zone of the spleen and exhibits 44% homology with the receptor identified in brain tissues (Matsuda et al. 1990), though this value rises to 68% in the transmembrane domains. Since multiple receptor subtypes exist, a consistent receptor nomenclature was adopted. The receptor nomenclature committee of IUPHAR recommended that the cannabinoid receptor be abbreviated as CB with a numerical subscript assigned according to order of discovery. Thus, the receptor isolated initially in brain tissue (Matsuda et al. 1990) is designated CB_1, while that identified in the spleen (Munro et al. 1993) is designated CB_2. Though only a limited number of cannabinoids were evaluated, based upon binding properties it was concluded that the CB_2 receptor was indeed cannabinoid. The cloning of this receptor is consistent with the findings of others (Kaminski et al. 1992) showing that the spleen contains a cannabinoid binding site as well as the requisite mRNA.

The sequence of the cannabinoid receptors falls into the growing category of G-protein coupled receptors, which share structural and functional homologies. Signal transduction of ligand-receptor binding occurs via GTP-binding and G-proteins. Despite the fact that three sites of glycosylation are predicted from the structure of the receptor, biochemical studies indicate that the cannabinoid receptor need not be glycosylated to decrease cAMP (Howlett et al. 1990b). Structurally all are predicted to possess seven transmembrane domains based upon the amino acid sequence. When the CB_1 receptor amino acid sequence was compared with that of 38 other G-protein coupled receptors, the cannabinoid receptor was found to be in a novel subgroup along with the ACTH and melanocortin receptors (MOUNTJOY et al. 1992). This subset of receptors: (1) lacks proline residues in the fourth and/or fifth membrane domains (where existence in G-protein coupled receptors is though to participate in forming a binding "pocket" by introduction of a bend into the linear nature of the α-helical structure); (2) lacks one or both of the cysteine residues (disrupting disulfide bond formation between the first and second extracellular loops); and (3) possesses amino acid residue homology of between 32% and 39%. Though the cannabinoid receptor shares 20% homology with the δ opioid receptor (EVANS et al. 1992), assigning relevance to this degree of homology is difficult. Since the opioid receptor belongs to a class of peptide-responsive receptors, it is conceivable that the cannabinoid receptor is also responsive to an (as yet unidentified) endogenous peptide. Regardless, knowledge of the conserved amino acids between receptors that bind different ligands provides suitable target amino acids for site-directed mutagenesis studies.

Molecular techniques also provide the means to examine the expression of mRNA. Cannabinoid receptor binding and mRNA levels were examined in whole brain homogenates prepared from mice that had been treated chronically with Δ^9-THC (ABOOD et al. 1993). No alterations in cannabinoid receptor mRNA or protein levels were found in whole brain homogenates, though the chronic treatment was sufficient to induce a 27-fold tolerance in one behavioral assay. However, it is possible that alterations might occur within distinct brain regions (OVIEDO et al. 1993), and such changes would not be apparent in whole brain preparations (ABOOD et al. 1993).

Cell lines transfected with the rat and human cannabinoid receptor clones have been investigated for their binding and signal transduction properties (FELDER et al. 1992). The affinity of [^{3}H]CP 55,940 was similar to other preparations. The number of sites in the cell line expressing the human cannabinoid receptor was comparable to that of rat cerebellum, while the expression of the rat receptor was lower. Interestingly, cannabinoid receptor-mediated inhibition of cAMP accumulation was significantly reduced in the cell line overexpressing the human receptor. The rank order of potency of 16 cannabinoids evaluated for both receptor affinity and adenylyl cyclase inhibition proved to be nearly indentical to that in an earlier report of receptor binding in rat brain and multiple behavioral effects (LITTLE and MARTIN 1991).

d) Molecular Modeling

Characterization of the interaction between the receptor and the ligand is crucial for understanding receptor activation, developing selective agonists, understanding antagonist actions, and distinguishing receptor subtypes. Two molecular modeling approaches can be useful in this regard. The first strategy involves modeling the receptor itself, which is quite difficult and therefore has received relatively little attention. The second strategy involves developing a pharmacophore which describes the three-dimensional steric and electrostatic properties of an agonist. Though the discovery of the AAI cannabinoids underscores the limitations of the empirical approach to drug design, one technique used to evaluate the structural determinants for ligand binding and biological activity is computer-assisted molecular modeling. Studies have focused on the role of the phenolic hydroxyl in possible ligand receptor interactions (Reggio et al. 1990; Semus and Martin 1990) and the importance of the C9 position substituent and the spatial orientation of the associated ring (Reggio et al. 1991, 1993). Use of comparative molecular field analysis to analyze pharmacological and binding data has produced a three-dimensional pharmacophore of the electrostatic and steric forces of cannabinoids capable of quantitating the variations in the potencies of a wide variety of cannabinoids (Thomas et al. 1991). Steric repulsion "behind" the ring associated with C9 and the double bond of Δ^9-THC was associated with decreased binding affinity and pharmacological potency. The steric bulk of a side chain (located at C3 of the phenolic ring) can be extended by adding up to a total of seven carbons, which increases affinity and potency. This model possessed reasonable predictive capabilities and accommodated the AAI cannabinoids. However, considerable refinement is needed before the emergence of a predictive model for either receptor subtypes or selective ligands. In general, these models have provided descriptive models of SARs without divulging new insights.

e) Cannabinoid Antagonist

The search for a cannabinoid antagonist has been the topic of a previous review (Martin et al. 1987). Historically, lack of a cannabinoid antagonist has hindered research progress, since antagonists have generally played major roles in the characterization of many receptor systems. Numerous marihuana constituents, along with weakly active or inactive cannabinoid analogs, have been evaluated for potential antagonist properties with relatively little success. Although there have been some intriguing observations with CBD, there is no convincing evidence it is a specific antagonist. One report indicates that 11-nor-Δ^9-THC-carboxylic acid is capable of antagonizing the cataleptic effects produced by Δ^9-THC (Burstein et al. 1987). However, this observation has not been replicated in other laboratories. Though some drugs can (sometimes partially) attenuate some of the effects of Δ^9-THC, most alterations produced by such agents apparently simply represent the net effect of drugs possessing opposite effects (e.g., amphetamine stimulation plus cannabinoid inhibition of motor activity) rather than specific antagonism.

Agents which bind irreversibly to receptors have proven to be useful in developing antagonists for several classes of drugs. Reports thus far indicate that nitrogen mustards CBD and Δ^8-THC lack agonist and antagonist effects (COMPTON et al. 1990b; LITTLE et al. 1987). By contrast, photoactivation of azido-analogs of Δ^8-THC results in irreversible binding to the cannabinoid receptor in vitro (BURSTEIN et al. 1991; CHARALAMBOUS et al. 1992), and 5'-azido-Δ^8-THC exhibited potent in vivo effects. Although photoactivatable analogs do not provide a means for producing antagonism in vivo, they suggest that a reactive group at the terminal position of the side chain may cause irreversible binding, which could produce delayed antagonism.

Receptor-specific cannabinoid antagonists have been reported in the AAI class of drugs (WARD et al. 1991). One AAI antagonist was capable of producing a rightward shift in the in vitro dose-response curve of various agonists. However, the antagonist was most effective against AAI agonists (shifts of 20-fold or more) but much less effective (approximately fivefold shift) against natural and synthetic cannabinoids. Additionally, the drug exhibited only moderate to weak affinity for the cannabinoid receptor and was not capable of blocking the effects of cannabinoids under in vivo conditions (COMPTON et al. 1992a).

However, a novel chemical structure (typifying a fourth subclass of cannabinoid structures besides the traditional, nonclassical, and AAI subclasses) has been described as a truly specific competitive cannabinoid receptor antagonist (RINALDI-CARMONA et al. 1994). This analog (SR141716A) is most closely related in structure to the AAI class of cannabinoids (both possessing a nitrogen-containing, five-member, heterocyclic ring), but instead of being a carboxy-aryl-substituted indole, like the AAI analogs, it is a carboxyamide-substituted pyrazole with phenyl ring substituents. Though data strongly suggest that the analog is the first specific competitive cannabinoid receptor antagonist to be effective in vivo, it has only been shown that SR141716A blocks the in vivo effects of WIN-55,212 (an AAI cannabinoid). The ability of this drug to block the effects of other cannabinoids must be demonstrated in light of the data presented on the AAI antagonists. Additionally, only antagonism of WIN-55,212-mediated hypothermia has been evaluated in terms of time course of action and specificity with respect to noncannabinoid hypothermic agents.

3. Second Messenger and Other Transduction Mechanisms

a) Adenylyl Cyclase

The role of cannabinoids in the modulation of cAMP levels in cell culture and in homogenates of brain regions has been widely demonstrated. Δ^9-THC decreased epinephrine-and prostaglandin-stimulated levels of cAMP in fibroblasts (KELLY and BUTCHER 1973) and decreased cAMP levels in the *Tetrahymena* (ZIMMERMAN et al. 1981) and in nonstimulated rat heart homogenates (LI and NG 1984). However, the effect in the fibrolast preparation was biphasic and a function of incubation time (KELLY and BUTCHER 1979), so pharmacological relevance was unclear. In contrast, the effects of cannabinoids on membrane fluidity in the liver and heart may alter coupling of glucagon to the G_s-protein leading to activation of adenylyl cyclase by glucagon in the liver and isoproterenol in the heart (Hillard et al. 1990). This effect was shown to be enantioselective, so it has been proposed that the receptor associated with the cannabinoids may utilize a phospholipid as part of the recognition site.

The findings that cannabinoids inhibited forskolin-stimulated adenylyl cyclase preceded the characterization of the receptor (Howlett 1985; Howlett and Fleming 1984). In neuroblastoma (N18TG2) or neuroblastoma X glioma (NG108-15) cell lines, cannabinoid-induced inhibition of cAMP formation has been consistent, reproducible, and independent of interaction with prostanoid, opioid, muscarinic, or adrenergic systems (Howlett 1984, 1985; Howlett and Fleming 1984). It also was not blocked by antagonists of other classical neurotransmitters (Bidaut-Russell and Howlett 1991; Devane et al. 1986; Howlett et al. 1992). Cannabinoid-induced inhibition of cAMP formation in NG108-15 cells was rapid and reversible (Dill and Howlett 1988l; Howlett 1985), occurred at low cannabinoid concentrations (Howlett et al. 1986), and was consistent with the SARs established for the cannabinoids (Howlett 1987; Howlett et al. 1990b; Howlett and Fleming 1984). The ability of cannabinoid analogs to inhibit adenylyl cyclase correlated well with their potency in several pharmacological assays, suggesting a cause-effect relationship (Howlett et al. 1990a).

Cannabinoids were reported to interact with a ribosylated membrane protein identified as the G_i-protein (Howlett et al. 1986). This result was subsequently reinforced by the finding that pertussis toxin attenuated cannabinoid effects on adenylyl cyclase (Howlett et al. 1988). Monovalent cations are recognized for their modulatory role in G-protein/receptor coupling (e.g., sodium) which is generally required for optimal inhibition of adenylyl cyclase by G_i/G_o-coupled receptors. In contrast, cannabinoid (and $GABA_B$) agonists inhibited adenylyl cyclase in a sodium-independent fashion in the cerebellum, but in a sodium-dependent fashion in the striatum (Pacheco et al. 1994). This differential effect was not due to either the receptor or the effector, so it is possible that different G-proteins are involved in these two brain regions.

Biochemical tolerance has been useful in studying receptor-regulated adenylyl cyclase activity. The cellular response to an agonist declines reversibly after continued exposure to drug. Exposure of N18TG2 neuroblastoma cells to Δ^9-THC attenuated cannabinoid-inhibited adenylyl cyclase activity without affecting cell morphology or growth (Dill and Howlett 1988). Cells pretreated for 24 h with $1 \mu M$ Δ^9-THC showed unaltered levels of basal cAMP, secretin-stimulated cAMP, and carbachol-inhibited cAMP, but Δ^9-THC produced only a 17% decrease (cf. 35% in controls) of cAMP accumulation. Thus, the desensitization was specific for the cannabinoid receptor-mediated response. The desensitization process was time- and dose-dependent and reversible.

In contrast to cell culture data, however, the modulation of forskolin-stimulated cAMP levels in mouse brain synaptosomes was not identical for all cannabinoids; Δ^9-THC and Δ^8-THC produced biphasic effects, while nonclassical cannabinoids only produced inhibition (LITTLE and MARTIN 1991). Additionally, levonantradol (which produces pronounced cannabinoid effects) did not alter cAMP levels, while CP-56,667 (the largely inactive enantiomer of CP-55,940) inhibited cAMP, which suggested little correlation between pharmacological activity and modulation of adenylyl cyclase. Inhibition of adenylyl cyclase activity was also observed for AAI cannabinoids in rat brain membranes (PACHECO et al. 1991).

Though in vitro studies are relatively consistent and in vivo studies have suggested that cannabinoid administration to rodents altered cAMP accumulation in brain, the effects were modest and frequently difficult to reproduce (MARTIN et al. 1994). The initial work in rodent brain indicates that the levels of cAMP in the brain are altered in a biphasic manner by cannabinoids in the mouse. The intraperitoneal (i.p.) administration of Δ^9-THC has been shown to increase cAMP in whole brain and brain regions, while higher doses decrease cAMP levels (DOLBY and KLEINSMITH 1974). These effects have been proposed to be correlated with the initial stimulatory effects of low doses of the cannabinoids, whereas the depressant effects were with the higher doses of the cannabinoids (DOLBY and KLEINSMITH 1977). Similarly, cannabinoids increased cAMP levels in other preparations, but the effect did not correlate well with the psychoactive potency of the cannabinoids (HILLARD and BLOOM 1983). In contrast, another study found that i.v. administration of cannabinoids did not alter cAMP concentrations in five brain regions of the mouse, while Δ^8-THC increased cAMP in only one brain region (ASKEW and HO 1974).

Despite evidence suggesting cannabinoid receptor/adenylyl cyclase association, it has been difficult to establish which pharmacological effects are mediated through this pathway. Most efforts have concentrated on demonstrating a role for adenylyl cyclase in cannabinoid-induced antinociception (HOWLETT et al. 1988). However, the affinity of cannabinoids at the CP-55,940 binding site and potency at inhibiting adenylyl cyclase (DEVANE et al. 1988) have been shown to be similar in rank order to the production of not only antinociception, but also hypothermia, spontaneous activity and catalepsy by the cannabinoids (COMPTON et al. 1993). Yet, pertussis toxin abolished the antinociceptive effects of cannabinoids, and i.t. administration of both forskolin and chloro-cAMP attenuated the antinociceptive effects of Δ^9-THC (WELCH et al. 1994). These results support a role for adenylyl cyclase in the actions of cannabinoids since these agents either elevate or mimic cAMP. It should be noted that these agents did not completely abolish the cannabinoid effects and that actions other than adenylyl cyclase may be implicated by pertussis toxin.

b) Calcium Ion Channels

There has been reasonable evidence supporing a role for cannabinoid modulation of neurotransmitter release (Dewey 1986). Calcium is the likely mediator of this action given its well characterized role in neurotransmitter release. Cannabinoids act to decrease the release of acetylcholine by decreasing the influx of presynaptic calcium (Kumbaraci and Nastuk 1980). Additionally, the effects of pertussis toxin on cannabinoid response could as easily be attributed to G-protein-linked ion channels as to adenylyl cyclase activity. Cannabinoids are known to decrease calcium uptake to several brain regions (Harris and Stokes 1982), although this effect does not correlate with psychoactivity. Direct measurement of the effects of cannabinoids on free intracellular calcium in brain tissue (using intracellular calcium indicators) has shown that depolarization-induced rises in intracellular calcium are attenuated by Δ^9-THC, though at micromolar concentrations (Martin et al. 1989). These concentrations are similar to those required for the alteration of neuronal transmission (Kumbaraci and Nastuk 1980), but higher than those required to block calcium uptake (Harris et al. Stokes 1982). In contrast, others (Okada et al. 1992) have reported that Δ^9-THC did not perturb calcium levels in rat brain. Yet, in mouse thymocytes Δ^9-THC has been shown to decrease concanavalin A-stimulated levels of free intracellular calcium by both inhibition of calcium influx and inhibition of intracellular mobilization of calcium. These authors proposed that such changes in calcium may explain the immune suppression observed with the cannabinoids (Yebra et al. 1992).

Very low concentrations of Δ^9-THC ($0.1\,\mathrm{n}M$) have been shown to enhance potassium-stimulated rises in intracellular calcium, while intermediate concentrations ($1-50\,\mathrm{n}M$) block potassium-stimulated rises in intracellular calcium. Electrophysiological studies in neuroblastoma cells indicated that $1-100\,\mathrm{n}M$ concentrations of several cannabinoids inhibited an Ω conotoxin-sensitive, high voltage-activated calcium channel. This effect was blocked by the administration of pertussis toxin and was independent of the formation of cAMP. Since the L-type calcium channel blocker nitrendipine failed to alter the cannabinoid effect, it was concluded that cannabinoids apparently interact with an N-type calcium channel. Such an effect would lead to a decrease in the release of neurotransmitters (Mackie and Hille 1992). Results from a similar study revealed that cannabinoids inhibit I_{Ca} current in neuroblastoma cells, but the effect was not dose-related suggesting lack of a receptor-mediated event. However, it was pertussis toxin- and Ω conotoxin-sensitive (Caulfield and Brown 1992). Thus, while cannabinoids have been shown to alter intracellular calcium, the role of the cannabinoid receptor in these events has been questioned. Additionally, in CHO cells cannabinoids were shown to induce a nonspecific release of intracellular calcium. Both the active $(-)$- and inactive $(+)$-enantiomers of the potent cannabinoid 11-OH-Δ^8-THC-DMH were able to release calcium in non-transfected and CB_1 transfected cells (Felder et al. 1992).

The interaction of the adenylyl cyclase system with intracellular calcium has also been documented (BROSTROM et al. 1978). cAMP has been shown to produce rises in free intracellular calcium in synaptosomes (OKADA et al. 1989; OLSON and WELCH 1991). Such interactions may lead to cellular events responsible for the release of neurotransmitters, such as the phosphorylation of calcium channels which increases calcium conductance (REUTER 1983). Thus, the modulation of intraceullar calcium by cannabinoids is possible either via their interaction with adenylyl cyclase or by a mechanism in-dependent of the formation of cAMP.

Although in vitro studies indicate a role for calcium in the effects of the cannabinoids, in vivo administration of various calcium channel modulators to mice has yielded results which indicate a lack of involvement of calcium directly in the antinociceptive effects of i.t. administered cannabinoids, whereas calcium modulation of i.c.v. administered cannabinoids is observed (WELCH et al. 1994). The antinociceptive effects of the cannabinoids (i.t.) were not altered directly by the administration of calcium or by other modulators such as nimodipine, verapamil, Ω conotoxin, thapsigargin, BAYK 8644, or ryanodine. In addition, cannabinoids administered i.t. were blocked by the calcium-gated potassium channel blocker, apamin, but not by blockers of any other potassium channels. These data indicate that the antinociceptive effects of the cannabinoids in the spinal cord are not mediated by calcium channels. Unlike the i.t. situation, the i.c.v. administration of cannabinoids results in antinociception which is blocked by i.c.v. admini-stration of calcium. In addition, thapsigargin (i.c.v.) blocks the effects of Δ^9-THC. Thus calcium modulation appears to play a role in the antinociceptive effects of cannabinoids in the brain. Apamin (i.c.v. or i.t.) fails to block the antinociceptive effects of i.c.v. administerred cannabinoids. Thus, the modulation of potassium channels by the cannabinoids may differ in the brain and in the spinal cord.

One other possible mechanism by which cannabinoids may decrease calcium entry is via interaction with the receptor-operated calcium channels stimulated by NMDA. Blockade of the NMDA-stimulated calcium channel has been described for $(+)$-11-OH-Δ^8-THC-DMH, an analog which is devoid of psychoactive effects of cannabimimetic properties in rodents (FEIGENBAUM et al. 1989).

c) Prostaglandins and Other Systems

Compelling evidence for the involvement of other second messenger systems in the pharmacological effects of cannabinoids does not exist, though there are many research avenues still open because cannabinoids appear to have some effect on almost any selected system or biochemical pathway (MARTIN 1986; MELLORS 1979). One example is the effects of cannabinoids on cellular ATPases (MARTIN 1986; PERTWEE 1988). Generally, the cannabinoids inhibit both cellular Na^+/K^+ ATPase and Mg^{2+}/Ca^{2+} ATPase. Thus, the effects of cannabinoids on calcium may, in part, be due to the alteration of Ca^{2+}

ATPase. activity. The effects of the cannabinoids on cellular energy via Na^+/K^+ ATPases disruption may explain the inhibition of neurotransmitter uptake. Most investigators have concluded that effects of the cannabinoids in ATPases may result from membrane perturbation or fluidization by the cannabinoids.

In several biological systems it has been shown that activation of receptors coupled to the formation of cyclic nucleotides results in a decrease in phosphoinositides (NISHIZUKA 1983, 1984). Δ^9-THC decreases the formation of *myo*-inositol trisphosphate (IP$_3$) in pancreatic islets (CHAUDRY et al. 1988). It is possible that the cannabinoids alter intracellular calcium and thus neuronal transmission by IP$_3$ formation. IP$_3$ formation has been shown to enhance the release of calcium from the endoplasmic reticulum, an organelle partially responsible for the buffering of intracellular calcium levels (NISHIZUKA 1983, 1984). However, there is no evidence that the effects of Δ^9-THC in brain or spinal cord are mediated through IP$_3$, though involvement in peripheral (e.g., cardiac) effects is still uncertain. While the binding of the cannabinoids within the cerebellum colocalized with that of forskolin, protein kinase C distribution was not localized in a similar pattern (HERKENHAM et al. 1991a). These studies support a role for cAMP rather than IP$_3$ in the actions of cannabinoids in the cerebellum. However, the interaction of the cannabinoids with IP$_3$ in other brain and spinal cord regions is not precluded.

Since, in the pituitary, cGMP enhances the formation of inositol phosphates (NAOR 1990), a possible interrelationship between IP$_3$ and cGMP formation has been hypothesized. Also, it has been shown that levonantradol (but not its inactive enantiomer dextronantradol) decreases basal and isoniazid-induced increases in cGMP in the cerebellum, possibly via an interaction with GABA (KOE et al. 1985; LEADER et al. 1981). In most systems the role of cGMP is unclear, although cGMP produces antinociceptive effects when injected into the brain of mice (VOCCI et al. 1978); therefore, it is possible the cannabinoids alter either IP$_3$ or cGMP formation in the production of antinociception. Though cGMP has also been linked to nitric oxide formation in the cerebellum, and cannabinoid mechanism of action pursued intensely in this brain region, there are no reports on the interaction of cannabinoids with nitric oxide.

Previous studies have suggested a role for cannabinoid agonists in arachidonic acid release and membrane phospholipid turnover. Δ^9-THC released arachidonic acid from mouse peritoneal cells and S49 cells (AUDETTE et al. 1991), and this effect was attenuated by pertussis toxin or cholera toxin. Thus, the release of arachidonic acid would appear to involve the G$_i$-protein. However, cell lines transfected with cannabinoid receptor have been evaluated recently for possible signal transduction systems (FELDER et al. 1992). Though CP-55,940 was able to release [^{3}H]arachidonic acid (at concentrations greater than 100 μM), it also did so in nontransfected CHO cells. Additionally, the inactive (+)-enantiomer of the potent cannabinoid agonist $11\text{-}OH\text{-}\Delta^8\text{-}THC\text{-}DMH$ was also able to stimulate [^{3}H]arachidonic acid release. These data indicated lack of involvement of the cannabinoid receptor.

The role of prostaglandins in the activity of cannabinoids is an area of research that has been previously reviewed (BURSTEIN 1992; MARTIN 1986). Though anandamide, the endogenous ligand for the cannabinoid receptor, has been shown to be an ethanolamine derivative of arachadonic acid (DEVANE et al. 1992b), the relationship of this product to others in the arachadonic acid cascade remains questionable. Several behavioral effects, in particular cataleptic and antinociceptive effects, of the cannabinoids have been proposed to be related to the formation of prostaglandins. In vitro cannabinoids have been shown to produce diverse effects on prostaglandin synthesis. Both inhibition (BURSTEIN et al. 1973; BURSTEIN and RAZ 1972; BURSTEIN et al. 1974; REICHMAN et al. 1987; SPRONCK et al. 1978) and stimulation of prostaglandin formation (BURSTEIN and HUNTER 1981; BURSTEIN et al. 1982, 1985; WHITE and TANSIK 1980) have been observed and are blocked by aspirin and mepacrine. $\Delta^9\text{-}THC$ has been shown to inhibit prostaglandin (PG)E$_1$ formation (HOWES and OSGOOD 1976) in rat brain. However, PGE$_1$ and $\Delta^9\text{-}THC$ act synergistically in the production of antinociception as well as cataleptic, anticonvulsant and sedative effects (BHATTACHARYA et al. 1980). Blockers of prostaglandin formation, such as aspirin and indomethacin, have been shown to modify the antinociceptive, cataleptic, and hypotensive effects of $\Delta^9\text{-}THC$ in rodents, supporting the notion that cannabinoids may increase the formation of prostaglandins. (BURSTEIN et al. 1982; DALTERIO et al. 1981; FAIRBAIRN and PICKENS 1979, 1980; JORAPUR et al. 1985). Similar findings have been reported to occur in humans (PEREZ-REYES et al. 1991), in whom some behavioral effects of the cannabinoids have been shown to be attenuated by indomethacin. Mice immunized against PGE$_2$ had reduced cataleptic effects (BURSTEIN et al. 1989; HUNTER et al. 1991). Since antibodies presumably could not enter the central nervous system, the effect was thought to be primarily peripheral. These results were in agreement with those indicating a rise in levels of PGE$_2$ and PGF$_{2\alpha}$ following administration of $\Delta^9\text{-}THC$ (BHATTACHARYA 1986). Since the binding of PGE$_2$ was decreased following $\Delta^9\text{-}THC$ administration, it appeared there were increased levels of the prostaglandin which decreased its binding (HUNTER et al. 1991). In contrast, $\Delta^9\text{-}THC$ has also been shown to inhibit the release of PGF$_{2\alpha}$ (RAFFEL et al. 1976) in rat brain.

4. Integration of Systems

a) Endogenous Cannabinoid System

Attempts to identify an endogenous ligand have resulted in the isolation of anandamide, an arachidonic acid derivative, from porcine brain which bound with high affinity to the cannabinoid receptor (DEVANE et al. 1992b).

Anandamide specifically bound to membranes from cells transfected with the cannabinoid receptor, but not to membranes from control nontransfected cells (VOGEL et al. 1993).

Additionally, anandamide inhibited forskolin-stimulated adenylyl cyclase in transfected cells (but not in control nontransfected cells), an effect which was blocked by pretreatment with pertussis toxin (VOGEL et al. 1993). Inhibition of adenylyl cyclase activity by anandamide in CHO cells expressing the human cannabinoid receptor was also observed and also blocked by pertussis toxin (FELDER et al. 1993). N-type calcium channels were inhibited by anandamide in N-18 neuroblastoma cells. Additionally, inhibition of N-type calcium channels was voltage-dependent and N-ethylmaleimide sensitive (MACKIE et al. 1993).

Anandamide was also shown to inhibit electrically stimulated contractions of mouse vas deferens much in the same fashion as Δ^9-THC. These effects were mediated via presynaptic actions on cholinergic neurons. Anandamide also reversed the stimulation of the miniature endplate potential firing frequency in the frog neuromuscular junction that was induced by hypertonic gluconate (VAN DER KLOOT 1994). Since the protein kinase A inhibitor Rp-cAMPS also blocks this stimulatory effect, it is possible the anandamide effect was mediated via protein kinase A. However, the increase in frequency produced by Sp-cAMPS (a protein kinase activator) was not attenuated by anandamide. Thus, anandamide inhibits the gluconate effect without altering protein kinase A activity, though apparently via calcium effects (see above).

Anandamide has also been reported to produce effects in the rat on the hypothalamic-pituitary-adrenal axis similar to those produced by Δ^9-THC (WEIDENFELD et al. 1994). Anandamide (i.c.v.) decreased CRF-41 levels in the median eminence and increased serum ACTH and corticosterone levels. These findings are consistent with the observations that cannabinoids exhibit anxiogenic properties (ONAIVI et al. 1990).

Preliminary studies in mice also indicated that anandamide shares some of the behavioral and other pharmacological effects of Δ^9-THC (FRIDE and MECHOULAM 1993). Other investigators (CRAWLEY et al. 1993) also found a similar reduction in spontaneous activity and body temperature in mice treated with anandamide. However, more detailed studies (SMITH et al. 1994) show that though anandamide and Δ^9-THC are very similar, there are also distinct differences. Of minor importance is the relatively short duration of action of anandamide, and the weak potency (anandamide is 4- to 20-fold less potent than Δ^9-THC). Interestingly, anandamide is largely inactive following i.p. administration, with the exception of the ability to produce profound sedative effects. Also, the antinociceptive properties of anandamide suggest a divergence from mechanisms for the production of other effects. The time course for anandamide-mediated antinociception is significantly longer than other effects and (unlike Δ^9-THC) is insensitive to administration of nor-BNI (see opiate interactions above).

Anandamide could function as an endogenous neurotransmitter or neuromodulator, since synthetic and metabolic pathways exist (DEUTSCH and CHIN 1993). Synthesis was demonstrated by incubating arachidonic acid and ethanolamine in the presence of rat brain homogenate. Anandamide was also synthesized in bovine brain fortified with arachidonate and ethanolamide (DEVANE and AXELROD 1994), with the level of synthesis being greatest in the hippocampus, intermediate (twofold lower) in the thalamus, striatum or frontal cortex, and lowest (five- to sixfold less) in the cerebellum, an area with the greatest receptor density. Based upon the fact that anandamide synthesis is enzyme CoA- and ATP-independent, it was proposed that synthesis occurred via a novel eicosanoid pathway (KRUSZKA and GROSS 1994). Anandamide was readily taken up by neuroblastoma or glioma cells and rapidly degraded by an amidase which can be blocked by phenylmethylsulfonyl fluoride (PMSF), a nonspecific peptidase and esterase inhibitor (DEUTSCH and CHIN 1993). The degradative enzyme resides in the membranes (DEUTSCH and CHIN 1993), which is corroborated by the fact that degradation occurs within ligand binding assays (ADAMS et al., in press; CHILDERS et al. 1994). Anandamide was also degraded by brain, liver, kidney and lung tissue, but not heart or muscle. There is also evidence that the metabolism of anandamide can be blocked by trifluoromethyl ketone, α-keto-ester and α-keto-amide analogs of anandamide by acting as transition state inhibitors (KOUTEK et al. 1994). Though separate enzymes appeared to be responsible for synthesis and degradation, since PMSF did not block synthesis (DEUTSCH and CHIN 1993), others have found that PMSF did inhibit synthesis (DEVANE and AXELROD 1994). Therefore, the question of multiple enzymes for anandamide synthesis and metabolism remains unanswered.

Anandamide may not be the only endogenous cannabinoid. A family of anandamides (similar structurally and physicochemically) may exist, since other endogenous unsaturated fatty acid ethanolamides (homo-γ-linolenylethanolamide and docosatetraenylethanolamide) have been isolated and also bind to the cannabinoid receptor (HANUS et al. 1993; MECHOULAM et al. 1994). Additionally, unlike the anandamides or glycerol derivatives, a more hydrophilic endogenous substance was described (EVANS et al. 1994), which could be released from neurons in a calcium-dependent fashion.

The last decade of progress in the cannabinoid field now supports the postulate that a cannabinoid neurochemical system exists. However, its role in the brain and its relationship to other neurochemical systems remains to be elucidated. Without direct evidence for a primary functional role, it would seem that the cannabinoid system is largely neuromodulatory, which is supported by the fact (RINALDI-CARMONA et al. 1994) that a putative cannabinoid antagonist (SR 141716A) administered alone appears to be devoid of typical cannabinoid effects in various rodent models (temperature, nociception, catalepsy, forced motor activity).

b) Spinal and Peripheral Cannabinoid Receptors

[^{3}H]CP-55,940 binds to the substantia gelatinosa of the spinal cord (Herkenham et al. 1990) at a level approximately 10% of that found in the substantia nigra, where maximal cannabinoid binding occurs. The substantia gelatinosa is responsible for the processing of pain transmission (Yaksh et al. 1988). Though the density of cannabinoid receptors is low relative to that of the brain, it is still much higher than that of substance P, which is a recognized transmitter involved in pain processing, in the dorsal horn of the spinal cord (Iverfeldt et al. 1988). Additionally, the substantia gelatinosa is also the principle location of the opioid receptors in the dorsal horn (Gamse et al. 1979). The colocalization of these two systems may be critical to the synergism observed following i.t. administration of inactive doses of cannabinoids and active doses of morphine (i.t. or i.c.v.) in the production of antinociception in mice (Smith and Martin 1992; Welch and Stevens 1992). Though parallel shifts in the morphine dose-response curve were produced by pretreatment with several cannabinoids, it is not clear that the response is mediated by a known cannabinoid receptor, since CP-55,940 was inactive in these procedures. However, it is possible that an as-yet-unidentified receptor might exist in spinal tissue, although describing the binding to this site would apparently require use of a radioligand other than CP-55,940.

The CB_2 receptor (discussed above) is structurally distinct from the brain CB_1 receptor (Munro et al. 1993) and has not been found in brain. The primary distinction between CB_1 and CB_2, besides anatomical location and primary structure, appears to be their affinity for CBN relative to that for Δ^9-THC. A review of the (brain CB_1) binding literature indicates that no single study has generated displacement data on both CBD and CBN to allow proper comparison to the Δ^9-THC value. Additionally, K_I values for all three of these analogs vary considerably between studies. These facts underscore the necessity of further characterization of CB_2 before concluding that its binding profile is distinguishable from that of CB_1. The functional role these receptors may play in the immune system is uncertain. However, the potential discovery of endogenous cannabinoids from peripheral tissue (see above) may suggest that existence of peripheral cannabinoid neuro-modulatory systems. Regardless, the existence of the CB_2 receptor suggests the possibility that yet other subtypes may exist.

c) Cardiovascular Mechanisms

The effects of cannabinoids on the vascular system appear to be mediated by altered autonomic control of both the heart and blood vessels (Adams et al. 1976; Benowitz et al. 1979; Jandhyala and Buckley 1977), and indeed the cannabinoids possess some anticholinergic properties which may contribute to this response (Drew and Miller 1974; Gascon and Peres 1973; Layman 1971; Rosell et al. 1976, 1979). Effects on heart rate have been linked to altered parasympathetic function of the vagus nerve (BENOWITZ et al. 1979; HOLLISTER 1986). In the dog, several acute cardiovascular and autonomic effects of Δ^9-THC were not observed following chronic Δ^9-THC treatment, suggesting tolerance. However, prolonged use may also induce pharmacological properties and/or mechanisms of action which may not exist in acute exposure (JANDHYALA and BUCKLEY 1977) and might mask previously observed events. Since some (but not all) cardiovascular effects were observed with nonpsychoactive drugs (ADAMS et al. 1977), the molecular mechanisms involved with these particular effects are probably not related to activation of CB_1 receptors in the brain. Additionally, since not all cardiovascular effects appear to be mediated by central and autonomic systems, some of the effects of cannabinoids would appear to be mediated by peripheral mechanisms. This last postulate is tentative, but the discovery of at least one peripheral receptor and one potential peripheral endogenous cannabinoid provide indirect support for this contention. However, cannabinoid receptors have not been detected in cardiac or vascular tissues, but it is possible such hypothetical sites represent a new subtype of receptor which is less sensitive to CP-55,940 and therefore not bound under currently used conditions or ligand concentrations, or that effects are mediated by an intermediate substance produced elsewhere in the periphery.

C. General Pharmacology

I. Pharmacokinetics

1. Absorption and Distribution

Δ^9-THC is absorbed rapidly and efficiently via the inhalation route. Detectable amounts of Δ^9-THC (7–18 ng/ml) have been measured following a single puff of marihuana smoke by individuals, and during a multiple puff session peak Δ^9-THC concentrations developed prior to the termination of smoking (HUESTIS et al. 1992b; PEREZ-REYES et al. 1981). Despite considerable intersubject variability, experienced individuals developed peak Δ^9-THC concentrations in excess of 100 ng/ml after smoking marihuana cigarettes (THC content 1.32 to 2.54%) (COCCHETTO et al. 1981; HUESTIS et al. 1992b; LEMBERGER et al. 1972b; OHLSSON et al. 1980; PEREZ-REYES et al. 1982). The initial increase in Δ^9-THC blood concentrations during smoking is followed by rapid redistribution to tissues. Subsequent release back into the circulation occurs slowly, which produces a prolonged elimination half-life.

Oral ingestion of Δ^9-THC or marihuana leads to the production of similar pharmacological effects as smoking, although substantial differences exist in the rate of onset of effects and in the amounts of cannabinoids appearing in blood. Following oral dosing with 15–20 mg of Δ^9-THC there

was a gradual increase in blood levels of Δ^9-THC over a period of 4–6 h
(WALL et al. 1983). Peak concentrations of Δ^9-THC were in the 10–15 ng/ml
range, while concurrent 11-hydroxy-Δ^9-THC concentrations were in the range
of 1–6 ng/ml. 11-*nor*-9-carboxy-Δ^9-THC concentrations were increased
approximately twofold over those observed following intravenous dosing of
Δ^9-THC.

Distribution of Δ^9-THC begins of occur immediately upon absorption.
Mean peak Δ^9-THC concentrations declined by 50% approximately 10 min
after the plateau was reached following smoking. Subsequently, concentrations declined much more slowly, but remained detectable for at least 4 h.
Much longer detection times for Δ^9-THC have been reported, particularly in
studies in which sensitive analytical methodologies were utilized. Concentrations of deuterium-labeled Δ^9-THC in plasma of chronic marihuana users
were detected for 13 days by GC/MS techniques (JOHANSSON et al. 1988).

2. Metabolism and Excretion

Following the rapid redistribution of Δ^9-THC to body tissues there is a slow
release from these tissues back to the circulatory system, which results in a
prolonged elimination half-life. Δ^9-THC is metabolized in humans by a
variety of oxidative routes which first produce hydroxylated metabolites,
followed by conversion to carboxylic acids, and subsequent excretion as
conjugates. The metabolite, 11-hydroxy-Δ^9-THC, is active (LEMBERGER et al.
1972a); however, it is formed in trace amounts when marihuana is smoked,
though greater amounts may be formed following oral ingestion. About
50% of a dose of Δ^9-THC is excreted in feces and 15% is excreted in urine
over a period of several days (WALL et al. 1983). The primary metabolite
excreted in urine is conjugated 11-*nor*-9-carboxy-Δ^9-THC. Blood concentrations of Δ^9-THC peak prior to drug-induced effects. The discrepany
between time course effects and cannabinoid blood concentrations, which
was first raised almost 50 years ago (LOEWE 1946), remains unsolved.

Marihuana plant material cooked in brownies and consumed by male
volunteers was studied to evaluate oral absorption (CONE et al. 1988).
Subjects scored significantly higher on behavioral measures after consumption
of brownies containing Δ^9-THC than placebo; however, the effects were
slow to appear and were variable. Urinalysis indicated that substantial
amounts of 11-*nor*-9-carboxy-Δ^9-THC were excreted in urine over a period
of 3–14 days.

The metabolism of Δ^9-THC to 11-hydroxy-Δ^9-THC and to 11-*nor*-9-
carboxy-Δ^9-THC occurs rapidly with peak blood concentrations of 11-
hydroxy-Δ^9-THC appearing shortly after peak Δ^9-THC concentrations
following either intravenous or oral administration (HUESTIS et al. 1992a;
WALL et al. 1983). Peak 11-nor-9-carboxy-Δ^9-THC concentrations appear
later (1–2 h) and decline slowly thereafter.

Due to complex distribution and elimination phases, a number of kinetic
models have been proposed to describe plasma Δ^9-THC data. Blood levels

for Δ^9-THC during the first 6 h after smoking have been adequately described by a triexponential function (BARNETT et al. 1982). Disposition of Δ^9-THC was described empirically as being represented by a two-compartment model with first order input from smoking. Others have utilized two- and four-compartment models to describe the disposition of Δ^9-THC administered intravenously. Half-life estimates for plasma Δ^9-THC range from 18 h to 4 days. Use of very sensitive assays usually results in longer half-lives and less variable Δ^9-THC clearance from blood that ranged from 650 to 1000 ml/min. Cannabinoids are excreted via bile and reabsorbed from the gastrointestinal tract, which likely contributes to their long half-life. Oral bioavailability of Δ^9-THC appears to be lower (6%–19%) than Δ^9-THC from smoked marihuana (14%–27%). Although several factors contribute to bioavailability, the experience of the smoker appears to play a key role. Subjects inhaling smoke from 4.5% Δ^9-THC marihuana cigarettes had a mean area-under-the-curve plasma concentration almost twice as high as that of subjects smoking 1.3% Δ^9-THC cigarettes (PEREZ-REYES 1985). The expected AUC ratio based on the relative potency of the two cigarettes was 3.6:1, whereas the observed AUC ratio was 1.8:1. This discrepancy led to speculation that smokers could sense the rate of appearance and intensity of their "high" and would titrate their intake accordingly. Heavy marihuana users smoked more efficiently (23%–27% bioavailability) than light smokers (10%–14% bioavailability) leading to the conclusion that the experienced smokers utilized a more adept smoking technique e.g., deeper inhalations (OHLSSON et al. 1985). In studies involving drug administration by intravenous infusion of Δ^9-THC and by smoked marihuana (LINDGREN et al. 1981; OHLSSON et al. 1982), there was a trend for heavy users to exhibit lower plasma concentrations than light users, but the differences were not statistically significant.

3. Relationship of THC Levels to Effects

Subjects begin to report behavioral effects after a single puff of marihuana smoke and these effects culminate at a time similar to or somewhat delayed with respect to blood Δ^9-THC concentrations (HUESTIS et al. 1992b). The delay between peak blood concentrations and peak drug effects is likely related to delays in penetration of the central nervous system and to subsequent redistribution of Δ^9-THC following rapid uptake by adipose tissues. The delay has been characterized as a counter-clockwise hysteresis between Δ^9-THC blood concentrations and drug (BARNETT et al. 1982, 1985; CONE and HUESTIS 1993). Prior to equilibrium, plasma concentrations increase rapidly while effects develop more slowly. Consequently, at early times after smoking marihuana, plasma concentrations are high while effects are low; whereas at later times, plasma concentrations may be low while effects become highly prominent. This time discordance between blood concentrations of Δ^9-THC and effects has led to conclusions that no meaningful relationships exist between blood concentration and effect (MASON and McBAY 1985; McBAY 1986).

II. Effects on Organ Systems

1. Brain

a) Electroencephalogram

Alterations in EEG recordings are found in both humans and animals, but interpretation of such data is difficult. It has been suggested that the subcortical spike activity might be related to motor manifestations of marihuana use (ROSENKRANTZ 1983). In animals, the areas most sensitive to the effects of the cannabinoids were the hippocampus, amygdala, and septal areas. Identical measures are not available in humans.

b) Cerebral Blood Flow and Glucose Metabolism

Normally cerebral blood flow (CBF) and cerebral metabolic rate (CMR) are closely coupled with brain activity. Drug-induced changes in CBF or CMR are likely to be representative of a change in brain function (MATHEW and WILSON 1993). Relatively little has been described concerning the effects of Δ^9-THC on these cerebral parameters (MATHEW and WILSON 1992). Acute Δ^9-THC generally increases CBF (MATHEW and WILSON 1992). A maximal, bilateral increase in CBF was observed 30 min following marihuana smoking, with greater increases observed in the frontal region and right hemisphere, though increases in both hemispheres correlated well with the degree of intoxication (MATHEW et al. 1992). This correlation suggests that stimulation, rather than an inhibition of neuronal activity, is principally responsible for the observed effects. A decrease in CBF observed in inexperienced marihuana smokers has been attributed, in part, to the anxiety response sometimes observed in first time users. Increased global CBF has been reported in animals receiving Δ^9-THC as well as an increase in cerebral blood velocity (related to increased capillary perfusion). Also, decreased CBF was observed in chronic heavy abusers, but no alterations were observed under chronic conditions of moderate or mild marihuana abuse. This attenuation could be interpreted as development of tolerance with the emergence of an exaggerated compensatory mechanism (MATHEW and WILSON 1992; VOLKOW and FOWLER 1993) or possibly of the unmasking of inhibitory actions once tolerance has developed to the stimulatory effects of Δ^9-THC.

Acute Δ^9-THC generally increases the CMR of glucose (MATHEW and WILSON 1992). However, Δ^9-THC has been shown to produce a biphasic response in limbic regions of animals (MARGULIES and HAMMER 1991). Effects on CMR in humans may be limited to specific areas of the brain such as the cerebellum and prefrontal cortex (VOLKOW and FOWLER 1993).

2. Immune System

a) Lymphoid Tissues

A thorough, and still relevant, review of the effects of marihuana and cannabinoids on the immune system is available (MUNSON and FEHR 1983). Though not definitive, the alteration of lymphoid organ weight is often considered an index of nonspecific immunosuppression. Δ^9-THC produced a reduction in thymus weight in monkeys, focal hemorrhages in rats (with no alteration in weight), and decreases in weight and cellularity in the mouse (MUNSON and FEHR 1983). Since the thymus provides immunocompetent lymphocytes to the secondary lymphoid organs, it seems plausible that marihuana negatively affects maturation of these cells in the developing individual (PROSS et al. 1992a). However, no data are available concerning thymic changes in humans.

There are no consistent effects observed on the spleen following marihuana or cannabinoid administration to various animal species (MUNSON and FEHR 1983). The most consistent results appear to be in the mouse model, in which treatments of 8 days or less induce a hypocellularity concomitant with organ weight loss. One study indicates that administration of the nonpsychoactive cannabinoids CBD and CBN produced a decrease in the white pulp of the spleen, suggesting a reduction in lymphocytes. Other results indicated that Δ^9-THC could either enhance or suppress aspects of the immune response, depending on the specific immune stimulants used and the specific parameter of immunity measured (PROSS et al. 1992c) as well as the age of the animal (NAKANO et al. 1993). However, no data are available concerning splenic changes in humans.

There are limited data on the effects of marihuana on lymph nodes (MUNSON and FEHR 1983). Despite the fact that proliferation of Ly2 (suppressor/cytotoxic) cells of splenic origin could be inhibited with low doses of Δ^9-THC, identical cells of lymph node origin were resistant to the suppressive effects, which illustrated the dependence of the immunomodulatory capability of Δ^9-THC on the organ source of lymphocytes (PROSS et al. 1992b). No data are available concerning changes in human lymph nodes.

b) Immune System Cells

Δ^9-THC produces a reversible inhibition of macrophage extrinsic anti-herpes activity, while producing no effect on macrophage intrinsic activity (CABRAL and VASQUEZ 1993). The suppressive effect of Δ^9-THC on extrinsic antiviral activity is reversible upon removal of the drug. Δ^9-THC did not alter virus uptake or replication within macrophage-like cells in culture. Other studies indicated that Δ^9-THC altered macrophage morphology, function, and motility. Studies in rodents indicated a potential decrease in motility, an increased ease of cytolysis, and variety of other, more minor, alterations (MUNSON and FEHR 1983). Δ^9-THC inhibited cell propagation and DNA synthesis, though the magnitude of these effects was dependent upon the number of cells in the culture and the protein content in the culture medium (TANG et al. 1992). As the cell number increased, the Δ^9-THC effect decreased. *Legionella* grew much better in macrophages treated with low doses of Δ^9-THC, though there was no change in the number or viability of the macrophages. Thus, it is apparent that Δ^9-THC has the ability to enhance the growth of the intracellular opportunistic pathogen *Legionella* that grows in A/J mouse macrophages (ARATA et al. 1992). There do not appear to be consistent changes in the total number of leukocytes in humans (MUNSON and FEHR 1983). However, high in vitro concentrations of both psychoactive and nonpsychoactive cannabinoids exhibit some immunosuppressive activity on leukocytes (MOLNAR et al. 1987).

Natural killer (NK) cell activity was reduced following exposure to cannabinoids. Δ^9-THC did not inhibit the binding to target cells of either cloned NK cells or freshly isolated mouse spleen cells, though killing capacity was restricted. Therefore, Δ^9-THC appears to directly inhibit NK cell cytolytic activity at a postbinding stage (period following adhesion of NK cells to target cells) (KAWAKAMI et al. 1988). Δ^9-THC treatment resulted in a suppression of splenic NK activity (KLEIN et al. 1987). Further experiments suggested that the psychoactive cannabinoids suppress NK cell function by interacting directly with the killer cells and disrupting events postbinding and during the programming for lysis (KLEIN et al. 1987). Δ^9-THC applied in vitro was toxic to human peripheral blood lymphocytes at high concentrations, but at lower concentrations still produced an inhibitory effect on NK activity against a human tumor cell line (SPECTER et al. 1986).

c) B Cells and Antibody Formation

There is no consistent change in B cell number in humans following cannabis administration (MUNSON and FEHR 1983). The proliferation of B cells in response to mitogens (bacterial LPS) was reduced, but no changes were observed in monkeys following other (pokeweed) mitogen treatment. Similar effects have not been demonstrated in humans (MUNSON and FEHR 1983). In rodents, cannabinoids inhibit IgG and IgM antibody secretion, a B cell function (MUNSON and FEHR 1983). Treatment of monkeys also resulted in a decrease in IgG and IgM, but treatment for 6 months was required. There are no consistent changes in human basal antibody production.

d) T Cells

Functional measures of T cells include the in vitro measurement of stimulation (blastogenesis and secretion of cytokines) by mitogens and the ability to kill allogeneic cells (MUNSON and FEHR 1983). In vitro measures to mitogenic stimulation have proven inconsistent. Also, the ability of T_{killer} cells to destroy allogeneic mastocytoma cells was decreased by Δ^9-THC treatment (MUNSON and FEHR 1983). Exposure to Δ^9-THC also resulted in suppression of concanavalin A-induced thymus cell proliferation, primarily evidenced in the single positive Ly2 (suppressor/cytotoxic) subpopulations (PROSS et al. 1992a). Though Δ^9-THC was found to suppress mitogen-induced proliferation, it also enhanced anti-CD3 antibody-induced proliferation (NAKANO et al. 1992). Δ^9-THC produced a suppression of Ly2 cell number following concanavalin A or phytohemagglutinin stimulation, but produced an increase of Ly2 cells following CD3 stimulation (NAKANO et al. 1992; PROSS et al. 1992c). However, it is clear that both age and organ source play a critical role in the generation of immunostimulation. This up-regulation of responsiveness was not seen in either lymph node cells of adult or young mice or in spleen cells of young mice, but was only observed on lymphocytes from adult spleens (NAKANO et al. 1993; PROSS et al. 1992b). Additionally, cototoxicity assays demonstrated that CTLs (cytotoxic T lymphocytes) from mice exposed to Δ^9-THC were deficient in anti-herpes virus (HSV1) cytolytic activity (FISCHER-STENGER et al. 1992). However, in vivo Δ^9-THC treatment had little effect on the number of T lymphocytes expressing the Lyt-2 (cytotoxic cell) or L3T4 (helper cell) antigens. CTL from drug-treated mice were able to bind specifically to the HSV1-infected targets, but in vivo Δ^9-THC treatment affected CTL cytoplasmic polarization toward the virus-infected target cell, and granule reorientation toward the effector cell-target cell interface (following cell conjugation) occurred at a lower frequency. These results suggest that Δ^9-THC elicits dysfunction in CTL by altering effector cell-target postconjugation events (FISCHER-STENGER et al. 1992). However, in humans the results are more variable.

e) Host Resistance

The immunomodulatory effects of marihuana include alterations in humoral, cell-mediated and innate immunity, and though most studies have shown immunosuppressive effects, there are reports that there may not be any deleterious effect or that some aspects of host immunity may be enhanced (YAHYA and WATSON 1987). Δ^9-THC or marihuana may reduced resistance to cancer growth and microbial pathogens in animals (YAHYA and WATSON 1987). In humans, studies have suggested potential links between episodes of marihuana abuse and increased infection by such organisms as those responsible for herpes and tuberculosis. These and similar anecdotal reports have not been corroborated (MUNSON and FEHR 1983). Studies conducted with appropriate control groups of humans have failed to indicate any decrease in resistance or any significant change in immunological responses related to T cell function. Evidence is generally supportive of some degree of immunosuppression only when in vitro studies are considered, and these have been flawed by the fact that most observations only occur at very high concentrations of Δ^9-THC (HOLLISTER 1988). When experimental studies have been conducted to more closely mimic the actual clinical or human abuse situations, the evidence has been less compelling for immunosuppression, decreased host resistance, or increased infection (HOLLISTER 1988).

3. Endocrine

a) Hypothalamic-Pituitary Hormones

In the rodent, acute administration of Δ^9-THC causes a decrease in the gonadotropins LH (luteinizing hormone) and FSH (follicular stimulating hormone). These effects appear to be mediated by disruption of the hypothalamic-pituitary system via alteration of dopaminergic, serotonergic, opioid and/or adrenergic controls of endocrine function (Fernández-Ruiz et al. 1992; Wenger et al. 1992). Decreased LH appears to be due to diminished release of LHRH (luteinizing hormone releasing hormone; also referred to as GnRH-gonadotropin releasing hormone), which was reported to accumulate in the hypothalamus (Rosenkrantz 1985), though some evidence suggests diminished biosynthesis of LHRH (Dewey 1986). One study suggests the effect of Δ^9-THC involves a mechanism which includes inhibitory actions within the preoptic-to-tuberal GnRH pathway (Tyrey 1992). Studies on the alteration of FSH levels in laboratory animals are somewhat inconclusive, though FSH levels have generally been found to be decreased (Rosenkrantz 1985). In humans, the acute alterations observed following marihuana or Δ^9-THC on gonadotropins are also somewhat unclear. Acute Δ^9-THC decreased LH, but did not change FSH, at the typically low doses abused by humans (Rosenkrantz 1985). Other evidence suggests either an increase, decrease, or no change in humans with either LH or FSH.

In rodents, the acute effect of marihuana or Δ^9-THC was reported to decrease prolactin (PRL) levels. Decreased PRL release in rodents appears to be due to diminished release of TRH (thyrotropin releasing hormone) (Rosenkrantz 1985), though it may also be altered by diminished GnRH. The reduction of PRL release following Δ^9-THC exposure, both in vivo and in vitro, might be elicited by a direct action of Δ^9-THC on the pituitary (Rodriguez de Fonseca et al. 1992a), though other data suggest that the initial site of action may not be in the region of the hypothalamus most intimately associated with pituitary function. Inhibition of the effects of Δ^9-THC on PRL (and ACTH, adrenocorticotropic hormone) by hypothalamic deafferentation suggests a more distant site for Δ^9-THC action (Tyrey 1984). In rats treated with estradiol, basal PRL levels were increased and a PRL surge occurred. However, estradiol stimulation of both basal and surge levels of PRL was significantly attenuated by concomitant Δ^9-THC treatment (Murphy et al. 1991). Lastly, though PRL was decreased in monkeys, in humans Δ^9-THC increased serum PRL levels following either oral or intravenous administration (Rosenkrantz 1985). No reports were found indicating whether lactation was altered in either humans of animals.

Growth hormone (GH) is generally reduced in both laboratory animals and humans (Rosenkrantz 1985). During a 9 h period following the acute administration of Δ^9-THC to rats, the episodic secretion of GH was suppressed in terms of mean plasma level, peak height, and integrated peak amplitude analyses. Although the physiological mechanisms involved in this response were undetermined, the data indicated that Δ^9-THC can inhibit the hypothalamic-pituitary control of normal episodic GH secretion (FALKENSTEIN and HOLLEY 1992). Interestingly, the effect appears to be biphasic, since higher doses of Δ^9-THC could induce an increase in GH in rats. However, increases in human GH have not been reported (ROSENKRANTZ 1985).

b) Gonadal Hormones

The general effect of Δ^9-THC marihuana on the gonadal hormone testosterone is to decrease serum levels in rodents, monkeys, and humans (ROSENKRANTZ 1985). However, it is important to note that these changes in humans occur at high oral or moderate intravenous doses. Many studies using low doses of marihuana administered via the inhalation route showed no acute change in testosterone. Inhibition of plasma testosterone may be due to a direct effect on synthesis in Leydig cells (BURSTEIN et al. 1978, 1979, 1980), although it appears that the Δ^9-THC-induced block of GnRH (gonadotropin releasing hormone) release results in lowered LH and FSH and subsequently reduced testosterone production by the Leydig cells of the testis (HARCLERODE 1984). Other results indicate that the nonpsychoactive cannabinoid CBD also suppresses hepatic testosterone oxidation at the 2α, 16α, and 17 positions through selective inhibition of a specific cytochrome P-450 in the adult male rat (NARIMATSU et al. 1988). Additionally, smoked marihuana condensate, Δ^9-THC, and CBN have been found to inhibit specific binding of dihydrotestosterone to the androgen receptor, but did so with dissociation constants in the range of $210-580\,nM$. While it is difficult to interpret the meaning of low dissociation constants, some of the anti-androgenic effects associated with marihuana use may, at least in part, be due to inhibition at the receptor level (PUROHIT et al. 1980).

Marihuana abuse during the time of established hormonal cycles may render human females anovulatory and produce delayed and smaller surges in estrogen and progesterone (ROSENKRANTZ 1985). However, data in female monkeys did not corroborate alterations in progesterone, though treatment for 1 year suggested a shortened luteal phase and either decreased or had no effect on serum estradiol and progesterone (ROSENKRANTZ 1985). Rat data suggest that Δ^9-THC is neither pro- nor antiestrogenic with respect to phase I responses (increased uterine macromolecular uptake within 6 h of estrogen administration), but in terms of phase II responses (hyperplasia and hypertrophy occurring 12–24 h following estrogen administration), Δ^9-THC was modestly pro-estrogenic in the progesterone-treated uterus, but was anti-estrogenic in the presence of estradiol. These estrogen agonistic/antagonistic effects of Δ^9-THC on uterine phase II responses did not adversely affect the process of implantation and decidualization (PARIA et al. 1992). Δ^9-THC antagonizes estradiol action on the anterior pituitary. Δ^9-THC also prevented the estradiol-induced increase in pituitary weight but had no effect on either the uterine or oviduct weight response (Paria et al. 1992).

c) Thyroid Hormones

Decreases in both T_3 (triiodothyronine) and T_4 (thyroxine) have been documented in rodents and appear to result from diminished TSH (thyroid stimulating hormone; thyrotropin) release. Diminished TSH levels appear to be due to the inhibition of its release. Additionally, there is a disruption of iodine uptake and release. In humans a similar disruption in iodine uptake and release was observed, but no change in levels of either T_3 or T_4. However, the subjects were chronic abusers (4–7 years) so the lack of an effect may be due to tolerance (Rosenkrantz 1985).

d) Glucocorticoid Hormones

Δ^9-THC can induce certain endocrine changes including stimulation of adrenocortical function (Dewey 1986). The effects of cannabinoids on glucocorticoid hormones and receptors (Eldridge and Landfield 1990; Rodriguez de Fonseca et al. 1991b), with special reference to the hippocampus, have been reviewed recently (Eldridge and Landfield 1992). In summary, cannabinoids stimulated adrenal corticosterone secretion either alone or in combination with physiological stressors (e.g., footshock) in laboratory animals. Δ^9-THC-induced increases in corticosterone appeared to be mediated by increased release of pituitary ACTH (Dewey 1986; Rosenkrantz 1985). Also, it has been suggested that normal functioning of the pituitary-adrenal axis requires a properly functioning hippocampus, which may suggest one region through which cannabinoid-mediated alterations could be induced (Eldridge and Landfield 1992). Additionally, Δ^9-THC appears to interact in a noncompetitive fashion with the type II corticosteroid-binding receptor in the hippocampus. Since Δ^9-THC was able to down-regulate this receptor in the hippocampus, it appears Δ^9-THC possesses at least partial agonist activity, suggesting that cannabinoids may disinhibit the negative feedback control of endogenous glucocorticoids (Eldridge and Landfield 1992). Additionally, corticosterone treatment appears to increase binding of the cannabinoid ligand CP-55,940 to the hippocampus but not the cerebellum (Eldridge and Landfield 1992). These findings suggest some specificity of the interaction between cannabinoids and glucocorticoids. Lastly, corticosterone was able to partially reverse the inhibition of binding of CP-55,490 induced by in vitro addition of GTP analogs (Eldridge and Landfield 1992). In humans, no effect on corticosterone level was produced by acute oral Δ^9-THC at doses pharmacologically relevant to human abuse patterns (Rosenkrantz 1985), nor under a variety of related clinical regimens. One of the most notable observations was that Δ^9-THC administration induced aging-like degenerative changes in rat brain similar to that resulting from elevated corticosterone (LANDFIELD et al. 1988).

It has now been demonstrated that an i.c.v. injection of anandamide ($50-150\mu g$/rat) significantly increases serum levels of ACTH and cortico-sterone in a dose-dependent manner and causes pronounced depletion of CRF-41 in the median eminence (WEIDENFELD et al. 1994). These data suggest that anandamide parallels Δ^9-THC in activating the hypothalamo-pituitary adrenal axis via mediation of a central mechanism which involves the secretion of CRF-41. It is of interest that the caudate-putamen of adrenalectonized rats contains 50% higher levels of mRNA for the can-nabinoid receptor than the controls. This increase could be counteracted by dexamethasone (MAILLEUX and VANDERHAEGHEN 1993b). Taken together with the findings of Weidenfeld and colleagues, it seems possible that the corticoid and anandamide systems could be mutually regulatory.

e) Reproduction

The accumulation of many reproduction studies in both animals and humans have produced conflicting results and widely varying conclusions with time, which may be due in great part to a combination of differences in experi-mental design and interspecies differences in drug tolerance (WENGER et al. 1992). However, Δ^9-THC has been described as a reproductive toxicant in both humans and animals in various studies (Basloch 1983). In animal studies, Δ^9-THC produces adverse effects on gametogenesis (both oogenesis and spermatogenesis), on embryogenesis (organogenesis and fetal development), and upon postnatal development (TUCHMANN-DUPLESSIS 1993). Conclusions that marihuana may be linked to infertility have been proposed, in part, due to data indicating large reductions in sperm concentrations following admin-istration of four to sixteen marihuana joints per week for a 4 week period (BUCHANAN and DAVIS 1984). However, as these reductions occurred in the absence of changes in FSH, LH, PRL, cortisol, and T_4 or testosterone, they would appear to be related to direct cellular effects rather than neuronal disruption. Other evidence suggesting alterations to the male reproductive tract include findings of oligospermia with Leydig and Sertoli cell dysfunc-tion, though there did not appear to be an associated sterility (ROSENKRANTZ 1985). Evidence suggesting gynecomastia remains controversial (ROSENKRANTZ 1985). Besides altered spermatogenesis, there may also be alterations in sex organ physical characteristics (BLOCH 1983; ROSENKRANTZ 1985), which appears to be the result of both direct actions on tissues and the indirect effect of reduced androgenic hormones. Though the effects of marihuana on fertility in men may involve the pituitary-hypothalamic axis, other effects seem to be produced via alteration of specific cells of the testis. Interestingly, the mechanism of action, at least on the Sertoli cells, appears to involve a pertussis toxin-independent pathway for the reduction of FSH-induced accumulation of cAMP (HEINDEL and KEITH 1989), suggesting cellular effects that are independent of the cannabinoid receptor described in brain tissue. Additional evidence that this response may not be mediated through the cannabinoid receptor includes the fact that Δ^9-THC does not alter forskolin stimulation of cAMP in the Sertoli cells and that cannabinoids devoid of psychoactivity also inhibit FSH-induced accumulation of cAMP (HEINDEL and KEITH 1989).

4. Cardiovascular

a) Blood Pressure

The effects of cannabinoids of cardiovascular function have been reviewed (DEWEY 1986; HOLLISTER 1986; TENNANT 1983). Generally, the hypotensive effect in animals appears to be centrally mediated, though some direct action upon the heart (SMILEY et al. 1976), or the nerve terminals of the heart, appears likely (GASCON and PERES 1973; JANDHYALA and BUCKLEY 1977). Also, the effect of Δ^9-THC on animals can be biphasic, with an initial vasoconstrictive phase and an associated increase in vascular pressure, followed by a period of hypotension (ADAMS et al. 1976). One difference between animal and human data is that the effect in humans generally appears to have been smaller in magnitude, though this may be due to the use of relatively small doses as well as postural considerations in humans (ROSENKRANTZ 1985). The blood pressure response in humans (HOLLISTER 1986; LEMBERGER et al. 1974; MALIT et al. 1975; WEISS et al. 1972) was minimal (though orthostatic hypotension occurred), while a clear hypotensive effect was observed in animals (CAVERO et al. 1973; DEWEY et al. 1972; PRADHAN 1984; WILLIAMS et al. 1973). However, no lasting effects on blood pressure have been described (TENNANT 1983). Inhibited vascular reflexes and decreased peripheral resistance have been reported (ROSENKRANTZ 1983). Incidentally, the time course of the psychoactive effect closely parallels the time course of the reddening of the conjunctiva, which is due to local vasodilation (HOLLISTER 1986).

b) Heart Rate and EKG

In animals, a bradycardia was the predominant effect observed following marihuana administration (TENNANT 1983), although biphasic responses have been described following administration of sufficiently low doses of Δ^9-THC (ROSENKRANTZ 1983). Yet, in humans tachycardia was produced (HOLLISTER 1986; LEMBERGER et al. 1974; LINTON et al. 1975). Interestingly, the time course of the tachycardia in humans closely parallels the time course of the psychoactive effects (HOLLISTER 1986). The tachycardia in humans appears to have been due to sympathetic stimulation in combination with parasympathetic inhibition (HOLLISTER 1986; ROSENKRANTZ 1983). The net effect of marihuana on humans was to increase myocardial work load, myocardial oxygen demand, and to decrease oxygen delivery (HOLLISTER

1986) which was also observed in animals (SMILEY et al. 1976). Smoking marihuana may dispose individuals to heart problems such as angina (ARONOW and CASSIDY 1975). Despite this, there is no evidence for significant changes in EKG, and field studies failed to disclose any abnormalities in heavy abusers (KOCHAR and HOSKO 1973; TENNANT 1983).

5. Gastrointestinal

In some humans a significant degree of diarrhea and related gastrointestinal upset (vomiting, cramps) occurred following marihuana abuse (TENNANT 1983). It was suggested that this may have been due to the fact that cannabinoids apparently decrease gastric acid secretion (GAHLINGER 1984), which could make the smoker more susceptible to infection by gastrointestinal bacteria and thereby produce an aggravated diarrhea. However, the mechanism of action is unclear, but it is possible that cannabinoids interfered with one or more of the neuronal controls of gastric secretion. In animal models, low doses of intravenous psychoactive cannabinoids exerted an inhibitory effect on GI transit and motility and slowed the rate of gastric emptying and small intestinal transit (SHOOK and BURKS 1989). Therefore, marihuana is capable of altering function of the GI tract, suggesting that low doses might produce mild constipation. However, it is possible that the humans suffering diarrhea and cramps ingested quantities larger than the relative amount evaluated in the animal nodels.

6. Renal

Renal toxicity has been observed following intravenous injection of cannabis extracts. Generally, renal insufficiency is not observed in humans, though there was at least one instance of such a report. Following consumption of a cannabis butter preparation an elderly man suffered constipation and urinary retention. The problem was sufficiently severe that the man required urethral catheterization. It was hypothesized that the mechanism of action might have been interference with peripheral cholinergic activity (TENNANT 1983). However, Δ^9-THC has also been shown to possess an antihistaminergic activity on the rabbit kidney, which seems to be a competitive antagonism at the histamine H1-receptor (TURKER et al. 1975). Also, evidence exists to suggest a Δ^9-THC-induced release of prostaglandin-like material from rabbit kidney (KAYMAKCALAN et al. 1975). The actions of cannabinoids on renal function has not received much attention in recent years.

III. Toxicity

1. Respiratory Effects

One effect of marihuana that is produced regardless of route of administration is the depression of the respiratory system, and this effect appears to be mediated by central mechanisms. Lethality in animals could readily be demonstrated following acute drug administration (ROSENKRANTZ 1983) and appears to have been due to respiratory depression (ROSENKRANTZ 1983), which was preceded by dyspnea and apnea (FORNEY and KIPLINGER 1971). These effects on respiration may have been due to an upward shift in the carbon dioxide set point as well as depression of the respiratory center in the medulla (DEWEY 1986). However, marihuana-induced deaths in humans, in the absence of any other drugs, are an unheard of event. Thus, lethality data for marihuana in humans is primarily anecdotal. However, the lethality of Δ^9-THC administered by various routes and formulations in different species indicates large differences between oral and intravenous lethality (ROSENKRANTZ 1983). Lastly, there was unpublished evidence to suggest a role for cardiac arrest in the production of lethality following high dose cannabinoid administration, but whether this was a result of respiratory depression or other factors was unclear and unsubstantiated (ROSENKRANTZ 1983).

2. Psychotic Episodes

The suggestion that Δ^9-THC induces psychopathologies (BARTOLUCCI et al. 1969; GEORGE 1970; TALBOTT and TEAGUE 1969) has been examined and a listing of medical literature associating marihuana with mental illness compiled (NAHAS 1993b; NAHAS and LATOUR 1992). However, attempts to identify a "cannabis psychosis" have been unsuccessful (DEWEY 1986; HOLLISTER 1986; TASCHNER 1983; THORNICROFT 1990), even in parts of the world where consumption of marihuana has previously been associated with admission to hospitals for psychiatric conditions (CHKILI and KTIOUET 1993; DEFER 1993).

The effects of marihuana on schizophrenic symptoms are widely recognized to be detrimental, yet approximately one-third of all schizophrenics continue to self-medicate with marihuana (NEGRETE 1993). Paranoid schizophrenics apparently recognize the worsening of symptomatology brought on by marihuana. Schizophrenics abusing marihuana have been reported to be more difficult to effectively treat, or their symptoms worsen even when appropriate neuroleptic levels were maintained (KNUDSEN and VILMAR 1984). Marihuana appears to consistently exacerbate the "positive" symptomatology of schizophrenia while producing inconsistent effects on "negative" symptoms. Patients who self-medicate with marihuana indicate their goal is to reduce negative symptoms.

The question of the causal relationship between abuse of marihuana and the development of schizophrenia has not been established, although some believe abuse leads to psychosis (ALLEBECK 1993; NEGRETE 1993). Those individuals abusing marihuana who also develop psychiatric problems suffer from rapid-onset schizophrenia and exhibit positive symptoms including auditory hallucinations and commenting voice (ALLEBECK 1993). Of those schizophrenics that previously abused marihuana, almost 70% developed psychosis after more than 1 year of marihuana abuse. Though the mental abnormalities and related conditions attributed to cannabis abuse exist, it does not appear as though the psychosis can be distinguished from that either: (a) induced by other drugs of abuse or (b) found as endogenous schizophrenia (TASCHNER 1983). It is difficult to point to any one drug as the causative agent since these individuals are polydrug abusers. Proper studies have not been performed to determine the relative risk of development of psychiatric problems within marihuana abusers compared to nonabusers. However, the relative risk would actually appear to be small given the widespread abuse of marihuana.

3. Neurochemical and Histological Effects

The potential of neurochemical and histological damage produced by cannabinoids has been evaluated in both rats and monkeys. These results have been reviewed previously (ALI et al. 1991; SLIKKER et al. 1992). Generally, 7 months after a 1 year period of inhalation exposure of male rhesus monkeys, there was no evidence of neurochemical, histological or electron microscopic alterations in hippocampal volume, neuronal size, number or length of CA3 pyramidal cell dendrites or synaptic connections. Though Δ^9-THC could not be construed to be neurotoxic to CA3 neurons in these monkeys, further studies in the CA1, dentate granule cells, and cerebellar granule cells were being conducted to rule out other potential neurotoxic effects which were suggested elsewhere (ELDRIDGE and LANDFIELD 1992; SCALLET et al. 1987). However, these largely negative results were obtained following a 1 year period of inhalation exposure of male rhesus monkeys (SLIKKER et al. 1992). It is quite possible that this period of treatment was too short to produce effects. There have also been attempts to monitor neurological changes in rats but the conditions (duration of exposure, marihuana vs THC, dose, etc.) have differed from those described above for monkeys. Administration of Δ^9-THC for a minimum period of 3 months was required before histochemical alterations were observed in the rat (ALI et al. 1991; SCALLET 1991; SCALLET et al. 1987). Comparatively, a 3 month period is a large portion (8%–10%) of the rat life span, and to obtain a similar exposure period in monkeys would require a 3 year exposure period and in humans would correspond to a 7 or 10 year period. A review of data in rats following lengths of Δ^9-THC administration of 3 months or greater indicated the formation in the CA3 region of the hippocampus of short, broken, axodendritic connections; a significant degree of extracellular space; and subcellular organelles (vesicles, mitochondria) were not separated from extracellular space by intact membranes (SCALLET 1991). Other observations included a smaller neuronal size and fewer synaptic densities in the CA3 region. Reduced neuronal density was observed in the CA1 stratum pyramidal cells as was an increase in the proportion of opaque material within the cytoplasm of astroglia. It is important to point out that the degree of histological change was greater in peripubertal animals than in young adults. Though it is entirely possible that these neurotoxic effects involved the cannabinoid receptor, it is important to realize that other possibilities exist. Additionally, it is possible that these structural changes resulted from indirect effects. The observed alterations could also have been produced by large increases in plasma corticosterone, which might have produced neurotoxic effects in the hippocampus via specific glucocorticoid receptors (SCALLET 1991).

IV. Tolerance

1. Animals

Specific cannabis-mediated effects to which tolerance develops in a variety of species have been reviewed elsewhere (HARRIS et al. 1977; JONES 1983). Generally, tolerance develops to some degree to all cannabis-induced effects. However, there are exceptions in which the degree of tolerance development is so slight as to be considered not of physiological importance. Tolerance development has been shown to occur in all species studied. Parameters to which tolerance develops include simple physiological indices and complex behaviors mediated via the central nervous system. Some of these parameters in laboratory animals include Δ^9-THC-induced anticonvulsant activity, catalepsy, depression of locomotor activity, hypothermia, hypotension, immunosuppression, static ataxia in dogs, and alteration of response rates and accuracy on schedule-controlled behaviors.

The degree of tolerance that can be developed to Δ^9-THC is quite high. A 100-fold development of tolerance has clearly been observed in pigeons, dogs, and rodents. However, some reports actually indicate lack of activity with doses following chronic treatment which were 300- or 6000-fold higher than those initially effective in producing an effect. Tolerance has also been shown to the lethal effect of Δ^9-THC in pigeons. Similarly, tolerance to the toxic effects of oral doses of Δ^9-THC as high as 250 mg/kg per day in rats has been reported.

The onset of tolerance can be very rapid or may require a prolonged treatment period. However, generally, only 1 week of daily administration is required to observe tolerance to most simple parameters measured in rodents, dogs, or monkeys. Examples of rapid onset of tolerance include rodent hypothermia and decreased locomotor activity. In these cases, a decrement in response can be observed 24–48 h later upon administration of a second dose of Δ^9-THC, with nearly complete tolerance observed after a third treatment. In monkeys, tolerance develops to the sedative properties of Δ^9-THC after 2 weeks of oral treatment, while tolerance to some excitatory components of behavior required 2 months of treatment prior to the development of tolerance. Thus, tolerance develops differentially in all species as a function of the parameter measured. It is also not necessary to treat animals on a frequent basis in order to develop tolerance. The administration interval of 7–9 days in the pigeon and 8 days in the dog has proven sufficient to maintain tolerance. Similarly, one Δ^9-THC treatment per week, for a period of 7 weeks, is sufficient to produce tolerance in pigeons to the suppressive effect of Δ^9-THC on response rate in positive reinforcement paradigms. Additionally, the tolerance developed using these kinds of treatment protocols is long-lasting. The tolerance development observed in the dog clearly was still present for at least 23 days. Though tolerance may be observed for a period of months in some parameters, the tolerance developed to other effects of Δ^9-THC have been shown to last for only up to 24 h.

2. Humans

Most investigators suggest that pronounced tolerance must occur prior to the development of physical dependence to a drug. There is convincing evidence of tolerance development to Δ^9-THC in humans (JONES et al. 1976). Tolerance developed to cannabis-induced increases in cardiovascular and autonomic functions, to decreased intraocular pressure, to sleep disturbances and sleep EEG, as well as mood and behavioral changes in those subjects receiving oral Δ^9-THC. It is not too surprising that there is less agreement with regard to the development of behavioral tolerance to cannabis. Psychological effects are highly complex and dependent upon many factors, not the least of which is the interaction between the subject and the environmental situation. In the studies in which high doses of Δ^9-THC have been employed, behavioral tolerance has been found. For example, studies with oral administration of Δ^9-THC revealed tolerance development to the subjective effects following a few days of 10 mg Δ^9-THC treatment administered several times each day (JONES 1983). Ten days of treatment with repetitive 30 mg doses of Δ^9-THC produced even greater tolerance to the behavioral effects. Tolerance to Δ^9-THC can best be summarized as relatively little tolerance development when the doses are small or infrequent and the drug exposure is of limited duration (HOLLISTER 1986). Tolerance clearly develops when individuals are exposed to high doses for a sustained period of time.

V. Dependence

1. Animals

The most robust demonstration of physical dependence in laboratory animals has been made with chronic administration of drugs possessing a relatively short half-life. The long half-life and resultant long duration of action of Δ^9-

THC precludes the rapid induction of a drug-free system necessary for producing readily observable withdrawal signs and symptoms. Generally, the chronic administration of a drug with a half-life of greater than 35 h tends not to be followed by a withdrawal syndrome upon abrupt cessation of abuse. With these considerations in mind, studies were conducted in monkeys by intravenously administering Δ^9-THC every 6 h, with increasingly larger doses for 14 days. Administration at the highest dose attained for 12 more days (36 day regimen) produced significant physiological effects during drug abstinence (KAYMAKCALAN 1973). Symptoms appeared 12 h after the last drug treatment, and continued for 5 days. Symptoms included anorexia, hyperirritability, aggressiveness, tremors and twitching, penile erection, and masturbation with ejaculation, as well as behaviors interpreted as hallucinations. However, it is not clear that the observed behaviors were in fact withdrawal, since Δ^9-THC was not clearly shown to reverse the effects. Similarly, after continuous intravenous infusion of Δ^9-THC (daily dose of 1.2 mg/kg) for 10 days, three of four monkeys suffered a disruption of schedule-controlled behavior (BEARDSLEY et al. 1986). Observers also noted that animals were aggressive and hyperactive during abstinence. Additionally, this withdrawal syndrome could be reversed by administration of Δ^9-THC. These studies may indeed suggest that cannabis is capable of producing dependence in animals, though in either experiment the symptoms were not severe.

Δ^9-THC produces a unique behavioral change in dogs first described as static ataxia (WALTON et al. 1937). The administration of effective doses of Δ^9-THC on a daily basis produced tolerance to this effect. Increasing the dose to very high levels did not overcome this tolerance. However, the administration of increasingly large doses of Δ^9-THC to dogs over 11 days did not produce withdrawal symptoms during an 8 day period of drug abstinence (McMILLAN and DEWEY 1972). Likewise, pigeons given daily intramuscular injections of very high doses of Δ^9-THC did not show withdrawal signs when the drug was removed (McMILLAN et al. 1970). Soon after the end of this treatment regimen there was a decrement in the operant behavior of the pigeons. However, this interruption of behavior was not felt to be an indication of withdrawal, since normal behavior was not reestablished when the drug was readministered.

It is, of course, impossible to measure psychological dependence in laboratory animals. Self-administration of drugs may be an indication of psychological dependence and/or abuse potential or craving. Yet, there are few reports which claim to have established experimental models in which animals self-administer Δ^9-THC or any of the majority of its analogs. The inability to maintain self-administration of Δ^9-THC was best shown when only a portion of the animals treated would self-administer Δ^9-THC after having had the drug administered to them for a long period of time prior to allowing the animal control of its drug supply (KAYMAKCALAN 1973). Despite the opportunity to self-administer Δ^9-THC to prevent possible symptoms of withdrawal, only a small portion of the monkeys self-administered during the abstinence period. Instead, when given a choice, some monkeys self-administered cocaine rather than Δ^9-THC. This choice suggests that, even when experiments are designed to enhance Δ^9-THC self-administration, the abuse potential and possible development of psychological dependence to Δ^9-THC is tremendously less than to cocaine. Δ^9-THC (as well as related analogs) did not substitute for drugs with strong reinforcing properties (CARNEY et al. 1977). This failure also suggests limited potential for development of physical cross-dependence as well as limited psychological dependence due to weak reinforcing properties.

2. Humans

It is well established that chronic heavy use of either cannabis or hashish does not result in a withdrawal syndrome with severe symptomatology. The number of well controlled studies on the development of psychological or physical dependence to cannabis in humans is much less than those in various animal species. However, cannabis has been used for centuries, and there are a considerable number of reports in the literature, regarding the long term use of this material. The occurrence of a psychological dependence, abuse liability, or craving is more probable than physical dependence. It has been difficult to draw conclusions from epidemiological data on the psychological dependence of marihuana given the plethora of social and legal factors that impact on the drug abuser. However, there are numerous case reports of psychological dependence to cannabis (JONES 1983).

The early evidence for "dependence" upon cannabis arose from uncontrolled clinical observations following cessation of chronic drug intake. Most reports originated in countries such as India, Greece or Jamaica where cannabis or hashish had been used for long periods and was much higher in potency than the material smoked in the U.S. There were very early reports that smokers suddenly deprived of cannabis became hyperirritable, experienced auditory and visual hallucinations, and masturbated incessantly for 3–5 weeks (FRASER 1949). Abstinence symptoms in Egyptian hashish smokers were characterized as dysphoria, hyperirritability and insomnia (SOUEIF 1976). South African smokers reportedly experienced anxiety, restlessness, nausea, and cramps when cannabis was suddenly unavailable. Not too surprisingly, the conclusions reached by different investigators vary considerably. However, there are some commonalities among the descriptions of cannabis withdrawal which include hyperirritability, tremors, sweating, auditory and visual hallucinations, dysphoria, anxiety, negativism, insomnia or abnormal sleep patterns.

Some of the symptoms of cannabis "withdrawal" that have been described in uncontrolled clinical studies have also been reported in more controlled experiments. In a very early study, subjects smoked an average of 17 marihuana cigarettes daily for 39 days and reported feeling "jittery" upon withdrawal, although observers were not able to detect any symptoms (Williams et al. 1946). Almost 30 years later, studies were conducted in which subjects were placed in a controlled environment and allowed them to smoke a self-determined number of marihuana cigarettes for a 21-day period (Greenberg et al. 1976; Mendelson et al. 1976). Upon termination of the smoking period, some subjects experienced rapid weight loss, decreased appetite, tremor, increased anxiety, hostility, decreased friendliness, etc. In another study, volunteers smoked marihuana for 64 days in a hospital. These subjects were allowed to self-medicate by smoking as many cigarettes as they wished, each containing 20 mg of Δ^9-THC (Cohen et al. 1976). Sleeplessness, anorexia, nausea and irritability developed after cessation of smoking. While similar conclusions can be drawn from all of the above studies, the results should be interpreted cautiously. These studies lacked placebo or double-blind controls, and attempts were not made to reverse the withdrawal symptoms by reinitiation of marihuana smoking. Additionally, there are always problems with confining individuals for long periods of time. It may well be that some of the subjects exhibited mood changes as a consequence of confinement. It should be kept in mind that the subjects were aware of the treatment schedule, and therefore could anticipate termination of drug administration. The issue of self-administration has both advantages and disadvantages. Self-administration may well provide the most realistic treatment regimen for marihuana users, and therefore cessation of such treatment would have relevance to the real world situation.

The development of tolerance and dependence to cannabis and Δ^9-THC has been examined under a more rigorous treatment paradigm (Jones 1983; Jones and Benowitz 1976; Jones et al. 1976, 1981). The premise was that if dependence did not result under these conditions, then it was highly unlikely to occur under less stringent conditions. Either Δ^9-THC or cannabis extract was administered orally to volunteers every 3 or 4 h, 24 h a day, for up to 21 days. The 30 mg dose of Δ^9-THC resulted in peak blood levels that were comparable to those obtained by smoking a marihuana cigarette. Cessation of treatment usually resulted in subjective effects which were first reported within 5–6 h after the last dose of Δ^9-THC. The most prominent and frequent symptoms were increased irritability and restlessness. Other prominent and somewhat variable symptoms were insomnia, anorexia, increased sweating and mild nausea. Objective changes included body weight loss, increased body temperature, and hand tremor. Both the subjective and objective changes could be diminished by smoking a marihuana cigarette or by readministration of oral Δ^9-THC, suggesting establishment of a withdrawal syndrome. The intensity of the effects observed was dependent upon the length of the treatment time and the dose of Δ^9-THC.

VI. Δ^9-THC During Pregnancy

1. Effect on Dams and Litters

Marihuana use in humans has been attributed to the low birth weight and small gestational size of infants (HATCH and BRACKEN 1986). In animals, Δ^9-THC during pregnancy increases the frequency of stillbirths (GAL and SHARPLESS 1984; HUTCHINGS et al. 1989b) and decreases litter size (WENGER et al. 1992). Resorption rates increased in mice but not rats following in utero exposure. A decrease in maternal food and water consumption occurred and led to decreased maternal weight gain, which may be the cause of various effects associated with prenatal exposure (ABEL 1985b). Prenatal exposure to cannabinoids led to a decrease in pup birth weight, which may be the only postnatal effect on offspring reliably demonstrated. In rodents, increased resorption of fetuses, perinatal death, and altered sex ratio (MORGAN et al. 1988) also affects the final characteristics of the litter (WENGER et al. 1992). Some of these effects have been attributed to altered LH, FSH, progesterone, placental steroid excretion and/or inhibited prostaglandin synthesis (DALTERIO et al. 1984; WENGER et al. 1992) In mice and rats low to moderate doses of Δ^9-THC did not affect the length of gestation, maternal viability, or maternal weight gain, though high doses of Δ^9-THC did prevent maternal weight gain (BLOCH 1983). The decreased production of milk following birth of the litter has been linked to high neonatal mortality. Diminished lactation appears to be due to disruption of prolactin release and disruption of hypothalamic neuronal controls (BLOCH 1983).

2. Developmental Toxicity

A great deal of effort has been expended on investigation of the effects of perinatal exposure to Δ^9-THC, largely based upon initial data suggesting the existence of various deficits in humans (O'CONNELL and FRIED 1991; TENNES et al. 1985) and with the anticipation of finding a definable syndrome equivalent to the fetal alcohol syndrome. For example, in humans, prenatal marihuana exposure was reported to be related to tremors, increased startle, and poorer habituation to visual stimuli of offspring (FRIED and MAKIN 1987). Investigations of minor physical abnormalities indicated that there was no correlation between the number of anomalies present in an individual and marihuana use, though two anomalies (true ocular hypertelorism and severe epicanthus) were found only among children of heavy users of cannabis (O'CONNELL and FRIED 1984). However, a survery of the literature indicates that prenatal exposure to cannabinoids does not produce malformations in humans and only does so in mice following exposure to high doses administered by the intraperitoneal route (ABEL 1985b). Long-term studies on postnatal effects have produced generally inconsistent results, which may be due to methodological flaws in experimental design (ABEL

1985b). Experiments that do not consider the confounding influences of maternal toxicity (prenatal and postnatal) are likely to yield a high rate of false-positive results, which has been observed in studies of cannabis that preceded current concerns for pair-feeding and surrogate fostering (ABEL 1985b; HUTCHINGS and DOW-EDWARDS 1991). Nearly all such studies found neurobehavioral effects that included changes in activity as well as impairments in learning and memory (HUTCHINGS and DOW-EDWARDS 1991). Thus, it is now generally concluded that there are no significant lasting effects that can be demonstrated on marihuana exposed offspring (HUTCHINGS and DOW-EDWARDS 1991). Transient decrements on rodent body growth (HUTCHINGS et al. 1989a) and brain potein synthesis (MORGAN et al. 1988) have been observed in neonates following perinatal marihuana exposure, but these effects appeared to be due to maternal toxicity (HUTCHINGS and DOW-EDWARDS 1991). When marihuana is administered via smoke inhalation a ventilation/perfusion imbalance is created and fetal oxygen availability limited (CLAPP et al. 1987), but the effect appears to be related to the 30% reduction in maternal respiration, suggesting any fetal effects are an indirect toxicity, and this decrement has not been related to any developmental toxicity.

3. Neural Development

The effects of perinatal cannabinoid exposure on development, with special emphasis on disruption of dopaminergic neurons of the nigrostriatal, mesolimbic, and tuberoinfundibular systems, has been reviewed (RODRIGUEZ DE FONSECA et al. 1991a, 1992a,b). Alterations in these systems have been suggested to result in changes in locomotor activity. However, numerous other studies failed to detect changes in motor and endocrine function (BRAKE et al. 1987; HUTCHINGS et al. 1989b). It is unclear whether any of these events on dopaminergic neurons are mediated by the cannabinoid CB1 receptor, especially considering the sexual dimorphism described. The dopaminergic effects of perinatal cannabinoids on males is more pronounced and prolonged than the effects observed in females (RODRIGUEZ DE FONSECA et al. 1992b). However, the presence of the cannabinoid receptor during the critical time of early development has been described (RODRIGUEZ DE FONSECA et al. 1993).

4. Teratogenicity

Though Δ^9-THC possesses some potential for production of teratogenic effects (DALTERIO 1986), such alterations were only observed after very high doses of Δ^9-THC administered specifically before the end of organogenesis (WENGER et al. 1992). Many other studies find little evidence to support this contention (ROSENKRANTZ et al. 1986), and clinical studies have also failed to resolve this issue (TUCHMANN-DUPLESSIS 1993; WENGER et al. 1992). A review of animal data has not provided convincing evidence of teratogenesis (GAL and SHARPLESS 1984; ROSENKRANTZ et al. 1986). Some data would suggest that marihuana abuse during pregnancy can induce fetal stress and hypoxia, and despite evidence suggesting enhanced startle or tremors in babies, a follow-up study at 1 year of age indicated no adverse mental or motor effects (GAL and SHARPLESS 1984). Fetotoxicity has been suggested (NAHAS 1993a). However, reviews by others (ROSENKRANTZ 1985), indicate there is a lack of solid evidence supporting embryotoxicity in women despite findings of increased resorption of fetuses and perinatal death in rodents (ROSENKRANTZ 1985; WENGER et al. 1992).

5. Fetotoxicity – Interactions with Ethanol

A review of the neurobehavioral and developmental effects of fetal drug exposure indicates that the drugs most commonly associated with an adverse developmental outcome are alcohol, anticonvulsants, narcotics, etc. (GAL and SHARPLESS 1984). However, the potential for an interaction between marihuana and other substances suggests many potential dangers. One potential danger in pregnancy is suggested by the fact that combination treatment with marihuana and alcohol, at doses that were inactive alone, produced complete fetal mortality in mice and a 73% fetal mortality in rats (ABEL 1985a). A superadditive effect was also suggested when alcohol (1 g/kg) and marihuana extract (50 or 100 mg/kg Δ^9-THC) were coadministered (ABEL and DINTCHEFF 1986).

D. Behavioral Pharmacology

I. Unlearned Behaviors/Ethology

1. General Comments

DEWEY (1986) has indicated that "little if any conclusive evidence has been presented which shows that the cannabinoids affect any peripheral system without working at least indirectly through the central nervous system (CNS)." Although cannabinoids have been shown to produce direct cellular actions on peripheral tissue, most effects of interest to researchers appear to involve a neural component of the CNS or autonomic system.

A variety of centrally mediated phenomena have been observed in mouse, rat, dog, rabbit and monkey and are reviewed elsewhere (DEWEY 1986; HOLLISTER 1986; RAZDAN 1986), but include measures of spontaneous and forced (rotorod) locomotion, hypothermia, immobility, antinociception, drug discrimination, static ataxia, anticonvulsant actions and operant behavioral measures. Also, hypersensitivity to auditory or tactile stimulation has been observed (FERRI et al. 1981). By defining the spectrum of activity (efficacy, potency, etc.) of naturally occurring cannabinoids in a series of procedures (MARTIN et al. 1987), it has been possible to determine whether new synthetic and structurally diverse chemical structures are cannabimimetic (COMPTON et al. 1992a,b; MARTIN et al. 1987).

2. Consummatory Behavior

It is well known that marihuana users consistently report an increased hunger during acute intoxication of the drug (HALIKAS et al. 1985). The numerous anecdotal accounts indicating that marihuana increases feeding behavior and body weight have suggested its potential thereapeutic use as an appetite stimulant for cancer or AIDS patients (PLASSE et al. 1991). In actuality, there is a scarcity of experimental evidence supporting this can-nabinoid action. In a previous review (MEYER 1978), the weight gain asso-ciated with marihuana smoking has been suggested to result from an increased appetite for sweets and carbohydrate-containing fruit drinks. Whether the self-reported appetite-enhancing effects of marihuana in humans is a direct drug effect or results from a complex interaction between the drug and social influences is an unresolved issue. Some research supports the latter explanation (FOLTIN et al. 1986). In an attempt to simulate a natural setting, human subjects were housed in a residential laboratory and allowed to smoke marihuana cigarettes prior to a private work period and during a social access period. A single active marihuana cigarette prior to the private work period had no effect on food intake. The administration of two or three active marihuana cigarettes during the social access period did not increase meal size but did increase daily caloric intake as between-meal snack items. Further research is needed to ascertain the mechanism by which cannabinoids enhance appetite.

In contrast, there is little experimental support in the animal literature for appetite enhancing effects of the cannabinoids. These compounds are typically reported to decrease food consumption and weight gain relative to the vehicle-treated subjects (ABEL and SCHIFF 1969; SOFIA and BARRY 1974). The initial anorectic effect of Δ^9-THC by either i.p. or i.v. route of admin-istration in rats, was found to diminish after 5 days of frequent administra-tion; however, daily weight gain remained suppressed compared to the controls (MICZEK and DIHIT 1980; VERBENE et al. 1980).

3. Motor Behavior

Comparison of changes in motor activity between animals and humans has not been easy (ROSENKRANTZ 1983). Human motor activity is highly variable, and greatly affected by prior drug exposure, psychosocial setting, cultural customs, etc. However, when high doses of cannabinoids are administered intravenously to humans a definite lethargy and sedation has been demonstrated which would seem to resemble animal results (ROSENKRANTZ 1983). In laboratory animals cannabinoids have been shown to elicit locomotor effects in a variety of tasks including the static ataxia test in dogs, alterations in spontaneous activity in mice, catalepsy in mice and rats, and impairment in the rotorod test (CONSROE and MECHOULAM 1987; LITTLE et al. 1988). The static ataxia test was one of the earlier behavioral tests employed to evaluate the psychoactivity of cannabinoids (WALTON et al. 1937). In this paradigm, a dog is administered an intravenous dose of cannabinoid and the degree to which the animal exhibits motor dysfunction is rated by observers. Although noncannabinoid substances also produce ataxia, this paradigm was useful in identifying psychoactivity in both naturally occurring and synthetic cannabinoids (MARTIN et al. 1976, 1984; WILSON and MAY 1974, 1979). Recently, the dog static ataxia test for the assessment of cannabinoids has been largely replaced by rodent models of motor behavior that also reliably predict psychoactivity.

One measure that is used to identify cannabinoid activity is the assessment of spontaneous locomotor activity. Cannabinoids generally lead to decreases in spontaneous locomotion (LITTLE et al. 1988). A variety of cannabinoids including the naturally occurring compounds as well as the synthetic agents from either the aminoalkylindole class (COMPTON et al. 1992a) or the bicyclic class (COMPTON et al. 1992b) have all been shown to produce hypomotility. Similarly, cannabinoids produce decreases in response rates under different schedules of reinforcement in a variety of species (CARNEY et al. 1979; ZUARDI and KARNIOL 1983). Several studies have demonstrated biphasic effects on spontaneous locomotion. Cannabinoids are known to produce hypomotility at medium to high dose and increases in activity after treatment with low doses. A similar phenomenon has been observed using the low rate differential reinforcement schedule of operant behavior in rats (HILTUNEN et al. 1989). The basis for the dissociation between the inhibitory and stimulatory effects of marihuana on locomotor activity may be related to its effects in different brain structures. However, the phenomenon is also found with other central depressants, such as the barbiturates and minor tranquilizers (HARRIS et al. 1966).

Another well established motor effect of marihuana is its propensity to cause animals to maintain a rigid posture or catalepsy. Cataleptic effects of cannabinoids have been assessed both the bar immobility test (FERRI et al. 1981) and the ring immobility test (PERTWEE 1972). Clearly, the extrapyramidal system seems to play an important role in these effects. Intracerebral administration of either Δ^9-THC or 11-OH-Δ^9-THC into the caudate putamen had a moderate cataleptic effect (GOUGH and OLLEY 1978). Although neither of these drugs produced catalepsy in the globus pallidus, intrapallidal injections of the potent analog 11-OH-Δ^8-THC-dimethylheptyl produced catalepsy (PERTWEE and WICKENS 1991).

4. Social Behavior

a) Motivation in Humans

The belief that "frequent use of marihuana by young adolescents can impede normal maturation and cause or contribute to an amotivational syndrome" has sometimes been expressed (SCHWARTZ et al. 1987; TUNVING 1987; WATANABE et al. 1984), but a controversy exists concerning the existence of an "amotivational syndrome" as associated with long-term marihuana abuse (HOLLISTER 1986; MAYKUT 1984; PAGE 1983; SOLOMONS and NEPPE 1989). An amotivational syndrome could generally be described as a condition of apathy, lethargy, a flattening of affect, and a lack of goal-oriented behavior. Attempts to verify the existence of such an effect in controlled humans studies or epidemiological studies in localities of great abuse have either failed to provide evidence of such a syndrome or observed other factors which could potentially produce the phenomenon observed or only found residual effects of acute Δ^9-THC administration (DEWEY 1986; FOLTIN et al. 1989, 1990; HOLLISTER 1986; MAYKUT 1984). Additionally, some changes that could be observed in an individual's character during long-term abuse of marihuana did not appear different from that produced by the abuse of any other licit or illicit drug (TASCHNER 1983). It seems likely that the lack of "motivation" in humans is more a function of drug abuse and psychosocial issues than of marihuana abuse per se.

b) Sensory and Other Effects in Animals

A 1 year period of repeated marihuana inhalation exposure in male rhesus monkeys appeared to reduce the "motivational" aspects of food reinforced responding in a progressive ratio protocol of an operant behavioral task (PAULE et al. 1992; SLIKKER et al. 1992). The general health of the animals was not compromised, though both short- and long-term treatment stressed animals significantly as evidenced by urinary cortisol output. Similarly, cessation produced a physiological stress response that could have been indicative of a "withdrawal" phenomenon. There were no residual behavioral effects of chronic marihuana treatment 7 months after the termination of treatments. Similar studies in rodents (SCALLET 1991) indicated altered performances in mazes, avoidance of footshock by motor activity, performance in memory tasks (in an eight-arm radial maze), deficits on differential reinforcement of a low lever-pressing response rate operant schedules, and decrements in rotorod performance.

One of the most notable cannabinoid effects is their ability to inhibit the perception of noxious sensory stimulation. Treatment with Δ^9-THC has been reported to decrease pain in patients suffering from neoplastic disease with a potency similar to that of codeine; however, the sedative and other intoxicating effects of the drug limited its clinical usefulness (NOYES et al. 1975).

In the nonhuman animal literature, the antinociceptive properties of cannabinoids have been demonstrated in a variety of pain assays including the tail-flick test (MARTIN 1985b), the hot plate test (FRIDE and MECHOULAM 1993; WELCH and STEVENS 1992), and the p-phenylquinone writhing test (FORMUKONG et al. 1988; HAUBRICH et al. 1990). Much of the research examining neuroanatomical and neurochemical mechanisms of cannabinoid-induced antinociception have employed the tail-flick test. Cannabinoids appear to produce antinociception through both spinal and supraspinal components of action because spinal transection attenuated but did not completely block the antinociceptive effects of intravenously administered cannabinoids (LICHTMAN and MARTIN 1991; SMITH and MARTIN 1992). In addition, spinal administration and i.c.v. administration of cannabinoids were found to produce antinociception in a variety of species (MARTIN et al. 1993; WELCH et al. 1994). Although those results indicate brain involvement, additional studies are required to elucidate the neural substrates of cannabinoid-induced antinociception. The occurrence of a higher concentration of cannabinoid receptors in the periaqueductal gray, a structure that plays an important role in antinociception, than other brainstem structures suggested its involvement in cannabinoid-induced antinociception. Consequently, intracerebral administration of CP-55,940 into the ventrolateral periaqueductal gray in the region of the dorsal raphe was found to elicit a potent antinociceptive effect, thus indicating its involvement in the antinociceptive effects of cannabinoids. Whether other brain areas also contribute to the antinociceptive effects of cannabinoids is an issue for additional research.

II. Conditioned Effects

1. Drug Discrimination

Despite the lack of methodological means to measure euphoria in animals (see self-administration below), the drug discrimination paradigm (BALSTER and PRESCOTT 1992) has been a very useful model to assess the intoxicating effects of cannabinoids. In this paradigm, nonhuman primates, rats, or pigeons are trained to make two different responses for reinforcement contingent upon whether the training drug or vehicle were administered (GOLD et al. 1992; JÄRBE and HILTUNEN 1987; WEISSMAN 1978). Once the subjects are able to discriminate successfully they can be administered other drugs to determine if these substances produce similar or different stimulus characteristics from the training drug. Drugs fround to generalize to Δ^9-THC in the drug discrimination paradigm have also been reported to be marihuana-like in humans or bind to the cannabinoid receptor. Cannabinoids that are known to possess distinct structures but nonetheless bind to the cannabinoid receptor, including the aminoalkylindole (COMPTON et al. 1992a) and bicyclic

(Gold et al. 1992) compounds, have been shown to substitute for Δ^9-THC. This paradigm elicited a high degree of specificity because substances belonging to other drug classes, including dopaminergic, benzodiazepine, opioid, cholinergic, and noradrenergic, do not reliably substitute for THC (Balster and Prescott 1992). The relative potencies of drugs that generalize to THC also exhibit similar binding affinity to the cannabinoid receptor.

2. Self-Administration

The facts that 60% of all young adults in the United States have used marijuana in their lifetime and more than 10% of this age group use it on a regular basis (Johnston et al. 1993) strongly suggest that this drug is a positive reinforcer. The self-administration paradigm in animals has been a valuable tool in predicting the abuse liabilities of drugs. However, there has been a general inability to obtain cannabinoid self-administration in non-human animals. The few published studies which employed this paradigm reported that Δ^9-THC was an ineffective reinforcer in rhesus monkeys (Harris et al. 1974). Because of the relatively delayed onset and the long duration of effect of cannabinoids and their rate decreasing effects, attempts were made to adapt the self-administration procedure by using a fixed interval schedule to circumvent response rate suppression (Mansbach et al., in press). Similar to studies employing the fixed ratio schedule, this attempt also failed to establish self-administration in laboratory animals. The apparent inability to establish cannabinoids as a reinforcer in the self-administration paradigm suggests a limitation in this model to predict the abuse liability of drugs in humans or that they have a low abuse liability.

Alternatively, cannabinoids have also been documented to elicit various aversive effects. In laboratory rodents, Δ^9-THC has been shown to act as an unconditioned stimulus in the taste aversion paradigm. Moreover, Δ^9-THC has been found to act as an anxiogenic agent in the elevated plus maze (Onaivi et al. 1990). Therefore, it may be that these apparent negative hedonic properties may mask the appetitive properties of cannabinoids and thus account for their failure to serve as positive reinforcers in the self-administration paradigm. Other research has demonstrated that cannabinoids act upon brain regions involved with reinforcement. Relatively high concentrations of cannabinoid receptors have been found in the nucleus accumbens (Herkenham et al. 1991b; Jansen et al. 1992; Thomas et al. 1992). Gardner and his colleagues have provided convincing evidence that cannabinoids produce effects upon the mesotelencephalic dopamine reward pathway similar to other rewarding drugs (Gardner and Lowinson 1991). Δ^9-THC was found to reduce the amount of electric current required for self-stimulation in the medial forebrain bundle (Gardner et al. 1988). In addition, systemic administration of Δ^9-THC was found to increase DA efflux in the nucleus accumbens (Chen et al. 1990). These effects are similar to those of other drugs which are reported to have positive hedonic effects in humans.

3. Performance, Memory and Learning

a) Intoxication and Performance Impairment

It seems safe to assume that the goal of most marihuana abusers is to attain a state of intoxication (CHAIT and ZACNY 1992; JONES 1971). The possible role of cannabinoids regarding the brain reward system has been summarized by others (GARDNER 1992; GARDNER and LOWINSON 1991). The euphoria coincides with adverse effects of behavioral toxicity, including alteration of motor control, sensory functions, and the cognitive process (NAHAS 1993a; NAHAS and LATOUR 1992). Impairment of both motor control and cognitive processes could easily lead to accidents and traffic fatalities (AUSSEDAT and NIZIOLEK-REINHARDT 1993). Nonvehicular accidents (SODERSTROM et al. 1993) have been linked to abuse of marihuana. However, the question asked should be: What is the relationship between marihuana consumption, blood or urine levels of drug, and the degree of incoordination or loss of function that is produced? (HOLLISTER et al. 1981; SODERSTROM et al. 1993). The resolution of this issue would more clearly substantiate the detrimental effects of marihuana abuse by establishing the causal relationship between the period of psychomotor disruption and in vivo levels of Δ^9-THC or metabolites, which has obvious medico-legal implications.

The impact of marihuana on task performance in humans has been reviewed extensively (CHAIT and PIERRI 1992; LEIRER et al. 1991, 1993). Although there are innumerable problems interpreting a large number of studies when a diversity of methods and approaches have been taken, the authors were able to draw several general conclusions. In summary, at moderate levels of intoxication, there is a weak correlation between the incidences of heart rate increases and levels of euphoria. Marihuana and Δ^9-THC adversely affect gross and simple motor ability (body sway as measured on a "wobble board" and hand tremor), as well as some psychomotor behaviors (rotary pursuit, digit symbol substitution test, reaction time in choice reaction time tasks, accuracy in divided attention tasks, sustained attention) while not adversely affecting other tasks (simple reaction time, hand-eye coordination). Interestingly, in some studies in which chronic abusers were evaluated, the results suggested, in comparison to similar studies not using chronic abusers, that a large degree of tolerance may develop in humans to some of these acute effects. In conclusion, similar to the situation with alcohol consumption, cannabis intoxication of an experienced abuser may be difficult to detect except in performance tasks for which they have had no previous training or in tasks requiring a great deal of skill and/or manual dexterity. However, cannabis intoxication in an inexperienced individual would be readily detectable, but not necessarily on all performance measures.

Cannabinoid-induced impairment of flying (LEIRER et al. 1991) and driving (HOLLISTER 1986; MOSKOWITZ 1985) have been documented. Each of these tasks presumably require a great deal of manual dexterity and undisrupted cognition and therefore might be expected to readily detect the intoxicating effects of any drugs. A review of the impaired flying studies (Leirer et al. 1993) suggests that individuals trained on computerized flight simulations perform less well than controls on five of the eight variables measured for up to 24 h after treatment. However, in a second more sophisticated experiment, the researchers failed to replicate those results. Yet, in a third study where the level of flight difficulty increased, and subjects were allowed less training on the simulation than in the first study, the global score (aggregate of six variables) for simulated flight was significantly altered at times up to 24 h. It is interesting that these latter studies did not attempt to replicate the detrimental effects of age and Δ^9-THC consumption on simulated flight, during which older "pilots" faired worse than their younger counterparts. The data suggested that either the level of impairment (though statistically significant) was not of functional relevance in terms of performance (at least in younger pilots) or the testing procedure was not appropriate for measuring impairment in humans. Also, it should be noted that impairment was not observed in all individuals (Chait and Pierri 1992).

There is little doubt that automobile accidents have been linked to intoxication of the driver by marihuana and a variety of other drugs, sometimes used in combination (Mason and McBay 1984; Moskowitz 1985). Co-abuse of marihuana with either alcohol (Wechsler et al. 1984) or with phencyclidine (Poklis et al. 1987) is common. However, it is also true that abuse of marihuana alone can disrupt driving performance if the task is of sufficient difficulty or the dose high enough. A summary of these results (Hollister 1986) suggests that intoxicating levels of alcohol impairs performance more than does marihuana. Unlike alcohol intoxication, not all driving measures were affected by marihuana and not all subjects were affected. The combination of alcohol with marihuana was more detrimental than either drug alone. Interestingly, when allowed to smoke marihuana until intoxicated, 94% of the individuals failed a roadside sobriety test 90 min after smoking, and 60% failed 150 min after smoking.

b) Memory and Time Perception

Δ^9-THC impairs memory and learning (Chait and Pierri 1992; Schwartz 1993), but results on specific evaluations are often inconsistent and test specific (Chait and Pierri 1992). The paradigms in which Δ^9-THC produces its greatest effects (10%–50% decrement) are in free recall tasks or short-term memory function (Chait and Pierri 1992). Some reviewers believe that data indicate long-term (possibly permanent) impairment of short-term memory in adolescent age chronic marihuana abusers (Schwartz 1993). It also appears that some individuals suffer no memory impairment at all and that as a group those with any type of learning disability are more affected than the exceptionally gifted student group (Schwartz 1993). Thus, the question could be asked: Are marihuana abusers unsuccessful students because they smoke cannabis, or do they smoke cannabis because they are underachievers? Preliminary data support the latter contention and also suggest that continued abuse of marihuana and other substances also involves other factors (JOHNSON 1988; JOHNSON and PANDINA 1991; LABOUVIE 1990).

A thorough review of the literature indicated that Δ^9-THC reliably alters the perception of time (CHAIT and PIERRI 1992). Subjects overestimated time elapsed relative to real clock time or experienced an increase in the subjective rate of time. Attempts to demonstrate other behavioral effects on mental function have not met with such certainty (CHAIT and PIERRI 1992). Mixed or inconsistent results have been obtained on the Stroop (color and word) test, mental arithmetic capability, and various "creativity" tasks, although significant effects of marihuana administration were observed on an embedded figures task (finding geometric figures within a more complex design) and on verbal output tests. Thus, psychomotor performance would be expected to the impaired if short-term memory and/or time perception were required for that task. Perhaps this is true and is reflected in driving or piloting studies, but evaluation of work productivity in groups of heavy marihuana abusers has indicated no decrement in performance (HOLLISTER 1986).

c) Memory in Animals

Cannabinoids have long been known to impair learning and memory in a variety of tasks in rodents (CARLINI et al. 1970), nonhuman primates (FERRARO and GRILLY 1973), and humans (ABEL 1971). In rats, Δ^9-THC has been found to disrupt memory as assessed in the delayed match-to-sample (DMTS) task (HEYSER et al. 1993), Lashley III maze (CARLINI et al. 1970), and the eight-arm radial-maze (LICHTMAN et al., in press; NAKAMURA et al. 1991). In nonhuman primates Δ^9-THC has been found to disrupt chaining behavior and the DMTS (RUPNIAK et al. 1991). In addition to Δ^9-THC, the structurally distinct synthetic compounds CP-55,940 and WIN-55,212-2 were also found to impair working memory in rats. Interestingly, anandamide failed to impair working memory in either the eight-arm radial maze or the delayed nonmatch to sample tasks (CRAWLEY et al. 1993). This difference between Δ^9-THC and the endgoenous ligand of cannabinoid in memory function underscores the other differences found between these compounds (SMITH et al. 1994).

The high concentration of cannabinoid receptors found in the hippocampus (HERKENHAM et al. 1991b; JANSEN et al. 1992; THOMAS et al. 1992) may mediate the disruptive effects of these drugs on cognition. Research from a convergence of in vitro and in vivo studies further implicates the involvement of the hippocampus in cannabinoid-induced memory impairment. Δ^9-THC applied to hippocampal tissue biphasically affected long-term potentiation (NOWICKY et al. 1987), a neural mechanism believed to play a prominent role in information storage in the brain. Long-term administration of Δ^9-THC decreased the concentration of synapses in the CA3 region of the hippocampus (SCALLET et al. 1987). Δ^9-THC-induced impairment in the DMTS task was associated with a specific decrease in hippocampal cell discharge (HEYSER et al. 1993). Direct evidence implicating hippocampal involvement was that intrahippocampal administration of CP-55,940 led to a dose-related increase in the number of errors committed in the eight-arm radial maze task. The effects of intrahippocampal CP-55,940 were apparently specific to cognition because no other cannabinoid pharmacological effects (e.g., antinociception, hypothermia, and catalepsy) were elicited. Δ^9-THC has been shown to alter cerebral metabolism in several brain regions including the hippocampus (MARGULIES and HAMMER 1991).

D. Conclusions

Marihuana remains one of the most widely abused substances throughout the world. Despite a wide range of pharmacological effects on most organ systems, the health consequences of chronic marihuana abuse are relatively mild when compared to those of most other abused substances. There is no question that marihuana is capable of producing impairment of performance in individuals while intoxicated. Memory and learning decrements are well documented under specific circumstances. Attempts to establish neurochemical and histological damage produced by cannabinoids have not resulted in definitive conclusions. However, the current data suggest that a rigorous treatment during a long exposure period will be required if permanent neurological deficits are produced.

Considerable progress has been made regarding the characterization of cannabinoid receptor subtypes in brain and peripheral tissues, and there has been some insight into second messenger systems. The discovery of endogenous cannabinoids and the characterization of their synthetic and metabolic enzymes provides the basic foundation for establishing an endogenous cannabinoid system. The recent development of a cannabinoid antagonist will greatly facilitate this undertaking. The physiological role of cannabinoids should emerge in the not too distant future.

Acknowledgements. Portions of this research were supported by NIDA grants DA-00490 and DA-03672.

References

Abel EL (1971) Retrieval of information after use of marihuana. Nature 231:58
Abel EL (1985a) Alcohol enhancement of marihuana-induced fetotoxicity. Teratology 31:35–40
Abel EL (1985b) Effects of prenatal exposure to cannabinoids. In: Pinkert TM (ed) Current research on the consequences of maternal drug abuse. US Government Printing Office, Washington DC, pp 20–35

Abel EL, Dintcheff BA (1986) Increased marihuana-induced fetotoxicity by a low dose of concomitant alcohol administration. J Stud Alcohol 47:440–443

Abel EL, Schiff BB (1969) Effects of the marihuana homologue, parahexyl, on food and water intake and curiousity in the rat. Psychon Sci 16:38

Abood ME, Sauss C, Fan F, Tilton CL, Martin BR (1993) Development of behavioral tolerance of Δ^9-THC without alteration of cannabinoid receptor binding or mRNA levels in the whole brain. Pharmacol Biochem Behav 46:575–579

Adams IB, Thomas BF, Compton DR, Razdan RK, Martin BR (in press) Evaluation of cannabinoid receptor binding and in vivo activities for anandamide analogs. J Pharmacol Exp Ther

Adams MD, Earnhardt JT, Dewey WL, Harris LS (1976) Vasoconstrictor actions of Δ^8- and Δ^9-tetrahydrocannabinol in the rat. J Pharmacol Exp Ther 196:649–656

Adams MD, Earnhardt JT, Martin BR, Harris LS, Dewey WL, Razdan RK (1977) A cannabinoid with cardiovascular activity but no overt bahavioral effects. Experientia 33:1204–1205

Adams R (1942) Marihuana. Harvey Lect 37:168–197

Ali SF, Newport GD, Scallet AC, Paule MG, Bailey JR, Slikker W (1991) Chronic marijuana smoke exposure in the rhesus monkey. IV. Neurochemical effects and comparison to acute and chronic exposure to Δ^9-tetrahydrocannabinol (THC) in rats. Pharmacol Biochem Behav 40:677–682

Allebeck P (1993) Schizophrenia and cannabis: cause-effect relationship? In: Nahas GG, Latour C (eds) Cannabis: physiopathology, epidemiology, detection. CRC Press, Boca Raton, pp 113–117

Arata S, Newton C, Klein T, Friedman H (1992) Enhanced growth of Legionella pneumophila in tetrahydrocannabinol-treated macrophages. Proc Soc Exp Biol Med 199:65–67

Aronow WS, Cassidy J (1975) Effect of smoking marihuana and of a high-nicotine cigarette on angina pectoris. Clin Pharmacol Ther 17(S):549–554

Askew WE, Ho BT (1974) The effects of tetrahydrocannabinols on cyclic AMP levels in rat brain areas. Experientia 30:879–880

Audette CA, Burstein SH, Doyle SA, Hunter SA (1991) G-protein mediation of cannabinoid-induced phospholipase activation. Pharmacol Biochem Behav 40: 559–563

Aussedat M, Niziolek-Reinhardt S (1993) Detection of cannabis and other drugs in 120 victims of road accidents. In: Nahas GG, Latour C (eds) Cannabis: physiopathology, epidemiology, detection. CRC Press, Boca Raton, pp 73–77

Balster RL, Prescott WR (1992) Δ^9-tetrahydrocannabinol discrimination in rats as a model for cannabis intoxication. Neurosci Biobehav Rev 16:55–62

Barnett C, Chiang C, Perez-Reyes M, Owens S (1982) Kinetic study of smoking marijuana. J Pharmacokinet Biopharm 10:495–506

Barnett G, Licko V, Thompson T (1985) Behavioral pharmacokinetics of marijuana. Psychopharmacology 85:51–56

Bartolucci G, Fryer L, Perris C, Shagass C (1969) Marijuana psychosis: a case report. Can Psychiatr Assoc J 14:77–79

Beardsley PM, Balster RL, Harris LS (1986) Dependence on tetrahydrocannabinol in rhesus monkeys. J Pharmacol Exp Ther 239:311–319

Benowitz NL, Rosenberg J, Rogers W, Bachman J, Jones RT (1979) Cardiovascular effects of intavenous Δ^9-tetrahydrocannabinol: autonomic nervous mechanisms. Clin Pharmacol Ther 25:440–446

Bhargava HN (1976) Effect of some cannabinoids on naloxone-precipitated abstinence in morphine-dependent mice. Psychopharmacology 49:267–270

Bhargava HN (1978) Time course of the effects of naturally occurring cannabinoids on morphine abstinence syndrome. Pharmacol Biochem Behav 8:7–11

Bhattacharya SK (1986) Δ^9-Tetrahydrocannabinol increases brain prostaglandins in the rat. Psychopharmacology 90:499

Bhattacharya SK, Ghosh P, Sanyal AK (1980) Effects of prostaglandins on some central pharmacological actions of cannabis. Indian J Med Res 71:955–960

Bidaut-Russell M, Howlett A (1988) Opioid and cannabinoid analgetics both inhibit cyclic AMP production in the rat striatum. In: Hamon M, Cros I, Meunier J (eds) Advances in the biosciences: proceedings of the international narcotics research conference. Pergamon, Oxford

Bidaut-Russell M, Howlett AC (1991) Cannabinoid receptor-regulated cyclic AMP accumulation in the rat striatum. J Neurochem 57:1769–1773

Bloch E (1983) Effects of marihuana and cannabinoids on reproduction, endocrine function, development, and chromosomes. In: Fehr KO, Kalant H (eds) Cannabis and health hazards: proceedings of an ARF/WHO scientific meeting on adverse health and behavioral consequences of cannabis use. Addiction Research Foundation, Toronto, pp 355–432

Bloom AS, Dewey WL (1978) A comparison of some pharmacological actions of morphine and Δ^9-tetrahydrocannabinol in the mouse. Psychopharmacology 57: 243–248

Brake SC, Hutchings DE, Morgan B, Lasalle E, Shi T (1987) Delta-9-tetrahydro-cannabinol during pregnancy in the rat: II. Effects on ontogeny of locomotor activity and nipple attachment in the offspring. Neurotoxicol Teratol 9:45–49

Brostrom MA, Brostrom CO, Breckenridge BM, Wolff DJ (1978) Calcium-dependent regulation of brain adenylate cyclase. Adv Cyclic Nucleotide Res 9:85–99

Buchanan JF, Davis LJ (1984) Drug-induced infertility. Drug Intell Clin Pharm 18:122–132

Burstein S (1992) Eicosanoids as mediators of cannabinoid action. In: Murphy L, Bartke A (eds) Marijuana/cannabinoids: neurobiology and Neurophysiology. CRC Press, Boca Raton, pp 73–91

Burstein S, Hunter SA (1981) Prostaglandins and cannabis – VIII. Elevation of phospholipase A_2 activity by cannabinoids in whole cells and subcellular preparations. J Clin Pharmacol 21:240S–248S

Burstein S, Raz A (1972) Inhibition of prostaglandin E2 biosynthesis by Δ^1-THC. Prostaglandins 2:369–374

Burstein S, Levin E, Varanelli C (1973) Prostaglandins and cannabis-II: inhibition of biosynthesis by the naturally occuring cannabinoids. Biochem Pharmacol 22:2905–2910

Burstein S, Varanelli C, Slade LT (1974) Prostaglandins and cannabis-III. Inhibition of biosynthesis by essential oil components of marihuana. Biochem Pharmacol 24:1053–1054

Burstein SH, Hunter SA, Shoupe TS, Taylor P (1978) Cannabinoid inhibition of testosterone synthesis by mouse Leydig cells. Res Commun Chem Pathol Pharmacol 19:557–560

Burstein S, Hunter SA, Shoupe TS (1979) Site of inhibition of leydig cell testosterone synthesis by Δ^9-tetrahydrocannabinol. Mol Pharmacol 15:633–640

Burstein S, Hunter SA, Sedor C (1980) Further studies on the inhibition of Leydig cell testosterone production by cannabinoids. Biochem Pharmacol 29:2153–2154

Burstein S, Ozman K, Burstein E, Palermo N, Smith E (1982) Prostaglandins and cannabis – XI: inhibition of Δ^1-tetrahydrocannabinol-induced hypotension by aspirin. Biochem Pharmacol 31:591–592

Burstein S, Hunter SA, Renzulli L (1985) Prostaglandins and cannabis XIV. Tolerance to the stimulatory actions of cannabinoids on arachidonate metabolism. J Pharmacol Exp Ther 235:87–91

Burstein S, Hunter SA, Latham V, Renzulli L (1987) A major metabolite of Δ^1-tetrahydrocannabinol reduces its cataleptic effect in mice. Experientia 43: 402–403

Burstein SH, Hull K, Hunter SA, Shilstone J (1989) Immunization against prostaglandins reduces Δ^1-THC-induced catalepsy in mice. Mol Pharmacol 35: 6–9

Burstein SH, Audette CA, Charalambous A, Doyle SA, Guo Y, Hunter SA, Makriyannis A (1991) Detection of cannabinoid receptors by photoaffinity labelling. Biochem Biophys Res Commun 176:492–497

Cabral GA, Vasquez R (1993) Δ^9-tetrahydrocannabinol suppresses macrophage extrinsic anti-herpesvirus activity. In: Nahas GG, Latour C (eds) Cannabis: physiopathology, epidemiology, detection. CRC Press, Boca Raton, pp 137–153

Carlini EA, Hamaoui A, Bieniek D, Korte F (1970) Effects of $(-)\Delta^9$-trans-tetrahydrocannabinol and a synthetic derivative on maze performance of rats. Pharmacology 4:359–368

Carney JM, Uwayday IM, Balster RL (1977) Evaluation of a suspension system for intravenous self-administration studies with water-insoluble compounds in the Rhesus monkey. Pharmacol Biochem Behav 7:357–364

Carney JM, Balster RL, Martin BR, Harris LS (1979) Effects of systemic and intraventricular administration of cannabinoids on schedule-controlled responding in the squirrel monkey. J Pharmacol Exp Ther 210:399–404

Caulfield MP, Brown DA (1992) Cannabinoid receptor agonists inhibit Ca current in NG108-15 neuroblastoma cells via a pertussis toxin-sensitive mechanism. Br J Pharmacol 106:231–232

Cavero I, Buckley JP, Bhagavan SJ (1973) Hemodynamic and myocardial effects of $(-)$-Δ^9-trans-tetrahydrocannabinol in anesthetized dogs. Eur J Pharmacol 24:243–251

Chait LD, Pierri J (1992) Effects of smoked marijuana on human performance: a critical review. In: Murphy L, Bartke A (eds) Marijuana/cannabinoids: neurobiology and neurophysiology. CRC Press, Boca Raton, pp 387–423

Chait LD, Zacny JP (1992) Reinforcing and subjective effects of oral Δ^9-THC and smoked marijuana in humans. Psychopharmacology 107:255–262

Charalambous A, Yan G, Houston DB, Howlett AC, Compton DR, Martin BR, Makriyannis A (1992) 5'-Azido-Δ^8-THC: A novel photoaffinity label for cannabinoid receptor. J Med Chem 35:3076–3079

Chaudry A, Thompson RH, Rubin RP, Laychock SG (1988) Relationship between Δ^9-tetrahydrocannabinol-induced arachidonic acid release and secretagogue-evoked phosphoinositide breakdown and Ca^{2+} mobilization of exocrine pancreas. Mol Pharmacol 34:543–548

Chen J, Paredes W, Lowinson JH, Gardner EL (1990) Δ^9-Tetrahydrocannabinol enhances presynaptic dopamine efflux in medial prefrontal cortex. Eur J Pharmacol 190:259–262

Chen J, Marmur R, Pulles A, Paredes W, Gardner E (1993) Ventral tegmental microinjection of Δ^9-tetrahydrocannabinol enhances ventral tegmental somato-dendritic dopamine levels but not forebrain dopamine levels: evidence for local neural action by marijuana's psychoactive ingredient. Brain Res 621:65–70

Chesher GB, Jackson DM (1985) The quasimorphine withdrawal syndrome: effect of cannabinol, cannabidiol and tetrahydrocannabinol. Pharmacol Biochem Behav 23:13–15

Chesher GB, Dahl CJ, Everingham M, Jackson DM, Merchant-Williams H, Starmer GA (1973) The effect of cannabinoids on intestinal motility and their antinociceptive effect in mice. Br J Pharmacol 49:588–594

Childers SR, Fleming L, Konkoy C, Marckel D, Pacheo M, Sexton T, Ward S (1992) Opioid and cannabinoid receptor inhibition of adenylyl cyclase in brain. Ann NY Acad Sci 654:33–51

Childers SR, Sexton T, Roy MB (1994) Effects of anandamide on cannabinoid receptors in rat brain membranes. Biochem Pharmacol 47:711–715

Chkili T, Ktiouet JE (1993) Prospective study of 104 psychiatric cases associated with cannabis use in a Moroccan medical center. In: Nahas GG, Latour C (eds) Cannabis: physiopathology, epidemiology, detection. CRC Press, Boca Raton, pp 101–104

Clapp JFI, Wesley M, Cooke R, Pekala R, Holstein C (1987) The effects of marijuana smoke on gas exchange in ovine pregnancy. Alcohol Drug Res 7:85–92

Cocchetto D, Owens S, Perez-Reyes M, Di Guiseppi S, Miller L (1981) Relationship between plasma delta-9-tetrahydrocannabinol concentration and pharmacologic effects in man. Psychopharmacology 75:158–164

Cohen S, Lessin P, Hahn PM, Tyrell ED (1976) A 94-day cannabis study. In: Braude MC, Szara S (eds) Pharmacology of marihuana. Raven, New York, pp 621–626

Compton DR, Dewey WL, Martin BR (1990a) Cannabis dependence and tolerance production. In: Erickson CK, Javors MA, Morgan WW (eds) Addiction potential of abused drugs and drug classes. Hayworth, Binghampton, pp 129–147

Compton DR, Little PJ, Martin BR, Gilman JW, Saha JK, Jorapur VS, Sard HP, Razdan RK (1990b) Synthesis and pharmacological evaluation of amino, azido, and nitrogen mustard analogues of 10-substituted cannabidiol and 11- or 12-substituted Δ^8-tetrahydrocannabinol. J Med Chem 33:1437–1443

Compton DR, Gold LH, Ward SJ, Balster RL, Martin BR (1992a) Aminoalkylindole analogs: cannabimimetic activity of a class of compounds structurally distinct from Δ^9-tetrahydrocannabinol. J Pharmacol Exp Ther 263:1118–1126

Compton DR, Johnson MR, Melvin LS, Martin BR (1992b) Pharmacological profile of a series of bicyclic cannabinoid analogs: classification as cannabimimetic agents. J Pharmacol Exp Ther 260:201–209

Compton DR, Rice KC, De Costa BR, Razdan RK, Melvin LS, Johnson MR, Martin BR (1993) Cannabinoid structure-activity relationships: correlation of receptor binding and in vivo activities. J Pharmacol Exp Ther 265:218–226

Cone E, Huestis M (1993) Relating blood concentrations of tetrahydrocannabinol and metabolites to pharmacologic effects and time of marihuana usage. Ther Drug Mon 15:527–532

Cone EJ, Johnson RE, Paul BD, Mell LD, Mitchell J (1988) Marijuana-laced brownies: behavioral effects, physiologic effects, and urinalysis in humans following ingestion. J Anal Toxicol 12:169–175

Consroe P, Mechoulam R (1987) Anticonvulsant and neurotoxic effects of tetrahydrocannabinol stereoisomers. In: Rapaka RS, Markiyannis A (eds) Structure-activity relationships of the cannabinoids. US Government Printing Office, Washington DC, pp 59–66

Crawley J, Corwin R, Robinson J, Felder C, Devane W, Axelrod J (1993) Anandamide, an endogenous ligand of the cannabinoid receptor, induces hypomotility and hypothermia in vivo in rodents. Pharmacol Biochem Behav 46:967–972

Dalterio SL (1986) Cannabinoid exposure: effects on development. Neurobehav Toxicol Teratol 8:345–352

Dalterio S, Bartke A, Harper MJK, Huffman R, Sweeney C (1981) Effects of cannabinoids and female exposure on the pituitary-testicular axis in mice: possible involvement of prostaglandins. Biol Reproduct 24:315–322

Dalterio S, Steger R, Mayfield D, Bartke A (1984) Early cannabinoid exposure influences neuroendocrine and reproductive functions in male mice: I. Prenatal exposure. Pharmacol Biochem Behav 20:107–113

Defer B (1993) Cannabis and schizophrenia: how causal a relationship? In: Nahas GG, Latour C (eds) Cannabis: physiopathology, epidemiology, detection. CRC Press, Boca Raton, pp 119–121

Deutsch DG, Chin SA (1993) Enzymatic synthesis and degradation of anadamide, a cannabinoid receptor agonist. Biochem Pharmacol 46:791–796

Devane W, Axelrod J (1994) Enzymatic synthesis of anandamide, an endogenous ligand for the cannabinoid receptor by brain membranes. Proc Natl Acad Sci USA 91:6698–6701

Devane WA, Spain JW, Coscia CJ, Howlett AC (1986) An assessment of the role of opioid receptors in the response to cannabimimetic drugs. J Neurochem 46:1929–1935

Devane WA, Dysarz FA, Johnson MR, Melvin LS, Howlett AC (1988) Determination and characterization of a cannabinoid receptor in rat brain. Mol Pharmacol 34:605–613

Devane WA, Breuer A, Sheskin T, Järbe TUC, Eisen MS, Mechoulam R (1992a) A novel probe for the cannabinoid receptor. J Med Chem 35:2065–2069

Devane WA, Hanus L, Breuer A, Pertwee RG, Stevenson LA, Griffin G, Gibson D, Mandelbaum A, Etinger A, Mechoulam R (1992b) Isolation and structure of a brain constituent that binds to the cannabinoid receptor. Science 258: 1946–1949

Dewey WL (1986) Cannabinoid pharmacology. Pharmacol Rev 38:151–178

Dewey WL, Jenkins J, Rourke T, Harris LS (1972) The effects of chronic administration of trans-Δ^9-tetrahydrocannabinol on behavior and the cardiovascular system. Arch Int Pharmacol Ther 198:118–131

Dill JA, Howlett AC (1988) Regulation of adenylate cyclase by chronic exposure to cannabimimetic drugs. J Pharmacol Exp Ther 244:1157–1163

Dolby TW, Kleinsmith LJ (1974) Effects of Δ^9-tetrahydrocannabinol on the levels of cyclic adenosine 3′,5′-monophosphate in mouse brain. Biochem Pharmacol 23:1817–1825

Dolby TW, Kleinsmith LJ (1977) Cannabinoid effects on adynalate cyclase and phosphodiesterase activities of mouse brain. Can J Physiol Pharmacol 55:934–942

Drew WG, Miller LL (1974) Cannabis: neural mechanisms and behavior – a theoretical review. Pharmacology 11:12–32

Eldridge JC, Landfield PW (1990) Cannabinoid interactions with glucocorticoid receptors in rat hippocampus. Brain Res 534:135–141

Eldridge JC, Landfield PW (1992) Cannabinoid-glucocorticoid interactions in the hippocampal region of the brain. In: Murphy L, Bartke A (eds) Marijuana/ cannabinoids: neurobiology and neurophysiology. CRC Press, Boca Raton, pp 93–117

ElSohly MA, Ross SA (1994) Quaterly report, NIDA potency monitoring project. Report no 50

Evans D, Lake JT, Johnson MR, Howlett A (1994) Endogenous cannabinoid receptor binding activity released from rat brain slices by depolarization. J Pharmacol Exp Ther 268:1271–1277

Evans DM, Johnson MR, Howlett AC (1992) Ca^{2+}-dependent release from rat brain of cannabinoid receptor binding activity. J Neurochem 58:780–782

Fairbairn JW, Pickens JT (1979) The oral activity of Δ^9-tetrahydrocannabinol and its dependence on prostaglandin E2. Br J Pharmacol 67:379–385

Fairbairn JW, Pickens JT (1980) The effect of conditions influencing endogenous prostaglandins on the activity of Δ^9-tetrahydrocannabinol in mice. Br J Pharmacol 69:491–493

Falkenstein BA, Holley DC (1992) Effect of acute intravenous administration of Δ^9-tetrahydrocannabinol on the episodic secretion of immunoassayable growth hormone in the rat. Life Sci 50:1109–1116

Feigenbaum JJ, Bergmann R, Richmond SA, Mechoulam R, Nadler V, Kloog Y, Sokolovsky M (1989) Nonpsychotropic cannabinoid acts as a functional N-methyl-D-aspartate receptor blocker. Proc Natl Acad Sci USA 86:9584–9587

Felder CC, Veluz JS, Williams HL, Briley EM, Matsuda LA (1992) Cannabinoid agonists stimulate both receptor- and non-receptor-mediated signal transduction pathways in cells transfected with and expressing cannabinoid receptor clones. Mol Pharmacol 42:838–845

Felder CC, Briley EM, Axelrod J, Simpson JT, Mackie K, Devane WA (1993) Anandamide, an endogenous cannabimimetic eicosanoid, binds to the cloned human cannabinoid receptor and stimulates receptor-mediated signal transduction. Proc Natl Acad Sci USA 90:7656–7660

Fernández-Ruiz JJ, de Fonseca FR, Navarro M, Ramos JA (1992) Maternal cannabinoid exposure and brain development: changes in the ontogeny of dopaminergic neurons. In: Murphy L, Bartke A (eds) Marijuana/cannabinoids: neurobiology and neurophysiology. CRC Press, Boca Raton, pp 119–164

Ferraro DP, Grilly DM (1973) Lack of tolerance to Δ^9-tetrahydrocannabinol in chimpanzees. Science 179:490–492

Ferri S, Costa G, Murari G, Panico AM, Rapisarda E, Speroni E, Arrigo RR (1981) Investigations on behavioral effects of an extract of Cannabis sativa L. in the rat. Psychopharmacology 75:144–147

Fischer-Stenger K, Updegrove AW, Cabral GA (1992) Δ^9-tetrahydrocannabinol decreases cytotoxic T lymphocyte activity to herpes simplex virus type 1-infected cells. Proc Soc Exp Biol Med 200:422–430

Foltin RW, Brady JV, Fischman MW (1986) Behavioral analysis of marijuana effects on food intake in humans. Pharmacol Biochem Behav 25:577–582

Foltin RW, Fischman MW, Brady JV, Kelly TH, Bernstein DJ, Nellis MJ (1989) Motivational effects of smoked marihuana: behavioral contingencies and high-probability recreational activities. Pharmacol Biochem Behav 34:871–877

Foltin RW, Fischman MW, Brady JV, Bernstein DJ, Nellis MJ, Kelly TH (1990) Marijuana and behavioral contingencies. Drug Dev Res 20:67–80

Formukong EA, Evans AT, Evans FJ (1988) Analgesic and antiinflammatory activity of constituents of Cannabis sativa L. Inflammation 12:361–371

Forney RB, Kiplinger GF (1971) Toxicology and pharmacology of marijuana. Ann NY Acad Sci 191:74–82

Fraser JD (1949) Withdrawal symptoms in cannabis-indica addicts. Lancet 257: 747–748

Fride E, Mechoulam R (1993) Pharmacological activity of the cannabinoid receptor agonist, anandamide, a brain constituent. Eur J Pharmacol 231:313–314

Fried PA, Makin JE (1987) Neonatal behavioral correlates of prenatal exposure to marihuana, cigarettes and alcohol in a low risk population. Neurotoxicol Teratol 9:1–7

Gahlinger PM (1984) Gastrointestinal illness and cannabis use in a rural Canadian community. J Psychoactive Drugs 16:263–265

Gal P, Sharpless MK (1984) Fetal drug exposure-behavioral teratogenesis. Drug Intell Clin Pharm 18:186–201

Gamse R, Holzer P, Lembeck F (1979) Indirect evidence for presynaptic location of opiate receptors in chemosensitive primary sensory neurones. Arch Pharmacol 308:281–285

Gardner EL (1992) Cannabinoid interaction with brain reward systems – the neurobiological basis of cannabinoid abuse. In: Murphy L, Bartke A (eds) Marijuana/cannabinoids: neurobiology and neurophysiology. CRC Press, Boca Raton, pp 275–335

Gardner EL, Lowinson JH (1991) Marijuana's interaction with brain reward systems – update 1991. Pharmacol Biochem Behav 40:571–580

Gardner EL, Paredes W, Smith D, Donner A, Milling C, Cohen A, Morrison D (1988) Facilitation of brain stimulation reward by Δ^9-tetrahydrocannabinol. Psychopharmacology 96:142–144

Gascon AL, Peres MT (1973) Effect of Δ^9- and Δ^8-tetrahydrocannabinol on the peripheral autonomic nervous system in vitro. Can J Physiol Pharmacol 51:12–21

Gennings C, Carter JWH, Martin BR (1994) Response-surface analysis of morphine sulfate and Δ^9-tetrahydrocannabinol interaction in mice. In: Lange N, Ryan L (eds) Case studies in biometry. Wiley, New York, pp 429–451

George HR (1970) Two psychotic epidodes associated with cannabis. Br J Addict 65:119–121

Gérard C, Mollereau C, Vassart G, Parmentier M (1990) Nucleotide sequence of a human cannabinoid receptor cDNA. Nucleic Acids Res 18:7142

Gilbert PE (1981) A comparison of THC, nantradol, nabilone, and morphine in the chronic spinal dog. J Clin Pharmacol 21:311S–319S

Gold L, Balster RL, Barrett RL, Britt DT, Martin BR (1992) A comparison of the discriminative stimulus properties of Δ^9-THC and CP-55 940 in rats and rhesus monkeys. J Pharmacol Exp Ther 262:479–486

Gough AL, Olley JE (1978) Catalepsy induced by intrastriatal injections of Δ^9-THC and 11-OH-Δ^9-THC in the rat. Neuropharmacology 17:137–144

Greenberg I, Mendelson JH, Kuehnle JC, Mello N, Babor TF (1976) Psychiatric and behavioral observations of casual and heavy marijuana users in a controlled research setting. Ann NY Acad Sci 282:72–84

Halikas JA, Weller RA, Morse CL, Hoffmann RG (1985) A longitudinal study of marijuana effects. Int J Addict 20:701–711

Hanus L, Gopher A, Almog S, Mechoulam R (1993) Two new unsaturated fatty acid ethanolamides in brain that bind to the cannabinoid receptor. J Med Chem 36:3032–3034

Harclerode J (1984) Endocrine effects of marijuana in the male: preclinical studies. In: Braude MC, Ludford JP (eds) Marijuana effects on the endocrine and reproductive systems. US Government Printing Office, Washington DC, pp 46–64

Harris RA, Stokes JA (1982) Cannabinoids inhibit calcium uptake by brain synaptosomes. J Neurosci 2:443–447

Harris LS, Pearl J, Aceto MD (1966) Similarities in the effects of barbiturates. Psychonom Sci 4:267–268

Harris R, Waters W, McLendon D (1974) Evaluation of reinforcing capability of Δ^9-tetrahydrocannabinol in rhesus monkeys. Psychopharmacology 37:23–29

Harris LS, Dewey WL, Razdan RK (1977) Cannabis: its chemistry, pharmacology, and toxicology. In: Martin WR (ed) Drug addiction. Springer, Berlin Heidelberg New York, pp 371–429 (Handbook of experimental pharmacology, vol 45/2)

Harris LS, Carchman RA, Martin BR (1978) Evidence for the existence of specific cannabinoid binding sites. Life Sci 22:1131–1138

Hatch EE, Bracken MB (1986) Effect of marijuana use in pregnancy on fetal growth. Am J Epidemiol 124:986–993

Haubrich DR, Ward SJ, Baizman E, Bell MR, Bradford J, Ferrari R, Miller M, Perrone M, Pierson AK, Saelens JK, Luttinger D (1990) Pharmacology of pravadoline: a new analgesic agent. J Pharmacol Exp Ther 255:511–522

Heindel JJ, Keith WB (1989) Specific inhibition of FSH-stimulated cAMP accumulation by Δ^9-tetrahydrocannabinol in cultures of rat Sertoli cells. Toxicol Appl Pharmacol 101:124–134

Herkenham M, Lynn AB, Little MD, Johnson MR, Melvin LS, DeCosta BR, Rice KC (1990) Cannabinoid receptor localization in the brain. Proc Natl Acad Sci USA 87:1932–1936

Herkenham M, Lynn AB, DeCosta BR, Richfield EK (1991a) Neuronal localization of cannabinoid receptors in the basal ganglia of the rat. Brain Res 547:267–274

Herkenham M, Lynn AB, Johnson MR, Melvin LS, de Costa BR, Rice KC (1991b) Characterization and localization of cannabinoid receptors in rat brain: a quantitative in vitro autoradiographic study. J Neurosci 11:563–583

Heyser CJ, Hampson RE, Deadwyler SA (1993) Effects of Δ^9-tetrahydrocannabinol on delayed match to sample performance in rats: alterations in short-term memory associated with changes in task specific firing of hippocampal cells. J Pharmacol Exp Ther 264:294–307

Hillard CJ, Bloom AS (1983) Possible role of prostaglandins in the effects of the cannabinoids on adenylate cyclase activity. Eur J Pharmacol 91:21–27

Hillard CJ, Pounds JJ, Boyer DR, Bloom AS (1990) Studies of the role of membrane lipid order in the effects of Δ^9-tetrahydrocannabinol on adenylate cyclase activation in heart. J Pharmacol Exp Ther 252:1075–1082

Hiltunen AJ, Jarbe TU, Kamkar MR, Archer T (1989) Behaviour in rats maintained by low differential reinforcement rate: effects of Δ^1-tetrahydrocannabinol, cannabinol and cannabidiol, alone and in combination. Neuropharmacology 28:183–189

Hine B (1985) Morphine and Δ^9-tetrahydrocannabinol: two-way cross tolerance for antinociceptive and heart rate responses in the rat. Psychopharmacology 87:34–38

Hine B, Friedman E, Torellio M, Gershon S (1975) Morphine-dependent rats: blockade of precipitated abstinence by tetrahydrocannabinol. Science 187: 443–445

Hollister LE (1986) Health aspects of cannabis. Pharmacol Rev 38:1–20

Hollister LE (1988) Marijuana and immunity. J Psychoactive Drugs 20:3–8

Hollister LE, Gillespie HK, Ohlsson A, Lindgren J-E, Wahlen A, Agurell S (1981) Do plasma concentrations of Δ^9-tetrahydrocannabinol reflect the degree of intoxication? J Clin Pharmacol 21:171S–177S

Howes JF, Osgood PF (1976) Cannabinoids and the inhibition of prostaglandin synthesis. In: Nahas GG (ed) Marihuana: chemistry, biochemistry and cellular effects. Springer, Berlin Heidelberg New York, pp 415–424

Howlett AC (1984) Inhibition of neuroblastoma adenylate cyclase by cannabinoid and nantradol compounds. Life Sci 35:1803–1810

Howlett AC (1985) Cannabinoid inhibition of adenylate cyclase. Biochemistry of the response in neuroblastoma cell membranes. Mol Pharmacol 27:429–436

Howlett AC (1987) Cannabinoid inhibition of adenylate cyclase: relative activity of constituents and metabolites of marihuana. Neuropharmacology 26:507–512

Howlett AC, Fleming RM (1984) Cannabinoid inhibition of adenylate cyclase. Pharmacology of the response in neuroblastoma cell membranes. Mol Pharmacol 26:532–538

Howlett AC, Qualy JM, Khachatrian LL (1986) Involvement of Gi in the inhibition of adenylate cyclase by cannabimimetic drugs. Mol Pharmacol 29:307–313

Howlett AC, Johnson MR, Melvin LS, Milne GM (1988) Nonclassical cannabinoid analgetics inhibit adenylate cyclase: development of a cannabinoid receptor model. Mol Pharmacol 33:297–302

Howlett AC, Bidaut-Russell M, Devane WA, Melvin LS, Johnson MR, Herkenham M (1990a) The cannabinoid receptor: biochemical, anatomical and behavioral characterization. TINS 13:420–423

Howlett AC, Champion TM, Wilken GH, Mechoulam R (1990b) Stereochemical effects of 11-OH-Δ^8-tetrahydrocannabinol-dimethylheptyl to inhibit adenylate cyclase and bind to the cannabinoid receptor. Neuropharmacology 29:161–165

Howlett AC, Evans DM, Houston DB (1992) The cannabinoid receptor. In: Murphy L, Bartke A (eds) Marijuana/cannabinoids: neurobiology and neurophysiology. CRC Press, Boca Raton, pp 35–72

Huestis MA, Henningfield JE, Cone EJ (1992a) Blood Cannabinoids: I. Absorption of THC and formation of 11-OH-THC and THCCOOH during and after smoking marijuana. J Anal Toxicol 16:276–282

Huestis MA, Sampson AH, Holicky BJ, Henningfield JE, Cone EJ (1992b) Characterization of the absorption phase of marijuana smoking. Clin Pharmacol Ther 52:31–41

Hunter SA, Audette CA, Burstein S (1991) Elevation of brain prostaglandin E_2 levels in rodents by Δ^1-THC. Prostaglandins Leukot Essent Fatty Acids 43:185

Hutchings DE, Brake SC, Morgan B (1989a) Animal studies of prenatal Δ^8-tetrahydrocannabinol: female embryolethality and effects on somatic and brain growth. Ann NY Acad Sci 562:133–144

Hutchings DE, Gamagaris Z, Miller N, Fico TA (1989b) The effects of prenatal exposure to Δ^9-tetrahydrocannabinol to the rest-activity cycle of the preweanling rat. Neurotoxicol Teratol 11:353–356

Hutchings DE, Dow-Edwards D (1991) Animal models of opiate, cocaine, and cannabis use. Clin Perinatol 18:1–22

Iverfeldt K, Ögren SO, Barfai T (1988) Substance P receptors in the rat spinal cord: the effect of GTP and of chronic antidepressant treatment. Acta Physiol Scand 132:175–179

Jandhyala BS, Buckley JP (1977) Autonomic and cardiovascular effects of chronic Δ^9-tetrahydrocannabinol administration in mongrel dogs. Res Commun Chem Pathol Pharmacol 16:593–607

Jansen EM, Haycock DA, Ward SJ, Seybold VS (1992) Distribution of cannabinoid receptors in rat brain determined with aminoalkylindoles. Brain Res 575:93–102

Järbe TU, Hiltunen AJ (1987) Cannabimimetic activity of cannabinol in rats and pigeons. Neuropharmacol 26:219–228

Johansson E, Agurell S, Hollister L, Halldin M (1988) Prolonged apparent half-life of delta-9-tetrahydrocannabinol in plasma of chronic marijuana users. J Pharm Pharmacol 40:374–375

Johson MR, Melvin LS (1986) The discovery of nonclassical cannabinoid analgetics. In: Mechoulam R (ed) Cannabinoids as therapeutic agents. CRC Press, Boca Raton, pp 121–144

Johnson V (1988) Adolescent alcohol and marijuana use: a longitudinal assessment of a social learning perspective. Am J Drug Alcohol Abuse 14:419–439

Johnson V, Pandina RJ (1991) Effects of the family environment on adolescent substance use, delinquency, and coping styles. Am J Drug Alcohol Abuse 17:71–88

Johnston LD, O'Malley PM, Bachman JG (1994) National survey results on drug use from the monitoring the future study, 1975–1993, National survey results on drug use from the monitoring the future study, 1975–1993, vol 1 and 2. US Department of Health and Human Services, Washington DC

Jones RT (1971) Marihuana-induced "high": influence of expectation, setting and previous drug experience. Pharmacol Rev 23:359–369

Jones RT (1983) Cannabis tolerance and dependence. In: Fehr KO, Kalant H (eds) Cannabis and health hazards. Addiction Research Foundation, Toronto, pp 617–689

Jones RT, Benowitz N (1976) The 30-day trip – clinical studies of cannabis tolerance and dependence. In: Braude MC, Szara S (eds) Pharmacology of marihuana. Raven, New York, pp 627–642

Jones RT, Benowitz N, Bachman J (1976) Clinical studies of cannabis tolerance and dependence. Ann NY Acad Sci 282:221–239

Jones RT, Benowitz NL, Herning RI (1981) Clinical relevance of cannabis tolerance and dependence. J Clin Pharmacol 21:143S–152S

Jorapur VS, Khalil ZH, Duffley RP, Razdan RK, Martin BR, Harris LS, Dewey WL (1985) Hashish: Synthesis and central nervous system activity of some novel analogues of cannabidiol and oxepin derivatives of Δ^9-tetrahydrocannabinol. J Med Chem 28:783–787

Kaminski NE, Abood ME, Kessler FK, Martin BR, Schatz AR (1992) Identification of a functionally relevant cannabinoid receptor on mouse spleen cells that is involved in cannabinoid-mediated immune modulation. Mol Pharmacol 42:736–742

Kawakami Y, Klein TW, Newton C, Djeu JY, Dennert G, Spector S, Friedman H (1988) Suppression by cannabinoids of a cloned cell line with natural killer cell activity. Proc Soc Exp Biol Med 187:355–359

Kaymakcalan S (1973) Tolerance to and dependence on cannabis. Bull Narc 25:39–47

Kaymakcalan S, Deneau GA (1972) Some pharmacologic properties of synthetic Δ^9-tetrahydrocannabinol. Acta Med Turc [Suppl] 1:5

Kaymakcalan S, Ercan ZS, Turker RK (1975) The evidence of the release of prostaglandin-like material from rabbit kidney and guinea-pig lung by (−)-trans-Δ^9-THC. J Pharm Pharmacol 27:564–568

Kelly LA, Butcher RW (1973) The effects of Δ^1-tetrahydrocannabinol on cyclic AMP levels in WI-38 fibroblasts. Biochim Biophysica Acta 1973:540–544

Kelly LA, Butcher RW (1979) Effects of Δ^1-tetrahydrocannabinol (THC) on cyclic AMP metabolism in cultured human fibroblasts. In: Abood LG, Sharp CW (eds) Membrane mechanisms of drugs of abuse. Liss, New York, pp 227–236

Klein TW, Newton C, Friedman H (1987) Inhibition of natural killer cell function by marijuana components. J Toxicol Environ Health 20:321–332

Knudsen P, Vilmar T (1984) Cannabis and neuroleptic agents in schizophrenia. Acta Psychiatr Scand 69:162–174

Kochar MS, Hosko MJ (1973) Electrocardiographic effects of marihuana. J Am Med Assoc 225:25–27

Koe BK, Milne GM, Weissman A, Johnson MR, Melvin LS (1985) Enhancement of brain [^{3}H]flunitrazepam binding and analgesic activity of synthetic cannabimimetics. Eur J Pharmacol 109:201–212

Koutek B, Prestwich GD, Howlett AC, Chin SA, Salehani D, Akhavan N, Deutsch DG (1994) Inhibitors of arachidonyl ethanolamide hydrolysis. J Biol Chem 269:22937–22940

Kruszka K, Gross R (1994) The ATP- and CoA-independent synthesis of arachidonoylethanolamide. J Biol Chem 269:14345–14348

Kumbaraci NM, Nastuk WL (1980) Effects of Δ^9-tetrahydrocannabinol on excitable membranes and neuromuscular transmission. Mol Pharmacol 17:344–349

Kuster JE, Stevenson JI, Ward SJ, D'Ambra TE, Haycock DA (1993) Aminoalkylindole binding in rat cerebellum: selective displacement by natural and synthetic cannabinoids. J Pharmacol Exp Ther 264:1352–1363

Labouvie EW (1990) Personality and alcohol and marihuana use: patterns of convergence in young adulthood. Int J Addict 25:237–252

Landfield PW, Cadwallader LB, Vinsant S (1988) Quantitative changes in hippocampal structure following long-term exposure to Δ^9-tetrahydrocannabinol: possible mediation by glucocorticoid systems. Brain Res 443:47

Layman JM (1971) Some actions of Δ^1-tetrahydrocannabinol and cannabidiol at cholinergic junctions. Proc Br Pharmacol Soc 41:379p

Leader JP, Koe BK, Weisman A (1981) GABA-like actions of levonantradol. J Clin Pharmacol 21:2625–2705

Leirer VO, Yesavage JA, Morrow DG (1991) Marijuana carry-over effects on aircraft pilot performance. Aviat Space Environ Med 62:221–227

Leirer VO, Yesavage JA, Morrow DG (1993) Marijuana carry-over effects on psychomotor performance: a chronicle of research. In: Nahas GG, Latour C (eds) Cannabis: physiopathology, epidemiology, detection. CRC Press, Boca Raton, pp 47–60

Lemberger L, Crabtree RE, Rowe HM (1972a) 11-Hydroxy-Δ^9-tetrahydrocannabinol: pharmacology, disposition, and metabolism of a major metabolite of marihuana in man. Science 177:62–64

Lemberger L, Weiss JL, Watanabe AM, Galanter IM, Wyatt RJ, Cardon PV (1972b) Δ^9-Tetrahydrocannabinol: temporal correlation of the psychologic effects and blood levels after various routes of administration. N Eng J Med 286:685–688

Lemberger L, McMahon R, Archer R, Matsumoto K, Rowe H (1974) Pharmacologic effects and physiologic disposition of delta^{6a,10a} dimethylheptyl tetrahydrocannabinol (DMHP) in man. Clin Pharmacol Ther 15:380–386

Li DMF, Ng CKM (1984) Effects of Δ^1- and Δ^6-tetrahydrocannabinol on the adynalate cyclase activity in ventricular tissue of the rat heart. Clin Exp Pharmacol Physiol 11:81–85

Lichtman AH, Martin BR (1991) Spinal and supraspinal mechanisms of cannabinoid-induced antinociception. J Pharmacol Exp Ther 258:517–523

Lichtman AH, Dimen KR, Martin BR (1995) Systemic or intrahippocampal cannabinoid administration impairs spatial memory in rats. Psychopharmacology 119:282–290

Lindgren J, Ohlsson A, Agurell S, Hollister L, Gillespie H (1981) Clinical effects and plasma levels of delta-9-tetrahydrocannabinol in heavy and light users of cannabis. Psychopharmacology 74:208–212

Linton PH, Kuechenmeister CA, White HB, Travis RP (1975) Marijuana: heart rate and EEG response. Res Commun Chem Pathol Pharmacol 10:201–214

Little PJ, Martin BR (1991) The effects of Δ^9-tetrahydrocannabinol and other cannabinoids on cAMP accumulation in synaptosomes. Life Sci 48:1133–1141

Little PJ, Kaplan NC, Martin BR (1987) Pharmacological profile of Δ^9-THC carbamate. Alcohol Drug Res 7:517–523

Little PJ, Compton DR, Johnson MR, Melvin LS, Martin BR (1988) Pharmacology and stereoselectivity of structurally novel cannabinoids in mice. J Pharmacol Exp Ther 247:1046–1051

Little PJ, Compton DR, Mechoulam R, Martin BR (1989) Stereochemical effects of 11-OH-dimethylheptyl-Δ^8-tetrahydrocannabinol. Pharmacol Biochem Behav 32:661–666

Loewe S (1946) The rate of disappearance of marihuana-active substance from the circulating blood. J Pharmacol Exp Ther 294–296

Lynn A, Herkenham M (1994) Localization of cannabinoid receptors and nonsaturable high-density cannabinoid binding sites in peripheral tissues of the rat: implications for receptor-mediated immune modulation by cannabinoids. J Pharmacol Exp Ther 268:1612–1623

Mackie K, Hille B (1992) Cannabinoids inhibit N-type calcium channels in neurobalstoma-glioma cells. Proc Natl Acad Sci USA 89:3825–3829

Mackie K, Devane W, Hille B (1993) Anandamide, an endogenous cannabinoid, inhibits calcium currents as a partial agonist in N18 neuroblastoma cells. Mol Pharmacol 44:498–503

Mailleux P, Vanderhaeghen J (1993a) Dopaminergic regulation of cannabinoid receptor mRNA levels in the rat caudate-putamen: an in situ hybridization study. J Neurochem 61:1705–1712

Mailleux P, Vanderhaeghen JJ (1993b) Glucocorticoid regulation of cannabinoid receptor messenger RNA levels in the rat caudate-putamen. An in situ hybridization study. Neurosci Lett 156:51–53

Malit LA, Johnstone RE, Bourke DI, Kulp RA, Klein V, Smith TC (1975) Intravenous Δ^9-tetrahydrocannabinol: effects on ventilatory control and cardiovascular dynamics. Anesthesiology 42:666–673

Mansbach RS, Nicholson KL, Martin BR, Balster RL (in press) Failure of Δ^9-tetrahydrocannabinol and CP 55,940 to maintain intravenous self-administration under a fixed-interval schedule in rhesus monkeys. Behav Pharmacol

Margulies JE, Hammer RP (1991) Δ^9-tetrahydrocannabinol alters cerebral metabolism in a biphasic, dose-dependent manner in rat brain. Eur J Pharmacol 202:373–378

Martin BR (1985a) Characterization of the antinociceptive activity of intravenously administered Δ^9-tetrahydrocannabinol in mice. In: Harvey DJ (ed) Marihuana '84 proceedings of the oxford symposium on cannabis. IRL Press, Oxford, pp 685–692

Martin BR (1985b) Structural requirements for cannabinoid-induced antinociceptive activity in mice. Life Sci 36:1523–1530

Martin BR (1986) Cellular effects of cannabinoids. Pharmacol Rev 38:45–74

Martin BR, Dewey WL, Harris LS, Beckner JS (1976) ^{3}H-Δ^9-tetrahydrocannabinol tissue and subcellular distribution in the central nervous system and distribution in peripheral organs of tolerant and nontolerant dogs. J Pharmacol Exp Ther 196:128–144

Martin BR, Kallman MJ, Kaempf GF, Harris LS, Dewey WL, Razdan RK (1984) Pharmacological potency of R- and S-3'-hydroxy-Δ^9-tetrahydrocannabinol: additional structural requirement for cannabinoid activity. Pharmacol Biochem Behav 21:61–65

Martin BR, Compton DR, Little PJ, Martin TJ, Beardsley PM (1987) Pharmacological evaluation of agonistic and antagonistic activity of cannabinoids. In: Rapaka RS, Makriyannis A (eds) Structure-activity relationships of cannabinoids. US Government Printing Office, Washington DC, pp 108–122

Martin BR, Howlett AS, Welch SP (1989) Cannabinoid action in the central nervous system. In: Harris LS (ed) Problems of drug dependence 1988. Proceedings of the 50th annual scientific meeting. US Government Printing Office, Washington DC, 275–283

Martin BR, Compton DR, Thomas BF, Prescott WR, Little PJ, Razdan RK, Johnson MR, Melvin LS, Mechoulam R, Ward SJ (1991) Behavioral, biochemical, and molecular modeling evaluations of cannabinoid analogs. Pharmacol Biochem Behav 40:471–478

Martin BR, Welch SP, Abood M (1994) Progress toward understanding the cannabinoid receptor and its second messenger systems. In: August JT, Anders MW, Murad F (eds) Advances in pharmacology. Academic, San Diego, pp 341–397

Martin WJ, Lai NK, Patrick SL, Tsou K, Walker JM (1993) Antinociceptive actions of cannabinoids following intraventricular administration in rats. Brain Res 629:300–304

Mason AP, McBay AJ (1984) Ethanol, marijuana, and other drug use in 600 drivers killed in single-vehicle crashes in North Carolina, 1978–1981. J Forensic Sci 29:987–1026

Mason AP, McBay AJ (1985) Cannabis: pharmacology and interpretation of effects. J Forensic Sci 30:615–631

Mathew RJ, Wilson WH (1992) The effects of marijuana on cerebral blood flow and metabolism. In: Murphy L, Bartke A (eds) Maijuana/cannabinoids: neurobiology and neurophysiology. CRC Press, Boca Raton, pp 337–386

Mathew RJ, Wilson WH (1993) Acute changes in cerebral blood flow after smoking marijuana. Life Sci 52:757–767

Mathew RJ, Wilson WH, Humphreys DF, Lowe JV, Wiethe KE (1992) Regional cerebral blood flow after marijuana smoking. J Cereb Blood Flow Metab 12:750–758

Matsuda LA, Lolait SJ, Brownstein MJ, Young AC, Bonner TI (1990) Structure of a cannabinoid receptor and functional expression of the cloned dDNA. Nature 346:561–564

Maykut MO (1984) Health consequences of acute and chronic marihuana use, health consequences of acute and chronic marihuana use. Pergamon, Oxford

McBay A (1986) Drug concentrations and traffic safety. Adv Alcohol Subst Abuse 2:51–59

McMillan DE, Dewey WL (1972) On the mechanism of tolerance to Δ^9-THC. In: Lewis MF (ed) Current research in marihuana. Academic, New York, pp 97–128

McMillan DE, Harris LS, Frankenheim JM, Kennedy JS (1970) L-Δ^9-transtetrahydrocannabinol in pigeons: tolerance to the behavioral effects. Science 169:501–503

McMillan DE, Dewey WL, Harris LS (1971) Characteristics of tetrahydrocannabinol tolerance. Ann NY Acad Sci 191:83–99

Mechoulam R, Lander N, Srebnik M, Zamir, I, Breuer A, Shalita B, Dikstein S, Carlini EA, Leite JR, Edery H, Porath G (1984) Recent advances in the use of cannabinoids as therapeutic agents. In: Agurell S, Dewey W, Willette (eds) The cannabinoids: chemical, pharmacologic and therapeutic aspects. Academic, New York, pp 777–793

Mechoulam R, Feigenbaum JJ, Lander N, Segal M, Jarbe TUC, Hiltunen AJ, Consroe P (1988) Enantiomeric cannabinoids: stereospecificity of psychotropic activity. Experientia 44:762–764

Mechoulam R, Hanus L, Ben-Shabat S, Fride E, Weidenfeld J (1994) The an-andamides, a family of endogenous cannabinoid ligands – chemical and biological studies. Neuropsychopharmacology 10:145S–145S

Mellors A (1979) Cannabinoids and membrane-bound enzymes. In: Nahas GG, Paton WDM (eds) Marihuana biological effects-analysis, metabolsim, cellular responses, reproduction and brain. Pergamon, Oxford, pp 329–342

Melvin LS, Johnson MR, Harbert CA, Milne GM, Weissman A (1984) A cannabinoid derived prototypical analgesic. J Med Chem 27:67–71

Mendelson JH, Babor TF, Kuehenle JC, Rossi AM, Berstein JG, Mello NK, Greenberg I (1976) Behavioral biologic aspects of marihuana use. Ann NY Acad Sci 282:186–210

Meyer RE (1978) Behavioral pharmacology of marihuana. In: Lipton MA, DiMascio A, Killam KF (eds) Psychopharmacology: a generation of progress. Raven, New York, pp 1639–1652

Miczek KA, Dihit BN (1980) Behavioral and biochemical effects of chronic Δ^9-tetrahyrocannabinol in rats. Psychopharmacology 67:195–202

Molnar J, Petri I, Berek I, Shoyama Y, Nishioka I (1987) The effects of cannabinoids and cannabispiro compounds on Escherichia coli adhesion to tissue culture cells and on leukocyte functions in vitro. Acta Microbiol Hung 34:233–240

Morgan B, Brake SC, Hutchings DE, Miller N, Gamagaris Z (1988) Δ^9-tetrahydrocannabinol during pregnancy in the rat: effects on development of RNA, DNA, and protein in offspring brain. Pharmacol Biochem Behav 31:365–369

Moskowitz H (1985) Marihuana and driving. Accid Anal Prevent 17:323–345

Mountjoy KG, Robbins LS, Mortrud MT, Cone RD (1992) The cloning of a family of genes that encode the melanocortin receptors. Science 257:1248–1251

Munro S, Thomas KL, Abu-Shaar M (1993) Molecular characterization of a peripheral receptor for cannabinoids. Nature 365:61–64

Munson AE, Fehr KO (1983) Immunological effects of cannabis. In: Fehr KO, Kalant H (eds) Cannabis and health hazards: proceedings of an ARF/WHO scientific meeting on adverse health and behavioral consequences of cannabis use. Addiction Research Foundation, Toronto, pp 257–354

Murphy LL, Newton SC, Dhali J, Chavez D (1991) Evidence for a direct anterior pituitary site of Δ^9-tetrahydrocannabinol action. Pharmacol Biochem Behav 40:603–607

Nahas G (1993a) General toxicity of cannabis. In: Nahas GG, Latour C (eds) Cannabis: physiopathology, epidemiology, detection. CRC Press, Boca Raton, pp 5–17

Nahas G (1993b) Historical outlook of the psychopathology of cannabis. In: Nahas GG, Latour C (eds) Cannabis: physiopathology, epidemiology, detection. CRC Press, Boca Raton, pp 95–99

Nahas G, Latour C (1992) The human toxicity of marijuana. Med J Aust 156:495–497

Nakamura EM, Da Silva EA, Concilio GV, Wilkinson DA, Masur J (1991) Reversible effects of acute and lot-term administration of Δ^9-tetrahydrocannabinol (THC) on memory in the rat. Drug Alcohol Depend 28:167–175

Nakano Y, Pross SH, Friedman H (1992) Modulation of interleukin 2 activity by Δ^9-tetrahydrocannabinol after stimulation with concanavalin A, phytohemagglutinin, or anti-CD3 antibody. Proc Soc Exp Biol Med 201:165–168

Nakano Y, Pross S, Friedman H (1993) Contrasting effect of Δ^9-tetrahydrocannabinol on IL-2 activity in spleen and lymph node cells of mice of different ages. Life Sci 52:41–51

Naor Z (1990) Cyclic GMP stimulates inostitol phosphate production in cultured pituitary cells: Possible implications to signal transduction. Biochem Biophys Res Commun 167:982–992

Narimatsu S, Watanabe K, Yamamoto I, Yoshimura H (1988) Mechanism for inhibitory effect of cannabidiol on microsomal testosterone oxidation in male rat liver. Drug Metab Dispos 16:880–889

Navarro M, Fernandez-Ruiz JJ, De Miguel R, Hernandez ML, Cebeira M, Ramos JA (1993) Motor disturbances induced by an acute dose of Δ^9-tetrahydrocannabinol: possible involvement of nigrostriatal dopaminergic alterations. Pharmacol Biochem Behav 45:291-298

Negrete JC (1993) Effects of cannabis on schizophrenia. In: Nahas GG, Latour C (eds) Cannabis: physiopathology, epidemiology, detection. CRC Press, Boca Raton, pp 105-112

Newman LM, Lutz MP, Gould MH, Domino EF (1974) Δ^9-tetrahydrocannabinol and ethyl alcohol: evidence for cross-tolerance in the rat. Science 175:1022-1023

Niemi WD (1979) Effect of Δ^9-tetrahydrocannabinol on synaptic transmission in the electric eel electroplaque. Res Commun Chem Pathol Pharmacol 25:537-546

Nishizuka Y (1983) Calcium, phospholipid turnover and transmembrane signalling. Philos Trans R Soc Lond 302:101-112

Nishizuka Y (1984) The role of protein kinase C in cell surface signal transduction and tumour promotion: Nature 308:693-697

Nowicky AV, Teyloer TJ, Vardaris RM (1987) The modulation of long-term potentiation by Δ^9-tetrahydrocannabinol in the rat hippocampus, in vitro. Brain Res Bull 19:663-672

Noyes JR, Brunk SF, Avery DH, Canter A (1975) The analgesic properties of Δ^9-tetrahydrocannabinol and codeine. Clin Pharmacol Ther 18:84-89

Nye JS, Seltzman HH, Pitt CG, Snyder SH (1984) Labelling of a cannabinoid binding site in brain with a [^{3}H]quarternary ammonium analogue of Δ^8-THC. In: Harvey DJ (ed) Marijuana '84: proceedings of the oxford symposium on cannabis. IRL Press, Oxford, pp 253-262

Nye JS, Seltzman HH, Pitt CG, Snyder SS (1985) High-affinity cannabinoid binding sites in brain membranes labeled with [^{3}H]-5'-trimethylammonium-Δ^8-tetrahydrocannabinol. J Pharmacol Exp Ther 234:784-791

O'Connell CM, Fried PA (1984) An investigation of prenatal cannabis exposure and minor physical anomalies in a low risk population. Neurobehav Toxicol Teratol 6:345-350

O'Connell CM, Fried PA (1991) Prenatal exposure to cannabis: a preliminary report of postnatal consequences in school-age children. Neurotoxicol Teratol 13: 631-639

Ohlsson A, Lindgren J-E, Wahlen A, Agurell S, Hollister LE, Gillespie HK (1980) Plasma Δ^9-tetrahydrocannabinol concentrations and effect after oral and intravenous administration and smoking. Clin Pharmacol Ther 28:409-416

Ohlsson A, Lindgren J-E, Wahlen A, Agurell, S, Hollister L, Gillespie H (1982) Single dose kinetics of deuterium labelled Δ^1-tetrahydrocannabinol in heavy and light cannabis users. Biomed Mass Spectrom 9:6-11

Ohlsson A, Agurell S, Londgren J, Gillespie H, Hollister L (1985) Pharmacokinetics studies of delta-1-tetra hydrocannabinol in man. In: Barnett G, Chiang C (eds) Pharmacokinetics and pharmacodynamics of psychoactive drugs. Biomedical, Stony Brook, pp 75-92

Okada M, Mine K, Fujiwara M (1989) Relationship of calcium and adenylate cyclase messenger systems in rat brain synaptosomes. Brain Res 501:23-31

Okada M, Urae A, Mine K, Shoyama Y, Iwasaki K, Fujiwara M (1992) The facilitating and suppressing effects of Δ^9-tetrahydrocannabinol on the rise in intrasynaptosomal Ca^{2+} concentration in rats. Neurosci Lett 140: 55-58

Olson KG, Welch SP (1991) The effects of dynorphin A (1-13) and U50 488H on free intracellular calcium in guinea pig cerebellum. Life Sci 48:575-581

Onaivi ES, Green MR, Martin BR (1990) Pharmacological characterization of cannabinoids in the elevated plus maze. J Pharmacol Exp Ther 253:1002-1009

Oviedo A, Glowa J, Herkenham M (1993) Chronic cannabinoid administration alters cannabinoid receptor binding in rat brain: a quantitative autoradiographic study. Brain Res 616:293-302

Pacheco M, Childers SR, Arnold R, Casiano F, Ward SJ (1991) Aminoalkylindoles: actions on specific G-protein-linked receptors. J Pharmacol Exp Ther 257: 170–183

Pacheco M, Ward S, Childers S (1994) Differential requirements of sodium for coupling of cannabinoid receptors to adenylyl cyclase in rat brain membranes. J Neurochem 62:1773–1782

Page JB (1983) The amotivational syndrome hypothesis and the Costa Rica study: relationships between methods and results. J Psychoactive Drugs 15:261–267

Paria BC, Kapur S, Dey SK (1992) Effects of 9-ene-tetrahydrocannabinol on uterine estrogenicity in the mouse. J Steroid Biochem Mol Biol 42:713–719

Paule MG, Allen RR, Bailey JR, Scallet AC, Ali SF, Brown RM, Slikker W (1992) Chronic marijuana smoke exposure in the rhesus monkey II: effects on progressive ratio and conditioned position responding. J Pharmacol Exp Ther 260:210–222

Perez-Reyes M (1985) Pharmacodynamics of certain drugs of abuse. In: Barnett G, Chiang C (eds) Pharmacokinetics and pharmacodynamics of psychoactive drugs. Biomedical Publications, pp 287–310

Perez-Reyes M, Owens S, Di Guiseppi S (1981) The clinical pharmacology and dynamics of marijuana cigarette smoking. J Clin Pharmacol 21:201S–207S

Perez-Reyes M, Di Guiseppi S, Davis KH, Schindler VH, C.E. C (1982) Comparison of effects of marihuana cigarettes of three different potencies. Clin Pharmacol Ther 31:617–624

Perez-Reyes M, Burstein SH, White WR, McDonald SA, Hicks RE (1991) Antagonism of marihuana effects by indomethacin in humans. Life Sci 48:507–515

Pertwee RG (1972) The ring test: a quantitative method for assessing the "cataleptic" effect of cannabis in mice. Br J Pharmacol 46:753–763

Pertwee RG (1988) The central neuropharmacology of psychotropic cannabinoids. Pharmacol Ther 36:189–261

Pertwee (1990) The central neuropharmacology of psychotropic cannabinoids. In: Balfour DJK (ed) Psychotropic drugs of abuse. Pergamon, New York, p 355

Pertwee R (1992) In vivo interactions between psychotropic cannabinoids and other drugs involving central and peripheral neurochemical mediators. In: Murphy L, Bartke A (eds) Marihuana/cannabinoids: neurobiology and neurophysiology. CRC Press, Boca Raton, pp 165–218

Pertwee RG, Wickens AP (1991) Enhancement by chlordiazepoxide of catalepsy induced in rats by intravenous or intrapallidal injections of enantiomeric cannabinoids. Neuropharmacology 30:237–244

Plasse TF, Gorter RW, Krasnow SH, Lane M, Shepard KV, Wadleigh RG (1991) Recent clinical experience with dronabinol. Pharmacol Biochem Behav 40: 695–700

Poklis A, Maginn D, Barr JL (1987) Drug findings in "driving under the influence of drugs" cases: a problem of illicit drug use. Drug Alcohol Depend 20:57–62

Pradhan SN (1984) Pharmacology of some synthetic tetrahydrocannabinols. Neurosci Biobehav Rev 8:369–385

Pross S, Nakano Y, Smith J, Widen R, Rodriguez A, Newton C, Friedman H (1992a) Suppressive effect of tetrahydrocannabinol on specific T cell subpopulations in the thymus. Thymus 19:97–104

Pross SH, Nakano Y, McHugh S, Widen R, Klein TW, Friedman H (1992b) Contrasting effects of THC on adult murine lymph node and spleen cell populations stimulated with mitogen or anti-CD3 antibody. Immunopharmacol Immunotoxicol 14:675–687

Pross SH, Nakano Y, Widen R, McHugh S, Newton CA, Klein TW, Friedman H (1992c) Differing effects of Δ^9-tetrahydrocannabinol (THC) on murine spleen cell populations dependent upon stimulators. Int J Immunopharmacol 14: 1019–1027

Purohit V, Ahluwahlia BS, Vigersky RA (1980) Marihuana inhibits dihydrotestosterone binding to the androgen receptor. Endocrinology 107:848–850

Raffel G, Clarenbach P, Peskar BA, Hertting G (1976) Synthesis and release of prostaglandins by rat brain synaptosomal fractions. J Neurochem 26:493–498

Razdan RK (1986) Structure-activity relationships in cannabinoids. Pharmacol Rev 38:75–149

Reggio PH, Seltzman HH, Compton DR, Prescott WR, Martin BR (1990) An investigation of the role of the phenolic hydroxyl in cannabinoid activity. Mol Pharmacol 38:854–862

Reggio PH, McGaughey GB, Odear DF, Seltzman HH, Compton DR, Martin BR (1991) A rational search for the separation of psychoactivity and analgesia in cannabinoids. Pharmacol Biochem Behav 40:479–486

Reggio PH, Panu AM, Miles S (1993) Characterization of a region of steric interference at the cannabinoid receptor using the active analog approach. J Med Chem 36:1761–1771

Reichman M, Nen W, Hokin LE (1987) Effects of Δ^9-tetrahydrocannabinol on prostaglandin formation in brain. Mol Pharmacol 32:686–690

Reuter H (1983) Calcium channel modulation by neurotransmitters, enzymes and drugs. Nature 301:569–574

Rinaldi-Carmona M, Barth F, Héaulme M, Shire D, Calandra B, Congy C, Martinez S, Maruani J, Néliat G, Caput D, Ferrara P, Soubrié P, Brelière JC, Le Fur G (1994) SR141716A, a potent and selective antagonist of the brain cannabinoid receptor. FEBS Lett 350:240–244

Rodriguez de Fonseca F, Cebeira M, Fernandez RJJ, Navarro M, Ramos JA (1991a) Effects of pre- and perinatal exposure to hashish extracts on the ontogeny of brain dopaminergic neurons. Neuroscience 43:713–723

Rodriguez de Fonseca F, Fernandez RJJ, Murphy L, Eldridge JC, Steger RW, Bartke A (1991b) Effects of Δ^9-tetrahydrocannabinol exposure on adrenal medullary function: evidence of an acute effect and development of tolerance in chronic treatments. Pharmacol Biochem Behav 40:593–598

Rodriguez de Fonseca F, Fernández-Ruiz JJ, Murphy LL, Cebeira M, Steger RW, Bartke A, Ramos JA (1992a) Acute effects of Δ^9-tetrahydrocannabinol on dopaminergic activity in several rat brain areas. Pharmacol Biochem Behav 42:269–275

Rodriguez de Fonseca F, Hernandez ML, de MR, Fernandez RJJ, Ramos JA (1992b) Early changes in the development of dopaminergic neurotransmission after maternal exposure to cannabinoids. Pharmacol Biochem Behav 41:469–474

Rodriguez de Fonseca F, Ramos JA, Bonnin A, Fernandez RJJ (1993) Presence of cannabinoid binding sites in the brain from early postnatal ages. Neuroreport 4:135–138

Rodriguez de Fonseca F, Gorriti M, Fernandez RJJ, Palomo T, Ramos JA (1994) Downregulation of rat brain cannabinoid binding sites after chronic Δ^9-tetrahydrocannabinol treatment. Pharmacol Biochem Behav 47:33–40

Rosell S, Agurell S, Martin B (1976) Effects of cannabinoids on isolated smooth muscle preparations. In: Nahas GG (ed) Marihuana: chemistry, biochemistry, and cellular effects. Springer, Berlin Heidelberg New York, pp 397–406

Rosell S, Bjorkroth U, Agurell S, Leander K, Ohlsson A, Martin B, Mechoulam R (1979) Relation between effects of cannabinoid derivatives on the twitch response of the isolated guinea-pig ileum and their psychotropic properties. In: Nahas GG, Paton WDM (eds) Marihuana biological effects-analysis, metabolism, cellular responses, reproduction and brain. Pergamon, Oxford, pp 63–70

Rosenkrantz H (1983) Cannabis, marihuana, and cannabinoid toxicological manifestations in man and animals. In: Fehr KO, Kalant H (eds) Cannabis and health hazards: proccedings of an ARF/WHO scientific meeting on adverse health and behavioral consequences of cannabis use. Addiction Research Foundation, Toronto, pp 91–176

Rosenkrantz H (1985) Cannabis components and responses of neuroendocrine-reproductive targets: An overview. In: Paton W, Nahas GG (eds) Marihuana

'84: proceedings of the Oxford symposium on cannabis. IRL Press, Oxford, pp 457–505

Rosenkrantz H, Grant RJ, Fleischman RW, Barker JR (1986) Marihuana-induced embryotoxicity in the rabbit. Fund Appl Toxicol 7:236–243

Roth SH, Williams PJ (1979) The non-specific membrane binding properties of Δ^9-tetrahydrocannabinol and the effects of various solubilizers. J Pharm Pharmacol 31:224–230

Rupniak NM, Samson NA, Steventon MJ, Iversen SD (1991) Induction of cognitive impairment by scopolamine and noncholinergic agents in rhesus monkeys. Life Sci 48:893–899

Sanders J, Jackson DM, Starmer GA (1979) Interactions among the cannabinoids in the antagonism of abdominal constriction response in the mouse. Psychopharmacology 61:281–285

Scallet AC (1991) Neurotoxicology of cannabis and THC: a review of chronic exposure studies in animals. Pharmacol Biochem Behav 40:671–676

Scallet AC, Uemura E, Andrews A, Ali SF, McMillan DE, Paule MG, Brown RM, Slikker W (1987) Morphometric studies of the rat hippocampus following chronic Δ^9-tetrahydrocannabinol (THC). Brain Res 436:193–198

Schwartz RH (1993) Chronic marihuana smoking and short-term memory impairment. In: Nahas GG, Latour C (eds) Cannabis: physiopathology, epidemiology, detection. CRC Press, Boca Raton, pp 61–71

Schwartz RH, Hoffmann NG, Jones R (1987) Behavioral, psychosocial, and academic correlates of marijuana usage in adolescence. A study of a cohort under treatment. Clin Pediatr (Phila) 26:264–270

Semus SF, Martin BR (1990) A computergraphic investigation into the pharmacological role of the THC-cannabinoid phenolic moiety. Life Sci 46:1781–1785

Shook JE, Burks TF (1989) Psychoactive cannabinoids reduce gastrointestinal propulsion and motility in rodents. J Pharmacol Exp Ther 249:444–449

Slikker W Jr, Paule MG, Ali SF, Scallet AC, Bailey JR (1992) Behavioral, neurochemical, and neurohistological effects of chronic marijuana smoke exposure in the nonhuman primate. In: Murphy L, Bartke A (eds) Marijuana/cannabinoids: neurobiology and neurophysiology. CRC Press, Boca Raton, pp 219–273

Smiley KA, Karler R, Turkanis SA (1976) Effects of cannabinoids on the perfused rat heart. Res Commun Chem Pathol Pharmacol 14:659–675

Smith PB, Martin BR (1992) Spinal mechanisms of Δ^9-tetrahydrocannabinol-induced analgesia. Brain Res 578:8–12

Smith PB, Welch SP, Martin BR (1993) nor-Binaltorphimine specifically inhibits Δ^9-tetrahydrocannabinol-induced antinociception in mice without altering other pharmacological effects. J Pharmacol Exp Ther

Smith PB, Compton DR, Welch SP, Razdan RK, Mechoulam R, Martin BR (1994) The pharmacological activity of anandamide, a putative endogenous cannabinoid, in mice. J Pharmacol Exp Ther 270:219–227

Soderstrom CA, Trifillis AL, Shankar BS, Clark WE, Cowley A (1993) Marijuana and alcohol use among 1023 trauma patients. In: Nahas GG, Latour C (eds) Cannabis: physiopathology, epidemiology, detection. CRC Press, Boca Raton, pp 79–92

Sofia RD, Barry H (1974) Acute and chronic effects of Δ^9-tetrahydrocannabinol on food intake by rats. Psychopharmacology 39:213–222

Sofia RD, Nalepa SD, Harakal JJ, Vassar HB (1973) Anti-edema and analgesic properties of Δ^9-tetrahydrocannabinol (THC). J Pharmacol Exp Ther 186:646–655

Solomons K, Neppe VM (1989) Cannabis – its clinical effects. S Afr Med J 76:102–104

Soueif MI (1976) Cannabis-type dependence: the psychology of chronic heavy consumption. Ann NY Acad Sci 282:121–125

Specter SC, Klein TW, Newton C, Mondragon M, Widen R, Friedman H (1986) Marijuana effects on immunity: suppression of human natural killer cell activity of Δ^9-tetrahydrocannabinol. Int J Immunopharmacol 8:741–745

Spronck HJW, Luteijn JM, Salemink CA, Nugteren DH (1978) Inhibition of prostaglandin biosynthesis by derivatives of olivetol formed under pyrolysis of cannabidiol. Biochem Pharmacol 27:607–608

Substance Abuse and Mental Health Services Administration (1994) Preliminary estimates from the 1993 National Household Survey on Drug Abuse. US Department of Health and Human Services, Washington DC (Advance report no 7)

Talbott JA, Teague JW (1969) Marihuana Psychosis. JAMA 210:299–302

Tang JL, Lancz G, Specter S, Bullock H (1992) Marijuana and immunity: tetrahydrocannabinol-mediated inhibition of growth and phagocytic activity of the murine macrophage cell line, P388D1. Int J Immunopharmacol 14:253–262

Taschner KL (1983) Psychopathology and differential diagnosis of so-called Cannabis psychoses. Fortschr Neurol Psychiatr 51:235–248

Tennant FS (1983) Clinical toxicology of cannabis use. In: Fehr KO, Kalant H (eds) Cannabis and health hazards: proceedings of an ARF/WHO scientific meeting on adverse health and behavioral consequences of cannabis use. Addiction Research Foundation, Toronto, pp 69–90

Tennes K, Avitable N, Blackard C, Boyles C, Hassoun B, Holmes L, Kreye M (1985) Marijuana: prenatal and postnatal exposure in the human. In: Pinkert TM (ed) Current research on the consequences of maternal drug abuse. US Government Printing Office, Washington DC, pp 48–60

Thomas BF, Compton DR, Martin BF, Semus SF (1991) Modeling the cannabinoid receptor: a three-dimensional quantitative structure-activity analysis. Mol Pharmacol 40:656–665

Thomas BF, Wei X, Martin BR (1992) Characterization and autoradiographic localization of the cannabinoid binding site in rat brain using $[^3H]11$-OH-Δ^9-THC-DMH. J Pharmacol Exp ther 263:1383–1390

Thornicroft G (1990) Cannabis and psychosis: is there epidemiological evidence for an association? Br J Psychiatry 157:25–33

Tramposch A, Sangdee C, Franz DN, Karler R, Turkanis SA (1981) Cannabinoid-induced enhancement and depression of cat monosynaptic reflexes. Neuropharmacology 20:617–621

Tuchmann-Duplessis H (1993) Effects of cannabis on reproduction. In: Nahas GG, Latour C (eds) Cannabis: physiopathology, epidemiology, detection. CRC Press, Boca Raton, pp 187–192

Tulunay FC, Ayhan IH, Portoghese PS, Takemori AE (1981) Antagonism by chlornaltrexamine of some effects of Δ^9-tetrahydrocannabinol in rats. Eur J Pharmacol 70:219–224

Tulunay FC, Ayhan IH, Sparber SB (1982) The effects of morphine and Δ^9-tetrahydrocannabinol on motor activity in rats. Psychopharmacology 78:358–360

Tunving K (1987) Psychiatric aspects of cannabis use in adolescents and young adults. Pediatrician 14:83–91

Turkanis SA, Karler R (1986) Effects of Δ^9-tetrahydrocannabinol, 11-hydroxy-Δ^9-tetrahydrocannabinol and cannabidiol on neuromuscular transmission in the frog. Neuropharmacology 25:1273–1278

Turker RK, Kaymakcalan S, Ercan ZS (1975) Antihistaminic action of (−)-trans-Δ^9-tetrahydrocannabinol. Arch Int Pharmacodyn Ther 214:254–262

Tyrey L (1984) Endocrine aspects of cannabinoid action in female subprimates: search for sites of action. In: Braude MC, Ludfor JP (eds) Marijuana effects on the endocrine and reproductive systems. US Government Printing Office, Washington DC, pp 65–81

Tyrey L (1992) Δ^9-tetrahydrocannabinol attenuates luteinizing hormone release induced by electrochemical stimulation of the medial preoptic area. Biol Reprod 47:262–267

Uphouse LA, Welch SP, Ellis EF, Embrey JP (1993) Antinociceptive activity of intrathecal ketoralac is blocked by the kappa-opioid antagonist, nor-binaltorphimine. Eur J Pharmacol 242:53–58

Van der Kloot W (1994) Anandamide, a naturally-occurring agonist of the cannabinoid receptor, blocks adenylate cyclase at the frog neuromuscular junction. Brain Res 649:181–184

Vaysse PJ, Gardner EL, Zukin RS (1987) Modulation of rat brain opioid receptors by cannabinoids. J Pharmacol Exp Ther 241:534–539

Verbene AJM, Taylor DA, Fennessy MR (1980) Withdrawal-like behaviour induced by inhibitors of biogenic amine reuptake in rats treated chronically with Δ^9-tetrahydrocannabinol. Psychopharmacology 68:261–267

Vocci FJ, Petty SK, Dewey WL (1978) Antinociceptive action of butyryl derivatives of guanosine 3':5'-cyclic monophosphate. J Pharmacol Exp Ther 207:892–898

Vogel Z, Barg J, Levy R, Saya D, Heldman E, Mechoulam R (1993) Anandamide, a brain endogenous compound, interacts specifically with cannabinoid receptors and inhibits adenylate cyclase. J Neurochem 61:352–355

Volkow ND, Fowler JS (1993) Use of positron emission tomography to study drugs of abuse. In: Nahas GG, Latour C (eds) Cannabis: physiopathology, epidemiology, detection. CRC Press, Boca Raton, pp 21–43

Wall ME, Sadler BM, Brine D, Taylor H, Perez-Reyes M (1983) Metabolism, disposition, and kinetics of Δ^9-tetrahydrocannabinol in men and women. Clin Pharmacol Ther 34:352–363

Walton RP, Martin LF, Keller JH (1937) The relative activity of various purified products obtained from American grown hashish. J Pharmacol Exp Ther 62: 239–251

Ward SJ, Baizman E, Bell M, Childers S, D'Ambra T, Eissenstat M, Estep K, Haycock D, Howlett A, Luttinger D, Miller M (1991) Aminoalkylindoles (AAIs): a new route to the cannabinoid receptor? In: Harris LS (ed) Problems of drug dependence 1990: proceedings of the 52nd annual scientific meeting. US Government Printing Office, Washington DC, pp 425–426

Watanabe N, Moroji T, Tada K, Aoki N (1984) A therapeutic trial of caerulein to a long-term heavy marihuana user with amotivational syndrome. Prog Neuropsychopharmacol Biol Psychiatry 8:419–421

Wechsler H, Rohman M, Kotch JB, Idelson RK (1984) Alcohol and other drug use and automobile safety: a survey of Boston-area teen-agers. J Sch Health 54: 201–203

Weidenfeld J, Feldman S, Mechoulam R (1994) The effect of the brain constituent anandamide, a cannabinoid receptor agonist, on the hypotholamo-pituitary-adrenal axis in the rat. Neuroendocrinology 59:110–112

Weiss JL, Watanabe AM, Lemberger L, Tamarkin NR, Cardon PV (1972) Cardiovascular effects of Δ^9-tetrahydrocannabinol in man. Clin Pharmacol Ther 13: 671–684

Weissman A (1978) Generalization of the discriminative stimulus properties of Δ^9-tetrahydrocannabinol to cannabinoids with therapeutic potential. In: Colpaert FC, Rosecrans JA (eds) Stimulus properties of drugs: ten years of progress. Elsevier/North-Holland Biomedical, Amsterdam, pp 99–122

Welch SP (1993) Blockade of cannabinoid-induced antinociception by nor-binaltorphimine, but not N,N-diallyl-tyrosine-aib-phenylalanine-leucine, ICI 174,864 or naloxone in mice. J Pharmacol Exp Ther 256:633–640

Welch SP, Stevens DL (1992) Antinociceptive activity of intrathecally administered cannabinoids alone, and in combination with morphine, in mice. J Pharmacol Exp Ther 262:10–18

Welch SP, Thomas C, Patrick GS (1994) Modulation of cannabinoid-induced antinociception following intracerebroventricular versus intrathecal administration to mice: possible mechanisms for interaction with morphine. J Pharmacol Exp Ther 272:310–331

Wenger T, Croix D, Tramu G, Leonardelli J (1992) Effects of Δ^9-tetrahydro cannabinol on pregnancy, puberty, and the neuroendocrine system. In: Murphy L, Bartke A (eds) Marijuana/cannabinoids: neurobiology and neurophysiology. CRC Press, Boca Raton, pp 539–560

Westlake TM, Howlett AC, Ali SF, Paule MG, Scallet AC, Slikker W Jr (1991) Chronic exposure to Δ^9-tetrahydrocannabinol fails to irreversibly alter brain cannabinoid receptors. Brain Res 544:145–149

White HL, Tansik RL (1980) Effects of Δ^9-tetrahydrocannabinol and cannabidiol on phospholipase and other enzymes regulating arachidonate metabolism. Prostagladins Med 4:409–411

Wickens AP, Pertwee RG (1993) Δ^9-Tetrahydrocannabinol and anandamide enhance the ability of muscimol to induce catalepsy in the globus pallidus of rats. Eur J Pharmacol 250:205–208

Williams EG, Himmelsbach CK, Wikler A, Ruble DC, Lloyd BJ Jr (1946) Studies on marihuana and pyrahexyl compound. Public Health Rep 61:1059–1083

Williams RB, Ng LKY, Lamprecht F, Roth K, Kopin IJ (1973) Δ^9-Tetrahydrocannabinol: A hypotensive effect in rats. Psychpharmacology 28:269–274

Wilson RS, May EL (1974) 9-Nor-Δ^8-tetrahydrocannabinol, a cannabinoid of metabolic interest. J Med Chem 17:475–476

Wilson RS, May EL (1975) Analgesic properties of the tetrahydrocannabinols, their metabolites, and analogs. J Med Chem 18:700–703

Wilson RS, May EL (1979) Some 9-hydroxycannabinoid-like compounds. Synthesis and evaluation of analgesic and behavioral properties. J Med Chem 22:886–888

Yahya MD, Watson RR (1987) Immunomodulation by morphine and marijuana. Life Sci 41:2503–2510

Yaksh TL (1981) The antinociceptive effects of intrathecally-administered levonantradol and desacetyllevonantradol in the rat. J Clin Pharmacol 21:3345–3405

Yaksh TL, Al-Rodhan NRF, Jensen TS (1988) In: Fields HL, Besson JM (eds) Sites of action of opiates in production of analgesia. Elsevier, Amsterdam, pp 371–394 (Progress in brain research, vol 77)

Yebra M, Klein TW, Friedman H (1992) Δ^9-tetrahydrocannabinol suppresses concanavalin A induced increase in cytoplasmic free calcium in mouse thymocytes. Life Sci 51:151–160

Zimmerman S, Zimmerman AM, Laurence H (1981) Effect of Δ^9-tetrahydrocannabinol on cyclic nucleotides in synchronously dividing Tetrahymena. Can J Biochem 59:489–493

Zuardi AW, Karniol IG (1983) Effects on variable-interval performance in rats of delta 9-tetrahydrocannabinol and cannabidiol, separately and in combination. Braz J Med Biol Res 16:141–146

CHAPTER 4
Cocaine*

M.W. Fischman and C.-E. Johanson

A. History and Epidemiology

Cocaine is the principal alkaloid of *Erythoxylon coca*, a shrub that grows in the Andean Highlands and the northwestern portions of the Amazon River in South America. The coca plant has been cultivated by the Indians in these areas for several thousand years, and the dried leaves are still used today. Those living in the Highland areas mix the leaves with lime or ash and chew or suck this combination. Addition of the alkaline ash promotes release of the cocaine by changing the pH of saliva. Amazonian Indians first pulverize the alkaline and coca leaf combination. They place this combination into their mouths, where it is mixed with saliva and then swallowed. Coca ingested in these ways has been used for religious, medicinal, and energizing purposes for centuries (see Johanson and Fischman 1989 for a more complete discussion of this history). Since significant cocaine plasma levels can be attained when coca leaves are chewed and sucked as described (Paly et al. 1980), the effects achieved by coca chewing are undoubtedly due to the actions of this principal constituent.

Cocal leaves, unlike tobacco, did not achieve immediate popularity in Europe following their discovery in the New World. This was most likely due to deterioration during the time required to transport them from South America, resulting in no discernible effect following consumption. However, after isolation of cocaine from the coca leaf by Niemann in the 1850s and uncritical endorsement by many political leaders and entertainers of the time, cocaine use became popular. In particular, it was used as an additive in wines and patent medicines. As physicians and researchers (e.g., Sigmund Freud, Karl Koller, William Halsted) in the late 1800s and early 1990s found more and more uses for this substance, coca and cocaine products were prescribed for a multiplicity of ills (Johanson and Fischman 1989). But, the increasing interest of the American Medical Association in raising the

---

*This chapter was prepared during the same period of time that another chapter on the same topic was being prepared by CEJ. That chapter will appear in *Psychopharmacology: The Fourth Generation of Progress* (edited by F.E. Bloom and D.J. Kupfer) and was co-authored by Charles R. Schuster. Portions of the two chapters overlap and all authors have agreed to this arrangement. In addition, the editors of both books have been informed.

standards of medical practice, the ready availability of cocaine powder, and the deleterious publicity about cocaine's effects led to the regulation of the manufacture and sale of patent medicines (Pure Food and Drug Act of 1906) and then to a registration requirement of those involved in the importation, manufacture, or distribution of opium or coca products (Harrison Narcotics Act of 1914). As a result of these measures and others, cocaine use in the United States was substantially reduced by the 1920s and remained relatively low until the early 1970s (see MUSTO 1992, for a more complete discussion of this history).

In 1974, 5.4 million people in the United States reported having used cocaine at least once in their lifetime (FISHBURNE et al. 1980). Three years later that number had increased to 9.8 million and in 1982 it was estimated at 21.5 million. Data from the National Institute on Drug Abuse (NIDA) National Household Surveys indicate that past-year and past-30-day prevalence rates peaked in 1985. The most recent data available from the 1992 survey[1] revealed the lowest prevalence rates since 1977. Unfortunately, although past-year and past-30-day use have declined substantially, past-week use has not. This more frequent use of cocaine remains at the same high level as in the immediately preceeding years. Further support for the lack of decline in problematic cocaine use are data on emergency room (ER) admissions associated with cocaine use (Drug Abuse Warning Network, DAWN). Twenty-eight percent of all emergency room admissions involving drugs were cocaine-related in 1992, compared with 9% of all admissions in 1985. Because there is considerable evidence that the compilation of ER data by the DAWN system provides an underestimate, the actual number of ER visits and deaths associated with cocaine use are likely much higher (POLLOCK et al. 1991). Despite this suspected underestimation, cocaine is still second only to alcohol in combination with other drugs in ER mentions.

Cocaine is currently placed in schedule II of the Comprehensive Drug Abuse Prevention and Control Act of 1970, indicating that it has acceptable medical use but a high potential for abuse as well. It is the only local anesthetic that is capable of producing significant vasoconstriction, and thus could be extremely useful in surgeries requiring decreased bleeding in order to more readily visualize the surgical field. Although initially used extensively in ophthalmology because of these desirable attributes, it caused sloughing of the corneal epithelium. Its use is now limited to topical application in the upper respiratory tract (RITCHIE and GREENE 1990). When used carefully, cocaine is an excellent local anesthetic with rapid induction of anesthesia, good vasoconstriction, and no signs of mucosal or nerve damage (BARASH 1977). This contrasts significantly with its toxicity when used illicitly.

---

[1] The National Institute on Drug Abuse, now part of the National Institutes of Health, no longer conduct the survey. It is conducted by Substance Abuse, Mental Health and Alcohol Administration.

B. General Pharmacology

I. Pharmacokinetics

Cocaine pharmacokinetics have been studied using a number of different routes: oral, intranasal, intravenous, and smoked. Although the general pharmacological effects of cocaine are similar regardless of its route of administration, it has been assumed that mode of administration contributes to the likelihood of abuse. The oral route of administration, studied experimentally by administering cocaine in gelatin capsules, is characterized by relatively slow absorption and peak levels that do not appear until approximately 65 min after ingestion (VAN DYKE et al. 1978). Nevertheless, cocaine taken by the oral route produces subjective effects that are similar to those experienced by the intranasal route in humans and has dependence-related behavioral effects similar to cocaine delivered by other routes in animals and humans (MEISCH et al. 1993; TANG and FALK 1987; VAN DYKE et al. 1978). Although not generally abused by this route, toxic and lethal effects have been reported following the rupture of cocaine-filled plastic bags swallowed by smugglers or dealers trying to avoid arrest (FISHBAIN and WETLI 1981).

Cocaine hydrochloride is readily absorbed from the nasal mucosa when it is inhaled into the nose as a powder (known as insufflation). Peak venous plasma levels are attained approximately 30 min later (JAVAID et al. 1978). Inhalation of 96 mg results in peak plasma levels of 150–200 ng/ml (JAVAID et al. 1983), and plasma levels remain elevated for a prolonged period of time due to local vasoconstriction of the nasal mucosa. Although insufflation has been the most common route for nonmedical administration, this is no longer the case in most major cities in the United States (NATIONAL INSTITUTE ON DRUG ABUSE 1992). In keeping with that change, the smoked and intravenous routes are associated with a higher frequency of ER mentions (23% for smoked, 20% for IV, and 12% for snorted) and number of people entering treatment for cocaine abuse or dependence.

Venous plasma levels of cocaine peak more rapidly after it is smoked or intravenously injected, reflecting a rapid delivery of drug to the brain. Although cocaine hydrochloride is used for intravenous injection, the base form ("free base" or "crack") is the form smoked because it vaporizes at a lower temperature than does cocaine hydrochloride. At the temperature required to vaporize the hydrochloride, the chemical structure of cocaine is destroyed. When cocaine is injected intravenously, peak venous plasma levels have been reported 4 min after injection, with levels to 250–300 ng/ml after a single injection of 32 mg. A similar rise in venous plasma levels to 200–250 ng/ml after smoking 50 mg cocaine base has been reported (FOLTIN and FISCHMAN 1991). Venous plasma levels are correlated with the smoked and intravenous doses administered and, regardless of the route of administration, cocaine's elimination half-life is approximately 40 min, although longer half-lives of 56–80 min have been reported, perhaps associated with considerably higher doses. There have been several reports estimating cocaine half-lives of 20–110 h in cocaine abusers with histories of high dose chronic use, suggesting that cocaine can accumulate, presumably in fatty tissues (CONE and WEDDINGTON 1989; WEISS 1988a). During abstinence, cocaine is likely released from these storage sites and excreted as cocaine and metabolites over a period of 2–3 weeks (CONE and WEEDINGTON 1989).

EVANS et al. (1993, 1994) compared venous and arterial plasma levels after smoked or intravenous administration of single doses of cocaine. When doses of cocaine base of 25 and 50 mg were smoked or doses of 16 and 32 mg were administered intravenously, arterial plasma levels reached a peak within 15 s whereas venous plasma levels did not reach their peak until after 3–6 min. This rapid rise of arterial cocaine levels may very well be related to the immediate "rush" reported by individuals using cocaine by these routes. Further, the peak arterial levels reached by either route of administration were approximately ten times higher than the peak venous plasma levels and reached similar levels after both routes. Thus, smoking cocaine base or intravenously injecting cocaine hydrochloride resulted in considerably greater levels of cocaine reaching the heart and presumably the brain than would have been predicted from prior data on venous plasma levels.

Cocaine is metabolized by plasma and liver cholinesterases into two principal water-soluble metabolites, benzoylecgonine and ecgonine methyl ester, which are excreted in the urine along with smaller amounts of ecgonine, norcocaine, and various hydroxylated products (VITTI and BONI 1985). Oxidative pathways account for considerably smaller amounts of cocaine metabolism. The principal metabolite found in plasma is benzoylecgonine, with less than 5% present as ecgonine methyl ester (ISENSCHMID et al. 1992). The major metabolites of cocaine can be found in urine for periods up to 36 h after the last administration of the drug, with reports of longer periods in extremely heavy users (WEISS 1988a). Cocaine can also be measured in saliva (CONE and WEDDINGTON 1989) and in hair (CONE et al. 1991). The meconium of infants born to women who have used cocaine during pregnancy has also been found to contain levels of cocaine and its metabolites (OSTREA et al. 1992). Breast milk of nursing cocaine-using mothers also contains cocaine (WIGGINS et al. 1989).

It has been estimated that between 60% and 90% of cocaine abusers use it in combination with alcohol (GRANT and HARFORD 1990; ROUNSAVILLE et al. 1991; WEISS et al. 1988). When cocaine is taken in combination with alcohol, another metabolite, cocaethylene, is formed. Recent research suggests that cocaethylene is a psychoactive substance with a pharmacological profile like that of cocaine (PEREZ-REYES and JEFFCOAT 1992), but it is more slowly metabolized. Cocaethylene is equipotent to cocaine in blocking the reuptake of dopamine but significantly less potent in blocking the reuptake of serotonin (BRADBERRY et al. 1993). In addition, the LD_{50} in mice for cocaethylene is significantly lower than that for cocaine (HEARN et al.

1991; KATZ et al. 1992). The combination of cocaine and alcohol also has been shown to produce greater changes in heart rate and blood pressure than either drug alone (FARRE et al. 1990; FOLTIN and FISCHMAN 1989; HIGGINS et al. 1990) as well as greater hepatotoxicity (BOYER and PETERSEN 1990). These experimental studies may help to explain the previously puzzling fact that for many cases of death apparently associated with cocaine use, very low blood levels of cocaine were reported (MITTLEMAN and WETLI 1984; WETLI and WRIGHT 1979). Because it is possible that these individuals had also consumed alcohol, the toxicity may have been largely due to the consequent formation of cocaethylene. This hypothesis is substantiated by recent studies (HEARN et al. 1991; HIME et al. 1991) of post mortem blood and tissue that have reported low levels of cocaine but high levels of cocaethylene, suggesting that the latter compound could have been responsible for the fatality.

An additional complication of the interaction between cocaine and alcohol is that alcohol increases the blood levels of intranasally administered cocaine (PEREZ-REYES and JEFFCOAT 1992). This may be due to alterations in the absorption of cocaine through the nasal mucosa, secondary to alcohol-induced vasodilation. Further, it appears that the formation of cocaethylene after a single inhalation of cocaine following alcohol consumption does not contribute to the physiological or subjective effects observed over the next 2 h. For instance, in the study by PEREZ-REYES and JEFFCOAT (1992), cocaine's subjective and physiological effects increased relatively rapidly after its administration and were returning to baseline during the period when cocaethylene levels in venous plasma were slowly increasing. However, it is possible that under conditions in which cocaine-alcohol combinations are taken repeatedly within a relatively short time period, cocaethylene levels could continue to rise and reach toxic levels in the absence of reported CNS effects that likely regulate continued self-administration.

II. Organ and System Toxicity

Although cocaine use has been associated with toxicity to almost every organ system of the body, the cardiovascular system appears to be most commonly affected (see reviews by BENOWITZ 1992; CHOW et al. 1990; HOLLANDER and HOFFMAN 1991). Myocardial ischemia and myocardial infarction have both been associated with cocaine use even in individuals with no evidence of abnormal coronary arteries (ISNER and CHOKSHI 1991; MINOR et al. 1991; PENTEL and HATSUKAMI 1994; SMITH et al. 1987; ZIMMERMAN et al. 1991). LANGE et al. (1989) demonstrated that a large number of individuals reporting cocaine-induced chest pain showed evidence of coronary vasoconstriction when administered relatively low doses of intranasal cocaine. Interestingly, this vasoconstriction was not temporally associated with further chest pain or myocardial ischemia. In addition, this cocaine-induced vasoconstriction occurs in both diseased and nondiseased artery segments but is significantly greater in the diseased segments (Brogan et al. 1991; Flores et al. 1990).

The mechanisms underlying cardiovascular toxicity are not well understood. These effects are undoubtedly complex because cocaine is a local anesthetic and also inhibits neuronal uptake of catecholamines. Cocaine's local anesthetic effects on the heart and blood vessels are likely to result in antiarrhythmic and vasodilatory actions, whereas effects at adrenergic, dopaminergic, and serotonergic synapses within the CNS are more likely to result in excitation, leading to seizure activity and increased peripheral sympathetic tone with accompanying tachycardia and vasoconstriction (Wilkerson 1988). Cocaine's sympathomimetic actions appear to predominate at lower doses, whereas its local anesthetic actions are more likely at higher doses (Herman and Vick 1987; Jain et al. 1987; Lew and Angus 1983; Stewart et al. 1963; Trendelenburg 1968). Thus, under conditions of rapid and complete absorption of a large dose, cocaine's local anesthetic actions are likely to predominate, resulting in decreased arterial blood pressure, decreased pacemaker activity, and myocardial depression. Experimental verification of these potential effects in humans are almost impossible to obtain. Although clinical studies involving controlled administration of cocaine in humans have failed to report significant numbers of adverse cardiovascular effects (Fischman et al. 1985; Frankenfield et al. 1994; Pentel and Hatsukami 1994; Perez-Reyes et al. 1982), the use of low, pure and infrequent doses and the careful selection of only healthy subjects all contribute to the low probability of noting adverse effects.

A puzzling aspect of cocaine's cardiovascular effects is the time lag between cocaine use and cocaine-associated angina pectoris or myocardial infarction. Although some cocaine users experience these symptoms within minutes, many do not have symptoms until hours after their last drug use. Brogan et al. (1992), studying intranasal cocaine effects in patients in a cardiac catheterization laboratory, found that cocaine causes recurrent coronary vasoconstriction, initially related to peak cocaine blood levels with later vasoconstriction temporally related to decreasing cocaine blood levels and an increase in concentration of its metabolites. It is possible that these metabolites might contribute to cocaine's cardiovascular toxicity, but no data exist to support this hypothesis. Alternatively, Majid et al. (1992) have suggested that the coronary vasospasm may not be primary in producing acute myocardial infarction. Instead, they hypothesize that platelet activation related to repeated cocaine use, by occluding and causing repeated spasm of small vessels, may play an important role in the development of cocaine-induced myocardial infarction.

In addition to cardiovascular changes, cocaine-related neurovascular complications, including cerebral parenchymal hemorrhages, subarachnoid hemorrhages, and ischemic cerebral manifestations, have been reported in response to all routes of administration (Jacobs et al. 1989). Pathological changes in the vasculature including arteriolar thickening, increased perivascular deposits of collagen and glycoprotein, and inflammatory cellular infiltrates are frequently present (CHOW et al. 1990), and it has been suggested that the combination of cocaine-induced systemic hypertension with these underlying vascular abnormalities may result in hemorraghic stroke (BRUST and RICHTER 1977; LICHTENFELD et al. 1984; SCHWARTZ and COHEN 1984). A large increase in systemic blood pressure has also been implicated in a case of acute rupture of the ascending aorta during cocaine intoxication (BARTH et al. 1986).

It has been suggested, based on a careful review of cocaine's cardiovascular toxicity, that more than one mechanism may be responsible for cocaine's deleterious effects on the myocardium (CHOW et al. 1990). It is possible that subgroups of the population have increased vulnerability to cocaine's cardiotoxic effects (e.g., via atherosclerotic changes). A different etiology may be responsible for the cocaine-related myocardial infarctions that appear to occur in users who have normal coronary arteries, as shown by angiography (ISNER et al. 1986). LEVINE et al. (1987) have suggested that enhanced sympathetic activity combined with blood pressure elevation could play a role, as could increased synaptic levels of serotonin, a potent vasoconstrictor for large and medium size arteries. There are also data suggesting that repeated cocaine use may decrease cerebral blood flow, especially in the frontal and temporal cortex (VOLKOW et al. 1987).

There have been recent reports showing an association between nontraumatic rhabdomyolisis and cocaine abuse. The mechanism responsible for this association is not known, but it is important clinically to recognize its possible occurrence for several reasons. These include the fact that chest pain associated with rhabdomyolisis can be mistaken for a myocardial infarction. Furthermore, untreated rhabdomyolisis may lead to fatal kidney failure (RUBIN and NEUGARTEN 1989).

Other neurologic complications, such as seizures, are also associated with cocaine use. This may be related to cocaine's hyperpyrexic actions (RITCHIE and GREENE 1990) or to cocaine's ability to "kindle" neurons that then results in a reduction in the seizure threshold during subsequent administration (POST et al. 1976). Seizure development can also occur in conjunction with ventral tachycardia and fibrillation (CREGLER and MARK 1986).

III. Fetal and Developmental Toxicity

Cocaine, norcocaine, and cocaethylene cross the placenta readily (SCHENKER et al. 1993), but it has been difficult to assess the incidence of neonatal exposure to cocaine in utero. Early clinical research findings describing cocaine's teratogenic effects (e.g., CHASNOFF et al. 1985) have not been replicated, and it is likely that many of the toxic effects attributed to maternal cocaine use, such as low birth weight, smaller head circumference, and a shorter gestational period, were a result of other factors. Women who abuse cocaine often are infected with sexually transmitted dieseases, do not obtain prenatal care, have poor nutrition, and abuse other drugs besides cocaine. It is difficult to separate cocaine's specific effects from the effects of these other risks associated with drug abuse (BEHNKE and EYLER 1993). In fact, as recently pointed out in a commentary with contributions from most of the major researchers in the field of cocaine teratology (HUTCHINGS 1993), cocaine is actually a relatively weak teratogen. In addition, LUTIGER and colleagues (1991) found few effects that could be attributed to maternal cocaine use in a meta-analysis of the relationship between use of cocaine during pregnancy and fetal development. Although only genitourinary tract malformations were consistently found to be associated with maternal cocaine use, more recent and better controlled studies have not replicated that finding (HUTCHINGS 1993). Further, the low birth weight seen in many neonates that are born to cocaine-using mothers can be significantly attenuated if the mother receives prenatal care (RACINE et al. 1993). Although animal research would seem to be the obvious answer for controlling many of the confounding variables found in human clinical research, the use of animal models raises its own set of difficulties. Questions related to species to be used, route, dose, and frequency of administration, as well as when in the gestational period to administer cocaine, adulterants commonly found in street drugs, and interactions with other drugs commonly used with cocaine, point to some of the complexities to be considered in any animal model.

In addition to structural abnormalities, researchers have also noted behavioral problems in infants of cocaine-exposed mothers. These have included increased irritability and reactivity, particularly in response to environmental challenges that might be considered stressful (ANDAY et al. 1989; MAGNANO et al. 1992). Researchers have also evaluated the behavioral development of human infants exposed to cocaine in utero as they have become older. COLES and PLATZMAN (1993) found few differences in toddlers on measures of global intellectual functioning, although ALESSANDRI et al. (1993) have shown that when the assessments are more complex and specific, there is evidence of cognitive and learning deficits in humans. Animal studies have also shown specific learning deficits as well as abnormal behavioral patterns following exposure to stressful situations (HEYSER et al. 1992; MOLINA et al. 1994; SPEAR et al. 1989a,b). However, it continues to be difficult to control for the following myriad of potential confounds or intervening variables emanating from the sampling scheme: (1) difficulty in defining and following a cohort; (2) defining, measuring, and verifying cocaine use; (3) the use of other drugs, prenatal care, and complications; (4) other maternal characteristics; (5) differences in socioeconomic level; (6) differences in age at which behavioral development is measured; and (7) postnatal environmental differences. The latter has been particularly ignored. NEUSPIEL (1994) reviewed several studies that used a particular scale, the Brazelton Neonatal Behavioral Assessment Scale, to study the effects of cocaine on infant and toddler behavior. The conclusions reached in these eight studies were often conflicting, and NEUSPIEL (1994) concluded that it would be difficult to attribute any noted developmental deficit specifically to cocaine use, although he was optimistic about future research using more refined methods.

IV. Behavioral Toxicity

A significant consequence of chronic cocaine use is the development of behavioral pathology in chronic cocaine abusers. Cocaine is generally taken in repeated dose cycles, and chronic intoxication can result in a psychosis characterized by paranoia, impaired reality testing, anxiety, a stereotyped compulsive repetitive pattern of behavior, and vivid visual, auditory, and tactile hallucinations, including delusions of insects crawling under the skin (JAFFE 1985; POST et al. 1976; SIEGEL 1978). It has also been reported that a psychosis-like syndrome was induced iatrogenically in a patient treated with a topical anesthetic containing 3 ml of 10% cocaine every 4 h (LESKO et al. 1982). In addition, less severe behavioral changes resulting from repeated cocaine use can include restlessness, irritability, hyperactivity, paranoid ideation, hypervigilance, impaired interpersonal relationships, and disturbances of eating and sleeping (GAWIN and ELLINWOOD 1988; SHERER et al. 1988). Although it is generally assumed that these symptoms related to chronic cocaine use dissipate within a few weeks after cessation of cocaine use, there have been reports of residual symptomatic and cognitive impairments (e.g., MANSCHRECK et al. 1987).

C. Neurobiology of Cocaine's Behavioral Effects

A major consequence of the cocaine epidemic and enhanced public concern about its abuse has been an acceleration in the number of investigations designed to increase the understanding of cocaine's CNS effects (BUDNEY et al. 1992). In particular, emphasis has been placed on understanding its reinforcing actions. Although it has been suspected for decades that dopamine (DA) plays a major role in the behavioral actions of cocaine, seminal research in the 1990s has provided even more difinitive evidence of DA's role in the behavioral actions of cocaine, in terms of receptor targets, neuroanatomical locus, and changes in neurotransmission. This research has also pointed to interactions with other neurochemical systems. Nevertheless, DA is viewed as so important in the behavioral actions of cocaine that the description of its mediation often is called the "dopamine hypothesis" (KUHAR et al. 1991). The implications of the DA hypothesis are extremely important as they suggest an approach for developing medications to treat cocaine dependence as well as to ameliorate changes in brain function that may occur as a consequence of chronic use. In addition, understanding the behavioral or reinforcing actions produced by cocaine can also help elucidate

CNS mechanisms of behavioral control by other reinforcers (Wise and Rompre 1989).

I. Receptor Targets

Studies in the 1980s demonstrated the existence of cocaine binding sites in rat striatum as well as monkey caudate-putamen, with affinities of cocaine and related compounds at these sites correlated with their potencies as DA reuptake blockers (Kennedy and Hanbauer 1983; Madras et al. 1989). Furthermore, there is a significant positive correlation between the potencies of cocaine and related compounds as DA reuptake blockers and their ability to maintain self-administration behavior (Bergman et al. 1989; Ritz et al. 1987). Although cocaine also blocks the reuptake of norepinephrine (NE) and serotonin (5-HT), significant correlations with self-administration were not found for these other transporter systems (Ritz et al. 1987). Thus, the action of cocaine at its binding sites that results in blockade of DA uptake appears to mediate the effects of cocaine that contribute to its abuse. Further, there are good arguments in favor of the conclusion that the cocaine "receptor" and the DA transporter are identical proteins. Cloning experiments have provided additional support. Transfection of COS cells, which do not take up DA or bind cocaine, with a single cDNA for DA uptake results in the development of the ability of these cells to both transport DA and bind cocaine (Shimada et al. 1991 but also see Madras et al. 1989). However, the existence of both a high- and a low-affinity cocaine binding site, at least in the striatum, is difficult to explain (Calligaro and Eldefrawi 1988; Madras et al. 1989). Other studies have shown that several DA reuptake inhibitors, such as high affinity analogs of cocaine, mazindol, and several GBR 12909 analogues, seem to bind to a common site and interact competitively, which leads to the parsimonious conclusion that they bind to the DA transporter (Carroll et al. 1992; Patel et al. 1992; Reith et al. 1992; Ritz et al. 1990a). Others have reported evidence, such as noncompetitive inhibition, consonant with the idea of multiple, perhaps overlapping, sites for various ligands and DA itself (Johnson et al. 1992; Rothman 1990; Rothman et al. 1991). Nevertheless, Grilli et al. (1991) have shown that the expression of cocaine binding sites and DA reuptake sites occurs simultaneously during in vitro cellular development. A final indication that cocaine binding sites and the DA transporter are related is that cocaine binding sites are distributed within the CNS in areas of high concentrations of DA nerve terminals. For instance, Kaufman et al. (1991) used [^{3}H]CFT as the radioligand in an autoradiographic study to map the distribution of cocaine binding sites in monkey brain. This congener of cocaine has the advantage of having a high affinity, low dissociation rate, and low levels of nonspecific binding relative to cocaine itself. As a result, even low concentrations of binding can be detected. Furthermore, its behavioral actions are virtually identical to cocaine (Spealman et al. 1991a,b).

Using this probe, KAUFMAN et al. (1991) found CFT binding sites to be distributed in DA-rich areas such as the caudate, putamen, and nucleus accumbens. In areas such as the substantia nigra and amygdala, where the density of DA and its transporter are low, binding sites were minimal. In areas with high concentrations of NE and 5-HT transporters, there were no indications of any binding sites. However, in areas such as the globus pallidus, which is known to have DA receptors, little binding was detected, a finding that needs clarification. Nevertheless, the strong correspondence between the distribution of CFT binding and DA-containing neurons provides evidence of the intimate relationship between the two "receptors." Similar conclusions were reached by KUHAR and colleagues using a different ligand ($[^{125}I]$RTI-55) in rats (BOJA et al. 1992; CLINE et al. 1992).

In addition to the identification of cocaine binding sites, the characteristics of the cocaine structure that are significant for its binding activity have also been described. The important structural features include a levorotatory configuration, a β-oriented substituent at C-2 and C-3, and a benzene ring at the C-3 carbon (RITZ et al. 1990b). The effects of changes in these structural features on binding are described by CARROLL et al. (1992). Several laboratories also have been involved in describing the characterisitics of the DA transporter protein. The transporter appears to be heterogeneous, with differences occurring in different brain regions (LEW et al. 1991; RITZ et al. 1992). These studies led to the cloning and expression of cDNA for the cocaine-sensitive DA receptor (KILTY et al. 1991; SHIMADA et al. 1991), a major breakthrough in furthering an understanding about how cocaine exerts its initial CNS effects. In particular, this finding provides an opportunity to elucidate the cellular mechanism of DA uptake and how it is altered by cocaine. An understanding of the sequence of cellular events would be useful for the development of therapeutic drugs. Another approach that appears promising for the development of therapeutic drugs is exemplified in a study by KITAYAMA et al. (1992). They assumed that only certain portions of the DA transporter are relevant to the binding of cocaine. If site-directed mutations of the transporter are made, clues to the relevant areas could then be deduced from further binding studies. The study found, in fact, that an aspartic acid residue lying within a particular region is necessary for both DA transport and cocaine binding, whereas other areas are only necessary for DA transport. The authors interpret this finding as holding promise for the development of cocaine antagonists that do not interfere with DA transport.

II. Sensitization

With chronic administration, the motor-activating effects of cocaine increase in magnitude. This resultant sensitization was an early observation and has been shown with a variety of psychomotor stimulants (ROBINSON 1988). Conditioning processes also appear to play an important role in sensitization

(POST et al. 1981). Because the DA systems mediating the motor effects of cocaine are related to those for its reinforcing effects (ROBINSON 1988), an understanding of sensitization has relevance for cocaine abuse (KALIVAS et al. 1993). Sensitization has also been proposed as a model for human psychosis (POST et al. 1976).

The development of in vivo methods that allow the monitoring of changes in extracellular DA levels and behavior simultaneously over time has made a major contribution in terms of increasing the knowledge base of the CNS actions of repeatedly administered cocaine that result in sensitization. These methods include microdialysis (IMPERATO and DI CHIARA 1984; UNGERSTEDT 1984) and voltammetry (JUSTICE 1987). Although these in vivo methods have enjoyed great popularity, they have certain limitations and might best be viewed as complementary to standard approaches involving the use of synaptosomal preparations, tissue slice preparations, and electrophysiological techniques.

UNGERSTEDT and colleagues (HURD et al. 1988; HURD and UNGERSTEDT 1989) used microdialysis to demonstrate that extracellular DA levels in several brain regions increased in a dose-dependent fashion following acute administration of cocaine at dose levels known to increase locomotor behavior. At the same time, CARBONI et al. (1989) showed that the increased DA levels following cocaine were more prominent in the nucleus accumbens, the area thought to mediate cocaine's reinforcing and locomotor effects (ROBERTS et al. 1980), than in the caudate. With the microdialysis technique, increases have also been reported in the ventral tegmental area (VTA) (BRADBERRY and ROTH 1989) and striatum (AKIMOTO et al. 1989). Although increases also occur in the prefrontal cortex, they only occur following relatively high doses and, in general, this area appears to be less affected at equal doses than the nucleus accumbens (MAISONNEUVE et al. 1990; MOGHADDAM and BUNNEY 1989). Subsequent research has focused on the effects of chronic administration. For instance, PETTIT et al. (1990) showed that the elevation of DA in response to cocaine was augmented in the nucleus accumbens after a repeated regimen of cocaine. In a subsequent study by NG et al. (1991), these findings were confirmed and extended using voltammetry techniques. These investigators showed that release of DA in the accumbens was augmented following 10 days of repeated cocaine. Since under the same regimen, tissue levels of DA are unchanged (KALIVAS et al. 1988) and synthesis is decreased (BROCK et al. 1990; TRULSON and ULISSEY 1987), the authors postulated that the increases in levels of DA efflux were due to a redistribution of DA stores into releasable pools. NG et al. (1991) also provided evidence that DA reuptake rate was increased with repeated cocaine administration, but the mechanism underlying this increase, such as increased numbers of reuptake sites, was not clear. Using the in vitro technique of examining ^{3}H-DA uptake in tissue slices, some researchers have also found evidence of an increase in DA reuptake rate following chronic cocaine administration (e.g., YI and JOHNSON 1990), whereas other

researchers have found persistent inhibition, at least in the nucleus accumbens (e.g., IZENWASSER and COX 1990). Thus, it is not clear at this time
how alterations in the reuptake mechanism change with time and contribute
to sensitization or any other behavioral effect.

Although there are important limitations of in vivo techniques, only
they can provide a direct demonstration of the relationship between increased
extracellular DA levels and locomotor activity in behaving animals, even
though early studies failed to take advantage of this attribute. KALIVAS and
DUFFY (1990) did measure both effects simultaneously in the same animals
and also found a correlation between changes in levels of extracellular DA
and motor activity, both of which increased over a 4-day regimen of repeated
treatment. However, there are other in vivo studies indicating that a high
correlation between extracellular levels of DA and behavioral sensitization
is not always observed. For instance, even in the study by KALIVAS and
DUFFY (1990), absolute levels of DA in the accumbens and motor activity
after the initial adminstration of cocaine were not highly correlated. Similar
results were found by HOOKS et al. (1992). In addition, doses of different
drugs that produce similar changes in motor behavior do not necessarily
produce similar increases in extracellular DA (KUCZENSKI and SEGAL 1988,
1992). Furthermore, a subsequent study by KALIVAS and DUFFY (1993a) that
examined more carefully the time course of the two effects found that
following a 5-day regimen of 15 mg/kg cocaine, behavioral sensitization in
response to an acute injection of the drug continued to be observed over at
least a 20-day period posttreatment. Although augmented levels of DA in
the nucleus accumbens in response to an injection of cocaine were observed
when animals were tested 10 and 20 days after the termination of treatment,
this augmentation was not observed 4 days posttreatment. To account for
the failure to find a continued correlation between augmented extracellular
DA levels and an augmented behavioral response at all time points, KALIVAS
and DUFFY (1993a) postulated that tolerance to the increased DA efflux had
occurred. This tolerance was evident immediately following the chronic
regimen but subsided with time. This hypothesis is corroborated in a study
by IMPERATO et al. (1992), in which levels of DA were measured in the
nucleus accumbens prior to cocaine administration during a 9 day regimen
of 10 mg/kg given twice daily. Initially, these basal levels were increased
relative to control but after approximately 6 days they were markedly
decreased and this decrease persisted for at least a week (IMPERATO et al.
1992). Similar results were found by KOOB and associates (WEISS et al.
1992b). They reported that basal DA levels were increased on day 1 following a 10-day regimen of 10 mg/kg or 30 mg/kg cocaine in rats but returned to
saline-treated levels by day 7. These investigators also noted that while
cocaine-induced DA was augmented on day 1 following the chronic regimen
relative to the control group, if increases were expressed as a percentage of
basal levels, which were also increased on day 1, the percentage increase
was less compared to chronically treated saline rats.

There appears to be some evidence that extracellular DA in the terminal fields in response to drug challenge is no longer increased in response to an injection of cocaine immediately after a chronic period of administration (KALIVAS and DUFFY 1993a). The mechanism for this decrease has not been elucidated, although speculation is rampant. For instance, KALIVAS and colleagues (KALIVAS and DUFFY 1993a; KALIVAS and STEWART 1991) suggested that the decrease (or absence of an increase) might be due to decreased synthesis, increased postsynaptic sensitivity, increased rate of reuptake, or alteration in the regulation of dopamine release from terminals. Furthermore, the mechanism underlying initiation of sensitization is difficult to understand in the absence of evidence of augmented extracellular DA in terminal fields. KALIVAS and DUFFY (1993a) offered the explanation that initiation rather than expression of sensitization may be mediated by other mechanisms. In a subsequent study, KALIVAS and DUFFY (1993b) provided additional evidence about the nature of this mechanism. Instead of measuring extracellular DA in terminal fields, dialysis probes were placed in the area of DA cell bodies (A10) of the VTA that project to the accumbens. They found that levels of extracellular DA in that area and locomotor activity in response to an injection of cocaine both were augmented 24h after the drug regimen was terminated whereas 13 days later only the locomotor response was still increased, at a time when increased levels of DA in terminal fields are seen. KALIVAS and STEWART (1991) have postulated that changes in DA processes related to sensitization could be explained by a variety of mechanisms that occur sequentially. Increased DA somatodendritic release in the VTA results in a decrease in inhibitory impulse-regulating somatodendritic D_2 autoreceptor sensitivity in the area of A10 cell bodies which in turn leads to increased neuronal firing and subsequent increases in extracellular DA in the nucleus accumbens.

Several studies using local administration of substances into the VTA have also provided evidence of the potential role of the VTA in the initiation of sensitization (but see DI CHIARA 1993). Injections of stimulants directly into the VTA result in augmented responses to systemically administered stimulants, whereas injections into the nucleus accumbens do not (KALIVAS and STEWART 1991; KALIVAS and WEBER 1988). Further, the administration of a DA antagonist into that area prevents the development of sensitization to stimulants (STEWART and VEZINA 1989). Acute local injections of cocaine decrease firing rates of VTA neurons because of increased levels of extracellular DA. These increased levels, which are due to cocaine's action in blocking DA reuptake, in turn decrease firing rate because of their action on inhibitory impulse-regulating somatodendritic autoreceptors. However, this effect diminishes with chronic administration (BRODIE and DUNWIDDIE 1990; LACEY et al. 1990) and it has been postulated that the subsensitivity of the VTA autoreceptors may involve reduced G-protein coupling (KALIVAS et al. 1992; NESTLER et al. 1990; STEKETEE and KALIVAS 1991; STEKETEE et al. 1991).

Using electrophysiological techniques, WHITE and colleagues (HENRY and WHITE 1992) have also shown that inhibitory impulse-regulating soma-todendritic D2 autoreceptors in the VTA become subsensitive with repeated cocaine administration and thus firing rate increases (ACKERMAN and WHITE 1990; HENRY et al. 1989). However, this subsensitivity, which results in these increased levels of neuronal firing, lasts less than 8 days (ACKERMAN and WHITE 1990) and thus may only be important in initiating sensitization, as suggested by KALIVAS and others. Additional evidence of increased neuronal firing of VTA neurons with chronic administration is the finding of increased tyrosine hydroxylase levels in the VTA but not in the nucleus accumbens (BEITNER-JOHNSON et al. 1992a; BEITNER-JOHNSON and NESTLER 1991). VTA neurons, but not neurons in other areas, have also been shown to have decreased numbers of neurofilament proteins. This decrease could alter VTA neuronal cytostructure and its functional integrity (BEITNER-JOHNSON et al. 1992a,b).

Electrophysiological studies by WHITE and colleagues have also shown that DA receptors in the nucleus accumbens become supersensitive to the effects of extracellular DA (HENRY et al. 1989). This increased sensitivity appears to be limited to D_1 receptors (HENRY and WHITE 1991). However, there is no evidence that this increased sensitivity is due to increased receptor D_1 density (KLEVEN et al. 1990b; PERIS et al. 1990), increased receptor affinity (MAYFIELD et al. 1992), or increased D_1 DA receptor-stimulated adenylate cyclase activity, although there is some evidence of increases in intracellular adenylate cyclase and cyclic AMP-dependent protein kinase activity in the nucleus accumbens but not in other areas following chronic cocaine administration (TERWILLIGER et al. 1991). Although HENRY and WHITE (1991) did not find changes in postsynaptic D_2 sensitivity, others have reported changes in D_2 binding in the nucleus accumbens (GOEDERS and KUHAR 1987; PERIS et al. 1990). Although the demonstrations of postsynaptic changes are appealing, the duration of these changes is problematic. For instance, the increase in D_2 binding is short-lived (PERIS et al. 1990) and D_1 receptor supersensitivity subsides after 2 months (HENRY and WHITE 1992).

Although most investigators would agree that DA systems play a major role in the development and maintenance of sensitization to the effects of cocaine, cocaine also blocks the reuptake of NE and 5HT (KOE 1976; RITZ et al. 1990b; ROSS and RENYI 1969). Cocaine binding sites have been localized on 5-HT terminals and the affinity of cocaine for these sites is greater than its affinity at DA sites (KOE 1976; REITH et al. 1983; RITZ et al. 1990b). Furthermore, the degree of inhibiton of 5-HT firing is greater than the inhibition of firing of DA neurons in the VTA (CUNNINGHAM et al. 1992a,b; PITTS and MARWAH 1987). Finally, cocaine also has been shown to bind to 5-HT$_3$ sites (KILPATRICK et al. 1987) and to decrease 5-HT synthesis (GALLOWAY 1990).

CUNNINGHAM and colleagues (1992b) have described changes in 5-HT systems correlated with behavioral sensitization. Reuptake blockade of 5-

HT in the dorsal raphe by acute administrations of cocaine results in decreased spontaneous 5-HT neuronal firing. This change appears to be mediated by an augmented inhibition resulting from increased stimulation of $5\text{-}HT_{1A}$ impulse-modulating autoreceptors (Cunningham and Lakoski 1988, 1990). Further, after 7 days of repeated administrations of cocaine that produce sensitization, firing rate inhibition in response to cocaine is augmented (Cunningham et al. 1992a,b). It may be that decreased neuronal firing would result in decreased release of 5-HT in neuronal projections to areas such as the VTA and nucleus accumbens. This in turn would diminish the inhibitory influence of 5-HT, furthering increasing DA neurotransmission (Cunningham et al. 1992a,b).

It is clear that the mechanism(s) underlying sensitization are complex and involve interactions among a variety of neurochemical pathways within the CNS. Kalivas and colleagues (1993) give a description of the entire circuitry that may be involved in both the initiation and maintenance of sensitization. To date there is not a clear picture of all the changes in various systems that occur subsequent to a chronic regimen, in part because researchers have used different chronic regimens, different techniques to measure changes, and different times of assessment. Each of the proposed mechanisms, changes in reuptake mechanisms, release, and receptor(s) sensitivity is supported by convincing empirical evidence. However, additional studies are clearly needed to differentiate and elucidate interactions among the potential explanations of the developmental course in the initiation and expression of sensitization. Furthermore, the influence of conditioning factors in this sequence of events is now just beginning to be considered. Nevertheless, a great deal has been learned with the advent of increased research interest in cocaine and new methodologies about the sequence of events that affect behavioral responses following continued cocaine treatment.

III. Reinforcement

Cocaine is a robust positive reinforcer and most previous studies have demonstrated that its reinforcing effects are mediated by mesolimbic/mesocortical dopaminergic neuronal systems (Johanson and Fischman 1989; Johanson and Schuster 1981; Koob and Bloom 1988; Young and Herling 1986). One line of evidence supporting a DA mediation is that other DA reuptake blockers and DA agonists support self-administration behavior (Roberts 1993; Weed et al. 1993; Wilson and Schuster 1976; Woolverton et al. 1984). Although these findings do not provide direct evidence of the mechanism underlying cocaine's reinforcing effects, they demonstrate that activation of DA receptors can maintain behavior. A second line of evidence is that both D_1 and D_2 DA antagonists modify cocaine self-administration (e.g., de Wit and Wise 1976; Koob et al. 1987). However, cocaine self-administration is not blocked by noradrenergic antagonists, such as phentolamine (DE WIT and WISE 1976). A third line of evidence is that depletions of DA produced by injections of the neurotoxin 6-OHDA into the meso-limbic/mesocortical dopaminergic neuronal pathway, including the VTA, nucleus accumbens, and ventral pallidum, attenuate cocaine self-administration, whereas depletions of NE do not (e.g., HUBNER and KOOB 1987; ROBERTS et al. 1977, 1980; ROBERTS and KOOB 1982). In many of these studies, however, the percent reduction in DA did not correlate with the extent of changes in self-administration behavior. Finally, direct injections of cocaine into the medial prefrontal cortex maintain self-administration behavior although direct injections into the nucleus accumbens or VTA do not (GOEDERS and SMITH 1983, 1993). These reinforcing effects are cor-related with decreases in DA utilization in frontal cortex and increases in the nucleus accumbens (GOEDERS and SMITH 1993). In addition, DA antagonists block the reinforcing effects of direct injections of cocaine. However, depletions of DA in the medial prefrontal cortex produced by the neurotoxin 6-OHDA have produced inconsistent effects. For instance, MARTIN-IVERSON et al. (1986) found no change in self-administration be-havior and GOEDERS and SMITH (1986) reported a suppression of intracranial cocaine self-administration, whereas in a study by SCHENK et al. (1991), the reinforcing effects of cocaine were increased following DA deple-tion in this area. In recent years new evidence supporting a DA mech-anism of cocaine's reinforcing effects has come from microdialysis studies. PETTIT and JUSTICE (1989) placed a microdialysis probe in the nucleus accumbens and showed that extracellular DA levels in that area were increased when cocaine was self-administered and levels were correlated with intake (PETTIT and JUSTICE 1991). HURD and colleagues (1989) re-ported that there was no longer a change in extracellular DA following cocaine self-administration with repeated exposure to cocaine during self-administration sessions (9–10 days). The authors speculated that a variety of mechanisms might mediate this presumed tolerance, including changes in uptake, release mechanisms, or postsynaptic receptor sensitivity (HURD et al. 1989, 1990). In summary, although the interpretation of many of the findings from self-administration studies that support the view that DA in mesolimbic/mesocortical pathways mediates cocaine's reinforcing effects is complex (see JOHANSON and FISCHMAN 1989 for a more thorough dis-cussion), most investigators, largely due to the confluence of evidence, have supported this conclusion and continue to examine the details of the mechanisms.

One approach in delineating the details is to separate the role of dif-ferent DA receptors in mediating the reinforcing effects of cocaine. Much of the earlier self-administration work using DA agonists and antagonists pointed to a prominent role for postsynaptic D_2 receptors (JOHANSON and FISCHMAN 1989), although the specificity of this role had been questioned (WOOLVERTON and VIRUS 1989). However, more recent studies have shown a potential role for D_1 receptors as well. For instance, although previously only D_2 agonists had been shown to support self-administration behavior (Woolverton et al. 1984), at least one D_1 agonist also maintains this behavior (Weed et al. 1993). Furthermore, SCH 23390, a relatively specific D_1 antagonist, blocks cocaine's reinforcing effects in rats when delivered systemically, as well as directly, into the nucleus accumbens (Koob et al. 1987; Maldonado et al. 1993). Another D_1 antagonist, SCH 39166, had a similar effect in squirrel monkeys (Bergman et al. 1990), although in the rhesus monkey SCH 23390 did not affect cocaine self-administration (Woolverton 1986). Studies in rats using more complex schedules have concluded that both D_1 and D_2 antagonists specifically decrease the reinforcing effects of cocaine (Hubner and Moreton 1991; Roberts et al. 1989). There is also evidence that D_3 receptors mediate the reinforcing effects of cocaine (Caine and Koob 1993). Although the relative contribution of the activation of each of these receptors, as well as others, is not yet known, it is clear that D_2 receptors alone are not responsible for the reinforcing effects of cocaine.

In addition to self-administration approaches, drug discrimination paradigms may provide complementary assessments of reinforcing effects, based upon the idea that both are related to the subjective effects of drugs in humans (Schuster and Johanson 1988). Johanson and Fischman (1989) reviewed many of the earlier studies that indicated that cocaine's discriminative stimulus (DS) effects are also mediated by DA. Recent studies support this conclusion and also provide evidence of multiple receptor subtype involvement. For instance, in a study using squirrel monkeys, Spealman et al. (1991b) demonstrated that cocaine analogs that were more potent than cocaine in binding to cocaine binding sites and in inhibiting uptake of DA were also more potent substitutes for cocaine as a DS. Although both D_1 and D_2 antagonists blocked cocaine's DS effects, neither D_1 nor D_2 agonists alone nor their combination completely substituted for cocaine. Similar results with antagonists were found in pigeons (Johanson and Barrett 1993). However, studies in rats and rhesus monkeys have not replicated entirely these findings (Kleven et al. 1990a; Witkin et al. 1991) and again, as with the findings from self-administration studies, the roles of different receptor subtypes are not completely understood. Furthermore, it is likely that as new DA receptor subtypes are described, they too may contribute in some manner to the reinforcing and discriminative stimulus effects of cocaine.

In addition to the further evidence of a role for DA in cocaine-maintained responding and discriminative stimulus effects, there have also been some recent studies implicating 5-HT. Loh and Roberts (1990) showed that depletions of 5-HT in the medial forebrain bundle or amygdala induced by injections of the neurotoxin, 5,7-dihydroxytryptamine increased the reinforcing effects of cocaine, as indicated by an increase in break-point under a progressive ratio schedule. Carroll et al. (1990) reported that rate of cocaine self-administration was reduced by fluoxetine, a 5-HT reuptake blocker. These investigators concluded that this change may have been due to either a decrease (blockade) or an increase in the reinforcing effects of cocaine. However, CUNNINGHAM and CALLAHAN (1991), using a drug discrimination paradigm, showed that fluoxetine shifted the cocaine dose-response function to the left, indicative of potentiation. In a drug discrimination study with pigeons, both fluoxetine and 8-OH-DPAT (a 5-HT$_{1A}$ agonist) partially substituted for, and the putative 5-HT$_{1A}$ antagonist NAN-190 partially blocked, the DS effects of cocaine (JOHANSON and BARRETT 1993).

Other neurotransmitters may also affect cocaine self-administration. For instance, injections of APV, a selective NMDA receptor antagonist, into the nucleus accumbens block the reinforcing effects of cocaine, and MK-801, a noncompetitive NMDA antagonist, also has been shown to interfere with the acquisition and maintenance of cocaine self-administration (PULVIRENTI et al. 1992; SCHENK et al. 1993). Because there is evidence of glutamatergic projections to the nucleus accumbens which interact with DA systems, these results indicate a modulatory role for glutamate in the reinforcing effects of cocaine via dopaminergic pathways (PULVIRENTI et al. 1992). It is also well known that endogenous opioid and DA systems interact in the CNS (KOOB and BLOOM 1988; WATSON et al. 1988). For instance, MELLO and colleagues (1989, 1990b) showed that buprenorphine, a mixed opioid agonist/antagonist, specifically reduced the self-administration of cocaine in rhesus monkeys both following an acute administration as well as during chronic administration. However, several investigators have questioned whether this effect is due to the antagonist effects of buprenorphine or to its agonist effects (CARROLL and LAC 1992; WINGER et al. 1992). Furthermore, laboratory research with humans suggests that buprenorphine can decrease cocaine self-administration by non opiate-dependent cocaine users (FOLTIN and FISCHMAN 1994). Research subjects given the opportunity to choose between cocaine (8, 16 or 32 mg/70 kg/injection) and tokens exchangeable for other reinforcers were pretreated with 0, 2, or 4 mg buprenorphine 60 min prior to their choice session. Pretreatment with buprenorphine significantly decreased choice of 16 or 32 mg/70 kg cocaine and increased token choice. The effects of cocaine in combination with buprenorphine were similar to the effects of morphine-cocaine combinations (FOLTIN and FISCHMAN 1992), again suggesting that the agonist effects of these low doses of buprenorphine might be responsible for the shift in cocaine choice in these non opiate-dependent research subjects.

Many of the self-administration studies designed to elucidate CNS mechanisms clearly have implications for the development of new medications for treatment. Such studies have also been done to determine mechanisms underlying the already demonstrated clinical utility of antidepressants and dopamine agonists. For instance, although there have been negative results reported (O'BRIEN et al. 1988; WEISS 1988b), GAWIN et al. (1989) showed the efficacy of desipramine for treating cocaine dependence in a carefully conducted clinical trial. In monkey self-administration studies, however, desipramine treatment produces inconsistent effects, ranging from decreases, no effect, or actual increases in self-administration (Kleven and Woolverton 1990; Mello et al. 1990a). In a human self-administration study, desipramine appeared to have no effect on self-administration behavior, although there were indications that it modified cocaine's subjective effects, decreasing some stimulant-related effects and increasing some generally aversive effects (Fischman and Foltin 1988). In contrast to the effects of desipramine in animal self-administration studies, bromocriptine, a direct D_2 DA agonist, which has also been suggested as a useful pharmacotherapy (Dackis and Gold 1985a), did suppress cocaine self-administration in a dose-dependent manner (Kleven and Woolverton 1990). However, due to the low specificity of the effect and evidence that the treatment enhanced cocaine's reinforcing and stimulatory effects, the authors were skeptical about its clinical utility.

IV. Medications Development

Dackis and Gold (1985b) hypothesized that the use of cocaine over extended periods of time led to a depletion in DA. Depletions of DA produced profound behavioral manifestations when cocaine use was terminated. A detailed description of this cocaine withdrawal syndrome was provided by Gawin and Kleber (1986), and they too believed that alterations in DA function were responsible. Therefore, it became geneally accepted that medications that would increase DA brain levels or activate DA receptors would reverse this withdrawal. Furthermore, it was hypothesized that craving was a manifestation of withdrawal and thus its amelioration would decrease the probability of relapse induced by craving (Dackis and Gold 1985b). Unfortunately, more recent studies (Satel et al. 1991; Weddington et al. 1990) conducted under controlled conditions have not replicated the findings of Gawin and Kleber (1986). In fact, these studies have only reported mild changes in mood and craving during withdrawal. It may be important to point out, however, that these studies were conducted in an in-patient setting where environmental influences, such as conditioned cues, might have been missing. In addition to the failure to reliably demonstrate a withdrawal syndrome, the idea that brain levels of DA are depleted after excessive cocaine abuse has weak support. Prolactin release, which is under central inhibitory DA control, has been reported to be increased initially following abstinence (Dackis and Gold 1985b; Mendelson et al. 1989). However, this finding was not replicated by Satel and colleagues (1991). Nevertheless, PET studies of human cocaine abusers have provided support for the role of DA dysfunction during cocaine withdrawal. Volkow and colleagues (1988) reported a decrease in cerebral blood flow in the prefrontal cortex in individuals who had previously abused cocaine. The authors suggested that this effect was due to changes in neuronal functioning due to neurotoxic changes in DA neuronal systems. A second study using [18F] fluorodeoxyglucose showed that glucose metabolism was increased in the orbitofrontal cortex and basal ganglia after cessation of cocaine use in humans (VOLKOW et al. 1991). This finding also indicates a decrease in DA activity. Furthermore, these levels had returned to normal 2–4 weeks after cessation when symptoms of withdrawal were largely dissipated.

Because it is extremely difficult to measure changes in DA function directly in humans and, in addition, it is impossible to control exposure dose, exposure duration, comorbid psychiatric conditions, and use of other types of drugs in human research participants, studies in experimental animal preparations are needed. Thus, animal studies have been conducted to determine whether there is irreversible damage to DA neuronal elements that might lead to decreased levels of DA following chronic cocaine administration. This search was stimulated in part by the findings that repeated administrations of methamphetamine (MA), another abused psychomotor stimulant, destroy DA nerve terminals, particularly in the caudate (SEIDEN and RICAURTE 1987). This selective destruction is associated with depletion of DA as well as decreases in tyrosine hydroxylase activity and reduced DA receptor binding. If these changes also occurred following chronic administration of cocaine, it would provide substantive support for DA dysfunction in humans. Although other MA-like drugs have also been shown to have similar long-lasting neurotoxic effects, it has not been shown that cocaine decreases tissue levels of DA or its metabolites (KLEVEN et al. 1988; YEH and DE SOUZA 1991). Changes do occur in D_1, D_2 and/or DA transporter binding site densities in both rats and monkeys (FARFEL et al. 1992; KLEVEN et al. 1990b; PERIS et al. 1990), although none of the studies tested animals longer than 2 weeks after the chronic regimen was terminated. In some cases, the effects were shown to be reversed after a relatively short time period (e.g., PERIS et al. 1990).

Although there is not evidence from animal studies of DA depletion, there has been recent data indicating a functional DA depletion. These data, which have been obtained using microdialysis techniques, show a decrease in basal levels of extracelluar DA following chronic cocaine at certain time periods post abstinence. For instance, although PARSONS et al. (1991) showed no change in basal DA levels immediately after a 10-day regimen of 20 mg/kg per day cocaine, after 10 days of abstinence these levels were significantly reduced, which the investigators attributed to reduced release. ROSSETTI et al. (1992), using a regimen of 15 mg/kg twice daily for 16 days, reported similar findings and also reported decreases in basal DA following other drugs of abuse. ROSSETTI et al. (1992) related these decreases in basal DA levels to the findings of MARKOU and KOOB (1991), who demonstrated that, following a period of self-administration, there was an increase in the threshold of intracranial self-stimulation. This increase and its duration were dose- and time-dependent. That increases in threshold are evidence of an aversive state is based upon the findings that drugs of abuse, which produce euphoria, including cocaine, decrease threshold (e.g., ESPOSITO et al. 1978). Other investigators have shown increased levels of basal DA immediately following a chronic regimen, which, however, are not maintained over time (Imperato et al. 1992; Weiss et al. 1992b). Nevertheless, when expressed as a percent of basal DA levels, the chronic cocaine regimen resulted in a diminished increase in DA levels in response to cocaine immediately after the chronic regimen (Weiss et al. 1992b). Another interesting finding is that, within hours following a short regimen of i.v. cocaine self-administration, there were decreases in basal DA levels (Weiss et al. 1992a) and these decreases were directly related to the duration of self-administration. The authors also relate these findings to the changes reported in threshold for intracranial stimulation. However, the results are somewhat difficult to interpret since, although basal DA levels were reduced relative to the levels of DA prior to the period of self-administration, baseline levels in the animals experienced in the self-administration paradigm were significantly greater than in control animals. Thus, the "decreased" levels were still greater than the levels in control rats.

Despite the difficulty interpreting many of the results in both animal and human studies, the possibility of DA dysfunction has provided the basis for the development of several medications for the treatment of cocaine abuse and their evaluation in clinical trials. The evidence that cocaine's reinforcing effects are mediated by DA has provided an additional impetus as well. Dopaminergic agonists and antagonists have been suggested as potential treatments for cocaine abuse much as opiate agonists and antagonists have been used in the treatment of opiate abuse. Direct and indirect dopamine agonists such as bromocriptine and amantadine have been reported to alleviate the symptoms of cocaine withdrawal (Dackis et al. 1987; Giannini et al. 1987; Tennant and Sagherian 1987), but a double-blind trial reported a 70% dropout rate due to bromocriptine's side effects (Giannini and Billett 1987). In addition, recent double-blind placebo-controlled trials (e.g., Weddington et al. 1991) have not confirmed any positive effects. The effectiveness of other DA agonists, such as methylphenidate or mazindol, in reducing cocaine use has not been demonstrated in controlled clinical trials. Although there is evidence from self-administration studies that DA antagonists decrease the reinforcing effects of cocaine, controlled clinical trials with antagonists have not been conducted. While there has been a report of success in an open trial with the DA blocker flupenthixol decanoate (Gawin et al. 1989), the double-blind clinical trial of this medication has not yet been reported. One of the few medications to show efficacy in promoting cocaine abstinence and maintaining treatment program attendance in a double-blind placebo-controlled study is the antidepressant desipramine (Gawin et al. 1989). Gawin et al. (1989) hypothesized that its effectiveness was based upon its ability to reverse the neurochemical changes produced by chronic exposure to cocaine, perhaps by blocking DA uptake. Other controlled trials carried out soon after concluded that desipramine was no more effective than placebo in promoting cocaine abstinence (Kosten et al. 1992). Although a meta-analysis of nine placebo-controlled studies of desipramine concluded that it promoted greater rates of abstinence than placebo (LEVIN and LEHMAN 1991), only two of the included reports were positive. In addition, one of these was an interim report and when it was finished the investigators concluded that the drug was no more effective than placebo (ARNDT et al. 1992; O'BRIEN et al. 1988). However, desipramine appears to promote cocaine abstinence (LEAL et al. 1994) or other positive outcome measures (ARNDT et al. 1994) significantly more than placebo if patients with diagnosed anti-social personality are deleted from the analysis. This suggests that pharmacological interventions for the treatment of cocaine abuse may have to be targeted to specific subpopulations of users (e.g., methylphenidate for those with Attention Deficit Disorder, lithium for cocaine abusers diagnosed with concurrent bipolar manic-depressive or cyclothymic disorders). In addition, a creative and potentially effective approach to treating cocaine abusers was recently reported by LANDRY et al. (1993), who, using myoclonal antibody techniques, have developed an artificial enzyme that inactivates cocaine as soon as it enters the blood stream by binding the cocaine and breaking it into two inactive metabolites. Thus, it has the potential for destroying much of the cocaine before it reaches the brain. However, this technique is still in an early test phase and it is not clear whether such an approach will ultimately prove practical.

Although the epidemic use of cocaine peaked between 1980 and 1985, the demand for treatment has been somewhat delayed and is still rising. With the increase in demand, there began a concerted effort to develop appropriate treatment strategies. Because of clinicians' experience with treating opioid dependence with medications and because of early speculations about DA depletion, initial treatment efforts were in the area of pharmacotherapy, as partially reviewed above. Although medications must be used within the context of a behavioral therapy in order to be maximally successful, by holding the latter constant, it should be possible to determine the efficacy of pharmacological agents to aid in treatment. Medications can modify cocaine use in a number of ways including decreasing craving, initiating abstinence, decreasing the positive effects of cocaine, substituting for cocaine, or blocking cocaine's effects. The general approach to evaluating a new medication, initiation of an open-label trial followed, if successful, with a controlled trial is expensive, time-consuming and, if successful, generally cannot elucidate the behavioral mechanism by which it was effective. Recent laboratory research has suggested that combining a cocaine choice/self-administration procedure with measures of cocaine's subjective and physiological effects might be an efficient predictor of the efficacy of a potential treatment medication (FISCHMAN and FOLTIN 1992). In addition, the combination of measures collected should provide useful information about their mechanisms of action. Using this model, as described above, FISCHMAN and colleagues (1990) reported that desipramine maintenance did not affect cocaine-taking bahavior but did appear to reduce reports of wanting cocaine and changed cocaine's profile of subjective effects such that some of its stimulant-like effects were reduced and some of its aversive effects were potentiated. Kleber (1988) suggested that desipramine provides a treatment "window of opportunity," and the laboratory data describing desipramine's effect suggest a possible mechanism for that "window." A more recent laboratory study evaluating buprenorphine in combination with cocaine has shown that cocaine choice/self-administration is reduced after buprenorphine pretreatment (Foltin and Fischman 1994), an effect that is probably related to buprenorphine's agonist effects. Hopefully, this approach will continue to be used before the initiation of expensive and time-consuming clinical trials, particularly given the low rate of success of the latter.

D. Final Comments

Although cocaine abuse continues to be a major public health problem, substantial advances in our understanding of the mechanisms underlying cocaine's dependence-producing effects have been made. Research into the cellular, physiological, and behavioral effects of cocaine suggest that, although the interactions among the various neuronal systems affected by cocaine are complex and poorly understood, the DA systems are clearly significant in mediating many of these effects. It is also clear that environmental influences modulate cocaine's effects in important ways, and any research into neuronal mechanisms must also involve the environmental determinants of cocaine's reinforcing effects if we are to be successful in developing effective pharmacological interventions for the treatment of cocaine abusers. Importantly, although such pharmacological interventions may well provide us with the "window of opportunity" required to work with cocaine abusers, the problem of cocaine abuse is a behavioral problem, and the combination of pharmacotherapy with behavioral and psychosocial interventions is our only hope for the successful treatment of cocaine-dependent individuals.

Acknowledgements. Manuscript preparation supported by U.S. Public Health Service Grants DA03818, DA06234 and DA03476 from the National Institute on Drug Abuse.

References

Ackerman JM, White FJ (1990) A10 somatodendritic dopamine autoreceptor sensitivity following withdrawal from repeated cocaine treatment. Neurosci Lett 117:181–187

Akimoto K, Hamamura T, Otsuki S (1989) Subchronic cocaine treatment enhances cocaine-induced dopamine efflux, studied by in vivo intracerebral dialysis. Brain Res 490:339–344

Alessandri SM, Sullivan MW, Imaizumi S, Lewis M (1993) Learning and emotional responsibity in cocaine-exposed infants. Dev Psychol 29:989–997

Anday EK, Cohen ME, Kelley NE, Leitner DS (1989) Effect of in utero cocaine exposure on startle and its modification. Dev Pharmacol Ther 12:137–145

Arndt IO, Dorozynsky L, Woody GE, McLellan AT, O'Brien CP (1992) Desipramine treatment of cocaine dependence in methadone-maintained patients. Arch Gen Psychiatry 49:888–893

Arndt IO, McLellan AT, Dorozynsky L, Woody GE, O'Brien CP (1994) Desipramine treatment of cocaine dependence: role of antisocial personality disorder. J Nerv Ment Dis 182:151–156

Barash PG (1977) Cocaine in clinical medicine. In: Petersen RC, Stillman RC (eds) Cocaine, 1977. US Government Printing Office, Washington DC, pp 193–200

Barth CW, Bray M, Roberts WC (1986) Rupture of the ascending aorta during cocaine intoxication. Am J Cardiol 57:496

Behnke M, Eyler FD (1993) The consequences of prenatal substance use for the developing fetus, newborn, and young child. In J Addict 28:1341–1391

Beitner-Johnson D, Nestler EJ (1991) Morphine and cocaine exert common chronic actions on tyrosine hydroxylase in dopaminergic brain reward regions. J Neurochem 57:344–347

Beitner-Johnson D, Guitart X, Nestler EJ (1992a) Common intracellular actions of chronic morphine and cocaine in dopaminergic brain reward regions. In: Kalivas PW, Samson HH (eds) The neurobiology of drug and alcohol addiction. New York Academy of Sciences, New York, pp 70–87

Beitner-Johnson D, Guitart X, Nestler EJ (1992b) Neurofilament proteins and the mesolimbic dopamine system: common regulation by chronic morphine and chronic cocaine in the rat ventral tegmental area. J Neurosci 12:2165–2176

Benowitz NL (1992) How toxic is cocaine? In: Edwards G (ed) Cocaine: scientific and social dimensions. Ciba Foundation Symposium. Wiley, Chichester, pp 125–148

Bergman J, Madras B, Johnson S, Spealman R (1989) Effects of cocaine and related drugs in nonhuman primates. J Pharmacol Exp Ther 251:150–155

Bergman J, Kamien J, Spealman R (1990) Antagonism of cocaine self-administration by selective dopamine D1 and D2 antagonists. Behav Pharmacol 1:355–363

Boja JW, Mitchell WM, Patel A, T.A. K, Carroll FI, Lewin AH, Abraham P, Kuhar MJ (1992) High-affinity binding of [125I]RTI-55 to dopamine and serotonin transporters in rat brain. Synapse 12:27–36

Boyer CS, Petersen DR (1990) Potentiation of cocaine-mediated hepatotoxicity by acute and chronic ethanol. Alcoholism 14:28–31

Bradberry C, Roth R (1989) Cocaine increases extracellular dopamine in rat nucleus accumbens and ventral tegmental area as shown by in vivo dialysis. Neurosci Lett 103:97–102

Bradberry CW, Nobiletti JB, Elsworth JD, Murphy B, Jatlow P, Roth RH (1993) Cocaine and cocaethylene: microdialysis comparison of brain drug levels and effects on dopamine and serotonin. J Neurochem 60:1429–1435

Brock JW, Ng JP, Justice JB (1990) Effect of chronic cocaine on dopamine synthesis in the nucleus accumbens as determined by microdialysis perfusion with NSD-1015. Neurosci Lett 117:234–239

Brodie MS, Dunwiddie TV (1990) Cocaine effects in the ventral tegmental area: evidence for an indirect dopaminergic mechanism of action. Naunyn Schmiedebergs Arch Pharmacol 342:660–665

Brogan WC, Lange RA, Kim AS, Moliterno DJ, Hillis LD (1991) Alleviation of cocaine-induced coronary vasoconstriction by nitroglycerin. J Am Coll Cardiol 18:581–586

Brogan WC, Lange RA, Glamann B, Hillis LD (1992) Recurrent coronary vasoconstriction caused by intranasal cocaine: possible role for medicine. Ann Intern Med 116:556–561

Brust JC, Richter RW (1977) Stroke associated with cocaine abuse? NY State J Med 77:1473–1475

Budney AJ, Higgins ST, Hughes JR, Bickel WK (1992) The scientific/clinical response to the cocaine epidemic – a MEDLINE search of the literature. Drug Alcohol Depend 30:143–149

Caine SG, Koob GF (1993) Modulation of cocaine self-administration in the rat through D-3 dopamine receptors. Science 260:1814–1816

Calligaro DO, Eldefrawi ME (1988) High affinity stereospecific binding of [^{1}H]cocaine in striatum and its relationship to the dopamine transporter. Membr Biochem 7:87–106

Carboni E, Imperato A, Perezzani L, Di Chiara G (1989) Amphetamine, cocaine, phencyclidine and nomifensine increase extracellular dopamine concentrations preferentially in the nucleus accumbens of freely moving rats. Neuroscience 28:653–661

Carroll ME, Lac ST (1992) Effects of buprenorphine on self-administration of cocaine and a nondrug reinforcer in rats. Psychopharmacology 106:439–446

Carroll ME, Lac ST, Acensio M, Kragh R (1990) Fluoxetine reduces intravenous cocaine self-administration in rats. Pharmacol Biochem Behav 35:237–244

Carroll FI, Lewin AH, Boja JW, Kuhar MJ (1992) Cocaine receptor: biochemical characterization and structure-activity relationships of cocaine analogues at the dopamine transporter. J Med Chem 35:969–981

Chasnoff IJ, Burns WJ, Schnoll SH, Burns KA (1985) Cocaine use in pregnancy. N Engl J Med 313:666–669

Chow JM, Menchen SL, Paul BD, Stein RJ (1990) Vascular changes in the nasal submucosa of chronic cocaine addicts. Am J Forensic Med Pathol 11:136–143

Cline EJ, Scheffel U, Boja JW, Mitchell WM, Carroll FI, Abraham P, Lewin AH, Kuhar MJ (1992) In vivo binding of [125I]RTI-55 to dopamine transporters: pharmacology and regional distribution with autoradiography. Synapse 12:37–46

Coles CD, Platzman KA (1993) Behavioral development in children prenatally exposed to drugs and alcohol. Int J Addict 28:1393–1433

Cone EJ, Weddington WW Jr (1989) Prolonged occurrence of cocaine in human saliva and urine after chronic use. J Anal Toxicol 13:65–68

Cone EJ, Yousefnejad D, Darwin WD, Maguire T (1991) Testing human hair for drugs of abuse. II. Identification of unique cocaine metabolites in hair of drug abusers and evaluation of decontamination procedures. J Anal Toxicol 15:250–255

Cregler LL, Mark H (1986) Medical complications of cocaine abuse. N Engl J Med 315:1495–1500

Cunningham KA, Callahan PM (1991) Monoamine reuptake inhibitors enhance the discriminative state induced by cocaine in the rat. Psychopharmacology 104:177–180

Cunningham KA, Lakoski JM (1988) Electrophysiological effects of cocaine and procaine on dorsal raphe serotonin neurons. Eur J Pharmacol 148:457–462

Cunningham KA, Lakoski JM (1990) The interaction of cocaine with serotonin dorsal raphe neurons: single unit extracellular recording studies. Neuropsychopharmacology 3:41–50

Cunningham KA, Paris JM, Goeders NE (1992a) Chronic cocaine enhances serotonin autoregulation and serotonin uptake binding. Synapse 11:112–123

Cunningham KA, Paris JM, Goeders NE (1992b) Serotonin neurotransmission in cocaine sensitization. In: Kalivas PW, Samson HH (eds) The neurobiology of drug and alcohol addiction. New York Academy of Sciences, New York, pp 117–127

Dackis CA, Gold MS (1985a) Bromocriptine as a treatment for cocaine abuse: the dopamine depletion hypothesis. Int J Psychiatry Med 15:125–135

Dackis CA, Gold MS (1985b) Pharmacological approaches to cocaine addiction. J Subst Abuse Treat 2:139–145

Dackis CA, Gold MS, Sweeney DR, Byron JPJ, Climko R (1987) Single-dose bromocriptine reverses cocaine craving. Psychiatry Res 20:261–264

de Wit H, Wise RA (1976) Blockade of cocaine reinforcement in rats with the dopamine receptor blocker pimozide, but not with the noradrenergic blockers phentolamine or phenoxybenzamine. Can J Psychol 31:195–203

Di Chiara G (1993) Searching for the hidden order in chaos. Commentary on Kalivas et al.: the pharmacology and neural circuitry of sensitization to psychostimulants. Behav Pharmacol 4:335–337

Esposito RU, Motola AHD, Kornetsky C (1978) Cocaine: acute effects on reinforcement thresholds for self-stimulation behavior to the medial forebrain bundle. Pharmacol Biochem Behav 8:437–439

Evans SM, Cone EJ, Marco AP, Henningfield JE (1993) A comparison of the arterial kinetics of smoked and intravenous cocaine. In: Harris L (ed) Problems of Drug Dependence, 1992. US Department of Health and Human Services, Rockville MD

Evans SM, Cone EJ, Henningfield JE (1994) Rapid arterial kinetics of intravenous and smoked cocaine: relationship to subjective and cardiovascular effects. Presented at the annual meeting of the College on Problems of Drug Dependence 56th annual scientific meeting, Palm Beach, Florida

Farfel GM, Kleven MS, Woolverton WL, Seiden LS, Perry BD (1992) Effects of repeated injections of cocaine on catecholamine receptor binding sites, dopamine transporter binding sites and behavior in rhesus monkey. Brain Res 578:235–243

Farre M, Llorente M, Ugena B, Lamax X, Cami J (1990) Interaction with ethanol. In: Harris L (ed) Problems of drug dependence 1990. Proceedings of the 52nd annual scientific meeting. The Committee on Problems of Drug Dependence. National Institute on Drug Abuse, Rockville

Fischman MW, Foltin RW (1988) The effects of desipramine maintenance on cocaine self-administration in humans. Psychopharmacology 96:S20

Fischman MW, Foltin RW (1992) Self-administration of cocaine by humans: a laboratory perspective. In: Bock G, Whelan J (eds) Cocaine: scientific and social dimensions. Ciba Foundation Symposium. Wiley, Chichester, pp 165–180

Fischman MW, Schuster CR, Javaid J, Hatano Y, Davis J (1985) Acute tolerance development to the cardiovascular and subjective effects of cocaine. J Pharmacol Exp Ther 235:677–682

Fischman MW, Foltin RW, Nestadt G, Pearlson GD (1990) Effects of desipramine maintenance on cocaine self-administration by humans. J Pharmacol Exp Ther 253:760–770

Fishbain DA, Wetli CV (1981) Cocaine intoxication, delirium and death in a body packer. Ann Emerg Med 10:531–532

Fishburne P, Abelson H, Cisin I (1980) The National Survey on Drug Abuse: main findings, 1979. Superintendent of Documents, US Government Printing Office, Washington DC

Flores ED, Lange RA, Cigarroa RG, Hillis LD (1990) Effect of cocaine on coronary artery dimensions in atheroschlerotic coronary artery disease: enhanced vasoconstriction at sites of significant stenoses. J Am Coll Cardiol 16:74–79

Foltin RW, Fischman MW (1989) Ethanol and cocaine interactions in humans: cardiovasular consequences. Pharmacol Biochem Behav 31:877–883

Foltin RW, Fischman MW (1991) Smoked and intravenous cocaine in humans: acute tolerance, cardiovascular and subjective effects. J Pharmacol Exp Ther 257: 247–261

Foltin RW, Fischman MW (1992) The cardiovascular and subjective effects of intravenous cocaine and morphine combinations in humans. J Pharmacol Exp Ther 261:623–632

Foltin RW, Fischman MW (1994) Effects of buprenorphine on the self-administration of cocaine by humans. Behav Pharm 5:79–89

Frankenfield DL, Lange WR, Weinhold LL, Contoreggi CS, Gorelick DA (1994) Risk factors for adverse cardiovascular events in cocaine-dependent research subjects. College on Problems of Drug Dependence, Proceedings of the 56th annual scientific meeting

Galloway MP (1990) Regulation of dopamine and serotonin synthesis by acute administration of cocaine. Synapse 6:63–72

Gawin F, Allen D, Hurnblestone B (1989) Outpatient treatment of "crack" cocaine smoking with flupenthixol decanoate. Arch Gen Psychiatry 46:322–325

Gawin FH, Ellinwood EH (1988) Cocaine and other stimulants: actions, abuse and treatment. N Engl J Med 318:1173–1182

Gawin FH, Kleber HD (1986) Abstinence symptomatology and psychiatric diagnosis in cocaine abusers. Clinical observations. Arch Gen Psychiatry 43:107–113

Gawin FH, Kleber HD, Byck R, Rounsaville BJ, Kosten TR, Jatlow PI, Morgan C (1989) Desipramine facilitation of initial cocaine abstinence. Arch Gen Psychiatry 46:117–121

Giannini AJ, Billett W (1987) Bromocriptine-desipramine protocol in treatment of cocaine addiction. J Clin Pharmacol 27:549–554

Giannini AJ, Baumgartel P, DiMarzio LR (1987) Bromocriptine therapy in cocaine withdrawal. J Clin Pharmacol 27:267–270

Goeders NE, Kuhar MJ (1987) Chronic cocaine administration induces opposite changes in dopamine receptors in the striatum and nucleus accumbens. Alcohol Drug Res 7:207–216

Goeders NE, Smith JE (1983) Cortical dopaminergic involvement in cocaine reinforcement. Science 221:773–775

Goeders NE, Smith JE (1986) Reinforcing properties of cocaine in the medial prefrontal cortex: primary action on presynaptic dopaminergic terminals. Pharmacol Biochem Behav 25:191–199

Goeders NE, Smith JE (1993) Intracranial cocaine self-administration into the medial prefrontal cortex increases dopamine turnover in the nucleus accumbens. J Pharmacol Exp Ther 265:592–600

Grant BF, Harford TC (1990) Concurrent and simultaneous use of alcohol with cocaine-results of a national survey. Drug Alcohol Depend 25:97–104

Grilli M, Wright AG, Hanbauer I (1991) Characterization of [³H]dopamine uptake sites and [³H]cocaine recognition sites in primary cultures of mesencephalic neurons during in vitro development. J Neurochem 56:2108–2115

Hearn WL, Rose S, Wagner J, Ciarleglios A, Mash DC (1991) Cocaethylene is more potent that cocaine in mediating lethality. Pharmaco Biochem Behav 3:531–533

Henry DJ, White FJ (1991) Repeated cocaine administration causes persistent enhancement of D1 dopamine receptor sensitivity within the rat nucleus accumbens. J Pharmacol Exp Ther 258:882–890

Henry DJ, White FJ (1992) Electrophysiological correlates of psychomotor stimulant-induced sensitization. In: Kalivas PW, Samson HH (eds) The neurobiology of drug and alcohol addiction. New York Academy of Sciences, New York

Henry DJ, Greene MA, White FJ (1989) Electrophysiological effects of cocaine in the mesoaccumbens dopamine system: repeated administration. J Pharmacol Exp Ther 251:833–839

Herman EH, Vick J (1987) A study of direct effect of cocaine on the heart. Fed Proc 46:1146

Heyser CJ, Miller JS, Spear NE, Spear LP (1992) Prenatal exposure to cocaine disrupts cocaine-induced conditioned place preference in rats. Neurotoxicol Teratol 14:57–64

Higgins ST, Bickel WK, Hughes JR, Lynn M, Capeless MA (1990) Behavioral and cardiovascular effects of cocaine and alcohol combinations in humans. In: Harris L (eds) Problems of drug dependence 1990. Proceedings of the 52nd annual scientific meeting. The Committee on Problems of Drug Dependence, National Institute on Drug Abuse, Rockville MD

Hime GW, Hearn WL, Rose S, Cofino J (1991) Analysis of cocaine and cocaethylene in blood and tissues by GD-NPD and GC-ion trap mass spectrometry. J Anal Toxicol 15:241–245

Hollander JE, Hoffman RS (1991) Cocaine-induced myocardial infarction: an analysis and review of the literature. J Emerg Med 10:169–177

Hooks MS, Jones GH, Smith AD, Neill DB, Justice JB (1992) Response to novelty predicts the locomotor and nucleus accumbens response to cocaine. Synapse 9:121–128

Hubner CB, Koob GF (1987) Ventral pallidal lesions produce decreases in cocaine and heroin self-administration in the rat. Neurosci Abstr 13:1717

Hubner CB, Moreton JE (1991) Effects of selective D1 and D2 dopamine antagonists on cocaine self-administration in the rat. Psychopharmacology 105:151–156

Hurd Y, Kehr J, Ungerstedt U (1988) In vivo microdialysis as a technique to monitor drug transport: correlation of extracellular cocaine levels and dopamine overflow in the rat brain. J Neurochem 51:1314–1316

Hurd YL, Ungerstedt U (1989) Cocaine: an in vivo microdialysis evaluation of its acute action on dopamine transmission in rat striatum. Synapse 3:48–54

Hurd YL, Weiss F, Anden N-E, Koob GF, Ungerstedt U (1989) Cocaine reinforcement and extracellular dopamine overflow in rat nucleus accumbens: an in vivo microdialysis study. Brain Res 498:199–203

Hurd YL, Weiss F, Koob G, Ungerstedt U (1990) The influence of cocaine self-administration on in vivo dopamine and acetylcholine neurotransmission in rat caudate-putamen. Neurosci Lett 109:227–233

Hutchings DE (1993) The puzzle of cocaine's effects following maternal use during pregnancy: are there reconcilable differences? Neurotoxicol Teratol 15:281–286

Imperato A, Di Chiara G (1984) Trans-striatal dialysis coupled to reverse phase high performance liquid chromatography with electrochemical detection: a new method for the study of the in vivo release of endogenous dopamine and metabolites. J Neurosci 4:966

Imperato A, Mele A, Scrocco MG, Puglisi-Allegra S (1992) Chronic cocaine alters limbic extracellular dopamine. Neurochemical basis for addiction. Eur J Pharmacol 212:299–300

Isenschmid DS, Fischman MW, Foltin RW, Caplan YH (1992) Concentration of cocaine and metabolites in plasma of humans following intravenous administration and smoking of cocaine. J Anal Tox 16:311–314

Isner JM, Chokshi SK (1991) Cardiac complications of cocaine abuse. Annu Rev Med 42:133–138

Isner JM, Estes NAM III, Thompson PD, Costanzo-Nordin MR, Subramanian R, Miller G, Katsas G, Sweeney K, Sturner WQ (1986) Acute cardiac events temporarily related to cocaine abuse. N Engl J Med 315:1438–1443

Izenwasser S, Cox BM (1990) Daily cocaine treatment produces a persistent reduction of [3H]dopamine uptake in vitro in rat nucleus accumbens but not in striatum. Brain Res 531:338–341

Jacobs IG, Roszler MH, Kelly JK, Klein MA, Kling GA (1989) Cocaine abuse: neurovascular complications. Radiology 170:223–227

Jaffe J (1985) Drug addiction and drug abuse. In: Gilman AG, Goodman LS, Rall TW, Murad F (eds) The pharmacological basis of therapeutics. Macmillan, New York, pp 532–581

Jain R, Gatti PJ, Visner M, Albrecht KG, Moront MG, Rackley CE, Gillis RA (1987) Effects of cocaine on cardiorespiratory function and on cardiovascular responses produced by bilateral carotid occlusion (BCO) and IV norepinephrine. Fed Proc 46:402

Javaid JI, Musa MN, Fischman MW, Schuster CR, Davis JM (1983) Kinetics of cocaine in humans after intravenous and intranasal administration. Biopharm Drug Dispos 4:9–18

Javaid JL, Fischman MW, Schuster CR, Dekirmenjian H, Davis JM (1978) Cocaine plasma concentration: relation to physiological and subjective effects in humans. Science 202:227–227

Johanson CE, Barrett JE (1993) The discriminative stimulus effects of cocaine in pigeons. J Pharmacol Exp Ther 267:1–8

Johanson CE, Fischman MF (1989) The pharmacology of cocaine related to its abuse. Pharmacol Rev 41:3–52

Johanson CE, Schuster CR (1981) Animal models of drug self-administration. In: Mello NK (ed) Advances in substance abuse: behavioral and biological research. JAI, Connecticut, pp 219–297

Johnson KM, Bergmann JS, Kozikowski AP (1992) Cocaine and dopamine differentially protect [3H]mazindol binding sites from aklylation by N-ethylmaleimide. Eur J Pharmacol 227:411–415

Justice JB (1987) Voltammetry in the neurosciences. Humana, Clifton

Kalivas PW, Duffy P (1990) Effect of acute and daily cocaine treatment on extracellular dopamine in the nucleus accumbens. Synapse 5:48–58

Kalivas PW, Duffy P (1993a) Time course of extracellular dopamine and behavioral sensitization to cocaine. I. Dopamine axon terminals. J Neurosci 13:266–275

Kalivas PW, Duffy P (1993b) Time course of extracellular dopamine and behavioral sensitization to cocaine. II. Dopamine perikarya. J Neurosci 13:276–284

Kalivas PW, Stewart J (1991) Dopamine transmission in the initiation and expression of drug- and stress-induced sensitization of motor activity. Brain Res Rev 16:223–244

Kalivas PW, Weber B (1988) Amphetamine injection into the ventral mesencephalon sensitizes rats to perpheral amphetamine and cocaine. J Pharmacol Exp Ther 245:1095–1102

Kalivas PW, Duffy P, DuMars LA, Skinner C (1988) Behavioral and neurochemical effects of acute and daily cocaine administration in rats. J Pharmacol Exp Ther 245:485–492

Kalivas PW, Striplin CD, Steketee JD, Klitenick MA, Duffy P (1992) Cellular mechanisms of behavioral sensitization to drugs of abuse. In: Kalivas PW, Samson HH (eds) The neurobiology of drug and alcohol addiction. New York Academy of Sciences, New York

Kalivas PW, Sorg BA, Hooks MS (1993) The pharmacology and neural circuitry of sensitization to psychostimulants. Behav Pharmacol 4:315–334

Katz JL, Terry P, Witkin J (1992) Comparative behavioral pharmacology and toxicology of cocaine and its ethanol-derived metabolite, cocaine ethyl-ester (cocaethylene). Life Sci 50:1351–1361

Kaufman MJ, Spealman RD, Madras BK (1991) Distribution of cocaine recognition sites in monkey brain: I. In vitro autoradiography with [3H]CFT. Synapse 9:177–187

Kennedy LT, Hanbauer I (1983) Sodium-sensitive cocaine binding to rat striatal membrane: possible relationship to dopamine uptake sites. J Neurochem 41:172–178

Kilpatrick GJ, Jones BJ, Tyers MB (1987) Identification and distribution of 5-HT3 receptors in rat brain using radioligand binding. Nature 330:746–748

Kilty JE, Lorang D, Amara SG (1991) Cloning and expression of a cocaine-sensitive rat dopamine transporter. Science 254:578–579

Kitayama S, Shimada S, Xu H, Markham L, Donovan DM, Uhl GR (1992) Dopamine transporter site-directed mutations differentially alter substrate transport and cocaine binding. Proc Natl Acad Sci USA 89:7782–7785

Kleber H (1988) Introduction-cocaine abuse: historical, epidemiological, and psychological prerspectives. J Clin Psychiatry 49:3–6

Kleven MS, Woolverton WL (1990) Effects of bromocriptine and desipramine on behavior maintained by cocaine or food presentation in rhesus monkeys. Psychopharmacology 101:208–213

Kleven MS, Woolverton WL, Seiden LS (1988) Lack of long-term monoamine depletions following continuous or repeated exposure to cocaine. Brain Res Bull 21:233–237

Kleven MS, Anthony EW, Woolverton WL (1990a) Pharmacological characterization of the discriminative stimulus effects of cocaine in rhesus monkeys. J Pharmacol Exp Ther 254:312–317

Kleven MS, Perry BD, Woolverton WL, Seiden LS (1990b) Effects of repeated injections of cocaine on D1 and D2 dopamine receptors in rat brain. Brain Res 532:265–270

Koe BK (1976) Molecular geometry of inhibitors of the uptake of catecholamines and serotonin in synaptosomal preparations of rat brain. J Pharmacol Exp Ther 199:649–661

Koob GF, Bloom FE (1988) Cellular and molecular mechanisms of drug dependence. Science 242:715–723

Koob GF, Le HT, Creese I (1987) The D1 dopamine receptor antagonist SCH 23390 increases cocaine self-administration in the rat. Neurosci Lett 79:315–20

Kosten RT, Morgan CM, Falcione J, Schottenfeld RS (1992) Pharmacotherapy for cocaine-abusing methadone-maintained patients using amantadine or desipramine. Arch Gen Psychiatry 49:894–898

Kuczenski R, Segal DS (1988) Psychomotor stimulant-induced sensitization: behavioral and neurochemical correlates. In: Kalivas PW, Barnes CD (eds) Sensitization in the nervous system. Telford, Caldwell, pp 175–206

Kuczenski R, Segal DS (1992) Differential effects of amphetamine and dopamine uptake blockers (cocaine, nomifensine) on caudate and accumbens dialysate dopamine and 3-methoxytyramine. J Pharmacol Exp Ther 262:1085–1094

Kuhar MJ, Ritz MC, Boja JW (1991) The dopamine hypothesis of the reinforcing properties of cocaine. Trends Neurosci 14:299–302

Lacey MG, Mercuri NB, North RA (1990) Actions of cocaine on rat dopaminergic neurones in vitro. Br J Pharmacol 99:731–735

Landry DW, Zhao K, Yang GX-Q, Glickman M, Georgiadis TM (1993) Antibody-catalyzed degradation of cocaine. Science 259:1899–1901

Lange RA, Cigarroa RG, Yancy CW, Willard JE, Popma JJ, Sills MN, McBride W, Kim AS, Hillis LD (1989) Cocaine-induced coronary-artery vasoconstriction. N Engl J Med 321:1557–1562

Leal J, Ziedonis D, Kosten T (1994) Antisocial personality disorder as a prognostic factor for pharmacotherapy of cocaine dependence. Drug Alcohol Depend 35:31–35

Lesko LM, Fischman MW, Javaid JI, Davis JM (1982) Iatrogenous cocaine psychosis. N Engl J Med 307:1153

Levin FR, Lehman AF (1991) Meta-analysis of desipramine as an adjunct in the treatment of cocaine addiction. J Clin Psychopharmacol 11:374–378

Levine SR, Washington JM, Moen M, Kieran SN, Junger S, Welch KMA (1987) Crack-associated stroke. Neurology 37:1092–1093

Lew MJ, Angus JA (1983) Disadvantages of cocaine as a neuronal uptake blocking agent: comparison with desipramine in guinea-pig right atrium. J Auton Pharmacol 3:61–71

Lew R, Vaughan R, Simantov R et al. (1991) Dopamine transporters in the nucleus accumbens and the striatum have different apparent molecular weights. Synapse 8:152–153

Lichtenfeld PJ, Rubin DB, Feldman RS (1984) Subarachnoid hemorrhage precipitated by cocaine snorting. Arch Neurol 41:223–224

Loh EA, Roberts DCS (1990) Break-points on a progressive ratio schedule reinforced by intravenous cocaine increase following depletion of forebrain serotonin. Psychopharmacology 101:262–266

Lutiger B, Graham K, Einarson TR, Koren G (1991) Relationshiop between gestational cocaine use and pregnancy outcome: a meta-analysis. Teratology 44:405–414

Madras BK, Fahey MA, Bergman J, Canfield DR, Spealman RD (1989) Effects of cocaine and related drugs in nonhuman primates. I. [3H]Cocaine binding sites in caudate-putamen. J Pharmacol Exp Ther 251:131–141

Magnano CL, Gardner JM, Karmel BZ (1992) Differences in salivary cortisol levels in cocaine-exposed and noncocaine-exposed NICU infants. Dev Psychobiol 25:93–103

Maisonneuve IM, Keller RW, Glick SD (1990) Similar effects of d-amphetamine and cocaine on extracellular dopamine levels in medial prefrontal cortex in rats. Brain Res 535:221–226

Majid PA, Cheirif JB, Rikey R, Sanders WE, Patel B, Zimmerman JL, Dellinger RP (1992) Does cocaine cause coronary vasospasm in chronic cocaine abusers? A study of coronary and systemic hemodynamics. Clin Cardiol 15:253–258

Maldonado R, Robledo P, Chover AJ, Caine SB, Koob GF (1993) D1 dopamine receptors in the nucleus accumbens modulate cocaine self-administration in the rat. Pharmacol Biochem Behav 45:239–242

Manschreck TC, Allen DF, Neville M (1987) Freebase psychosis: cases from a Bahamian epidemic of cocaine abuse. Compr Psychiatry 28:555–564

Markou A, Koob GF (1991) Postcocaine anhedonia: an animal model of cocaine withdrawal. Neuropsychopharmacology 4:17–26

Martin-Iverson MT, Szostak C, Fibiger HC (1986) 6-Hydroxydopamine lesions of the medial prefrontal cortex fail to influence intravenous self-administration of cocaine. Psychopharmacology 88:310–314

Mayfield RD, Larson G, Zahniser NR (1992) Cocaine-induced behavioral sensitization and D_1 dopamine receptor function in rat nucleus accumbens and striatum. Brian Res 573:331–335

Meisch RA, Bell SM, Lemaire GA (1993) Orally self-administered cocaine in rhesus monkeys – transition from negative or neutral behavioral effects to positive reinforcing effects. Drug Alcohol Depend 32:143–158

Mello NK, Mendelson JH, Bree MP, Lukas SE (1989) Buprenorphine suppresses cocaine self-administration by rhesus monkeys. Science 245:859–862

Mello NK, Lukas SE, Bree MP, Mendelson JH (1990a) Desipramine effects on cocaine self-administration by rhesus monkeys. Drug Alcohol Depend 26: 103–116

Mello NK, Mendelson JH, Bree MP, Lukas S (1990b) Buprenorphine and naltrexone effects on cocaine self-administration by rhesus monkeys. J Pharmacol Exp Ther 254(3):926–939

Mendelson JH, Mello NK, Teoh SK, J.E, Cochin J (1989) Cocaine effects on pulsatile secretion of anterior pituitary, gonadal, and adrenal hormones. J Clin Endocrinol Metab 69:1256–1260

Minor RL, Scott BD, Brown DD, Winniford MD (1991) Cocaine-induced myocardial infarction in patients with normal coronary arteries. Ann Intern Med 115:797–806

Mittleman RE, Wetli CV (1984) Death caused by recreational cocaine use. An update. JAMA 252:1889–1893

Moghaddam B, Bunney BS (1989) Differential effect of cocaine on extracellular dopamine levels in rat prefrontal cortex and nucleus accumbens: comparison to amphetamine. Synapse 4:156–161

Molina VA, Wagner JM, Spear LP (1994) The behavioral response to stress is altered in adult rats exposed prenatally to cocaine. Physiol Behav 55:941–945

Musto DF (1992) Cocaine's history, especially the American experience. In: Edwards G (ed) Cocaine: scientific and social dimensions. Wiley Chichester

National Institute on Drug Abuse (1992) Epidemiologic trends in drug abuse. In Community Epidemiology Work Group, publication number 93-3560 (575 pp). National Institutes of Health, Washington DC

Nestler EJ, Termwilliger RZ, Walker JR, Sevarino KA, Duman RS (1990) Chronic cocaine treatment descreases levels of the G protein subunits G_{ia} and G_{oa} in discrete regions of rat brain. J Neurochem 55:1079–1082

Neuspiel DR (1994) Overview: behavior in cocaine-exposed infants and children: Association versus causality. Drug Alcohol Depend 36:101–108

Ng JP, Hubert GW, Justice JB (1991) Increased stimulated release and uptake of dopamine in nucleus accumbens after repeated cocaine adminstration as measured by in vivo voltammetry. J Neurochem 56:1485–1492

O'Brien CP, Childress AR, Arndt IO, McLellan AT, Woody GE, Maany I (1988) Pharmacological and behavioral treatments of cocaine dependence: controlled studies. J Clin Psychiatry 49 [Suppl]:17–22

Ostrea EM, Brady M, Gause S, Raymundo AL, Stevens M (1992) Drug screening of newborns by meconium analysis: a large-scale, prospective, epidemiologic study. Pediatrics 89:107–113

Paly D, Jatlow P, Van Dyke C, Cabieses F, Byck R (1980) Plasma levels of cocaine in native Peruvian coca chewers. In: Jeri FR (ed) Cocaine, 1980, proceedings of the Interamerican seminar on medical and sociological aspects of coca and cocaine. Pacific, Lima, pp 86–89

Parsons LH, Smith AD, Justice JB (1991) Basal extracellular dopamine is decreased in the rat neuclues accumbens during abstinence from chronic cocaine. Synapse 9:60–65

Patel A, Boja J, Lever J, Lew R, Simantov R, Carroll FI, Lewin AH, Philip A, Gao Y, Kuhar MJ (1992) A cocaine analog and a GBR analog label the same protein in rat striatal membranes. Brain Res 576:173–174

Pentel PR, Hatsukami D (1994) 12-lead and continuous ECG recordings of subjects during inpatient administration of smoked cocaine. Drug Alcohol Depend 35:107–116

Perez-Reyes M, Jeffcoat AR (1992) Ethanol/cocaine interaction: cocaine and cocaethylene plasma concentrations and their relationship to subjective and cardiovascular effects. Life Sci 51:553–563

Perez-Reyes M, Di Guiseppi S, Ondrusek G, Jeffcoat AR, Cook CE (1982) Freebase cocaine smoking. Clin Pharmacol Ther 32:459–465

Peris J, Boyson SJ, Cass WA, Curella P, Dwoskin L, Larson G, Lin L-H, Yasuda R, Zahniser N (1990) Persistence of neurochemical changes in dopamine systems after repeated cocaine administration. J Pharmacol Exp Ther 253:38–44

Pettit H, Justice J Jr (1989) Dopamine in the nucleus accumbens during cocaine self-administration as studied by in vivo microdialysis. Pharmcol Biochem Behav 34:899–904

Pettit HO, Justice JB (1991) Effect of dose on cocaine self-administration behavior and dopmaine levels in the nucleus accumbens. Brain Res 539:94–102

Pettit HO, Pan H-T, Parsons LH, Justice JB (1990) Extracellular concentrations of cocaine and dopamine are enhanced during chronic cocaine administration. J Neurochem 55:798–804

Pitts DK, Marwah J (1987) Cocaine modulation of central monoaminergic neurotransmission. Pharmacol Biochem Behav 26:453–461

Pollock DA, Holmgreen P, Lui K-J, Kirk ML (1991) Discrepancies in the reported frequency of cocaine-related deaths, United States, 1983 through 1988. JAMA 266:2233–2237

Post RM, Kopanda RT, Black KE (1976) Progressive effects of cocaine on behavior and central amine metabolism in rhesus monkeys: relationship to kindling and psychosis. Biol Psychiatry 11:403–419

Post RM, Lickfeld A, Squillace KM, Contel NR (1981) Drug-environment interaction: context dependency of cocaine induced behavioral sensitization. Life Sci 28:755–760

Pulvirenti L, Maldonado-Lopez R, Koob GF (1992) MNDA receptors in the nucleus accumbens modulate intravenous cocaine but not heroin self-administration in the rat. Brain Res 594:327–330

Racine A, Joyce T, Anderson R (1993) The association between prenatal care and birth weight among women exposed to cocaine in New York City. JAMA 270:13

Reith MEA, Sershen H, Allen DL, Lajtha A (1983) A portion of [^{3}H] cocaine binding in brain is associated with serotonergic neurons. Mol Pharmacol 23:600–606

Reith MEA, de Costa B, Rice KC, Jacobson AE (1992) Evidence for mutually exclusive binding of cocaine, BTCP, GBR 12935, and dopamine to the dopamine transporter. Eur J Pharmacol 227:417–425

Ritchie JM, Greene NM (1990) Local anesthetics. In: Gilman AG, Rall TW, Nies AS, Taylor P (eds) Goodman and Gilman's the pharmacological basis of therapeutics. Pergamon, New York, pp 311–331

Ritz MC, Lamb RJ, Goldberg SR, Kuhar MJ (1987) Cocaine receptors on dopamine transporters are related to self-administration of cocaine. Science 237:1219–23

Ritz MC, Boja JW, Grigoriadis D, Zaczek R, Carroll FI, Lewis AH, Kuhar MJ (1990a) [^{3}H]WIN 35,065-2: a ligand for cocaine receptors in striatum. J Neurochem 55:1556–1562

Ritz MC, Cone EJ, Kuhar MJ (1990b) Cocaine inhibition of ligand binding at dopamine, norepinephrine and serotonin transporters: a structure-activity study. Life Sci 46:635–645

Ritz MC, Kuhar MJ, George FR (1992) Molecular mechanisms associated with cocaine effects: Possible relationships with effects of ethanol. In: Galanter M (ed) Recent development in alcoholism, vol 10: alcohol and cocaine: similarities and differences. Plenum, New York, pp 273–302

Roberts DCS (1993) Self-administration of GBR 12909 on a fixed ratio and progressive ratio schedule in rats. Psychopharmacology 111:202–206

Roberts DCS, Koob GF (1982) Disruption of cocaine self-administration following 6-hydroxydopamine lesions of the ventral tegmental area in rats. Pharmacol Biochem Behav 17:901–904

Roberts DCS, Corcoran ME, Fibiger HC (1977) On the role of ascending catecholaminergic systems in intravenous self-administration of cocaine. Pharmacol Biochem Behav 6:615–620

Roberts DCS, Koob GF, Klonoff P, Fibiger HC (1980) Extinction and recovery of cocaine self-administration following 6-hydroxydopamine lesions of the nucleus accumbens. Pharmacol Biochem Behav 12:781–787

Roberts DCS, Loh EA, Vickers G (1989) Self-administration of cocaine on a progressive ratio shedule in rats: dose-response relationship and effect of haloperidol pretreatment. Psychopharmacology 97:535–538

Robinson TE (1988) Stimulant drugs and stress: Factors influencing individual differences in the susceptibility to sensitization. In: Kalivas PW, Barnes CD (eds) Sensitization in the nervous systems. Telford, Caldwell, pp 145–173

Ross SB, Renyi AL (1969) Inhibition of the uptake of tritiated 5-hydroxytryptamine in brain tissue. Eur J Pharmacol 7:270–277

Rossetti ZL, Hmaidan Y, Gessa GL (1992) Marked inhibition of mesolimbic dopamine release: a common feature of ethanol, morphine, cocaine and amphetamine abstinence in rats. Eur J Pharmacol 221:227–234

Rothman RB (1990) High affinity dopamine reuptake inhibitors as potential cocaine antagonists: a strategy for drug development. Life Sci 46:17

Rothman RB, Mele A, Reid AA, Akunne HC, Greig N, Thurkauf A, DeCosta BR, Rice KC, Pert A (1991) GBR 12909 antagonizes the ability of cocaine to elevate extracellular levels of dopamine. Pharmacol Biochem Behav 40:387

Rounsaville BJ, Foley-Anton S, Carroll K, Budde D, Pursoff BA, Gawin F (1991) Psychiatric diagnosis of treatment-seeking cocaine abusers. Arch Gen Psychiatry 48:43–51

Rubin RB, Neugarten J (1989) Cocaine rhabdomyolysis masquerading as myocardial ischemia. Am J Med 86:551–553

Satel SL, Price LH, Palumbo JM, McDougle CJ, Krystal JH, Gawin G, Charney DS, Heninger GR, Kleber HK (1991) Clinical phenomenology and neurobiology of cocaine abstinence: a prospective inpatient study. Am J Psychiatry 148:1712–1716

Schenk S, Horger BA, Peltier R, Shelton K (1991) Supersensitivity to the reinforcing effects of cocaine following 6-hydroxydopamine lesions to the medial prefrontal cortex in rats. Brain Res 543:227–235

Schenk S, Valadez A, McNamara C, House DT, Higley D, Bankson MG, Gibbs S, Horger BA (1993) Development and expression of sensitization to cocaine's reinforcing properties: role of NMDA receptors. Psychopharmacology 111:332–338

Schenker S, Yang YQ, Johnson RF, Downing JW, Schenken RS, Henderson GI, King TS (1993) The transfer of cocaine and its metabolites across the term human placenta. Clin Pharmacol Ther 53:329–339

Schuster CR, Johanson CE (1988) Relationship between the discriminative stimulus properties and subjective effects of drugs. In: Colpaert FC, Balster RL (eds) Transduction mechanisms of drug stimuli. Springer, Berlin Heidelberg New York, pp 161–175 (Psychopharmacology series, vol 4)

Schwartz KA, Cohen JA (1984) Subarachnoid hemorrhage precipitated by cocaine snorting. Arch Neurol 41:704

Seiden LS, Ricaurte GA (1987) Neurotoxicity of methamphetamine and related drugs. In: Meltzer HY (ed) Psychophamacology: the third generation of progress. Raven, New York

Sherer MA, Kumor KM, Cone EJ, Jaffe JH (1988) Suspiciousness induced by four-hour intravenous infusions of cocaine. Arch Gen Psychiatry 45:673–677

Shimada S, Kitayama S, Lin C-L, Patel A, Nanthakumar E, Gregor P, Kuhar M, Uhl G (1991) Cloning and expression of a cocaine-sensitive dopamine transporter complementary DNA. Science 254:576–577

Siegel RK (1978) Cocaine hallucinations. Am J Psychiatry 135:309–314

Smith HWB, Liberman HA, Broady SL, Battey LL, Donohue BC, Morris DC (1987) Acute myocardial infarction temporally related to cocaine use. Ann Intern Med 107:13–18

Spealman RD, Bergman J, Madras BK (1991a) Self-administration of the high-affinity cocaine analog 2ba-carbomethoxy-3b-(4-fluorophenyl)tropane. Pharmacol Biochem Behav 39:1011–1013

Spealman RD, Bergman J, Madras BK, Melia KF (1991b) Discriminative stimulus effects of cocaine in squirrel monkeys: involvement of dopamine receptor sub-types. J Pharmacol Exp Ther 258:945–953

Spear LP, Kirstein CL, Bell J, Yoottanasumpun V, Greenbaum R, O'Shea J, Hoffman H, Spear NE (1989a) Effects of prenatal cocaine exposure on behavior during the early postnatal period. Neurotoxicol Teratol 11:57–63

Spear LP, Kirstein CL, Frambes NA (1989b) Cocaine effects on the developing central nervous system: behavioral, psychopharmacological, and neurochemical studies. Ann NY Acad Sci 526:290–307

Steketee JD, Kalivas PW (1991) Sensitization to psychostimulants and stress after injection of pertussis toxin into the A10 dopamine region. J Pharmacol Exp Ther 259:916–924

Steketee JD, Striplin CD, Murray TF, Kalivas PW (1991) Possible role for G-proteins in behavioral sensitization to cocaine. Brain Res 545:287–291

Stewart DM, Rogers WP, Hahaffey JE, Witherspoon S, Woods EF (1963) Effect of local anesthetics on the cardiovascular system of the dog. Anesthesiology 24:620–624

Stewart J, Vezina P (1989) Microinjections of SCH-23390 into the ventral tegmental area and substantia nigra pars reticulata attenuate the development of sensitization to the locomotor activating effects of systemic amphetamine. Brain Res 495:401–406

Tang M, Falk JL (1987) Oral self-administration of cocaine: chronic excessive intake by schedule induction. Pharmacol Biochem Behav 28:517–519

Tennant FS, Sagherian AA (1987) Double-blind comparison of amantadine and bromocriptine for ambulatory withdrawal from cocaine dependence. Arch Intern Med 147:109–112

Terwilliger RZ, Beitner-Johnson D, Sevarino KA, Crain SM, Nestler EJ (1991) A general role for adaptations in G-proteins and the cyclic AMP system in mediating the chronic actions of morphine and cocaine on neuronal function. Brain Res 548:100–110

Trendelenburg U (1968) The effect of cocaine on the pacemaker of isolated guinea-pig atria. J Pharmacol Exp Ther 161:222–231

Trulson ME, Ulissey JJ (1987) Chronic cocaine administration decreases dopamine synthesis rate and increases [³H]spiroperidol binding in rat brain. Brain Res Bull 19:35–38

Ungerstedt U (1984) Measurement of neurotransmitter release by intracranial dialysis. In: Marsden CA (ed) Measurement of neurotransmitter release in vivo. Wiley, London, pp 81–105

Van Dyke C, Jatlow P, Ungerer J, Barash PG, Byck R (1978) Oral cocaine: plasma concentrations and central effects. Science 200:211–213

Vitti TG, Boni RL (1985) Metabolism of cocaine. In: Barnett G, Chiang CN (eds) Pharmacokinetics and pharmacodynamics of psychoactive drugs: a research monograph. Biomedical, Foster City, pp 427–440

Volkow ND, Gould KL, Mullani N, Adler S, Krajewski K (1987) Changes in cerebral blood flow of chronic cocaine users. J Cereb Blood Flow Metab 7 [Suppl]:S292

Volkow ND, Mullani N, Gould KL, Adler S, Krajewski K (1988) Cerebral blood flow in chronic cocaine users: a study with positron emission tomography. Br J Psychiatry 152:641–648

Volkow ND, Fowler JS, Wolf AP, Hitzemann R, Dewey S, Bendriem B, Alpert R, Hoff A (1991) Changes in brain glucose metabolism in cocaine dependence and withdrawal. Am J Psychiatry 148:621–626

Watson SJ, Trujillo KA, Herman JP, Akil H (1988) Neuroanatomical and neuro- chemical substrates of drug-seeking behavior: overview and future directions. In: Goldstein A (ed) Molecular and cellular aspects of the drug addictions. Springer, Berlin Heidelberg New York, pp 29–91

Weddington W, Brown BS, Haertzen CA, Cone EJ, Dax EM, Herning RI, Michaelson BS (1990) Changes in mood, craving, and sleep during short-term abstinence reported by male cocaine addicts: A controlled, residential study. Arch Gen Psychiatry 47:861–868

Weddington W, Brown BS, Haertzen CA, Hess JM, Mahaffey JR, Kolar AF, Jeffe JH (1991) Comparison of amantadine and desipramine combined with psy- chotherapy for treatment of cocaine dependence. Am J Drug Alcohol Abuse 17:137–152

Weed MR, Vanover KE, Woolverton WL (1993) Reinforcing effect of the D1 dopamine agonist SKF 81297 in rhesus monkeys. Psychopharmacology 113:51–52

Weiss F, Markou A, Lorang MT, Koob GF (1992a) Basal extracellular dopamine levels in the nucleus accumbens are decreased during cocaine withdrawal after unlimited-access self-administration. Brain Res 593:314–318

Weiss F, Paulus MP, Lorang MR, Koob GF (1992b) Increases in extracellular dopamine in the nucleus accumbens by cocaine are inversely related to basal levels: Effects of acute and repeated administration. J Neurosci 12:4372–4380

Weiss RD (1988a) Protracted elimination of cocaine metabolites in long-term, high- dose cocaine abusers. Am J Med 85:879–880

Weiss RD (1988b) Relapse to cocaine abuse after initiating desipramine treatment. J Am Med Assoc 260:2545–2546

Weiss RD, Mirin SM, Griffin ML, Michael JL (1988) Psychopathology in cocaine abusers: changing trends. J Nerv Ment Dis 176:719–725

Wetli CV, Wright RK (1979) Death caused by recreational cocaine use. JAMA 241:2519–2522

Wiggins RC, Rolsten C, Ruiz B, Davis CM (1989) Pharmacokinetics of cocaine: basic studies of route, dosage, pregnancy, and lactation. Neurotoxicology 10: 367–382

Wilkerson RD (1988) Cardiovascular toxicity of cocaine. In: Clouet D, Ashgar K, Brown R (eds) Mechanisms of cocaine abuse and toxicity. US Government Printing Office, Washington DC, pp 304–324

Wilson MC, Schuster CR (1976) Mazindol self-administration in the rhesus monkey. Pharmacol Biochem Behav 4:207–210

Winger G, Skjoldager P, Woods JH (1992) Effects of buprenorphine and other opioid agonists and antagonists on alfentanil- and cocaine-reinforced responding in rhesus monkeys. J Pharmacol Exp Ther 261:311–317

Wise R, Rompre P (1989) Brain dopamine and reward. Annu Rev Psychol, 40: 191–225

Witkin JM, Nichols DE, Terry P, Katz JL (1991) Behavioral effects of selective dopaminergic compounds in rats discriminating cocaine injection. J Pharmacol Exp Ther 257:706–713

Woolverton WL (1986) Effects of a D1 and a D2 dopamine antagonist on the self-administration of cocaine and piribedil by rhesus monkeys. Pharmacol Biochem Behav 24:531–535

Woolverton WL, Virus RM (1989) The effects of a D1 and a D2 dopamine antagonist on behavior maintained by cocaine or food. Pharmacol Biochem Behav 32:691–697

Woolverton WL, Goldberg LI, Ginos JE (1984) Intravenous self-administration of dopamine receptor agonists by rhesus monkeys. J Pharmacol Exp Ther 230: 678–683

Yeh SY, De Souza EB (1991) Lack of neurochemical evidence for neurotoxic effects of repeated cocaine administration in rats on brain monoamine neurons. Drug Alcohol Depend 27:51–61

Yi S-J, Johnson KM (1990) Effects of acute and chronic administration of cocaine on striatal uptake, compartmentalization and release of [^{3}H]dopamine. Neuropharmacology 29:475–486

Young AM, Herling S (1986) Drugs as reinforcers: studies in laboratory animals. In: Goldberg SR, Stolerman IP (eds) Behavioral analysis of drug dependence. Academic, Orlando, pp 9–67

Zimmerman JL, Dellinger RP, Majid PA (1991) Cocaine associated chest pain. Ann Emerg Med 20:611–615

CHAPTER 5

Opioid Analgesics

L. Dykstra

A. Background

Shortly after the petals fall from the flower of the poppy plant, the remaining
seed capsule is slit open with a sharp knife, rendering a milky sap which,
when dried, yields the substance known as opium. Although people have
known since ancient times that they could relieve their pain and feel pleasure
by eating or smoking this substance, it was not until the early 1800s that the
substance responsible for opium's effects was isolated and given the name
morphine, after Morpheus, the Greek god of dreams. Morphine's alkaline
properties were different from those of other plant-derived products, thus its
isolation from opium marked the beginning of alkaloid chemistry. Since
then other alkaloids have been isolated from opium, including papaverine,
thebaine and codeine (Wasacz 1981).

Not long after its isolation from opium, morphine was introduced into
medical practice and used widely to treat ailments as diverse as asthma,
diarrhea, diabetes and ulcers. Although some misuse of morphine occurred
because of its easy availability in patent medicines, the development of the
hypodermic needle as a means of administering morphine during the Civil
War was a major contributor to its abuse. Concern about opioid abuse
increased on into the 1990s, and in 1914 the Harrison Narcotic Act was
enacted to outlaw the purchase of opioid preparations over the counter and
monitor the treatment of opioid-dependent individuals (Brecher 1972;
Wasacz 1981). Despite these restrictions, opioid abuse continues today and
estimates from the 1992 National Household Survey indicate that approximately 1.8 million people in the United States have used illegal opioids such
as heroin in their lifetime. In addition, opioid abuse has particularly serious
consequences today because of the high prevalence of AIDS among intravenous heroin users.

Interest in the opioids centers on two very important pharmacological
characteristics: (1) their analgesic effects and (2) their tendency to produce
tolerance and physical dependence upon chronic administration. As a result
of the close association between their analgesic effects and the development
of tolerance and dependence, scientists have attempted to develop analgesics
that might possess the beneficial analgesic effects of the opioids without
some of their unwanted effects, in particular effects such as tolerance and

dependence (FRASER and HARRIS 1967). A number of opioid analgesics have been identified in this search; however, none of them is without some degree of tolerance and/or dependence potential. Nevertheless, compounds with a lesser degree of dependence than that of morphine have been identified and a great deal of progress has been made in limiting the problems of dependence in clinical settings (FOLEY 1989; HILL et al. 1991; MELZACK 1990). Indeed, the Federal Agency for Health Care Policy and Research recently suggested that opioids are underutilized in clinical settings because of unjust concerns about their abuse potential.

The opioid analgesics include a large number of synthetic and semisynthetic compounds with morphine being the prototype compound (see Table 1 for a list of those opioids most commonly used clinically). Among these is heroin, whose effects are essentially identical to those of morphine except for a faster onset of action and greater potency. Since heroin is rapidly metabolized to morphine, morphine is probably responsible for most of heroin's actions. Codeine is another important opioid. Although it is less efficacious than morphine as an analgesic, it is particularly active when administered by the oral route and is very effective as a cough suppressant. Other opioids of interest include fentanyl and related compounds, all of which have shorter durations of action than morphine and therefore are useful in brief surgical or orthopedic manipulations. Methadone and *l-α*-acetylmethadol (LAAM), which have an extended duration of action and are active by the oral route, are particularly useful in the treatment of opioid dependence. The structures of morphine and a few other representative opioids are shown in Fig. 1.

Although the various synthetic and semisynthetic opioids differ from morphine in terms of potency, rate of absorption or duration of action, as a group they have many effects in common with morphine and therefore are

Table 1. Commonly used opioid analgesics

Generic name	Brand name	Analgesic dose[a]	Duration (h)
Morphine	Morphine	10	4–5
Heroin	Heroin	5	4–5
Hydromorphone	Dilaudid	1.3	2–3
Codeine	Codeine	130	3–4
Methadone	Dolophine	10	4–5
Meperidine	Demerol	75	3–4
Propoxyphene	Darvon	65 (oral)	4–6
Fentanyl	Sublimaze	0.1	1–2
Pentazocine	Talwin	10–30	4–6
Nalbuphine	Nubain	10	4–6
Butorphanol	Stadol	2	4–6
Buprenorphine	Buprenex	0.4	4–5

Data adapted from JAFFE and MARTIN (1990).
[a] Doses administered by intramuscular or subcutaneous route, except as noted.

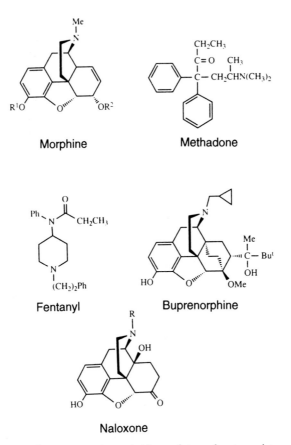

Fig. 1. Structures of representative opioid agonists and antagonists

classified together as *opioid agonists*. Moreover, opioid analgesics can be differentiated from other classes of analgesics by the fact that their effects are readily reversed by *opioid antagonists* such as naloxone and naltrexone. The structure of naloxone is also shown in Fig. 1. The opioid antagonists reverse (or antagonize) the effects of opioid agonists and are particularly useful in reversing the respiratory depressant effects of the opioid agonists. In individuals dependent on opioids, administration of an opioid antagonist immediately produces (i.e., precipitates) the signs and symptoms of withdrawal.

Another interesting group of opioids is the *mixed action opioid agonist/ antagonists*. This group of opioids produces opioid agonist effects under some conditions and opioid antagonist effects under other conditions. In general, these compounds produce a lesser degree of analgesia as well as a lesser degree of dependence than morphine and are thought to produce their effects by interacting with more than one opioid receptor type (see

Sect. B.I.2). Therefore, they are also called *partial agonists*. Several important compounds fall within this category, including nalbuphine and butorphanol; the benzomorphans pentazocine and cyclazocine; and the oripavine derivative buprenorphine.

B. Cellular and Molecular Effects

I. Opioid Receptors

1. Early Discoveries

It is now well-known that the opioids exert their effects via a system of highly selective receptors. Evidence for this notion came, first of all, from the discovery in 1973 of high-affinity sites that bind opioid compounds in a stereoselective and saturable manner (Terenius 1973; Simon et al. 1973; Pert and Snyder 1973). Importantly, nonopioids do not bind to these sites. Moreover, the affinity of various opioids for these sites correlates very well with their potencies in a variety of pharmacological assays. One of those assays is the guinea pig ileum (GPI) bioassay, in which the longitudinal muscle of the guinea pig ileum is dissected out from the ileum and attached to an apparatus that measures its tension. When the GPI is stimulated electrically, acetylcholine is released and the longitudinal muscle contracts. The opioids inhibit the release of acetylcholine and thereby inhibit the muscle contraction. Importantly, the potency of the opioids in inhibiting contraction of the GPI correlates with their affinity of opioid receptors (Creese and Snyder 1975). Moreover, since opioid receptors in the GPI are similar to those in the CNS, activity in the GPI generally correlates with activity in other assays of opioid activity. For example, there is a high correlation between opioid potency in the GPI and in assays for analgesic activity, reinforcing efficacy and suppression of withdrawal (Pearl et al. 1968; Young et al. 1981; Magnan et al. 1982).

Additional evidence to support the receptor-mediated action of the opioids came from the discovery, in 1975, of two endogenous opioid-like substances in the brain, methionine (met) and leucine (leu) enkephalin. Both were shown to produce dose-dependent, naloxone-reversible inhibition of responses in the guinea pig ileum and the mouse vas deferens, another bioassay for opioid activity (Hughes et al. 1975; Lord et al. 1977). On the basis of these and other findings, investigators proposed the existence of an endogenous opioid system through which morphine and other opioids produced their effects.

2. Opioid Receptor Multiplicity

Although the opioids share a number of pharmacological effects, there are also significant differences between them. These differences were first noted in a series of classic studies conducted by MARTIN and colleagues (1976) in the chronic spinal dog. Briefly, MARTIN found that morphine produced analgesia. respiratory depression and a well-defined withdrawal syndrome following termination of chronic intake. Withdrawal was suppressed by the subsequent administration of morphine and by a number of other opioid agonists. In contrast, ketocyclazocine, an opioid of the benzomorphan class, produced a different type of analgesia than that of morphine, did not decrease respiration and was not effective in suppressing signs of morphine withdrawal.

These observations led MARTIN to postulate the existence of more than one type of opioid receptor; (1) the mu opioid receptor which was selective for morphine-like drugs; (2) the kappa opioid receptor, which was selective for ketocyclazocine-like drugs; and (3) a sigma opioid receptor which was selective for another class of drugs represented by N-allylnormetazocine. Although sigma receptors presently are not considered opioid receptors, MARTIN's classification of mu and kappa opioid receptor types is well accepted today. Moreover, the sequences of both the mu (CHEN et al. 1993; FUKUDA et al. 1993; WANG et al. 1993) and the kappa (LI et al. 1993; MENG et al. 1993; YASUDA et al. 1993) receptor types have been successfully cloned.

There is a great deal of evidence to support the existence of at least one other subtype, the delta receptor, and other opioid receptors have also been proposed (LORD et al. 1977; CHANG and CUATRECASAS 1979; SMITH and SIMON 1980). The initial hypothesis about the existence of delta receptors was advanced by KOSTERLITZ and colleagues after they noted that the relative potency of opioid alkaloids such as morphine and the enkephalins differed in two bioassays commonly used to examine opioid activity, the GPI and the mouse vas deferens (LORD et al. 1977). On the basis of these differences, they postulated that morphine-like opioid alkaloids produce their effects through mu opioid receptors whereas the enkephalins produce their effects through a different system of opioid receptors, called delta receptors. Although less is known about the effects of delta agonists than of mu or kappa agonists, the recent success in cloning the delta opioid receptor gene promises to expand our understanding of the function of delta opioid receptors (EVANS et al. 1992; KIEFFER et al. 1992).

In addition to the mu, kappa and delta opioid receptor types, several opioid receptor subtypes have been proposed. These include mu_1 and mu_2 receptor subtypes (WOLOZIN and PASTERNAK 1981; PASTERNAK 1986), at least two kappa receptor subtypes (ATTALI et al. 1982; ZUKIN et al. 1988; DEVLIN and SHOEMAKER 1990; HORAN et al. 1991), and $delta_1$ and $delta_2$ receptor subtypes (SOFUOGLO et al. 1991; MATTIA et al. 1991; JIANG et al. 1991).

The classification of multiple opioid receptor types is also supported by the discovery of distinct families of opioid peptides which correspond with the multiple categories of opioid receptors (JAFFE and MARTIN 1990). These include the enkephalins, the endorphins and the dynorphins. Each of these families is derived from a genetically distinct precursor polypeptide. Proopiomelanocortin is the precursor for the opioid peptide β-endorphin. Pro-enkephalin is the precursor for the enkephalins and pro-dynorphin is the precursor for dynorphin. These endogenous peptides have preferences for specific receptor types, but they are not receptor-specific. For example, peptides from pro-dynorphin bind predominantly to kappa sites; however, they have some affinity for mu and delta sites as well (Akil et al. 1984).

3. Selective Opioid Agonists and Antagonists

In making inferences about action at the various opioid receptors, investigators often compare the effects of opioid agonists and antagonists thought to have selective action at specific opioid receptor types. In this context, several compounds have been identified that display opioid receptor selectivity and these are listed in Table 2. Morphine is listed here as the prototypic mu compound more for historical reasons than by virtue of its selectivity since it also has some affinity for kappa and delta receptors (Takemori and Portoghese 1987). The opioid peptide DAMGO ([D-Ala^2, NMePhe^4, Gly-ol^5]) and the nonpeptide sufentanil are currently considered some of the most selective mu agonists available. Other commonly used selective mu agonists are PL017, l-methadone, fentanyl, and levorphanol (Zimmerman and Leander 1990).

Table 2 also lists antagonists with selectivity for the μ, κ and σ opioid receptors. Mu antagonists include the irreversible antagonist β-FNA (β-

Table 2. Receptor selective agonists and antagonists

Receptor	Agonist	Antagonist
Mu (μ)	Morphine[a] DAMGO Sufentanil	β-FNA Naloxone, naltrexone[b]
Kappa (κ)	U50,488 U69,593 C1977	nor-BNI
Delta (δ)	DPDPE Deltorphin	ICI 174864 Naltrindole

Abbreviations: DAMGO: [D-Ala^2, NMePhe^4, Gly-ol^5] enkephalin; β-FNA: β-funaltrexamine; U69,593: {(5a,7a,8b)-(−)-N-methyl-N-(7-(a-pyrrolidinyl)-1-oxas-piro-(4,5)dec-8-yl) benzeneacetamine}; U50,488: {trans-3,4-dichloro-N-methyl-N[2-(1-pyrrolidinyl)cyclo-hexyl] benzeneacetamide methanesulfonate hydrate}; nor-BNI: nor-binaltorphimine; DPDPE: [D-Pen^2, D-Pen^5] enkephalin; deltorphin: [^{3}H] D-Ala^2]deltorphin II ((5R)-(5a,7a,8b)-N-methyl-N-[7-(1-pyrrolidinyl)-1-oxaspiro [4,5]dec-8-yl]-4-benzofuranacetamide monohydrochl-oride); ICI 174864: diallyl-Tyr-Aib-Aib-Phe-Leu.
[a] Also has some activity at kappa and delta receptors.
[b] Selectivity based on concentration.

funaltrexamine) (PORTOGHESE et al. 1980; WARD et al. 1982) and naloxone and naltrexone. Although naloxone and naltrexone bind to kappa and delta opioid receptors as well as to mu receptors, they have approximately ten times greater affinity at the mu receptor (MAGNAN et al. 1982). Therefore, they are often used to infer mu receptor selectivity by comparing the dose of naloxone or naltrexone necessary to antagonize the effects of a mu agonist vs the dose necessary to antagonize the effects of kappa or delta opioid agonists. The pA_2 technique for estimating the differential potency of an antagonist to reverse the effects of an agonist (SCHILD 1947) has been particularly useful for characterizing action at different opioid receptor types (TAKEMORI et al. 1969; DYKSTRA et al. 1987b).

Within the kappa category, U69,593, U50,488, CI977 and the endogenous peptide dynorphin are usually considered the most selective kappa agonists and nor-binaltorphimine (nor-BNI) the most selective kappa antagonist (LAHTI et al. 1985; PORTOGHESE et al. 1987; TAKEMORI et al. 1988; HUNTER et al. 1990). Within the delta category, the peptide agonist [D-Pen^2, D-Pen^5]enkephalin (DPDPE) has a well-established, delta-selective profile with approximately 100–300 times greater affinity for delta than for mu and kappa opioid receptors (MOSBERG et al. 1983). In addition, the antagonists ICI 174864 and naltrindole are considered the most selective delta antagonists (ZIMMERMAN and LEANDER 1990).

Although there is some evidence that the different types of opioid receptors are independent, other studies suggested that some opioid receptor types may be linked with each other such that administration of one type of opioid agonist might alter the activity of another. Early evidence indicated that the endogenous delta ligands leu and met-enkephalin could modulate morphine-induced analgesia (VAUGHT and TAKEMORI 1970; BARRETT and VAUGHT 1982). More recently, investigators have shown that the selective delta agonist DPDPE modulates morphine's analgesic effects (HEYMAN et al. 1989a,b). Findings such as these support hypotheses advanced by ROTHMAN and colleagues that delta agonists may bind to two different binding sites, one that is part of a mu-delta receptor complex and another which binds only delta agonists (BOWEN et al. 1981; ROTHMAN et al. 1985).

It is also important to note that the action of individual opioids is not necessarily mediated by just one receptor type. As a group, the mixed-action opioid agonist/antagonists are thought to produce their opioid-like effects by binding to more than one opioid receptor type. Differences in their spectra of action relates both to differences in their selectivity for given opioid receptor types as well as their efficacy at these different receptors. For example, a number of kappa agonists have considerable activity as mu antagonists (MILLER et al. 1986; SHELDON et al. 1987; CRAFT and DYKSTRA 1992a,b). Similarly, several mu agonists, including morphine, are active at kappa and delta receptors (TAKEMORI and PORTOGHESE 1987; CRAFT and DYKSTRA 1993). The partial mu agonist buprenorphine also has a well-established profile as a kappa antagonists (SU 1985; NEGUS and DYKSTRA 1988; LEANDER 1988).

4. Distribution

Opioid receptors are found throughout the CNS including the spinal cord and in the innervation of certain smooth muscle systems of all vertebrates and of some invertebrates. Autoradiographic studies of the distribution of opioid receptors in the monkey, rat and human brain indicate that opioid receptors exist in different regions of the CNS. Regions rich in opioid binding sites include the limbic system, with particularly high levels in the amygdala, hypothalamus, thalamus, nucleus accumbens and caudate as well as pathways associated with the perception of pain such as the periaqueductal gray. Within the spinal cord, a large number of opioid receptor sites are fond in the substantia gelatiosa (Mansour et al. 1988; Kuhar et al. 1973).

The various opioid receptor types differ in their regional distributions and species specificity (Robson et al. 1985). For example, the GPI contains primarily mu and kappa receptors, whereas the mouse vas deferens is thought to have a very high proportion of delta receptors in addition to mu and kappa receptors (Lord et al. 1977). Within the CNS, kappa receptors are generally more prominent in the monkey than in the rat (Mansour et al. 1988). Moreover, within a given organism, there is a distribution of opioid receptor types. For example, in rats mu opioid binding is observed at several levels of the CNS, whereas the distribution of delta receptors is more restricted and occurs predominantly in forebrain structures (Mansour et al. 1988).

II. Postreceptor Events

Although the action of the opioids may be explained by their initial binding with opioid receptors, this interaction activates second messenger systems, leading to subsequent alterations in numerous neurotransmitter systems. Currently, evidence indicates that opioid receptors belong to the large family of membrane-bound receptors that interact with G-proteins and use cyclic AMP as one of their second messengers. Indeed, all three types of opioid receptor are though to be coupled to G-proteins in at least some tissues (Di Chiara and North 1992). As a result, stimulation of opioid receptors inhibits adenylate cyclase activity, leading to a dose-dependent decrease in the synthesis of cyclic AMP (Loh and Smith 1990; Childers 1991). These effects are readily antagonized by opioid antagonists such as naloxone.

Experiments with NG108-15 neuroblastoma-glioma cells have shown that opioid-induced inhibition of adenylate cyclase is attenuated when cells are treated chronically with opioid agonists. Moreover, the addition of naloxone to cells that have been treated chronically with morphine produces an increase in adenylate cyclase activity above control levels. On the basis of these effects, some investigators have suggested that this system might provide a model of tolerance and dependence (Gmerek 1988; Childers 1991).

Investigators working with a variety of preparations have also shown that mu and delta opioid receptors are coupled to voltage- and/or calcium-dependent potassium channels. As a result, mu and delta agonists increase potassium conductance, shorten action potential duration and consequently reduce calcium conductance whereas kappa agonists appear to inhibit voltage-dependent calcium channels directly (WERZ and MacDONALD 1983, 1985; WERZ et al. 1987; NORTH 1986; NORTH et al. 1987). These effects also appear to be mediated by G-proteins and can be inhibited by pertussis toxin. In either case, the end result is a decrease in the rate of neural discharge and a consequent decrease in the amount of neurotransmitter released. As a result, opioids alter neurotransmission within several neurotransmitter systems. For example, modulation of dopaminergic neurotransmission is thought to be a major effect of the opioids (MULDER et al. 1988). The role of noradrenergic, dopaminergic and serotonergic neurotransmitter systems in opioid-induced effects such as analgesia, reinforcement, tolerance and dependence is discussed in subsequent sections.

C. General Pharmacology

I. Pharmacokinetics and Metabolism

Opioids enter the body in several ways. They are readily absorbed from the gastrointestinal tract, the nasal mucosa or the lungs. They are also active when administered via intravenous, intramuscular or subcutaneous injection. Once in the bloodstream, the opioids are distributed throughout the body with accumulation in the kidney, lung, liver, spleen, digestive tract and muscle as well as the brain. Although most opioids penetrate the brain, they differ in the ease with which penetration takes place. For example, heroin is very lipid-soluble and readily enters the brain, whereas morphine is less lipid-soluble and therefore enters the brain more slowly. The opioid peptides do not cross the blood-brain barrier and thus are only active when administered directly into the brain. The opioids are rapidly metabolized in the liver with the major pathway for detoxification being conjugation with glucuronic acid. The conjugated product is then rapidly excreted by the kidney. For example, 90% of a dose of morphine is excreted within a day of administration (JAFFE and MARTIN 1990).

II. General Effects

As with all drugs, the effects of the opioids are dose-dependent. At small to moderate doses, morphine produces drowsiness, decreases in anxiety, pain relief and often feelings of euphoria. Nausea and vomiting sometimes occur due to stimulation of the chemoreceptor trigger zone. In addition, the opioids are particularly effective in suppressing the cough reflex. At higher doses, the mu opioids depress respiratory centers and produce a shallow respiratory response (see JAFFE and MARTIN 1990 for a review). Although mu opioids generally depress respiration at high doses, the kappa opioids are less likely to produce respiratory depression, even at analgesic doses (HOWELL et al. 1988).

Opioids have little effect on the cardiovascular system although they may reduce blood pressure slightly, due to dilation of the peripheral blood vessels. Opioids usually produce constriction of the pupils (miosis) is people; but in certain other species, such as monkeys and cats, the opioids produce mydriasis. The opioids also have prominent effects on the GI tract. They decrease motility in the stomach, small intestine and the large intestine and increase tone. As a consequence, the passage of the contents is slowed and a greater amount of water is absorbed from the GI tract. Constipation often occurs and this can be a serious complication in individuals taking opioids for long periods of time. The opioids affect neuroendocrine function, inhibiting the release of hypothalamic releasing factors and pituitary hormones. The opioids also modify urinary output with mu opioids generally producing antidiuretic effects and kappa agonists producing diuresis (MARTIN 1984; LEANDER et al. 1987; JAFFE and MARTIN 1990).

III. Immune Function

Clinical studies suggest that increases in disease susceptibility among intravenous opioid users may be related to alterations in immune status (DONAHOE et al. 1986). With the advent of AIDS and its spread through intravenous drug use, there is a great deal of interest in the direct effects of opioid administration on immune status as well as how the various states of dependence, withdrawal and alterations in nutritional status that accompany opioid use may affect immune status.

In rodents, acute morphine treatment inhibits phagocytic functions (TUBARO et al. 1983), depresses the natural killer (NK) cell activity of spleen-, blood- and bone marrow-derived cells (SHAVIT et al. 1986, 1987; BAYER et al. 1990) and suppresses the proliferative response of blood lymphocytes to mitogen stimulation (BAYER et al. 1990). Moreover, these effects are dose-dependent, and naltrexone-reversible (LYSLE et al. 1993). Chronic exposure to morphine also suppresses a number of immune responses, including in vitro measures of proliferation of both T and B lymphocytes to mitogenic stimuli (BRYANT et al. 1987, 1988) and in vivo measures of delayed-type hypersensitivity (BRYANT and ROUDEBUSH 1990). There is also evidence that chronic administration of opioids results in compromised immune function in humans. Heroin-dependent individuals show a reduction in immune determinants such as T cell function (WYBRAN et al. 1979; McDONOUGH et al. 1980; DONAHOE et al. 1986) and NK cell activity (NOVICK et al. 1989). Interestingly, reports indicate that abnormalities in immune function in heroin users may normalize on methadone maintenance (NOVICK et al. 1989).

The mechanism by which morphine exerts these effects is still unclear. Morphine may alter immune status by acting directly on opioid receptors present on the surface of immune cells (WYBRAN et al. 1979; SIBINGA and GOLDSTEIN 1988). It is also possible that some of morphine's immuno-modulatory effects involve central opioid receptors. For example, micro-injections of morphine directly into the periaqueductal gray region of the rat brain depresses splenic NK cell activity and this depression is reversed by naltrexone (WEBER and PERT 1989). Additional investigations suggest that morphine's effects may be mediated by noradrenergic systems (FECHO et al. 1993). Clearly, further research is required to resolve these issues.

IV. Analgesia

1. Clinical Observations

The primary clinical use of the opioids is to relieve severe pain, in particular pain associated with heart attacks, serious injury, postoperative discomfort and terminal illnesses such as cancer. The opioids are vastly superior to all other drugs as strong analgesics because pain relief occurs without loss of consciousness, although drowsiness is common. Moreover, pain relief is selective, in that other sensory modalities such as vision and touch are not altered. When a 10 mg dose of morphine is given to nontolerant patients who are experiencing pain, they report that the pain is less intense and what pain they do experience does not bother them as much. Thus, morphine affects both the sensation of pain and the suffering associated with pain (JAFFE and MARTIN 1990).

2. Mechanism of Action of Mu Agonists

The opioids also alter sensitivity to painful stimuli in animals. Since it is impossible to determine an animal's affective response to pain, the effects of opioids are determined by exposing animals to potentially painful stimuli such as heat or electric shock and measuring either the time it takes them to respond to these stimuli or the intensity at which they respond. If the time it takes the animal to respond to the stimulus is prolonged following drug administration and if it can be determined that the prolongation is not due simply to sedation or motor incapacitation, then it is presumed that the drug altered perception of the painful stimulus. Alterations in responding under these conditions are referred to as *antinociception* and are generally very predictive of a drug's ability to produce analgesia in people.

It is generally believed that the effects of systemic administration of morphine reflect an interaction between direct action in the brain and spinal cord (YAKSH and RUDY 1977, 1978; YAKSH and NOUEIHED 1985) and indirect alteration of descending systems of pain modulation (BASBAUM and FIELDS 1984; HAMMOND 1986). For example, opioid administration is thought to stimulate opioid receptors in the spinal dorsal horn, which in turn inhibit the release of substance P from primary afferent nerves that are involved in the transmission of pain (JESSELL and IVERSEN 1977; SUAREZ-ROCA and MAIXNER 1992). The opioids have also been shown to act in the brainstem raphe region, the thalamus, the periaqueductal gray, the limbic system and the somatosensory cortex (MILLAN 1986). All of these regions contain opioid receptors and abundant amounts of endogenous opioids (MANSOUR et al. 1988). Thus, it has been hypothesized that drugs such as morphine produce their effects by interacting with mu opioid receptor systems in these areas.

Evidence to support the involvement of mu opioid receptors in morphine's effects comes from investigations showing that the potency of opioid agonists in producing antinociception parallels their potency in binding to mu opioid receptors (PEARL et al. 1968). Additional evidence comes from studies showing that the analgesic effects of mu opioids are antagonized by the mu-selective antagonist β-FNA as well as by doses of naloxone that are selective for mu-opioid receptors (TAKEMORI et al. 1969; WARD and TAKEMORI 1983; SCHMAUSS and YAKSH 1984; DYKSTRA et al. 1987b; ZIMMERMAN et al. 1987; DYKSTRA and MASSIE 1988; MJANGER and YAKSH 1991).

It is also agreed that opioid-induced antinociception involves interactions with the serotonergic and noradrenergic systems, by way of activation of descending systems of pain modulation (BASBAUM and FIELDS 1984; YAKSH 1985; HAMMOND 1986). Microinjection of morphine into the ventricles or into various sites in the brain increases the release of both norepinephrine and serotonin. Moreover, morphine's antinociceptive effects can be blocked by noradrenergic and serotonergic antagonists (YAKSH 1979; WIGDOR and WILCOX 1987) and potentiated by drugs which inhibit the uptake of norepinephrine or serotonin (LARSEN and ARNT 1984). There is also evidence to support a role for other receptor systems such as histamine and adenosine in morphine's effects (GOGAS et al. 1989; SWEENEY et al. 1987, 1991; DELANDER et al. 1992).

3. Kappa and Delta Agonists

Although opioids of the mu type are the most commonly used analgesics, opioids of the kappa and delta types also have analgesic effects. In animals, kappa agonists produce antinociception following systemic, intracerebral and spinal administration (TYERS 1980; VONVOIGHTLANDER et al. 1983; SCHMAUSS and YAKSH 1984; WARD and TAKEMORI 1983; PORRECA et al. 1984, 1987b; DYKSTRA et al. 1987a; MILLAN 1989). In general, the antinociceptive effects of the kappa agonists are antagonized by doses of naloxone or naltrexone larger than those that antagonize the effects of mu agonists (MILLAN 1989; MILLAN et al. 1989). Interestingly, the antinociceptive effects of the kappa agonists are not antagonized by the mu-selective antagonist β-

FNA (DYKSTRA et al. 1987b; ZIMMERMAN et al. 1987; NEGUS and DYKSTRA 1988) nor do they show cross-tolerance with mu agonists (CRAFT and DYKSTRA 1990); nevertheless, they are antagonized by kappa selective antagonists such as nor-BNI (TAKEMORI et al. 1988). Taken together, these findings suggest that analgesia following administration of kappa agonists is mediated by kappa opioid receptors rather than by mu opioid receptors.

The analgesic effects of the delta opioid agonists are not as well-documented as those of the kappa agonists, partly because the first delta agonists examined for analgesic activity were rapidly metabolized and lacked clear delta specificity. More recently, however, chemists have developed delta agonists such as DPDPE (MOSBERG et al. 1983) that are more stable and selective, and these produce antinociception in animals when administered intracerebrally as well as intrathecally (PORRECA et al. 1984, 1987a,b). Moreover, their effects are reversed with selective delta antagonists (CALCAGNETTI and HOLTZMAN 1991; DROWER et al. 1991). There is also some evidence that delta agonists can modulate the antinociceptive effects of mu agonists. For example, recent experiments have shown that doses of the delta agonist DPDPE that do not produce antinociception alone can increase the potency of morphine. Since these effects can be blocked by delta antagonists, it has been suggested that the modulation is delta-mediated (VAUGHT and TAKEMORI 1979; JIANG et al. 1990; PORRECA et al. 1992; MALMBERG and YAKSH 1992).

V. Tolerance

When the opioids are given repeatedly, larger doses are required to achieve a given effect, and the opioid dose-effect curve moves rightward, suggestive of tolerance. Tolerance does not develop uniformly to all actions of the opioids and depends on the dose administered and the frequency of administration. In general, tolerance occurs to opioid-induced respiratory depression, analgesia, euphoria, sedation, but not to effects on the pupil or the GI tract. In addition, cough suppression does not show a great deal of tolerance (JAFFE and MARTIN 1990).

In general, opioids show cross-tolerance with each other. That is, if an organism develops tolerance to morphine upon repeated administration, tolerance will extend to other morphine-like opioids. Interestingly, opioids that act at kappa opioid receptors generally do not show cross-tolerance with morphine or other mu agonists, providing further evidence that these opioids produce their effects by interacting with different opioid receptor types. The lack of cross-tolerance between mu and kappa agonists is seen under both in vitro (SCHULZ et al. 1981a,b) and in vivo conditions (MARTIN et al. 1976; VONVOIGHTLANDER et al. 1983; GMEREK et al. 1987; CRAFT and DYKSTRA 1990).

Although pharmacokinetic changes may mediate some aspects of opioid-induced tolerance, mechanisms such a receptor down-regulation, desensitization and alterations in opioid receptor signal transduction are thought to play the major role in opioid tolerance. Recent evidence indicates that the decreases in adenylate cyclase which follow acute administration of the opioids are attenuated during chronic exposure to opioids. Although the precise mechanism whereby this occurs are not presently understood, current hypotheses suggest that chronic opioid administration leads to an uncoupling of the opioid receptor-adenylate cyclase complex and that the development of tolerance is accompanied by an up-regulation of G-proteins (Law et al. 1982; Christie et al. 1987; Nestler and Tallman 1988; Nestler et al. 1989; Johnson and Fleming 1989).

VI. Physical Dependence

Discontinuation of morphine after long periods of administration results in physiological and psychological disturbances indicative of a withdrawal syndrome. In people, this syndrome includes sweating, increased respiratory and heart rate as well as restlessness, body aches, insomnia, nausea, irritability and anxiety. These symptoms reach their peak about 48–72h after discontinuation of chronic intake and run their course within 7–10 days. In all cases, withdrawal is readily suppressed by administration of an opioid. The severity of these symptoms depends on the length and intensity of the chronic opioid treatment schedule as well as whether withdrawal is natural (following termination of opioid administration) or precipitated by an opioid antagonist (Martin 1984; Jaffe 1990). A similar set of symptoms occurs following long term administration of other mu agonists, although the time course and severity may differ. For example, withdrawal following termination of buprenorphine is less intense than that observed with morphine (Fudala et al. 1989).

Withdrawal is also observed in animals that have been treated chronically with opioid agonists. The classic signs of withdrawal from morphine in nonhuman primates are very similar to those observed in people and include abdominal defense reactions, vomiting and excessive vocalization (Woods and Gmerek 1985). In rodents, withdrawal behavior consists of hyperactivity, weight loss, diarrhea, wet-dog shakes in rats and stereotypic jumping in mice (Wei et al. 1973). All of these signs can be suppressed by the administration of opioid agonists. Investigations in animals have also shown that partial opioid agonists such as butorphanol, nalbuphine and buprenorphine produce withdrawal of the morphine type, although the signs tend to be milder (Jacob et al. 1979; Woods and Winger 1987).

Withdrawal following chronic administration of mu agonists can also be observed in specific sites in the central nervous system. One area of particular interest is the locus coeruleus, which has high levels of norepinephrine and a large number of opioid binding sites. Electrophysiological studies indicate that acute administration of opioids decreases spontaneous discharge rates in the locus coeruleus. Tolerance develops to this effect upon chronic administration and spontaneous discharge rates are increased when withdrawal is precipitated with naloxone (RASMUSSEN et al. 1990; VALENTINO and WEHBY 1989). Interestingly, α_2 adrenergic agonists such as clonidine block these withdrawal symptoms (AGHAJANIAN 1978). These data have led to the hypothesis that noradrenergic hyperactivity in the locus coeruleus mediates opiate withdrawal. Since morphine dependent rats show increased adenylate cyclase activity, it has been suggested that adenylate cyclase up-regulation may be involved in this effect.

Although chronic administration of kappa agonists produces a form of dependence, it differs from that seen with mu agonists in a number of ways. Kappa-induced withdrawal in primates is characterized by hyperactivity, unusual tongue movements, yawning, scratching and grooming. Interestingly, kappa-induced withdrawal can be suppressed by other kappa agonists, but not by mu agonists. Conversely, signs of withdrawal from morphine-like agonists can be suppressed by another mu agonist, but not by kappa agonists (GMEREK et al. 1987).

D. Behavioral Pharmacology

A great deal of our understanding about the abuse potential of the opioids comes from the field of behavioral pharmacology. Investigations in this area have shown that the opioids are potent reinforcers, maintaining robust self-administration behavior in a number of species. The opioids also possess distinct discriminative stimulus properties in both animals and people, and these properties are thought to be closely related to their abuse potential. Moreover, many of the effects of the opioids are readily conditioned, a finding that has particular relevance for understanding the maintenance of drug-seeking behavior and relapse following periods of abstinence. The opioids also alter behaviors that are not directly related to their abuse potential; nevertheless, alterations in these behaviors represent some of the consequences of drug usage. Of particular interest along these lines are the effects of the opioids on consummatory behavior, motor activity and processes related to learning and memory.

I. Reinforcing Effects

1. Self-Administration and Place Preference

One very prominent behavioral characteristic of morphine and related opioid agonists is their ability to function as positive reinforcers. The most commonly used procedure for investigating the reinforcing properties of opioids in animals is the self-administration procedure, in which rats or monkeys are prepared with in-dwelling venous catheters to deliver drug solutions and delivery of the drug solution is made contingent on the completion of one or more responses (Thompson and Schuster 1964). A number of mu opioid agonists (e.g., morphine, fentanyl, heroin, meperidine, methadone, levorphanol, codeine and etorphine) function as reinforcers under these conditions (Young et al. 1981; Woods et al. 1982). Moreover, many of the drugs classified as partial opioid agonists such as butorphanol, nalbuphine and buprenorphine also maintain self-administration behavior, however, rates of responding maintained by the partial agonists are often lower than those maintained by full agonists (Mello et al. 1981; Young et al. 1984; Lukas et al. 1986). In contrast to the mu opioid agonists, opioids that have a kappa-like profile generally do not maintain self-administration behavior (Woods and Winger 1987). Although the reinforcing effects of delta agonists have not been examined with intravenous self-administration procedures, there is evidence from other behavioral paradigms that opioids of the delta type can function as reinforcers when they are administered directly into the brain (see below).

Mu opioids function as reinforcers under a variety of conditions. Very early studies showed that if monkeys are given unlimited access to morphine, they will self-administer increasing levels of morphine after which intake stabilizes at a fairly constant level (Deneau et al. 1969). Opioids are also self-administered under more limited access conditions (e.g., when they are available for brief periods once or twice a day) as well as when the drug is delivered under various schedules of reinforcement. For example, morphine can be made available under a fixed ratio schedule of reinforcement which requires the animal to make a specific number of responses in order to receive drug. If saline is substituted for morphine or if an opioid antagonist is administered after animals have self-administered morphine for several weeks, signs of withdrawal occur and these can be suppressed by readministering morphine or by administering other mu agonists (Thompson and Schuster 1964; Young et al. 1981).

Another very important aspect of opioid self-administration is its dose-dependent nature. The rate at which an animal responds to an opioid is a function of the dose of the drug available. As shown in Fig. 2, low doses of a variety of mu agonists maintain low rates of self-administration, intermediate doses maintain maximal rates whereas higher doses markedly reduce rates of self-administration, presumably because of the direct rate-suppressing effects of the drug itself (Young et al. 1981).

Opioid self-administration procedures have also been used to examine the potential of certain pharmacological treatments to reduce opioid self-administration. For example, buprenorphine, a partial mu agonist with a lesser degree of dependence than morphine, has been shown to decrease opioid self-administration in monkeys (Mello et al. 1983; Winger et al. 1992). On the basis of these findings, it has been suggested that buprenorphine might be useful in the treatment of heroin dependence, a suggestion that has been verified in a variety of clinical situations (Mello et al. 1982; Bickel et al. 1988; Kosten and Kleber 1988).

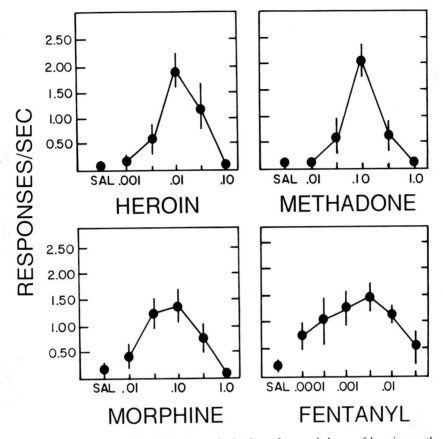

Fig. 2. Rates of responding following substitution of several doses of heroin, methadone, morphine or fentanyl in monkeys responding under an FR 30 schedule of codeine presentation. *Ordinate*, rates of responding in responses/sec. *Abscissa*, injection dose (mg/kg), log scale. Each point represents the mean ± SEM of the average response rate during two substitution sessions in each of three monkeys. Points at *SAL* represent the mean ± SEM of the average rate for each monkey during two saline substitution sessions. (Adapted from YOUNG et al. 1981)

Another procedure that is commonly used to investigate the reinforcing properties of opioid agonists is the place preference procedure (STOLERMAN 1992). In this procedure, administration of a drug such as morphine is paired with a distinct environment, and vehicle injection is paired with a different environment. Following several conditioning sessions in which drug administration is paired with the distinct environment, the distribution of time spent in the two environments is determined. If rats spend more time in the drug-paired environment, then the drug is said to produce a place preference, serving as a positive reinforcer. Conversely, if rats spend less time in the drug-paired environment, then the drug is said to produce a place aversion.

Data from place conditioning procedures support the hypothesis that mu and delta agonists, but not kappa agonists, possess positive reinforcing effects. For example, morphine, fentanyl, sufentanil and the delta peptide DPDPE all produce place preferences, whereas kappa agonists produce place aversions (VAN DER KOOY et al. 1982; MUCHA and HERZ 1985; SHIPPENBERG et al. 1987; BALS-KUBIK et al. 1993). These findings with the kappa agonists are in keeping with studies in people indicating that kappa agonists produce unpleasant, dysphoric effect (PFEIFFER et al. 1986; KUMOR et al. 1986).

2. Neurobiological Mechanisms

Additional studies have attempted to elucidate the neurobiological mechanisms underlying opioid reinforcement. It has been proposed that the reinforcing effects of opioid agonists are primarily mediated by mu opioid receptors. Evidence supporting this hypothesis comes from experiments showing that opioid self-administration is altered by mu-preferring antagonists (KOOB et al. 1984; VACCARINO et al. 1985; BERTALMIO and WOODS 1989; NEGUS et al. 1993). As shown in Fig. 3, the mu-preferring opioid antagonist quadazocine produces parallel rightward shifts in the self-administration dose-effect curve for the mu agonist alfentanil, and these shifts occur within a dose range that suggests mu-mediated action.

Several sites within the central nervous system have been implicated in opioid reinforcement, however, the mesocorticolimbic dopaminergic system has received most attention (KOOB 1992). This system originates in the ventral tegmental area (VTA) and projects to the nucleus accumbens (NA), frontal cortex, amygdala and septal areas. It contains both mu and delta

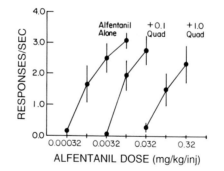

Fig. 3. Rates of responding following substitution of several doses of alfentanil alone and of alfentanil in combination with two doses of the opioid antagonist, quadazocine. Data were obtained by substituting alfentanil for codeine in monkeys responding under an FR 30 schedule of codeine presentation. *Ordinate*, rate of responding in responses/s. *Abscissa*, injection dose of alfentanil (mg/kg), log scale. Each point represents the mean ± SEM of the average response rate in six monkeys. (Adapted from BERTALMIO and WOODS 1989)

opioid receptors as well as cells containing endogenous opioid peptides (MANSOUR et al. 1988).

Several strategies are used to examine the role of the mesolimbic dopamine system in opioid reinforcement. First, opioid agonists are micro-injected directly into areas of the mesolimbic dopamine system and their reinforcing effects observed. Microinjection of both mu and delta agonists into either the VTA or the NA produces conditioned place preferences and/or self-administration in rats, and these effects can be antagonized by naloxone (BOZARTH and WISE 1981; GOEDERS et al. 1984; SHIPPENDBERG et al. 1987; BALS-KUBIK et al. 1993). On the basis of these findings, other investigators have reasoned that if activation of the mesolimbic dopamine system is necessary for opioid reinforcement, then dopaminergic antagonists should block opioid reinforcement. There is some evidence that dopamine antagonists can attenuate opioid reinforcement and that opioid reinforce-ment can be altered by selective lesions of the mesolimbic dopamine system; however, these findings are not always consistent and await further inves-tigation (SPYRAKI et al. 1983; PETTIT et al. 1984; LEONE and DI CHIARA 1987; DWORKIN et al. 1988; see NEGUS and DYKSTRA 1989 and KOOB and BLOOM 1988 for more thorough reviews of this literature).

II. Discriminative Stimulus Characteristics

Animals can be readily trained to discriminate between the presence and absence of a variety of drugs. Since these unique discriminative stimulus effects may be related to the subjective effects that people experience when taking these drugs, it is thought that an understanding of a drug's discri-minative stimulus properties might provide insights into their abuse potential (SCHUSTER et al. 1981). One common procedure used to examine the discri-minative stimulus properties of the opioids involves training animals to respond on one lever when an opioid agonist such as morphine is ad-ministered and on another lever when morphine vehicle is administered. Once training has taken place with a given dose of morphine (called the training dose) and the animals have met some criterion (e.g., 80% or more of their responses consistently occur on the correct lever), other doses of morphine are administered and dose-effect curves reflecting morphine's discriminative stimulus properties are generated. It has been shown that morphine's discriminative stimulus effects are shared by several opioids of the mu type, but not by inactive opioid isomers nor by nonopioid compounds such as amphetamine and pentobarbital (Fig. 4). This type of pharmaco-logical specificity has been observed under a wide range of conditions and in a variety of species, including rats, monkeys and pigeons (SHANNON and HOLTZMAN 1976; SCHAEFER and HOLTZMAN 1977; HOLTZMAN 1983; PICKER and DYKSTRA 1987). Partial opioid agonists such as nalbuphine also substitute for morphine in the drug discrimination procedure; however, the dose of morphine used as a training stimulus is an important determinant of sub-

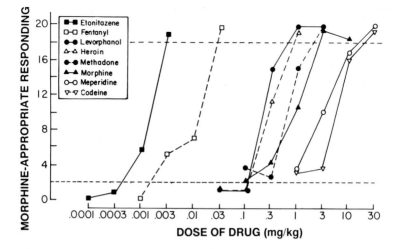

Fig. 4. Substitution curves for several mu opioid agonists in rats trained to discriminate between saline and 3.0 mg/kg of morphine. Durgs were administered s.c. 30 min before the start of a session. *Ordinate*, number of trials in which rats responded on the morphine-appropriate level (out of a total of 20 trials). *Abscissa*, dose of each drug (mg/kg), log scale. The *lower dashed line* shows the level of morphine-appropriate responding following saline and the *upper dashed line* shows the level of morphine-appropriate responding following 3.0 mg/kg of morphine. Each point is the mean number of trials on the morphine-appropriate level based on one observation in each of 37 (morphine curve) or 4–5 rats. (Adapted from Holtzman 1983)

stitution by partial agonists (Young et al. 1992; Picker et al. 1992). It has also been shown that people can discriminate between various opioids and saline (Jasinski 1977; Preston et al. 1987; Bickle et al. 1989; Kamien et al. 1993) and the effects observed are similar to those seen in animals.

The discriminative stimulus properties of morphine and other opioids of the mu type appear to be related to action at mu opioid receptors. First of all, it has been shown that the potency of various opioids to substitute for morphine in animals that have been trained to discriminate between morphine and saline correlates with their affinity for mu opioid receptors as measured in a variety of binding assays (Holtzman and Locke 1988). Moreover, the discriminative stimulus effects of mu agonists can be antagonized by mu-preferring opioid antagonists in a competitive manner, resulting in displacement of the dose-effect curve to the right (Shannon and Holtzman 1976; Bertalmio and Woods 1987). Investigations with selective mu antagonists such as β-FNA also suggest that the discriminative stimulus effects of morphine-like drugs are mu-mediated (Locke and Holtzman 1986a; France and Woods 1987). Finally, tolerance has been shown to develop to morphine's discriminative stimulus effects, with cross-tolerance extending to other mu agonists (Young et al. 1991). Taken together, these findings provide evidence that the discriminative stimulus properties of morphine reflect action at mu opioid receptors.

Yet another interesting aspect of the discriminative stimulus effects of the opioids is revealed during morphine withdrawal. In a series of studies, HOLTZMAN and colleagues have shown that rats dependent on morphine can be trained to discriminate between an injection of the opioid antagonist naltrexone (which precipitates withdrawal) and saline. Interestingly, the discriminative stimulus effects of naltrexone in morphine-dependent rats parallels another measure of opioid withdrawal in rats, namely the loss of body weight. Moreover, the discriminative stimulus effects of withdrawal can be reversed by a number of mu opioid agonists, but not by the inactive isomers nor by nonopioid drugs such as diazepam, haloperidol and pentobarbital (HOLTZMAN 1985; GELLERT and HOLTZMAN 1979). Similar effects are seen in morphine-dependent monkeys in which the discriminative stimulus effects of naltrexone correlate closely with signs of opioid withdrawal and are reversed by subsequent administration of opioid agonists (FRANCE and WOODS 1989).

Although morphine's discriminative effects are shared by other mu opioid agonists, kappa opioid agonists generally do not share morphine's discriminative stimulus effects. Nevertheless, animals can be trained to discriminate between a kappa agonist and saline and the discriminative stimulus properties of kappa agonists are shared by other kappa agonists and antagonized by opioid antagonists. Data from experiments with naltrexone and the mu-selective antagonist β-FNA suggest that the discriminative stimulus effects of kappa agonists are not mediated by mu opioid receptors and most likely involve action at kappa opioid receptors (DYKSTRA et al. 1987a,b; PICKER and DYKSTRA 1987; BERTALMIO and WOODS 1987; FRANCE and WOODS 1990). These differences are consistent with the distinct profiles observed for mu and kappa agonists under other conditions.

III. Schedule-Controlled Behavior

In most investigations of schedule-controlled behavior, animals perform a simple response such as a lever press or a key peck in order to receive food or water. Delivery of these events usually occurs according to some schedule. In general, both full and partial mu agonists, as well as opioids of the kappa type, decrease responding maintained by different schedules of food presentation in a variety of species (McMILLAN and MORSE 1967; GOLDBERG et al. 1976, 1981; OLIVETO et al. 1991). To date very little information is available about the effects of selective delta agonists on schedule-controlled responding; however, at least one study has shown that the delta peptide DPDPE produces transient increases in rates of schedule-controlled responding at low doses and transient decreases at high doses (ADAMS and HOLTZMAN 1991).

It is important to note that these decreases in schedule-controlled responding are not unique to opioids, since a variety of drugs decrease rates of responding maintained by food presentation. What distinguishes the opioids from other drugs is the fact that their rate-decreasing effects can be antagonized by opioid antagonists. Opioid antagonists such as naloxone produce dose-dependent, parallel shifts to the right in the morphine dose-effects curve for rate of responding over a dose range that is selective for mu opioid receptors (Goldberg et al. 1981; McMillan et al. 1970). Moreover, morphine's effects on schedule-controlled responding are antagonized by the mu-selective antagonist β-FNA (France and Woods 1985).

Repeated administration of opioid agonists typically results in an attenuation of their effects on schedule-controlled responding, i.e., tolerance develops. For example, tolerance develops to morphine's rate-decreasing effects in animals responding under a schedule of food presentation, as evidenced by shifts to the right in the morphine dose-effects curve following chronic morphine administration (Heifetz and McMillan 1971; Adams and Holtzman 1990b). Moreover, morphine-tolerant animals are cross-tolerant to the effects of other mu opioid agonists and to delta agonists, but are not cross-tolerant to the effects of kappa agonists (Doty et al. 1989; Craft et al. 1989; Picker et al. 1990; Adams and Holtzman 1991).

One very interesting aspect of opioid tolerance is that it can be modified by the behavioral conditions accompanying drug administration. In one experiment, Sannerud and Young (1986) examined the effects of repeated morphine adminsitration in two groups of rats responding under a schedule of food presentation. The separate groups of rats were administered 10 mg/kg of morphine every day either prior to the experimental session or after the daily session had terminated. Tolerance developed in rats that received an equal amount of morphine after the session. This suggests that the opportunity to perform a task in the presence of the drug can contribute to the development of tolerance. Thus, both behavioral and pharmacological variables are involved in morphine tolerance. This is in keeping with other reports that behavioral and environmental variables play an important role in tolerance to morphine's analgesic effects (See Sect. D.IV).

Schedule-controlled responding also provides a very sensitive measure of opioid antagonist activity. When doses of the opioid antagonist naltrexone that do not alter rates of responding in morphine-naive rats are administered to rats receiving morphine chronically, these same doses of naltrexone decrease rates of responding markedly. Indeed, investigators have shown that the ED_{50} for naltrexone's rate-decreasing effects in morphine-treated rats and monkeys can be several log units lower than in untreated animals (France and Morse 1989; Adams and Holtzman 1990a; Oliveto et al. 1991). This increased sensitivity to opioid antagonists in morphine-maintained animals is often used as a measure of dependence.

In summary, examinations of the opioids on schedule-controlled responding indicate that the effects are dose-dependent, shared by several opioid agonists and readily antagonized by an opioid antagonist. In addition, tolerance to opioid agonists and sensitivity to opioid antagonists has been shown to develop in animals treated chronically with opioids. Moreover, although both mu and kappa agonists decrease rates of schedule-controlled behavior, cross-tolerance studies differentiate their action.

IV. Conditioned Drug Effects

Environmental events that are associated with the adminsitration of a drug can subsequently evoke some of the effects associated with delivery of that drug (i.e., conditioning). One of the most convincing demonstrations of opioid conditioning is that of conditioned withdrawal. In a classic experiment by GOLDBERG and SCHUSTER (1967), monkeys were treated chronically with morphine and trained to press a lever under a fixed ratio schedule of food presentation. After responding was stable, a tone was presented followed by an injection of an opioid antagonist that immediately precipitated signs of withdrawal. After several experimental sessions in which the tone was paired with the delivery of the opioid antagonist, presentation of the tone alone produced withdrawal signs, including decreases in rates of responding, bradycardia, emesis and excessive salivation.

Another classic demonstration of the conditioning of opioid effects is that observed during the development of tolerance to morphine's analgesic effects. In an important series of studies, SIEGEL (1976) showed that tolerance to morphine's analgesic effects was only displayed when animals were tested in the same environment in which they had previously received a chronic regimen of morphine injections. Rats that received the same regimen of morphine injections but were tested in a different environment did not show tolerance. Since Siegel's initial report, conditioned tolerance to morphine's analgesic effects has been demonstrated under a variety of conditions (TIFFANY and MAUDE-GRIFFIN 1988).

These findings have important implications for understanding the variables that play a role in the maintenance of drug-seeking behavior and in the relapse to drug-taking behavior following periods of abstinence (see SCHUSTER 1986 and O'BRIEN et al. 1986 for reviews). Findings such as these have also expanded our understanding of some of the long-term consequences of drug administration, since conditioned drug effects have been shown to persist long after drug administration has terminated.

V. Learning and Memory

Consideration of drug effects on learning and memory involves two general types of procedures. In one type, drug administration occurs prior to training of a new task and effects on acquisition are observed. In a second type of procedure, drug administration occurs after a task has been learned (post-training) and the effects on retention of the task are observed. There is extensive evidence that opioid agonists impair retnetion of a task when they are administered after training has taken place. These effects are dose- and time-dependent, antagonized by opioid antagonists and observed under a variety of conditions (McGaugh 1989). Studies of opioid antagonists also support a role for the opioid system in the regulation of memory since retention is enhanced by posttraining administration of opiate antagonists (Gallagher et al. 1983).

In contrast to effects observed for the opioids on retention of a task, investigations of opioid effects on the acquisition of new behaviors do not reveal a consistent pattern of effects across species or procedures. For example, neither mu nor kappa agonists alter acquisition under a procedure in which monkeys acquire a new sequence of responses during each experimental session (Moerschbaecher and Thompson 1983; Moerschbaecher et al. 1987). However, there is some evidence that mu opioid agonists do alter response accuracy in rats under some conditions (Moerschbaecher et al. 1984). Thus, clear conclusions about the action of opioid agonists on acquisition await futher investigation.

VI. Unlearned Behaviors

In addition to the behaviors discussed above, many of which play a prominent role in our understanding of opioid abuse, the opioids also alter other behaviors such as consummatory behavior and motor activity. The effects of opioids on consummatory behavior have been examined in several different species. On the basis of these investigations, it is generally agreed that opioid agonists increase food and water intake whereas opioid antagonists decrease intake (Holtzman 1974; Sanger and McCarthy 1981; Morley et al. 1983; Cooper et al. 1988). Although the precise mechanism whereby the opioids alter these behaviors is not clear, there is evidence that the effect is centrally mediated and most likely involves the hypothalamus (Gosnell 1987). It also is not clear whether one particular opioid receptor subtype is involved in opioid-induced alterations in food and water intake. Indeed, alterations in feeding behavior occur following administration of mu, kappa and delta opioid agonists and blockade of both mu and kappa opioid receptors alters certain types of feeding (Morley et al. 1983; Levine et al. 1990; Ukai and Holtzman 1988).

The opioids also have very prominent effects on motor activity and these effects may differ depending on the species and the type of opioid agonist examined. In people, low doses of mu opioids have minimal effects on motor activity; however, very high doses can produce rigidity. Low doses of morphine and other mu-like compounds produce naltrexone-reversible increases in motor activity in rats whereas higher doses first decrease and then increase motor activity. In general, mu agonists only produce sedation in primates. By contrast, the prominent effects of kappa agonists are sedation in both rodents and primates (IWAMOTO 1981; LOCKE and HOLTZMAN 1986b; GMEREK et al. 1987).

Since morphine's effects on locomotor activity cna be altered by various manipulations of the dopamine system (IWAMOTO 1981) and since the injection of morphine directly into sites rich in dopamine also increases locomotor activity, it has been hypothesized that dopamine systems are involved in opioid-induced alterations in motor activity (KALIVAS et al. 1983). It is also interesting that morphine-induced alterations in motor activity do not show tolerance following repeated adminsitration. Instead sensitization often occurs and usually persists for several months. Investigations of opioid-induced sensitization indicate that dopamine mechanisms play a major role in this process as well (KALIVAS and DUFFY 1987).

E. Summary

In summary, opioid analgesics such as morphine display a broad range of activities with predominant effects including analgesia at both spinal and supraspinal sites, respiratory depression, cough suppression and altered GI function. The abuse of the opioids is thought to be related to their reinforcing and discriminative stimulus effects and to their tendency to produce tolerance and dependence upon repeated administration. The action of the opioids is mediated by interaction with at least three different types of opioid receptors, mu, kappa and delta, and is readily reversed by the administration of opioid antagonists. Action at opioid receptors alters G-protein-linked second messenger systems as well as the activity of potassium and calcium channels, leading to alterations in the function of a number of different neurotransmitter systems. The number and type of opioid compounds are diverse, including agonists with selectivity for each of the opioid receptor types, mixed action opioids with agonist activity at more than one receptor type and opioids with antagonist activity. These diverse compounds cna display effects very different from those of the prototypic opioid analgesic morphine.

References

Adams JU, Holtzman SG (1990a) Pharmacologic characterization of the sensitization to the rate-decreasing effects of naltrexone induced by acute opioid pretreatment in rats. J Pharmacol Exp Ther 253:483–489

Adams JU, Holtzman SG (1990b) Tolerance and dependence after continuous morphine infusion from osmotic pumps measured by operant responding in rats. Psychopharmacology 100:451–458

Adams JU, Holtzman SG (1991) Effects of receptor-selective opioids on operant behavior in morphine-treated and untreated rats. Pharmacol Biochem Behav 38:195–200

Aghajanian GK (1978) Tolerance of locus coeruleus neurones to morphine and suppression of withdrawal response by clonidine. Nature 267:186–188

Akil H, Watson SJ, Young E, Lewis ME, Khachaturian H, Walker JM (1984) Endogenous opioids: biology and function. Annu Rev Neurosci 7:223–255

Attali B, Gouardères C, Mazarguil H, Audigier Y, Cros J (1982) Evidence for multiple "kappa" binding sites by use of opioid peptides in the guineapig lumbosacral spinal cord. Neuropeptides 3:53–64

Bals-Kubik R, Ableitner A, Herz A, Shippenberg TS (1993) Neuroanatomical sites mediating the motivational effects of opioids as mapped by the conditioned place preference paradigm in rats. J Pharmacol Exp Ther 264:489–495

Barrett RW, Vaught JL (1982) The effects of receptor selective opioid peptides on morphine-induced analgesia. Eur J Pharmacol 80:427–430

Basbaum AI, Fields HL (1984) Endogenous pain control systems: brainstem spinal pathways and endorphin circuitry. Annu Rev Neurosci 7:309–338

Bayer BM, Daussin S, Hernandez M, Irvin L (1990) Morphine inhibition of lymphocyte activity is mediated by an opioid-dependent mechanism. Neuropharm 29:369–374

Bertalmio AJ, Woods JH (1987) Differentiation between mu and kappa receptor-mediated effects in opioid drug discrimination: apparent pA_2 analysis. J Pharmacol Exp Ther 243:591–597

Bertalmio AJ, Woods JH (1989) Reinforcing effect of alfentanil is mediated by mu opioid receptors: apparent pA_2 analysis. J Pharmacol Exp Ther 251:455–460

Bickel WK, Stitzer ML, Bigelow GE, Liebson IA, Jasinski DR, Johnson RE (1988) Buprenorphine: dose-related blockade of opioid challenge effects in opioid dependent humans. J Pharmacol Exp Ther 247:47–53

Bickel WK, Bigelow GE, Preston KL, Liebson IA (1989) Opioid drug discrimination in humans: stability, specificity and relation to self-reported drug effect. J Pharmacol Exp Ther 251:1053–1063

Bowen WD, Gentlemen S, Herkenham M, Pert CB (1981) Interconverting mu and delta forms of the opiate receptors in rat striatal patches. Proc Natl Acad Sci USA 78:4818–4822

Bozarth MA, Wise RA (1981) Intracranial self-administration of morphine into the ventral tegmental area in rats. Life Sci 28:551–555

Brecher EM (1972) Licit and illicit drugs. Little Brown, Toronto

Bryant HU, Roudebush RE (1990) Suppressive effects of morphine pellet implants on in vivo parameters of immune function. J Pharmacol Exp Ther 255:410–414

Bryant HU, Bernton EW, Holaday JW (1987) Immunosuppressive effects of chronic morphine treatment in mice. Life Sci 41:1731–1738

Bryant HU, Bernton EW, Holaday JW (1988) Morphine pellet-induced immunomodulation in mice: temporal relationships. J Pharmacol Exp Ther 245:913–920

Calcagnetti DJ, Holtzman SG (1991) Delta opioid antagonist, naltrindole, selectively blocks analgesia induced by DPDPE but not DAGO or Morphine. Pharmacol Biochem Behav 38:185–190

Chang K-J, Cuatrecasas P (1979) Multiple opiate receptors: enkaphalin and morphine bind to receptors of different specificity. J Biol Chem 254:2610–2618

Chen Y, Mestek A, Liu J, Hurley JA, Yu L (1993) Molecular cloning and functional expression of a μ-opioid receptor from rat brain. Mol Pharmacol 44:8–12

Childers SR (1991) Opioid receptor-coupled second messenger systems. Life Sci 48:1991–2003

Christie MJ, Williams JT, North RA (1987) Cellular mechanism of opioid tolerance: studies in single brain neurons. Mol Pharmacol 32:633–638

Cooper SJ, Jackson A, Kirkham TC, Turkish S (1988) Endorphins, opiates and food intake. In: Rodgers RJ, Cooper SJ (eds) Endorphins, opiates and behavioural processes. Wiley, New York, pp 143–186

Craft RM, Dykstra LA (1990) Differential cross-tolerance to opioids in squirrel monkeys responding under a shock titration schedule. J Pharmacol Exp Ther 252:945–952

Craft RM, Dykstra LA (1992a) Agonist and antagonist activity of kappa opioids in the squirrel monkey: I. Antinociception and urine output. J Pharmacol Exp Ther 260:327–333

Craft RM, Dykstra LA (1992b) Agonist and antagonist activity of kappa opioids in the squirrel monkey: II. Effect of chronic morphine treatment. J Pharmacol Exp Ther 260:334–342

Craft RM, Dykstra LA (1993) Morphine antagonizes U50,488's effects in squirrel monkey shock titration procedure. Eur J Pharmacol 234:199–207

Craft RM, Picker MJ, Dykstra LA (1989) Differential cross-tolerance to opioid agonists in morphine-tolerant pigeons responding under a schedule of food presentation. J Pharmacol Exp Ther 249:386–393

Creese I, Snyder SH (1975) Receptor binding and pharmacological activity of opiates in the guinea-pig intestine. J Pharmacol Exp Ther 194:205–219

DeLander GE, Mosberg HI, Porreca F (1992) Involvement of adenosine in antinociception produced by spinal or supraspinal receptor-selective opioid agonists: dissociation from gastrointestinal effects in mice. J Pharmacol Exp Ther 263:1097–1104

Deneau G, Yanagita T, Seevers MH (1969) Self-administration of psychoactive substances by the monkey. Psychopharmacology 16:30–48

Devlin T, Shoemaker WJ (1990) Characterization of kappa opioid binding using dynorphin A_{1-13} and U69 593 in the rat brain. J Pharmacol Exp Ther 253:749–759

Di Chiara GD, North RA (1992) Neurobiology of opiate abuse. Trends Pharmacol Sci 13:185–193

Donahoe RM, Nicholson JKA, Madden JJ, Donahoe F, Shafer DA, Gordon D, Bokos P, Falek A (1986) Coordinate and independent effects of heroin, cocaine and alcohol abuse on T-cell E-rosette formation and antigenic marker expression. Clin Immunol Immunopathol 41:254–264

Doty P, Picker MJ, Dykstra LA (1989) Differential cross-tolerance to opioid agonists in morphine-tolerant squirrel monkeys responding under a schedule of food presentation. Eur J Pharmacol 174:171–180

Drower EJ, Stapelfeld A, Rafferty MF, de Costa BR, Rice KC, Hammond DL (1991) Selective antagonism by naltrindole of the antinociceptive effects of the delta opioid agonist cyclic[D-Penicillamine²-D-Penicillamine⁵]enkephalin in the rat. J Pharmacol Exp Ther 259:725–731

Dworkin SI, Guerin GF, Goeders NE, Smith JE (1988) Kainic acid lesions of the nucleus accumbens selectively attenuate morphine self-administration. Pharmacol Biochem Behav 29:175–181

Dykstra LA, Massie CA (1988) Antagonism of the analgesic effects of mu and kappa opioid agonists in the squirrel monkey. J Pharmacol Exp Ther 246:813–821

Dykstra LA, Gmerek DE, Winger G, Woods JH (1987a) Kappa opioids in rhesus monkeys. I. Diuresis, sedation, analgesia and discriminative stimulus effects. J Pharmacol Exp Ther 242:413–420

Dykstra LA, Gmerek DE, Winger G, Woods JH (1987b) Kappa opioids in rhesus monkeys. II. Analysis of the antagonistic actions of quadazocine and β-funaltrexamine. J Pharmacol Exp Ther 242:421–427

Evans CJ, Keith DE Jr, Morrison H, Magendzo K, Edwards RH (1992) Cloning of a delta opioid receptor by functional expression. Science 258:1952–1955

Fecho K, Dykstra LA, Lysle DT (1993) Evidence for β-adrenergic receptor involvement in the immunomodulatory effects of morphine. J Pharmacol Exp Ther 265:1079–1087

Foley KM (1989) The rational use of analgesics in the management of patients with acute and chronic pain. Clin Neurosurg 35:360–384

France CP, Morse WH (1989) Pharmacological characterization of supersensitivity to naltrexone in squirrel monkeys. J Pharmacol Exp Ther 250:928–936

France CP, Woods JH (1985) Antagonistic and rate-suppressing effects of opioid antagonists in the pigeon. J Pharmacol Exp Ther 235:442–447

France CP, Woods JH (1987) β-funaltrexamine antagonizes the discriminative stimulus effects of morphine but not naltrexone in pigeons. Psychopharmacology 91:213–216

France CP, Woods JH (1989) Discriminative stimulus effects of naltrexone in morphine-treated rhesus monkeys. J Pharmacol Exp Ther 250:937–943

France CP, Woods JH (1990) Discriminative stimulus effects of opioid agonists in morphine-dependent pigeons. J Pharmacol Exp Ther 254:626–632

Fraser HF, Harris LS (1967) Narcotic and narcotic antagonist analgesics. Annu Rev Pharmacol 7:277–300

Fudala PJ, Johnson RE, Bunker E (1989) Abrupt withdrawal of buprenorphine following chronic administration. Clin Pharmacol Ther 45:186

Fukuda K, Kato S, Mori K, Nishi M, Takeshima H (1993) Primary structures and expression from cDNAs of rat opioid receptor δ- and μ-subtypes. Fed Eur Biochem Soc 327:311–314

Gallagher M, King RA, Young NB (1983) Opiate antagonists improve spatial memory. Science 221:975–976

Gellert VF, Holtzman SG (1979) Discriminative stimulus effects of naltrexone in the morphine-dependent rat. J Pharmacol Exp Ther 211:596–605

Gmerek DE, Dykstra LA, Woods JH (1987) Kappa opioids in rhesus monkeys. III. Dependence associated with chronic administration. J Pharmacol Exp Ther 242:428–436

Gmerek DE (1988) Physiological dependence on opioids. In: Rodgers RJ, Cooper SJ (eds) Endorphins, opiates and behavioural processes. Wiley, New York, pp 25–52

Gogas KR, Hough LB, Eberle NB, Lyon RA, Glick SD, Ward SJ, Young RC, Parsons ME (1989) A role for histamine and H_2-receptors in opioid antinociception. J Pharmacol Exp Ther 250:476–484

Goeders NE, Lane JD, Smith JE (1984) Self-administration of methionine enkephalin into the nucleus accumbens. Pharmacol Biochem Behav 20:451–455

Goldberg SR, Schuster CR (1967) Conditioned suppression by a stimulus associated with nalorphine in morphine-dependent monkeys. J Exp Anal Behav 10:235–242

Goldberg SR, Morse WH, Goldberg DM (1976) Some behavioral effects of morphine, naloxone and nalorphine in the squirrel monkey and the pigeon. J Pharmacol Exp Ther 196:625–636

Goldberg SR, Morse WH, Goldberg DM (1981) Acute and chronic effects of naltrexone and naloxone on schedule-controlled behavior of squirrel monkeys and pigeons. J Pharmacol Exp Ther 216:500–509

Gosnell BA (1987) Central structures involved in opioid-induced feeding. Fed Proc 46:163–167

Hammond DL (1986) Control systems for nociceptive afferent processing: the descending inhibitory pathways. In: Yaksh TL (ed) Spinal afferent processing. Plenum, New York, pp 391–416

Heifetz SA, McMillan DE (1971) Development of behavioral tolerance to morphine and methadone using the schedule-controlled behavior of the pigeon. Psychopharmacologia 19:40–52

Heyman JS, Vaught JL, Mosberg HI, Haaseth RC, Porreca F (1989a) Modulation of m-mediated antinociception by δ agonists in the mouse: selective potentiation of morphine and normorphine by [D-Pen^2,D-Pen^5]enkephalin. Eur J Pharmacol 165:1–10

Heyman JS, Jiang Q, Rothman RB, Mosberg HI, Porreca F (1989b) Modulation of m-mediated antinociception by δ agonists: characterization with antagonists. Eur J Pharmacol 169:43–52

Hill HF, Mackie AM, Coda BA (1991) Patient-controlled analgesic infusion. Adv Pain Res Ther 18:507–521

Holtzman SG (1974) Behavioral effects of separate and combined administration of naloxone and d-amphetamine. J Pharmacol Exp Ther 189:51–60

Holtzman SG (1983) Discriminative stimulus properties of opioid agonists and antagonists. In: Cooper SJ (ed) Theory in psychopharmacology, vol II. Academic, New York, pp 2–45

Holtzman SG (1985) Discriminative stimulus effects of morphine withdrawal in the dependent rat: suppression by opiate and nonopiate drugs. J Pharmacol Exp Ther 233:80–86

Holtzman SG, Locke KW (1988) Neural mechanisms of drug stimuli: experimental approaches. In: Colpaert FC, Balster RL (eds) Psychopharmacology series 4: transduction mechanisms of drug stimuli. Springer, Berlin Heidelberg New York, pp 139–153

Horan P, de Costa BR, Rice KC, Porreca F (1991) Differential antagonism of U69,593- and bremazocine-induced antinociception by (−)-UPHIT: evidence of kappa opioid receptor multiplicity in mice. J Pharmacol Exp Ther 257:1154–1161

Howell LL, Bergman J, Morse WH (1988) Effects of levorphanol and several kappa-selective opioids on respiration and behavior in rhesus monkeys. J Pharmacol Exp Ther 245:364–372

Hughes J, Smith TW, Kosterlitz HW, Fothergil LA, Morgan BA, Morris HR (1975) Identification of two related pentapeptides from brain with potent opiate agonist activity. Nature 258:577–579

Hunter JC, Leighton GE, Meecham KG, Boyle SJ, Horwell DC, Rees DC, Hughes J (1990) CI-977, a novel and selective agonist for the κ-opioid receptor. Br J Pharmacol 101:183–189

Iwamoto ET (1981) Locomotor activity and antinociception after putative mu, kappa and sigma opioid receptor agonists in the rat: influence of dopaminergic agonists and antagonists. J Pharmacol Exp Ther 217:451–460

Jacob JJC, Michaud GM, Tremblay EC (1979) Mixed agonist-antagonist opiates and physical dependence. Br J Clin Pharmacol 7:291–296

Jaffe JH (1990) Drug addiction and drug abuse. In: Gilman AG, Rall TW, Nies AS, Taylor P (eds) Goodman and Gilman's the pharmacological basis of therapeutics. Pergamon, New York, pp 552–573

Jaffe JH, Martin WR (1990) Opioid analgesics and antagonists. In: Gilman AG, Rall TW, Nies AS, Taylor P (eds) Goodman and Gilman's the pharmacological basis of therapeutics. Pergamon, New York, pp 485–521

Jasinski DR (1977) Assessment of the abuse potentiality of morphine-like drugs. In: Martin WR (ed) Drug Addiction I. Morphine, sedative-hypnotic and alcohol dependence. Springer, Berlin Heidelberg New York, pp 197–258 (Handbook of experimental pharmacology, vol 45)

Jessell TM, Iverson LL (1977) Opiate analgesics inhibit substance P release from rat trigeminal nucleus. Nature 268:549–551

Jiang Q, Mosberg HI, Porreca F (1990) Modulation of the potency and efficacy of mu-mediated antinociception by delta agonists in the mouse. J Pharmacol Exp Ther 254:683–689

Jiang Q, Takemori AE, Sultana M, Portoghese PS, Bowen WD, Mosberg HI, Porreca F (1991) Differential antagonism of opioid delta antinociception by [D-Ala^2,Leu^5,Cys^6]enkephalin and naltrindole 5′-isothiocyanate: evidence for delta receptor subtypes. J Pharmacol Exp Ther 257:1069–1075

Johnson SM, Fleming WW (1989) Mechanisms of cellular adaptive sensitivity changes: applications to opioid tolerance and dependence. Pharmacol Rev 41:435–488

Kalivas PW, Duffy P (1987) Sensitization to repeated morphine injection in the rat: possible involvement of A10 dopamine neurons. J Pharmacol Exp Ther 241:204–212

Kalivas PW, Widerlöv E, Stanley D, Breese GR, Prange AJ Jr (1983) Enkephalin action on the mesolimbic system: a dopamine-dependent and a dopamine-independent increase in locomotor activity. J Pharmacol Exp Ther 227:229–237

Kamien JB, Bickel WK, Hughes JR, Higgins ST, Smith BJ (1993) Drug discrimination by humans compared to nonhumans: current status and future directions. Psychopharmacology 111:259–270

Kieffer BL, Befort K, Gaveriaux-Ruff C, Hirth CG (1992) The δ-opioid receptor: isolation of a cDNA by expression cloning and pharmacological characterization. Proc Natl Acad Sci USA 89:12048–12052

Koob GF (1992) Drugs of abuse: anatomy, pharmacology and function of reward pathways. Trends Pharmacol Sci 13:177–184

Koob GF, Bloom FE (1988) Cellular and molecular mechanisms of drug dependence. Science 242:715–723

Koob GF, Pettit HO, Ettenberg A, Bloom FE (1984) Effects of opiate antagonists and their quaternary derivatives on heroin self-administration in the rat. J Pharmacol Exp Ther 229:481–486

Kosten TR, Kleber HD (1988) Buprenorphine detoxification from opioid dependence: a pilot study. Life Sci 42:635–641

Kuhar MJ, Pert CB, Snyder SH (1973) Regional distribution of opiate receptor binding in monkey and human brain. Nature 245:447–450

Kumor KM, Haertzen CA, Johnson RE, Kocher T, Jasinski D (1986) Human psychopharmacology of ketocyclazocine as compared with cyclazocine, morphine and placebo. J Pharmacol Exp Ther 238:960–968

Lahti RA, Mickelson MM, McCall JM, VonVoigtlander PF (1985) [^{3}H]U-69593 A highly selective ligand for the opioid κ receptor. Eur J Pharmacol 109:281–284

Larson J-J, Arnt J (1984) Spinal 5-HT or NA uptake inhibition potentiates supraspinal morphine antinociception in rats. Acta Pharmacol Toxicol 54:72–75

Law PV, Hom DS, Loh HH (1982) Loss of opiate receptor activity in neuroblastoma X glioma NG 108-15 hybrid cells after chronic opiate treatment. Mol Pharmacol 22:1–4

Leander JD (1988) Buprenorphine is a potent kappa opioid receptor antagonist in pigeons and mice. Eur J Pharmacol 151:457–461

Leander JD, Hart JC, Zerbe RL (1987) Kappa agonist-induced diuresis: evidence for stereoselectivity, strain differences, independence of hydration variables and a result of decreased plasma vasopressin levels. J Pharmacol Exp Ther 242:33–39

Leone P, Di Chiara G (1987) Blockade of D-1 receptors by SCH 23390 antagonizes morphine- and amphetamine-induced place preferences conditioning. Eur J Pharmacol 135:251–254

Levine AS, Grace M, Billington CJ, Portoghese PS (1990) Norbinaltorphimine decreases deprivation and opioid-induced feeding. Brain Res 534:60–64

Li S, Zhu J, Chen C, Chien Y-W, Deriel JK (1993) Molecular cloning and expression of a rat κ opioid receptor. Biochem J 295:629–633

Locke KW, Holtzman SG (1986a) Behavioral effects of opioid peptides selective for mu or delta receptors. I. Morphine-like discriminative stimulus effects. J Pharmacol Exp Ther 238:990–996

Locke KW, Holtzman SG (1986b) Behavioral effects of opioid peptides selective for mu or delta receptors. II. Locomotor activity in nondependent and morphine-dependent rats. J Pharmacol Exp Ther 238:997–1003

Loh HH, Smith AP (1990) Molecular characterization of opioid receptors. Ann Rev Pharmacol Toxicol 30:123–147

Lord JAH, Waterfield AA, Hughes J, Kosterlitz HW (1977) Endogenous opioid peptides: multiple agonists and receptors. Nature 267:495–499

Lukas SE, Brady JV, Griffiths RR (1986) Comparison of opioid self-injection and disruption of schedule-controlled performance in the baboon. J Pharmacol Exp Ther 238:924–931

Lysle DT, Coussons ME, Watts VJ, Bennett EH, Dykstra LA (1993) Morphine-induced alterations of immune status: dose dependency, compartment specificity, and antagonism by naltrexone. J Pharmacol Exp Ther 265:1071–1078

Magnan J, Paterson SJ, Tavani A, Kosterlitz HW (1982) The binding spectrum of narcotic analgesic drugs with different agonist and antagonist properties. Arch Pharmacol 319:197–205

Malmberg AB, Yaksh TL (1992) Isobolographic and dose-response analyses of the interaction between intrathecal mu and delta agonists: effects of naltrindole and its benzofuran analog (NTB). J Pharmacol Exp Ther 263:264–275

Mansour A, Khachaturian H, Lewis ME, Akil H, Watson SJ (1988) Anatomy of CNS opioid receptors. Trends Neurosci 11:308–314

Martin WR (1984) Pharmacology of opioids. Pharmacol Rev 35:283–323

Martin WR, Eades CG, Thompson JA, Huppler RE, Gilbert PE (1976) The effects of morphine- and nalorphine-like drugs in the non-dependent and morphine-dependent chronic spinal dog. J Pharmacol Exp Ther 197:517–532

Mattia A, Vanderah T, Mosberg HI, Porreca F (1991) Lack of antinociceptive cross-tolerance between [D-Pen^2,D-Pen^5]enkephalin and [D-Ala^2]deltorphin II in mice: evidence for delta receptor subtypes. J Pharmacol Exp Ther 258:583–587

McDonough RJ, Madden JJ, Falek A, Shafer DA, Pline M, Gordon D, Bokos P, Kuehnle JC, Mendelson J (1980) Alteration of T and null lymphocyte frequencies in the peripheral blood of human opiate addicts: in vivo evidence for opiate receptor sites on T lymphocytes. J Immunol 125:2539–2543

McGaugh HL (1989) Involvement of hormonal and neuromodulatory systems in the regulation of memory storage. Annu Rev Neurosci 12:255–287

McMillan DE, Morse WH (1967) Some effects of morphine and morphine antagonists on schedule-controlled behavior. J Pharmacol Exp Ther 157:175–184

McMillan DE, Wolf PS, Carchman RA (1970) Antagonism of the behavioral effects of morphine and methadone by narcotic antagonists in the pigeon. J Pharmacol Exp Ther 175:443–458

Mello NK, Mendelson JH, Kuehnle JC (1981) Buprenorphine self-administration by rhesus monkey. Pharmacol Biochem Behav 15:215–225

Mello NK, Mendelson JH, Kuehnle JC (1982) Buprenorphine effects on human heroin self-administration: an operant analysis. J Pharmacol Exp Ther 223:30–39

Mello NK, Bree MP, Mendelson JH (1983) Comparison of buprenorphine and methadone effects on opiate self-administration in primates. J Pharmacol Exp Ther 225:378–386

Meng F, Xie G-X, Thompson RC, Mansour A, Goldstein A, Watson SJ, Akil H (1993) Cloning and pharmacological characterization of a rat κ opioid receptor. Proc Natl Acad Sci USA 90:9954–9958

Melzack R (1990) The tragedy of needless pain. Sci Am 262:27–33

Millan MJ (1986) Multiple opioid systems and pain. Pain 27:303–347

Millan MJ (1989) Kappa-opioid receptor-mediated antinociception in the rat. I. Comparative actions of mu- and kappa-opioids against noxious thermal, pressure and electrical stimuli. J Pharmacol Exp Ther 251:334–341

Millan MJ, Czlonkowski A, Lipkowski A, Herz A (1989) Kappa-opioid receptor-mediated antinocicpetion in the rat. II. Supraspinal in addition to spinal sites of action. J Pharmacol Exp Ther 251:342–350

Miller L, Shaw JS, Whiting EM (1986) The contribution of intrinsic activity to the action of opioids in vitro. Br J Pharmacol 87:595–601

Mjanger E, Yaksh TL (1991) Characteristics of dose-dependent antagonism by β-funaltrexamine of the antinociceptive effects of intrathecal mu agonists. J Pharmacol Exp Ther 258:544–550

Moerschbaecher JM, Thompson DM (1983) Differential effects of prototype opioid agonists on the acquisition of conditional discriminations in monkeys. J Pharmacol Exp Ther 226:738–748

Moerschbaecher JM, Mastropolo, J Winsauer PJ, Thompson DM (1984) Effects opioids on accuracy of a fixed-ratio discrimination in monkeys and rats. J Pharmacol Exp Ther 230:541–549

Moerschbaecher JM, Brocklehurst C, Devia C, Faust WB (1987) Effects of kappa agonists and dexoxadrol on the acquisition of conditional discriminations in monkeys. J Pharmacol Exp Ther 243:737–744

Morley JE, Levine AS, Yim GK, Lowy MT (1983) Opioid modulation of appetite. Neurosci Biobehav Rev 7:281–305

Mosberg HI, Hurst R, Hruby VJ, Gee K, Yamamura HI, Galligan JJ, Burks TF (1983) Bio-penicillamine enkephalins possess highly improved specificity toward δ opioid receptors. Proc Natl Acad Sci USA 80:5871–5874

Mucha RJ, Herz A (1985) Motivational properties of kappa and mu opioid receptor agonists studied with place and taste preference conditioning. Psychopharmacology 86:274

Mulder AH, Frankhuyzen AL, Schoffelmeer ANM (1988) Modulation by opioid peptides of dopaminergic neurotransmission at the pre- and postsynaptic level. In: Illes P, Farsang C (eds) Regulatory roles of opioid peptides. VCH, Weinheim, pp 268–281

Negus SS, Dykstra LA (1988) κ antagonist properties of buprenorphine in the shock titration procedure. Eur J Pharmacol 156:77–86

Negus SS, Dykstra LA (1989) Neural substrates mediating the reinforcing properties of opioid analgesics. In: Watson RR (ed) Biochemistry and physiology of substance abuse, vol I. CRC Press, Boca Raton, pp 211–242

Negus SS, Henriksen SJ, Mattox SR, Pasternak GW, Portoghese PS, Takemori AE, Weinger MB, Koob GF (1993) Effects of antagonists selective for mu, delta and kappa opioid receptors on the reinforcing effects of heroin in rats. J Pharmacol Exp Ther 265:1245–1252

Nestler EJ, Tallman JF (1988) Chronic morphine treatment increases cyclic AMP-dependent protein kinase activity in the rat locus coeruleus. Mol Pharmacol 33:127–132

Nestler EJ, Erdos JJ, Terwilliger R, Duman RS, Tallman JF (1989) Regulation of G proteins by chronic morphine in the rat locus coeruleus. Brain Res 476:230–239

North RA (1986) Opioid receptor types and membrane ion channels. Trends Neurobiol Sci 114–117

North RA, Williams JT, Surprenant A, Christie MJ (1987) m and δ receptors belong to a family of receptors that are coupled to potassium channels. Proc Natl Acad Sci USA 5487–5491

Novick DM, Ochshorn M, Ghali V, Croxson TS, Mercer WD, Chiorazzi N, Kreek MJ (1989) Natural killer cell activity and lymphocyte subsets in parenteral heroin abusers and long-term methadone maintenance patients. J Pharmacol Exp Ther 250:606–610

O'Brien CP, Ehrman RN, Ternes JW (1986) Classical conditioning in human opioid dependence. In: Goldberg SR, Stolerman IP (eds) Behavioral analysis of drug dependence. Academic, New York, pp 329–356

Oliveto AH, Picker MJ, Dykstra LA (1991) Acute and chronic morphine administration: effects of mixed-action opioids in rats and squirrel monkeys responding under a schedule of food presentation. J Pharmacol Exp Ther 257:8–18

Pasternak GW (1986) Multiple m opiate receptors: biochemical and pharmacological evidence for multiplicity. Biochem Pharmacol 35:361–364

Pearl J, Aceto MD, Harris LS (1968) Prevention of writhing and other effects of narcotics and narcotic antagonists in mice. J Pharmacol Exp Ther 160:217–230

Pert CB, Snyder SH (1973) Opiate receptor: demonstration in nervous tissue. Science 179:1011–1014

Pettit HO, Ettenberg A, Bloom FE, Koob GF (1984) Destruction of dopamine in the nucleus accumbens selectively attenuates cocaine but not heroin self-administration in rats. Psychopharmacology 84:167–173

Pfeiffer A, Brantl V, Herz A, Emrich HM (1986) Psychotomimesis mediated by κ-opiate receptors. Science 233:774–776

Picker MJ, Dykstra LA (1987) Comparison of the discriminative stimulus properties of U50488 and morphine in pigeons. J Pharmacol Exp Ther 243:938–945

Picker MJ, Negus SS, Powell KR (1990) Differential cross-tolerance to mu and kappa opioid agonists in morphine-tolerant rats responding under a schedule of food presentation. Psychopharmacology 103:129–135

Picker MJ, Craft RM, Negus SS, Powell KR, Mattox SR, Jones SR, Hargrove BK, Dykstra LA (1992) Intermediate efficacy mu opioids: examination of their morphine-like stimulus effects and response rate-decreasing effects in morphine-tolerant rats. J Pharmacol Exp Ther 263:668–681

Porreca F, Mosberg HI, Hurst R, Hruby VJ, Burks TF (1984) Roles of mu, delta and kappa opioid receptors in spinal and supraspinal mediation of gastrointestinal transit effects and hot-plate analgesia in the mouse. J Pharmacol Exp Ther 230:341–348

Porreca F, Heyman JS, Mosberg HI, Omnaas JR, Vaught JL (1987a) Role of mu and delta receptors in the supraspinal and spinal analgesic effects of [D-Pen^2, D-Pen^5]enkephalin in the mouse. J Pharmacol Exp Ther 241:393–400

Porreca F, Mosberg HI, Omnaas JR, Burks TF, Cowan A (1987b) Supraspinal and spinal potency of selective opioid agonists in the mouse writhing test. J Pharmacol Exp Ther 240:890–894

Porreca F, Takemori AE, Sultana M, Portoghese PS, Bowen WD, Mosberg HI (1992) Modulation of mu-mediated antinociception in the mouse involves opiod delta-2 receptors. J Pharmacol Exp Ther 263:147–152

Portoghese PS, Larson DL, Sayre LM, Fries DS, Takemori AE (1980) A novel opioid receptor site directed alkylating agent with irreversible narcotic antagonistic and reversible agonistic activities. J Med Chem 23:233–234

Portoghese PS, Lipkowski AW, Takemori AE (1987) Binaltorphimine and nor-binaltorphimine, potent and selective κ-opioid receptor antagonists. Life Sci 45:1287–1292

Preston KL, Bigelow GE, Bickel W, Liebson IA (1987) Three-choice drug discrimination in opioid-dependent humans: hydromorphone, naloxone and saline. J Pharmacol Exp Ther 243:1002–1009

Rasmussen K, Beitner-Johnson DB, Krystal JH, Aghajanian GK, Nestler EJ (1990) Opiate withdrawal and the rat locus coeruleus: behavioral, electrophysiological, and biochemical correlates. J Neurosci 10:2308–2317

Robson LE, Gillan MGC, Kosterlitz HW (1985) Species differences in the concentrations and distributions of opioid binding sites. Eur J Pharmacol 112:65–71

Rotherman RB, Danks JA, Jacobson AE, Burke TR Jr, Rice KC (1985) Leucine enkephalin noncompetitively inhibits the binding of ^{3}H-naloxone to the opiate mu-recognition site: Evidence for delta-mu binding site interactions in vitro. Neuropeptides 6:351–364

Sanger DJ, McCarthy PS (1981) Increased food and water intake produced in rats by opiate receptor agonists. Psychopharmacology 74:217–220

Sannerud CA, Young AM (1986) Modification of morphine tolerance by behavioral variables. J Pharmacol Exp Ther 237:75–81

Schaefer GJ, Holtzman SG (1977) Discriminative effects of morphine in the squirrel monkey. J Pharmacol Exp Ther 201:67–75

Schild HO (1947) Drug antagonism. Br J Pharmacol Chemother 2:189–206

Schmauss C, Yaksh TL (1984) In vivo studies on spinal opiate receptor systems mediating antinociception. II. Pharmacological profiles suggesting a differential association of mu, delta and kappa receptors with visceral chemical and cutaneous thermal stimuli in the rat. J Pharmacol Exp Ther 228:1–12

Schulz R, Wüster M, Rubini P, Herz A (1981a) Differentiation of opiate receptors in the brain by the selective development of tolerance. Pharmacol Biochem Behav 14:75–79

Schulz R, Wüster M, Rubini P, Herz A (1981b) Functional opiate receptors in the guinea-pig ileum: Their differentiation by means of selective tolerance development. J Pharmacol Exp Ther 219:547–550

Schuster CR (1986) Implications of laboratory research for the treatment of drug dependence. In: Goldberg SR, Stolerman IP (eds) Behavioral analysis of drug dependence. Academic, New York, pp 357–385

Schuster CR, Fischman MW, Johanson CE (1981) Internal stimulus control and subjective effects of drugs. In: Thompson T, Johanson CE (eds) Behavioral pharmacology of human drug dependence. NIDA Monogr 37:116–129

Shannon HE, Holtzman SG (1976) Evaluation of the discriminative effects of morphine in the rat. J Pharmacol Exp Ther 198:54–65

Shavit Y, DePaulis A, Martin FC, Terman GW, Pechnick RN, Zane CJ, Gale RP, Liebeskind JC (1986) Involvement of brain opiate receptors in the immune-suppressive effect of morphine. Proc Natl Acad Sci USA 83:7114–7117

Shavit Y, Martin FC, Yirmiya R, Ben-Eliyahu S, Terman GW, Weiner H, Gale RP, Liebeskind JC (1987) Effects of a single adminsitration of morphine or footshcock stress on natural killer cell cytotoxicity. Brain Behav Immun 1:318–328

Sheldon RJ, Nuan L, Porreca F (1987) Mu antagonist properties of kappa agonists in a model of rat urinary bladder motility in vivo. J Pharmacol Exp Ther 243:234–240

Shippenberg TS, Bals-Kubik R, Herz A (1987) Motivational properties of opioids; evidence that an activation of δ-receptors mediates reinforcement processes. Brain Res 436:234–239

Sibinga NES, Goldstein A (1988) Opioid peptides and opioid receptors in cells of the immune system. Annu Rev Immunol 6:219–249

Siegel S (1976) Morphine analgesic tolerance: its situation specificity supports a pavlovian conditioning model. Science 193:323–325

Simon EJ, Hiller JM, Edelman I (1973) Stereospecific binding of the potent narcotic analgesic [^{3}H]$_2$-etorphine to rat-brain homogenate. Proc Natl Acad Sci USA 70:1947–1949

Smith JR, Simon EJ (1980) Selective protection of stereospecific enkephalin and opiate binding against inactivation by N-ethyklmaleimide: evidence for two classes of opiate receptors. Proc Natl Acad Sci USA 77:281–284

Sofuoglu M, Portoghese PS, Takemori AE (1991) Differential antagonism of delta opioid agonsits by naltrindole and its benzofuran analog (NTB) in mice; Evidence for delta opioid receptor subtypes. J Pharmacol Exp Ther 257:676–680

Spyraki C, Fibiger HC, Phillips AG (1983) Attenuation of heroin reward in rats by disruption of the mesolimbic dopamine system. Psychopharmacology 79:278–283

Stolerman I (1992) Drugs of abuse: behavioural principles, methods and terms, Trends Pharmacol Sci 13:170–176

Su TP (1985) Further demonstration of kappa opioid binding sites in the brain: evidence for heterogeneity. J Pharmacol Exp Ther 232:144–148

Suarez-Roca H, Maixner W (1992) Morphine produces a multiphasic effect on the release of substance P frim rat trigeminal nucleus slices by activating different opioid receptor subtypes. Brain Res 579:195–203

Sweeney MI, White TD, Sawynok J (1987) Involvement of adenosine in the spinal antinociceptive effects of morphine and noradrenaline. J Pharmacol Exp Ther 243:657–665

Sweeney MI, White TD, Sawynok J (1991) Intracerebroventricular morphine releases adenosine and adenosine 3′, 5′-cyclic monophosphate from the spinal cord via a serotonergic mechanism. J Pharmacol Exp Ther 259:1013–1018

Takemori AE, Kupferberg HJ, Miller JW (1969) Quantitative studies of the antagonism of morphine by nalorphine and naloxone. J Pharmacol Exp Ther 169:39–45

Takemori AE, Portoghese PS (1987) Evidence for the interaction of morphine with kappa and delta opioid receptors to induce analgesia in β-funaltrexamine-treated mice. J Pharmacol Exp Ther 243:91–94

Takemori AE, Ho BY, Naeseth JS, Portoghese PS (1988) Nor-binaltorphimine, a highly selective kappa-opioid antagonist in analgesic and receptor binding assays. J Pharmacol Exp Ther 246:255–258

Terenius L (1973) Stereospecific interaction between narcotic analgesics and a synaptic plasma membrane fraction of rat cerebral cortex. Acta Pharmacol Toxicol 32:317–320

Thompson T, Schuster CR (1964) Morphine self-administration, food-reinforced, and avoidance behaviors in rhesus monkeys. Psychopharmacologia 5:87–94

Tiffany ST, Maude-Griffin PM (1988) Tolerance to morphine in the rat: associative and nonassociative effects. Behav Neurosci 102:534–543

Tubaro E, Borelli G, Croce C, Cavallo G, Santiangeli C (1983) Effect of morphine on resistance to infection. J Infect Dis 148:656–666

Tyers MB (1980) A classification of opiate receptors that mediate antinociception in animals. Br J Pharmacol 69:503–512

Ukai M, Holtzman SG (1988) Effects of β-funaltrexamine on ingestive behaviors in the rat. Eur J Pharmacol 153:161–165

Vaccarino FJ, Pettit HO, Bloom FE, Koob GF (1985) Effects of intracerebroventricular administration of methyl naloxonium, chloride on heroin self-administration in the rat. Pharmacol Biochem Behav 23:495–498

Valentino RJ, Wehby RG (1989) Locus ceruleus discharge characteristics of morphine-dependent rats: effects of naltrexone. Brain Res 488:126–134

Van Der Kooy D, Mucha RF, O'Shaughnessy M, Bucenlieks P (1982) Reinforcing effects of brain microinjections of morphine revealed by conditioned place preference. Brain Res 243:107

Vaught JL, Takemori AE (1979) Differential effects of leucine and methionine enkephalin on morphine-induced analgesia, acute tolerance and dependence. J Pharmacol Exp Ther 208:86–90

VonVoightlander PF, Lahti RA, Ludens JH (1983) U-50,488: a selective and structurally novel non-mu (kappa) opioid agonist. J Pharmacol Exp Ther 224:7–12

Wang JB, Imai Y, Eppler CM, Gregor P, Spivak CE, Uhl GR (1993) μ opiate receptor: cDNA cloning and expression. Proc Natl Acad Sci USA 90:10230–10234

Ward SJ, Takemori AE (1983) Relative involvement of mu, kappa and delta receptor mechanisms in opiate-mediated antinociception in mice. J Pharmacol Exp Ther 224:525–530

Ward SJ, Portoghese PS, Takemori AE (1982) Pharmacological characterization in vivo of the novel opiate, β-funaltrexamine. J Pharmacol Exp Ther 220:494–498

Wasacz J (1981) Natural and synthetic narcotic drugs. Am Sci 69:318–324

Weber RJ, Pert A (1989) The periaqueductal gray matter mediates opiate-induced immunosuppression. Science 245:188–190

Wei E, Loh HH, Way EL (1973) Quantitative aspects of precipitated abstinence in morphine-dependent rats. J Pharmacol exp Ther 184:398–403

Werz MA, MacDonald RL (1983) Opioid peptides with differential affinity for mu and delta receptors decrease sensory neuron calcium-dependent action potentials. J Pharmacol Exp Ther 227:394–402

Werz MA, MacDonald RL (1985) Dynorphin and neoendorphin peptides decrease dorsal root ganglion neuron calcium-dependent action potential duration. J Pharmacol Exp Ther 234:49–56

Werz MA, Grega DS, Macdonald RL (1987) Actions of mu, delta and kappa opioid agonists and antagonists on mouse primary afferent neurons in culture. J Pharmacol Exp Ther 243:258–263

Wigdor S, Wilcox GL (1987) Central and systemic morphine-induced antinociception in mice: contribution of descending serotonergic and noradrenergic pathways. J Pharmacol Exp Ther 242:90–95

Winger G, Skjoldager P, Woods JH (1992) Effects of buprenorphine and other opioid agonists and antagonists on alfentanil- and cocaine-reinforced responding in rhesus monkeys. J Pharmacol Exp Ther 261:311–317

Wolozin BL, Pasternak GW (1981) Classification of multiple morphine and enkephalin binding sites in the central nervous system. Proc Natl Acad Sci USA 78:6181–6185

Woods JH, Gmerek DE (1985) Substitution and primary dependence studies in animals. Drug Alcohol Depend 14:233–247

Woods JH, Winger G (1987) Opioids, receptors and abuse liability. In: Meltzer HY (ed) Psychopharmacology: the third generation of progress. Raven, New York, pp 1555–1564

Woods JH, Young AM, Herling S (1982) Classification of narcotics on the basis of their reinforcing, discriminative, and antagonist effects in rhesus monkeys. Fed Proc 41:221–227

Wybran J, Appelboom T, Famaey JP, Govaerts A (1979) Suggestive evidence for receptors for morphine and methionine-enkephalin on normal human blood T-lymphocytes. J Immunol 123:1068–1070

Yaksh TA (1979) Direct evidence that spinal serotonin and noradrenaline terminals mediate the spinal antinociceptive effects of morphine in the periaqueductal gray. Brain Res 160:180–185

Yaksh TA (1985) Pharmacology of spinal adrenergic systems which modulate spinal nociceptive processing. Pharmacol Biochem Behav 22:845–858

Yaksh TA, Rudy TA (1977) Studies on the direct spinal action of narcotics in the production of analgesia in the rat. J Pharmacol Exp Ther 202:411–428

Yaksh TA, Rudy TA (1978) Narcotic analgetics: CNS sites and mechanisms of action as revealed by intracerebral injection techniques. Pain 4:299–359

Yaksh TA, Noueihed R (1985) The physiology and pharmacology of spinal opiates. Annu Rev Pharmacol Toxicol 25:433–462

Yasuda K, Raynor K, Kong H, Breder CD, Takeda J, Reisine T, Bell GI (1993) Cloning and functional comparison of κ and δ opioid receptors from mouse brain. Proc Natl Acad Sci USA 90:6736–6740

Young AM, Swain HH, Woods JH (1981) Comparison of opioid agonists in maintaining responding and in suppressing morphine withdrawal in rhesus monkeys. Psychopharmacology 74:329–335

Young AM, Stephens KR, Hein DW, Woods JH (1984) Reinforcing and discriminative stimulus properties of mixed agonist-antagonist opioids. J Pharmacol Exp Ther 229:118–126

Young AM, Kapitsopoulos G, Makhay MM (1991) Tolerance to morphine-like stimulus effects of mu opioid agonists. J Pharmacol Exp Ther 257:795–805

Young AM, Masaki MA, Geula C (1992) Discriminative stimulus effects of morphine: effects of training dose on agonist and antagonsit effects of mu opioids. J Pharmacol Exp Ther 261:246–257

Zimmerman DM, Leander JD (1990) Selective opioid receptor agonists and antagonists: research tools and potential therapeutic agents. J Med Chem 33:895–902

Zimmerman DM, Leander JD, Reel JK, Hynes MD (1987) Use of β-funaltrexamine to determine mu opioid receptor involvement in the analgesic activity of various opioid ligands. J Pharmacol Exp Ther 241:374–378

Zukin RS, Eghbali M, Olive D, Unterwald EM, Tempel A (1988) Characterization and visualization of rat and guinea pig brain κ opioid receptors: evidence for κ_1 and κ_2 opioid receptors. Proc Natl Acad Sci USA 85:4061–4065

CHAPTER 6

Phencyclidine: A Drug of Abuse and a Tool for Neuroscience Research

R.L. BALSTER and J. WILLETTS

A. Introduction and History

Phencyclidine (PCP; 1-(1-phenylcyclohexyl)piperidine hydrochloride) was launched as an intravenous anesthetic under the trade name Sernyl by Parke, Davis and Company in the late 1950s. Unlike other general anesthetics, phencyclidine produced unconsciousness, absence of pain perception and amnesia without suppressing vital reflexes; the so-called dissociative anesthetic state induced by phencyclidine differed qualitatively from anesthesia induced by depressant drugs and volatile anesthetics (DOMINO 1964). However, unwanted psychotomimetic effects, often resembling acute schizophrenic episodes, were associated with emergence from phencyclidine-induced anesthesia and the drug was withdrawn from clinical development (GREIFENSTEIN et al. 1958; DOMINO 1964).

Phencyclidine is easily synthesized and emerged as a drug of abuse soon after its clinical debut. Along with a number of related compounds from the arylcycloalkylamine series, phencyclidine is regulated in the United States under the Controlled Substance Act; a number of active phencyclidine analogs may yet appear as "designer drugs" (JERRARD 1990). Phencyclidine is the drug of choice for many abusers, particularly in certain urban areas of the United States (THOMBS 1989; GORELICK and BALSTER 1995).

Phencyclidine has appealed to a wide audience as a research tool in neuroscience, not only from the perspective of its abuse liability, but also from the perspective of developing new therapeutic agents. Phencyclidine's favorable profile as an anesthetic led to the search for related compounds with a diminished potential for psychotomimetic effects. Ketamine, a closely related but less potent analog of phencyclidine with a shorter duration of action, is used throughout the world in some surgical or diagnostic procedures in humans. Conversely, the psychotomimetic effects observed initially in the clinical population and subsequently in abusers of the drug led to the hypothesis that phencyclidine intoxication served as a model for schizophrenia (see JAVITT 1987). The finding that phencyclidine is an antagonist of excitatory amino acids acting through the N-methyl-D-aspartate (NMDA) receptor (reviewed by LODGE and JOHNSON 1990) and the involvement of this receptor in seizure activity, long-term potentiation, excitotoxicity and neuronal damage following ischemia led to the search for phencyclidine-like drugs as antiepileptic and neuroprotective agents. While the therapeutic fruits of these labors have yet to be harvested, a great deal has consequently been learned about the requirements for phencyclidine-like biological activity both within the arylcycloakylamine series as well as in structurally unrelated compounds. In addition, basic scientific knowledge of the neurobehavioral pharmacology of phencyclidine has contributed greatly to the rapid progress being made in understanding excitatory amino acid neurotransmission and its role in brain function and disease.

B. Overview of Pharmacological Profile

I. Humans

Few studies have systematically addressed dose-related acute effects of phencyclidine in humans; indeed, no experimental laboratory studies of phencyclidine intoxication have been carried out in humans for more than 30 years. This puts phencyclidine in a somewhat unique position among drugs of abuse in that no reference data from standard assessments of subjective states are available for the compound. Thus, there is often considerable confusion as to the nature of phencyclidine-like effects. Its subjective effects include perceptual distortions, dissociation from the environment, and disordered thinking. It also produces impairment of coordinated movement. However, phencyclidine is not usually classified as an hallucinogen since its effects differ considerably from those of classical hallucinogens such as the indolamines and substituted amphetamines (Gorelick and Balster 1995).

Phencyclidine has variously been known as "angel dust," "PeaCe Pill," "wack" or "wacky" among users. Smoking appears to be the preferred mode of administration, since the dose level can be more easily titrated than by injection or inhalation routes. Phencyclidine is often mixed with parsley, tobacco or marijuana for smoking or is dissolved and then placed on a cigarette (Wesson and Washburn 1990). It has also been mixed with various forms of cocaine for combined use. Pyrolysis products of phencyclidine have been recognized, but their effects have not been well characterized (Cook and Jeffcoat 1990; Martin and Boni 1990).

Phencyclidine's acute toxicity presents as a range of clinical signs and symptoms (Gorelick and Balster 1995) including disorientation, confusion, impaired memory and judgement, and labile and inappropriate affect without other evidence of psychosis. Signs of phencyclidine intoxication include: increased muscle tone, tremor, brisk deep tendon reflexes, nystagmus, ataxia, moderate elevation of heart rate and blood pressure, diaphoresis, lacrimation, and increased bronchial and salivary secretions. Seizures, hyperthermia, intracranial hemorrhage, apnea and acute rhabdomyolysis have also been associated with its use. In patients presenting with light or deep coma, the symptoms of delirium may persist for several days during recovery. It is unknown exactly what percentage of phencyclidine users exhibit psychotic episodes attributable to phencyclidine use and whether these symptoms are observed only in certain users who have underlying predisposing conditions necessary for their expression.

Phencyclidine has an almost folkloric reputation for inducing violent behavior and "superhuman" strength. However, in a retrospective study of 81 clinical reports of phencyclidine toxicity in humans, BRECHER et al. (1988) found a lack of corroborating evidence to support this view: Criteria including the presence of phencyclidine (in the absence of other drugs) in bodily fluids or post mortem tissue as well as the violent behavior being reported by persons other than the user were met in only three of the hundreds of patients described in the reports. WISH (1986) also reported that arrestees with phencyclidine use confirmed by positive urine samples were less likely than others to have been involved in aggressive crimes. Thus, phencyclidine abuse may be no more likely to be associated with violence and aggression than abuse of alcohol and other drugs. Although there is no scientific evidence that phencyclidine increases muscular strength, it is likely that phencyclidine intoxication may render users insensitive to their own injuries or unmindful of their own or others' safety. Thereby, phencyclidine users may pose a formidable risk, and law enforcement personnel are understandably on guard against these potentially dangerous situations. However, it should not be assumed that most people who abuse phencyclidine will become violent nor should every inexplicable act of violence be attributed to phencyclidine abuse.

An unknown percentage of phencyclidine users progress to chronic use. Although tolerance and physical dependence do not appear to be prominent features of human phencyclidine abuse, abusers do report difficulty stopping phencyclidine use (GORELICK and WILKINS 1989). The specific motivations for continued phencyclidine use are unclear and likely to vary with the individual. Clinical reports of withdrawal effects after discontinuation of phencyclidine abuse exist (TENNANT et al. 1981), but a withdrawal syndrome is not a component of the DSM-IV diagnosis of phencyclidine use disorders. Little has been published on methods which could be used for phencyclidine detoxification and relapse prevention in chronic abusers (DAGHESTANI and SCHNOLL 1989).

Intrauterine exposure to phencyclidine by maternal use during pregnancy has been reported to result in fetal growth retardation, precipitation of labor, signs of fetal distress and abnormal neonatal neurological findings (CHASNOFF et al. 1986; AHMAD 1987; GOLDEN et al. 1987; TABOR et al. 1990). It is unclear, however, to what extent the general lifestyle associated with illicit phencyclidine use in the mother during pregnancy and beyond may contribute to effects observed in the neonate. The signs and symptoms described for phencyclidine-exposed neonates are different from the withdrawal symptoms typically described for narcotic-exposed infants, but more closely resemble those manifested in cocaine-exposed neonates (Rahbar et al. 1993). Despite the abnormal behaviors reported in the newborn period, mental and psychomotor development in phencyclidine-exposed babies appears to be no different than that of control infants, at least through 2 years of age (Chasnoff et al. 1986; Wachsman et al. 1989).

II. Animals

Phencyclidine administration in animals produces a variety of effects. Some of these are unique, others resemble those of classical stimulants such as amphetamines or classical depressants such as barbiturates. Some of the grossly observable effects resemble those seen in intoxicated humans. In nonhuman primates, for example, dose-related effects are observed to include salivation, lacrimation, nystagmus, ataxia and anesthesia with little respiratory depression.

The amphetamine-like stimulant effects of phencyclidine have been reviewed by Balster (1987). These effects, seen more prominently in rodents than primates, include sympathetic nervous system stimulation, increases in spontaneous motor activity, ipsilateral rotation in substantia nigra-lesioned rats, suppression of plasma prolactin levels and increases in low rates of scheduled-controlled operant behavior. Phencyclidine also enhances effects of amphetamine-like drugs. Stereotyped behavior is also produced, although it differs from that observed with amphetamine and in rodents includes a characteristic head-weaving (Sturgeon et al. 1979; Koek et al. 1987). Like dopaminergic agonists, phencyclidine disrupts prepulse inhibition of the startle reflex in rodents (Mansbach and Geyer 1989), a model of sensorimotor gating deficits often associated with schizophrenia (Judd et al. 1992). The ability of dopaminergic agonists to disrupt prepulse inhibition is consistent with theories implicating alterations in dopaminergic neurotransmission in this disorder, though dopaminergic mechanisms may not be responsible for these effects of phencyclidine (Keith et al. 1991).

Phencyclidine's depressant effects, notably motor incoordination, muscle relaxation and anticonvulsant effects (Hayes and Balster 1985; Leander et al. 1988; Church and Lodge 1990), resemble those produced by barbiturates, benzodiazepines and ethanol (Balster and Wessinger 1983). Phencyclidine also enhances the toxic (Chait and Balster 1978a; Boren and Consroe 1981), anesthetic (Daniel 1989) and behavioral (Woolverton and Balster 1981; Thompson and Moerschbaecher 1982) effects of depressant drugs. Like other depressants, phencyclidine has activity in various animal models predictive of anxiolytic activity (Chait et al. 1981; Porter et al. 1989; McMillan et al. 1991). However, its effects are generally less robust than those of the benzodiazepines and are not apparent in all species (reviewed by Wiley and Balster 1992, 1993).

Antinociceptive effects of phencyclidine and ketamine, occurring through a nonopioid mechanism, are observed in a variety of animal tests (Smith et al. 1985; FRANCE et al. 1989). While these effects occur at subanesthetic doses, they are generally accompanied by clear behavioral effects, as appears to be the case in humans.

C. Cellular Mechanisms of Action

By far the most influential finding in phencyclidine research in the last decade or so was the discovery that phencyclidine affects NMDA receptor-mediated excitatory amino acid neurotransmission (reviewed by LODGE and JOHNSON 1990). Phencyclidine also affects a number of other neurotransmitter systems, including the dopaminergic, nicotinic and muscarinic cholinergic systems, and has somewhat more poorly understood interactions with noradrenergic and serotonergic neurotransmission (reviewed by JOHNSON 1987; JOHNSON and JONES 1990; JOHNSON et al. 1993). Phencyclidine also functions as an inhibitor of brain nitric oxide synthetase (OSAWA and DAVILA 1993) and blocks potassium channels (ALBUQUERQUE et al. 1981; KOKOZ et al. 1994). Thus, phencyclidine is a somewhat promiscuous compound and determining which of its neuropharmacological effects are responsible for which of its pharmacological, behavioral and psychological effects has been the subject of extensive investigation.

I. The Phencyclidine Receptor

The best characterized cellular effect of phencyclidine is its interaction with a specific binding site in neural tissues. This site was first identified in 1979 when two separate groups (ZUKIN and ZUKIN 1979; VINCENT et al. 1979), using radioligand binding techniques, identified a unique binding site for phencyclidine and other arylcycloalkylamines which satisfied many of the biochemical and pharmacological criteria for a physiologically relevant receptor. There are many reviews describing the pharmacological specificity and other characteristics of this phencyclidine receptor (JOHNSON 1987; ZUKIN and ZUKIN 1988). Actions at this site now are clearly known to be an important neural basis for the production of behavioral effects of phencyclidine-like drugs relevant to their abuse (BALSTER and WILLETTS 1988). This conclusion is primarily based on the excellent concordance between binding site affinity and the potency and efficacy of arylcycloalkylamines and other compounds described below for the production of phencyclidine-like behavioral effects in animal studies.

Much is known about the absolute molecular characteristics of arylcycloalkylamines which are necessary for affinity for the phencyclidine receptor (KAMENKA and CHICHEPORTICHE 1988; THURKAUF et al. 1988; JACOBSON et al. 1992). In addition, some of the most useful tools to explore the neuropharmacology of phencyclidine are compounds which structurally differ from the arylcyclohexylamines but bind to the phencyclidine receptor and share significant aspects of phencyclidine's neuropharmacological and behavioral effects. For example, the 1,3-substituted dioxolanes etoxadrol and dexoxadrol, which produce a nearly identical profile of effects to that of phencyclidine (Brady et al. 1982), have been very useful for studying the stereoselectivity of phencyclidine-like effects and for modeling the phencyclidine pharmacophore (Jacobson et al. 1987; Thurkauf et al. 1988). More recently, dizocilpine (MK-801), which was developed years ago as a possible anticonvulsant (Clineschmidt et al. 1982), has been shown to be a potent ligand for phencyclidine sites of action in the brain and is now used extensively as a selective probe for the phencyclidine receptor (Wong et al. 1986). Even more potent analogs of dizocilpine have now been synthesized (Gee et al. 1993), and photoaffinity ligands for the phencyclidine receptor have been prepared from this series (Linders et al. 1993).

II. Phencyclidine and the *Sigma* Binding Site

An historically important phencyclidine-like drug is *N*-allylnormetazocine (NANM), a compound which has been the basis for much confusion in the research literature. NANM, also known as SKF-10047, was selected by Martin et al. (1976) as the prototypic *sigma* opioid agonist based upon its unique profile of pharmacological effects in the dog. It now appears almost certain, given the similarities in the cellular and pharmacological actions of NANM and phencyclidine, that its *sigma*-like profile of effects is produced by NANM's actions on phencyclidine receptors, which were historically termed "phencyclidine/*sigma* opioid receptors." There is now little evidence that any subtype of opiate receptor plays a role in the phencyclidine-like effects of NANM. The dextrorotatory isomers of NANM and many other 6,7-benzomorphans such as cyclazocine lack classical opioid effects whereas the levorotatory isomers possess varying degrees of *mu* or *kappa* opioid agonist or antagonist activity. However, the levorotatory forms of the benzomorphans often differ relatively little from their dextrorotary counterparts in affinity for phencyclidine receptors. Thus, the expression of the in vivo pharmacological effects of the racemates of benzomorphans such as NANM and cyclazocine are complexly determined by their ratio of phencyclidine-like and opioid effects (Kumor et al. 1986; Slifer and Balster 1988; Pfeiffer et al. 1986; Vaupel and Cone 1991).

Great confusion was introduced into the literature when radioligand binding studies revealed that [^{3}H]*d*-NANM also binds with high affinity to another site clearly different from that of the historical phencyclidine/*sigma* receptor (Martin et al. 1984; Sircar et al. 1986; Tam 1983; Zukin et al. 1984). While this site was termed a *sigma* receptor (Quirion et al. 1987), there is little evidence that it has anything to do with mediating the shared pharmacological profiles of phencyclidine and the benzomorphans discussed above (Balster and Willetts 1988). Indeed, other compounds with high affinity for the *sigma* site, including (+)-3-ppp (Largent et al. 1984), di-tolyl guanidine (WEBER et al. 1986) and certain putative novel antipsychotics (LARGENT et al. 1988) do not share phencyclidine's or NANM's profile of behavioral effects (BALSTER 1989; HOLTZMAN 1989). Thus, these high-affinity *sigma* sites are not responsible for the behavioral and psychological effects of phencyclidine. The physiological relevance of this site is currently unclear. Some evidence implicates the *sigma* site in motor behavior and in the modulation of phosphoinositide metabolism; other evidence suggests that it may be an intracellular protein such as an enzyme (WALKER et al. 1990). There is also evidence for heterogeneity of *sigma* sites (ITZHAK and STEIN 1990; WALKER et al. 1990; QUIRION et al. 1992). Investigators need to be very clear to distinguish phencyclidine receptor pharmacology from the pharmacology of the high-affinity *sigma* binding sites, the role of which in mediating psychotomimetic effects of any drug class remains unclear (MUSACCHIO 1990).

III. Phencyclidine and the NMDA Receptor Complex

Although the importance of the phencyclidine receptor to understanding the neural basis for phencyclidine abuse was very well understood by the early 1980s, the physiological function of this site became clearer following the discovery that phencyclidine antagonized activation of cat spinal neurons by NMDA (LODGE and ANIS 1982; ANIS et al. 1983). NMDA was being used at that time to selectively activate a subtype of excitatory amino acid receptor (WATKINS and EVANS 1981). Phencyclidine and NMDA receptors were found to have similar regional distributions in the central nervous system; they are intimately associated, but distinct, sites (JOHNSON and SNELL 1986; MARAGOS et al. 1986, 1988; MONAGHAN and COTMAN 1986). Since this time, a number of other sites have been discovered to be associated with the NMDA receptor (Fig. 1), which probably represents one of the most complex and unusual receptor systems discovered to date (WONG and KEMP 1991).

Phencyclidine is a noncompetitive NMDA antagonist (SNELL and JOHNSON 1985; HARRISON and SIMMONDS 1985) acting at the level of an ion channel (HONEY et al. 1985; SNELL et al. 1987; FAGG 1987; JAVITT et al. 1987; JOHNSON et al. 1993; KLOOG et al. 1988). NMDA receptor-associated ion channels display an unusual voltage sensitivity and blocking action by magnesium ions which is relieved by membrane depolarization, allowing influx of calcium into the cell (see review by ASCHER and NOWAK 1987). This use dependency of NMDA receptors provides a means by which they may participate in neuroplasticity and in the destabilization of membranes during convulsions; excessive calcium flux resulting from NMDA receptor activation during neural injury also appears to play an important role in excitotoxicity (OLNEY 1990).

Another important feature of the NMDA receptor is the large number of allosteric modulatory sites on the receptor complex. Besides the phencyclidine site in the channel, there is a well-characterized strychnine-insensitive

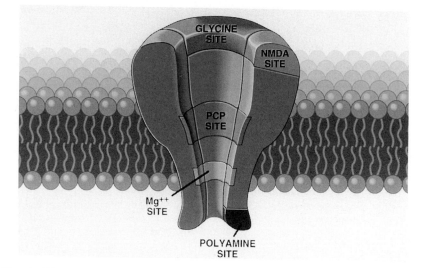

Fig. 1. Diagram of the *N*-methyl-D-aspartate (*NMDA*) receptor complex showing modulatory sites for phencyclidine (*PCP*), glycine, magnesium and polyamines

glycine co-agonist site (JOHNSON and ASCHER 1987; MONAGHAN et al. 1989). Glycine facilitates channel opening by agonists acting at the NMDA site (JOHNSON and ASCHER 1987; BONHAUS et al. 1987; REYNOLDS et al. 1987; WONG et al. 1987). Competitive antagonists at the glycine site can function as noncompetitive NMDA antagonists (PALFREYMAN and BARON 1991). There is also a polyamine regulatory site (RANSOM and STEC 1988; REYNOLDS and MILLER 1989) and an additional site at which divalent cations such as zinc may also regulate NMDA receptor function (REYNOLDS and MILLER 1988). The development of compounds with selective agonist or antagonist actions at each of these sites has provided a means of studying the pharmacology of the NMDA receptor (WATKINS et al. 1990; BIGGE 1993; JOHNSON et al. 1993).

The application of molecular biology techniques to the study of the NMDA receptor has provided much additional information on its structure and function (KUSHNER et al. 1993; HOLLMANN and HEINEMANN 1994). The pharmacology of NMDA receptors expressed in *Xenopus* oocytes following injection of crude extracts of exogenous mRNA from rat brain is similar to that observed in neuronal NMDA receptors (VERDOORN et al. 1987; KUSHNER et al. 1988; KLECKNER and DINGLEDINE 1988; LERMA et al. 1989, 1990). NMDA and phencyclidine receptors are co-expressed, though in certain cell lines (i.e., neuroblastoma-brain hybrid cell line NCB-20), the NMDA site is not functionally activated. Nonetheless, it is expressed and functionally activated when mRNA from these cell lines is injected into oocytes (LERMA et al. 1989). This indicates that, in the NCB-20 cells, the expression of functional NMDA receptors may be subject to regulation or be defective. It is unclear at this time whether such deficiency or regulatory mechanisms occur in vivo, although it is interesting to note that there is also biochemical evidence indicating that some phencyclidine binding sites are not associated with NMDA receptors (WOOD and RAO 1989; RAO et al. 1990a; SUN and LARSON 1993).

The recent cloning of the NMDA receptor will contribute substantially to our knowledge of the functioning of this receptor system (MORIYOSHI et al. 1991). Studies have shown that, like other ligand-gated ion channel receptors, the NMDA receptor is composed of several subunits (NAKANISHI 1992; HOLLMANN and HEINEMANN 1994). At present, two major families of these subunit proteins have been identified, referred to in the rat as the NMDAR1 and NMDAR2 subunits. The NMDAR1 subunit is found in eight different isoforms generated by alternative RNA splicing, and four distinct subtypes of NMDAR2 subunits encoded by separate genes have been identified (MONYER et al. 1992; SUGIHARA et al. 1992). Expression studies suggest that an NMDAR1 subunit is essential for the function of the NMDA receptor (NAKANISHI 1992; HOLLMANN and HEINEMANN 1994) and it can combine with different subtypes of NMDAR2 subunits to form different heteromeric configurations with different functional properties (MONYER et al. 1992; ISHII et al. 1993). As of this writing, genes for several of these subunits have now been mapped in the rat (KURAMOTO et al. 1994) and the genomic location of the NMDAR1 subunit is known for humans (KARP et al. 1993). Rapid progress can be expected in further characterization of the molecular biology of the NMDA receptor complex leading to new opportunities for functional studies. For example, it has recently been shown that treatment of rats with an antisense oligodeoxynucleotide to the complementary DNA coding for the NMDAR1 subunit reduced neurotoxicity elicited by NMDA or by focal ischemic infarction (WAHLESTEDT et al. 1993). This adds support to the already substantial evidence for the involvement of NMDA receptors in neurotoxicity. There is also evidence emerging for NMDA receptor heterogeneity based on different combinations of subunits expressed in different brain regions (SUGIHARA et al. 1992). Drugs with selectivity for specific subunit combinations have been identified (WILLIAMS 1993).

IV. Behavioral Correlates of Phencyclidine/NMDA Interactions

It is now apparent that the previously identified phencyclidine receptor exists in the channel of the NMDA ionophore since there is an excellent correlation between the potencies of arylcyclohexylamines and other phencyclidine-like drugs as NMDA antagonists and their activities at the phencyclidine receptor site (LODGE and JOHNSON 1990). By implication, this also suggests that NMDA antagonism probably plays an important role in producing behavioral and psychological effects of phencyclidine relevant to its abuse (WILLETTS et al. 1990). Consistent with a role for NMDA receptors in epilepsy and neuronal injury, phencyclidine and other NMDA antagonists are anticonvulsant in a variety of models and in a number of species (ROGAWSKI and PORTER 1990); neuroprotective effects of phencyclidine and other noncompetitive NMDA antagonists have also been shown in vitro and in vivo (BOAST 1988; FOSTER et al. 1988; OLNEY 1990).

The already-mentioned effects of phencyclidine in animal models of anxiety are mimicked by other NMDA antagonists such as ketamine and dizocilpine. Increases in punished behaviors are observed in rats and pigeons, although as with phencyclidine, the magnitude and range of effective doses is typically less than is seen with benzodiazepines under comparable conditions (e.g., BRANDÃO et al. 1980; CHAIT et al. 1981; PORTER et al. 1989; McMILLAN et al. 1991; SANGER and JOLY 1991) and antipunishment effects are typically not produced in squirrel monkeys (CLINESCHMIDT et al. 1982; MANSBACH et al. 1991). However, dizocilpine is active in nonpunishment procedures such as the elevated plus-maze (CORBETT and DUNN 1991; SHARMA and KULKARNI 1991), separation-induced vocalization (WINSLOW et al. 1990) and antipredator defensiveness (BLANCHARD et al. 1992). Other types of NMDA antagonists also are generally active in rodents (KOEK and COLPAERT 1991; WILEY et al. 1992; WILLETTS et al. 1993; FAIMAN et al. 1994), suggesting a possible role for the NMDA receptor in anxiety disorders and the possibility that NMDA antagonists may be developed as anxiolytics (WILEY and BALSTER 1992, 1993; BARRETT and WILLETTS 1994).

Based upon potency relationships among phencyclidine-like drugs, it would appear that NMDA antagonism also serves as the basis for their antinociceptive effects. This is consistent with the increasing evidence for a role for NMDA receptors in pain perception (SKILLING et al. 1988; DICKENSON and AYDAR 1991; KLEPSTAD et al. 1990; VACCARINO et al. 1993). The generally poor separation of antinociceptive from direct behavioral effects of phencyclidine-like compounds, however, suggests they may not be clinically useful analgesics (FRANCE et al. 1989).

Although we have been stressing the similarities between phencyclidine-like NMDA antagonists, it should also be pointed out that in some studies, differences in the behavioral effects of dizocilpine and phencyclidine or ketamine are apparent (MARQUIS et al. 1989a; BEARDSLEY et al. 1990; LEHMANNMASTER and GEYER 1991; MAREN et al. 1991; FRANCE et al. 1991). The relevance of phencyclidine's interactions with other neurotransmitter systems or cellular fuctions in this respect are unclear. Nevertheless, it is likely that subtle differences in selectivity of action for various types of ligands having affinity for the the phencyclidine receptor, such as phencyclidine and dizocilpine, may underlie subtle differences in their behavioral or psychological effects. Additionally, different heteromeric NMDA receptor channels expressed in *Xenopus* oocytes display differing sensitivities to dizocilpine, although there is little variability in the sensitivities of the differing channels to phencyclidine, ketamine or NANM (YAMAKURA et al.

1993). Again, such differing sensitivities, if indeed they translate to mammalian systems, may give rise to differences in the behavioral effects of phencyclidine and other NMDA antagonists which have affinity for the channel site. Another very plausible, if less provocative, explanation for differences in behavioral effects of these compounds may lie in their differing pharmacokinetic properties (WOODS and WINGER 1991).

V. Phencyclidine and Dopaminergic Effects – Direct or Indirect Interactions?

There is a little doubt that phencyclidine can also serve as an indirect-acting dopaminergic agonist in many model systems, particularly in rodent species (JOHNSON and JONES 1990). What is in doubt is whether this is primarily an effect of phencyclidine acting on sites within dopaminergic synapses. Phencyclidine itself has little affinity for postsynaptic dopamine receptors although it does have some affinity for the dopamine transporter (AKUNNE et al. 1991; KUHAR et al. 1990; ROTHMAN et al. 1989; reviewed by JOHNSON and JONES 1990). It is not a potent dopamine releasing agent. It is also possible that phencyclidine's dopaminergic effects may result as an indirect consequence of its activation of presynaptic dopaminergic neurons. This may be through its primary actions at the NMDA receptor complex, but there is also evidence that phencyclidine's dopaminergic effects may arise from an action on voltage-dependent potassium channels (MOUNT et al. 1990).

An important line of evidence suggesting that phencyclidine's direct actions at dopamine synapses may not be responsible for its amphetamine-like behavioral effects comes from studies of dizocilpine (SNELL et al. 1988). This compound produces most, if not all, of the amphetamine-like effects of phencyclidine (CLINESCHMIDT et al. 1982; DIMPFEL and SPULER 1990; TRICKLEBANK et al. 1989), and, like phencyclidine, increases extracellular dopamine levels in the prefrontal cortex after local injection (HONDO et al. 1994), yet it lacks meaningful affinity for the dopamine transporter (SNELL et al. 1988; RAO et al. 1990b). Its potency as an amphetamine-like stimulant is much better predicted by its high affinity for the phencyclidine site on the NMDA receptor (LILJEQUIST et al. 1991). In addition, the affinity of phencyclidine for the dopamine transporter is considerably less than its affinity for the site on the NMDA receptor. This suggests that the dopaminergic effects of phencyclidine and other phencyclidine-like drugs, seen in vivo and in many functional in vitro assays, may be secondary to NMDA antagonism. Indeed, there are many studies showing functional interrelationships between glutamatergic and dopaminergic systems in brain (OVERTON and CLARK 1992; ZHANG et al. 1992; JOHNSON et al. 1993).

The cellular basis for the dopaminergic actions of phencyclidine is important to understand because of its relevance to the possible relationship between phencyclidine and schizophrenia. Consequently, this question is a matter of considerable current research interest (Zhang et al. 1992; French et al. 1993; Bristow et al. 1993; Baunez et al. 1994; Hondo et al. 1994). There is also increasing evidence that interactions between glutamatergic and dopaminergic systems may play a role in the neural basis of drug reinforcement and that glutamate antagonists may be able to attenuate effects of drugs of abuse related to their reinforcing properties (Bespalov et al. 1994; Pulvirenti et al. 1994). One of the interesting phencyclidine analogs to be the subject of recent scientific investigation is *N*-[1-(2-benzo(b)thiophenyl)] cyclohexylpiperidine (BTCP). BTCP has little activity at phencyclidine sites but is a potent catecholamine uptake inhibitor with a cocaine-like profile of behavioral effects (Vignon et al. 1988; Koek et al. 1989). This compound is proving to be useful in characterizing some of the dopaminergic aspects of phencyclidine's neuropharmacology independent of its actions as an NMDA antagonist.

D. Behavioral Pharmacology of Phencyclidine Abuse

I. Discriminative Stimulus Effects

Phencyclidine's unique effects are probably best revealed by studies of its discriminative stimulus effects (Balster 1991). These effects have been well-characterized in a number of species and much is known about the neuro-pharmacological basis for phencyclidine discrimination. Animals can be readily trained to discriminate phencyclidine, or another phencyclidine-like compound such as ketamine, from saline. To briefly summarize over a decade of studies on phencyclidine discrimination, it has been found that the discriminative stimulus effects of phencyclidine are shared fully only by compounds having affinity for the phencyclidine receptor. Furthermore, the relative potencies of such compounds in producing phencyclidine-like effects agree with their relative potencies as phencyclidine-site selective NMDA antagonists (see reviews by Balster 1987, 1991; Balster and Willetts 1988; Willetts et al. 1990; Woods et al. 1991). The exception to this may be what are referred to as low-affinity channel blockers, which may have less ability to produce direct phencyclidine-like effects on behavior while still effective when channels are opened in pathological conditions (Rogawski et al. 1991). This possibility needs additional study, however, since at least one low-affinity blocker, memantine, does have phencyclidine-like discriminative stimulus effects in rats (Sanger et al. 1992). Compounds acting through other neurotransmitter systems, including other stimulants, depressants and hallucinogens, neither substitute for phencyclidine nor antagonize phencyclidine discrimination (Browne 1986; Poling et al. 1979; Shannon 1982; Beardsley and Balster 1988), indicating that systems independent of NMDA receptors are not involved or not essential in mediating the discriminative stimulus effects of phencyclidine.

Interestingly, NMDA antagonists which do not have affinity for the phencyclidine site typically do not fully substitute for phencyclidine or ketamine (reviewed by WILLETTS et al. 1990). Similarly, phencyclidine-like drugs produce at best only partial substitution in animals trained to discriminate other types of NMDA antagonists (WILLETTS et al. 1989; GOLD and BALSTER 1993; SINGH et al. 1990; WILEY and BALSTER 1994). It has been suggested that biodispositional differences alone may account for such differences between these different classes of NMDA antagonists (WOODS and WINGER 1991). However, it is also likely that these differences represent truly dissimilar effects irrespective of differing biodisposition. For example, dizocilpine has a slower onset of action than phencyclidine but its discriminative stimulus effects are phencyclidine-like (BEARDSLEY et al. 1990; KOEK et al. 1988; WILLETTS and BALSTER 1988).

There are no such direct comparisons of the effects of phencyclidine and other NMDA antagonists in humans. Thus, inferences about phencyclidine-like effects of other NMDA antagonists which have been given to humans are difficult to verify without direct comparisons under carefully controlled laboratory conditions. Nevertheless, dizocilpine, ketamine, NANM and other NMDA channel blockers such as dexoxadrol and dextrorphan that have been tested in humans produce disturbing psychological effects (SIEGEL 1978; LASAGNA and PEARSON 1965; KEATS and TELFORD 1964; ISBELL and FRASER 1953; KRYSTAL et al. 1994; TROUPIN et al. 1986). In addition, a number of arylcyclohexylamine analogs of phencyclidine such as TCP and PCE, though not directly studied in humans, have appeared as "designer drugs" of abuse. Taken together, it appears that data support the hypothesis that the ability of drugs to produce phencyclidine-like discriminative stimulus effects is predictive of their potential to produce disturbing subjective effects and abuse liability in humans. Although there is some evidence from limited clinical testing of competitive NMDA antagonists that they can produce some disturbing psychological effects, it is unclear at this time if these effects are truly phencyclidine-like (SVEINBJORNSDOTTIR et al. 1993).

II. Self-Administration

Phencyclidine is self-administered by all species in which it has been studied (e.g., BALSTER et al. 1973; CARROLL and MEISCH 1980; LUKAS et al. 1984; CARROLL 1987; RISNER 1982). Responding is also maintained by other phencyclidine-site selective NMDA antagonists, including ketamine and other arylcycloalkylamines, NANM, dizocilpine, dexoxadrol and etoxadrol (BRADY et al. 1982; SLIFER and BALSTER 1983; LUKAS et al. 1984; KOEK et al. 1988; BEARDSLEY et al. 1990). The conditions for self-administration may be more important for some compounds than for others; for example, dizocilpine is not self-administered by monkeys with recent histories of cocaine self-administration (BEARDSLEY et al. 1990) but is self-administered by monkeys with recent histories of phencylidine or ketamine self-administration (KOEK et al. 1988; Beardsley et al. 1990). There are few data on the self-administration of other types of NMDA antagonists, but those studies that have been described have indicated that competitive antagonists are not self-administered (Winger et al. 1991; Willetts et al. 1992).

As with a number of other classes of drug reinforcers, the self-administration of phencyclidine-like drugs by animals can be modified by changing food deprivation conditions (Carroll and Meisch 1984), the response requirement for injection (Marquis et al. 1989b) and the availability of alternative reinforcers (Carroll 1985). The sensitivity of phencyclidine self-administration to testing conditions suggests that animal studies can be used to explore the complex interactions between pharmacology and context that underlie individual differences in vulnerability to phencyclidine abuse (Carroll 1993).

III. Tolerance and Dependence

Tolerance to the behavioral effects of phencyclidine in animals has been shown in a number of studies (Balster and Woolverton 1981). When effects on learned operant behavior are examined, only two- to fourfold shifts in dose-effect curves are seen when moderate, behaviorally active doses are given repeatedly (e.g., Chait and Balster 1978b; Brocco et al. 1983; Wenger 1983). This magnitude of tolerance is less than that generally seen with other classes of drugs of abuse such as the opioids, cannabinoids and sympathomimetic stimulants. When unlearned behaviors such as feeding or catalepsy are measured, somewhat larger degrees of tolerance can be seen when maximally tolerated doses of phencyclidine are given (Balster and Woolverton 1980; Lu et al. 1992). When behavioral effects of phencyclidine that are related to dopaminergic activation such as motor activity increases and stereotypy are studied, sensitization can occur with repeated dosing (Greenberg and Segal 1986; Xu and Domino 1994). Cross-tolerance has been seen among phencyclidine-like drugs (Wenger 1983) and between phencyclidine and other types of NMDA antagonists (Lu et al. 1992). Evidence suggests that most tolerance is pharmacodynamic rather than biodispositional and that learning factors play an important role in the development and persistence of tolerance (Woolverton and Balster 1979) and its situational specificity (Smith 1991).

There is evidence from animal studies that phencyclidine can produce physical dependence (Spain and Klingman 1985). A grossly obsrvable withdrawal syndrome including tremors, oculomotor hyperactivity, bruxism, fearfulness, vocalizations, diarrhea and, in some animals, emesis and convulsions was seen in rhesus monkeys whose self-administration of very large daily doses of phencyclidine was discontinued (Balster and Woolverton 1980). While such an easily observable syndrome is not observed at lower doses of phencyclidine, subtle changes in learned behavior are seen following cessation of repeated phencyclidine administration (Slifer et al. 1984;

CARROLL 1987; BEARDSLEY and BALSTER 1987; WESSINGER 1987; MASSEY and WESSINGER 1990b; CARROLL and CARMONA 1991; WESSINGER and Owens 1991). While the neural basis of this dependence is unknown, some evidence points toward cross-dependence between phencyclidine and other phencyclidine-like NMDA antagonists such as ketamine and d-NANM. However, consistent changes in phencyclidine/NMDA receptor regulation with repeated dosing have not been observed (MANALLACK et al. 1989; MASSEY and WESSINGER 1990a; WEISSMAN et al. 1991). Thus, the extent to which NMDA receptors are involved in this phenomenon requires further investigation. Nonetheless, more systematic study of the factors which might lead to dependence in human abusers and the role of dependence in continuation of phencyclidine use appears warranted.

IV. Modification of Drug Tolerance and Dependence

Although repeated phencyclidine administration can produce tolerance and dependence, it is interesting that when phencyclidine or other phencyclidine-like NMDA antagonists are given repeatedly in combination with other drugs of abuse they can block or reduce changes in sensitivity and/or dependence development to these drugs. Most of this research has focused on the modification of opioid tolerance and dependence (TRUJILLO and AKIL 1991, 1994; MAREK et al. 1991; BEN-ELIYAHU et al. 1992; LUTFY et al. 1993), but prevention of tolerance to effects of ethanol and other depressant drugs has also been reported (KHANNA et al. 1991; FILE and FERNANDES 1994). In addition, sensitization to the effects of amphetamines, cocaine and nicotine can be blocked by phencyclidine-like drugs (KARLER et al. 1989; STEWART and DRUHAN 1993; SHOAIB and STOLERMAN 1992). Although most studies have used the phencyclidine-site selective NMDA antagonist dizocilpine, there is also evidence that competitive NMDA antagonists may also be able to block tolerance development (TISEO and INTURRISI 1993). Although the neural basis by which NMDA antagonists interfere with tolerance and dependence development is not known, it is possible that their interference with NMDA receptor-mediated neuroadaptive processes is involved. The ability of NMDA antagonists to interfere with learning and memory, as discussed below, is consistent with this hypothesis.

V. Learning and Memory

The membrane physiology of NMDA receptor ion channels suggests that they may play an important role in integration of cell firing and neuronal plasticity. In particular, the voltage-dependent magnesium block of the channel means that the functionality of NMDA receptor activation is activity-dependent. Consistent with this is the evidence that NMDA receptors are critically involved in long-term potentiation in the hippocampus, an extensively studied form of activity-dependent synaptic plasticity (Collingridge and Singer 1990). It follows from this that NMDA antagonists could be expected to disrupt learning and memory functions just as they do long-term potentiation; the amnestic effects of phencyclidine in humans supports this contention.

The effects of phencyclidine and other NMDA antagonists on learning and memory have been extensively studied in a wide variety of animal models (Morris et al. 1986; Ward et al. 1990; McLamb et al. 1990; Itoh et al. 1991; Woods et al. 1991). However, it is often difficult to separate true effects on learning and memory from performance disruption caused by the drug. In the repeated acquisition procedure, whereby the acquisition, retention and performance of a task by an animal can be independently measured, phencyclidine and dizocilpine consistently disrupt retention and acquisition at doses lower than those that disrupt performance; less consistent effects are observed with competitive NMDA antagonists (Moerschbaecher 1992). It is not known if the learning and memory impairment produced by NMDA antagonists in animal studies has implications for neuropsychological sequelae in phencyclidine abusers, but these effects are interesting from the standpoint of the basic neurobiology of neuronal and behavioral adaptation.

E. Concluding Remarks

As NMDA antagonists begin to be considered for various therapeutic uses, an important question to be addressed is whether they all will produce phencyclidine-like behavioral/psychological side effects and have phencyc-lidine-like abuse liability. As discussed above, based on results of animal testing and the limited data from human testing, it appears that NMDA antagonists that act at the phencyclidine site in the channel are very likely to produce phencyclidine-like side effects in humans. It is important from a drug development perspective, however, that such effects be considered in the context of the drug's therapeutic use and in relation to the therapeutic index for producing beneficial effects vs side effects. For some indications, such as neuroprotection after stroke or trauma, phencyclidine-like side effects of NMDA antagonists may be of little concern. It should be remembered that ketamine has been used clinically for over 30 years.

In addition to the phencyclidine site, there are a number of other modu-latory sites on the NMDA receptor where drugs can exert NMDA antagonist effects. Evidence is emerging that different classes of site-selective NMDA antagonists produce different profiles of neurochemical and behavioral effects (French et al. 1992, 1993; Willetts et al. 1990; Woods et al. 1991; Ginski and Witkin 1994). In addition, competitive NMDA antagonists have a better therapeutic index for separation of clinically useful effects from direct effects on behavior (Willetts et al. 1990; De Sarro and De Sarro 1993). However, there is evidence that, under pathological conditions of NMDA receptor activation such as that produced by kindling, fewer differences are seen in the behavioral effects of competitive antagonists and phencyclidine-like noncompetitive antagonists than in normal animals (Löscher and Honack 1991). This implies that in patients the side effect profile of competitive antagonists may be less favorable than predicted from studies in normals. In any case, it is clear that important differences can be seen in the behavioral effects of phencyclidine-like and competitive NMDA antagonists. These differences may be important in defining the therapeutic efficacy and side effect profiles of these two classes of NMDA antagonists.

Results with glycine site and polyamine site NMDA antagonists are even more promising. Although systemically effective drugs for these sites are only now becoming available for study, results so far show that they may produce behavioral effects that are even more dissimilar from those of phencyclidine than are obtained with competitive antagonists (Singh et al. 1990; Koek and Colpaert 1992). The glycine site low-affinity partial agonist (+)-HA-966, which functions as an NMDA antagonist under many circumstances, has also been shown to lack the dopaminergic effects seen with phencyclidine-like antagonists and can even block dopaminergic activation in mesolimbic areas (Bristow et al. 1993). Polyamine site and/or subtype-selective NMDA antagonists, such as ifenprodil and eliprodil, also appear to be completely devoid of phencyclidine-like behavioral effects in animals (Jackson and Sanger 1988; Balster et al. 1994). It may be that the therapeutic effects and phencyclidine-like effects of NMDA antagonists can be successfully separated. It should not be surprising that behavioral effects of different classes of site-selective NMDA antagonists can be shown to differ when one considers the differences in behavioral effects seen with GABA agonists acting at different sites on the GABA receptor complex. By understanding the physiological consequences of interaction of compounds acting at various sites on the NMDA receptor, we may better understand the contribution to this site may make to the unique effects of phencyclidine in humans.

Acknowledgements. Preparation of this review was supported in part by the National Institute on Drug Abuse, grant DA-01442.

References

Ahmad G (1987) Abuse of phencyclidine (PCP) A laboratory experience. J Toxicol Clin Toxicol 25:341–346

Akunne HC, Reid AA, Thurkauf A, Jacobson AE, De Costa BR, Rice KC, Heyes MP, Rothman RB (1991) [^{3}H]-[2-(2-Thienyl)cyclohexyl] piperidine labels two high-affinity binding sites in human cortex: further evidence for phencyclidine binding sites associated with the biogenic amine reuptake complex. Synapse 8:289–300

Albuquerque EX, Aguayo LG, Warnick JE, Weinstein H, Glick SD, Maayani S, Ickowicz RK, Blaustein MP (1981) The behavioral effects of phencyclidines may be due to their blockade of potassium channels. Proc Natl Acad Sci U S A 78:7792–7796

Anis NA, Berry SC, Burton NR, Lodge D (1983) The dissociative anaesthetics, ketamine and phencyclidine, selectively reduce excitation of central mammalian neurones by N-methyl-aspartate. Br J Pharmacol 79:565–575

Ascher P, Nowak L (1987) Electrophysiological studies of NMDA receptors. Trends Neurosci 10:284–288

Balster RL (1987) The behavioral pharmacology of phencyclidine. In: Meltzer HY (ed) Psychopharmacology: the third generation of progress. Raven, New York, pp 1573–1579

Balster RL (1989) Substitution and antagonism in rats trained to discriminate (+)-N-allylnormetazocine from saline. J Pharmacol Exp Ther 249:794–796

Balster RL (1991) Discriminative stimulus properties of phencyclidine and other NMDA antagonists. In: Glennon RA, Järbe TUC, Frankenheim J (eds) Drug discrimination: applications to drug abuse research. National Institute on Drug Abuse Research Monograph Series 116. DHHS publication no (ADM) 92-1878. US Government Printing Office, Washington, pp 163–180

Balster RL, Wessinger WD (1983) Central nervous system depressant effects of phencyclidine. In: Kamenka J-M, Domino EF, Geneste P (eds) Phencyclidine and related arylcyclohexylamines: present and future applications. NPP, Ann Arbor, pp 291–309

Balster RL, Willetts J (1988) Receptor mediation of the discriminative stimulus properties of phencyclidine and *sigma*-opioid agonists. In: Colpaert FC, Balster RL (eds) Transduction mechanisms of drug stimuli. Springer, Berlin Heidelberg New York, pp 122–135

Balster RL, Woolverton WL (1980) Continuous access phencyclidine self-administration by rhesus monkeys leading to physical dependence. Psychopharmacology 70:5–10

Balster RL, Woolverton WL (1981) Tolerance and dependence to phencyclidine. In: Domino EF (ed) PCP (Phencyclidine): historical and current perspectives. NPP, Ann Arbor, pp 293–306

Balster RL, Johanson CE, Harris RT, Schuster CR (1973) Phencyclidine self-administration in the rhesus monkey. Pharmacol Biochem Behav 1:167–172

Balster RL, Nicholson KL, Sanger DJ (1994) Evaluation of the reinforcing effects of eliprodil in rhesus monkeys and its discriminative stimulus effects in rats. Drug Alcohol Depend 35:211–216

Barrett JE, Willetts J (1994) Conflict procedures as pharmacological screens for anxiolytic drugs. In: Paloma T, Archer T (eds) Strategies for studying brain disorders, vol 1: depressive, anxiety and drug abuse disorders. Farrand, London, pp 179–196

Baunez C, Nieoullon A, Almaric M (1994) N-methyl-D-aspartate receptor blockade impairs behavioural performance of rats in a reaction time task: new evidence for glutamatergic-dopaminergic interactions in the striatum. Neuroscience 61: 521–531

Beardsley PM, Balster RL (1987) Behavioral dependence upon phencyclidine and ketamine in the rat. J Pharmacol Exp Ther 242:203–211

Beardsley PM, Balster RL (1988) Evaluation of antagonists of the discriminative stimulus and response rate effects of phencyclidine. J Pharmacol Exp Ther 244:34–40

Beardsley PM, Hayes BA, Balster RL (1990) The self-administration of MK-801 can depend upon drug-reinforcement history and its discriminative stimulus properties are phencyclidine-like in rhesus monkeys. J Pharmacol Exp Ther 252:953–959

Ben-Eliyahu S, Marek P, Vaccarino AL, Mogil JS, Sternberg WF, Liebeskind JC (1992) The NMDA receptor antagonist MK-801 prevents long-lasting non-associative morphine tolerance in the rat. Brain Res 575:304–308

Bespalov A, Dumpis M, Piotrovsky L, Zvartau E (1994) Excitatory amino acid receptor antagonist kynurenic acid attenuates rewarding potential of morphine. Eur J Pharmacol 264:233–239

Bigge CF (1993) Structural requirements for the development of potent N-methyl-D-aspartic acid (NMDA) receptor antagonists. Biochem Pharmacol 45:1547–1561

Blanchard DC, Blanchard RJ, Corobrez ADP, Veniegas R, Rodgers RJ, Shepherd JK (1992) MK-801 produces a reduction in anxiety-related antipredator aggressiveness in male and female rats and a gender-dependent increase in locomotor behavior. Psychopharmacology 108:352–362

Boast CA (1988) Neuroprotection after brain ischemia: role of competitive N-methyl-D-aspartate antagonists. In: Cavalheiro EA, Lehmann J, Turski L (eds) Frontiers in excitatory amino acid research. Liss, New York, pp 691–698

Bonhaus DW, Burge BC, McNamara JO (1987) Biochemical evidence that glycine allosterically regulates an NMDA receptor-coupled ion channel. Eur J Pharmacol 142:489–490

Boren JL, Consroe P (1981) Phencyclidine (PCP) and ethanol: potentiation of lethality and sleep time with combined administration in rats. Neurobehav Toxicol Teratol 3:11–14

Brady KT, Woolverton WL, Balster RL (1982) Discriminative stimulus and reinforcing properties of etoxadrol and dexoxadrol in monkeys. J Pharmacol Exp Ther 220:56–62

Brandão ML, Fontes JCS, Graeff FG (1980) Facilitatory effect of ketamine on punished behavior. Pharmacol Biochem Behav 13:1–4

Brecher M, Wang BW, Wong H, Morgan JP (1988) Phencyclidine and violence: clinical and legal issues. J Clin Psychopharmacol 8:397–401

Bristow LJ, Hutson PH, Thorn HL, Tricklebank MD (1993) The glycine/NMDA receptor antagonist, R-(+)-HA-966, blocks activation of the mesolimbic dopaminergic system induced by phencyclidine and dizocilpine (MK-801) in rodents. Br J Pharmacol, 108:1156–1163

Brocco MJ, Rastogi SK, McMillan DE (1983) Effects of chronic phencyclidine administration on the schedule-controlled behavior in rats. J Pharmacol Exp Ther 226:449–454

Browne RG (1986) Discriminative stimulus properties of PCP mimetics. In: Clouet DH (ed) Phencyclidine: an update. Natinal Institute on Drug Abuse Research Monograph 64 (DHHS publication no ADM 86-1443). US Government Printing Office, Washington DC, pp 134–147

Carroll ME (1985) Concurrent phencyclidine and saccharin access: presentation of an alternative reinforcer reduces drug intake. J Exp Anal Behav 4:131–144

Carroll ME (1987) A quantitative assessment of phencyclidine dependence produced by oral self-administration in rhesus monkeys. J Pharmacol Exp Ther 242:405–412

Carroll ME (1993) The economic context of drug and non-drug reinforcers affects acquisition and maintenance of drug-reinforced behavior and withdrawal effects. Drug Alcohol Depend 33:201–210

Carroll ME, Carmona G (1991) Effects of food FR and food deprivation on disruptions in food-maintained performance of monkeys during phencyclidine withdrawal. Psychopharmacology 104:143–149

Carroll ME, Meisch R (1980) Oral phencyclidine (PCP) self-administration in rhesus monkeys: effects of feeding conditins. J Pharmacol Exp Ther 214:339–346

Carroll ME, Meisch RA (1984) Increased drug-reinforced behavior due to food deprivation. In: Thompson T, Dews PB, Barrett JE (eds) Advances in behavioral pharmacology, vol 4. Academic, New York, pp 44–88

Chait LD, Balster RL (1978a) Interaction between phencyclidine and pentobarbital in several species of laboratory animals. Commun Psychopharmacol 2:351–356

Chait LD, Balster RL (1978b) The effects of acute and chronic phencyclidine on schedule-controlled behavior in the squirrel monkey. J Pharmacol Exper Ther 204:77–87

Chait LD, Wenger GR, McMillan DE (1981) Effects of phencyclidine and ketamine on punished and unpunished responding by pigeons. Pharmacol Biochem Behav 15:145–148

Chasnoff IJ, Burns KA, Burns WJ, Schnoll SH (1986) Prenatal drug exposure: effects on neonatal and infant growth and development. Neurobehav Toxicol Teratol 8:357–362

Church J, Lodge D (1990) Anticonvulsant actions of phencyclidine receptor ligands: correlation with N-methyl-D-aspartate antagonism in vivo. Gen Pharmacol 21: 165–170

Clineschmidt BV, Williams M, Witoslawski JJ, Bunting PR, Risley EA, Totaro JA (1982) Restoration of shock-suppressed behavior by treatment with (+)-5-methyl-10-11-dihydro-5H-dibenzo[a,d] cyclohepten-5,10-imine (MK-801), a substance with potent anticonvulsant, central sympathomimetic, and apparent anxiolytic properties. Drug Dev Res 2:147–163

Collingridge GL, Singer W (1990) Excitatory amino acids and synaptic plasticity. Trends Pharmacol Sci 8:290–296

Cook CE, Jeffcoat AR (1990) Pyrolytic degradation of heroin, phencyclidine and cocaine: identification of products and some observations on their metabolism. In: Chang CN, Hawks RL (eds) Research findings on smoking of abused substances. National Institute on Drug Abuse Research Monograph 99. DHHS publication no (ADM)90-1690. US Government Printing Office, Washington DC, pp 97–120

Corbett R, Dunn RW (1991) Effects of HA-966 on conflict, social interaction, and plus maze behavior. Drug Dev Res 24:201–205

Daghestani AN, Schnoll SH (1989) Phencyclidine abuse and dependence. In: Treatments of psychiatric disorders, vol 2. A task force report of the American Psychiatric Association. Washington DC: American Psychiatric Association, pp 1209–1218

Daniel LC (1989) The noncompetitive N-methyl-D-aspartate antagonists, MK-801, phencyclidine and ketamine, increase the potency of general anesthetics. Pharmacol Biochem Behav 36:111–115

De Sarro GB, De Sarro A (1993) Anticonvulsant properties of non-competitive antagonists of the N-methyl-D-aspartate receptor in genetically epilepsy-prone rats: comparison with CPPene. Neuropharmacology 32:51–68

Dickenson AH, Aydar E (1991) Antagonism at the glycine site on the MMDA receptor reduces spinal nociception in the rat. Neurosci Lett 121:263–266

Dimpfel W, Spuler M (1990) Dizocilpine (MK-801), ketamine and phencyclidine: low doses affect brain field potentials in the freely moving rat in the same way as activation of dopaminergic transmissin. Psychopharmacology 101:317–323

Domino EF (1964) Neurobiology of phencyclidine (Sernyl). Int Rev Neurobiol 6:303–347

Fagg GE (1987) Phencyclidine and related drugs bind to the activated N-methyl-D-aspartate receptor-channel complex in rat brain membranes. Neurosci Lett 76: 221–227

Faiman CP, Viu E, Skolnick P, Trullas R (1994) Differential effects of compounds that act at strychnine-insensitive glycine receptors in a punishment procedure. J Pharmacol Exp Ther 270:528–533

File SE, Fernandes C (1994) Dizocilpine prevents the development of tolerance to the sedative effects of diazepam in rats. Pharmacol Biochem Behav 47:823–826

Foster AC, Gill R, Woodruff GN, Iversen LI (1988) Non-competitive NMDA receptor antagonists and ischaemia-induced neuronal degeneration. In: Cavalheiro EA, Lehmann J, Turski L (eds) Frontiers in excitatory amino acid research. Liss, New York, pp 707–714

France CP, Snyder AM, Woods JH (1989) Analgesic effects of phencyclidine-like drugs in rhesus monkeys. J Pharmacol Exp Ther 250:197–201

France CP, Moerschbaecher JM, Woods JH (1991) MK-801 and related compounds in monkeys: discriminative stimulus effects and effects on a conditional discrimination. J Pharmacol Exp Ther 257:727–734

French ED, Lin JY, Simms D (1992) Characterization of possible mechanisms by which phencyclidine (PCP) and PCP-like drugs alter the activity of A_{10} dopamine neurons: electrophysiological and behavioral studies. In: Kamenka J-M, Domino EF (eds) Multiple sigma and PCP receptor ligands: mechanisms for neuromodulation and neuroprotection? NPP, Ann Arbor, pp 445–457

French ED, Mura A, Wang T (1993) MK-801, phencyclidine (PCP), and PCP-like drugs increase burst firing in rat A10 dopamine neurons: comparison to competitive NMDA antagonists. Synapse 13:108–116

Gee KR, Barmettler P, Rhodes MR, McBurney RN, Reddy NL, Hu LY, Cotter RE, Hamilton PN, Weber E, Keana JF (1993) 10,5-(Iminomethano)-10,11-dihydro-5H-dibenzo[a,d]cycloheptene and derivatives. Potent PCP receptor ligands. J Med Chem 36:1938–1946

Ginski MJ, Witkin JM (1994) Sensitive and rapid behavioral defferentatation of N-methyl-D-aspartate receptor antagonists. Psychopharmacology 114:573–582

Gold LH, Balster RL (1993) Effects of NMDA receptor antagonists in squirrel monkeys trained to discriminate the competitive NMDA receptor antagonist NPC 12626 from saline. Eur J Pharmacol 230:285–292

Golden NL, Kuhnert BR, Sokol RJ, Martier S, Williams T (1987) Neonatal manifestations of maternal phencyclidine exposure. J Perinat Med 15:185–191

Gorelick DA, Wilkins JN (1989) Inpatient treatment of PCP abusers and users. Am J Drug Alcohol Abuse 15:1–12

Gorelick DA, Balster RL (1995) Phencyclidine (PCP). In: Bloom FE, Kupfer DJ (eds) Psychopharmacology: the fourth generation of progress. Raven, New York, pp 1767–1776

Greifenstein FD, Yoshitake J, DeVauit M, Gajewska JE (1958) A study of 1-arylcyclohexylamines for anesthesia. Anesth Analg 37:283

Greenberg BD, Segal DS (1986) Evidence for multiple opiate receptor involvement in different phencyclidine-induced unconditioned behaviors in rats. Psychopharmacology 88:44–53

Harrison NL, Simmonds MA (1985) Quantitative studies on some antagonists of N-methyl-D-aspartate in slices of rat cerebral cortex. Br J Pharmacol 84:381–391

Hayes BA, Balster RL (1985) Anticonvulsant properties of phencyclidine-like drugs in mice. Eur J Pharmacol 117:121–125

Hollmann M, Heinemann S (1994) Cloned glutamate receptors. Annu Rev Neurosci 17:31–108

Holtzman SG (1989) Opioid and phencyclidine-like discriminative stimulus effects of ditolylguanidine, a selective sigma ligand. J Pharmacol Exp Ther 248:1054–1062

Hondo H, Yonezawa Y, Nakahara T, Nakamura K, Hirano M, Uchimura H, Tashiro N (1994) Effect of phencyclidine on dopamine release in the rat prefrontal cortex; an in vivo microdialysis study. Brain Res 633:337–342

Honey CR, Miljkovic Z, MacDonald JF (1985) Ketamine and phencyclidine cause a voltage-dependent block of responses to L-aspartic acid. Neurosci Lett 61:135–139

Isbell H, Fraser HF (1953) Actions and addiction liabilities of dromoran derivatives in man. J Pharmacol Exp Ther 107:524–530

Ishii T, Moriyoshi K, Sugihara H, Sakurada K, Kadotani H, Yokoi M, Akazawa C, Shigemoto R, Mizuno N, Masu M, Nakanishi S (1993) Molecular characterization of the family of the N-methyl-D-aspartate receptor subunits. J Biol Chem 268: 2836–2843

Itoh J, Nabeshima T, Kameyama T (1991) Utility of an elevated plus-maze for dissociation of amnesic and behavior effects of drugs in mice. Eur J Pharmacol 194:71–76

Itzhak Y, Stein I (1990) Sigma-binding sites in the brain: an emerging concept for multiple sites and their relevance for psychiatric disorders. Life Sci 47:1073–1081

Jackson A, Sanger DJ (1988) Is the discriminative stimulus produced by phencyclidine due to an interaction with N-methyl-D-aspartate receptors? Psychopharmacology 96:87–92

Jacobson AE, Harrison EA Jr, Mattson MV, Rafferty MF, Rice KC, Woods JH, Winger G, Soloman RE, Lessor RA, Silverton JV (1987) Enantiomeric and diastereomeric dioxadrols: behavioral, biochemical and chemical determination of the configuration necessary for phencyclidine-like properties. J Pharmacol Exp Ther 243:110–117

Jacobson AE, Linders JTM, Mattson MV, George C, Iorio MA (1992) The 1-(1-phenyl-(2-, 3-, 4-methylcyclohexyl)piperidines revisited: synthesis, stereochemistry, absolute configuration, computer assisted modeling and biological effects. In: Kamenka J-M, Domino EF (eds) Multiple sigma and PCP receptor ligands: mechanisms for neuromodulation and neuroprotection? NPP, Ann Arbor, pp 61–74

Javitt DC (1987) Negative schizophrenic symptomatology and the PCP (phencyclidine) model of schizophrenia. Hillside J Clin Psychiatry 9:12–35

Javitt DC, Jotkowitz A, Sircar R, Zukin SR (1987) Non-competitive regulation of phencyclidine/sigma receptors by the N-methyl-D-aspartate antagonist, d-(−)-2-amino-5-phosphonovaleric acid. Neurosci Lett 78:193–198

Jerrard DA (1990) "Designer drugs" – a current perspective. J Emerg Med 8:733–741

Johnson JW, Ascher P (1987) Glycine potentiates the NMDA response in cultured mouse brain neurones. Nature 325:529–531

Johnson KM Jr (1987) Neurochemistry and neurophysiology of phencyclidine. In: Meltzer HY (ed) Psychopharmacology: the third generation of progress. Raven, New York, pp 1581–1588

Johnson KM, Jones SM (1990) Neuropharmacology of phencyclidine: basic mechanisms and therapeutic potential. Annu Rev Pharmacol Toxicol 30:707–750

Johnson KM, Snell LD (1986) Involvement of dopaminergic, cholinergic and glutamatergic mechanisms in the action of phencyclidine-like drugs. In: Clouet DH (ed) Phencyclidine: an update. National Institute on Drug Abuse Research Monograph 64 (DHHS publication no ADM 86-1443). US Government Printing Office, Washington DC, pp 52–66

Johnson KM, Snell LD, Sacaan AI, Jones SM (1993) Pharmacologic regulation of the NMDA receptor-ionophore complex. In: De Souza EB, Clouet D, London ED, (eds) Sigma, PCP and NMDA receptors. National Institute on Drug Abuse Research Monograph Series 133 (DHHS publication no 93-3587). US Government Printing Office, Washington DC, pp 13–39

Judd LL, McAdams L, Budnick B, Braff DL (1992) Sensory gating deficits in schizophrenia: new results. Am J Psychiatry 149:488–493

Kamenka J-M, Chicheportiche R (1988) The conformational adaptation of the phencyclidine molecular pattern to the lipophilicity of its surrourdings. In: Domino EF, Kamenka J-M (eds) Sigma and phencyclidine-like compounds as molecular probes in biology NPP, Ann Arbor, pp 1–10

Karler R, Calder LD, Chaudhry IA, Turkanis SA (1989) Blockade of reverse tolerance to cocaine and amphetamine by MK-801. Life Sci 45:599–606

Karp SJ, Masu M, Eki T, Ozawa K, Nakanishi S (1993) Molecular cloning and chromosomal localization of the key subunit of the human N-methyl-D-aspartate receptor. J Biol Chem 268:3728–3733

Keats AS, Telford J (1964) Narcotic antagonists as analgesics. Clinical aspects. In: Gould RF (ed) Molecular modification of drug design. American Chemical Society, Washington DC, pp 170–176

Keith VA, Mansbach RS, Geyer MA (1991) Failure of haloperidol to block the effects of phencyclidine and dizocilpine on prepulse inhibition of startle. Biol Psychiatry 30:557–566

Khanna JM, Wu PH, Winer J, Kalant H (1991) NMDA antagonist inhibits rapid tolerance to ethanol. Brain Res Bull 25:643–645

Kleckner NW, Dingledine R (1988) Requirements for glycine in activation of NMDA-receptors expressed in Xenopus oocytes. Science 241:835–837

Klepstad P, Maurset A, Moberg ER, Aye I (1990) Evidence of a role for NMDA receptors in pain perception. Eur J Pharmacol 187:513–518

Kloog Y, Haring R, Sokolovsky M (1988) Kinetic characterization of the phencyclidine-N-methyl-D-aspartate receptor interaction: Evidence for a steric blockade of the channel. Biochemistry 27:843–848

Koek W, Colpaert FC (1992) N-methyl-D-aspartate antagonism and phencyclidine-like activity: behavioral effects of glycine site ligands. In: Kamenka J-M, Domino EF (eds) Multiple sigma and PCP receptor ligands: mechanisms for neuromodulation and neuroprotection? NPP, Ann Arbor, pp 665–671

Koek W, Woods JH, Ornstein P (1987) A simple and rapid method for assessing similarities among directly observable effects of drugs: PCP-like effects of 2-amino-5-phosphonovalerate in rats. Psychopharmacology 91:297–304

Koek W, Woods JH, Winger GD (1988) MK-801, a proposed noncompetitive antagonist of excitatory amino acid neurotransmission, produces phencyclidine-like behavioral effects in pigeons, rats and rhesus monkeys. J Pharmacol Exp Ther 245:969–974

Koek W, Colpaert FC, Woods JH, Kamenka J-M (1989) The phencyclidine (PCP) analog N-(1-(2-benzo(B)thiophenyl) cyclohexyl) piperidine shares cocaine-like but not other characteristic behavioral effects with PCP, ketamine and MK-801. J Pharmacol Exp Ther 250:1019–1027

Kokoz YM, Alekseev AE, Povsun AA, Korystova AF, Peres-Saad H (1994) Anesthetic phencyclidine, blocker of the ATP-sensitive potassium channels. FEBS Lett 337:277–280

Krystal JH, Karper LP, Seibyl JP, Freeman GK, Delaney R, Bremner JD, Heninger GR, Bowers MB Jr, Charney DS (1994) Subanesthetic effects of the noncompetitive NMDA antagonist, ketamine, in humans. Psychotomimetic, perceptual, cognitive and neuroendocrine responses. Arch Gen Psychiatry 51:199–214

Kuhar MJ, Boja JW, Cone EJ (1990) Phencyclidine binding to striatal cocaine receptors. Neuropharmacology 29:293–297

Kumor KM, Haertzen CA, Johnson RE, Kocher T, Jasinski D (1986) Human psychopharmacology of ketocyclazocine as compared with cyclazocine, morphine and placebo. J Pharmacol Exp Ther 238:960–968

Kuramoto T, Maihara T, Masu M, Nakanishi S, Serikawa T (1994) Gene mapping of NMDA receptors and metabotropic glutamate receptors in the rat (Rattus norvegicus). Genomics 19:358–361

Kushner L, Lerma J, Zukin RS, Bennett MVL (1988) Coexpression of N-methyl-D-aspartate and phencyclidine receptor in Xenopus oocytes injected with rat brain mRNA. Proc Natl Acad Sci USA 85:3250–3254

Kushner L, Bennett MVL, Zukin RS (1993) Molecular biology of PCP and NMDA receptors. In: De Souza EB, Clouet D, London ED (eds) Sigma, PCP and NMDA receptors. National Institute on Drug Abuse Research Monograph Series 133 (DHHS publication no 93-3587). US Government Printing Office, Washington DC, pp 159–183

Largent BL, Gundlach AL, Snyder SH (1984) Psychotomimetic opiate receptors labeled and visualized with (+)-[^{3}H]3-(3-hydroxyphenyl)-N-(1-propyl)piperidine. Proc Natl Acad Sci USA 81:4983–4987

Largent BL, Wikström, Snowman AM, Snyder SH (1988) Novel antipsychotic drugs share high affinity for sigma receptors. Eur J Pharmacol 155:345–347

Lasagna L, Pearson JW (1965) Analgesic and psychotomimetic properties of dexoxadrol. Proc Soc Exp Biol Med 118:353–354

Leander JD, Rathbun RC, Zimmerman DM (1988) Anticonvulsant effects of phencyclidine-like drugs: relation to N-methyl-D-aspartate antagonism. Brain Res 454:368–372

Lehmannmaster VD, Geyer MA (1991) Spatial and temporal patterning distinguishes the locomotor activating effects of dizocilpine and phencyclidine in rats. Neuropharmacology 30:629–636

Lerma J, Kushner L, Spray DC, Bennett MVL, Zukin RS (1989) mRNA from NCB-20 cells encodes the N-methyl-D-aspartate/phencyclidine receptor: a Xenopus oocyte study. Proc Natl Acad Sci USA 86:1708–1711

Lerma L, Zukin RS, Bennett MVL (1990) Glycine decreases desensitization of N-methyl-D-aspartate receptors expressed in Xenopus oocytes and is required for NMDA responses. Proc Natl Acad Sci USA 87:2354–2358

Liljequist S, Ossowska K, Grabowska-Andén M, Andén N-E (1991) Effect of the NMDA receptor antagonist, MK-801, on locomotor activity and on the metabolism of dopamine in various brain areas of mice. Eur J Pharmacol 195:55–61

Linders JTM, Monn JA, Mattson MV, George C, Jacobson AE, Rice KA (1993) Sythesis and binding properties of MK-801 isothiocyanates; (+)-3-isothiocyanato-5-methyl-10,11-dihydro-5H-dizenzo[a,d]cyclohepten-5-,10-imine hydrochloride: a new, potent and selective electrophilic affinity ligand for the NMDA receptor-coupled phencyclidine binding site. J Med Chem 36:2499–2407

Lodge D, Anis NA (1982) Effects of phencyclidine on excitatory amino acid activation of spinal interneurones in the cat. Eur J Pharmacol 77:203–204

Lodge D, Johnson KM (1990) Noncompetitive excitatory amino acid receptor antagonists. Trends Pharmacol Sci 11:81–86

Löscher W, Honack D (1991) The novel competitive N-methyl-D-aspartate (NMDA) antagonist CGP 37849 preferentially induces phencyclidine-like behavioral effects in kindled rats: attenuation by manipulation of dopamine, alpha-1 and serotonin-1_A receptors. J Pharmacol Exp Ther 257:1146–1153

Lu Y, France CP, Woods JH (1992) Tolerance to the cataleptogenic effect of the N-methyl-D-aspartate (NMDA) receptor antagonists in pigeons: cross tolerance between PCP-like compounds and competitive NMDA antagonists. J Pharmacol Exp Ther 263:499–504

Lutfy K, Hurlbut DE, Weber E (1993) Blockade of morphine-induced analgesia and tolerance in mice by MK-801. Brain Res 616:83–88

Lukas SE, Griffiths RR, Brady JV, Wurster RM (1984) Phencyclidine-analogue self-injection by the baboon. Psychopharmacology 83:316–320

Manallack DT, Lodge D, Beart PM (1989) Subchronic administration of MK-801 in the rat decreases cortical binding of [^{3}H]D-AP5, suggesting down-regulation of the cortical N-methyl-D-aspartate receptors. Neuroscience 30:87–94

Mansbach RS, Geyer MA (1989) Effects of phencyclidine and phencyclidine biologs on sensorimotor gating in the rat. Neuropsychopharmacology 2:299–308

Mansbach RS, Willetts J, Jortani SA, Balster RL (1991) NMDA antagonists: lack of an antipunishment effect in squirrel monkeys. Pharmacol Biochem Behav 39:977–981

Maragos WF, Chu DCM, Greenamyre JT, Penney JB, Young AB (1986) High correlation between the localization of [^{3}H] TCP binding and NMDA receptors. Eur J Pharmacol 123:173–174

Maragos WF, Penney JB, Young AB (1988) Anatomic correlation of NMDA and [^{3}H] TCP-labelled receptors in rat brain. J Neurosci 8:493–501

Marek P, Ben-Eliyahu S, Gold M, Liebeskind JC (1991) Excitatory amino acid antagonists (kynurenic acid and NK-801) attenuate the development of morphine tolerance in the rat. Brain Res 547:77–81

Maren S, Baudry M, Thompson RF (1991) Differential effects of ketamine and Mk-801 on the induction of long-term potentiation. Neuroreport 2:239–242

Marquis KL, Paquette NC, Gussio RP, Moreton JE (1989a) Comparative electroencephalographic and behavioral effects of phencyclidine, (+)-SKF-10,047 and MK-801 in rats. J Pharmacol Exp Ther 251:1104–1112

Marquis KL, Webb MG, Moreton JE (1989b) Effects of fixed-ratio size and dose on phencyclidine self-administration by rats. Psychopharmacology 97:179–182

Martin BR, Boni J (1990) Pyrolysis and inhalation studies with phencyclidine and cocaine. In: Chang CN, Hawks RL (eds) Research findings on smoking of abused substances. National Institute on Drug Abuse Research Monograph 99. DHHS publication no (ADM)90-1690. US Government Printing Office, Washington DC, pp 141–158

Martin BR, Katzen JS, Woods JA, Tripathi HL, Harris LS, May EL (1984) Stereoisomers of [^{3}H]-N-allylnormetazocine bind to different sites in mouse brain. J Pharmacol Exp Ther 231:539–544

Martin WR, Eades CG, Thompson JA, Huppler RE, Gilbert PE (1976) The effects of morphine- and nalorphine-like drugs in the nondependent and morphine-dependent chronic spinal dog. J Pharmacol Exp Ther 197:517–532

Massey BM, Wessinger WD (1990a) Alterations in rat brain [³H]-TCP binding following chronic phencyclidine administration. Life Sci 47:139–143

Massey BM, Wessinger WD (1990b) Effects of terminating chronic phencyclidine on schedule-controlled behavior in rats. Pharmacol Biochem Behav 36:117–121

McLamb RL, Williams LR, Nanry KP, Wilson WA, Tilson HA (1990) MK-801 impedes the acquisition of spatial memory task in rats. Pharmacol Biochem Behav 37:41–45

McMillan DE, Hardwick WC, de Costa BR, Rice KC (1991) Effects of drugs that bind to PCP and sigma receptors on punished respoding. J Pharmacol Exp Ther 258:1015–1018

Moerschbaecher JM (1992) The role of excitatory amino acids in learning and memory. In: Simon RP (ed) Excitatory amino acids. Fidia Research Foundation Symposium series, vol 9. Thieme Medical, New York, pp 211–214

Monaghan DT, Cotman CW (1986) Identification and properties of NMDA receptors in rat brain synaptic plasma membranes. Proc Natl Acad Sci U S A 83:7532–7536

Monaghan DT, Bridges RJ, Cotman CW (1989) The excitatory amino acid receptors: their classes, pharmacology, and distinct properties in the function of the central nervous system. Annu Rev Pharmacol Toxicol 29:365–402

Monyer H, Sprengel R, Schoepfer R, Herb A, Higuchi M, Lomeli H, Burnashev N, Sakmann B, Seeburg PH (1992) Heteromeric NMDA receptors: molecular and functional distinction of subtypes. Science 256:1217–1221

Moriyoshi K, Masu M, Ishii T, Shigemoto R, Mizuno N, Nakashini S (1991) Molecular cloning and characterization of the rat NMDA receptor. Nature 354:31–37

Morris RGM, Anderson E, Lynch G, Baudry M (1986) Selective impairment of learning and blockade of long-term potentiation by an N-methyl-D-asprtate receptor antagonist, AP-5. Nature 319:774–776

Mount H, Boksa P, Chaudieu I, Quirion R (1990) Phencyclidine and related compounds evoke [³H] dopamine release from rat mesencephalic cell cultures by a mechanism indepedent of the phencyclidine receptor, sigma-binding site, or dopamine uptake site. Can J Physiol Pharmacol 68:1200–1206

Musacchio JM (1990) The psychotomimetic effects of opiates and the σ receptor. Neuropsychopharmacology 3:191–200

Nakanishi S (1992) Molecular diversity of glutamate receptors and implications for brain function. Science 258:597–603

Olney JW (1990) Excitotoxic amino acids and neuropsychiatric disorders. Annu Rev Pharmacol Toxicol 30:47–71

Osawa Y, Davila JC (1993) Phencyclidine, a psychotomimetic agent and drug of abuse, is a suicide inhibitor of brain nitric oxide synthase. Biochem Biophys Res Commun 194:1435–1439

Overton P, Clark D (1992) Electrophysiological evidence that intrastriatally administered N-methyl-D-aspartate augments striatal dopamine tone in the rat. J Neurotransmission 4:1–14

Palfreyman MG, Baron B (1991) Non-competitive NMDA antagonists, acting at the glycine site. In: Meldrum BS (ed) Excitatory amino acid antagonists. Blackwell Scientific, Oxford, pp 101–129

Pfeiffer A, Brantl V, Herz A, Emrich HM (1986) Psychotomimesis mediated by κ opiate receptors. Science 233:774–776

Poling AD, White FJ, Appel JB (1979) Discriminative stimulus properties of phencyclidine. Neuropharmacology 18:459–463

Porter JH, Wiley JL, Balster RL (1989) Effects of phencyclidine-like drugs on punished behavior in rats. J Pharmacol Exp Ther 248:997–1002

Pulvirenti L, Maldonado-Lope R, Koob GF (1994) NMDA receptors in the nucleus accumbens modulate intraenous cocaine but not heroin self-administration. Brain Res 594:327

Quirion R, Chicheportiche R, Contreras PC, Johnson KM, Lodge D, Tam SW, Woods JH, Zukin SR (1987) Classification and nomenclature of phencyclidine and *sigma* receptor sites. Trends Neurosci 10:444–446

Quirion R, Bowen WD, Itzhak Y, Junien J-L, Musacchio JM, Rothman RB, Su T-P, Tam SW, Taylor DP (1992) Classification of *sigma* binding sites: a proposal. In: Kamenka J-M, Domino EF (eds) Multiple sigma and PCP receptor ligands: mechanisms for neuromodulation and neuroprotection? NPP, Ann Arbor, pp 959–965

Rahbar F, Fomufod A, White D, Westney LS (1993) Impact of intrauterine exposure to phencyclidine (PCP) and cocaine on neonates. J Natl Med Assoc 85:349–352

Ransom RW, Stec NL (1988) Cooperative modulation of [^{3}H]MK-801 binding to the N-methyl-D-aspartate receptor ion-channel complex by L-glutamate, glycine and polyamines. J Neurochem 51:830–836

Rao TS, Contreras PC, Wood PL (1990a) Are *N*-methyl-D-aspartate (NMDA) and phencyclidine (PCP) receptors always functionally coupled to each other? Neurochem Int 1:1–8

Rao TS, Kim HC, Lehmann J, Martin LL, Wood PL (1990b) Selective activation of dopaminergic pathways in the mesocortex by compounds that act at the phencyclidine (PCP) binding site: tentative evidence for PCP recognition sites not coupled to N-methyl-D-aspartate (NMDA) recetors. Neuropharmacology 29:225–230

Reynolds IJ, Miller RJ (1988) Multiple sites for the regulation of the N-methyl-D-aspartate receptor. Mol Pharmacol 33:581–584

Reynolds IJ, Miller RJ (1989) Ifenprodil is a novel type of N-methyl-D-aspartate antagonist: interaction with polyamines. Mol Pharmacol 36:758–765

Reynolds IJ, Murphy SN, Miller RJ (1987) [^{3}H]-labelled MK-801 binding to the excitatory amino acid receptor complex from rat brain is enhanced by glycine. Proc Natl Acad Sci U S A 84:7744–7748

Risner ME (1982) Intravenous self-administration of phencyclidine and related compounds in the dog. J Pharmacol Exp Ther 221:637–643

Rogawski MA, Porter RJ (1990) Antiepileptic drugs: pharmacological mechanisms and clinical efficacy with consideration of promising developmental stage compounds. Pharmacol Rev 42:223–286

Rogawski MA, Yamaguchi S, Jones SM, Rice KC, Thurkauf A, Monn JA (1991) Anticonvulsant activity of the low affinity uncompetitive *N*-methyl-D-aspartate antagonist (+)-5-aminocarbyonyl-10,11-dihydro-5H-dibenz[a,d]cyclopeen-5,10-imine: comparison with the structural analogs dizocilpine (MK-801) and carbamazepine. J Pharmacol Exp Ther 259:33–37

Rothman RB, Reid AA, Monn JA, Jacobson AE, Rice KC (1989) The psychotomimetic drug phencyclidine labels 2 high affinity binding sites in guinea pig brain: Evidence for N-methyl-D-aspartate coupled and dopamine reuptake carrier-associated phencyclidine binding sites. Mol Pharmacol 36:887–896

Sanger DJ, Joly D (1991) The effects of NMDA antagonists on punished exploration in mice. Behav Pharmacol 2:57–63

Sanger DJ, Terry P, Katz JL (1992) Memantine has phencyclidine-like but not cocaine-like discriminative stimulus effects in rats. Behav Pharmacol 3:265–268

Shannon HE (1982) Pharmacological analysis of the phencyclidine-like discriminative stimulus properties of narcotic deriatives in rats. J Pharmacol Exp Ther 222:146–151

Sharma AC, Kulkarni SK (1991) MK-801 produces antianxiety effect in elevated plus-maze in mice. Drug Dev Res 22:251–258

Shoaib M, Stolerman IP (1992) MK-801 attenuates behavioural adaptation to chronic nicotine administration in rats. Br J Pharmacol 105:514–515

Siegel RK (1978) Phencyclidine and ketamine intoxication: A study of four populations of recreational users. In: Petersen RC, Stillman RC (eds) Phencyclidine (PCP) abuse: an appraisal. National Institute on Drug Abuse Research monograph 21.

DHEW publication no (ADM) 78-728. US Government Printing Office, Washington DC, pp 119–147

Singh L, Menzies R, Tricklebank MD (1990) The discriminative stimulus properties of (+)-HA-966, an antagonist at the glycine/N-methyl-D-aspartate receptor. Eur J Pharmacol 186:129–132

Sircar R, Nichtenhauser R, Ieni JR, Zukin SR (1986) Characterization and autoradiographic visualization of (+)-[³H]SKF10,047 binding in rat and mouse brain: further evidence for phencyclidine/"sigma opiate" receptor commonality. J Pharmacol Exp Ther 247:681–588

Skilling SR, Smullin DH, Larson AA (1988) Extracellular amino acid concentrations in the dorsal spinal cord of freely moving rats following veratridine and nociceptive stimulation. J Neurochem 51:127–132

Slifer BL, Balster RL (1983) Reinforcing properties of stereoisomers of the putative sigma agonists N-allylnormetazocine and cyclazocine in rhesus monkeys. J Pharmacol Exp Ther 225:522–528

Slifer BL, Balster RL (1988) Phencyclidine-like discriminative stimulus effects of the stereoisomers of alpha- and beta-cyclazocine in rats. J Pharmacol Exp Ther 244:606–612

Slifer BL, Balster RL, Woolverton WL (1984) Behavioral dependence produced by continuous phencyclidine infusion in rhesus monkeys. J Pharmacol Exp Ther 230:399–406

Smith DJ, Perrotti JM, Mausell AL, Monroe PJ (1985) Ketamine analgesia is not related to an opiate action in the periaqueductal gray region of the rat brain. Pain 21:253–265

Smith JB (1991) Situational specificity of tolerance to effects of phencyclidine on responding of rats under fixed-ratio and spaced-responding schedules. Psychopharmacology 103:121–128

Snell LD, Johnson KM (1985) Antagonism of N-methyl-D-aspartate-induced transmitter release in the rat striatum by phencyclidine-like drugs and its relationship to turning behavior. J Pharmacol Exp Ther 235:50–57

Snell LD, Jones SM, Johnson KM (1987) Inhibition of N-methyl-D-aspartate-induced hippocampal [³H] norepinephrine release by phencyclidine is dependent upon potassium concentration. Neurosci Lett 78:333–337

Snell, LD, Yi S-J, Johnson KM (1988) Comparison of the effects of MK-801 and phencyclidine on catecholamine uptake and NMDA-induced norepinephrine release. Eur J Pharmacol 145:223–226

Spain JW, Klingman GI (1985) Continuous intravenous infusion of phencyclidine in unrestrained rats results in the rapid induction of tolerance and physical dependence. J Pharmacol Exp Ther 234:415–424

Stewart J, Druhan JP (1993) Development of both conditioning and sensitization of the behavioral activating effects of amphetamine is blocked by the non-competitive NMDA receptor antagonist, MK-801. Psychopharmacology 110:125–132

Sturgeon RD, Fessler RG, Meltzer HY (1979) Behavioral rating scales for assessing phencyclidine-induced locomotor activity, stereotyped behavior and ataxia in rats. Eur J Pharmacol 59:169–179

Sugihara H, Moriyoshi K, Ishii T, Masu M, Nakanishi S (1992) Structures and properties of seven isoforms of NMDA receptors generated by alternative splicing. Biochem Biophys Res Comm 185:826–832

Sun X, Larson AA (1993) MK-801 and phencyclidine act at phencyclidine sites that are not linked to N-methyl-D-aspartate activity to inhibit behavioral sensitization to kainate. Neuroscience 54:773–779

Sveinbjornsdottir S, Sander JWAS, Upton D, Thompson PJ, Patsalos PN, Hirt D, Emre M, Lowe D, Duncan JS (1993) The excitatory amino acid antagonist D-CPP-ene (SDZ EAA-494) in patients with epilepsy. Epilepsy Res 16:165–174

Tabor BL, Smith-Wallace T, Yonekura ML (1990) Perinatal outcome associated with PCP versus cocaine use. Am J Drug Alcohol Abuse 16:337–348

Tam SW (1983) Naloxone-inaccessible *sigma* receptor in rat central nervous system. Proc Natl Acad Sci U S A 80:6703–6707

Tennant FS Jr, Rawson, RA, McCann M (1981) Withdrawal from chronic phencyclidine (PCP) depedence with desipramine. Am J Psychiatry 138:845–847

Thombs DL (1989) A review of PCP abuse trends and perceptions. Publ Health Rep 104:325–328

Thompson DM, Morerschbaecher JM (1982) Phencyclidine in combination with pentobarbital: supra-additive effects on complex operant behavior in patas monkeys. Pharmacol Biochem Behav 16:159–165

Thurkauf A, Zenk PC, Balster RL, May EL, George C, Carroll FI, Mascarella SW, Rice KC, Jacobson AE, Mattson MV (1988) Synthesis, absolute configuration, and molecular modeling study of etoxadrol, a potent phencyclidine-like agonist. J Med Chem 31:2257–2263

Tiseo PJ, Inturrisi CE (1993) Attenuation and reversal of morphine tolerance by the competitive N-methyl-D-aspartate receptor antagonist, LY274614. J Pharmacol Exp Ther 264:1090–1096

Tricklebank MD, Singh L, Oles RJ, Preston C, Iversen SD (1989) The behavioural effects of MK-801: a comparison with antagonists acting non-competitively and competitively at the NMDA receptor. Eur J Pharmacol 167:127–135

Troupin AS, Mendius JR, Cheng F, Risinge MW (1986) MK-801. In: Meldrum B, Porter R (eds) New anticonvulsant drugs. Libby, London, pp 191–201

Trujillo KA, Akil H (1991) Inhibition of opiate tolerance by non-competitive N-methyl-D-aspartate receptor antagonists. Brain Res 633:178–188

Vaccarino AL, Marek P, Kest B, Weber E, Keana JFW, Liebeskind JC (1993) NMDA receptor antagonists, MK-801 and ACEA-1011, prevent the development of tonic pain following subcutaneous formalin. Brain Res 615:331–334

Vaupel DB, Cone EJ (1991) Pharmacodynamic and pharmacokinetic actions of ketocyclazocine enantiomers in the dog: absence of *sigma*-like or phencyclidine-like activity. J Pharmacol Exp Ther 256:211–221

Verdoorn TA, Kleckner NW, Dingledine R (1987) Rat brain N-methyl-D-aspartate receptors expressed in Xenopus oocytes. Science 238:1114–1116

Vignon J, Pinet V, Cerruti C, Kamenka J-M, Chicheportiche R (1988) [^{3}H]N-[1-(2-benz(b)thiophenyl)-cyclohexyl]piperidine([^{3}H]BTCP): a new phencyclidine analog selective for the dopamine uptake complex. Eur J Pharmacol 148:427–436

Vincent JP, Kartalovski B, Geneste P, Kamenka JM, Lazdunski M (1979) Interaction of phencyclidine ("angel dust") with a specific receptor in rat brain membranes. Proc Natl Acad Sci USA 76:4678–4682

Wachsman L, Scheutz S, Chan LS, Wingert WA (1989) What hapens to babies exposed to phencyclidine (PCP) in utero? Am J Drug Alcohol Abuse 15:31–39

Wahlestedt C, Golanov E, Yamamoto S, Yee F, Ericson H, Yoo H, Inturrisi CE, Reis DJ (1993) Antisense oligonucleotides to NMDA-R1 receptor channel protect cortical neurons from excitotoxicity and reduce focal ischemic infarctions. Nature 363:260–264

Walker JM, Bowen WD, Walker FO, Matsumoto RR, de Costa R, Rice KC (1990) *Sigma* receptors: biology and function. Pharmacol Rev 42:355–402

Ward L, Mason SE, Abraham WC (1990) Effects of the NMDA antagonists CPP and MK-801 on radial arm maze performance in rats. Pharmacol Biochem Behav 35:785–790

Watkins JC, Evans RH (1981) Excitatory amino acid transmitters. Annu Rev Pharmacol Toxicol 21:165–204

Watkins JC, Krosgaard-Larson P, Honoré T (1990) Structure-activity relationships in the development of excitatory amino acid receptor agonists and competitive antagonists. Trends Pharmacol Sci 11:25–33

Weber E, Sonders M, Quarum J, McLean S, Pou S, Keana JFW (1986) 1,3-Di(2-[5-^{3}H]tolyl)guanidine: a selective ligand that labels σ-type receptors for psychotomimetic opiates and antipsychotic drugs. Proc Natl Acad Sci USA 83:8784–8788

Weissman AD, Casanova MF, Kleinman JE, DeSouza EB (1991) PCP and *sigma* receptors in brain are not altered after repeated exposure to PCP in humans. Neuropsychopharmacology 4:95–102

Wenger G (1982) Tolerance to phencyclidine in pigeons: cross-tolerance to ketamine. J Pharmacol Exp Ther 255:646–652

Wessinger WD (1987) Behavioral dependence on phencyclidine in rats. Life Sci 41:355–360

Wessinger WD, Owens SM (1991) Phencyclidine dependence: the relationship of dose and serum concentrations to operant behavioral effects. J Pharmacol Exp Ther 258:207–215

Wesson DR, Washburn P (1990) Current patterns of drug abuse that involve smoking. In: Chang CN, Hawks RL (eds) Research findings on smoking of abused substances. National Institute on Drug Abuse Research monograph 99. DHHS publication no (ADM) 90-1690. US Government Printing Office, Washington DC, pp 5–11

Wiley JL, Balster RL (1992) Preclinical evaluation of N-methyl-D-aspartate antagonists for antianxiety effects: a review. In: Kamenka J-M, Domino EF (eds) Multiple sigma and PCP receptor ligands: mechanisms for neuromodulation and neuroprotection? NPP, Ann Arbor, pp 801–815

Wiley JL, Balster RL (1993) NMDA antagonists: a novel class of anxiolytics? In: Hamon M, Ollat H, Thiébot M-H (eds) Anxiety: neurobiology, clinic and therapeutic perspectives. Libbey Eurotext/Colloque INSERM 232, Montrouge, pp 177–184

Wiley J, Balster RL (1994) Effects of competitive and noncompetitive N-methyl-D-aspartate (NMDA) antagonists in squirrel monkeys trained to discriminate D-CPPene (SDZ EAA 494) from vehicle. Psychopharmacology 116:266–272

Wiley JL, Porter JH, Compton AD, Balster RL (1992) Antipunishment effects of acute and repeated administration of phencyclidine and NPC 12626 in rats. Life Sci 50:1519–1528

Willetts J, Bobelis DJ, Balster RL (1989) Drug discrimination based on the competitive N-methyl-D-aspartate antagonist, NPC 12626. Psychopharmacology 99:458–462

Willetts J, Balster RL, Leander JD (1990) The behavioral pharmacology of NMDA receptor antagonists. Trends Pharmacol Sci 11:423–428

Willetts J, Morse WH, Lee-Parritz D (1992) Behavioral pharmacology of NMDA antagonists. In: Simon RP (ed) Excitatory amino acids. Fidia Research Foundation symposium series, vol 9. Thieme Medical, New York, pp 203–210

Willetts J, Clissold DB, Hartman TL, Brandsgaard RR, Hamilton GS, Ferkany JW (1993) Behavioral pharmacology of NPC 17742, a competitive N-methyl-D-aspartate antagonist. J Pharmacol Exp Ther 265:1055–1062

Williams K (1993) Ifendopril discriminates between subtypes of the *N*-methyl-D-aspartate receptor: selectivity and mechanisms at recombinant heteromeric receptors. Mol Pharmacol 44:851–859

Winger G, France CP, Woods JH (1991) Intravenous self-injection in rhesus In: Meldrum BS, Moroni F, Simon RP (eds) Excitatory amino acids. Fidia Research Foundation symposium series, vol 5. Raven, New York, pp 539–545

Winslow JT, Insel TR, Trullas R, Skolnick P (1990) Rat pup isolation calls are reduced by functional antagonists of the NMDA receptor complex. Eur J Pharmacol 190:11–21

Wish ED (1986) PCP and crime: just another illicit drug? In: Clouet DH (ed) Pheycyclidine: an update. National Institute on Drug Abuse Research monograph 64 (DHHS publication no ADM86-1443). US Government Printing Office, Washington DC, pp 174–189

Wong EHF, Kemp JA (1991) Sites for antagonism on the N-methyl-D-aspartate receptor channel complex. Annu Rev Pharmacol Toxicol 31:401–425

Wong EHF, Kemp JA, Preistley T, Knight AR, Woodruff GN, Iversen LL (1986) The anticonvulsant MK-801 is a potent N-methyl-D-aspartate antagonist. Proc Natl Acad Sci USA 83:7104–7108

Wong EHF, Knight AR, Ransom R (1987) Glycine modulates [³H]MK-801 binding
    to the NMDA receptor in rat brain. Eur J Pharmacol 142:487–488
Wood PL, Rao TS (1989) NMDA-coupled and uncoupled forms of the PCP receptor
    – Preliminary in vivo evidence for PCP receptor subtypes. Psychiatry 13:519–523
Woods JH, Winger G (1991) Phencyclidine (PCP) and related substances. In: Drug
    abuse and drug abuse research. Third triennial report to Congress from the
    Secretary, Department of Health and Human Services. DHHS publication no
    (ADM)91-1708 US Government Printing Office, Washington DC, pp 145–159
Woods JH, Koek W, France CP, Moerschbaecher Jm (1991) Behavioral effects of
    NMDA antagonists. In: Meldrum BS (ed) Excitatory amino acid antagonists.
    Blackwell Scientific, Oxford, pp 237–264
Woolverton WL, Balster RL (1979) Tolerance to the behavioral effects of phencyc-
    lidine: importance of behavioral and pharmacological variables. Psychophar-
    macology 64:19–24
Woolverton WL, Balster RL (1981) Effects of combinations of phencyclidine and
    pentobarbital on fixed-interval performance in rhesus monkeys. J Pharmacol
    Exp Ther 217:611–618
Xu X, Domino EF (1994) Phencyclidine-induced behavioral sensitization. Pharmacol
    Biochem Behav 47:603–608
Yamakura T, Mori H, Masaki H, Shimoji K, Mishina M (1993) Different sensitivities of
    NMDA receptor channel subtypes to non-competitive antagonists. Neuroreport
    4:687–690
Zhang J, Chiodo LA, Freeman AS (1992) Electrophysiological effects of MK-801 on
    rat nigrostriatal and mesoaccumbal dopaminergic neurons. Brain Res 590:153–163
Zukin SR, Zukin RL (1979) Specific [³H]phencyclidine binding in rat central nervous
    system. Proc Natl Acad Sci USA 76:5372–5376
Zukin RS, Zukin SR (1988) Phencyclidine, σ and NMDA receptors: emerging
    concepts. In: Domino EF, Kamenka J-M (eds) Sigma and phencyclidine-like
    compounds as molecular probes in biology. NPP, Ann Arbor, pp 407–424
Zukin SR, Brady KT, Slifer BL, Balster RL (1984) Behavioral and biochemical
    stereoselectivity of σ opiate/PCP receptor. Brain Res 294:174–177

CHAPTER 7

Benzodiazepine Discontinuation Syndromes: Clinical and Experimental Aspects

L.G. MILLER and D.J. GREENBLATT

A. Clinical Aspects of Benzodiazepine Discontinuation

Early reports of benzodiazepine discontinuation syndromes suggested that these effects occurred at high doses, and that discontinuation syndromes were uncommon or even rare in patients receiving therapeutic doses (WOODS et al. 1987). Subsequent studies indicated that discontinuation effects could occur in patients receiving therapeutic doses of several benzodiazepines, although the incidence of significant discontinuation syndromes remained uncertain (PETURSSON and LADER 1981). More recent studies provide a broad range of estimates for the occurrence of significant discontinuation effects, from 5% to 35% in patients receiving benzodiazepines for at least 1 month (SHADER and GREENBLATT 1993).

I. Risk Factors for Benzodiazepine Discontinuation Effects

1. Benzodiazepine Compounds

More than 40 benzodiazepines are in therapeutic use worldwide. Of the 15 most commonly used benzodiazepines (alprazolam, bromazepam, chlordiazepoxide, clobazam, clonazepam, diazepam, flunitrazepam, flurazepam, lorazepam, lormetazepam, midazolam, nitrazepam, oxazepam, temazepam, triazolam), discontinuation effects have been reported after chronic use of each compound, with the possible exception of midazolam, which is primarily used in acute or subacute settings (MARRIOTT and TYRER 1993). Whether individual compounds are more commonly associated with discontinuation effects remains controversial. It is likely that short half-life benzodiazepines are associated with more frequent discontinuation effects due to their pharmacokinetics: in a once daily dosing regimen, levels of these compounds are more likely to fall to undetectable concentrations (BUSTO et al. 1986). In contrast, long half-life benzodiazepines such as diazepam and especially flurazepam are less likely to be associated with discontinuation effects due to prolonged levels of the parent drug or active metabolites (GREENBLATT et al. 1990b). Largely anecdotal evidence suggests that alprazolam is particularly associated with discontinuation effects (GREENBLATT and WRIGHT 1993). However, correlative pharmacokinetic/pharmacodynamic studies support the effects of dose and duration rather than a specific effect of alprazolam (GREENBLATT et al. 1993). Finally, it has been hypothesized that high potency benzodiazepines are more likely to be associated with discontinuation effects, presumably due to high receptor affinity (TYRER 1993). This hypothesis is not supported by neurochemical evidence (see below).

2. Dose

Although it has been generally assumed that the likelihood of discontinuation effects increases with dose, evidence for this association is limited (MURPHY and TYRER 1991). Recent studies of patients receiving chronic alprazolam indicate that discontinuation effects increase at higher doses and further suggest the presence of a threshold under which discontinuation effects are much less likely (RICKELS et al. 1990; GREENBLATT et al. 1993).

3. Duration

It is generally understood that longer duration of treatment predisposes to benzodiazepine effects (RICKELS et al. 1983; COVI et al. 1973). However, in one study, longer duration of treatment was associated with slightly fewer discontinuation effects (HOLTON et al. 1992).

4. Personality Type

There is a considerable literature concerning the interaction of personality type and benzodiazepine discontinuation. This issue will not be reviewed here. The bulk of current evidence supports an association between "dependent" personality type and the likelihood of discontinuation effects (e.g., TYRER 1985).

II. Clinical Benzodiazepine Discontinuation Effects

The discontinuation or abstinence syndrome after benzodiazepine use is variable in nature, severity and duration. In general, there are four possible resuts after discontinuation (SHADER and GREENBLATT 1993): (1) no effects; (2) relapse, with return to premorbid symptoms; (3) rebound, generally characterized by symptoms similar to premorbid symptoms but more intense; and (4) withdrawal, generally characterized by symptoms of sympathetic discharge and associated premorbid symptoms.

In rare cases, discontinuation can lead to full tonic-clonic seizures and even death. Much more commonly, a true withdrawal reaction is characterized by such symptoms as sensitivity to light and sound, headaches, palpitations, sweating and dysphoria.

The clinical approach to benzodiazepine discontinuation has been reviewed elsewhere (WOODS et al. 1988). In most patients, gradual reduction of dose over several weeks is sufficient to prevent severe discontinuation effects. Other approaches involve use of pharmacologic adjuncts, such as β-blockers, carbamazepine, clonidine, and antidepressants, but evidence is limited concerning the efficacy of these interventions (Woods et al. 1992). Use of buspirone and related compounds is probably not efficacious.

B. Laboratory Aspects of Benzodiazepine Discontinuation

I. Behavioral Pharmacology

Chronic benzodiazepine administration has been associated with the development of tolerance to virtually all behavioral effects of benzodiazepines (File and Andrews 1993). Whether tolerance to the anxiolytic effect occurs remains uncertain, in part due to limitations in animal models of anxiolysis (Greenblatt et al. 1990a). Similarly, the relation between tolerance and discontinuation syndromes remains uncertain, although most evidence indicates a link between these two phenomena. For example, courses of benzodiazepines which do not induce tolerance, either due to dosage or duration, are not associated with discontinuation effects (Miller 1991). Courses of some benzodiazepine partial agonists also do not produce tolerance or discontinuation syndromes.

Several pharmacologic interventions have been reported to alter benzodiazepine discontinuation syndromes. Several investigators reported that the benzodiazepine antagonist flumazenil, administered either chronically or on a single occasion during benzodiazepine agonist administration, or approximately 24 h after discontinuation, prevented the development of withdrawal effects in rodents and primates (Lamb and Griffiths 1985; Gallager et al. 1986; Baldwin and File 1988). Similar results were reported in rodents with the use of the benzodiazepine inverse agonist FG-7142 (Hitchcott et al. 1992).

Nonbenzodiazepines have also been reported to limit benzodiazepine discontinuation effects. A calcium channel antagonist, verapamil, was reported to prevent discontinuation effects in diazepam-treated rodents. Partial effects were also reported with the GABAb agonist baclofen (File et al. 1991), the tricyclic antidepressant tianeptine (File and Andrews 1993), buspirone (File and Andrews 1991), carbamazepine (Galpern et al. 1991), and the "peripheral" benzodiazepine receptor antagonist PK 11195 (Byrnes et al. 1993).

Although a single unifying hypothesis is not suggested by these data, the results indicate that benzodiazepine discontinuation is most likely to: (1) be related to the development of tolerance, and perhaps occurs only in the setting of some degree of tolerance; (2) depend upon benzodiazepine/GABAa receptor configuration; (3) involve multiple neurotransmitter systems beyond the GABAa system.

II. Neurochemical Effects

The effects of chronic benzodiazepine administration have been studied extensively in animal models since the identification of the benzodiazepine binding site over 15 years ago. Initial studies were contradictory, in part due to differences in drug, dose and route of administration. Subsequent studies yielded at least a partial picture of the neurochemical alterations associated with benzodiazepine tolerance and dependence.

Investigations of chronic benzodiazepine effects have been performed primarily in rodents and as such have been subject to pharmacokinetic limitations: rapid clearance of benzodiazepines in rodents prevents the development of chronic blood and tissue drug concentrations. The use of sustained release techniques for drug delivery has obivated this problem (Miller et al. 1987). A second limitation has been the use of in vitro assays for receptor binding and function. It is now clear that results obtained in vitro differ in some cases from those obtained ex vivo and routinely from data obtained in vivo (Schoch et al. 1993).

Initial in vivo studies using sustained release drug delivery indicated both tolerance and a behavioral discontinuation syndrome in mice (Miller et al. 1988a,b). These behavioral effects were associated temporally with benzodiazepine receptor down-regulation in several brain regions followed by receptor up-regulation after discontinuation. GABAa receptor function was similarly affected. Analogous results were obtained for the benzodiazepine agonists clonazepam and alprazolam (Miller et al. 1989a; Lopez et al. 1990; Galpern et al. 1991). In contrast, a benzodiazepine antagonist (flumazenil) and an inverse agonist (FG 7142) produced opposite results. In addition, a partial agonist compound, bretazenil, produced neither tolerance nor receptor changes (Miller et al. 1989b; Miller et al. 1990; Pritchard et al. 1991).

These data indicate that benzodiazepine tolerance and discontinuation effects have correlates in benzodiazepine receptor alterations and GABAa receptor function. The molecular basis for these effects remains uncertain. Changes in benzodiazepine receptor synthetic rates have been hypothesized to account for the "overshoot" in receptors after discontinuation, and some evidence supports this idea (Miller et al. 1991). Several studies indicate a reduction in several GABAa receptor subunit mRNAs associated with chronic benzodiazepine administration, but this effect occurs after the receptor changes noted above (Kang and Miller 1992). Thus, it is unlikely that receptor effects merely follow receptor concentrations, in turn dictated by mRNA levels. It is possible that changes in the distribution of GABAa receptor subunits occur, which then affect receptor responsiveness; little evidence is available concerning this hypothesis (Gallager and Primus 1993).

Recent cloning and sequencing of the human GABAa $\alpha 1$ subunit indicates the presence of a benzodiazepine responsive region in the upstream sequence

(KANG et al. 1994). Further study should indicate whether benzodiazepines exert a direct effect on gene transcription, perhaps accounting for changes in receptors associated with tolerance and drug discontinuation.

C. Conclusion

Chronic administration of benzodiazepines can produce a discontinuation syndrome in humans and experimental animals. This syndrome is characterized by transient exacerbation of premorbid symptoms (rebound) or by symptoms of sympathetic activity. Discontinuation effects can occur at therapeutic doses; effects appear to be more likely at higher doses. Contribution of duration of treatment is variable, but prolonged exposure appears to be a risk factor. Animal models indicate that discontinuation syndromes are related to the development of tolerance to benzodiazepine effects. Changes in GABAa/benzodiazepine receptor structure and function appear to be associated with discontinuation effects, although the molecular nature of these alterations has not yet been defined.

References

Baldwin HA, File SE (1988) Reversal of increased anxiety during benzodiazepine withdrawal: evidence for an anxiogenic endogenous ligand for the benzodiazepine receptor. Brain Res Bull 20:603–606

Busto U, Sellers EM, Naranjo CA, Cappell H, Sanchez-Craig M, Sykora K (1986) Withdrawal reaction after long-term therapeutic use of benzodiazepines. N Engl J Med 315:854–859

Byrnes JJ, Miller LG, Perkins K, Greenblatt DJ, Shader RI (1993) Chronic benzodiazepine administration XI. Concurrent administration of PK11195 attenuates lorazepam discontinuation effects. Neuropsychopharmacology 8:267–273

Covi L, Lipman RS, Pattison JH (1973) Length of treatment with anxiolytic sedatives and response to their sudden withdrawal. Acta Psychiatr Scand 49:51–64

File SE, Andrews N (1991) Low but not high doses of buspirone reduce the anxiogenic effects of diazepam withdrawal. Psychopharmacology 105:578–582

File SE, Mabbutt PS, Andrews N (1991) Diazepam withdrawal responses measured in the social interaction test of anxiety and their reversal by baclofen. Psychopharmacology 104:62–66

File SE, Andrews N (1993) Benzodiazepine withdrawal: behavioral pharmacology and neurochemical changes. Biochem Soc Symp 59:997–1106

Gallager DW, Heninger K, Heninger G (1986) Periodic benzodiazepine antagonist administration prevents benzodiazepine withdrawal symptoms in primates. Eur J Pharmacol 132:31–38

Gallager DW, Primus RJ (1993) Benzodiazepine tolerance and dependence: GABAa receptor complex locus of change. Biochem Soc Symp 59:135–151

Galpern WR, Lumpkin M, Greenblatt DJ, Shader RI, Miller LG (1991) Chronic benzodiazepine administration. VII. Behavioral tolerance and withdrawal and receptor alterations associated with clonazepam administration. Psychopharmacology 104:225–230

Greenblatt DJ, Wright CE (1993) Clinical pharmacokinetics of alprazolam. Clin Pharmacokinet 24:453–471

Greenblatt DJ, Miller LG, Shader RI (1990a) Neurochemical and pharmacokinetic correlates of the clinical action of benzodiazepine hypnotic drugs. Am J Med 88 [Suppl 3A]:18s–24s

Greenblatt DJ, Miller LG, Shader RI (1990b) Benzodiazepine discontinuation syndromes. J Psychiatr Res 24 [Suppl 2]:73–79

Greenblatt DJ, Harmatz JS, Shader RI (1993) Plasma alprazolam concentrations: relation to efficacy and side effects in the treatment of panic disorder. Arch Gen Psychiatry 50:715–722

Hitchcott PK, Zharkovsky A, File SE (1992) Concurrent treatment with verapamil prevents diazepam withdrawal-induced anxiety, in the absence of altered calcium flux in cortical synaptosomes. Neuropharmacology 31:55–60

Holton A, Riley P, Tyrer P (1992) Factors predicting long-term outcome after chronic benzodiazepine therapy. J Affect Disord 24:245–252

Kang I, Miller LG (1992) A quantitative assay for $GABA_A$ receptor subunit mRNAs using the polymerase chain reaction. Mol Neuropharmacol 2:249–254

Kang I, Lindquist DG, Kinane TB, Ercolani L, Pritchard GA, Miller LG (1994) Isolation and characterization of the promoter of the human GABAa receptor alpha1 subunit gene. J Neurochem 62:1643–1646

Lamb RJ, Griffiths RR (1985) Effects of repeated RO 15-1788 administration in benzodiazepine-dependent baboons. Eur J Pharmacol 110:257–261

Lopez F, Miller LG, Greenblatt DJ, Chesley S, Schatzki A, Shader RI (1990) Chronic administration of benzodiazepines. V. Rapid onset of behavioral and neurochemical alterations after discontinuation of alprazolam. Neuropharmacology 29:237–241

Marriott S, Tyrer P (1993) Benzodiazepine dependence: avoidance and withdrawal. Drug Saf 9:93–103

Miller LG (1991) Chronic benzodiazepine administration: from the patient to the gene. J Clin Pharmacol 31:492–495

Miller LG, Greenblatt DJ, Shader RI (1987) Benzodiazepine receptor binding: influence of physiologic and pharmacologic factors. Biopharm Drug Dispos 8:103–114

Miller LG, Greenblatt DJ, Barnhill JG, Shader RI (1988a) Chronic benzodiazepine administration. I. Tolerance is associated with benzodiazepine receptor downregulation and decreased gamma-aminobutyric acid$_A$ receptor function. J Pharmacol Exp Ther 246:170–176

Miller LG, Greenblatt DJ, Roy RB, Summer WR, Shader RI (1988b) Chronic benzodiazepine administration: II. Discontinuation syndrome is associated with upregulation of gamma-aminobutyric acid$_A$ receptor complex binding and function. J Pharmacol Exp Ther 246:177–182

Miller LG, Woolverton S, Greenblatt DJ, Lopez F, Shader RI (1989a) Chronic benzodiazepine administration. IV. Rapid development of tolerance and receptor downregulation associated with alprazolam administration. Biochem Pharmacol 38:3773–3777

Miller LG, Greenblatt DJ, Roy RB, Gaver A, Lopez F, Shader RI (1989b) Chronic benzodiazepine administration. III. Upregulation of gamma-aminobutyric acid$_A$ receptor binding and function associated with chronic benzodiazepine antagonist administration. J Pharmacol Exp Ther 248:1096–1101

Miller LG, Galpern WR, Greenblatt DJ, Lumpkin M, Shader RI (1990) Chronic benzodiazepine administration. VI. A partial agonist produces behavioral effects without tolerance or receptor alterations. J Pharmacol Exp Ther 254:33–38

Miller LG, Lumpkin M, Greenblatt DJ, Shader RI (1991) Accelerated benzodiazepine recovery following lorazepam discontinuation. FASEB J 5:93–97

Murphy SM, Tyrer P (1991) A double-blind comparison of the effects of gradual withdrawal of lorazepam, diazepam and bromazepam in benzodiazepine dependence. Br J Psychiatry 158:511–516

Peturrson H, Lader MH (1981) Withdrawal from long-term benzodiazepine treatment. Br Med J 283:643–645

Pritchard GA, Galpern WR, Lumpkin M, Miller LG (1991) Chronic benzodiazepine administration. VIII. Receptor upregulation produced by chronic exposure to the inverse agonist FG-7142. J Pharmacol Exp Ther 258:280–285

Rickels K, Case GW, Downing RW, Winokur A (1983) Long-term diazepam therapy and clinical outcome. JAMA 250:767–771

Rickels K, Schweizer E, Case WG, Greenblatt DJ (1990) Long-term therapeutic use of benzodiaepines. I. Effects of abrupt discontinuation. Arch Gen Psychiatry 47:899–907

Schoch P, Moreau JL, Martin JR, Haefely WE (1993) Aspects of benzodiazepine receptor structure and function with releance to drug tolerance and dependence. Biochem Soc Symp 59:121–134

Shader RI, Greenblatt DJ (1993) Use of benzodiazepines in anxiety disorders. N Engl J Med 328:1398–1405

Tyrer P (1985) Neurosis divisible. Lancet 1:685–688

Tyrer P (1993) Withdrawal from hypnotic drugs. BMJ 305:706–708

Woods JH, Katz JL, Winger G (1987) Abuse liability of benzodiazepines. Pharmacol Rev 39:251–419

Woods JH, Katz JL, Winger G (1988) Use and abuse of benzodiazepines: issues relevant to prescribing. JAMA 260:3476–3480

Woods JH, Katz JL, Winger G (1992) Benzodiazepines: use, abuse and consequences. Pharmacol Rev 44:151–347

CHAPTER 8
Nicotine

J.E. HENNINGFIELD, R.M. KEENAN, and P.B.S. CLARKE

A. Introduction

Tobacco use is a form of drug dependence, pharmacologically mediated by the actions of nicotine at central and peripheral receptors. Nonpharmacological factors are also important in determining patterns of tobacco use and the prevalence of use but it is the pharmacologic factors which define tobacco use as a form of psychoactive drug dependence and which will be the focus of this review. The dependence process involves psychoactive and reinforcing effects of nicotine, tolerance and physical dependence to nicotine, as well as effects produced by nicotine that some users feel is useful if not indispensable. The neuropharmacological underpinnings of the dependence process have been the subject of considerable research in recent years and may pave the way towards more effective treatments. The goal of this review is to provide an overview of major findings and contemporary theory regarding the dependence producing actions of nicotine so as to provide a useful guide to further research. Our approach will be to provide an overview of the clinical phenomena of nicotine dependence and the general pharmacology of nicotine as well as to highlight recent research on the cellular and molecular basis of these effects.

B. Abuse and Dependence in Humans

I. Epidemiology

There are more than 1 billion cigarette smokers worldwide and of these, approximately one half billion will be killed by tobacco (CONNOLLY and CHEN 1993; PETO et al. 1994). In fact, approximately one half of continuing smokers will die prematurely as a result or their tobacco use (PETO et al. 1994). This will amount to approximately 3 million deaths per year throughout the 1990s and 10 million deaths per year by the 2020s. About 400 000 deaths per year occur in the U.S. and this will remain generally stable if current smoking trends persist (MCGINNIS and FOEGE 1993).

It has been estimated that more than one in three people who continue to smoke will die prematurely because of their tobacco exposure and that approximately 20% of all deaths occurring each year in the United States are attributable to cigarette smoking; however, cigarette smoking caused mortality can be significantly reduced by cessation of smoking at any age (US DHHS 1990). The causes of death in order of incidence are cardiovascular disease (43%); all forms of tobacco-caused cancer (36%), respiratory diseases (20%), and all other smoking-caused deaths (1%) (Centers for Disease Control 1993). Cigarette smoking is an important contributor to the fact that the three primary causes of mortality are similar for men and women: heart disease, cancer, and stroke.

The spread of nicotine dependence follows the course of an infectious disease, with transmission being largely by person to person exposure to cigarettes. Frequently, this is augmented by tobacco company promotions, free sampling campaigns and other marketing efforts. The absence of active public health and educational campaigns against smoking apparently leaves populations highly vulnerable to developing dependence. For example, Russell (1990) has reported that a survey of adults in the Great Britain in the early 1960s indicated that 94% of those who smoked more than three cigarettes became "long term regular smokers." More recently collected data in the United States and Great Britain suggest that between 30% and 50% of people who try smoking cigarettes escalate to regular patterns of use (McNeill 1991; Henningfield et al. 1991; Hirschman et al. 1984). Consistent with these observations, data from the 1991 National Health Survey indicate that approximately 70% of adolescents tried smoking, whereas approximately 25% were smoking every day of the 30 days prior to the survey. Since it is unlikely that cigarettes have become substantially less addictive or that people have become biologically less vulnerable in two to three decades, it would appear that educational efforts and social factors and smoking policies have decreased the likelihood of progression from use to dependence.

The United States provides an interesting case history for why continued research on the pharmacology of nicotine is needed. The prevalence of smoking declined form 42% of adults in 1965 to approximately 25% in 1990. Unfortunately, there has been no decline since 1990, either in adult prevalence or in smoking prevalence by the primary source of adult smokers, i.e., adolescent and preadolescent smokers (Giovino et al. 1995). Approximately one half of young smokers try to quit smoking by age 18 and approximately one third of all smokers attempt to quit each year by quitting for at least one day. It is not clear how many more people attempt to quit but relapse within 24 h and thus are not even counted as attempted quitters. It is clear that only about 2.5% of all smokers become nonsmokers each year, leaving much room for improvement.

The foregoing statistics illustrate the important intersection of population based public health efforts and individually driven treatment efforts. The population targeted efforts are presumably most important in pressuring people to quit smoking and in providing support to those attempting to sustain abstinence; however, the limitations on treatment efficacy, acceptability and availability mean that the vast majority of quitting efforts are short-lived. Biomedical research may be the key to new treatments that could reverse this situation.

II. Clinical Aspects and Pathophysiology of Nicotine Dependence

The pathophysiological consequences of tobacco smoke exposure include destructive tissue effects that contribute to pulmonary disease, cellular changes that contribute to cancer, and the neuronal changes produced by nicotine exposure and reinforcing effects of nicotine that lead to addiction. Without the pathophysiologic changes leading to nicotine addiction, there would not be enormous public health toll and cost to our health care system because most people would be more readily able to stop smoking or at least to smoke at substantially lower levels. That goal is the expressed desire of the vast majority of smokers, but only a minority of these smokers ever achieve lasting abstinence. Those who do achieve lasting abstinence greatly reduce their risk of premature death and significantly lengthen their lives (US DHHS 1990). The pathophysiology of the nicotine dependence process has been reviewed in detail elsewhere and will only be summarized here (see HENNINGFIELD et al. 1995a).

A variety of surveys and clinical studies indicate that the vast majority of cigarette smokers are nicotine-dependent, an observation consistent with the remarkably short-lived success of more than 90% of most self-managed cessation attempts. In the United States, approximately 90% of cigarette smokers smoke more than five cigarettes per day, and the majority have tried to reduce their tobacco intake or report various symptoms of dependence (HENNINGFIELD et al. 1990). By contrast, among people who had consumed five or more alcoholic drinks in the past 30 days only 17% reported feeling a need to drink or dependence; among people who had used cocaine in the previous year, only 16% had used it in the previous week with only less than 10% feeling they needed the drug or were dependent (HENNINGFIELD 1992).

The tobacco dependence process, like other pathogenically induced disorders, involves host or individual factors, environmental factors, and the level of exposure to the pathogen itself. Initiation is often mediated by a variety of social and cultural factors, much as other forms of psychoactive drug use. However, over time, the reinforcing effects of the drug strengthen and the individual's control over use is lessened. Whereas a variety of social and other factors may continue to be operative, cigarette addiction is powerfully and critically driven by the positively and negatively reinforcing actions of nicotine, as will be discussed further on in this review.

Nicotine dependence shares many critical features in common with other drug dependencies (SCHUSTER et al., this volume). The role of nicotine in the use of tobacco products is functionally similar to the role of cocaine, morphine and ethanol in the use of coca and opiate-derived products and alcoholic beverages, respectively. While certain commonalties in the effects of these drugs lead to their being categorized similarly, as dependence producing psychoactive drugs that lead to harmful use and abuse, the pharmacology of these substances differs in many respects (HENNINGFIELD et al. 1995a). Like other drug dependencies, nicotine dependence is a "progressive," "chronic", "relapsing" disorder. In the United States, the mean age of onset of cigarette smoking is 13–14 years (US DHHS 1994; GALLUP 1992). It is older in some populations in the U.S. (e.g., African Americans) and in many countries where access to cigarettes to children is highly restricted. No age level appears to confer invulnerability to dependence; however, the level of dependence to nicotine observed in adults has been found to be inversely related to the age of initiation of smoking when measured by diagnostic criteria of the American Psychiatric Association (BERSLAU et al. 1992).

The progression of increased tobacco intake is accompanied by the development of tolerance and physiological dependence. In the United States, the average cigarette smoker consumes approximately 20 cigarettes per day (Centers for Disease Control 1994). The actual number of cigarettes smoked per day appears to be influenced by factors such as cigarette nicotine delivery, price, the restrictions on smoking (US DHHS 1988; BURNS et al. 1992; SWEANOR et al. 1992). Thus, cigarette smokers with less dispensable income might smoke a smaller number of cigarettes but extract more nicotine per cigarette, thereby sustaining a high dependence level. Such considerations are important in cross-cultural and cross-national comparisons of smoking rates.

Tobacco use tends to be chronic with many people attempting to quit but persistent remission only occurring in the minority of cases. In fact, studies by RAVIES and KANDEL (1987) showed that whereas the use of cocaine, heroin and alcohol all declined sharply with age, the progression of cigarette smoking was slowed but not reversed as individuals aged. The chronic phase of the addictive process is highly resistant to substantial modification. For example, efforts to reduce tobacco smoke and nicotine exposure by smoking cigarettes with lower nicotine delivery ratings or to smoke fewer cigarettes are usually partially or completely thwarted by compensatory changes in how the cigarettes are smoked (BENOWITZ et al. 1983; BENOWITZ and JACOB 1984; KOZLOWSKI 1981, 1982; US DHHS 1988).

Efforts to achieve and sustain abstinence are generally short lived with the majority of individuals who quit on their own relapsing within a few days (HUGHES et al. 1992). Providing some level of medical support extends the mean period of remission by a week or more (KOTTKE et al. 1988), and providing nicotine replacement in addition to adjunctive therapy can extend the mean duration of remission by 6 months or more (FOULDS et al. 1993).

III. Tolerance and Physical Dependence

Tolerance refers to the decreased responsiveness to the same dose of the drug as a function of repeated drug exposure (JAFFE 1985; US DHHS 1988).

It is often demonstrated by the observation that increased doses of a drug are required to obtain pharmacologic effects formerly produced by lower doses. Nicotine tolerance appears to be substantially acquired during youth as people progress from a few cigarettes upon initial exposure to higher levels (US DHHS 1988; McNEIL et al. 1989). Administering nicotine to a tobacco deprived smoker can substantially increase heart rate and euphoria measures and decrease knee reflex strength (US DHHS 1988). However, with repeated doses, heart rate stabilizes at a level intermediate to that produced by the first dose and that occurring when nicotine deprived. In addition, subjective effects are minimal, and the knee reflex may appear normal (US DHHS 1988; DOMINO and VON BAUMGARTEN 1969; SWEDBERG et al. 1990). Tolerance to a variety of the behavioral, physiologic, and subjective effects of nicotine has been stuided (US DHHS 1988). There are several physiologic mechanisms of nicotine tolerance including decreased responsiveness to the drug at the site of drug action and possibly increased metabolism in chronic smokers (US DHHS 1988). It is possible that the development of tolerance is related to up-regulation of nicotine receptors, as will be discussed in Sect. C.

Tolerance resulting from nicotine exposure has been measured using various pharmacodynamic assays. For example, PERKINS et al. (1991a) demonstrated that acute tolerance to the pressor effect of intranasal nicotine develops rapidly in smokers and nonsmokers. In a series of investigations in smokers and nonsmokers, repeated exposure to the same dose of nasal nicotine administered over several days produced progressively decreased subjective, physiological and behavioral responses over time (PERKINS et al. 1990a,b, 1991a,b).

Like morphine and alcohol, chronic nicotine administration leads to physiologic or physical dependence such that abrupt abstinence is accompanied by a syndrome of signs and symptoms. The severity of the withdrawal syndrome can range from unpleasant to debilitating (US DHHS 1988). The clinical course has been described in detail elsewhere (AMERICAN PSYCHIATRIC ASSOCIATION 1987). In brief, the syndrome includes increased craving, anxiety, irritability, appetite, and decreased cognitive capabilites and heart rate. Onset is within approximately 8 h after the last cigarette; the symptoms peak within the first few days, then subside over the next few weeks. Symptoms may persist for months or more in some individuals. The magnitude of the withdrawal syndrome is directly related to the level of nicotine dependence, and measured by cotinine concentration or the Fagerstrom Tolerance Questionnaire score (FAGERSTROM and SCHNEIDER 1989; HEATHERTON et al. 1991), although there is considerable variability within and across individuals (US DHHS 1988).

The tobacco withdrawal syndrome is pharmacologically mediated by nicotine deprivation, although behavioral conditioning factors are certainly important (HENNINGFIELD and NEMETH-COSLETT 1988; US DHHS 1988). Nicotine withdrawal has been observed in abstinent cigarette smokers, smokeless tobacco users, and chronic users of nicotine gum (HATSUKAMI et al. 1987, 1991). Furthermore, many abstinent smokers complain about cognitive deficits. Because the tobacco withdrawal syndrome is associated with moderate to severe levels of physical/psychological discomfort, it is difficult for tobacco users to abstain from cigarette use for any extended period (see HUGHES et al. 1990). It is a commonly held belief that people continue to use tobacco and fail at cessation in order to avoid any or all of the above-mentioned withdrawal symptoms associated with abstinence (STOLERMAN 1991). One of the most important and probably least understood features of the nicotine withdrawal syndrome is "craving for tobacco." Cravings and urges to smoke cigarettes and/or use smokeless tobacco have been described as major obstacles confronting tobacco users attempting to quit. Craving for tobacco has been identified as one of the most prominent symptoms of nicotine withdrawal (see HUGHES et al. 1990; TIFFANY and DROBES 1990, 1991). Abstinent cigarette smokers report that craving for cigarettes is the most troublesome symptom they experience over the first month of quitting (WEST et al. 1989), and there is evidence that the intensity of urges and cravings associated with cigarette smoking is comparable in magnitude with the craving associated with other addictive disorders (KOZLOWSKI et al. 1989). Animal models of nicotine dependence have also been developed as will be discussed further on in this review (Sect. BIV).

C. General Pharmacology

Nicotine and its metabolites exert a multitude of pharmacologic effects, including reinforcement of tobacco use. Understanding these effects is critical to understanding the underpinnings of the dependence process.

I. Chemistry and Pharmacokinetics

1. Absorption

Nicotine is a tertiary amine existing in both isomeric forms, but tobacco contains only the levo-rotary form, (S)-nicotine, which has greater pharmacologic activity (POOL et al. 1985). It is a water- and lipid-soluble weak base with an 8.0 index of ionization (LIDE 1991). Thus, the nicotine present in the mildly alkaline smoke of cigars, pipes, chewing tobacco and snuff, is readily absorbed across mucosal membranes of the mouth and nose (RUSSELL et al. 1980; SVENSSON 1987). Cigarette smoke, in contrast, is mildly acidic and is absorbed in the lungs, where the massive area of distribution renders pH less important (GORI et al. 1986). Smokeless tobacco products vary widely in their pH levels and this appears to be the primary means by which nicotine dosage is controlled by tobacco manufacturers (HENNINGFIELD et al. 1995; DJORDJEVIC et al. 1995).

Studying nicotine dose-response relationships with commercially marketed cigarettes is complicated by the fact that cigarettes contain a range of

6–11 mg of nicotine of which up to 40% is bioavailable to the cigarette smoker (BENOWITZ and HENNINGFIELD 1994; HENNINGFIELD et al. 1994). There is little direct relationship between advertised cigarette nicotine yield levels and absorbed nicotine when people smoke cigarettes ad libitum (HENNINGFIELD et al. 1994; US DHHS 1988). Thus, the intent of researchers to manipulate nicotine dose by simply giving volunteers cigarettes with varying estimated yields may have been thwarted by subtle changes in smoking topography which allowed the volunteer to obtain lower than expected doses from the cigarettes rated most highly, and higher than expected doses for the low rated cigarettes.

The tobacco cigarette is the most toxic and addictive widely used vehicle for nicotine delivery. Nicotine is volatilized at the tip of a burning cigarette where it is carried by particulate matter ("tar" droplets) deep into the lungs with inspired air. The approximately 800°C pyrolysis at the tip of the cigarette is also the source of carbon monoxide and many other toxicologically significant products.

Nicotine is rapidly absorbed through tissue barriers and pulmonary alveoli following smoke inhalation into the lungs. Intravenous and inhaled nicotine produce an almost instantaneous nicotine bolus in arterial blood pumped from the heart which is delivered to the brain within 10 s (US DHHS 1988). Available data from studies of nicotine distribution in arterial blood (HENNINGFIELD et al. 1990, 1993) and studies of the absorption of radiolabeled nicotine (ARMITAGE et al. 1975) suggest that absorption characteristics are similar to those of gases such as oxygen which are exchanged in the lung from inspired air to venous blood (KETY 1951). Thus, smoke inhalation produces arterial boli which may be ten times more concentrated than the levels meausred in venous blood (HENNINGFIELD et al. 1990, 1993). A similar phenomenon of arterial boli occurs when cocaine is smoked, adding to the addictiveness and toxicity of "crack" cocaine (EVANS et al. 1995).

Following oral smokeless tobacco use, venous blood nicotine concentration peaks in 15 min (BENOWITZ et al. 1988). Time to peak blood nicotine concentration may be 30 min or longer following nicotine gum use (BENOWITZ et al. 1988) and several hours for transdermally delivered nicotine (PALMER et al. 1992).

a. Nicotine Delivery Kinetics May Determine Effects of Alternate Nicotine Delivery Systems

It is now clear that the vehicle of nicotine delivery is a determinant of not only toxicity, but also of the nature and magnitude of the effects produced by nicotine due to its control over the nicotine concentrations in the blood over time. Addictive and cardioactive effects are generally directly related to the speed of delivery of nicotine and other drugs (HENNINGFIELD and KEENAN 1993; BENOWITZ 1993). For example, smoke inhalation essentially mimics the effects of a rapid intravenous injection and exposes the heart, brain, and fetus to high concentrations which dissipate within a few minutes. Conversely, nicotine polacrilex gum and transdermal patch systems are of low abuse liability and weak cardiovascular effects, in part because rapid absorption of high dose is not possible from either system. Moreover, release of nicotine from polacrilex requires a substantial work effort (chewing) (HENNINGFIELD and KEENAN 1993; US DHHS 1988). Smokeless tobacco systems similarly vary widely in their impact in accordance with their pH (FREEDMAN 1994; HENNINGFIELD et al. 1995; DJORDJEVIC et al. 1995). The delivery system also helps to determine the nature and quantity of other toxic substances to which the user is exposed.

These issues are not unique to nicotine. The drug dosage form affects the acceptability of a medication to patients, and compliance with instructions for medication use: similarly the drug dosage form is a determinant of the addiction potential of substances of abuse (FARRE and CAMI 1991; SELLERS et al. 1991; US DHHS 1988). The drug delivery system determines ease and convenience of use as well as the speed and amount of drug that is absorbed. For example, tobacco and coca leaves are rarely swallowed and addiction to swallowed formulations are uncommon, presumably because the bioavailability of the nicotine and cocaine is so poor via the gastrointestinal system; furthermore, cocaine or nicotine absorbed via this route do not produce the rapid onsetting and offsetting effects which characterize the most powerfully addicting drugs and drug forms (HENNINGFIELD and KEENAN 1993; JAFFE 1985; US DHHS 1988).

2. Distribution

Following absorption of nicotine into the body, about two thirds of the drug is present in the ionized form. Less than 5% of nicotine is bound to plasma proteins. Thus, nicotine is widely distributed in all body tissues and organs with the highest concentrations observed in brain, lungs, heart, kidneys, adrenal glands, liver and spleen. The steady state volume of distribution for nicotine is 180 liters which indicates extensive distribution. Following intravenous nicotine administration, the blood nicotine concentration declines rapidly due to uptake by the various organ tissues of the body. Nicotine freely crosses the placenta and has been found in amniotic fluid and breast milk of cigarette smokers.

3. Systemic Metabolism

In the earliest work on nicotine metabolism, LAUTENBACH (1886–1887) macerated liver tissue from dogs and rabbits with a nicotine solution. The tissue extract was then injected into dogs with no symptoms of nicotine poisoning noted. Conversely, the tissue extract of nicotine macerated with kidney produced death in all animals following injection. LAUTENBACH hypothesized that nicotine detoxification was occurring in the liver, but not the kidney. Subsequent research has demonstrated that nicotine undergoes extensive metabolism, primarily in the liver, but also to a lesser extent in lung and brain (US DHHS 1988). Systemic metabolism of nicotine produces several compounds in differing amounts and biologic activity. Cotinine, the major metabolite of nicotine (about 85% of ingested nicotine), is produced via a two-step oxidation in the liver. Cotinine is then further metabolized to several other compounds including cotinine-N-oxide, norcotinine, and hydroxycotinine, with only 17% of cotinine excreted unchanged in urine. Nicotine is also metabolized to nornicotine and nicotine-N-oxide.

For most routes of administration, the $t_{1/2}$ of nicotine is 2 h following an initial redistribution "half-life" phase of about 20 min (BENOWITZ 1988). The $t1/2$ of transdermally delivered nicotine is 4 h (PALMER et al. 1992), probably reflecting the continued release of some nicotine absorbed by the dermal tissues at the site of application. Using radiolabeled nicotine administered to cigarette smokers, the average nicotine intake per cigarette (FTC yield 1.1 mg) was estimated to be 2.3 mg, while the bioavailability of oral nicotine capsules and transdermally delivered nicotine were 44% and 82%, respectively (BENOWITZ et al. 1991a,b).

4. Brain Metabolic Activity

Efforts to understand the effects of nicotine on the regional metabolism of glucose emanated from studies of the distribution of nicotine binding sites in the brain. The 2-deoyx-D-[1–14 C]glucose method (SOKOLOFF et al. 1977) has been used to map the areas of the brain that respond metabolically to acute and chronic nicotine administration. The acute administration of nicotine in rats leads to a significant increase in cerebral glucose utilization (LONDON et al. 1985a,b, 1988; GRUNWALD et al. 1987) that is related to the distribution of nicotine binding sites (LONDON et al. 1985b). After chronic (twice daily, 10 days nicotine treatment) acute stimulatory effects of nicotine on glucose metabolism were still obtained. In the lateral geniculate and the superior colliculus tolerance to nicotine-induced stimulation was reported; sensitization was never seen (LONDON 1990). Chronic nicotine infusion (GRUNWALD et al. 1987) did not change the characteristic nicotine-induced stimulatory effect in most brain ares, but in the lateral geniculate body an increased response was reported. In general, animal stuides indicate that nicotine increases glucose utilization and the effect persists after chronic exposure.

In contrast to the findings with animals, that nicotine increases glucose utilization in certain regions of the brain, preliminary clinical data from a study of smokers and nonsmokers showed that intravenous nicotine decreased glucose metabolism in both groups of subjects (STAPLETON et al. 1992). Euphorigenic doses of abused drugs including cocaine (LONDON et al. 1990a) and morphine (LONDON et al. 1990b) also decreased cerebral glucose metabolism in humans. In line with a model of drug-induced euphoria (SWERDLOW and KOOB 1987), it has been proposed that a decrease in cortical glucose metabolism may reflect changes in dopaminergic modulation of the nucleus accumbens causing a decrease in thalamic activity (LONDON 1990).

5. Drug Interactions

Cigarette smoking is known to alter the metabolism of many medications used in the treatment of various medical disorders (US DHHS 1988). As a consequence, the effect of nicotine on hepatic drug metabolism has received significant experimental attention. In clinical practice, higher doses of a broad range of medications are given to cigarette smokers than to nonsmokers and tobacco cessation can result in increased plasma levels of these medications (BENOWITZ 1988; KYEREMATEN et al. 1983; KYEREMATEN and VESELL 1991; SEVNSSON 1987).

Nicotine replacement medications may not produce the same level of hepatic induction that cigarette smoke does. For example, one study found that theophylline half-life increased by 36% within 1 week of smoking cessation, but there was no apparent effect of nicotine polacrilex on theophylline metabolism (LEE et al. 1987). Consistent with these observations, caffeine concentrations can increase by more than 250% when people quit smoking, mainly due to decreased metabolism (BENOWITZ et al. 1989). Certain adverse effects of acute caffeine overdose are similar to those of nicotine withdrawal (e.g., anxiety and sleep disturbance) (SACHS and BENOWITZ 1988). These findings suggest that patients in treatment for nicotine addiction should be warned not to increase their caffeine intake, and possibly to decrease it by about one third. Follow-up evaluations of patient status should include questions about caffeine and all other drug intake to determine if intake should be adjusted. With respect to the methylxanthines, it is also worth noting that acute caffeine abstinence results in its own withdrawal syndrome (SILVERMAN et al. 1992) that might complicate simultaneous tobacco withdrawal. Furthermore, some caffeine intake may produce a small degree of relief of nicotine withdrawal symptoms (COHEN et al. 1994; OLIVETO et al. 1991).

Although the mechanism is not understood, chronic use of a variety of dependence producing drugs is positively associated with higher levels of tobacco use and dependence (KOZLOWSKI et al. 1993). Acute exposure to alcohol (GRIFFITHS et al. 1976; HENNINGFIELD et al. 1984; NIL et al. 1984; MELLO et al. 1980; KEENAN et al. 1990), pentobarbital (HENNINGFIELD et al. 1984), opioids (MELLO et al. 1980a; CHAIT and GRIFFITHS 1984), and amphetamine (HENNINGFIELD and GRIFFITHS 1981; CHAIT and GRIFFITHS 1983) increase cigarette smoking and tobacco intake. Caffeine and marijuana produce weak and unreliable effects on smoking across studies (CHAIT and GRIFFITHS 1983; NIL et al. 1984; KOZLOWSKI 1976; OSSIP and EPSTEIN 1981; MELLO et al. 1980; NEMETH-COSLETT et al. 1986a). Acute administration of the centrally distributed ganglionic blocker mecamylamine produced increased cigarette smoking (STOLERMAN et al. 1973; NEMETH-COSLETT et al. 1986b), whereas the noncentrally acting blocker pentolinium did not (STOLERMAN et al. 1973). The opioid antagonist naloxone produced a small reduction of smoking in one study (KARRAS and KANE 1980) and no effect over a wide dose range in another study (NEMETH-COSLETT et al. 1986a).

II. Pharmacodynamics

Nicotine acts at nicotinic receptors and has mixed pharmacologic effects depending upon the dose, the interval since prior administration, and the time following administration at which the measurement is taken. Thus, the pharmacodynamic effects of nicotine vary widely and, in part, are determined by the conditions under which it is administered. This section will review some of these effecs.

1. Cardiovascular Effects

Upon acute administration, nicotine produces potent pressor effects in animals and humans (US DHHS 1988). Nicotine has been shown to produce dose-related increases in heart rate and blood pressure which last over several minutes and then diminish. With each subsequent nicotine administration, the peak cardiovascular effect is diminished. This is an example of a phenomenon known as tachyphylaxis (i.e., acute tolerance). Within the cardiovascular system, tachyphylaxis is protective against nicotine toxicity. The mechanism of tachyphylaxis is not well understood, but it may be related to neurotransmitter depletion or to the formation of antagonistic-like metabolites and/or receptor desensitization.

Cigarette smokers who chronically abuse nicotine have a higher resting heart rate than nonsmokers. Furthermore, past research has shown that cigarette smokers have modestly, but significantly, lower blood pressure than nonsmokers (LARSON et al. 1961). Within a large group of cigarette smokers, BENOWITZ and SHARP (1989) noted a significant inverse correlation between serum cotinine concentration and blood pressure and suggested that this relationship could be secondary to the pharmacologic effects of cotinine and/or nicotine exposure.

While the mechanism of action underlying nicotine's effect of decreasing blood pressure is unknown, data regarding the effect of nicotine and cotinine on blood pressure have been accrued. For example, BORZELLECA et al. (1962) found that intravenous nicotine base (0.125 mg/kg) induced a potent pressor response in anesthetized dogs, whereas equivalent doses of intravenous cotinine produced no response. Conversely, at higher doses (10–500 mg/kg), intravenous cotinine transiently lowered blood pressure in a dose-dependent manner. Cotinine's depressor effect was not appreciably altered in either the decerebrate or spinal dog preparation, suggesting that the pharmacologic mechanism of action was not dependent upon intact central nervous system function. Pretreatment with atropine (a cholinergic muscarinic receptor antagonist) or diphenhydramine (a nonspecific histaminic receptor antagonist) had no effect on the depressor activity. Hence, cotinine's depressor effect was not mediated by cholinergic muscarinic and/or histaminic receptors. The authors concluded that the depressor effect of cotinine was secondary to the induction of vascular smooth muscle relaxation. Another interesting result of this experiment was that high dose cotinine pretreatment completely antagonized the potent pressor effects of intravenous nicotine. Consequently, high doses of cotinine in the presence of nicotine may serve as a pharmacologic nicotinic antagonist either directly or indirectly within the cardiovascular system.

Cigarette smoking is the greatest preventable risk factor for cardiovascular disease. Carbon monoxide exposure and the rapid high dose delivery of nicotine produced by cigarette smoking inhalation appear to be major determinants of this increased risk (see US DHHS 1983). In addition, nicotine and cotinine influence prostaglandin biosynthesis of prostacyclin (PGI2), which could have adverse cardiovascular consequences. Decreased levels of PGI2 are thought to increase atherogenesis progression. Using horse aorta and platelet preparations, the effects of nicotine and cotinine on the biosynthesis of PGI2 in vitro were examined (CHAHINE et al. 1990, 1991). Nicotine significantly decreased PGI2 synthetase activity which led to a decreased PGI2 concentration, while cotinine stimulated PGI2 synthetase activity resulting in increased PGI2 levels. Moreover, cotinine significantly attenuated the inhibitory effect of nicotine on PGI2 synthetase activity. Thus, nicotine's effect on PGI2 levels is thought play a role in the atherosclerosis noted in smokers. Furthermore, cotinine may serve to defend the cardiovascular system from atherosclerosis by antagonizing the effects of nicotine resulting in decreased PGI2 levels within the arterial system.

2. Electroencephalograph Effects

The spontaneous electroencephalogram (EEG) recorded from scalp electrodes is a convenient, noninvasive measure of drug action in the brain. GOLDING (1988) reported that tobacco but not sham smoking increased power in the β-band, reduced α and θ activity and had no effect on δ power; α frequency increased after smoking. Intravenous nicotine increased α power in discrete bursts that were correlated with subject-reported euphoria (LUKAS et al. 1990). KNOTT and VENABLES (1977) reported that smoking caused a decrease in α and θ power and increased β and α frequencies.

In several experiments, the EEG effects of nicotine have been measured in smokers deprived of tobacco. Nicotine adminstration in the form of smoked tobacco increased EEG α frequency (ULETT and ITIL 1969) and decreased α and θ power (HERNING et al. 1983). The EEG consequences of overnight (PICKWORTH et al. 1986) and extended (PICKWORTH et al. 1989) abstinence were reversed by nicotine polacrilex and these effects were prevented by pretreatment with mecamylamine (PICKWORTH et al. 1988), Furthermore, the EEG effects of overnight abstinence were accompanied by a slowing of cognitive performance that was also reversed by nicotine gum (SNYDER and HENNINGFIELD 1989). ROBINSON et al. (1992) reported that a low nicotine yield cigarette decreased δ power and increased β power but a cigarette from which the nicotine had been extracted caused no significant changes in the EEG.

The EEG synchrony that persistently follows nicotine deprivation in smokers has been temporally linked to changes in performance and subjective complaints of inability to concentrate (CONRIN 1980; EDWARDS and WARBURTON 1983). REVELL (1988) found that performance changes occur on a puff-by-puff basis, and KNOTT (1988) reported that EEG changes associated with arousal also occur while smoking a single cigarette. In most EEG studies, however, the EEG arousal effects are only apparent when the EEG is collected in a low arousal situation. For example, PICKWORTH et al. (1986, 1989) found that the EEG effects of nicotine and the effects of nicotine abstinence were most apparent in subjects at rest with eyes closed. In situations in which the subjects concentrated on a task or simply opened their eyes, the EEG effects of nicotine were diminished, confirming the importance of behavioral context in the effects of EEG effects of nicotine.

III. Systemic Effects

1. Immunologic Effects

Nicotine may act to regulate endocrine systems resulting in decreased immune competency. Increased glucocorticoid secretion resulting from adrenergic stimulation of the adrenal gland produces immune suppression through a reduction in thymus-dependent responses (MUNCK et al. 1984). However, nicotine-induced autonomic activation and the release of other pituitary hormones may serve to counteract the immunosuppressive actions of the glucocorticoids. For example, cholinergic stimulation itself may increase immune function (ATWEH et al. 1984). Vasopressin release increases the mitogenic activity of thymocytes (WHITFIELD et al. 1970) and enhances lymphocyte production of interferon (JOHNSON and TORRES 1985). Nicotine-induced endorphin release enhances lymphocytic function (WYBRAN 1985a,b). Thus, like morphine (a tumorogenic and immunosuppressive agent), endorphin may have inhibitory and/or facilitatory effects on the immune system (SHAVIT et al. 1985). Prolactin is an immunomodulatory hormone and the nicotine-induced decrease in its release may be immunosuppressive (see FUXE et al. 1989). A decrease in natural killer cell activity of monkey spleen cells, and a decrease white blood cell response to concanvalin (SOPORI et al. 1985) were reported after exposure of the animals to high doses of cigarette smoke. In view of the hypothesis that natural killer cells may play a protective role in the immune defense against cancer, reductions in natural killer cell activity could lead to increased susceptibility for many malignant diseases (see FUXE et al. 1989).

2. Hormonal Effects

Female cigarette smokers experience more infertility than do non-smokers. In fact, women who smoke cigarettes have lower endogenous estrogen levels than nonsmokers (US DHHS 1988). Barbieri et al. (1986) determined that nicotine inhibits estrogen formation without affecting progesterone biosynthesis in human choriocarcinoma cell cultures and term placental microsomes. Nicotine prevented the conversion of androstenedione to estradiol in a dose-dependent fashion. The observed effect on estrogen formation was mediated through aromatase enzyme inhibition. These findings may explain, in part, the decreased estrogen levels observed in women who smoke cigarettes and potentially the difference in fertility between smokers and nonsmokers.

Male cigarette smokers are also prone to sex hormone abnormalities compared to nonsmokers. Meikle et al. (1988) investigated the effects of nicotine on testosterone metabolism using isolated canine prostate cells. Nicotine competitively inhibited 3-α-hydroxysteroid dehydrogenase (3-α-HSD) activity which is involved in the enzymatic metabolism of 5-α-dihydrotestosterone (5-α-DHT) and results in altered testosterone production. Using the Leydig cells of the rat testis, Yeh et al. (1989) found that nicotine competitively inhibited multiple steps in testosterone biosynthesis. Patterson et al. (1990) performed similar androgen biosynthesis research with nicotine on mouse Leydig cells. Again, nicotine inhibited LH-stimulated testosterone production. The mechanism of inhibition was thought to involve calcium or a calcium-mediated metabolic pathway. These data demonstrate the ability of nicotine to alter testosterone biosynthesis and metabolism. The observed alterations in testosterone biosynthesis could impact upon androgen action in several tissues including skin, prostate or testis.

Nicotine has been observed to affect other endocrine systems. In a study of nicotine on pancreatic functioning, Tjälve and Popov (1973) found that nicotine at low concentrations stimulated and at high concentrations inhibited glucose-induced insulin secretion in isolated rabbit pancreas. The differential effects of nicotine on insulin secretion at all doses were attenuated by atropine and/or hexamethonium pretreatment. Thus, the effects of nicotine on insulin secretion are mediated through mechanisms involving both muscarinic and nicotinic cholinergic receptors. Humans, dogs and rabbits that are chronically exposed to nicotine have increased plasma levels of thyroid hormones. In male rats, however, nicotine decreases the release of thyroid hormones (see Fuxe et al. 1989) and this effect is more evident after intermittent than after chronic nicotine exposure.

3. Toxicity

Nicotine is a potent, highly toxic substance which has been routinely used as a pesticide for many years. In rats, the LD_{50} for intraperitoneally administered nicotine is 10 mg/kg (Kitamura 1958). Nicotine derives its toxicity and potential lethality from its ability to act as a potent cholinergic nicotinic receptor antagonist which inhibits the activity of the sympathetic and par-asympathetic nervous systems. In spite of its potential toxicity, nicotine poisoning deaths are rare, with none reported to the U.S. poison control centers in recent years. This is despite the fact that in addition to the several million replacement prescriptions written each year, more than 25 billion packages of cigarettes are distributed, each of which contains approximately 160 mg nicotine. The reasons include the poor bioavailability of orally ingested nicotine and the remarkable degree of tolerance that begins to develop even during a single nicotine use episode (BENOWITZ 1993).

Nicotine is psychotoxic at high doses and even at doses readily supplied by one cigarette in nontolerant persons, who may exhibit severe behavioral disruption, confusion and intoxication (US DHHS 1988, Appendix B; HENNINGFIELD and HEISHMAN 1995). In fact, certain South American popu-lations have used tobacco ritualistically to produce altered states of con-sciousness including hallucinations (WILBERT 1987; JARVIK 1995). Intoxication is rare in tolerant smokers, however, and tobacco associated antisocial behavior appears to be a more prominent feature of nicotine deprivation than of the acute effects of nicotine.

IV. Abuse Liability and Dependence Potential

1. Discriminative Effects

Human studies of the subjective effects of several forms of nicotine delivery and animal studies of the discriminative effects of parenterally administered nicotine have demonstrated that nicotine's psychoactive effects partially generalize to prototypic drugs of abuse. For example, nicotine produces dose-related discriminable effects in animals and subjective responses in humans that generalize more to stimulants such as amphetamine and cocaine than to depressants and opioids (US DHHS 1988). These effects can be blocked or at least reduced by pretreatment with centrally acting nicotinic antagonists (e.g., mecamylamine) but not by antagonists which do not pene-trate the central nervous system (e.g., pentolinium, hexamethonium) (STOLERMAN 1991). Similarly, partial blockade of nicotine's effects by in-creasing doses of mecamylamine led smokers to rate the delivered tobacco smoke as weaker and to show a preference for higher nicotine concentrations of smoke (ROSE et al. 1989). This effect and other effects on brain neuro-transmitters that might mediate the psychoactive effects of nicotine will be discussed in Sect. C of this chapter.

Psychoactive drugs which produce elevations in scores on drug-liking scales and other indices of mood elevation are often abused by humans and self-administered by animals (JASINSKI 1977; GRIFFITHS et al. 1980). Nicotine has been shown to produce such elevations in mood in humans when given by intravenous injection (JONES et al. 1978; HENNINGFIELD et al. 1985; KEENAN et al. 1994a) by nasal nicotine administration (PERKINS et al. 1992,

1994; Sutherland et al. 1992a,b), and by cigarette smoke administration (Henningfield et al. 1985; Pomerleau and Pomerleau 1992). Delivery of nicotine via the slow release polacrilex and transdermal medications produces little if any such elevations in mood (Henningfield and Keenan 1993). Interestingly, nicotine may produce discriminative effects which would facilitate its ability to modify behavior at very low doses and not produce distinct changes in mood (Perkins et al. 1994). Nicotine and other drugs abused by humans increase extrasynaptic dopamine concentrations in the mesolimbic system of the brain in rats (Di Chiara and Imperato 1988). Additional findings regarding the literature on the discriminative and psychoactive effects of nicotine have been reviewed in greater detail elsewhere (Rosencrans and Meltzer 1981; Henningfield 1984; Stolerman and Reavill 1989; Stolerman 1991).

An emerging research issue of relevance to developing medications for treating tobacco dependence is the role of nicotine's metabolites in mediating its dependence producing effects. For example, whereas the psychoactivity of nicotine has been well characterized in animals and tobacco users, there has been much less study of the psychoactivity, or pharmacology in general, of nicotine's metabolites. Cotinine has received the most attention, with studies demonstrating that intravenous and oral cotinine produces subjective and other behavioral effects in abstinent cigarette smokers at blood levels similar to those achieved through daily smoking (Benowitz et al. 1983; Keenan et al. 1994b, 1995; Schuh et al. 1995). Studies of continine's behavioral effects in animals have demonstrated dose-related changes in arrousal (Yamamoto and Domino 1965), food-reinforced rates of responding (Risner et al. 1985; Goldberg et al. 1983), and nicotine discrimination (Takada et al. 1989; Goldberg et al. 1981a). These data suggest that cotinine and possibly other nicotine metabolites have significant pharmacologic activity which needs further elucidation in preclinical and clinical investigation.

2. Reinforcing Effects

The potential of nicotine to serve as a reinforcer and thereby strengthen behaviors leading to its ingestion has been explored and confirmed in a variety of animal and human models (Swedberg et al. 1990). Although the reinforcing effects of intravenous nicotine were demonstrated by Deneau and Inoki in 1967, it was not until the research of Goldberg et al. (1981b) that a robust nonhuman primate model of nicotine self-administration was established. Henningfield and colleagues studied human nicotine self-administration in the early 1980s (Henningfield et al. 1984; Henningfield and Goldberg 1983). This work was extended by Corrigall (1991a,b) who used procedures analogous to those of Goldberg to develop a rat model of nicotine self-administration. In brief, it has now been demonstrated that, in the absence of the tobacco vehicle or the taste and other sensory effects of tobacco use, nicotine can serve as a positive reinforcer for five animal species in addition to humans (SWEDBERG et al. 1990).

Both the CORRIGALL and GOLDBERG procedures utilized small, rapidly delivered doses of intermittently available nicotine; in addition, the GOLDBERG procedure added paired stimuli, which further increased the amount of behavior sustained ultimately by nicotine injections and provide paired environmental stimuli with the injections. These procedures appear to model key features of cigarette-delivered nicotine and led to the establishment of strong reinforcing effects in animals. Cigarettes also enable people to easily provide themselves with rapid delivery of small intermittent doses and salient sensory stimuli. The importance of rate of delivery of nicotine to the reinforcing effects of the substance has not been adequately studied; however, it appears that rate of delivery is directly related to the subjective effects of the nicotine delivery system (HENNINGFIELD and KEENAN 1993). For example, whereas rapid intravenous injections or cigarette smoke inhalation produce psychoactive effects that may be pleasurable, slow infusions or delivery by the transdermal systems produce little if any discriminable psychoactive response (HENNINGFIELD and KEENAN 1993), and blunted or eliminated physiologic responses (BENOWITZ 1988; PALMER et al. 1992). Analogously, BALSTER and SCHUSTER (1973) found that the reinforcing effects of cocaine in rhesus monkeys were directly related to infusion rate, and DE WIT, BODKER and AMBRE (1992) showed that the subjective effects of pentobarbital in humans were directly related to the rate of drug administration. The role of infusion rate as a determinant of the reinforcing and other effects of nicotine is clearly a factor meriting further study, because this appears to be one means by which both the qualitative and the quantitative effects of nicotine are determined. Additional findings regarding the reinforcing effects of nicotine have been reviewed in detail elsewhere (GOLDBERG and HENNINGFIELD 1988; SWEDBERG et al. 1990; CORRIGALL 1991b).

3. Physical Dependence

As discussed earlier, nicotine produces physical dependence such that acute deprivation leads to a well defined withdrawal syndrome (AMERICAN PSYCHIATRIC ASSOCIATION 1994; HUGHES et al. 1990). Several promising animal models of various features of nicotine withdrawal have also been developed in animals that have been chronically maintained on nicotine for 10 days or longer. For example, a rat drug discrimination model revealed that nicotine deprivation produces an interoceptive state that generalizes to the presumably anxiogenic cue produced by pentylenetetrazol administration (HARRIS et al. 1986).

Two models of behavioral performance revealed the behaviorally disrupting effects of acute nicotine deprivation (CARROLL et al. 1989; CORRIGALL et al. 1988a). Another model of clinical relevance found that rats showed increased sensorimotor reactivity in an auditory startle paradigm (HELTON et al. 1993). The animal models should be useful for drug development research and for exploring the mechanisms of nicotine withdrawal (MAULTSBY et al. 1991). For example, it appears that both mecamylamine (MALIN et al. 1994) and naloxone (MALIN et al. 1993) administration can precipitate EEG and/or behavioral signs of withdrawal in rats maintained on nicotine, thus, suggesting the involvement of opioid and nicotinic components in nicotine dependence.

D. Neuropharmacology

The neuronal mechanisms underlying the pharmacological effects of nicotine have been the subject of intense study in recent years. This had led to a broader understanding of how nicotine produces receptor up-regulation, differentially activates receptor subpopulations, and releases a variety of neurohormones – actions that may contribute to the resultant dependence disorder. Recent and important neuropharmacological research on nicotine which have led to these conclusions will be summarized below.

I. Nicotinic Receptors

1. Receptor Diversity

Virtually all the known actions of nicotine appear to be mediated by nicotinic receptors (CLARKE 1987a; US DHHS 1988). It is believed that most, and perhaps all, nicotinic receptors are acetylcholine (ACh)-gated cation channels, permitting the influx of sodium and to some extent calcium. Nicotinic receptors form an extended family, as revealed principally by molecular genetic approaches (SARGENT 1993). All such receptors are probably composed of five protein subunits arranged around a central ionophore (ANAND et al. 1991; COOPER et al. 1991). Each subunit protein is encoded by a distinct gene; neuronal receptors appear to consist only of α subunits (that provide the principal binding sites for agonists) and β ("structural") subunits. To date, at least five α and three β subunit genes have been shown to be expressed in the central or peripheral nervous systems (SARGENT 1993).

Assuming that each neuronal receptor subtype is made up of a unique permutation of α and β subunits, there is considerable potential for receptor diversity. However, expression studies in *Xenopus* oocytes indicate that not all subunit combinations can function as nicotinic receptors (DENERIS et al. 1991). There are also indications from studies employing in situ hybridization histochemistry or protein isolation that certain receptor subunits may be preferentially coexpressed and coassemble to form functional receptors (WADA et al. 1989; WHITING et al. 1991). However, the prevalence of most nicotine ACh receptor (nAChR) subunits cannot be measured at the protein level because specific and quantifiable probes are lacking.

Receptors for nicotine are widely distributed in the CNS and in the periphery. In the CNS, nicotinic receptors are present in virtually every brain region (CLARKE and PERT 1985) and in the spinal cord (GILLBERG et al. 1988). Peripheral receptors are located mainly on autonomic ganglion cells, adrenal chromaffin cells, primary sensory neurons, and on skeletal muscle fibers. Doses of nicotine relevant to smoking exert actions in the CNS and at several peripheral sites, but muscle nAChRs appear unaffected at such doses.

Radioligand binding studies have identified two principal populations of nAChRs in mammalian brain, labeled with nanomolar affinity by ^{3}H-agonists and ^{125}I-labeled α-bungarotoxin, respectively (CLARKE et al. 1985b). The former are associated with receptors containing $\alpha4$ and $\beta2$ subunits (WHITING and LINDSTROM 1986; SWANSON et al. 1987; WHITING et al. 1991; FLORES et al. 1992), the latter with receptors containing $\alpha7$ subunits (SCHOEPFER et al. 1990; COUTURIER et al. 1990; SEGUELA et al. 1993). A third, less prevalent, subtype of putative receptors that may contain $\alpha3$ subunits has been identified by ^{125}I-labeled neuronal bungarotoxin binding in rat brain (SCHULZ et al. 1991). Nicotinic receptors in the brain that contain $\alpha4$ and $\beta2$ subunits are widely believed to represent important pharmacological targets for behaviorally active doses of nicotine that might be relevant to smoking (CLARKE 1987a). However, the evidence is rather indirect and largely correlational (LONDON et al. 1988; BENWELL et al. 1988; GRADY et al. 1992). The possibility therefore remains that other receptor subtypes play a greater role in behavioral responses to nicotine.

2. Receptor Regulation

Chronic in vivo treatment with nicotine, in sufficient doses, leads to an increase in [^{3}H]nicotine and ^{125}I-labeled α-bungarotoxin binding site density in rodent brain (MARKS et al. 1983a; SCHWARTZ and KELLAR 1983; EL-BIZRI and CLARKE 1994b). This effect has been termed "paradoxical" because chronic agonist treatment typically leads to a down-regulation of receptors. It has been proposed that nicotine up-regulates its receptors by acting as a "time-averaged" antagonist (HULIHAN-GIBLIN et al. 1990). Thus, nicotine treatment would produce a transition of the receptor to a desensitized state or to a more persistent "inactivated" state (MARKS et al. 1983b; SCHWARTZ and KELLAR 1985). Although there is ample evidence that nicotine can indeed produce acute and chronic tolerance in the CNS (LAPCHAK et al. 1989a; HULIHAN-GIBLIN et al. 1990; BENWELL et al. 1994; MARKS et al. 1993), there is some evidence to suggest that loss of receptor function per se may not be the stimulus leading to up-regulation (EL-BIZRI and CLARKE 1994b).

II. Cellular Mechanisms

At the cellular level, nicotine can be considered as an excitatory agent, since, when nicotinic receptors are present, acute administration tends to result in depolarization. If this occurs at the somatodendritic level, cell firing may increase, whereas acute depolarization of the nerve terminal tends to increase transmitter release.

With prolonged or repeated administration, tolerance may occur. Tolerance to nicotine may reflect a number of cellular processes with different dose relationships and time courses. Nevertheless, it is useful to distinguish acute tolerance (also known as tachyphylaxis), which occurs over a period of seconds to hours, from chronic tolerance, which may last for several weeks or longer. Acute tolerance could conceivably occur, for example, through receptor desensitization (Grady et al. 1994) or as a result of inactivation of voltage-gated ion channels. Chronic exposure to high doses of nicotine can result in a loss of sensitivity to nicotine lasting a number of days (Hulihan-Giblin et al. 1990; Lapchak et al. 1989a). The mechanisms underlying chronic tolerance remain largely unexplored.

Tolerance to nicotine can also be induced by concomitantly administered substances. Animals chronically exposed to ethanol exhibit a lessened response to the effects of nicotine thereby demonstrating cross-tolerance (Collins 1990). In mice, chronic exogenous corticosterone exposure decreases the sensitivity and density of cholinergic nicotinic receptors, which mimics tolerance to nicotine (Pauly et al. 1990a,b).

1. Effects of Nicotine on Cell Firing

A large number of electrophysiological studies have demonstrated that nicotine can alter neuronal firing in many brain regions (Clarke 1990). Most of this work is based upon extracellular recordings of single neurons in anesthetized animals and on the local application of drugs, given either by microiontophoresis or by pressure ejection. In only a few recent studies has it been possible to demonstrate unequivocally the occurrence of direct drug actions on the neuron under study. The sole dependent measure in almost all studies has been neuronal firing rate; drug-induced changes in the pattern of firing have received little attention (Grenhoff et al. 1986; Tung et al. 1989).

The preponderant electrophysiological effect of nicotine reported in the literature is one of excitation, but there are numerous examples of inhibitory effects. However, among those reports likely to reflect a direct drug action, inhibition has only been clearly shown in two brain areas – the cerebellum (De la Garza et al. 1987a,b) and dorsolateral septum (Wong and Gallagher 1989, 1991) of the rat. The latter action appears to be mediated by a calcium-dependent potassium conductance. The receptor subtype(s) associated with these inhibitory actions is not known, nor is it clear how widespread such actions might be in the brain.

2. Effect of Nicotine on Nerve Terminals

Nicotine has been reported to modulate the release of a number of brain transmitters in vitro. In some cases, direct presynaptic actions have been demonstrated, through the use of superfused synaptosomes or of brain slices superfused with tetrodotoxin. Several neurotransmitters and brain areas have been examined. Direct actions of nicotine on the nerve terminal have generally been found to be stimulatory. This is not surprising, since it appears that most, and perhaps all, nicotinic receptors in the brain function as ligand-gated cation channels. Neurochemical evidence for nicotine-induced depolarization of isolated nerve terminals has been provided (HILLARD and POUNDS 1991); however, such a depolarizing action may reduce rather than increase *impulse-dependent* transmitter release.

3. Effects of Nicotine on Transmitter Release in the Whole Animal

For numerous reasons, the effects of a drug on a given neurotransmitter system in vivo may differ markedly from those seen in isolated tissues in vitro. All attempts to demonstrate drug-induced transmitter release in the living animal are by their nature indirect and thus introduce problems of interpretation. In some early studies, tissue levels of transmitter were found to be decreased by nicotine, but this is at best an insensitive guide to release and usually required massive doses of the drug. Drug-induced changes in metabolite/transmitter ratios have also been reported; since considerable intraneuronal metabolism can take place, it is not clear what these changes represent. Inhibition of synthetic enzymes prior to nicotine administration can provide a measure of transmitter utilization, but here the relationship to release is presumed rather than demonstrated. More recently, in vivo microdialysis has provided a method for sampling the extracellular milieu in the brain region of interest. This approach is potentially problematic, not least because it usually leads to a local depletion of transmitter and, quite possibly, of other unidentified neuromodulators. Finally, in vivo electro-chemistry can provide a less invasive means of measuring transmitters that are electroactive (such as dopamine, noradrenaline and 5-HT); the challenge with this approach is to demonstrate chemical selectivity.

III. Neuronal Activity and Mechanisms of Reinforcement

The search for mechanisms underlying the reinforcing actions of nicotine has been largely confined to the brain. There are two principal reasons for this. Firstly, cigarette smoking in human subjects was found early on to be influenced by acute administration of the centrally active nicotinic antagonist mecamylamine, but not by the nicotinic antagonist pentolinium, which does not readily pass into the CNS (STOLERMAN et al. 1973). Secondly, virtually all of the behavioral effects of nicotine in animals have, where tested, been found to result from direct central effects of the drug (CLARKE 1987b). Of particular relevance, the voluntary self-administration of intravenous nicotine is reduced by prior central administration of the quasi-irreversible nicotinic antagonist chlorisondamine (CORRIGALL et al. 1992). Clearly, nicotine is likely to exert many actions in the brain and not all will contribute a reinforcing effect. In the following discussion, emphasis is placed on neuronal systems that may be relevant to nicotine dependence in humans.

1. Mesolimbic Dopamine

Several drugs of abuse appear to derive their reinforcing properties in animals from their ability to stimulate the mesolimbic dopamine (DA) system (Wise and Hoffman 1992). This neuronal system is widely thought to represent part of the reinforcement circuitry of the brain. Convergent evidence is now reviewed which indicates that it is also a target for nicotine.

Mesolimbic dopamine neurons form a pathway that ascends in parallel with the adjacent nigrostriatal dopamine system. Both neuronal populations express nAChRs on their cell bodies and/or dendrites and on terminals (Clarke and Pert 1985). They also appear to synthesize the same types of nAChR subunit (Wada et al. 1989; Wada et al. 1990). In anesthetized rats, systemic administration of nicotine increased cell firing of nigrostriatal DA neurons via a central mechanism (Clarke et al. 1985a); burst firing was also facilitated (Grenhoff et al. 1986). Nicotine did not, however, stimulate mesolimbic DA neurons in rats under general anesthesia (Grenhoff et al. 1986; Mereu et al. 1987); a stimulant action did occur in locally anesthetized, paralyzed animals (Mereu et al. 1987), but this action is likely to have been complicated by the stress imposed on the animals. However, intracellular recordings in rat midbrain slices have shown that nicotine (10 and $100\,\mu M$) depolarizes presumed mesolimbic DA cells via a direct action mediated by nAChRs (Calabresi et al. 1989). Nicotine-induced excitation of mesolimbic DA neurons has also been reported to occur in vitro at considerably lower concentrations (EC_{50} approximately $170\,nM$), more clearly relevant to cigarette smoking (Brodie 1991). Unlike nicotine-evoked DA release from synaptosomes (see below), this action did not manifest desensitization upon prolonged application of agonist.

Of all transmitters in the brain whose release is modulated by nicotine, DA has been the most extensively investigated in vitro. The majority of studies have used synaptosomes prepared from rat caudate-putamen, representing the nigrostriatal terminal field. DA release has also been demonstrated in other species (mice and cats), in brain slice preparations, and in the nucleus accumbens, which represents the principal terminal field of the mesolimbic DA system.

At least two DA releasing actions of nicotine have been identified in the striatum in vitro (Westfall et al. 1987). At high concentrations that are highly unlikely to occur in vivo (e.g., $1\,mM$), nicotine-induced DA release is associated with a "tyramine-like" effect. However, at lower concentrations, evoked release is concentration-dependent, mimicked by other nicotinic agonists, stereoselective, blocked by a number of nicotinic receptor antagonists, and dependent on external calcium (Giorguieff-Chesselet et al. 1979; Rapier et al. 1988, 1990; Grady et al. 1992; El-Bizri and Clarke

1994a). This second action of nicotine is typically associated with an EC50 of around $1\,\mu M$, suggesting that it may be pharmacologically relevant in vivo. However, it is also susceptible to profound desensitization in vitro (RAPIER et al. 1988; GRADY et al. 1994). Thus, declining effects have been seen with brief nicotine ($1\mu M$) challenges given as much as 30 min apart (RAPIER et al. 1988), and a near-total attenuation has been noted after continuous super-fusion with low nanomolar concentrations of the agonist (GRADY et al. 1994; ROWELL and HILLEBRAND 1994). Tests with other secretagogues suggest that desensitization occurs at the nicotinic receptor rather than on the trans-mitter release mechanism. It should also be noted that nicotine, in con-centrations as low as $1\,nM$, has been reported to inhibit [^{3}H]DA uptake in rat striatal tissue (IZENWASSER et al. 1991). This action appeared to be receptor-mediated but was probably not mediated by a direct action on DA terminals.

Convergent evidence indicates that acute administration of nicotine to freely moving rats can increase DA release in both the nigrostriatal and mesolimbic systems. In this regard, the mesolimbic system appears the more susceptible to doses of nicotine that might be relevant to tobacco smoking. Indeed, a fivefold difference in sensitivity has been reported between the two projections (IMPERATO et al. 1986). In another study, nicotine (0.1–0.4 mg/kg sc) increased DA utilization in a dose-dependent, stereoselective manner; significant effects were observed at 0.2 and 0.4 mg/kg and were confined to mesolimbic terminal fields (CLARKE et al. 1988).

The possible development of tolerance or sensitization has also been examined in rats implanted with microdialysis probes in the nucleus ac-cumbens. In one such study (DAMSMA et al. 1989), DA release evoked by nicotine (0.35 mg/kg sc) was not significantly altered by administration of the same dose 1 h before; the data suggest that some tolerance may nevertheless have occurred. When release of mesolimbic DA was repeatedly evoked by giving nicotine (0.35 mg/kg sc) once daily, nicotine-evoked release was comparable in size to that obtained in drug-naive animals. However, other investigators have reported that daily nicotine injections lead to a sensitized DA response under similar conditions (BENWELL and BALFOUR 1992). In contrast, chronic continuous infusion of nicotine abolished the release of nucleus accumbens DA evoked by acute nicotine challenge, as measured by intracerebral dialysis (BENWELL et al. 1994). The occurrence of receptor desensitization is likely to have contributed to the latter result.

Nicotinic receptors are present at both the somatodendritic and terminal levels of mesolimbic DA neurons (see above). In the whole animal, local application of nicotine into cell body or terminal regions can result in activation (LICHTENSTEIGER et al. 1982; MIFSUD et al. 1989). However, two studies suggest that DA release induced by *systemic* nicotine administration results mainly or exclusively from a somatodendritic action. In the first, nicotine-evoked DA release was found to be impulse-dependent, as shown by the intra-accumbens infusion of tetrodotoxin via the dialysis probe

(BENWELL et al. 1993). In the second study, accumbens DA release resulting from systemic nicotine administration was attenuated by local infusion of the nicotinic antagonist into the mesolimbic cell body region but not into the nucleus accumbens (NISELL et al. 1994).

It is not known why systemic nicotine stimulates the mesolimbic DA system more readily than the nigrostriatal system. It is also unclear why somatodendritic nAChRs appear more important than terminal-located receptors. However, it may be significant that, in vitro, somatodendritic nAChRs located on mesolimbic DA neurons appear to desensitize much less readily than nAChRs located on terminals of the nigrostriatal pathway (BRODIE 1991; RAPIER et al. 1988; GRADY et al. 1994).

2. Dorsal Noradrenergic Bundle

Neurons arising from the locus coeruleus (LC) form the dorsal noradrenergic bundle, providing a widespread innervation of much of the forebrain, including the hippocampus and cerebral cortex. Many functions have been proposed for this neuronal population, including arousal and vigilance. Noradrenergic (NA) neurons orginating in the LC have also been implicated in the physical aspects of opiate withdrawal (NESTLER 1992).

Nicotinic receptors are present at both the somatodendritic and terminal levels of NA neurons. Nicotine appears to excite NA cell bodies of the LC via two mechanisms. One excitatory effect occurs via a peripheral action and is discussed below. An additional, centrally derived excitation has been observed at higher doses (ENGBERG and HAJOS 1994). In vitro electrophysiological experiments have implicated a direct action that is susceptible to prolonged desensitization (EGAN and NORTH 1986).

Nicotine stimulates the release of NA in a number of isolated preparations and brain areas in vitro (BALFOUR 1982). However, not infrequently, the concentrations of nicotine studied have been outside the pharmacological range. In superfused hippocampal slices, nicotine (10 and $100\,\mu M$) produced a small and transient increase in [^{3}H]NA release; this action was significantly attenuated by mecamylamine ($10\,\mu M$) (SNELL and JOHNSON 1989). Nicotine has proved to be of greater potency in, synaptosomal preparations. possibly because the agonist reaches the receptors more rapidly, with less opportunity for desensitization to occur. In synaptosomes prepared from rat hippocampus or hypothalamus and incubated with [^{3}H]NA, nicotine (5 and $50\,\mu M$) increased both uptake and release of the neurotransmitter (BALFOUR 1973). Subsequently, [^{3}H]NA release evoked from a hypothalamic synaptosomal preparation by nicotine (1 and $10\,\mu M$) was shown to be mecamylamine-sensitive, indicative of a probable receptor-mediated action (YOSHIDA et al. 1980). In the whole animal, nicotine acutely increases NA release. As with DA release, it appears that NA release promoted by systemic nicotine results mainly or exclusively from a somatodendritic action (MITCHELL 1993).

3. Thalamocortical Projections

Thalamocortical relay neurons process and convey sensory signals to the neocortex. Drugs that modulate this flow of information may conceivably affect such central processes as arousal, attention, and memory. There are reports from animal and human studies suggesting that nicotine can affect all three processes (LEVIN 1992; JONES et al. 1992). The rat thalamus contains abundant [³H]nicotine binding sites and nAChR-related mRNA, particularly in the anterior nuclei and in "specific" thalamic relay nuclei (WADA et al. 1989; CLARKE et al. 1984). As far as known, a similar pattern exists in human thalamus (RUBBOLI et al. 1994; PERRY et al. 1989; ADEM et al. 1988). Electrophysiological studies have identified direct excitatory actions of nicotine on thalamocortical relay neurons in the dorsolateral and medial geniculate nuclei (ANDERSEN and CURTIS 1964; McCORMICK and PRINCE 1987a). The association of [³H]nicotine binding sites with the corresponding projection areas in cortex (PRUSKY et al. 1987; CLARKE 1991) suggests that nicotine may modulate thalamocortical transmission at the somatodendritic and terminal levels.

4. Habenulo-Interpeduncular System

The medial habenula and its principal projection target, the interpeduncular nucleus (IPN), represent a major target for nicotine in the brain. The relevance of this pathway to drug dependence is not clear. Both structures possess very dense [³H]nicotine binding and abundant nAChR-related mRNAs in the rat (WADA et al. 1989; CLARKE et al. 1984). Many neurons within these nuclei are directly excited by nicotine and nicotinic agonists (BROWN et al. 1983; McCORMICK and PRINCE 1987b; MULLE and CHANGEUX 1990; MULLE et al. 1991), and presynaptic modulation by nicotine has also been demonstrated within the IPN (BROWN et al. 1984; MULLE et al. 1991). In vivo, the IPN evinces one of the largest increases in expression of the immediate early gene c-*fos* after systemic administration of nicotine (PANG et al. 1993).

5. 5-HT Release

Serotoninergic (5-HT) afferents from the midbrain raphé nuclei innervate much of the neuraxis. Serotoninergic interactions with the ascending DA systems have been noted at different anatomical sites (HAGAN et al. 1993), and there are indications that 5-HT tone modulates the reinforcing effects of cocaine (RICHARDSON and ROBERTS 1991). Nicotinic receptor mRNA has been detected in raphe nuclei (WADA et al. 1989, 1990), but to date no electrophysiological data are available that might point to a direct somatodendritic action of nicotine. Lesion evidence suggests that some nicotinic receptors reside on 5-HT terminals in rat striatum and hypothalamus (SCHWARTZ et al. 1984), but there have been a few attempts to demonstrate presynaptic control of 5-HT release by nicotine in the brain (HERY et al. 1977; BALFOUR 1973; BECQUET et al. 1988; FUXE et al. 1979). Although the majority of these results suggest that nicotine exerts little if any direct effect, the procedures used to data may not have been optimal for observing such effects. Although nicotine can clearly alter 5-HT utilization when administered in vivo, the mechanisms involved remain obscure (BALFOUR 1982).

The nicotine metabolite cotinine also has neuropharmacologic activity in the serotonergic system (ESSMAN 1973; FUXE et al. 1979). ESSMAN (1973) examined the influence of cotinine and nicotine on 5-HT activity in the cerebral cortex, mesencephalon and diencephalon of the rat brain. Nicotine slightly decreased the 5-HT turnover rate in all brain regions, while cotinine markedly increased 5-HT turnover rate (approximately six- to eightfold) in the mesencephalon and diencephalon without significantly affecting activity in the cerebral cortex. In other work, cotinine increased spontaneous release of 5-HT and inhibited neuronal 5-HT uptake and release from rat brain neurons, thereby increasing 5-HT turnover, while nicotine had no systematic effect (FUXE et al. 1979). Further, cotinine and nicotine given concomitantly with α-propyldopacetamide, a drug which depletes 5-HT in the brain, reduced whole brain 5-HT depletion compared to saline. Pretreatment with mecamylamine had no effect on the 5-HT depletion reduction induced by cotinine or nicotine. It was concluded that the observed effects following nicotine administration were likely due to the presence of the nicotine metabolite cotinine. These studies demonstrate that the serotonergic system is acutely altered by the administration of cotinine and this could be a significant neuropharmacologic mechanism of action for cotinine in the CNS. Further, the majority of the observed pharmacologic effect related to the alteration of 5-HT activity of cotinine probably occurs within the mesencephalon and diencephalon.

6. Amino Acid Neurotransmitter Release

Although the amino acids, glutamate and γ-amino-butyric acid (GABA), are probably the most prevalent excitatory and inhibitory neurotransmitters in the brain, few studies have addressed the possiblity of a presynaptic control of release via nicotinic receptors. Nicotine was reported to inhibit glutamate release in rat striatal slices, an effect attributed to an indirect action mediated by interneurons (GODUHKIN et al. 1984). In contrast, direct nicotine-evoked [^{3}H]GABA release has been observed in superfused hippocampal synaptosomes; the EC_{50} was approximately $5\,\mu M$ and the effect was sensitive to nicotinic receptor blockers (WONNACOTT et al. 1989).

7. Acetylcholine Release

Nicotine has been reported to stimulate ACh release in several brain areas via a direct presynaptic action. Almost all studies to date have utilized [^{3}H] choline as a precursor, which is incorporated into the newly synthesized

ACh release pool. In mouse cerebral cortex synaptosomes, nicotine ($0.1\,\mu M$ and greater) increased ACh release in a concentration-dependent manner; this effect was mimicked by the nicotinic agonist DMPP, abolished by the antagonist hexamethonium, and was attenuated by removal of external calcium (ROWELL and WINKLER 1984). In rat cortical synaptosomes, in contrast, nicotine did not evoke ACh release (MEYER et al. 1987); however, the acetylcholinesterase inhibitor eserine was present in these experiments and has antagonistic activity at nAChRs (CLARKE et al. 1994).

Nicotinic modulation of ACh release has also been studied in brain slices from cerebral cortex and hippocampus (ARAUJO et al. 1988). Significant stimulatory effects of nicotine were consistently observed at 0.1 and $1\,\mu M$. Nicotinic-evoked release was blocked by the nicotinic receptor antagonists DHBE and d-tubocurarine and was insensitive to tetrodotoxin. Similar findings have been obtained in rat cerebellum (LAPCHAK et al. 1989b). In contrast, nicotine-evoked ACh release was not detectable in rat striatum using the same approach (ARAUJO et al. 1988). In guinea pig cortical slices, nicotine-evoked ACh efflux was observed but was prevented by tetrodotoxin, indicating an indirect effect (BEANI et al. 1985). Whether this represents a species difference or whether procedural differences are responsible is not clear. There are also reports of nicotine-induced cortical ACh release in freely moving animals (QUIRION et al. 1994), but as yet the mechanisms involved have not been elucidated.

8. Other Measures of Neuronal Activity

a. 2-Deoxyglucose Uptake

Cellular uptake of radiolabeled 2-deoxyglucose (2-DG) has been widely used as a measure of neuronal activity in the rat brain. In all studies involving nicotine, animals have been partially immobilized during testing; this is an important caveat, in view of evidence that responses to nicotine can be attenuated by stress and/or circulating corticosteroids (PAULY et al. 1990a,b; CAGGIULA et al. 1993).

Acute systemic administration of nicotine produced a dose-dependent, mecamylamine-sensitive increase in 2-DG uptake in many brain areas in rats (GRUNWALD et al. 1987, 1988; LONDON et al. 1988; McNAMARA et al. 1990). Typically, brain areas showing the greatest nicotine effect were those rich in [^{3}H]nicotine binding sites rather than ^{125}I-labeled α-bungarotoxin. The effects of chronic nicotine treatment have also been investigated. In one study, the acute response to an injection of nicotine (0.3 mg/kg sc) was blunted in several brain regions by prior subchronic nicotine administration (1 mg/kg sc bid for 11 days); no sensitization was seen (LONDON 1990). Since the nicotine pretreatment resulted in significant residual plasma nicotine levels (12 ng/ml) at the time of testing, it is not clear whether the tolerance observed should be considered to be acute or chronic. In another study, nicotine (or saline) was infused chronically by osmotic minipump for 2 weeks, resulting in plasma nicotine levels two to three times those typical of human cigarette smokers (Grunwald et al. 1988). 2-DG uptake, measured near the end of this period, was significantly increased in several brain areas, with a similar trend in many others. The pattern of effects resembled that obtained with an acute infusion of nicotine which achieved similar plasma levels. After 1 day of nicotine withdrawal, glucose utilization returned essentially to control levels (Schröck and Kuschinsky 1991).

b. C-fos *Expression*

Expression of the immediate early gene c-*fos* provides another measure of neuronal activation in the brain and can be mapped neuroanatomically in brain sections by in situ hybridization or immunocytochemistry. The pattern of nicotine-induced c-*fos* expression in rat brain (Ren and Sagar 1992; Pang et al. 1993) differs markedly from known nAChR distributions and does not resemble the distribution of nicotine-induced 2-DG uptake. This disparity is particularly striking in the ventral tegmental area, where it is the non-dopaminergic neurons that principally express c-*fos* (Pang et al. 1993).

9. Peripheral Effects

Whilst it is believed that the primary reinforcing actions of nicotine are largely of central origin, several observations suggest that peripheral actions of the drug may also play a role. Such actions may contribute to the primary reinforcing effects of nicotine, or act as secondary reinforcers by association.

Actions of nicotine in the mouth and respiratory tract may be particularly important. First, the immediate satisfaction of cigarette smoking is promoted by sensory cues that can be blunted by local anesthesia of the respiratory tract (Rose et al. 1985). Second, nicotine appears to be the most important component in tobacco smoke that causes airway irritation in humans (Lee et al. 1993). Third, nicotine can produce a burning sensation in the mouth which is blocked by mecamylamine (Jarvik and Assil 1988). Fourth, nicotinic receptors have been shown to reside on sensory nerve endings in several tissues (Suwandi and Bevan 1966; Tanelian 1991; Juan 1982). Fifth, mecamylamine can reduce subjective ratings of cigarette strength and harshness (Rose et al. 1989). Sixth, sensory cues associated with cigarette smoke appear to modulate smoke intake and craving (Rose et al. 1993). Lastly, inhalation of black pepper vapor, which irritates the respiratory airways, or of denicotinized cigarettes, can produce acute reduction of cigarette craving and certain other withdrawal symptoms (Butschky et al. 1995; Rose and Behm 1994).

Once nicotine has entered the bloodstream, it may exert additional peripheral effects that affect CNS activity. This is suggested by electrophysiologic experiments conducted in anaesthetized rats. In these studies, intravenous administration of nicotine resulted in an almost instantaneous and transient excitation of neurons located in the substantia nigra pars reticulata (CLARKE et al. 1985) or LC (HAJOS and ENGBERG 1988; ENGBERG and HAJOS 1994). These effects could be attenuated or blocked by peripherally active nicotinic antagonists. In one study, nicotine was found to be effective in a dose as low as $2\,\mu g/kg$, approximately equivalent to a single puff of tobacco smoke (CLARKE et al. 1985). It appears that nicotine may activate primary sensory afferents to produce these indirect effects in the brain (HAJOS and ENGBERG 1988).

E. Implications for Medications Development and Conclusions

The fundamental reason that tobacco dependence is a major concern secondary to no other form of drug abuse is that tobacco kills far more people than all other addictive drugs combined. In fact, the rates of tobacco-caused morbidity and mortality are accelerating rapidly world wide even though they are stabilizing in the United States. Simply put, this is because, except for a few countries such as Australia, Canada, the United States, and Sweden, demand reduction efforts (i.e., treatment and prevention) are not keeping pace with supply side efforts. Part of the problem is similar to that with other addictions, which is that current knowledge and treatment approaches which could help are not broadly available. Equally important, however, is that current treatment approaches leave much room for improvement with respect ot efficacy, acceptability to tobacco users, and cost. Improvements along these dimensions will require enhanced understanding of the pharmacology of nicotine and the pathophysiology of the dependence process. Such information should facilitate the development and administration of treatments which are selectively targeted to the clusters of symptoms that appear to distinguish various subpopulations of smokers.

The development of nicotine replacement medications in the 1980s and 1990s illustrated the important public health potential of medications for treating nicotine dependence as well as the limits of those medications. Many patients who were refractory to other approaches were able to give up smoking and sustain abstinence, and abstinence rates for those on the medications were often double those for persons treated with placebo (FIORE et al. 1992; SILAGY et al. 1994; HENNINGFIELD 1994). However, the fact remains that, in most trials, the majority of persons treated resume smoking within days or weeks and show little evidence that the medication provided any meaningful benefit. Future research needs to determine which patients might benefit from repeated replacement therapy, higher doses, or fundamentally different approaches. Several avenues of research and development appear particularly promising. Identification of receptor subpopulations in the CNS that mediate various features of nicotine dependence is fundamental. Such information will facilitate the development of more effective and selective nicotinic agonists, antagonists, and mixed agonist-antagonists, as well as nonnicotinic acting agents. In this regard, it is important to encourage approaches which are currently limited by lack of acceptable medications (e.g., antagonists), when the approach, in principle, could be quite useful (e.g., Henningfield 1984; Clarke 1991).

Further information on the pathophysiologic underpinnings of nicotine dependence could also be critical in determining if there are subpopulations of people who are particularly vulnerable to nicotine dependence and may require prophylactic therapies. It is also plausible that some nicotine-dependent people might require extended, or even life long, administration of medications to sustain productive lives in the absence of tobacco delivered nicotine. Further, identification of similarities and differences in the mediation of nicotine dependence with other forms of drug dependence will also undoubtedly continue to contribute to the development of more effective behavioral and pharmacologic approaches to treatment. Thus, nicotine dependence researchers are clearly faced by many important challenges that could contribute significantly to our ability to prevent and treat this form of drug dependence.

References

Adem A, Jossan SS, d'Argy R, Brandt I, Winblad B, Nordberg A (1988) Distribution of nicotinic receptors in human thalamus as visualized by 3H-nicotine and 3H-acetylcholine receptor autoradiography. J Neural Transm 73:77–83

American Psychiatric Association (1987) Diagnostic and statistical manual of mental disorders, 3rd edn (revised). American Psychiatric Association, Washington

American Psychiatric Association (1994) Diagnostic and statistical manual of mental disorders. 4th edn. American Psychiatric Association, Washington DC

Anand R, Conroy WG, Schoepfer R, Whiting P, Lindstrom J (1991) Neuronal nicotinic acetylcholine receptors expressed in Xenopus oocytes have a pentameric quaternary structure J Biol Chem 266:11192–11198

Andersen P, Curtis DR (1964) The excitation of thalamic neurones by acetylcholine. Acta Physiol Scand 61:85–99

Araujo DM, Lapchak PA, Collier B, Quirion R (1988) Characterization of N-[3H]methylcarbamylcholine binding sites and effect of N-methylcarbamylcholine on acetylcholine release in rat brain. J Neurochem 51:292–299

Armitage AK, Dollery CT, George CF, Houseman TH, Lewis PJ, Turner DM (1975) Absorption and metabolism of nicotine from cigarettes. Br Med J 4: 313–316

Atweh SF, Grayhack MS, Richman DP (1984) A cholinergic receptor site on murine lymphocytes with novel binding characteristics. Life Sci 35:2459–2469

Balfour DJ (1982) The effects of nicotine on brain neurotransmitter systems. Pharmacol Ther 16:269–282

Balfour DJK (1973) Effects of nicotine on the uptake and retention of ^{14}C-noradrenaline and ^{14}C-5-hydroxytyramine by rat brain homogenates. Eur J Pharmacol 23:19–26

Balster RL, Schuster CR (1973) Fixed-interval schedule of cocaine reinforcement: effect of dose and infusion duration. J Exp Anal Behav 20:119–129

Barbieri RL, Gochberg J, Ryan KT (1986) Nicotine, cotinine, and anabasine inhibit aromatase in human trophoblast in vitro. J Clin Invest 77:1727–1733

Beani L, Bianchi C, Nilsson L, Nordberg A, Romanelli L, Sivilotti L (1985) The effect of nicotine and cytisine on 3H-acetylcholine release from cortical slices of guinea-pig brain. Naunyn Schmiedebergs Arch Pharmacol 331:293–296

Becquet D, Faudon M, Hery F (1988) In vivo evidence for acetylcholine control of serotonin release in the cat caudate nucleus: influence of halothane anaesthesia. Neuroscience 27:819–826

Benowitz NL (1988) Pharmacologic aspects of cigarette smoking and nicotine addiction. N Engl J Med 319:1318–1330

Benowitz NL (1993) Nicotine replacement therapy: What has been accomplished-can we do better? Drugs 45:157–170

Benowitz NL, Jacob P III (1984) Daily intake of nicotine during cigarette smoking. Clin Pharm Therap 35:499–504

Benowitz NL, Sharp DS (1989) Inverse relation between serum cotinine concentration and blood pressure in cigarette smokers. Circulation 80:1309–1312

Benowitz NL, Henningfield JE (1994) Establishing a nicotine threshold for addiction. N Engl J Med 331:123–125

Benowitz NL, Kuyt F, Jacob P, Jones RT, Osman AL (1983) Cotinine disposition and effects. Clin Pharmacol Ther 34:604–611

Benowitz NL, Porchet H, Sheiner L, Jacob P III (1988) Nicotine absorption and cardiovascular effects with smokeless tobacco use: comparison with cigarettes and nicotine gum. Clin Pharmacol Ther 44:23–28

Benowitz NL, Hall SM, Modin G (1989) Persistent increase in caffeine concentrations in people who stop smoking. Br Med J 298:1075–1076

Benowitz NL, Chan K, Denaro CP, Jacob P III (1991a) Stable isotope method for studying transdermal drug absorption: the nicotine patch. Clin Pharmacol Ther 50:286–293

Benowitz NL, Jacob P III, Denaro C, Jenkins R (1991b) Stable isotope studies of nicotine kinetics and bioavailability. Clin Pharmacol Ther 49:270–277

Benwell MEM, Balfour DJK (1992) The effects of acute and repeated nicotine treatment on nucleus accumbens dopamine and locomotor activity. Br J Pharmacol 105:849–856

Benwell MEM, Balfour DJ, Anderson JM (1988) Evidence that tobacco smoking increases the density of $(-)$-[^{3}H]nicotine binding sites in human brain. J Neurochem 50:1243–1247

Benwell MEM, Balfour DJK, Lucchi HM (1993) Influence of tetrodotoxin and calcium on changes in extracellular dopamine levels evoked by systemic nicotine. Psychopharmacology (Berl) 112:467–474

Benwell MEM, Balfour DJK, Khadra LF (1994) Studies on the influence of nicotine infusions on mesolimbic dopamine and locomotor responses to nicotine. J Clin Invest 72:233–239

Borzelleca JF, Bowman ER, Mckennis H (1962) Studies on the respiratory and cardiovascular effects of $(-)$-cotinine. J Pharmacol Exp Ther 137:313–318

Breslau N, Kilbey MM, Andreski P (1992) Nicotine withdrawal symptoms and psychiatric disorders: Findings from an epidemiologic study of young adults. Am J Psychiatry 149:464–469

Brodie MS (1991) Low concentrations of nicotine increase the firing rate of neurons of the rat ventral tegmental area in vitro. In: Adlkofer F, Thurau K (eds) Effects of nicotine on biological systems. Birkhäuser, Basel, p 373

Brown DA, Docherty RJ, Halliwell JV (1983) Chemical transmission in the rat interpeduncular nucleus in vitro. J Physiol (Lond) 341:655–670

Brown DA, Docherty RJ, Halliwell JV (1984) The action of cholinomimetic substances on impulse conduction in the habenulointerpedunculopontine pathway of the rat. J Physiol (Lond) 353:101–109

Burns DM, Axelrad R, Bal D, Carol J, Davis RM, Myers ML, Pinney JM, Rigotti NA, Shopland DR (1992) Report of the tobacco policy research study group on smoke-free indoor air policies. Tobacco Control 1 Suppl:S14–S18

Butschky MF, Bailey D, Henningfield JE, Pickworth WE (1995) Smoking without nicotine delivery decreases withdrawal in 12-hour abstinent smokers. Pharmacol Biochem Behav 50:91–96

Caggiula AR, Epstein LH, Antelman SM, Saylor S, Knopf S, Perkins KA, Stiller R (1993) Acute stress or corticosterone administration reduces responsiveness to nicotine: implications for a mechanism of conditioned tolerance. Psychopharmacology (Berl) 111:499–507

Calabresi P, Lacey MG, North RA (1989) Nicotinic excitation of rat ventral tegmental neurones in vitro studied by intracellular recording. Br J Pharmacol 98:135–140

Carroll ME, Lac ST, Asencio M, Keenan RM (1989) Nicotine dependence in rats. Life Sci 45:1381–1388

Centers for Disease Control and Prevention (1993) Cigarette smoking-attributable mortality and years of potential life lost – United States. In: Chronic disease and health promotion. Adapted from the MMWR. tobacco topics 1990–1993. U.S. Department of Health and Human Services, Public Health Service, Center for Disease Control and Prevention, National Center for Chronic Disease Prevention and Health Promotion, pp 77–81

Centers for Disease Control (1994) Surveillance for selected tobacco-use behaviors– United States, 1900–1994. MMWR CDC Surveill Summ 43:11–15

Chahine R, Calderone A, Navarro-Delmasure C (1990) The in vitro effects of nicotine and cotinine on prostacyclin and thromboxane biosynthesis. Prostaglandins Leukot Essent Fatty Acids 40:261–266

Chahine R, Kaiser R, Pham Huu Chanh A (1991) Effects of nicotine and cotinine on prostacyclin biosynthesis. In: Adlkofer F, Thurau, K (eds) Effect of nicotine on biological systms. Birkhäuser, Basel, p 171

Chait LD, Griffiths RR (1983) Effects of caffeine on cigarette smoking and subjective response. Clin Pharmacol Ther 34:612–622

Chait LD, Griffiths RR (1984) Effects of methadone on human cigarette smoking and subjective ratings. J Pharmacol Exp Ther 229:636–640

Clarke PBS (1987a) Recent progress in identifying nicotinic cholinoceptors in mammalian brain. Trends Pharmacol Sci 8:32–35

Clarke PBS (1987b) Nicotine and smoking: a perspective from animal studies. Psychopharmacology (Berl) 92:135–143

Clarke PBS (1990) The central pharmacology of nicotine: electrophysiological approaches. In: Wonnacott S, Russell MAH, Stolerman IP (eds) Nicotine psychopharmacology: molecular, cellular, and behavioural aspects. Oxford University Press, Oxford, p 158

Clarke PBS (1991) Nicotinic receptors in rat cerebral cortex are associated with thalamocortical afferents. Soc Neurosci Abstr 17:384.18

Clarke PBS, Pert CB, Pert A (1984) Autoradiographic distribution of nicotine receptors in rat brain. Brain Res 323:390–395

Clarke PBS, Pert A (1985) Autoradiographic evidence for nicotine receptors on nigrostriatal and mesolimbic dopaminergic neurons. Brain Res 348:355–358

Clarke PBS, Hommer DW, Pert A, Skirboll LR (1985a) Electrophysiological actions of nicotine on substantia nigra single units. Br J Pharmacol 85:827–835

Clarke PBS, Schwartz RD, Paul SM, Pert CB, and Pert A (1985b) Nicotinic binding in rat brain: autoradiographic comparison of ^{3}H-acetylcholine, ^{3}H-nicotine and ^{125}I-alpha-bungarotoxin. J Neurosci 5:1307–1315

Clarke PBS, Fu DS, Jakubovic A, Fibiger HC (1988) Evidence that mesolimbic dopaminergic activation underlies the locomotor stimulant action of nicotine in rats. J Pharmacol Exp Ther 246:701–708

Clarke PBS, Reuben M, El-Bizri H (1994) Blockade of nicotinic responses by physostigmine, tacrine and other cholinesterase inhibitors in rat striatum. Br J Pharmacol 111:695–702

Cohen C, Pickworth WB, Bunker EB, Henningfield JE (1994) Caffeine antagonizes EEG effects of tobacco withdrawal. Pharmacol Biochem Behav 40:919–926

Collins AC (1990) Interactions of ethanol and nicotine at the receptor level. Recent Dev Alcohol 8:221–231

Connolly G, Chen T (1993) International health and tobacco use. In: Houston TP (ed) Tobacco use: an American crisis. American Medical Association, Chicago, p 72

Conrin J (1980) The EEG effects of tobacco smoking – a review. Clin Electroencephalogr 11:180–187

Cooper E, Couturier S, Ballivet M (1991) Pentameric structure and subunit stoichiometry of a neuronal nicotinic acetylcholine receptor. Nature 350:235–238

Corrigall WA (1991a) A rodent model for nicotine self-administration. In: Boulton AA, Baker GB, Wu PH (eds) Animal models of drug addiction. Humana, Totawa, pp 315–344 (Neuromethods, vol 24)

Corrigall WA (1991b) Understanding brain mechanisms in nicotine reinforcement. Br J Addict 86:507–510

Corrigall WA, Herling S, Coen KM (1989) Evidence for a behavioral deficit during withdrawal from chronic nicotine treatment. Pharmacol Biochem Behav 33: 559–562

Corrigall WA, Franklin KBJ, Coen KM, Clarke PBS (1992) The mesolimbic dopaminergic system is implicated in the reinforcing effects of nicotine. Psychopharmacology (Berl) 107:285–289

Couturier S, Bertrand D, Matter JM, Hernandez MC, Bertrand S, Millar N, Valera S, Barkas T, Ballivet M (1990) A neuronal nicotinic acetylcholine receptor subunit ($\alpha 7$) is developmentally regulated and forms a homo-oligomeric channel blocked by α-BTX. Neuron 5:847–856

Damsma G, Day J, Fibiger HC (1989) Lack of tolerance to nicotine-induced dopamine release in the nucleus accumbens. Eur J Pharmacol 168:363–368

De la Garza R, Bickford-Wimer PC, Hoffer BJ, Freedman R (1987a) Heterogeneity of nicotine actions in the rat cerebellum: an in vivo electrophysiologic study. J Pharmacol Exp Ther 240:689–695

De la Garza R, McGuire TJ, Freedman R, Hoffer BJ (1987b) The electrophysiological effects of nicotine in the rat cerebellum: evidence for direct postsynaptic actions. Neurosci Lett 80:303–308

De Wit H, Bodker B, Ambre J (1992) Rate of increase of plasma drug level influences subjective response in humans. Psychopharmacology (Berl) 107: 352–358

Deneau GA, Inoki R (1967) Nicotine self-administration in monkeys. Ann NY Acad Sci 142:277–279

Deneris ES, Connolly J, Rogers SW, Duvoisin R (1991) Pharmacological and functional diversity of neuronal nicotinic acetylcholine receptors. Trends Pharmacol Sci 12:34–40

Di Chiara G, Imperato A (1988) Drugs abused by humans preferentially increase synaptic dopamine concentrations in the mesolimbic system of freely moving rats. Proc Natl Acad Sci USA 85:5274–5278

Djordjevic MV, Hoffmann D, Glynn T, Connolly GN (1995) US commercial brands of moist snuff, 1994: assessment of nicotine, moisture, and pH. Tobacco Control 4:62–66

Domino EF, von Baumgarten AM (1969) Tobacco cigarette smoking and patellar reflex depression. Clin Pharmacol Ther 10:72–79

Edwards JA, Warburton D (1983) Smoking, nicotine and electrocortical activity. Pharmacol Ther 19:147–164

Egan TM, North RA (1986) Actions of acetylcholine and nicotine on rat locus coeruleus neurons in vitro. Neuroscience 19:565–571

El-Bizri H, Clarke PBS (1994a) Blockade of nicotinic receptor-mediated release of dopamine from striatal synaptosomes by chlorisondamine and other nicotinic antagonists administered in vitro. Br J Pharmacol 111:406–413

El-Bizri H, Clarke PBS (1994b) Regulation of nicotinic receptors in rat brain following quasi-irreversible nicotinic blockade by chlorisondamine and chronic treatment with nicotine. Br J Pharmacol 113:917–925

Engberg G, Hajos M (1994) Nicotine-induced activation of locus coeruleus neurons–
    an analysis of peripheral versus central induction. Naunyn Schmiedebergs Arch
    Pharmacol 349:443–446
Essman WB (1973) Nicotine-related neurochemical changes: some implications for
    motivational mechanisms and differences. In: Dunn WL (ed) Smoking behavior:
    motives and incentives. Winston, Washington, p 51
Evans SM, Cone EJ, Henningfield JE (1995) Rapid arterial kinetics of intravenous
    and smoked cocaine; Relationship to subjective and cardiovascular effects.
    NIDA Research Monograph 153. US Dept of Health and Human Services,
    Public Health Service. National Institutes of Health, Rockville (NIH publication
    no. 95-3883)
Fagerstrom KO, Schneider NG (1989) Measuring nicotine dependence in tobacco
    smoking: a review of the Fagerstrom Tolerance Questionnaire. J Behav Med
    12:159–182
Farre M, Cami J (1991) Pharmacokinetic considerations in abuse liability evaluation.
    Br J Addict 86:1601–1606
Fiore MC, Jorenby DE, Baker TB, Kenford SL (1992) Tobacco dependence and the
    nicotine patch: clinical guidelines for effective use. JAMA 268:2687–2694
Flores CM, Rogers SW, Pabreza LA, Wolfe BB, Kellar KJ (1992) A subtype of
    nicotinic cholinergic receptor in rat brain is composed of $\alpha 4$ and $\beta 2$ subunits and
    is up-regulated by chronic nicotine treatment. Mol Pharmacol 41:31–37
Foulds J, Stapleton J, Hayward M, Russell MAH, Feyerabend C, Fleming T,
    Costello J (1993) Transdermal nicotine patches with low-intensity support to aid
    smoking cessation in outpatients in a general hospital. Arch Fam Med 2:417–423
Freedman AM (1994) How tobacco giant doctors snuff brands to boost their "kick".
    Wall Street Journal 26 Oct; A1
Fuxe K, Everitt BJ, Hökfelt T (1979) On the action of nicotine and cotinine on
    central 5-hydroxytryptamine neurons. Pharmacol Biochem Behav 10:671–677
Fuxe K, Andersson K, Eneroth P, Harfstrand A, Agnati LF (1989) Neuroendocrine
    actions of nicotine and of exposure to cigarette smoke: Medical implications.
    Neuroendocrinology 13:19–41
Gallup GH (1992) Teen-age attitudes and behavior concerning tobacco. The George
    H. Gallup International Institute. GII 9104. Princeton, NJ
Gillberg PG, d'Argy R, Aquilonius SM (1988) Autoradiographic distribution of ^{3}H-
    acetylcholine binding sites in the cervical spinal cord of man and some other
    species. Neurosci Lett 90:197–202
Giorguieff-Chesselet MF, Kemel ML, Wandscheer D, Glowinski J (1979) Regulation
    of dopamine release by presynaptic nicotinic receptors in rat striatal slices: effect
    of nicotine in a low concentration. Life Sci 25:1257–1262
Giovino GA, Henningfield JE, Tomar SL, Escobedo LG, Slade J (1995) Epidemiology
    of tobacco use and dependence. Epidemiol Rev (in press)
Godukhin OV, Budantsev AY, Selifanova OV, Agapova VN (1984) Effect of cho-
    linomimetics on the release and uptake of L-3H glutamic acid in rat neostriatum.
    Cell Mol Neurobiol 4:117–124
Golding JF (1988) Effects of cigarette smoking on resting EEG, visual evoked
    potentials and photic driving. Pharmacol Biochem Behav 29:23–32
Goldberg SR, Kelleher RT, Goldberg DM (1981a) Fixed-ratio responding under
    second-order schedules of food presentation or cocaine injection. J Pharm
    Exper Ther 218:271–281
Goldberg SR, Spealman RD, Goldberg DM (1981b) Persistent behavior at high rates
    maintained by intravenous nicotine self-adminstration. Science 214:573–575
Goldberg SR, Henningfield JE (1988) Reinforcing effects of nicotine in humans and
    experimental animals responding under intermittent schedules of IV drug injec-
    tion. Pharmacol Biochem Behav 30:227–234
Goldberg SR, Risner ME, Stolerman IP, Reavill C, Garcha HS (1989) Nicotine and
    some related compounds: effects on schedule-controlled behavior and discrimi-
    native properties in rats. Psychopharmacology (Berl) 97:295–302

Gori GB, Benowitz NL, Lynch CJ (1986) Mouth versus deep airways absorption of nicotine in cigarette smokers. Pharmacol Biochem Behav 25:1181–1184

Grady SR, Marks MJ, Wonnacott S, Collins AC (1992) Characterization of nicotinic receptor-mediated [^{3}H]dopamine release from synaptosomes prepared from mouse striatum. J Neurochem 59:848–856

Grady SR, Marks MJ, Collins AC (1994) Desensitization of nicotine-stimulated [^{3}H]dopamine release from mouse striatal synaptosomes. J Neurochem 62: 1390–1398

Grenhoff J, Aston-Jones G, Svensson TH (1986) Nicotinic effects on the firing pattern of midbrain dopamine neurons. Acta Physiol Scand 128:351–358

Griffiths RR, Bigelow GE, Liebson I (1976) Facilitation of human tobacco self-administration by ethanol: A behavioral analysis. J Exp Anal Behav 25:279–292

Griffiths R, Bigelow GE, Liebson I, Kaliszak JE (1980) Drug preference in humans: Double-blind choice comparison of pentobarbital, diazepam and placebo. J Pharmacol Exp Ther 215:649–661

Grunwald F, Schrock H, Kuschinsky W (1987) The effect of an acute nicotine infusion on the local cerebral glucose utilization of the awake rat. Brain Res 400:232–238

Grunwald F, Schrock H, Theilen H, Biber A, Kuschinsky W (1988) Local cerebral glucose utilization of the awake rat during chronic administration of nicotine. Brain Res 456:350–356

Hagan RM, Kilpatrick GJ, Tyers MB (1993) Interactions between 5-HT$_3$ receptors and cerebral dopamine function: implications for the treatment of schizophrenia and psychoactive substance abuse. Psychopharmacology (Berl) 112 Suppl:S68–S75

Hajos M, Engberg G (1988) Role of primary sensory neurons in the central effects of nicotine. Psychopharmacology (Berl) 94:468–470

Harris CM, Emmett-Oglesby MW, Robinson NG, Lal H (1986) Withdrawal from chronic nicotine substitutes partially for the interoceptive stimulus produced by pentylenetetrazol (PTZ). Psychopharmacology (Berl) 90:85–89

Hatsukami DK, Gust SW, Keenan RM (1987) Physiological and subjective changes from smokeless tobacco withdrawal. Clin Pharmacol Ther 41:103–107

Hatsukami DK, Skoog K, Huber M, Hughes JR (1991) Signs and symptoms from nicotine gum abstinence. Psychopharmacology (Berl) 104:496–504

Heatherton TF, Kozlowski LT, Frecker RC, Fagerstrom KO (1991) The Fagerstrom Test for nicotine dependence: a revision of the Fagerstrom Tolerance Questionnaire. Br J Addict 86:1119–1127

Helton DR, Modlin DL, Tizzano JP, Rasmussen K (1993) Nicotine withdrawal: a behavioral assessment using schedule controlled responding, locomotor activity, and sensorimotor reactivity. Psychopharmacology (Berl) 113:205–210

Henningfield JE (1984) Pharmacologic basis and treatment of cigarette smoking. J Clin Psychiatry 45:24–34

Henningfield JE (1992) Occasional drug use: Comparing nicotine with other addictive drugs. Tobacco Control 1:161–162

Henningfield JE (1994) Do nicotine replacement medications work? A unique standard for nicotine. Addiction 89:434–436

Henningfield JE, Griffiths RR (1981) Cigarette smoking and subjective response: Effects of d-amphetamine. Clin Pharm Ther 30:497–505

Henningfield JE, Goldberg SR (1983) Nicotine as a reinforcer in human subjects and laboratory animals. Pharmacol Biochem Behav 19:989–992

Henningfield JE, Chait LD, Griffiths RR (1984) Effects of ethanol on cigarette smoking by volunteers without histories of alcoholism. Psychopharmacology 82:1–5

Henningfield JE, Miyasato K, Jasinski DR (1985) Abuse liability and pharmacodynamic characteristics of intravenous and inhaled nicotine. J Pharmacol Exp Ther 234:1–12

Henningfield JE, Nemeth-Coslett R (1988) Nicotine dependence, interface between tobacco and tobacco-related disease. Chest 93:37S–55S

Henningfield JE, London ED, Benowitz NL (1990) Arterio-venous differences in plasma concentration of nicotine after cigarette smoking. JAMA 263:2049–2050

Henningfield JE, Cohen C, Slade JD (1991) Is nicotine more addictive than cocaine? Br J Addict 86:565–569

Henningfield JE, Keenan RM (1993) Nicotine delivery kinetics and abuse liability. J Consul Clin Psychol 61:743–750

Henningfield JE, Stapleton JM, Benowitz NL, Grayson RF, London ED (1993) Higher levels of nicotine in arterial than in venous blood after cigarette smoking. Drug Alcohol Depend 33:23–29

Henningfield JE, Kozlowski LT, Benowitz NL (1994) A proposal to develop meaningful labeling for cigarettes. JAMA 272:312–314

Henningfield JE, Heishman SJ (1995) The addictive role of nicotine in tobacco use. Psychopharmacology 117:11–13

Henningfield JE, Schuh LM, Jarvik ME (1995a) Pathophysiology of tobacco dependence. In: Bloom FE, Kupfer DJ (eds) Psychopharmacology: the fourth generation of progress. Raven, New York, pp 1715–1729

Henningfield JE, Radzius A, Cone EJ (1995b) Estimation of available nicotine content of six smokeless tobacco products. Tobacco Control (in press)

Herning RI, Pickworth WB (1985) Nicotine gum improved stimulus processing during tobacco withdrawal. Psychohysiology 22:595

Herning RI, Jones RT, Bachman J (1983) EEG changes during tobacco withdrawal. Psychophysiology 20:507–512

Hery F, Bourgoin S, Hamon M, Ternaux JP, Glowinski J (1977) Control of the release of newly synthetized 3H-5-hydroxytryptamine by nicotinic and muscarinic receptors in rat hypothalamic slices. Naunyn Schmiedebergs Arch Pharmacol 296:91–97

Hillard CJ, Pounds JJ (1991) [^{3}H]Tetraphenylphosphonium accumulation in cerebral cortical synaptosomes as a measure of nicotine-induced changes in membrane potential. J Pharmacol Exp Ther 259:1118–1123

Hirschman RS, Leventhal H, Glynn K (1984) The development of smoking behavior: Conceptualization and supportive cross-sectional survey data. J Appl Social Psychol 14(3):184–206

Hughes JR, Higgins ST, Hatsukami DK (1990) Effects of abstinence from tobacco: a critical review. In: Kozlowski LT, Annis H, Cappell HD, Glaser F, Goodstadt M, Israel Y, Kalant H, Sellers EM, Vingilis J (eds) Research advances in alcohol and drug problems, vol 10. Plenum, New York, p 318

Hughes JR, Gulliver SB, Fenwick JW, Valliere WA, Cruser K, Pepper S, Shea P, Solomon LJ, Flynn BS (1992) Smoking cessation among self-quitters. Health Psychol 11:331–334

Hulihan-Giblin BA, Lumpkin MD, Kellar K (1990) Effects of chronic administration of nicotine on prolactin release in the rat: inactivation of prolactin response by repeated injections of nicotine. J Pharmacol Exp Ther 252:21–25

Imperato A, Mulas A, Di Chiara G (1986) Nicotine preferentially stimulates dopamine release in the limbic system of freely moving rats. Eur J Pharmacol 132:337–338

Izenwasser S, Jacocks HM, Rosenberger JG, Cox BM (1991) Nicotine indirectly inhibits [^{3}H]dopamine uptake at concentrations that do not directly promote [^{3}H]dopamine release in rat striatum. J Neurochem 56:603–610

Jaffe JH (1985) Drug addiction and drug abuse. In: Gilman AG, Goodman LS, Rall TW, Murad F (eds) Goodman and Gilman's the pharmacologic basis of therapeutics, 7th edn. MacMillan, New York, pp 532–581

Jarvik ME, Assil KM (1988) Mecamylamine blocks the burning sensation of nicotine on the tongue. Chem Senses 13:213–217

Jarvik ME (1995) The addictive role of nicotine in tobacco use: commentary. Psychopharmacology 117:18–20

Jasinski DR (1977) Assessment of the abuse potentiality of morphinelike drugs (methods used in man). In: Martin WR (ed) Handbook of experimental pharmacology, vol 45. Springer, Berlin Heidelberg New York, pp 197–249

Johnson HM, Torres BA (1985) Regulation of lymphokine production by arginine vasopressin and oxytocin: modulation of lymphocyte function by neurohypophyseal hormones. J Immunol 135:773–775

Jones GMM, Sahakian BJ, Levy R, Warburton DM, Gray JA (1992) Effects of acute subcutaneous nicotine on attention, information processing and short-term memory in Alzheimer's disease. Psychopharmacology (Berl) 108:485–494

Jones RT, Farrell TR III, Herning RI (1978) Tobacco smoking and nicotine tolerance. NIDA Res Monogr 20:202–208

Juan H (1982) Nicotinic nociceptors on perivascular sensory nerve endings. Pain 12:259–264

Karras A, Kane JM (1980) Naloxone reduces cigarette smoking. Life Sci 27:1541–1545

Keenan RM, Hatsukami DK, Pickens RW, Gust SW, Strelow LJ (1990) The relationship between chronic ethanol exposure and cigarette smoking in the laboratory and the natural environment. Psychopharmacology 100:77–83

Keenan RM, Jenkins AJ, Cone EJ, Henningfield JE (1994a) Smoked and iv nicotine, cocaine and heroin have similar abuse liability. J Addict Dis 13:259

Keenan RM, Hatsukami DK, Pentel P, Thompson T, Grillo MA (1994b) Pharmacodynamic effects of cotinine in abstinent cigarette smokers. Clin Pharmacol Ther 55:581–590

Keenan RM, Hatsukami DK, Pentel P, Thompson T, Grillo MA (1995) Reply to Mr. Foulds. Clin Pharmacol Ther 56:95–97

Kety SS (1951) The theory and application of the exchange of inert gas at the lungs and tissues. Pharmacol Rev 3:1–41

Kitamura T (1958) Studies on the pharmacological properties of nicotine related derivatives. Especially about the pharmacology of nornicotine. Folia Pharmacol Jpn 54:825–837

Knott VJ, Venables PH (1977) EEG alpha correlates of nonsmokers, smokers, smoking and smoking deprivation. Psychophysiology 14:150–156

Knott VJ (1988) Dynamic EEG changes during cigarette smoking. Neuropsychobiology 19:54–60

Kottke TE, Battista RN, DeFriese GH, Brekke ML (1988) Attributes of successful smoking cessation interventions in medical practice: a meta-analysis of 39 controlled trials. JAMA 259:2883–2889

Kozlowski LT (1976) Effects of caffeine consumption of nicotine consumption. Psychopharmacology 47:165–168

Kozlowski LT (1981) Tar and nicotine delivery of cigarettes: what a difference a puff makes. JAMA 245:158–159

Kozlowski LT (1982) The determinates of tobacco use: cigarette smoking in the context of other forms of tobacco use. Can J Pub Health 73:236–241

Kozlowski LT, Wilkinson DA, Skinner W, Kent C, Franklin T, Pope M (1989) Comparing tobacco cigarette dependence with other drug dependencies. JAMA 261:898–901

Kozlowski LT, Henningfield JE, Keenan RM, Lei H, Leigh G, Jelinek LC, Pope MA, Heartzen CA (1993) Patterns of alcohol, cigarette, and caffeine and other drug use in two drug abusing populations. J Subst Abuse Treat 10:171–179

Kyerematen GA, Dvorchik BH, Vesell ES (1983) Influence of different forms of tobacco intake on nicotine elimination in man. Pharmacology 26:205–209

Kyerematen GA, Vesell ES (1991) Metabolism of nicotine. Drug Metab Rev 23:3–41

Lapchak PA, Araujo DM, Quirion R, Collier B (1989a) Effect of chronic nicotine treatment on nicotinic autoreceptor function and N-3Hmethylcarbamylcholine binding sites in the rat brain. J Neurochem 52:483–491

Lapchak PA, Araujo DM, Quirion R, Collier B (1989b) Presynaptic cholinergic mechanisms in the rat cerebellum: evidence for nicotinic, but not muscarinic autoreceptors. J Neurochem 53:1843–1851

Larson PS, Haag HB, Silvette H (1961) Tobacco: experimental and clinical studies. Williams and Wilkins, Baltimore

Lautenbach BF (1886/1887) On a new function of the liver. Philadelphia Medical Times 7:387–394

Lee BL, Benowitz NL, Jacob P (1987) Cigarette abstinence, nicotine gum, and theophylline disposition. Ann Intern Med 106:553–555

Lee L-Y, Gerhardstein DC, Wang AL, Burki NK (1993) Nicotine is responsible for airway irritation evoked by cigarette smoke inhalation in men. J Appl Physiol 75:1955–1961

Levin ED (1992) Nicotinic systems and cognitive function. Psychopharmacology (Berl) 108:417–431

Lichtensteiger W, Hefti F, Felix D, Huwyler T, Melamed E, Schlumpf M (1982) Stimulation of nigrostriatal dopamine neurones by nicotine. Neuropharmacology 21:963–968

Lide DR (1991) CRC handbook of chemistry and physics, 72nd edn. CRC, Boca Raton

London ED (1990) Effects of nicotine on cerebral metabolism. In: Boch G, Marsh J (eds) The biology of nicotine dependence. Wiley, New York, p 131

London ED, Connolly RJ, Szikszay M, Wamsley JK (1985a) Distribution of cerebral metabolic effects of nicotine in the rat. Eur J Pharmacol 110:391

London ED, Waller SB, Wamsley JK (1985b) Autoradiographic localization of ^{3}H nicotine binding sites in the rat brain. Neurosci Lett 53:179–184

London ED, Connolly RJ, Szikszay M, Wamsley JK, Dam M (1988) Effects of nicotine on local cerebral glucose utilization in the rat. J Neurosci 8:3920–3928

London ED, Cascella NG, Wong DF, Phillips RL, Dannals RF, Links JM, Herning R, Grayson R, Jaffe JH, Wagner HN Jr (1990a) Cocaine-induced reduction of glucose utilization in human brain. Arch Gen Psychiaty 47:567–574

London ED, Broussolle EPM, Links JM, Wong DF, Cascella NG, Dannals DF, Sano M, Herning RI, Snyder FR, Rippetoe SK, Toung TJK, Jaffe JH, Wagner HN Jr (1990b) Morphine-induced metabolic changes in human brain: studies with positron emission tomography and ^{18}fluorine fluorodeoxyglucose. Arch Gen Psychiatry 47:73–81

Lukas SE, Mendelson JH, Amass L, Benedikt R (1990) Behavioral and EEG studies of acute cocaine administration: comparisons with morphine, amphetamine, pentobarbital, nicotine, ethanol and marijuana. NIDA Res Monogr 95:146–151

Malin DH, Lake JR, Carter VA, Cunningham JS, Wilson OB (1993) Naloxone precipitates nicotine abtinence syndrome in the rat. Psychopharmacology (Berl) 112:339–342

Malin DH, Lake JR, Carter VA, Cunningham JS, Hebert KM, Conrad DL, Wilson OB (1994) The nicotinic antagonist mecamylamine precipitates nicotine abstinence syndrome in the rat. Psychopharmacology 115:180–184

Marks MJ, Burch JB, Collins AC (1983a) Effects of chronic nicotine infusion on tolerance development and nicotinic receptors. J Pharmacol Exp Ther 226:817–825

Marks MJ, Burch JB, Collins AC (1983b) Genetics of nicotine response in four inbred strains of mice. J Pharmacol Exp Ther 226:291–302

Marks MJ, Grady SR, Collins AC (1993) Downregulation of nicotinic receptor function after chronic nicotine infusion. J Pharmacol Exp Ther 266:1268–1276

Maultsby PN, Lake JR, Cortes CW, Lanier JG, Roberts LK, Wilson OB, Malin DH (1991) Aminal model of nicotine dependence and abstinence syndrome. Soc Neurosci Abstr 17:1251

McCormick DA, Prince DA (1987a) Acetylcholine causes rapid nicotinic excitation in the medial habenular nucleus of guinea pig in vitro. J Neurosci 7: 742–752

McCormick DA, Prince DA (1987b) Actions of acetylcholine in the guinea-pig and cat medial and lateral geniculate nuclei, in vitro. J Physiol (Lond) 392:147–165

McGinnis JM, Foege WH (1993) Actual causes of death in the United States. JAMA 270:2207–2212

McNamara D, Larson DM, Rapoport SI, Soncrant TT (1990) Preferential metabolic activation of subcortical brain areaas by acute administration of nicotine to rats. J Cereb Blood Flow Metab 10:48–56

McNeill A (1991) The development of dependence on smoking in clildren. Br J Addict 86:589–592

McNeill AD, Jarvis MJ, Stapleton JA, West RJ, Bryant A, (1989) Nicotine intake in young smokers: longitudinal study of saliva cotinine concentrations. Am J Pub Health 79:172–175

Mello NK, Mendelson JH, Sellers ML, Kuehnle JC (1980) Effects of heroin self-administration on cigarette smoking. Psychopharmacology 67:45–52

Meikle AW, Liu XH, Taylor GN, Stringham JD (1988) Nicotine and cotinine effects on 3-alpha-hydroxysteroid dehydrogenase in canine prostate. Life Sci 43: 1845–1850

Mereu G, Yoon KW, Boi V, Gessa GL, Naes L, Westfall TC (1987) Preferential stimulation of ventral tegmental area dopaminergic neurons by nicotine. Eur J Pharmacol 141:395–399

Meyer EM, Arendash GW, Judkins JH, Ying L, Wade C, Kem WR (1987) Effects of nucleus basalis lesions on the muscarinic and nicotinic modulation of [^{3}H] acetylcholine release in the rat cerebral cortex. J Neurochem 49:1758–1762

Mifsud JC, Hernandez L, Hoebel BG (1989) Nicotine infused into the nucleus accumbens increase synaptic dopamine as measured by in vivo microdialysis. Brain Res 478:365–367

Mitchell SN (1993) Role of the locus coeruleus in the noradrenergic response to a systemic administration of nicotne. Neuropharmacology 32:937–949

Mulle C, Changeux J-P (1990) A novel type of nicotinic receptor in the rat central nervous system characterized by patch-clamp techniques. J Neurosci 10:169–175

Mulle C, Vidal C, Benoit P, Changeux J-P (1991) Existence of different subtypes of nicotinic acetylcholine receptors in the rat habenulo-interpeduncular system. J Neurosci 11:2588–2597

Munck A, Guyre PM, Holbrook NJ (1984) Physiological function of glucocorticoids in stress and their relation to pharmacological actions. Endocr Rev 5:25–44

Nemeth-Coslett R, Henningfield JE, O'Keefe MK, Griffiths RR (1986a) Effects of marijuana smoking on subjective ratings and tobacco smoking. Pharmacol Biochem Behav 25:659–665

Nemeth-Coslett R, Henningfield JE, O'Keefe MK, Griffiths RR (1986b) Effects of mecamylamine on human cigarette smoking and subjective ratings. Psycho-pharmacology 88:420–425

Nestler EJ (1992) Molecular mechainsms of drug addiction. J Neurosci 12:2439–2450

Nil R, Buzzi R, Battig K (1984) Effects of single doses of alcohol and caffeine on cigarette smoke puffing behavior. Pharmacol Biochem Behav 20:583–590

Nisell M, Nomikos GGF, Svensson TH (1994) Systemic nicotine-induced dopamine release in the rat nucleus accumbens is regulated by nicotinic receptors in the ventral tegmental area. Synapse 16:36–44

Oliveto AH, Hughes JR, Terry SY (1991) Effects of caffeine on tobacco withdrawal. Clin Pharmacol Ther 50:157–164

Ossip, DJ, Epstein LH (1981) Relative effects of nicotine and coffee on cigarette smoking. Addict Behav 6:35–39

Palmer KJ, Buckley MM, Faulds D (1992) Transdermal nicotine: a review of its pharmacodynamic and pharmacokinetic properties, and therapeutic efficacy as an aid to smoking cessation. Drugs 44:498–529

Pang Y, Kiba H, Jayaraman A (1993) Acute nicotine injections induce c-fos mostly in non-dopaminergic neurons of the midbrain of the rat. Brain Res Mol Brain Res 20:162–170

Patterson TR, Stringham JD, Meikle AW (1990) Nicotine and cotinine inhibit steroidogenesis in mouse Leydig cells. Life Sci 46:265–272

Pauly JR, Grün EU, Collins AC (1990a) Chronic corticosterone administration modulates nicotine sensitivity and brain nicotinic receptor binding in C3H mice. Psychopharmacology (Berl) 101:310–316

Pauly JR, Ullman EA, Collins AC (1990b) Strain differences in adrenalectomy-induced alterations in nicotine sensitivity in the mouse. Pharmacol Biochem Behav 35:171–179

Perkins KA, Epstein LH, Stiller RL, Sexton JE, Debske TD, Jacob RG (1990a) Behavioral performance effects of nicotine in smokers and nonsmokers. Pharmacol Briochem Behav 37:11–15

Perkins KA, Epstein LH, Stiller RL, Sexton JE, Fernstrom MH, Jacob RG, Solberg R (1990b) Metabolic effects of nicotine after consumption of a meal in smokers and nonsmokers. Am J Clin Nutr 52:228–233

Perkins KA, Epstein LH, Stiller RL, Fernstrom MH, Sexton JE, Jacob RG, Solberg R (1991a) Acute effects of nicotine on hunger and caloric intake in smokers and nonsmokers. Psychopharmacology (Berl) 103:103–109

Perkins KA, Stiller RL, Jennings JR (1991b) Acute tolerance to the cardiovascular effects of nicotine. Drug Alcohol Depend 29:77–85

Perkins KA, Grobe JE, Epstein LH, Caggiula AR, Stiller RL (1992) Effects of nicotine on subjective arousal may be dependent on baseline subjective state. J Subst Abuse 4:131–141

Perkins KA DiMarco A, Grobe JE, Scierka A, Stiller RL (1994) Nicotine discrimination in male and female smokers. Psychopharmacology 116:407–413

Perry EK, Smith CJ, Perry RH, Whitford C, Johnson M, Birdsall NJ (1989) Regional distribution of muscarinic and nicotinic cholinergic receptor binding activities in the human brain. J Chem Neuroanat 2:189–199

Peto R, Lopez AD, Boreham J, Thun M, Heath C Jr (1994) Mortality from smoking in developed countries 1950–2000. Oxford University Press, New York

Pickworth WB, Herning RI, Henningfield JE (1986) Electroencephalographic effects of nicotine chewing gum in humans. Pharmacol Biochem Behav 25:879–882

Pickworth WB, Herning RI, Henningfield JE (1988) Mecamylamine reduces some EEG effects of nicotine chewing gum in humans. Pharmacol Biochem Behav 30:149–153

Pickworth WB, Herning RI, Henningfield JE (1989) Spontaneous EEG changes during tobacco abstinence and nicotine substitution in human volunteers. J Pharmacol Exp Ther 251:976–982

Pomerleau CS, Pomerleau OF (1992) Euphoriant effects of nicotine in smokers. Phychopharmacology 108:460–465

Pool WF, Crooks PA (1985) Biotransformation of primary nicotine metabolites: In vivo metabolism of R-(+)-[14 C-NCH3]-N-methylnicotinium ion in the guinea-pig. Drug Metab Disposit 13:578–581

Prusky GT, Shaw C, Cynader MS (1987) Nicotine receptors are located on lateral geniculate nucleus terminals in cat visual cortex. Brain Res 412:131–138

Quirion R, Richard J, Wilson A (1994) Muscarinic and nicotinic modulation of cortical acetylcholine release monitored by in vivo microdialysis in freely moving adult rats. Synapse 17:92–100

Rapier C, Lunt GG, Wonnacott S (1988) Stereoselective nicotine-induced release of dopamine from striatal synaptosomes: concentration dependence and repetitive stimulation. J Neurochem 50:1123–1130

Rapier C, Lunt GG, Wonnacott S (1990) Nicotinic modulation of ^{3}H-dopamine release from striatal synaptosomes: pharmacological characteristation. J Neurochem 54:937–945

Raveis VH, Kandel DB (1987) Changes in drug behavior from the middle to the late twenties: Initiation, persistence, and cessation of use. Am J Pub Health 77:607–611

Ren T, Sagar SM (1992) Induction of c-fos immunostaining in the rat brain after the systemic administration of nicotine. Brain Res Bull 29:589–597

Revell AD (1988) Smoking and performance – a puff-by-puff analysis. Psychopharmacology (Berl) 96:563–565

Richardson NR, Roberts DCS (1991) Fluoxetine pretreatment reduces breaking points on a progressive ratio schedule reinforced by intravenous cocaine self-administration in the rat. Life Sci 49:833–840

Risner MR, Goldberg SR, Prada JA, Cone EJ (1985) Effects of nicotine, cocaine, and some of their metabolites on schedule-controlled responding by beagle dogs and squirrel monkeys. J Pharmacol Exp Ther 234:113–119

Robinson JH, Pritchard WS, Davis RA (1992) Psychopharmacological effects of smoking a cigarette with typical "tar" and carbon monoxide yields but minimal nicotine. Psychopharmacology (Berl) 108:466–472

Rose JE, Behm FM (1994) Inhalation of vapor from black pepper extract reduces smoking withdrawal symptoms. Drug Alcohol Depend 34:225–229

Rose JE, Tashkin DP, Ertle A, Zinser MC, Lafr R (1985) Sensory blockade of smoking satisfaction. Pharmacol Biochem Behav 23:289–293

Rose JE, Sampson A, Levin ED, Henningfield JE (1989) Mecamylamine increases nicotine preference and attenuates nicotine discrimination. Pharmacol Biochem Behav 32:933–938

Rose JE, Behm FM, Levin ED (1993) Role of nicotine dose and sensory cues in the regulation of smoke intake. Pharmacol Biochem Behav 44:891–900

Rosencrans JA, Meltzer LT (1981) Central sites and mechanisms of action of nicotine. Neurosci Biobehav Rev 5:497–501

Rowell PP, Winkler DL (1984) Nicotinic stimulation of [3h]acetylcholine release from mouse cerebral cortical synaptosomes. J Neurochem 43:1593–1598

Rowell PP, Hillebrand JA (1994) Characterization of nicotine-induced desensitization of evoked dopamine release from rat striatal synaptosomes. J Neurochem 63:561–569

Rubboli F, Court JA, Sala C, Morris C, Perry E, Clementi F (1994) Distribution of neuronal nicotinic receptor subunits in human brain. Neurochem Int 25:69–71

Russell MAH, Jarvis MJ, Feyerabend C (1980) A new age for snuff? Lancet 474–475

Russell MAH (1990) The nicotine addiction trap: A 40-year sentence for four cigarettes. Br J Addict 85:293–300

Sachs D, Benowitz N (1988) The nicotine withdrawal syndrome: Nicotine absence or caffeine excess? In: Harris LS (ed) Problems of drug dependence, 1988, NIDA research monograph 90, U.S. Government Printing Office, Washington DC, p 38 (DHHS publication no. (ADM) 89-1605)

Sargent PB (1993) The diversity of neuronal nicotinic acetylcholine receptors. Annu Rev Neurosci 16:403–443

Schoepfer R, Conroy WG, Whiting P, Gore M, Lindstrom J (1990) α-4 bungarotoxin binding protein cDNAs reveal subtypes of this branch of the ligand-gated ion channel gene superfamily. Neuron 5:35–48

Schröck H, Kuschinsky W (1991) Effects of nicotine withdrawal on the local cerebral glucose utilization in conscious rats. Brain Res 545:234–238

Schuh LM, Henningfield JE, Pickworth WB, Rothman R, Ohuoha D, Keenan RM (1995) Pharmacodynamic effects of cotinine. Annual Meeting of the Society for Research on Nicotine and Tobacco, San Diego

Schulz DW, Loring RH, Aizenman E, Zigmond RE (1991) Autoradiographic localization of putative nicotinic receptors in the rat brain using ^{125}I-neuronal bungarotoxin. J Neurosci 11:287–297

Schwartz RD, Kellar KJ (1983) Nicotinic cholinergic receptor binding sites in the brain: regulation in vivo. Science 220:214–216

Schwartz RD, Kellar KJ (1985) In vivo regulation of ^{3}H-acetylcholine recognition sites in brain by nicotinic cholinergic drugs. J Neurochem 45:427–433

Schwartz RD, Lehmann J, Kellar KJ (1984) Presynaptic nicotinic cholingeric receptors labeled by ^{3}H-acetylcholine on catecholamine and serotonin axons in brain. J Neurochem 42:1495–1498

Séguéla P, Wadiche J, Dineley-Miller K, Dani JA, Patrick JW (1993) Molecular cloning, functional properties, and distribution of rat brain α7: A nicotinic cation channel highly permeable to calcium. J Neurosci 13:596–604

Sellers EM, Otton SV, Busto UE (1991) Drug metabolism and interactions in abuse liability assessment. Brit J Addict 86:1607–1614

Shavit T, Terman GW, Martin FC, Lewis JW, Liebeskind JC, Gale RR (1985) Stress, opioid peptides, the immune system and cancer. J Immunol 135:834s–837s

Silagy C, Mant D, Fowler G, Lodge M (1994) Meta-analysis on efficacy of nicotine replacement therapies in smoking cessation. Lancet 343:139–142

Silverman K, Evans SM, Strain EC, Grifiths RR (1992) Withdrawal syndrome after the double-blind cessation of caffeine consumption. N Engl J Med 327:1109–1114

Snell LD, Johnson KM (1989) Effects of nicotinic agonists and antagonists on N-methyl-D-aspartate-induced 3H-norepinephrine release and 3H-(1-1-(2-thienyl) cyclohexyl-piperidine) binding in rat hippocampus. Synapse 3:129–135

Snyder FR, Henningfield JE (1989) Effects of nicotine administration following 12 h of tobacco deprivation: assessment on computerized performance tasks. Psychopharmacology (Berl): 97:17–22

Sokoloff L, Reivich M, Kennedy C, DesRosiers MH, Patlak CS, Petigrew KD, Sakurad O, Shinohara M (1977) The [^{14}C] deoxyglucose method for the measurement of local cerebral glucose utilization: theory, procedure, and normal values in the conscious and anesthetized albino rat. J Neurochem 28:897–916

Sopori ML, Gairola CC, DeLucia A, Bryant L, Cherian S (1985) Immune responsiveness of monkeys exposed chronically to cigarette smoke. Clin Immunol Immunopathol 36:338–344

Stapleton JM, Henningfield JE, Wong DF, Phillips RL, Gilson SF, Grayson RF, Dannals RF, London ED (1992) Effects of nicotine on cerebral metabolism and subjective responses in human volunteers. Soc Neurosci Abst 18:1074

Stolerman IP (1991) Behavioural pharmacology of nicotine: multiple mechanisms. Br J Addict 86:533–536

Stolerman IP, Reavill C (1989) Primary cholinergic and indirect dopaminergic mediation of behavioural effects of nicotine. Prog Brain Res 79:227–237

Stolerman IP, Goldfarb T, Fink R, Jarvik ME (1973) Influencing cigarette smoking with nicotine antagonists. Psychopharmacologia (Berl) 28:247–259

Sutherland G, Russell MAH, Stapleton J, Feyerbend C, Ferno O (1992a) Nasal nicotine spray: A rapid nicotine delivery system. Psychopharmacology 108: 512–518

Sutherland G, Stapleton JA, Russell MAH, Jarvis MJ, Hajek P, Belcher M, Feyerabend C (1992b) Randomised controlled trial of nasal nicotine spray in smoking cessation. Lancet 340:324–329

Suwandi IS, Bevan JA (1966) Antagonism of lobeline by ganglion-blocking agents at afferent nerve endings. J Pharmacol Exp Ther 153:1–7

Svensson CK (1987) Clinical pharmacokinetics of nicotine. Clin Pharmacokinet 12:30–40

Swanson LW, Simmons DM, Whiting PJ, Lindstrom J (1987) Immunohistochemical localization of neuronal nicotinic receptors in the rodent central nervous system. J Neurosci 7:3334–3342

Sweanor D, Ballin S, Corcoran RD, Davis A, Deasy K, Ferrence RG, Lahey R, Lucido S, Nethery WJ, Wasserman J (1992) Report of the tobacco policy research study group on tobacco pricing and taxation in the United States. Tobacco Control 1 Suppl:S31–S36

Swedberg MDB, Henningfield JE, Goldberg SR (1990) Nicotine dependency: animal studies. In: Wonnacott S, Russell MAH, Stolerman IP (eds) Nicotine psychopharmacology: molecular, cellular and behavioural aspects. Oxford University Press, Oxford, p 38

Swerdlow NR, Koob GF (1987) Dopamine, schizophrenia, mania, and depression: toward a unified hypothesis of cortico-striato-pallido-thalamic function. Behav Brain Sci 10:197–245

Takada K, Swedberg MD, Goldberg SR, Katz JL (1989) Discriminative stimulus effects of intravenous I-nicotine and nicotine analogs or metabolites in squirrel monkeys. Psychopharmacology (Berl) 99:208–212

Tanelian DL (1991) Cholinergic activation of population of corneal afferent nerves. Exp Brain Res 86:414–420

Tiffany ST, Drobes DJ (1990) Imagery and smoking urges: the manipulation of affective content. Addict Behave 15:531–539

Tiffany ST, Drobes DJ (1991) The development and initial validation of a questionnaire of smoking urges. Br J Addict 86:1467–1476
Tjälve H, Popov D (1973) Effect of nicotine and nicotine metabolites on insulin
  secretion from rabbit pancreas pieces. Endocrinology 92:1343–1348
Tung CS, Ugedo L, Grenhoff J, Engberg G, Svensson TH (1989) Peripheral induction
  of burst firing in locus coeruleus neurons by nicotine mediated via excitatory
  amino acids. Synapse 4:313–318
Ulett JA, Itil TM (1969) Quantitative electroencephalogram in smoking and smoking
  deprivation. Science 164:969–970
US Department of Health and Human Services (1983) The health consequences of
  smoking: cardiovascular disease. A report of the Surgeon General. Public Health
  Service, Office on Smoking and Health, Rockville
US Department of Health and Human Services (1988) The health consequences of
  smoking: nicotine addiction. A report of the Surgeon General. Public Health
  Service, Office on Smoking and Health, Rockville
US Department of Health and Human Services (1990) The health benefits of smoking
  cessation. Public Health Service, Centers for Disease Control Center for chronic
  Disease Prevention and Health Promotion Office on Smoking and Health,
  Rockville DHHS publication no. (CDC) 90-8416
US Department of Health and Human Services (1994) Preventing tobacco use
  among young people: a report of the Surgeon General. US Department of
  Health and Human Services. Public Health Service, Centers for Disease Control,
  National Center for Chronic Disease Prevention and Health Promotion, Office
  on Smoking and Health, Rockville
Wada E, Wada K, Boulter J, Deneris E, Heinemann S, Patrick J, Swanson LW
  (1989) Distribution of $\alpha2$, $\alpha3$, $\alpha4$, and beta 2 neuronal nicotinic receptor subunit
  mRNAs in the central nervous system: a hybridization histochemical study in
  the rat. J Comp Neurol 284:314–335
Wada E, McKinnon D, Heinemann S, Patrick J, Swanson LW (1990) The distribution
  of mRNA encoded by a new member of the neuronal nicotinic acetylcholine
  receptor gene family ($\alpha5$) in the rat central nerous system. Brain Res 526:45–53
West R, Hajek P, Belcher M (1989) Time course of cigarette withdrawal symptoms
  while using nicotine gum. Psychopharmacology (Berl) 99:143–145
Westfall TC (1974) Effect of nicotine and other drugs on the release of 3H-norepine-
  phrine and 3H-dopamine from rat brain slices. Neuropharmacology 13:693–
  700
Westfall TC, Perry H, Vickery L (1987) Mechanisms of nicotine regulation of
  dopamine. In: Martin WR, Van Loon GR, Iwamoto ET, Davis L (eds) Tobacco
  smoking and nicotine. Plenum, New York, p 209
Whitfield JP, MacManus JP, Gillan DJ (1970) The possible mediation by cyclic-AMP
  of the stimulation of thymocyte proliferation by vasopressin and the inhibition of
  this mitogenic action by thyrocalcitonin. J Cell Physiol 76:65–76
Whiting P, Lindstrom J (1986) Pharmacological properties of immuno-isolated
  neuronal nicotinic receptors. J Neurosci 6:3061–3069
Whiting P, Schoepfer R, Lindstrom J, Priestley T (1991) Structural and pharmacological characterization of the major brain nicotinic acetylcholine receptor subtype
  stably expressed in mouse fibroblasts. Mol Pharmacol 40:463–472
Wilbert J (1987) Tobacco and shamanisms in South America. Yale University Press,
  New Haven
Wise RA, Hoffman DC (1992) Localization of drug reward mechanisms by intracranial
  injections. Synapse 10:247–263
Wong LA, Gallagher JP (1989) A direct nicotinic receptor-mediated inhibition
  recorded intracellularly in vitro. Nature 341:439–444
Wong LA, Gallagher JP (1991) Pharmacology of nicotinic receptor-mediated inhibition
  in rat dorsolateral septal neurones. J Physiol (Lond) 436:325–346
Wonnacott S, Irons J, Rapier C, Thorne B, Lunt GG (1989) Presynaptic modulation
  of transmitter release by nicotine receptors. Prog Brain Res 79:157–163

Wybran J (1985a) Enkephalins and endorphins: activation molecules for the immune system and natural killer activity. Neuropeptides 5:371–374

Wybran J (1985b) Enkephalins and endorphins as modifiers of the immune system: Present and future. FEd Proc 44:92–94

Yamamoto K, Domino EF (1965) Nicotine-induced EEG and behavioral arousal. Int J Neuropharmacol 4:359–373

Yeh J, Barbieri RJ, Friedman AJ (1989) Nicotine and cotinine inhibit rat testes androgen biosynthesis in vitro. J Steroid Biochem 33:627–630

Yoshida K, Kato Y, Imura H (1980) Nicotine-induced release of noradrenaline from hypothalamic synaptosomes. Brain Res 182:361–368

CHAPTER 9

Caffeine Reinforcement, Discrimination, Tolerance and Physical Dependence in Laboratory Animals and Humans

R.R. GRIFFITHS and G.K. MUMFORD

A. Introduction

Caffeine is an excellent model compound for understanding drugs of abuse/ dependence (HOLTZMAN 1990). Historically, caffeine use dates back hundreds, possibly thousands, of years (GRAHAM 1984). Caffeine use spread worldwide from its initially constrained geographical origins, despite recurring efforts to restrict or eliminate its use motivated on moral, economic, political, religious or medical grounds (AUSTIN 1979). Currently, caffeine use is almost universal, with more than 80% of adults in North America regularly consuming behaviorally active doses of caffeine (GILBERT 1976; GRAHAM 1978; HUGHES et al. 1993b). The broad generality of caffeine self-administration is reflected in the facts that, worldwide, caffeine consumption occurs in markedly different vehicles (e.g., drinking of coffee, tea, maté, soft drinks; chewing of kola nuts; consumption of cocoa and guarana products) and in widely different, but culturally well-integrated, social contexts (e.g., the coffee break in the United States; tea time in the United Kingdom; kola nut chewing in Nigeria). The dependence-producing nature of caffeine is reflected in the experience of most people who know someone who has expressed a "need" for a cup of coffee or who claimed to be "addicted" to coffee or their morning caffeinated beverage of choice. Consistent with this anecdotal experience, a recent random-digit telephone survey found that about 17% of current caffeine users met the DSM-III-R psychiatric criteria for being moderately or severely drug dependent on caffeine (HUGHES et al. 1993b). Ultimately, the decision about whether to label caffeine as a drug of abuse is a social-political judgement that will be influenced by culturally accepted norms and perceived health risks, analogous to recent debate in the United States and Europe about whether to consider nicotine as a drug of abuse/ dependence (ROBINSON and PRITCHARD 1992a,b; WEST 1992; HUGHES 1993) or debate in South America about whether to consider oral cocaine use (via chewing or coca tea) as drug abuse (LLOSA 1993).

The outcome of the debate with caffeine is irrelevant to the scientific usefulness of caffeine as a model for understanding the behavioral pharmacological processes that underlie chronically self-administered drugs. Although caffeine produces relatively subtle behavioral effects in both laboratory animals and humans, it nonetheless produces most of the behavioral effects that historically have been considered definitional of a drug of abuse/dependence. The purpose of this chapter is to review a rapidly emerging literature in laboratory animals and humans concerning four behavioral pharmacological effects that caffeine shares with classic drugs of abuse/dependence: reinforcing effects, discriminative/subjective effects, tolerance and physical dependence. The present chapter supplements two other reviews (GRIFFITHS and WOODSON 1988a; GRIFFITHS and MUMFORD 1995) which provide more detailed analyses of selected aspects of the behavioral pharmacology of caffeine.

B. Reinforcing Effects of Caffeine in Laboratory Animals

Reinforcing efficacy of a drug refers to the relative effectiveness in establishing or maintaining behavior on which the delivery of the drug is dependent (GRIFFITHS et al. 1979). Intravenous self-injection in laboratory animals has a high degree of face validity and is often regarded by specialists as providing the most direct and unequivocal assessment of a drug's reinforcing effect (BOZARTH 1987). With this procedure, animals are given access to a lever, responding on which results in a drug injection. The ability of the injection to reinforce behavior is assessed by examining the establishment or maintenance of responding.

As reviewed elsewhere (GRIFFITHS and MUMFORD 1995), nine self-injection studies have examined whether caffeine can function as a reinforcer. Of the seven studies which demonstrated caffeine self-injection, four showed self-injection in all animals (DENEAU et al. 1969; GRIFFITHS et al. 1979; DWORKIN et al. 1993, GRIFFITHS, SANNERUD and KAMINSKI, cited in Table 1 in GRIFFITHS and MUMFORD 1995) while three (SCHUSTER et al. 1969; ATKINSON and ENSSLEN 1976; COLLINS et al. 1984) showed that only a subset of animals (25%–33%) self-injected caffeine. A sporadic pattern of caffeine self-injection, which is characterized by periods of relatively high rates of intake alternating irregularly with periods of low intake, has been reported in three studies with nonhuman primates which examined self-injection over an extended period of consecutive days (DENEAU et al. 1969; GRIFFITHS et al. 1979; GRIFFITHS and MUMFORD 1995).

These intravenous self-injection studies show that caffeine can function as a reinforcer under some conditions. However, the inconsistent results across animals and studies contrast with the results reported with the classic abused stimulants (e.g., amphetamine and cocaine) which have more consistently been shown to maintain intravenous self-injection across a wide range of species and conditions (GRIFFITHS et al. 1979). The variation in results with caffeine is analogous to that which has been reported in self-injection studies with nicotine: nicotine has not reliably maintained self-injection across animals and studies (GOLDBERG and HENNINGFIELD 1988; DWORKIN et al. 1993).

C. Reinforcing Effects of Caffeine in Humans

As in the animal drug self-administration laboratory, procedures have been developed in the human laboratory for assessing the reinforcing effects of drugs. As reviewed in more detail elsewhere (GRIFFITHS and WOODSON 1988c; GRIFFITHS and MUMFORD 1995), 11 studies provide unequivocal evidence of the reinforcing effects of caffeine (GRIFFITHS et al. 1986a,b; GRIFFITHS and WOODSON 1988b; GRIFFITHS et al. 1989; HUGHES et al. 1991, 1992a,b; OLIVETO et al. 1991, 1992b; EVANS et al. 1994; SILVERMAN et al. 1994). These studies demonstrated caffeine reinforcement under double-blind conditions using various subject populations (moderate and heavy caffeine users with and without histories of alcohol or drug abuse), using a variety of different methodological approaches (variations on both choice and ad libitum self-administration procedures), when caffeine was available in different vehicles (coffee, soda or capsules), when subjects did and did not have immediate past histories of chronic caffeine exposure, and in the context of different behavioral requirements after drug ingestion (vigilance vs relaxation activities).

Caffeine reinforcement has been demonstrated in about 45% of normal subjects with histories of moderate and heavy caffeine use. Caffeine reinforcement appears to be an inverted U-shaped function of dose, with high doses sometimes producing significant caffeine avoidance (GRIFFITHS and MUMFORD 1995).

As reviewed elsewhere (GRIFFITHS and MUMFORD 1995), studies have repeatedly demonstrated that qualitative ratings of subjective effects have covaried with measures of reinforcement or choice (GRIFFITHS et al. 1986a,b, 1989; GRIFFITHS and WOODSON 1988b; STERN et al. 1989; EVANS and GRIFFITHS 1992; HUGHES et al. 1993a) An example is provided from a choice study (EVANS and GRIFFITHS 1992) which assessed the subjective effects of placebo and caffeine on forced-exposure days preceding choice days. When the subjective effect data were retrospectively categorized into caffeine choosers and nonchoosers, a face-valid profile of changes in subjective effects emerged: (1) choosers showed "positive" subjective effects of caffeine relative to placebo (e.g., increased alert, content, energetic, liking); (2) nonchoosers showed "negative" effects of caffeine relative to placebo (e.g., increased anxiety, mood disturbance, jittery); and (3) choosers showed "negative" effects of placebo (e.g., increased headache, fatigue).

An implication of the findings that demonstrate a relationship between abstinence-associated headache, fatigue and drowsiness and choice/reinforcement is that physical dependence may potentiate the reinforcing effects of caffeine. In this regard, a retrospective analysis (HUGHES et al. 1993a) of data from four previous studies involving choice between caffeinated coffee (100 mg/cup) vs decaffeinated coffee showed that subjects who reliably reported more headache with decaffeinated than caffeinated coffee were 2.6 times more likely to show reliable choice of caffeinated coffee. To date, experimental studies directly attempting to demonstrate a potentiation of caffeine reinforcement by providing a history of chronic caffeine administration have provided mixed results (Griffiths et al. 1986a; Evans and Griffiths 1992).

D. Discriminative Stimulus Effects of Caffeine in Laboratory Animals

Most centrally acting drugs produce interoceptive stimuli that can be used to train different behavioral responses. Under typical conditions, subjects are trained (e.g., with food reinforcement or shock avoidance) to make one response (e.g., press left lever) after administration of drug and another response (e.g., press right lever) after administration of vehicle. When subjects reach a criterion level of discriminative responding under training conditions, similarities between the stimulus produced by the training drug and the stimulus produced by a novel drug can be assessed in test sessions in which responding is not differentially reinforced. The extent to which the novel drug occasions drug-appropriate responding (i.e., generalization) provides a measure of the similarity between the training stimulus and the test stimulus.

This section is not intended to be a comprehensive review of all studies in which caffeine has served as a test stimulus. Rather it will focus on the 15 studies in which caffeine served as the training stimulus (Overton and Batta 1977; Carney and Christensen 1980; Modrow et al. 1981a,b; Winter 1981; Overton 1982; Carney et al. 1985; Holloway et al. 1985b; Modrow and Holloway 1985; Holtzman 1986, 1987; Mumford and Holtzman 1991; Holloway et al. 1992; Mariathasan and Stolerman 1992; Gauvin et al. 1993) or as part of a training stimulus complex (Mariathasan and Stolerman 1992; Gauvin et al. 1993). Caffeine doses of 10–125 mg/kg have been used as training stimuli. Caffeine discrimination was acquired at markedly different rates across studies (range 13–93 sessions for acquisition). As has been the case with other training drugs (Overton 1982), the rate of acquisition appeared to be dose-dependent, with higher training doses producing more rapid acquisition (Overton 1982; Mumford and Holtzman 1991). All of the subjects have been rats and the training dose of caffeine has been administered parenterally with only one exception (Carney and Christensen 1980).

There is mounting evidence that the discriminative stimulus effects of low caffeine training doses may be related to the behavioral stimulant effects of caffeine (Holtzman 1986; Mumford and Holtzman 1991; Mariathasan and Stolerman 1992). Low and intermediate doses of caffeine (10–30 mg/kg) lie on the ascending limb of the caffeine locomotor stimulant dose-effect curve. Furthermore, many other behavioral stimulants occasion caffeine-appropriate responding at low caffeine training doses (Table 1). This broad

Table 1. Caffeine discrimination in laboratory animals: psychomotor stimulants, but not a variety of other drugs, occasion low-dose caffeine-appropriate responding[a]

Drugs which occasion low-dose caffeine-appropriate responding	Drugs which do not occasion low-dose caffeine-appropriate responding
d-Amphetamine	BCCE
Apomorphine	Ethylketocyclazocine
Cocaine	Fenfluramine
CGS 15943	IBMX
Diethylpropion	Mescaline
Mazindol	Papaverine
Methylphenidate	Pentylenetetrazol
Phendimetrazine	Phencyclidine
Theophylline	Phentolamine
	Yohimbine

[a] Caffeine training dose was 10 mg/kg. Data are adapted from MUMFORD and HOLTZMAN (1991).

pattern of generalization might suggest that the discrimination lacks pharmacological specificity. However, a number of compounds from various pharmacological classes do not occasion caffeine-appropriate responding at low caffeine training doses (Table 1), thus demonstrating the specificity of low-dose caffeine interoceptive stimuli.

Although there is good specificity at low caffeine doses, there appears to be asymmetry in the pattern of cross-generalization between low/intermediate doses of caffeine and other behavioral stimulants. With but two exceptions (JONES et al. 1980; MODROW and HOLLOWAY 1985), caffeine has not been shown to produce training-drug appropriate responding under test conditions in which other behavioral stimulants have served as training drugs. Thus, caffeine alone is insufficient to produce criterion level amphetamine- (KUHN et al. 1974; HOLLOWAY et al. 1985a; ROSEN et al. 1986), methamphetamine- (ANDO and YANAGITA 1992), cocaine- (HARLAND et al. 1989; SILVERMAN and SCHULTZ 1989; GAUVIN et al. 1989, 1990), procaine- (SILVERMAN and SCHULTZ 1989), cathinone- (GOUDIE et al. 1986), phenylpropanolamine- (MARIATHASAN and STOLERMAN 1992; GAUVIN et al. 1993) or ephedrine- (GAUVIN et al. 1993) appropriate responding.

Interestingly, although caffeine does not occasion drug-appropriate responding in animals trained with other behavioral stimulants alone, caffeine can augment the discriminative effects of subthreshold doses of the training drug in amphetamine- (SCHECHTER 1977), cocaine- (HARLAND et al. 1989) and cathinone-trained rats (SCHECHTER 1989). In addition, caffeine/methylephedrine mixtures have been shown to produce criterion level methamphetamine-appropriate (ANDO and YANAGITA 1992) responding. Further, in amphetamine- or cocaine-trained rats, selected mixtures of caffeine, ephedrine

and phenylpropanolamine were shown to produce criterion level amphe-
tamine- (Holloway et al. 1985a) and cocaine-appropriate (Gauvin et al.
1989) responding.

The discriminative stimulus effects of high caffeine doses appear to be
qualitatively different than the discriminative stimulus effects of low caffeine
training doses (Mumford and Holtzman 1991). This change in the discriminative stimuli associated with caffeine appears to occur as a graded
function of training dose. Figure 1 shows that there was an inverse correlation
between caffeine training dose and the extent to which cocaine, a prototypic
psychomotor stimulant, engendered caffeine-appropriate responding. Among
a large number of xanthine and nonxanthine compounds tested, only theophylline has been shown to occasion criterion level drug-appropriate responding in rats trained with high doses of caffeine (Winter 1981; Mumford
and Holtzman 1991). Thus, the discriminative effects of high doses of
caffeine are extremely specific.

I. Pharmacologic Mechanisms in Drug Discrimination

Many of the behavioral effects of caffeine in rodents have been linked to
antagonism of endogenous adenosine. Although several adenosine antagonists

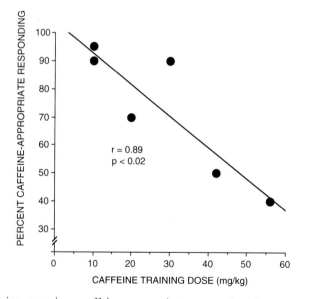

Fig. 1. Cocaine occasions caffeine-appropriate responding in rats as an inverse
function of caffeine training dose. Maximal percentage caffeine-appropriate responding
engendered by cocaine is presented as a function of caffeine training dose. Each data
point represents the results from a different training group across four studies in rats
(10 and 56 mg/kg, Mumford and Holtzman 1991; 20 mg/kg, Mariathasan and
Stolerman 1992; 10 and 30 mg/kg, Holtzman 1986; 42 mg/kg, Gauvin et al. 1993)

occasion caffeine-appropriate responding in rats trained to discriminate low doses of caffeine, caffeine-appropriate responding occurs only with those adenosine antagonists (i.e., CGS 15943, theophylline, 1,7-dimethylxanthine and β-hydroxyethyltheophylline) that share behavioral stimulant effects with these low caffeine training doses (SNYDER et al. 1981; CARNEY et al. 1985; FINN and HOLTZMAN 1987; HOLTZMAN 1991; MUMFORD and HOLTZMAN 1991). Adenosine antagonists such as theobromine and 8-chlorotheophylline, which are weak or ineffective in stimulating locomotor activity (SNYDER et al. 1981), and IBMX, which produces only behavioral depression (SNYDER et al. 1981), fail to produce caffeine-appropriate responding (CARNEY et al. 1985; HOLTZMAN 1986; MUMFORD and HOLTZMAN 1991). In addition, although caffeine and other adenosine antagonists block the discriminative stimulus effects of adenosine agonists (COFFIN and CARNEY 1983; SPENCER and LAL 1983; SPEALMAN and COFFIN 1988), adenosine agonists do not (HOLLOWAY et al. 1985b), or only partially (HOLTZMAN 1986), antagonize the discriminative stimulus effects of caffeine. This partial blockade occurs only at high doses of adenosine agonists and is not completely surmounted by increasing the dose of caffeine, suggesting a functional rather than a receptor-mediated antagonism (HOLTZMAN 1986).

Caffeine also inhibits cyclic nucleotide phosphodiesterase and it has been suggested that this cellular action might also contribute to the discriminative stimulus effects of caffeine. Given that the drugs that occasion low-dose caffeine-appropriate responding are all behavioral stimulants (discussed above) and that most selective phosphodiesterase inhibitors are behavioral depressants (CHOI et al. 1988; HOWELL 1993), it seems unlikely that phosphodiesterase inhibition would be involved in low-dose caffeine discriminative effects. The only low-dose caffeine discrimination study to test a phosphodiesterase inhibitor devoid of adenosine receptor activity (papaverine) failed to show caffeine-appropriate responding (MUMFORD and HOLTZMAN 1991). There are mixed results implicating phosphodiesterase inhibition in the discriminative effects of higher caffeine doses. One study showed that two phosphodiesterase inhibitors (Ro 20-1724 and papaverine) failed to occasion drug-appropriate responding at high caffeine training doses (MUMFORD and HOLTZMAN 1991). However, in another study, papaverine was shown to occasion caffeine-appropriate responding in rats trained to discriminate an intermediate dose of caffeine (HOLLOWAY et al. 1985b). A final study showed that a high dose of caffeine occasioned drug-appropriate responding in rats trained to discriminate the type IV selective phosphodiesterase inhibitor rolipram (YAMAMOTO et al. 1987). Thus, the role of phosphodiesterase inhibition in mediating the discriminative stimulus effects of high doses of caffeine remains open to interpretation.

Caffeine may affect catecholamine neurotransmitter systems and a variety of direct and indirect catecholamine agonists occasion caffeine-appropriate responding. A selective α_1 adrenergic receptor agonist (St 587) occasioned caffeine-appropriate responding (HOLTZMAN 1986). The discriminative effects of this α agonist and those of the caffeine training dose were blocked by both α (phentolamine, prazosin and yohimbine) and β (propranalol) adrenergic receptor antagonists and the antagonism of caffeine by phentolamine was surmountable, resulting in a rightward shift of the caffeine dose-response curve (HOLZMAN 1986). Other drugs including directly acting dopamine receptor agonists (apomorphine), catecholamine uptake inhibitors (cocaine) and agents that enhance release of newly synthesized (d-amphetamine) and stored (methylphenidate) catecholamines have been shown to occasion caffeine-appropriate responding (HOLTZMAN 1986, 1987; MUMFORD and HOLTZMAN 1991; MARIATHASAN and STOLERMAN 1992).

E. Subjective and Discriminative Stimulus Effects of Caffeine in Humans

Subjective effects of a drug usually refer to drug-induced changes in an individual's experiences or feelings that are not accessible to independent verification by an observer. Reasonably sophisticated methods for assessing human subjective effects of drugs were first developed about 40 years ago (BEECHER 1959; JAFFE and JAFFE 1989) and have been extensively used to evaluate caffeine. Typically, these methods involve having subjects self-rate on questionnaires their moods, feelings or behaviors after double-blind administration of drug.

Adaptation of drug discrimination methods, originally developed in the animal laboratory, for human research has largely occurred over the last 10 years and has been applied to caffeine only in the last several years (KAMIEN et al. 1993). In the drug discrimination paradigm, subjects are trained to respond differentially in the presence of different drug conditions that are administered under double-blind conditions. In the typical, two-response drug vs placebo procedure, subjects are reinforced (usually with money) for making one response (e.g., a correct verbal or written drug identification response: "I received Drug A") after double-blind administration of one drug condition, and an alternative response (e.g., "I received Drug B") after the other drug condition. Since the relationship between discriminative and subjective effects has often been of interest, most discrimination studies have included some questionnaire measures of subjective effects.

Although the measurement of subjective effects and discriminative stimulus effects are methodologically independent operations which could theoretically provide totally independent data, research across a range of compounds including caffeine has demonstrated a rather good correspondence between subjective and discriminative effects (PRESTON and BIGELOW 1991).

I. Qualitative Subjective Effects of Caffeine

Although regular users of caffeine-containing foods commonly report a profile of desirable subjective effects such as increased feelings of alertness

Table 2. Low and intermediate doses of caffeine (18–178 mg) produce a variety of "positive" subjective effects[a]

Subjective rating	Effect	References[b]
Well-being	+	3, 4, 5
Energy/active/vigor	+	1, 2, 3, 4, 5, 6
Alert/clear-headed	+	2, 3, 4, 5, 6
Concentration	+	3, 5
Self-confidence	+	3, 4, 5
Motivation for work	+	3, 4, 5, 6
Desire to talk/social	+	3, 4, 5
Imaginative	+	1
Efficiency	+	1, 2
Sleepy	−	3, 4, 5, 6
Muzzy/not clear-headed	−	3, 4, 5

[a] Results of group statistical analyses comparing caffeine and placebo.
[b] References: 1, LEATHWOOD and POLLET (1983); 2, LIEBERMAN et al. (1987); 3, GRIFFITHS et al. (1990a); 4, SILVERMAN and GRIFFITHS (1992); 5, MUMFORD et al. (1994); 6, SILVERMAN et al. (1994).

and well-being after consuming caffeine (GOLDSTEIN and KAIZER 1969), the reliable demonstration of such effects in the laboratory using rigorous double-blind procedures has been elusive. As reviewed in more detail elsewhere (GRIFFITHS and MUMFORD 1995), it is now clear that the qualitative subjective effects of caffeine are dose-dependent. High dietary doses of caffeine (200–800 mg) produce a predominately "dysphoric" profile of subjective effects characterized by increases in anxiety, nervousness or jittery (e.g., RAPOPORT et al. 1981; CHAIT and GRIFFITHS 1983; CHARNEY et al. 1984; LOKE 1988; MATTILA et al. 1988; EVANS and GRIFFITHS 1991a; OLIVETO et al. 1993). In contrast, relatively low caffeine doses (e.g., 20–200 mg), particularly when tested under conditions of caffeine deprivation or total abstinence, produce a profile of predominately positive subjective effects after caffeine administration (LEATHWOOD and POLLET 1983; LIEBERMAN et al. 1987; GRIFFITHS et al. 1990a; SILVERMAN and GRIFFITHS 1992; MUMFORD et al. 1994; SILVERMAN et al. 1994). Table 2 shows the types of positive, subjective effect changes produced by low and intermediate doses of caffeine. That studies have also shown such effects in subjects who were maintained on an otherwise caffeine-free diet (SILVERMAN and GRIFFITHS 1992; MUMFORD et al. 1994; SILVERMAN et al. 1994) indicates that physical dependence is not a necessary condition for demonstrating positive subjective effects of caffeine.

As discussed elsewhere (GRIFFITHS and MUMFORD 1995) there seem to be at least three factors that increase the likelihood of demonstrating such positive subjective effects of caffeine: (1) testing in caffeine deprivation or total abstinence; (2) testing low caffeine doses; and (3) testing under conditions (or in populations) in which caffeine functions as a reinforcer.

When caffeine produces positive changes in subjective effects, the profile of these changes (e.g., increases in well-being, energy/active; alert; concentration; self-confidence; motivation for work; desire to talk to people) is remarkably similar to that produced by *d*-amphetamine and cocaine (Foltin and Fischman 1991). A major difference in the subjective effect profile appears to be that caffeine is more likely to produce dysphoria/anxiety with increases in dose that *d*-amphetamine and cocaine.

As reviewed elsewhere (Griffiths and Mumford 1995), there is some evidence suggesting a dissociation between the positive/energy effect and the dysphoric/anxiogenic effect of caffeine. Also, results of questionnaire and experimental studies suggest that panic disorder patients may be particularly vulnerable to the anxiety- and panic-producing effects of caffeine.

II. Demonstration of Caffeine Discrimination

Using procedures that have demonstrated the discriminative stimulus effects of stimulants, opioids and sedatives (Preston and Bigelow 1991; Kamien et al. 1993), seven studies have demonstrated that a caffeine vs placebo discrimination can be established in a majority of subjects (Griffiths et al. 1990a; Evans and Griffiths 1991a; Silverman and Griffiths 1992; Oliveto et al. 1992a, 1993; Mumford et al. 1994; Silverman et al. 1994). All studies involved the double-blind administration of caffeine and placebo in capsules, with sessions conducted 3–7 days per week. In all studies, subjects were told that they were attempting to discriminate between two different drug conditions and, except for two studies (Griffiths et al. 1990a; Mumford et al. 1994), subjects were not informed about the exact nature of the two drug conditions (e.g., they were told that their training drugs would be selected from a variety of sedative or stimulant compounds). In all studies, subjects were required to abstain from dietary sources of caffeine at least overnight (Evans and Griffiths 1991a) and, in some instances, throughout the study (Griffiths et al. 1990a; Mumford et al. 1994).

III. Acquisition of Caffeine Discrimination

More than 80% of subjects acquired the caffeine vs placebo discrimination in the seven discrimination studies reported to date. The speed of acquisition of the discrimination varied widely across studies and across subjects, from as few as six sessions (Evans and Griffiths 1991a) to as many as 60 sessions (Silverman and Griffiths 1992). Doses at which the initial discrimination was acquired have ranged between 100 mg (Griffiths et al. 1990a) and 320 mg (Oliveto et al. 1992a). Unexpectedly, the results of one study suggested that when the training dose was administered daily, a 320 mg vs placebo discrimination was more difficult to acquire than a 178 mg vs placebo discrimination (Silverman and Griffiths 1992), possibly due to tolerance. Once the discrimination has been acquired, it has generally been quite stable. For example, one study showed no evidence for either increasing or decreasing trends in discrimination accuracy over as many as 129 sessions spanning a period of almost 9 months (EVANS and GRIFFITHS 1991a).

IV. Thresholds for Caffeine Discrimination

Identification of caffeine has repeatedly been shown to be an increasing function of the caffeine dose. After training a caffeine vs placebo discrimination at intermediate doses (200–320 mg) caffeine produced dose-related (50–600 mg) increases in correct identifications (≥80%), with no subjects correctly identifying 50 or 56 mg, two of nine subjects correctly identifying 100 mg, and most subjects correctly identifying caffeine doses at or above the training dose (EVANS and GRIFFITHS 1991a; OLIVETO et al. 1992a, 1993).

These thresholds for caffeine discrimination are generally higher than those obtained in studies that explicitly trained discrimination of progressively lower caffeine doses (GRIFFITHS et al. 1990a; SILVERMAN and GRIFFITHS 1992; MUMFORD et al. 1994). The lowest dose discriminated varied widely across the subjects, ranging from 1.8 to 178 mg, with about 70% of discriminating subjects detecting 56 mg or less and about 35% detecting 18 mg or less. These results document effects of caffeine at doses lower than those previously recognized to be behaviorally active.

V. Cross-Generalization with Other Drugs

Several studies have been conducted which provide some limited information about the discriminative stimulus effects of caffeine relative to other drugs. A study by CHAIT and JOHANSON (1988) showed that caffeine (100 and 300 mg in nondeprived subjects) did not occasion predominately drug-appropriate responding in subjects trained to discriminate 10 mg d-amphetamine from placebo. Two studies tested novel drug conditions in subjects trained to discriminate between 320 mg/70 kg caffeine and placebo (OLIVETO et al. 1992a, 1993). These studies showed that theophylline (56–320 mg/70 kg), a methylxanthine structurally similar to caffeine, produced 100% caffeine-appropriate responding in each of four subjects; however, this effect did not occur in a consistent dose-related manner across subjects; Methylphenidate (10–56 mg/70 kg), a psychomotor stimulant, produced caffeine-appropriate responding at the highest dose in three of four subjects. In contrast, the sedative triazolam (0.1–0.56 mg/70 kg) and the anxiolytic buspirone (1–32 mg/70 kg) both produced predominately placebo-appropriate responding.

VI. Relationship Between Caffeine Discriminative and Subjective Effects

When subjects who have been trained on a caffeine vs placebo discrimination are asked to report the basis on which they believe they make the discrimination (i.e., cues), they generally describe changes in mood or behavioral dispositions. Concordant with the dose-related qualitative changes in the subjective effect profile of caffeine previously discussed, subjects trained at intermediate doses of 200–320 mg tend to report more dysphoric negative subjective effects (e.g., anxious, jittery) as a basis for discriminating the caffeine condition (EVANS and GRIFFITHS 1991a); in contrast, subjects trained at lower doses tend to report more positive effects (e.g., well-being, energy) as the basis of the caffeine discrimination (GRIFFITHS et al. 1990a; SILVERMAN and GRIFFITHS 1992; MUMFORD et al., 1994). In two studies, headache and tired were reported as cues for the placebo condition, suggesting that caffeine withdrawal may be relevant to making the caffeine vs placebo discrimination under some conditions (GRIFFITHS et al. 1990a; EVANS and GRIFFITHS 1991a).

Despite subjects' perception that discrimination performance is based on subjective effect changes, it is not possible to conclude that there is a causative relationship between discriminative and subjective effects. However, data from a variety of studies demonstrate an impressive covariation between these measures.

Examination of individual subject data has shown a good relationship between those subjective dimensions reported as cues for the discrimination and those that were shown to be significantly influenced by the drug conditions (GRIFFITHS et al. 1990a; SILVERMAN and GRIFFITHS 1992). In one study, for example, 83% of subjective dimensions that were rated as being "very" important to making the discrimination were also associated with statistically significant caffeine vs placebo differences on standard mood rating scales; in contrast only 4% of subjective dimensions rated as being "not at all" important were statistically significant (GRIFFITHS et al. 1990a).

Studies which have trained progressively lower doses of caffeine provide additional evidence for the correspondence between discrimination and subjective effects. One study, which concurrently assessed discrimination and subjective effects, showed that, within subjects, significant subjective effects occurred at doses of caffeine that were discriminated, but not at doses for which there was no evidence of discriminative control (SILVERMAN and GRIFFITHS 1992). In two studies which assessed subjective effects after determining caffeine discrimination thresholds, the lowest dose discriminated significantly correlated with the number of significant subjective effects, the rating of magnitude of drug effect, and the speed of onset of drug effect rating (GRIFFITHS et al. 1990a; MUMFORD et al. 1994).

Finally, the covariation of discrimination and subjective effects was further illustrated in a study which showed that, at the caffeine training dose, the magnitude of subjective effects differed depending on whether the discrimination was correct or incorrect (OLIVETO et al. 1992a).

VII. Relationship Between Discrimination and Physical Dependence

Although withdrawal symptoms of headache and tired were reported as cues for identifying placebo in a caffeine vs placebo discrimination in two studies (GRIFFITHS et al. 1990a; EVANS and GRIFFITHS 1991a), there is no direct evidence demonstrating that caffeine physical dependence facilitates acquisition or increases sensitivity to caffeine discrimination. Studies that have trained a caffeine vs placebo discrimination under conditions in which caffeine physical dependence was unlikely to have occurred (SILVERMAN and GRIFFITHS 1992; MUMFORD et al. 1994) indicate that caffeine physical dependence is not a necessary condition for caffeine discrimination.

VIII. Relationship Between Discrimination and Reinforcement

No research has directly addressed the relationship between caffeine discrimination and caffeine reinforcement. However, the results of one recent study suggest that a history of caffeine discrimination may be important in demonstrating caffeine reinforcement (SILVERMAN et al. 1994). This study examined caffeine choice in eight subjects who had previously acquired a 100 mg caffeine vs placebo discrimination. During subsequent sessions, subjects ingested letter-coded capsules (letter codes established during discrimination training) containing 100 mg caffeine or placebo and then engaged in a vigilance or relaxation activity. In subsequent choice conditions, all subjects demonstrated significant caffeine reinforcement in one or more conditions when the vigilance activity was scheduled. The proportion of subjects showing caffeine reinforcement was greater than that reported in previous studies in normal subjects, suggesting that acquisition of caffeine discrimination may facilitate the subsequent demonstration of caffeine reinforcement.

F. Tolerance to Caffeine in Laboratory Animals

Tolerance refers to an acquired change in responsiveness of an individual as a result of exposure to drug such that an increased dose of drug is necessary to produce the same degree of response, or that less effect is produced by the same dose of drug. The development of tolerance is most unambiguously concluded from studies which compare full dose-response curves in the presence and absence of the tolerance-inducing manipulation (KALANT et al. 1971).

This section summarizes results of 15 studies which clearly demonstrate caffeine tolerance across different species and a range of experimental conditions. Caffeine tolerance has been demonstrated in mice (AHLIJANIAN and TAKEMORI 1986), rats (WAYNER et al. 1976; CARNEY 1982; CHOU et al. 1985; FINN and HOLTZMAN 1986, 1987, 1988; HOLTZMAN 1983, 1987, 1991; HOLTZMAN and FINN 1988; MUMFORD et al. 1988; YASUHARA and LEVY 1988;

Holtzman et al. 1991) and squirrel monkeys (Katz and Goldberg 1987). Caffeine has been administered chronically via i.p., i.m. and i.v. injections or po via adulteration of the animal's drinking water. The frequency of caffeine dosing leading to the development of tolerance has ranged from once every other day (Wayner et al. 1976) to several times daily over periods of days or weeks (Holtzman 1983). Tolerance has been demonstrated following chronic daily administration of a wide range of doses from as low as 10 mg/kg per day (Chou et al. 1985) to as high as 222 mg/kg per day (Ahlijanian and Takemori 1986).

Tolerance has been shown to develop to various caffeine effects, including the effects of caffeine on reticular neuron unit activity (Chou et al. 1985), schedule-controlled responding maintained by presentation of food (Katz and Goldberg 1987; Holtzman and Finn 1988), water (Wayner et al. 1976) and electric shock (Katz and Goldberg 1987), reinforcement thresholds for electrical brain stimulation (Mumford et al. 1988), locomotor activity (Holtzman 1983; Chou et al. 1985; Ahlijanian and Takemori 1986; Finn and Holtzman 1986, 1987, 1988; Holtzman 1991; Holtzman et al. 1991), caffeine-induced seizure thresholds (Yasuhara and Levy 1988) and discriminative responding in caffeine-trained animals (Holtzman 1987).

The time course of the development of tolerance to caffeine has varied across studies and is likely dependent on the variable investigated. In the case of caffeine-induced seizures, caffeine tolerance appears to develop within minutes of a caffeine pretreatment such that brain caffeine concentrations required to produce seizures are significantly elevated in caffeine-pretreated rats relative to saline-pretreated controls (Yasuhara and Levy 1988). Tolerance to locomotor stimulant effects also appears to occur relatively rapidly in contrast to tolerance to discriminative stimulus effects which develop more slowly, as described in more detail below.

Caffeine tolerance has been most widely studied with regard to its locomotor stimulant effect in rats. Studies have shown that such tolerance is rapid (effects occur within a single cumulative dosing session, Holtzman and Finn 1988), is usually insurmountable (the caffeine dose-response curve is displaced downward, Holtzman 1983; Ahlijanian and Taskemori 1986) and shows cross-tolerance to other methylxanthines (Finn and Holtzman 1987, 1988), but not to other nonmethylxanthine psychomotor stimulants such as d-amphetamine and methylphenidate (Holtzman 1983; Finn and Holtzman 1987).

Interestingly, the characteristics of tolerance development to the discriminative stimulus effects of caffeine in rats differ from those for locomotor stimulant effects. Tolerance to the discriminative stimulus effects of caffeine in rats appears to develop more slowly (i.e., it is not apparent within a single session, Holtzman 1987) and is characterized by a parallel rightward shift of the caffeine dose-response curve (Holtzman 1987). Also, rats tolerant to the discriminative stimulus effects of caffeine are cross-tolerant to the caffeine-like discriminative stimulus effects of methylphenidate (Holtzman 1987).

Although more complete parametric comparison studies are needed, the differences in tolerance development to discriminative stimulus and locomotor stimulant effects of caffeine suggest that different pharmacological mechanisms may be operative.

I. Role of Adenosine in Tolerance

Although caffeine tolerance to locomotor activity has been most widely studied, the mechanism(s) underlying such tolerance remain unclear. Two recent studies involving chronic caffeine treatment and withdrawal in mice provided ambiguous results because chronic caffeine treatment was associated with the emergence of locomotor depression rather than a diminution of locomotor stimulation (KAPLAN et al. 1993; NIKODIJEVIC et al. 1993). Such behavioral depression may represent caffeine toxicity rather than caffeine tolerance. The observations that caffeine is a competitive antagonist of adenosine, an endogenous neuromodulator (SNYDER 1985), and that chronic caffeine produces increases (usually 10%–30%) in the number of brain adenosine receptors (NEHLIG et al. 1992; DALY 1993) has led to speculation that such receptor up-regulation represents a mechanism underlying caffeine tolerance. However, this simple explanation is not consistent with the fact that tolerance to locomotor stimulant effects can be insurmountable (HOLTZMAN 1983; AHLIJANIAN and TAKEMORI 1986; FINN and HOLTZMAN 1986, 1987, 1988; HOLTZMAN 1991; HOLTZMAN et al. 1991). Also, there is no clear precedent for loss of potency of a competitive antagonist after chronic treatment with the antagonist. Consistent with this, caffeine retains full antagonist potency at adenosine receptors following a chronic caffeine dosing regimen that has been shown to produce insurmountable tolerance to the stimulant effects of caffeine (KATZ and GOLDBERG 1987; HOLTZMAN et al. 1991). Finally, on theoretical grounds, there is no reason to expect that a change in receptor number should affect the potency of a competitive antagonist (HOLTZMAN et al. 1991). Changes in a variety of other neurotransmitter receptors have been demonstrated following chronic caffeine administration (SHI et al. 1993a). Future investigations should determine whether such changes play a critical role in the behavioral effects associated with chronic caffeine administration.

G. Tolerance to Caffeine in Humans

As in the animal laboratory, tolerance development to caffeine has been clearly demonstrated in humans; however, detailed quantitative knowledge is quite fragmentary. As reviewed in more detail elsewhere (GRIFFITHS and MUMFORD 1995), 11 studies on tolerance to caffeine in humans have been reported.

Tolerance to the subjective effects of caffeine has been demonstrated only in one study (EVANS and GRIFFITHS 1992), in which two groups of subjects received either caffeine (300 mg t.i.d.) or placebo (t.i.d.) for 18 consecutive days. During the last 14 days of chronic dosing, the caffeine and placebo groups did not differ meaningfully on ratings of mood and subjective effects. Furthermore, after chronic dosing, compared to placebo, caffeine (300 mg b.i.d.) produced significant subjective effects (including increases in tension-anxiety, jittery/nervous/shaky, active/stimulated/energetic and strength of drug effect) in the chronic placebo group but not in the chronic caffeine group, suggesting the development of "complete" tolerance (i.e., caffeine effects no longer different from placebo) at these doses.

Two studies provided direct experimental evidence for caffeine tolerance to sleep disruption by demonstrating decreases in caffeine-induced disruption of objective measures of sleep after caffeine dosing of 250 mg b.i.d. for 2 days (ZWYGHUIZEN-DOORENBOS et al. 1990) or 400 mg t.i.d. for 7 days (BONNET and ARAND 1992). By day 7 in the latter study, a number of sleep measures (e.g., total sleep time, sleep efficiency, number of awakenings) were no longer different from baseline, suggesting the development of complete tolerance at these doses.

In addition to the three studies described above demonstrating tolerance to centrally mediated caffeine effects, there is good evidence that repeated daily caffeine administration produces decreased responsiveness to physiological effects of caffeine, including diuresis, parotid gland salivation, increased metabolic rate (oxygen consumption), increased blood pressure, increased plasma norepinephrine and epinephrine, and increased plasma renin activity (see GRIFFITHS and MUMFORD 1995). Several studies (ROBERTSON et al. 1981; AMMON et al. 1983; DENARO et al. 1991) have demonstrated complete tolerance to blood pressure and other cardiovascular and physiological responses with repeated daily caffeine administration (e.g., tolerance to 250 mg t.i.d. in 1–4 days, ROBERTSON et al. 1981).

Beyond the several demonstrations of complete tolerance development at high caffeine doses described above, there is little quantitative knowledge about the parameters which determine caffeine tolerance in humans. As with the developement of tolerance to most drugs, the degree of tolerance development to caffeine can be expected to depend on the caffeine dose, the dose frequency, the number of doses and the individual's elimination rate (SHI et al. 1993b). As in the infrahuman research, there is some indication that the rate and/or extent of tolerance development differ across different measures (DENARO et al. 1991; BONNET and ARAND 1992).

H. Caffeine Physical Dependence in Laboratory Animals

Physical dependence is manifested by time-limited biochemical, physiological and behavioral disruptions (i.e., a withdrawal syndrome) upon termination of chronic or repeated drug adminstration. To date, there have been nine reports of caffeine withdrawal in laboratory animals, most of which have documented substantial behavioral disruptions following cessation of chronic caffeine dosing (e.g., 50%–80% reductions in spontaneous locomotor activity (HOLTZMAN 1983; FINN and HOLTZMAN 1986); 20%–50% reductions in operant responding (CARNEY 1982; MUMFORD et al. 1988). These studies have examined caffeine withdrawal in rats (BOYD et al. 1965; VITIELLO and WOODS 1977; CARNEY 1982; HOLTZMAN 1983; FINN and HOLTZMAN 1986; HOLTZMAN and FINN 1988; MUMFORD et al. 1988), cats (SINTON and PETITJEAN 1989) and monkeys (CARROLL et al. 1989).

The manifestations of caffeine withdrawal have included decreases in locomotor activity (BOYD et al. 1965; HOLTZMAN 1983; FINN and HOLTZMAN 1986), decreases in operant behavior (CARNEY 1982; MUMFORD et al. 1988; CARROLL et al. 1989), decreases in reinforcement threshold for electrical brain stimulation (MUMFORD et al. 1988), increases in the ratio of time spent in slow wave sleep stages I and II (SINTON and PETITJEAN 1989), and avoidance of a preferred flavor when that flavor was paired with caffeine abstinence (VITIELLO and WOODS 1977).

The frequency of caffeine dosing leading to the development of dependence has ranged from once daily (CARNEY 1982) to several times daily (HOLTZMAN 1983). Physical dependence has been demonstrated following chronic daily doses ranging from 6 mg/kg per day (VITIELLO and WOODS 1977) to 190 mg/kg per day (BOYD et al. 1965). The dosing parameters necessary for the production of caffeine physical dependence have been most clearly defined in studies of locomotor activity (HOLTZMAN 1983; FINN and HOLTZMAN 1986). The severity of caffeine withdrawal is dependent upon the caffeine maintenance dose (HOLTZMAN 1983; FINN and HOLTZMAN 1986). Rats consuming 19 or 36 mg/kg per day caffeine for 6 weeks showed no change in locomotor activity (FINN and HOLTZMAN 1986), whereas animals consuming 67 mg/kg per day showed decreased locomotor activity following substitution of water for caffeine (FINN and HOLTZMAN 1986). This decrease lasted about 2 days and was maximal on the first day of withdrawal (activity levels were about 50% of prewithdrawal levels). Larger decreases in locomotor activity were reported in rats treated with higher doses of caffeine for longer periods of time (11 weeks of exposure to caffeine; approximately 160 mg/kg per day during the last 7 weeks (HOLTZMAN 1983). This decrease lasted 4 days and was maximal on the second day of withdrawal (activity levels were about 20% of prewithdrawal levels). There have been no studies to determine the minimum duration of caffeine exposure necessary to produce physical dependence; however, a significant decrease in operant responding has been demonstrated following as few as 10 days of chronic dosing (MUMFORD et al. 1988).

The latency to onset of caffeine withdrawal effects has not been evaluated rigorously but appears to occur within 24 h. Peak withdrawal effects occur from 24 to 48 h following termination of chronic caffeine dosing (VITIELLO and WOODS 1977; CARNEY 1982; HOLTZMAN 1983; FINN and HOLTZMAN 1986; MUMFORD et al. 1988; CARROLL et al. 1989; SINTON and PETITJEAN 1989). The

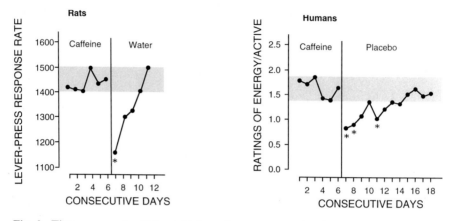

Fig. 2. Time course of caffeine withdrawal over consecutive days in rats and humans. The *left panel* presents data from a study in nine rats in which operant responding was maintained by electrical self-stimulation of the brain before and after substitution of water for caffeine (averaging 72 mg/kg per day) in drinking water (Mumford et al. 1988). The *right panel* presents data from a study in four withdrawal-sensitive subjects who rated "energy/active" before and after double-blind substitution of placebo capsules for caffeine capsules (100 mg/day) (Griffiths et al. 1990b). In both panels, *shaded areas* show the range of means from the last 6 days of a chronic caffeine administration period and *asterisks* indicate which withdrawal days are significantly different from control. The figure shows that, in both studies, withdrawal effects were maximal on the first day and tended to progressively decrease over consecutive days

left panel of Fig. 2 presents data from an illustrative study (Mumford et al. 1988) showing the time course of the suppression of lever-press responding maintained by electrical stimulation of the brain in a group of rats in which water was substituted for a caffeine solution after a period of chronic caffeine consumption. Although most withdrawal changes have been short-lived (e.g., one to several days), behavioral manifestations of withdrawal have been shown to last as long as 30 days following termination of chronic caffeine dosing (Sinton and Petitjean 1989).

I. Role of Adenosine in Physical Dependence

The pharmacological and physiological mechanisms underlying caffeine physical dependence need considerable further elaboration. Accumulating evidence, however, suggests a role for the endogenous neuromodulator adenosine. Caffeine is a competitive antagonist of adenosine (Snyder 1985) and chronic caffeine treatment has been reported to increase the number of brain adenosine receptors (Daly 1993), to shift brain A_1 adenosine receptors into a high affinity state (Green and Stiles 1986), and to increase functional sensitivity to adenosine (von Borstel et al. 1983; Ahlijanian and Takemori 1986; Green and Stiles 1986; Biaggioni et al. 1991). The increased functional tissue sensitivity to endogenous adenosine (e.g., in vascular and neural tissue of brain) has also been proposed as a mechanism for caffeine withdrawal headache and fatigue (von Borstel et al. 1983; Hirsh 1984).

I. Caffeine Physical Dependence in Humans

As in the animal laboratory, physical dependence manifested by time-limited biochemical, physiological and behavioral disruptions (i.e., a withdrawal syndrome) upon termination of chronic or repeated drug administration has been clearly demonstrated in humans. As has been reviewed elsewhere in more detail (Griffiths and Woodson 1988a; Griffiths and Mumford 1995) 53 case reports and human experimental studies of caffeine withdrawal have been reported.

The most frequently reported withdrawal symptom is headache (also cerebral fullness), which is characterized as being gradual in development, diffuse, throbbing and sometimes severe (Griffiths and Woodson 1988a). Other symptoms, in roughly decreasing order of prominence are drowsiness (e.g., increased sleepiness and yawning; decreased energy and alertness), increased work difficulty (decreased motivation for tasks/work; impaired concentration), decreased feelings of well-being/content (decreased self-confidence; increased irritability) decreased social/friendliness/talkativeness, flu-like feelings (muscle aches/stiffness; hot or cold spells; heavy feelings in arms or legs; nausea), and blurred vision (Griffiths and Woodson 1988a; Griffiths et al. 1990b; Silverman et al. 1992). In addition to these symptoms, composite scales of depression and anxiety may be elevated (Silverman et al. 1992) and psychomotor performance may be impaired (Griffiths and Woodson 1988a; Rizzo et al. 1988; Silverman et al. 1992). The occurrence of headache as a withdrawal symptom does not necessarily correlate with the occurrence of other symptoms (e.g., tiredness), suggesting that other signs and symptoms are not merely epiphenomena of headache (Griffiths and Woodson 1988a; Griffiths et al. 1990b).

The severity of caffeine withdrawal is an increasing function of caffeine maintenance dose (Griffiths and Woodson 1988a; Evans and Griffiths 1991b). When symptoms of caffeine withdrawal occur, the severity can vary from mild to extreme. At its worst, caffeine withdrawal is incompatible with normal functioning and is sometimes totally incapacitating (Griffiths and Woodson 1988a; Silverman et al. 1992).

The incidence of caffeine withdrawal is an increasing function of caffeine maintenance dose (Griffiths and Woodson 1988a; Galletly et al. 1989; Fennelly et al. 1991; Weber et al. 1993). The best estimates of the incidence of caffeine withdrawal in the general population come from a recent survey study and an experimental study. A recent random-digit telephone survey in Vermont showed that, among current users of caffeine who reported that they had abstained from caffeine for ≥ 24 h, 27% reported withdrawal headaches when they abstained (Hughes 1992; Hughes et al. 1993b). The experimental study (Silverman et al. 1992) involved 62 individuals from the general community with a distribution of caffeine intake similar to the general population in the United States (mean caffeine intake of 235 mg/day). The study involved a double-blind, approximately 48 h, caffeine abstinence trial under conditions which obscured that the purpose of the study was to investigate caffeine. During caffeine withdrawal 52% reported moderate or severe headache and 8%–11% showed abnormally high scores on standardized depression, anxiety and fatigue scales. The incidence of headache observed from the survey and experimental study in the general population (27%–52%) is in the range of that observed in several other recent studies conducted in special subject populations (19%–57%) (Griffiths et al. 1990b; van Dusseldorp and Katan 1990; Hughes et al. 1993a).

Although the incidence and severity of caffeine withdrawal are an increasing function of caffeine dose, two recent studies have shown that caffeine withdrawal can occur after relatively long-term administration of caffeine doses as low as 100 mg/day (Griffiths et al. 1990b; Evans and Griffiths 1991b).

The caffeine withdrawal syndrome follows an orderly time course (Griffiths and Woodson 1988a). Onset has been usually reported to occur 12–24 h after terminating caffeine intake, although onset as late as 36 h has been documented (Griffiths et al. 1990b). Peak withdrawal intensity has generally been described as occurring 20–48 h after abstinence. The duration of caffeine withdrawal has most often been described as ranging between 2 days and 1 week, although longer durations have been occasionally noted (Griffiths and Woodson 1988a; Griffiths et al. 1990b; van Dusseldorp and Katan 1990; Evans and Griffiths 1992). The right panel of Fig. 2 presents data from an illustrative study (Griffiths et al. 1990b) showing the time course of changes in ratings of energy/active in a group of human subjects in whom placebo capsules were substituted for caffeine after a period of chronic caffeine consumption.

Physiological mechanisms underlying caffeine withdrawal remain uncertain, although some studies suggest that increased blood volume, possibly adenosine-mediated, may be involved with caffeine withdrawal headache (von Borstel et al. 1983; Mathew and Wilson 1985).

J. Conclusions

This chapter reviews the rapidly developing literature in laboratory animals and humans concerning four behavioral pharmacological effects that caffeine shares with classic drugs of abuse/dependence: reinforcing effects, discriminative/subjective effects, tolerance and physical dependence. It is now clear from research with both laboratory animals and with humans that caffeine can function as a reinforcer under some conditions. The results of intravenous caffeine self-injection studies in nonhuman primaes and rats indicate that caffeine, like nicotine, functions as a reinforcer under a more limited range of conditions than classic psychomotor stimulant drugs of abuse. The results of self-administration and choice studies in humans clearly demonstrate the reinforcing effects of low and moderate doses of caffeine and suggest that such effects may be potentiated by caffeine physical dependence. The discriminative stimulus and subjective effects of caffeine have been demonstrated to be qualitatively different at low vs high doses in both laboratory animals and humans, with low-moderate doses, but not high doses, of caffeine showing substantial overlap with classic psychomotor stimulants. Tolerance to several behavioral effects of caffeine in laboratory animals and to the subjective and sleep-disrupting effects of caffeine in humans has been demonstrated, although parametric human studies are nonexistent. Physical dependence, as reflected in time-limited disruption of behavior in laboratory animals and a clinically significant withdrawal syndrome in humans, has been clearly demonstrated. It is widely speculated that the behavioral effects of caffeine are mediated primarily via antagonism of adenosine, an endogenous neuromodulator. Although there is some evidence that the psychomotor stimulant (data not reviewed) and physical dependence effects of caffeine are adenosine-mediated, data implicating adenosine in caffeine tolerance and discriminative effects are equivocal. Further studies of caffeine will not only advance our understanding of the most widely self-administered drug in the world, but should provide valuable insights into behavioral-pharmacological mechanisms underlying the drug abuse/dependence process.

Acknowledgements. Preparation of this review was supported, in part, by National Institute on Drug Abuse grants R01 DA01147 and RO1 DA03889. The authors thank Dr. Mary Abreu for helpful comments on the manuscript.

References

Ahlijanian MK, Takemori AE (1986) Cross-tolerance studies between caffeine and $(-)$-N^6-(Phenylisopropyl)-adenosine (PIA) in mice. Life Sci 38:577–588

Ammon HPT, Bieck PR, Mandalaz D, Verspohl EJ (1983) Adaptation of blood pressure to continuous heavy coffee drinking in young volunteers. A double-blind crossover study. Br J Clin Pharmacol 15:701–706

Ando K, Yanagita T (1992) Effects of an antitussive mixture and its consituents in rats discriminating methamphetamine from saline. Pharmacol Biochem Behav 41:783–788

Atkinson J, Ensslen M (1976) Self-administration of caffeine by the rat. Arzneimit-telforschung 26:2059–2061

Austin GA (1979) Perspectives on the history of psychoactive substance use. In: NIDA research issues 24, DHEW publication no (ADM) 79-810. US Government Printing Office, Washington DC, pp 50–66

Beecher HK (1959) Measurement of subjective responses: quantitative effects of drugs. Oxford University Press, New York

Biaggioni I, Paul S, Puckett A, Arzubiaga C (1991) Caffeine and theophylline as adenosine receptor antagonists in humans. J Pharmacol Exp Ther 258:588–593

Bonnet MH, Arand DL (1992) Caffeine use as a model of acute and chronic insomnia. Sleep 15:526–536

Boyd EM, Dolman M, Knight LM, Sheppard EP (1965) The chronic oral toxicity of caffeine. Can J Physiol Pharmacol 43:995–1007

Bozarth MA (1987) Methods of assessing the reinforcing properties of abused drugs. Springer, Berlin Heidelberg New York

Carney JM (1982) Effects of caffeine, theophylline and theobromine on scheduled controlled responding in rats. Br J Pharmacol 75:451–454

Carney JM, Christensen HD (1980) Discriminative stimulus properties of caffeine: studies using pure and natural products. Pharmacol Biochem Behav 13:313

Carney JM, Holloway FA, Modrow HE (1985) Discriminative stimulus properties of methylxanthines and their metabolites in rats. Life Sci 36:913–920

Carroll ME, Hagen EW, Asencio M, Brauer LH (1989) Behavioral dependence on caffeine and phencyclidine in rhesus monkeys: interactive effects. Pharmacol Biochem Behav 31:927–932

Chait LD, Griffiths RR (1983) Effects of caffeine on cigarette smoking behavior and subjective response. Clin Pharmacol Ther 34:612–622

Chait LD, Johanson CE (1988) Discriminative stimulus effects of caffeine and benzphetamine in amphetamine-trained volunteers. Psychopharmacology (Berl) 96:302–308

Charney DS, Galloway MP, Heninger GR (1984) The effects of caffeine on plasma MHPG, subjective anxiety, autonomic symptoms and blood pressure in healthy humans. Life Sci 35:135–144

Choi OH, Shamim MT, Padgett WL, Daly JW (1988) Caffeine and theophylline analogues: correlation of behavioral effects with activity as adenosine receptor antagonists and as phosphodiesterase inhibitors. Life Sci 43:387–398

Chou DT, Khan S, Forde J, Hirsh KR (1985) Caffeine tolerance: behavioral, electrophysiological and neurochemical evidence. Life Sci 36:2347–2358

Coffin VL, Carney JM (1983) Behavioral pharmacology of adenosine analogs. In: Daly JW, Kuroda Y, Phillis JW, Shimizu H, Ui M (eds) Physiology and pharmacology of adenosine derivatives. Raven, New York, pp 267–274

Collins RJ, Weeks JR, Cooper MM, Good PI, Russell RR (1984) Prediction of abuse liability of drugs using IV self-administration by rats. Psychopharmacology (Berl) 82:6–13

Daly JW (1993) Mechanism of action of caffeine. In: Garattini S (ed) Caffeine, coffee and health. Raven, New York, pp 97–150

Denaro CP, Brown CR, Jacob P III, Benowitz NL (1991) Effects of caffeine with repeated dosing. Eur J Clin Pharmacol 40:273–278

Deneau G, Yanagita T, Seevers MH (1969) Self-administration of psychoactive substances by the monkey: a measure of psychological dependence. Psychopharmacologia 16:30–48

Dworkin SI, Vrana SL, Broadbent J, Robinson JH (1993) Comparing the effects of nicotine, caffeine, methylphenidate and cocaine. Med Chem Res 2:593–602

Evans SM, Griffiths RR (1991a) Dose-related caffeine discrimination in normal volunteers: individual differences in subjective effects and self-reported cues. Behav Pharmacol 2:345–356

Evans SM, Griffiths RR (1991b) Low-dose caffeine physical dependence in normal subjects: dose-related effects. In: Harris LS (ed) Problems of drug dependence 1990. US Government Printing Office, Washington DC, p 446

Evans SM, Griffiths RR (1992) Caffeine tolerance and choice in humans. Psychopharmacology (Berl) 108:51–59

Evans SM, Critchfield TS, Griffiths RR (1994) Caffeine reinforcement demonstrated in a majority of moderate caffeine users. Behav Pharmacol 5:231–238

Fennelly M, Galletly DC, Puride GI (1991) Is caffeine withdrawal the mechanism of postoperative headache? Anesth Analg 72:449–453

Finn IB, Holtzman SG (1986) Tolerance to caffeine-induced stimulation of locomotor activity in rats. J Pharmacol Exp Ther 238:542–546

Finn IB, Holtzman SG (1987) Pharmacologic specificity of tolerance to caffeine-induced stimulation of locomotor activity. Psychopharmacology (Berl) 93:428–434

Finn IB, Holtzman SG (1988) Tolerance and cross-tolerance to theophylline-induced stimulation of locomotor activity in rats. Life Sci 42:2475–2482

Foltin RW, Fischman MW (1991) Assessment of abuse liability of stimulant drugs in humans: a methodological survey. Drug Alcohol Depend 28:3–48

Galletly DC, Fennelly M, Whitwam JG (1989) Does caffeine withdrawal contribute to postanaesthetic morbidity? Lancet 10:1335

Gauvin DV, Harland RD, Michaelis RC, Holloway FA (1989) Caffeine-phenylethylamine combinations mimic the cocaine discriminative cue. Life Sci 44:67–73

Gauvin DV, Criado JR, Moore KR, Holloway FA (1990) Potentiation of cocaine's discriminative effects by caffeine: a time-effect analysis. Pharmacol Biochem Behav 36:195–197

Gauvin DV, Moore KR, Youngblood BD, Holloway FA (1993) The discriminative stimulus properties of legal, over-the-counter stimulants administered singly and in binary and ternary combinations. Psychopharmacology (Berl) 110:309–319

Gilbert RM (1976) Caffeine as a drug of abuse. In: Gibbins RJ, Israel Y, Kalant H, Popham RE, Schmidt W, Smart RG (eds) Research advances in alcohol and drug problems, vol 3. Wiley, New York, pp 49–176

Goldberg SR, Henningfield JE (1988) Reinforcing effects of nicotine in humans and experimental animals responding under intermittent schedules of IV drug injection. Pharmacol Biochem Behav 30:227–234

Goldstein A, Kaizer S (1969) Psychoactive effects of caffeine in man. III. A questionnaire survey of coffee drinking and its effects in a group of housewives. Clin Pharmacol Ther 10:477–488

Goudie AJ, Atkinson J, West CR (1986) Discriminative properties of the psychostimulant dl-cathinone in a two lever operant task: lack of evidence for dopaminergic mediation. Neuropharmacology 25:85–94

Graham DM (1978) Caffeine – its identity, dietary sources, intake and biological effects. Nutr Rev 36:97–102

Graham HN (1984) Tea: the plant and its manufacture; chemistry and consumption of the beverage. In: Spiller GA (ed) The methylxanthine beverages and foods: chemistry, consumption and health effects. Liss, New York, pp 29–74

Green RM, Stiles GL (1986) Chronic caffeine ingestion sensitizes the A1 adenosine receptor-adenylate cyclase system in rat cerebral cortex. J Clin Invest 77:222–227

Griffiths RR, Woodson PP (1988a) Caffeine physical dependence: a review of human and laboratory animal studies. Psychopharmacology (Berl) 94:437–451

Griffiths RR, Woodson PP (1988b) Reinforcing effects of caffeine in humans. J Pharmacol Exp Ther 246:21–29

Griffiths RR, Woodson PP (1988c) Reinforcing properties of caffeine: studies in humans and laboratory animals. Pharmacol Biochem Behav 29:419–427

Griffiths RR, Brady JV, Bradford LD (1979) Predicting the abuse liability of drugs with animal drug self-administration procedures: psychomotor stimulants and hallucinogens. In: Thompson T, Dews PB (eds) Advances in behavioral pharmacology, vol 2. Academic, New York, pp 163–208

Griffiths RR, Bigelow GE, Liebson IA (1986a) Human coffee drinking: reinforcing and physical dependence producing effects of caffeine. J Pharmacol Exp Ther 239:416–425

Griffiths RR, Bigelow GE, Liebson IA, O'Keeffe M, O'Leary D, Russ N (1986b) Human coffee drinking: manipulation of concentration and caffeine dose. J Exp Anal Behav 45:133–148

Griffiths RR, Bigelow GE, Liebson IA (1989) Reinforcing effects of caffeine in coffee and capsules. J Exp Anal Behav 52:127–140

Griffiths RR, Evans SM, Heishman SJ, Preston KL, Sannerud CA, Wolf B, Woodson PP (1990a) Low-dose caffeine discrimination in humans. J Pharmacol Exp Ther 252:970–978

Griffiths RR, Evans SM, Heishman SJ, Preston KL, Sannerud CA, Wolf B, Woodson PP (1990b) Low-dose caffeine physical dependence in humans. J Pharmacol Exp Ther 255:1123–1132

Griffiths RR, Mumford GK (1995) Caffeine: a drug of abuse? In: Bloom FE, Kupfer DJ (eds) Psychopharmacology: the fourth generation of progress. Raven, New York, pp 1699–1713

Harland RD, Gauvin DV, Michaelis RC, Carney JM, Seale TW, Holloway FA (1989) Behavioral interaction between cocaine and caffeine: a drug discrimination analysis in rats. Pharmacol Biochem Behav 32:1017–1023

Hirsh K (1984) Central nervous system pharmacology of the dietary methylxanthines. In: Spiller GA (ed) The methylxanthine beverages and foods: chemistry, consumption, and health effects. Liss, New York, pp 235–301

Holloway FA, Michaelis RC, Huerta PL (1985a) Caffeine-phenylethylamine combinations mimic the amphetamine discriminative cue. Life Sci 36:723–730

Holloway FA, Modrow HE, Michaelis RC (1985b) Methylxanthine discrimination in the rat: possible benzodiazepine and adenosine mechanisms. Pharmacol Biochem Behav 22:815–824

Holloway FA, Goulden KL, Gauvin DV (1992) Masking of the chlordiazepoxide (CDP) cue by caffeine (CAF) in a three-choice CDP-SAL-CAF drug discrimination task. Soc Neurosci Abstr 18:813

Holtzman SG (1983) Complete, reversible, drug-specific tolerance to stimulation of locomotor activity by caffeine. Life Sci 33:779–787

Holtzman SG (1986) Discriminative stimulus properties of caffeine in the rat: noradrenergic mediation. J Pharmacol Exp Ther 239:706–714

Holtzman SG (1987) Discriminative stimulus effects of caffeine: tolerance and cross-tolerance with methylphenidate. Life Sci 40:381–389

Holtzman SG (1990) Caffeine as a model drug of abuse. Trends Pharmacol Sci 11:355–356

Holtzman SG (1991) CGS 15943, a nonxanthine adenosine receptor antagonist: effects on locomotor activity of nontolerant and caffeine-tolerant rats. Life Sci 49:1563–1570

Holtzman SG, Finn IB (1988) Tolerance to behavioral effects of caffeine in rats. Pharmacol Biochem Behav 29:411–418

Holtzman SG, Mante S, Minneman KP (1991) Role of adenosine receptors in caffeine tolerance. J Pharmacol Exp Ther 256:62–68

Howell LL (1993) Comparative effects of caffeine and selective phosphodiesterase inhibitors on respiration and behavior in rhesus monkeys. J Pharmacol Exp Ther 266:894–903

Hughes JR (1992) Clinical importance of caffeine withdrawal. N Engl J Med 327(16):1160–1161

Hughes JR (1993) Smoking is a drug dependence: a reply to Robinson and Pritchard. Psychopharmacology (Berl) 113:282–283

Hughes JR, Higgins ST, Bickel WK, Hunt WK, Fenwick JW, Gulliver SB, Mireault GC (1991) Caffeine self-administration, withdrawal, and adverse effects among coffee drinkers. Arch Gen Psychiatry 48:611–617

Hughes JR, Hunt WK, Higgins ST, Bickel WK, Fenwick JW, Pepper SL (1992a) Effect of dose on the ability of caffeine to serve as a reinforcer in humans. Behav Pharmacol 3:211–218

Hughes JR, Oliveto AH, Bickel WK, Higgins ST, Valliere W (1992b) Caffeine self-administration and withdrawal in soda drinkers, J Addict Dis 4:178

Hughes JR, Oliveto AH, Bickel WK, Higgins ST, Badger GJ (1993a) Caffeine self-administration and withdrawal: incidence, individual differences and interrelationships. Drug Alcohol Depend 32:239–246

Hughes JR, Oliveto AH, Helzer JE, Bickel WK, Higgins ST (1993b) Indications of caffeine dependence in a population-based sample. In: Harris LS (ed) Problems of drug dependence, 1992. US Government Printing Office, Washington DC, p 194

Jaffe JH, Jaffe FK (1989) Historical perspectives on the use of subjective effects measures in assessing the abuse potential of drugs. In: Fischman MW, Mello NK (eds) Testing for abuse liability of drugs. NIDA monograph no 92, DHHS publication no (ADM) 89-1613. US Government Printing Office, Washington DC, pp 43–72

Jones CN, Howard JL, McBennett ST (1980) Stimulus properties of antidepressants in the rat. Psychopharmacology (Berl) 67:111–118

Kalant H, LeBlanc AE, Gibbins RJ (1971) Tolerance to, and dependence on, some non-opiate psychotropic drugs. Pharmacol Rev 23:135–191

Kamien JB, Bickel WK, Hughes JR, Higgins ST, Smith B (1993) Drug discrimination by humans compared to nonhumans: current status and future directions. Psychopharmacology (Berl) 111:259–270

Kaplan GB, Greenblatt DJ, Kent MA, Cotreau-Bibbo MM (1993) Caffeine treatment and withdrawal in mice: relationships between dosage, concentrations, locomotor activity and A1 adenosine receptor binding. J Pharmacol Exp Ther 266:1563–1572

Katz JL, Goldberg SR (1987) Psychomotor stimulant effects of caffeine alone and in combination with an adenosine analog in the squirrel monkey. J Pharmacol Exp Ther 242:179–187

Kuhn DM, Appel JB, Greenberg I (1974) An analysis of some discriminative properties of d-amphetamine. Psychopharmacologia 39:57–66

Leathwood PD, Pollet P (1983) Diet-induced mood changes in normal populations. J Psychiatr Res 17:147–154

Lieberman HR, Wurtman RJ, Emde GG, Coviella ILG (1987) The effects of caffeine and aspirin on mood and performance. J Clin Psychopharmacol 7: 315–320

Llosa T (1993) Special issue on oral cocaine in humans. Clín Adicc Quím (J Chem Addict) 4:1–26

Loke WH (1988) Effects of caffeine on mood and memory. Physiol Behav 44:367–372

Mariathasan EA, Stolerman IP (1992) Drug discrimination studies in rats with caffeine and phenylpropanolamine administered separately and as mixtures. Psychopharmacology (Berl) 109:99–106

Mathew RJ, Wilson WH (1985) Caffeine consumption, withdrawal and cerebral blood flow. Headache 25:305–309

Mattila J, Seppala T, Mattila MJ (1988) Anxiogenic effect of yohimbine in healthy subjects: comparison with caffeine and antagonism by clonidine and diazepam. Int Clin Psychopharmacol 3:215–229

Modrow HE, Holloway FA (1985) Drug discrimination and cross generalization between two methylxanthines. Pharmacol Biochem Behav 23:425–429

Modrow HE, Holloway FA, Carney JM (1981a) Caffeine discrimination in the rat. Pharmacol Biochem Behav 14:683–688

Modrow HE, Holloway FA, Christensen HD, Carney JM (1981b) Relationship between caffeine discrimination and caffeine plasma levels. Pharmacol Biochem Behav 15:323–325

Mumford GK, Holtzman SG (1991) Qualitative differences in the discriminative stimulus effects of low and high doses of caffeine in the rat. J Pharmacol Exp Ther 258:857–865

Mumford GK, Neill DB, Holtzman SG (1988) Caffeine elevates reinforcement threshold for electrical brain stimulation: tolerance and withdrawal changes. Brain Res 459:163–167

Mumford GK, Evans SM, Kaminski BJ, Preston KL, Sannerud CA, Silverman K, Griffiths RR (1994) Discrminative stimulus and subjective effects of theobromine and caffeine in humans. Psychopharmacology (Berl) 115:1–8

Nehlig A, Daval J-L, Debry G (1992) Caffeine and the central nervous system: mechanisms of action, biochemical, metabolic and psychostimulant effects. Brain Res Brain Res Rev 17:139–170

Nikodijevic O, Jacobson KA, Daly JW (1993) Locomotor activity in mice during chronic treatment with caffeine and withdrawal. Pharmacol Biochem Behav 44:199–216

Oliveto AH, Hughes JR, Pepper SL, Bickel WK, Higgins ST (1991) Low doses of caffeine can serve as reinforcers in humans. In: Harris LS (ed) Problems of druig dependence, 1990. US Government Printing Office, Washington DC, p 442

Oliveto AH, Bickel WK, Hughes JR, Shea PJ, Higgins ST, Fenwick JW (1992a) Caffeine drug discrimination in humans: acquisition, specificity and correlation with self-reports. J Pharmacol Exp Ther 261:885–894

Oliveto AH, Hughes JR, Higgins ST, Bickel WK, Pepper SL, Shea PJ, Fenwich JW (1992b) Forced-choice versus free-choice procedures: caffeine self-administration in humans. Psychopharmacology (Berl) 109:85–91

Oliveto AH, Bickel WK, Hughes JR, Terry SY, Higgins ST, Badger GJ (1993) Pharmacological specificity of the caffeine discriminative stimulus in humans: effects of theophylline, methylphenidate and buspirone. Behav Pharmacol 4: 237–246

Overton DA [1982] Comparison of the degree of discriminability of various drugs using the T-maze drug discrimination paradigm. Psychopharmacology (Berl) 76:385–395

Overton DA, Batta SK (1977) Relationship between abuse liability of drugs and their degree of discriminability in the rat. In: Thompson T, Unna KR (eds) Predicting dependence liablility of stimulant and depressant drugs. University Park Press, Baltimore, pp 125–135

Preston KL, Bigelow GE (1991) Subjective and discriminative effects of drugs. Behav Pharmacol 2:293–313

Rapoport JL, Jensvold M, Elkins R, Buchsbaum MS, Weingartner H, Ludlow C, Zahn TP, Berg CJ, Neims AH (1981) Behavioral and cognitive effects of caffeine in boys and adult males. J Nerv Ment Dis 169:726–732

Rizzo AA, Stamps LE, Fehr LA (1988) Effects of caffeine withdrawal on motor performance and heart rate changes. Int J Psychophysiol 6:9–14

Robertson D, Wade D, Workman R, Woosley RL, Oates JA (1981) Tolerance to the humoral and hemodynamic effects of caffeine in man. J Clin Invest 67: 1111–1117

Robinson JH, Pritchard WS (1992a) The meaning of addiction: reply to West. Psychopharmacology (Berl) 108:411–416

Robinson JH, Pritchard WS (1992b) The role of nicotine in tobacco use. Psychopharmacology (Berl) 108:397–407

Rosen JB, Young AM, Beuthin FC, Louis-Ferdinand RT (1986) Discriminative stimulus properties of amphetamine and other stimulants in lead-exposed and normal rats. Pharmacol Biochem Behav 24:211–215

Schechter MD (1977) Caffeine potentiation of amphetamine: implications for hyperkinesis therapy. Pharmacol Biochem Behav 6:359–361

Schechter MD (1989) Potentiation of cathinone by caffeine and nikethamide. Pharmacol Biochem Behav 33:299–301

Schuster CR, Woods JH, Seevers MH (1969) Self-administration of central stimulants by the monkey. In: Sjoqvist F, Tottie M (eds) Abuse of central stimulants. Raven, New York, pp 339–347

Shi D, Nikodijevic O, Jacobson KA, Daly JW (1993a) Chronic caffeine alters the density of adenosine, adrenergic, cholinergic, GABA, and serotonin receptors and calcium channels in mouse brain. Cell Mol Neurobiol 13:247–261

Shi J, Benowitz NL, Denaro CP, Sheiner LB (1993b) Pharmacokinetic-pharmacodynamic modeling of caffeine: tolerance to pressor effects. Clin Pharmacol Ther 53:6–14

Silverman PB, Schultz KA (1989) Comparison of cocaine and procaine discriminative stimuli. Drug Dev Res 16:427–433

Silverman K, Evans SM, Strain EC, Griffiths RR (1992) Withdrawal syndrome after the double-blind cessation of caffeine consumption. N Engl J Med 327:1109–1114

Silverman K, Griffiths RR (1992) Low-dose caffeine discrimination and self-reported mood effects in normal volunteers. J Exp Anal Behav 57:91–107

Silverman K, Mumford GK, Griffiths RR (1994) Enhancing caffeine reinforcement by behavioral requirements following drug ingestion. Psychopharmacology (Berl) 114:424–432

Sinton CM, Petitjean F (1989) The influence of chronic caffeine administration on sleep parameters in the cat. Pharmacol Biochem Behav 32:459–462

Snyder SH (1985) Adenosine as a neuromodulator. Annu Rev Neurosci 8:103–124

Snyder SH, Katims JJ, Annau Z, Bruns RF, Daly JW (1981) Adenosine receptors and behavioral actions of methylxanthines. Proc Natl Acad Sci USA 78: 3260–3264

Spealman RD, Coffin VL (1988) Discriminative-stimulus effects of adenosine analogs: mediation by adenosine A_2 receptors. J Pharmacol Exp Ther 246:610–618

Spencer DG Jr, Lal H (1983) Discriminative stimulus properties of 1-phenylisopropyl adenosine: blockade by caffeine and generalization to 2-chloroadenosine. Life Sci 32:2329–2333

Stern KN, Chait LD, Johanson CE (1989) Reinforcing and subjective effects of caffeine in normal human volunteers. Psychopharmacology (Berl) 98:81–88

van Dusseldorp M, Katan MB (1990) Headache caused by caffeine withdrawal among moderate coffee drinkers switched from ordinary to decaffeinated coffee: a 12 week double blind trial. Br Med J 300:1558–1559

Vitiello MV, Woods SC (1977) Evidence for withdrawal from caffeine by rats. Pharmacol Biochem Behav 6:553–555

von Borstel RW, Wurtman RJ, Conlay LA (1983) Chronic caffeine consumption potentiates the hypotensive action of circulating adenosine. Life Sci 32:1151–1158

Wayner MJ, Jolicoeur FB, Rondeau DB, Barone FC (1976) Effects of acute and chronic administration of caffeine on schedule dependent and schedule induced behavior. Pharmacol Biochem Behav 5:343–348

Weber JG, Ereth MH, Danielson DR (1993) Perioperative ingestion of caffeine and postoperative headache. Mayo Clin Proc 68:842–845

West R (1992) Nicotine addiction: a re-analysis of the arguments. Psychopharmacology (Berl) 108:408–410

Winter JC (1981) Caffeine-induced stimulus control. Pharmacol Biochem Behav 15:157–159

Yamamoto T, Miyamoto K, Ueki S (1987) Rolipram as a discriminative stimuli: transfer to phosphodiesterase inhibitors. Jpn J Pharmacol 43:165–171

Yasuhara M, Levy G (1988) Rapid development of functional tolerance to caffeine-induced seizures in rats. Proc Soc Exp Biol Med 188:185–190

Zwyghuizen-Doorenbos A, Roehrs TA, Lipschutz L, Timms V, Roth T (1990) Effects of caffeine on alertness. Psychopharmacology (Berl) 100:36–39

CHAPTER 10

Classical Hallucinogens*

R. A. GLENNON

A. Introduction

Hallucinogenic agents, although typically nonaddicting, constitute a notable class of abused substances. Not only do hallucinogens possess a rich and fascinating folkloric history, but they constitute one of the oldest known classes of pharmacologically active agents. The earliest written account of psychoactive agents dates back 3500 years, and there is evidence that hallucinogenic plants were in use as long ago as 8000 B.C. The first mention of the use of hallucinogenic crude plant products in the New World was made, appropriately enough, by a member of the Columbus expedition. For a treatment of the historical and anthropological aspects of hallucinogens, see EFRON et al. (1967), FURST (1976) and HARNER (1973).

Although a small handful of substances has gained widespread name recognition – for example, the term "hallucinogen" often conjures up notions of drugs such as LSD and mescaline – there are perhaps hundreds of agents and naturally occurring substances that may be called hallucinogenic. Lately, some new designer drugs (i.e., controlled substance analogs) have been shown to possess hallucinogen-related structures. Thus, investigations with hallucinogens not only aid our understanding of hallucinogenic agents per se but can also have a bearing on these novel designer drugs. Apart from drug abuse considerations, hallucinogens may also be useful in the development of neuropsychiatric agents. Because certain hallucinogens act as $5-HT_2$ serotonin (5-HT) agonists, and because $5-HT_2$ antagonists may be effective in the treatment of schizophrenia, depression, and anxiety, antagonism of hallucinogen-induced effects in animals is becoming increasingly useful for the identification of novel serotonin antagonists.

Structurally, classical hallucinogens (see Sect. B for definitions) are commonly divided into two categories: those possessing an indolylalkylamine nucleus and those possessing a phenylalkylamine moiety (Fig. 1). The indolylalkylamines may be further divided into the: (1) simple N-substituted tryptamines, e.g., *N,N*-dimethyltryptamine (DMT), 5-methoxy DMT

---

* Portions of this manuscript were presented at a NIDA Technical Review on Classical Hallucinogens held in Bethesda, MD, in June, 1992, which forms the basis of an upcoming NIDA monograph.

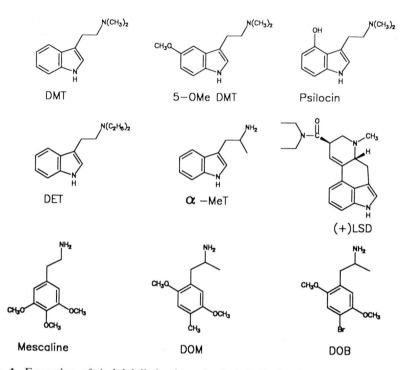

Fig. 1. Examples of indolylalkylamine classical hallucinogens: *N,N*-dimethyltry-ptamine (*DMT*, 5-methoxy DMT, psilocin, *N,N*-diethyltryptamine (*DET*), α-methyltrypta-mine (*α-MeT*), and (+)lysergic acid diethylamide (*LSD*); and of phenylalkylamine classical hallucinogens: mescaline, 1-(2,5-dimethoxy-4-methyl-phenyl)-2-aminopropane (*DOM*), and 1-(4-bromo-2,5-dimethoxyphenyl)-2-aminopropane (*DOB*)

(5-OMe DMT) psilocin (4-OH DMT), or *N,N*-diethyltryptamine, (DET), (2) α-methyltryptamines (α-MeT) e.g., 5-methoxy α-MeT, (3) ergolines e.g., (+)lysergic acid diethylamide (LSD), and (4) β-carbolines, e.g., harmaline-related alkaloids. The phenylalkylamines may be subdivided according to whether or not they possess an α-methyl group on the alkyl chain that separates the aromatic ring from the terminal amine; phenylal-kylamines without the α-methyl group are termed phenylethylamines, e.g., mescaline, whereas those with the α-methyl group are termed phenylisopro-pylamines or phenylaminopropanes, e.g., 1-(2,5-dimethoxy-4X-phenyl)-2-aminopropanes, where X can be methyl, bromo or iodo, i.e., DOM, DOB, and DOI, respectively. For a general review of hallucinogens, see Brimblecombe and Pinder (1975); for a discussion of classical hal-lucinogens, see Nichols and Glennon (1984) and Shulgin and Shulgin (1991). The present chapter deals exclusively with classical hallucinogens and structurally related agents.

B. Definitions

The term "classical hallucinogen" is being employed with increasing frequency. The purpose of this is to distinguish between different classes of agents that produce hallucinogenic effects. This is an important consideration for certain investigations, such as those aimed at elucidating mechanisms of action. For example, chronic administration (but, normally, not administration of single doses) of amphetamine can produce hallucinations (i.e., "amphetamine psychosis"); hallucinatory episodes may accompany the toxic delirium associated with the use of certain nonhallucinogenic agents; and the intoxication produced by phencyclidine (PCP) or certain opiates seems to be different from that produced by, for example, LSD. Each of these hallucinogenic effects is probably mechanistically distinct. Unfortunately, to date, there is no definition that adequately and accurately describes the actions of these agents. HOLLISTER (1968) proposed a set of criteria that should be met in order for an agent to be considered hallucinogenic: (1) in proportion to other effects, changes in thought, perception and mood should predominate, (2) intellectual or memory inpairment should be minimal, (3) stupor, narcosis, or excessive stimulation should not be an integral effect, (4) autonomic nervous system side effects should be minimal, and (5) addictive craving should be absent. Classical hallucinogens typically possess an indolealkylamine or phenylalkylamine moiety; as will be described below, they also bind at 5-HT$_2$ receptors and produce stimulus effects similar to those of DOM. However, specific inclusion criteria have yet to be defined.

Even within the category of agents called classical hallucinogens, it is recognized that not all agents necessarily produce identical effects. Indeed, not all classical hallucinogens produce hallucinations. Although hallucinogens may produce visual, and to a lesser extent, auditory hallucinations, they also produce closed-eye imagery, synesthesia, and other perceptual alterations. In fact, it has been said that a dose of a given agent may produce different effects in the same individual upon different occasions of administration (NARANJO 1973) and that the human subject is as much a contributor to the final definition of a drug's action as is the drug itself (SHULGIN and SHULGIN 1991). Thus, clearly, there exist some differences in effect. How can these differences be rationalized? There are several likely explanations: (1) effects may be dose- or route-dependent, (2) somatic or other side effects may contribute substantially to the observed differences, (3) the agents may not constitute a mechanistically homogeneous group of compounds, and/or (4) the classical hallucinogens may act via similar nonidentical mechanisms that share a common mechanistic component. Because there is some evidence for similarity of effect, the last explanation, i.e., the *common component hypothesis*, provides a framework for mechanistic investigations (GLENNON 1984). Identification of a common mechanistic component may be important to our understanding of these agents. Agents that produce dissimilar effects are likely acting via a different mechanism; such agents may need to be

categorized separately, and such categorization may influence future treatment modalities. This is particularly true of designer drugs that contain
structural elements similar to hallucinogens and, at the same time, possess
certain other structural features common to other classes of drugs of abuse
(e.g., stimulants).

C. Human vs Nonhuman Investigations

Hallucinogenic agents have been investigated using both human and
nonhuman subjects. Only human subjects can accurately assess and describe
the subjective effects of these agents. Relatively few hallucinogens have
been examined in humans, and even fewer have been examined under the
rigorous conditions normally accorded novel therapeutic agents undergoing
clinical trials. Legal constraints have discouraged new clinical studies. In
addition, some of the currently available data frequently cited in the literature were derived from studies that used small subject populations or that
did not examine multiple doses of drug (i.e., complete dose response studies
were not conducted). Indeed, some agents that are thought to be inactive
have been examined only at relatively low doses; the results of more recent
animal studies suggest that such agents might be active in humans if significantly higher doses are examined. Some other problematic studies have
involved patients suffering from various mental disorders, and yet other data
may be from anecdotal reports. Investigations involving nonhuman subjects
are much more common and allow the examination of greater numbers of
agents in large numbers of subjects, using multiple doses of drug and
carefully controlled conditions. However, there are obvious limitations to
this approach, the most prominent being that it is unknown if animals
experience subjective effects identical to those experienced by humans; that
is, do animals hallucinate? Thus, investigations in this field are fraught with
a variety of problems, not the least of which is: Do we rely on human data
regardless of source or quality, or do we put some faith in the results of
animal studies?

   Animal studies tend to fall into two categories: investigative or observa-
tional studies and interpretive studies. The former simply categorizes the
actions of known hallucinogens in animal subjects (e.g., effect on physiological functions and behavior) in an attempt to catalog their pharmacological effects without further interpretation. The latter addresses possible
mechanisms involved in the production of these effects. Mechanistic interpretation must necessarily be conservative, and identified mechanisms may
or may not be related to the hallucinogenic activity of the agents under
investigation.

   Another type of investigation involving animals is the development of
animal models to identify and study novel hallucinogens. Such studies begin
with examination of known hallucinogens to determine what effects are

common to a series of agents but absent upon administration of "inactive" agents. Once such an effect has been identified, the model is, ideally, challenged with other hallucinogens, nonhallucinogens, and ultimately with novel agents. Of course, it can never be assured that novel agents identified in this manner will be hallucinogenic until they have been evaluated in human subjects. Nevertheless, robust and reliable animals models can be valuable for further mechanistic investigations by allowing experimentation not appropriate (or allowed) in humans. Here also, greater numbers and doses of agents can be evaluated in relatively large subject populations.

Thus, studies involving human and nonhuman subjects have their own peculiar limitations, advantages and disadvantages. The ideal situation would likely be investigations involving both types of subjects.

D. Models and Assay Methods for Hallucinogenic Activity

I. Animal Models

Over the years, numerous animals models of hallucinogenic activity, or at least animal assays that might be useful for examining hallucinogenic agents, have been described (reviewed in GLENNON 1992). Animal models are of two types: behavioral and nonbehavioral. The behavioral models are further divided into *analog models* and *assay models* (STOFF et al. 1978); these models have also been referred to as isomorphic models and parallel models, respectively (JACOBS and TRULSON 1978). Analog models are correlational; that is, they rely on some drug-induced animal behavior for which there is a human counterpart (e.g., exploratory behavior). Assay models are inferential; that is, there need not be a relationship between the animal and human behavior so long as test drugs produce a dose-related effect that parallels human hallucinogenic potency. It has been whimsically suggested that if hallucinogens elicited tail-biting behavior in rodents in a dose-dependent manner with a potency that parallels human hallucinogenic potency, then tail-biting could be a useful assay model of hallucinogenic activity (STOFF et al. 1978). Nonbehavioral animal models may be of an analog or assay nature but simply rely on effects that are not necessarily behavioral (e.g., contraction of isolated muscle strip in a tissue bath).

Some common explicit animal models include: (1) the serotonin syndrome, (2) the ear-scratch reflex or scratch reflex stereotypy, (3) the head-twitch response, (4) rabbit hyperthermia, (5) limb-flick behavior in cats or limb-jerk in monkeys, (6) the startle reflex, (7) investigatory behavior, (8) disruption of fixed-ratio responding; the so-called hallucinogenic pause, and (9) drug discrimination using animals trained to standard hallucinogens (reviewed in GLENNON 1992). Combinations of these and other assays have been employed as test batteries (e.g., see OTIS et al. 1978; STOFF et al. 1978) with the hope that a combination of tests might prove more reliable than any single animal model. To date, however, there is no foolproof animal model that allows reliable predictions of hallucinogenic activity. That is not to say that the use of animal models is not worthwhile; indeed, animal models have significantly enhanced our understanding of hallucinogenic agents. They have allowed, for example, the formulation and evaluation of a multitude of hypotheses, and there are instances in which novel agents have been found active in animal models and only later have the results been verified in humans. Unfortunately, each model has resulted in some false positives (i.e., has identified an agent, known to be inactive in humans, as being potentially hallucinogenic) and/or false negatives (i.e., has identified a known hallucinogen as being potentially inactive).

A major problem that has haunted animal models of hallucinogenic activity/potency is the existence of what have been termed *enigmatic agents*. Certain agents are continually identified by various animal models as being "active" when in fact there is little (or no) supporting human data. These enigmatic agents fall into three broad categories. First, there are agents known to lack hallucinogenic activity in humans when administered in a single dose; amphetamine, an example of such an agent, is active in several animal models (e.g., rabbit hyperthermia). Second, there are agents that are generally regarded as lacking hallucinogenic properties and that may even be widely used therapeutically, but for which there are scattered accounts of hallucinogenic episodes in humans. Lisuride is typical of this type of enigmatic agent. Third, there are those agents for which human data are very limited or nonexistent; quipazine, for example, is active in many, if not most, animal models. It is this last category of agents that is most troublesome. Until additional clinical studies are conducted, it can never be known with certainty if these types of agents are truly without hallucinogenic effects. Nevertheless, these agents should continue to be used in future studies with animals in order to challenge new models and also to gain additional insight into the agents themselves.

II. Nonanimal Models

Nonanimal techniques have been employed to investigate hallucinogenic agents and, in particular, the structure-activity relationships of such agents. These may be classified as *stochastic interaction models*, *conservative molecule models*, and *mechanistic models* (KIER and GLENNON 1978). Stochastic interaction models are simulations of drug-receptor interactions in the absence of any understanding of the receptor involved; that is, interactions between an active drug molecule and some hypothetical receptor feature. The conservative molecule approach is an investigation of the structural influence (e.g., physicochemical or quantum chemical properties) of active agents on hallucinogenic activity; the approach is amechanistic and simply attempts to correlate hallucinogenic activity/potency with chemical structure. The mechanistic model is similar to the conservative molecule approach except that it allows development of quantitative relationships between properties of drugs and pharmacological activities that have potential mechanistic relevance (e.g., the influence of lipophilicity on receptor affinity for a series of active agents). Although these models may possess some predictive and/or mechanistic value, each requires animal or human data for initial input and, as such, cannot be considered a substitute for animal models.

One of the more exciting techniques to be recently explored is the modeling of drug-receptor interactions using graphics models of neurotransmitter receptors. Because the precise three-dimensional structures of neurotransmitter receptors are unknown at this time, different models, and indeed different hypothetical modes of drug-receptor interaction, are possible (reviewed in WESTKAEMPER and GLENNON 1991). Thus, these models will require validation by empirical methods such as site-directed mutagenesis and/or ligand binding utilizing chimeric receptors. Such studies will be further described in Sect. H.

III. The Drug Discrimination Paradigm

The drug discrimination paradigm (reviewed in GLENNON 1991) was listed above in Sect. D.I along with other animal models of hallucinogenic activity. In fact, it has never been claimed that the paradigm is a model of hallucinogenic activity. Rather, it is a drug detection procedure in which animals that have been trained to discriminate a given training drug from vehicle are essentially asked whether they recognize a novel agent as producing stimulus effects similar to those produced by the training drug. The paradigm is of general applicability and its utility is not limited to the investigation of hallucinogenic agents. Nevertheless, it has proven quite successful in qualitatively and quantitatively identifying hallucinogenic agents. Examples of each major category of classical hallucinogens (with the exception of the β-carbolines) have been used as training drugs in drug discrimination studies, and (+)LSD, 5-OMe DMT, mescaline, and DOM, representative of the four major subclasses of classical hallucinogens, have seen the most extensive application. Stimulus generalization occurs among all four of these agents regardless of which is used as the training drug. Animals trained to one of these agents do not recognize nonclassical hallucinogens such as phencyclidine, opiates, and other agents that may occasionally produce hallucinogenic effects. This is one reason why such agents have been classified under the common heading of classical hallucinogens; and this fact has some bearing on the above mentioned suggestion that classical hallucinogens, although perhaps capable of producing slightly different effects, seem able to produce a common effect. This hypothesis has been further tested using rats trained to discriminate DOM from vehicle. A large number of agents, including tryptamine hallucinogens, α-emthyltryptamines, ergolines, phenylethylamines, phenylisopropylamines, and β-carbolines have now been examined. To date, there have been no reports of false negatives, but several false positives have been noted: lisuride and quipazine, but not amphetamine, produce DOM-like stimulus effects. Structure-activity relationships have been formulated, mechanistic studies have been conducted, and for a series of agents (encompassing examples of each category of classical hallucinogens) for which human data are available there is a significant correlation ($r > 0.9$) between discrimination-derived ED50 values and human hallucinogenic potencies (reviewed in GLENNON 1991). It was the use of DOM-trained animals that originally led to the *5-HT$_2$ hypothesis* of hallucinogen action; this will be further discussed in Sect. F.II.

Another widely used training drug is (+)LSD (e.g., see CUNNINGHAM and APPEL 1987). Although the LSD stimulus generalizes to members of other categories of classical hallucinogens, its stimulus properties may involve multiple neurotransmitter mechanisms including dopaminergic and α_1-adrenergic mechanisms (MEERT et al. 1989; KOEK et al. 1992). False positives have been noted with (+)LSD as training drug. For example, lisuride, quipazine, but not amphetamine, produce (+)LSD-like stimulus effects. Interestingly, however, although animals trained to discriminate (+)LSD from vehicle recognize lisuride, animals can be trained to discriminate (+)LSD from lisuride (CUNNINGHAM and APPEL 1987). These results suggest that lisuride and (+)LSD produce sufficiently common effects to allow them to be recognized by the same animals but sufficiently different effects that allow them to be discriminated.

The actions of DOM, LSD, and structurally related agents, are at least partly related to a 5-HT$_2$ agonist mechanism (see Sect. F.II). Because there is evidence that 5-HT$_2$ antagonists may constitute novel classes of antipsychotic agents, antidepressants, and anxiolytics, antagonism of hallucinogen-induced effects in animals is being used to identify new 5-HT$_2$ antagonists. In particular, the drug discrimination paradigm with animals trained to discriminate various hallucinogens is being increasingly used for this purpose (GLENNON 1991; KOEK et al. 1992; MEERT et al. 1989; MEERT 1991).

E. Structure-Activity Relationships

Structure-activity relationships (SARs) have been formulated for the various classes of classical hallucinogens in order to: (1) determine how structural features influence activity/potency, (2) classify which agents possess membership in the family of agents referred to as classical hallucinogens, and (3) investigate mechanisms of action. Certain hallucinogens have been investigated in greater detail than others. For example, relatively little is known about the ergolines and β-carbolines, whereas much more is known about the indolylalkylamines and phenylalkylamines. The SARs of the latter two categories of agents in producing hallucinogenic activity and in producing activity in various animal models have been described in detail

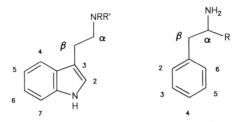

Fig. 2. General structures and numbering system for the indolylalkylamine (*left*) and phenylalkylamine (*right*) classical hallucinogens

(NICHOLS and GLENNON 1984). Where both human and animal data are available, the results are surprisingly consistent. Some of the important and consistent findings will be summarized here. See Fig. 2 for the chemical structures and numbering systems.

I. Indolylalkylamines

Primary amine analogs (i.e., derivatives with unsubstituted terminal amine groups) and most secondary amine analogs (i.e., analogs with one small alkyl substituent) are generally inactive. This may be related to problems associated with their inability to penetrate the blood-brain barrier and/or due to their rapid metabolism. N,N-Dialkylamine derivatives are typically active. Derivatives of DMT and DET are the most common in occurrence. Generally, there is little potency difference between DMT analogs and their DET counterparts. Further increasing the size of the alkyl substituents tends to decrease potency. Certain secondary amine analogs (in which the substituent is large enough to enhance lipophilicity or hinder metabolism) are also active.

α-Methyl substitution enhances the potency of primary amine analogs. In general, α-methyl analogs are several times more potent than the corresponding DMT counterparts. Otherwise, the SARs of the α-methyl-tryptamines seem to parallel those of the N,N-dialkyltryptamines. Introduction of the α-methyl group creates a chiral center; that is, two optical isomers are now possible. In the few cases in which individual optical isomers have been examined, the S(+) isomers are several times more potent than their R(−) enantiomers; nevertheless, both isomers possess activity.

Aromatic substitution can either increase or decrease the potency of DMT and α-methyltryptamine. In general, substitution at the 4- or 5-position by a methoxy group increases potency whereas similar substitution at the 6- or 7-position either decreases potency or abolishes activity. Hydroxylation at the 4-position results in active agents whereas hydroxylation at the 5-, 6-, or 7-positions abolishes activity. Two of the best investigated agents include 4-

OH DMT and 5-OMe DMT. Bufotenine (5-OH DMT or *N,N*-dimethyl 5-HT) has been reported to be both hallucinogenic and nonhallucinogenic; for a recent brief review, see McCLEOD and SITARAM (1985).

II. Phenylalkylamines

Although very few nonprimary amine analogs have been investigated in humans, in general the primary amines are more potent than their corresponding secondary or tertiary amines. In animal studies, even N-monomethylation can decrease potency by an order of magnitude.

α-Methyl-substituted derivatives (i.e., phenylisopropylamines) are usually two to ten times more potent than their corresponding unsubstituted counterparts (i.e., phenylethylamines). As with the indolylalkylamines, this may be related to a greater difficulty in penetrating the blood-brain barrier by, or to more rapid metabolism of, the phenylethylamines. With the phenylisopropylamines, the $R(-)$ isomers are usually several-fold more potent than their $S(+)$ enantiomers. Thus, whereas stereochemistry seems to play only a small role, the enantioselectivity for indolylalkylamine hallucinogens and phenylisopropylamine hallucinogens is reversed.

Aromatic substitutents have a very significant effect on activity. In the phenylisopropylamine series, in which the terminal amine and aromatic ring are unsubstituted, the agent (i.e., amphetamine) is inactive as a hallucinogen. Monomethoxy substitution typically results in agents that are not hallucinogenic, although these agents are not necessarily centrally inactive. Dimethoxy substitution can result in hallucinogenic agents depending upon the position of the methoxy groups. The best investigated of these is the 2,5-dimethoxy analog (2,5-DMA). The 2,4-DMA may also be active. Of the trimethoxy analogs (TMAs), 2,4,5-TMA is the best investigated. Several of the other possible TMAs are also active. With 2,5-DMA, further substitution at the 4-position leads to the most potent agents. For example, the 4-methyl analog is DOM and the 4-bromo analog is DOB. These general SARs are consistent with those of the phenylethylamines except, as mentioned above, the phenylethylamines are several-fold less potent than their phenylisopropylamine counterparts. See SHULGIN and SHULGIN (1991) for a comprehensive treatment of the role of substituent effects in the actions of phenylalkylamine hallucinogens in human subjects.

F. Mechanism of Action of Classical Hallucinogens

I. The 5-HT Hypothesis

Shortly after the discovery of the neurotransmitter 5-HT in the late 1940s and the realization that both 5-HT and LSD possess and indolylalkylamine nucleus, it was postulated that hallucinogens might act via a serotonergic mechanism (reviewed in WOOLLEY 1962). However, the structures of the phenylalkylamine hallucinogens bear a greater structural resemblance to catecholaminergic neurotransmitters such as epinephrine, norepinephrine and dopamine than to 5-HT. Thus, initially, it was not unreasonable to assume that catecholaminergic neurotransmitters might be involved in the actions of the phenylalkylamine hallucinogens. Indeed, even LSD has a profound effect on each of these (and other) neurotransmitter systems. Historically, however, there is little empirical support for involvement of most of these neurotransmitters in the common actions of the classical hallucinogens. In contrast, ever since its discovery, 5-HT has been consistently implicated in the actions of hallucinogens.

With the discovery of two distinct populations of peripheral 5-HT receptors (D receptors and M receptors), controversy arose during the 1950s: Do multiple populations of 5-HT receptors exist in human brain? Do hallucinogens act at a subpopulation of 5-HT receptors? If so, do they act as 5-HT agonists or antagonists? During the 1980s and early 1990s, numerous populations of 5-HT receptors were identified: 5-HT_{1A}, 5-HT_{1B}, 5-HT_{1C} (now 5-HT_{2C}), $5\text{-HT}_{1D\alpha}$, $5\text{-HT}_{1D\beta}$, 5-HT_{1E}, 5-HT_{1F}, 5-HT_2 (now 5-HT_{2A}), 5-HT_{2F} (now 5-HT_{2B}), 5-HT_3, and 5-HT_4. Some of the most recently described 5-HT receptors include 5-HT_5 and 5-HT_6. This multitude of 5-HT receptors only served to further confound the issue. Most of the 5-HT_1 (and probably 5-HT_4) receptors belong to a G-protein coupled superfamily of receptors involving an adenylate cyclase second messenger system. 5-HT_2 and 5-HT_{1C} receptors also belong to this family, but are linked to a phosphoinositol (PI) second messenger systems. Due to their significant transmembrane sequence homology and due to similarities in second messenger systems, 5-HT_{1C} receptors are now termed 5-HT_{2C} receptors. The originally defined 5-HT_2 receptors are currently referred to as 5-HT_{2A} receptors. For the same reasons, and to fill the gap between 5-HT_{2A} and 5-HT_{2C}, 5-HT_{2F} receptors were renamed 5-HT_{2B} receptors. Thus, the 5-HT_2 receptor family consists of three distinct subtypes: 5-HT_{2A}, 5-HT_{2B}, and 5-HT_{2C}. In the remainder of this chapter, the term "5-HT_2" will be used in a generic sense, or to refer to the entire family of 5-HT_2 receptors as a class. If specific information is available on a particular population of 5-HT_2 receptor, the more precise subpopulation name will be used. 5-HT_3 receptors are distinct in being ligand-gated ion channel receptors.

II. The 5-HT_2 Hypothesis

The issue now becomes even more complicated: Which population(s) of 5-HT receptors is involved in the actions of classical hallucinogens? Based on the finding that both the discriminative stimulus effects of DOM and DOM stimulus generalization to (+)LSD, 5-methoxy DMT, and mescaline could be potently antagonized by 5-HT_2 antagonists, it was proposed that the classical hallucinogens act as agonists at 5-HT_2 receptors (GLENNON et al.

1983). In support of this hypothesis, the binding of various hallucinogens at different populations of 5-HT receptors using radioligand binding techniques indicated a common affinity only for 5-HT_2, specifically 5-HT_{2A}, receptors (GLENNON et al. 1984). Indolylalkylamine hallucinogens are fairly nonselective and bind with high affinity at multiple populations of 5-HT receptors; ergolines such as LSD are without significant affinity for 5-HT_3 and 5-HT_4 receptors. In contrast, the phenylisopropylamine hallucinogens such as DOM, DOB, and DOI were found to bind rather selectively at 5-HT_{2A} receptors. Furthermore, there was a significant correlation ($r > 0.9$) between 5-HT_{2A} receptor affinity and both discrimination-derived ED_{50} values and human hallucinogenic potencies (reviewed in GLENNON 1990). For the first time, there was now evidence for a *5-HT$_2$ hypothesis* of hallucinogenic activity.

However, over the years, several serious problems have arisen regarding the 5-HT_2 hypothesis: (1) Do hallucinogens act as 5-HT_2 agonists or antagonists? (2) Are there subpopulations of 5-HT_2 receptors with which hallucinogens might interact differently? It might be noted that the 5-HT_2 hypothesis was first proposed prior to the discovery of 5-HT_{2B} and 5-HT_{2C} receptors. (3) May some other population of 5-HT receptors (instead of 5-HT_2) be involved in the actions of hallucinogens?

On the basis of drug discrimination studies, it was originally suggested that classical hallucinogens act as 5-HT_2 agonists; however, PIERCE and PEROUTKA (1988) challenged this concept, suggesting that hallucinogens, particularly (+)LSD, act as 5-HT_2 antagonists. This issue has been re-examined and it would appear that hallucinogens are *not* 5-HT_2 antagonists: classical hallucinogens are agonists, or at least partial agonists, at 5-HT_2 receptors (reviewed in GLENNON 1990). Certain hallucinogens, including LSD, may possess a low intrinsic activity; thus, when administered in combination with a full agonist, such agents might occasionally appear to behave as antagonists in some pharmacological assays. However, there is no evidence that 5-HT_2 antagonists (a number of which are currently in clinical trials) such as ketanserin, ritanserin, risperidone, and others produce hallucinogenic effects in humans.

Is it possible that subpopulations of 5-HT_2 receptors exist? In the *two-state hypothesis*, LYON et al. (1987) proposed that 5-HT_2 (now 5-HT_{2A}) receptors exist in two affinity states: a low-affinity state and an agonist high-affinity state. [^{3}H]Ketanserin, a 5-HT_2 antagonist, labels both states of the receptor. Yet the high-affinity state represents only 5%–20% of the binding. By analogy to the adrenergic system, a radiolabeled agonist ligand should label predominantly the high-affinity state of the receptor. Consequently, in an attempt to identify a candidate agonist for labeling, SAR investigation was conducted with DOB. The contribution of each structural feature of DOB to binding was examined and additional analogs were prepared. With the exception of the iodo derivative DOI, no analog was found to bind with higher affinity than DOB, and, at the same time, retain agonist character as measured using the drug discrimination paradigm. Subsequently, [^{3}H]DOB was synthesized and used as a radioligand in binding studies. Although antagonists were found to bind with comparable affinity at [^{3}H]ketanserin- and [^{3}H]DOB-labeled 5-HT$_2$ sites, agonists were shown to bind with upwards of 50-fold higher affinity at [^{3}H]DOB- vs [^{3}H]ketanserin-labeled sites. 5-HT, for example, binds with 100-fold higher affinity and R(−)DOB with 60-fold higher affinity at [^{3}H]DOB- vs [^{3}H]ketanserin-labeled sites. Thus, it was proposed that [^{3}H]DOB labels the high-affinity (i.e., 5-HT$_{2H}$) agonist state of 5-HT$_2$ receptors (reviewed in GLENNON 1990). Certain DOB analogs which were found to bind with higher affinity than DOB at [^{3}H]ketanserin-labeled sites, but which lacked agonist activity in the drug discrimination paradigm, were subsequently shown to possess nearly identical affinity, but not enhanced affinity, for [^{3}H]DOB-labeled sites. One such example is the 4-(3-phenylpropyl) analog of 2,5-DMA (DOPP). Based on the fact that DOPP binds with similar affinity at [^{3}H]ketanserin- and [^{3}H]DOB-labeled sites (K_i = 10 and 17 nM, respectively), it would seem likely that DOPP is a 5-HT$_2$ antagonist. Indeed, DOPP has been demonstrated to act as an antagonist of 5-HT at 5-HT$_2$ receptors linked to phosphoinositide hydrolysis (GLENNON et al. 1992a).

PIERCE and PEROUTKA (1989) later conducted related investigations using [^{77}Br]DOB and proposed an alternative explanation: two different populations (i.e., subpopulations) of 5-HT$_2$, i.e., 5-HT$_{2A}$, receptors (the *two-site hypothesis*) exist. It was suggested that radiolabeled phenylalkylamines label what they termed 5-HT$_{2A}$ receptors whereas [^{3}H]ketanserin labels 5-HT$_{2B}$ receptors. (These 5-HT$_{2A}$ and 5-HT$_{2B}$ receptors should not be confused with those described in Sect. F.I.) Part of their reasoning was based on the finding that [^{77}Br]DOB-labeled sites are absent in bovine cortex. However, because the extent of [^{3}H]ketanserin binding is lower in bovine tissue than in rat tissue and because 5-HT$_{2H}$ sites represent only a small fraction of this value, it was argued that 5-HT$_{2H}$ binding might not be observed with radioligands of low specific activity. The development of high specificactivity [^{125}I]DOI allowed the identification of 5-HT$_{2H}$ sites in bovine brain (GLENNON et al. 1988; TEITLER et al. 1990). Furthermore, the results of recent cloning studies favor the two-state concept in that a single 5-HT$_2$ receptor is expressed that behaves in a manner reminiscent of a two-state receptor population (reviewed in WEINSHANK et al. 1992). It might be noted, however, that the possibility of two different (overlapping) binding domains has not yet been excluded; that is, agonists and antagonists may bind in a slightly different manner at the same population (or state) of 5-HT$_2$ receptors.

III. The 5-HT$_{1C}$ Hypothesis

Finally, there is the issue of involvement of other (or additional) populations of 5-HT receptors in the actions of hallucinogens. Shortly after the 5-HT$_2$

hypothesis was proposed in 1983, PAZOS and coworkers (1984) described their discovery of 5-HT$_{1C}$ receptors. Like 5-HT$_2$ receptors, 5-HT$_{1C}$ receptors are linked to a phosphoinositol second messenger system. Furthermore, both receptors have now been cloned, and within the transmembrane domains there is about an 80% amino acid sequence homology (WEINSHANK et al. 1992). To underscore the similarity between these two types of receptors, they have been recently renamed 5-HT$_{2A}$ and 5-HT$_{2C}$ receptors. A reexamination of the binding of hallucinogens at these receptors found little difference between their 5-HT$_2$ and 5-HT$_{1C}$ (now 5-HT$_{2A}$ and 5-HT$_{2C}$) affinities (TITELER et al. 1988). Indeed, later studies have shown less than a tenfold difference in receptor affinity for a large series of phenylalkylamine derivatives (GLENNON et al. 1992b). As with 5-HT$_{2A}$ receptor affinities, 5-HT$_{2C}$ affinities are also correlated both with discrimination-derived ED$_{50}$ values and human hallucinogenic potencies. In addition BURRIS and SANDERS-BUSH (1988) reported that DOM acts as a 5-HT$_{2C}$ agonist. Furthermore, the 5-HT$_2$ hypothesis was based, in part, on the finding that 5-HT$_2$ antagonists (such as ketanserin and pirenperone) antagonize the stimulus effects of hallucinogens; it is now recognized that these 5-HT$_2$ antagonists are more accurately described as 5-HT$_{2A}$/5-HT$_{2C}$ antagonists. Thus, the likelihood exists that 5-HT$_{2A}$ and/or 5-HT$_{2C}$ receptors are involved in the actions of hallucinogenic agents.

IV. 5-HT$_{2A}$ vs 5-HT$_{2C}$ Receptors

It is quite difficult to ascribe a specific role for 5-HT$_{2C}$ vs 5-HT$_{2A}$ receptors in the mechanism of action of hallucinogens due to the lack of agents that display selectivity for one of these populations of receptors over the other. Nearly all agents that bind at 5-HT$_{2A}$ receptors bind at 5-HT$_{2C}$ receptors. However, there are a few agents that might offer hope in resolving this problem.

1. Studies with Selective Antagonists

The dopamine/5-HT$_{1A}$ antagonist spiperone binds with approximately 500-fold selectivity for 5-HT$_{2A}$ vs 5-HT$_{2C}$ receptors. Attempts to antagonize the stimulus effects of DOM using various doses of spiperone, with the intention that in might be more difficult to antagonize the DOM stimulus if the stimulus was 5-HT$_{2C}$-mediated, resulted in inconclusive results due to the severe disruptive effects of low doses of spiperone in combination with DOM (GLENNON 1991). However, recent stimulus antagonism studies with a new spiperone-related analog AMI-193, which displays 2000-fold selectivity for 5-HT$_{2A}$ vs 5-HT$_{2C}$ sites, suggest that the stimulus effects of DOM are primarily 5-HT$_{2A}$-mediated (ISMAIEL et al. 1993). AMI-193 dose dependently antagonized the stimulus effects of DOM without the disruption associated with spiperone (Fig. 3). Although these results point to a 5-HT$_{2A}$

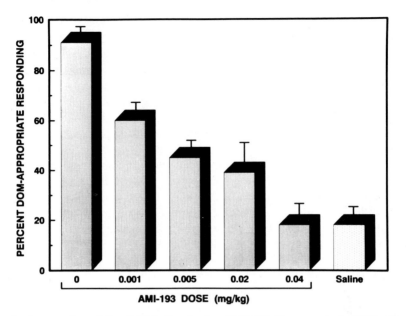

Fig. 3. Antagonism of the DOM stimulus by AMI-193 (Ismaiel et al. 1993). Doses of AMI-193 were administered 45 min prior to 1.0 mg/kg of the training drug DOM. Saline vehicle, administered alone, produced 18% DOM-appropriate responding

mechanism, a combined $5\text{-HT}_{2A}/5\text{-HT}_{2C}$ mechanism cannot be ruled out. That is, if DOM produces a compound stimulus involving both mechanisms, interference with one of the two mechanisms may be sufficient for stimulus antagonism. This problem cannot be satisfactorily resolved until a selective 5-HT_{2C} antagonist has been identified. Schreiber et al. (1994) have recently reported that the 5-HT_{2A}-selective antagonist MDL 100,907 or $R(+)\alpha$-(2,3-dimethoxyphenyl)-1-[2-(4-fluoropheny)ethyl]-4-piperidine-methanol, but not the newly developed 5-HT_{2C}-selective antagonist SB 200,646 or N-(1-methyl-5-indolyl)-N'-(3-pyridyl) urea, is capable of attenuating the stimulus effects of DOI in rats. These findings support earlier suggestions that DOX-type agents, where X = Me, Br, I, act primarily as 5-HT_{2A} agonists in producing their stimulus effects.

2. Studies with Lisuride

Sanders-Bush (1994) has recently reported that whereas lisuride is a pure 5-HT_{2C} antagonist, it behaves as a partial agonist at 5-HT_{2A} receptors. The argument has been made that, because lisuride is not hallucinogenic, if hallucinogens act via a 5-HT_{2A} mechanism, they are probably *not* acting as 5-HT_{2A} agonists (Sanders-Bush 1994). This argument is not without merit and raises serious questions concerning the 5-HT_2 hypothesis. However, lisuride is one of the enigmatic agents that has been shown to be active in animal models (e.g., GLENNON 1992; WHITE 1986) and for which there is some evidence of hallucinogenic activity in humans. Most human reports have come from patients who have been receiving lisuride for the treatment of Parkinson's disease; and in at least one case (TODES 1986) the LSD-like effects of lisuride have been argued to involve a dopaminergic rather than serotonergic mechanism (HOROWSKI 1986). However, it is possible that the emetic or other side effects of lisuride preclude use of high enough doses to observe any hallucinogenic effects in normal subjects. Thus, it cannot be argued with any confidence that lisuride is, in fact, a hallucinogen. Interestingly, however, when using rats trained to discriminate DOM from saline, lisuride seems to behave as a partial agonist. That is, at low doses lisuride produces DOM-like stimulus effects; however, given in combination with DOM, lower doses of lisuride attenuate the DOM stimulus effects (GLENNON 1994). Thus, these studies would further argue that the stimulus effects of DOM involve a 5-HT_{2A} mechanism.

3. Studies with TFMPP

1-(3-Trifluoromethylphenyl)piperzine (TFMPP) is a nonselective serotonergic agent that binds at multiple populations of 5-HT receptors. Evidence suggests that TFMPP is a 5-HT_{2C} agonist but a 5-HT_{2A} antagonist (or, at best, a 5-HT_{2A} partial agonist). Administration of TFMPP to animals trained to discriminate DOM from saline failed to result in stimulus generalization. In a parallel study, administration of DOM (or DOI) to animals trained to discriminate TFMPP from vehicle resulted only in partial generalization, followed, at slightly higher DOM (or DOI) doses, by disruption of behavior. In a third series of experiments, doses of DOM were administered in combination with $0.2\,\text{mg/kg}$ of TFMPP (ED_{50} dose $= 0.17\,\text{mg/kg}$) to rats trained to discriminate $0.5\,\text{mg/kg}$ of TFMPP from vehicle. The rationale was that lower (i.e., nondisruptive) doses of DOM should potentiate the effect of the near ED_{50} dose of TFMPP if both agents act via a common mechanism. The results were somewhat surprising in that low doses of DOM, rather than potentiating the effect, actually antagonized the effect of TFMPP. The results of these studies suggest that DOM and TFMPP produce stimulus effects that are dissimilar. However, a recent report on the complexity of the mechanisms underlying the TFMPP stimulus (HERNDON et al. 1992) make interpretation of these results rather difficult.

4. Radioligand Binding Studies

Although the stimulus effects of DOM may be primarily 5-HT_{2A} mediated, the issue remains as to whether hallucinogens produce their hallucinogenic effects in humans via a 5-HT_{2A} or 5-HT_{2C} mechanism. As mentioned above, human hallucinogenic potency correlates significantly both with 5-HT_{2A} and with 5-HT_{2C} affinity. This issue was recently reexamined (GLENNON and TEITLER, unpublished findings). Using the human potencies published by

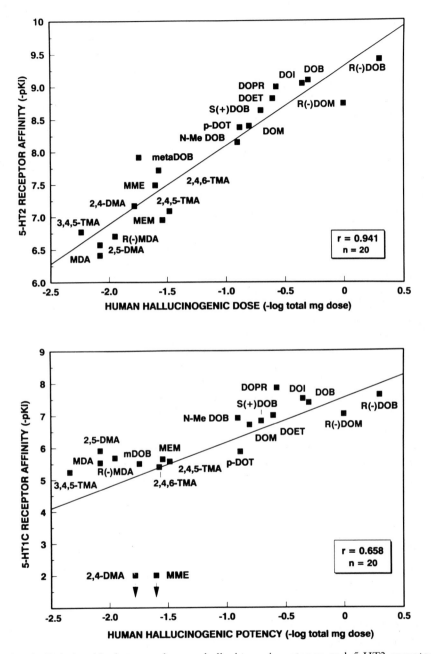

Fig. 4. Relationship between human hallucinogenic potency and 5-HT2 receptor affinity (*upper panel*) and 5-HT1C receptor affinity (*lower panel*) for 20 hallucinogens. Human hallucinogenic potency data are from SHULGIN and SHULGIN (1991)

SHULGIN and SHULGIN (1991), human potency (arithmetic means were used where dose ranges were cited) was found to correlate significantly ($r = 0.941$; $n = 20$) with 5-HT$_{2A}$ receptor affinity (Fig. 4). Potency is also highly correlated with 5-HT$_{2C}$ affinity ($r = 0.886$; $n = 18$) if two agents (i.e., 2,4-DMA and MME) are deleted from the correlation. However, if all 20 agents are included in the latter correlation, the correlation coefficient drops to $r = 0.658$ (Fig. 4). That is, there appears to be a better correlation between human potency and 5-HT$_{2A}$ binding data than with 5-HT$_{2C}$ binding data. It should be noted, however, as described above (Sect. C), that some human studies have not examined multiple doses of drug and/or have not involved large subject populations. Unfortunately, 2,4-DMA and MME are among such agents. Therefore, while these results are intriguing, and provide leads for future studies, additional investigation is certainly warranted. At this time, it cannot be concluded that human hallucinogenic activity involves a 5-HT$_{2A}$ rather than, or in addition to, a 5-HT$_{2C}$ mechanism. The existence of a third member of the 5-HT$_2$ family, 5-HT$_{2B}$ receptors, further confounds this issue. Recent studies (NELSON et al. 1994) have shown that although they display higher affinity for 5-HT$_{2A}$ and 5-HT$_{2C}$ receptors, hallucinogens nonetheless bind at human 5-HT$_{2B}$ receptors. Furthermore, for 17 agents, there was a significant ($r > 0.9$) correlation between 5-HT$_{2B}$ affinity and both 5-HT$_{2A}$ and 5-HT$_{2C}$ affinity.

V. Involvement of Other 5-HT Receptors

We and others have recently reported that there may be functional interactions between different populations of 5-HT receptors such that action at one may modulate activation of another (reviewed in GLENNON et al. 1991). Thus, interaction of an agonist at one population of 5-HT receptors may modulate the effect of the interaction of a second agonist at a different population of 5-HT receptors. For instance, indolylalkylamine hallucinogens are nonselective serotonergic agents. What is the effect of a nonselective agonist that can interact at more than one population of receptors at the same time? What about a nonselective agent that is an agonist at one population and an antagonist at another? Experiments necessary to sort out these types of interactions could be rather labor intensive and their interpretation quite complicated. Nonetheless, this, too, offers direction for future work. For example, it is possible that the nonselective nature of certain agents (e.g., 5-HT, administered as such or as the 5-HT precursor tryptophan) that can interact at multiple populations of 5-HT receptors results in the agents being nonhallucinogenic because they can "turn off" certain receptor-mediated events that might have otherwise resulted in hallucinogenic activity. Do some of the enigmatic agents act in a similar manner? That is, enigmatic agents may give false positives in a particular animal model or assay that primarily involves only a single receptor population (e.g., radioligand binding); but in other instances, in which multiple populations of receptors are present, interaction of these nonselective agents at various populations may modulate or abolish this effect and the agents might appear inactive.

The DOM stimulus does not generalize to the 5-HT$_{1A}$ agonist 8-OH DPAT nor does an 8-OH DPAT stimulus generalize to DOM. Furthermore, DOM does not bind at 5-HT$_{1A}$ receptors nor does 8-OH DPAT bind with significant affinity at 5-HT$_2$ receptors. More recently, we have demonstrated that very low doses of 8-OH DPAT amplify the stimulus effects of DOM in DOM-trained rats. For example, animals given 0.05 mg/kg of 8-OH DPAT in combination with the ED50 dose of DOM behave as if they have received the training dose of DOM (i.e., stimulus generalization occurs upon administration of the ED$_{50}$ dose of DOM) (see inset, Fig. 5). Furthermore, pretreatment of animals with 0.05 mg/kg of 8-OH DPAT results in a leftward shift of the DOM dose-response curve (Fig. 5). These results would seem to suggest that low doses of the 5-HT$_{1A}$-selective agonist influence the stimulus effects of DOM. Additional studies are required to further understand the details of this interaction.

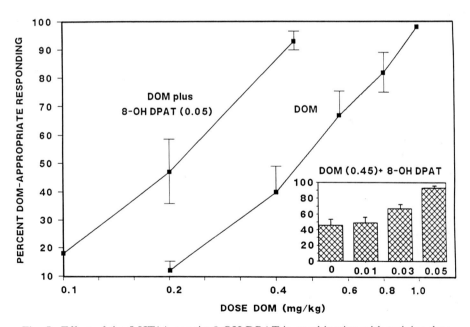

Fig. 5. Effect of the 5-HT1A agonist 8-OH DPAT in combination with training drug DOM (1 mg/kg) in rats trained to discriminate DOM from saline. Dose-response curve for DOM (*right*) and for DOM in animals pretreated with 0.05 mg/kg of 8-OH DPAT (*left*). *Inset* shows effect of different doses of 8-OH DPAT administered combination with the ED$_{50}$ dose (0.45 mg/kg) of DOM. Data previously reported (Glennon 1991)

Similar studies have been conducted with 5-HT$_3$ agents (GLENNON 1994). For example, very low doses of the 5-HT$_3$ antagonist zacopride attenuate the stimulus effects of DOM even though zacopride does not bind at 5-HT$_2$ receptors and DOM does not bind at 5-HT$_3$ receptors. A dose of 0.001 mg/kg of zacopride in combination with the training dose of DOM results in about 30% DOM-appropriate responding; higher doses appear to have less of an attenuating effect. The 5-HT$_3$ (partial) agonist meta-chlorophenylbiguanide (mCPBG) also has an unusual effect on the DOM stimulus. A dose of 0.5 mg/kg of mCPBG seems to attenuate the stimulus effects of 1 mg/kg of DOM; higher doses have less of an effect. However, administered alone, mCPBG seems to result in partial generalization (higher doses producing disruption of behavior). We have conducted parallel studies using rats trained to discriminate the structurally related agent MDMA from saline. Zacopride and LY 278584 (at doses of between 0.0003 and 0.001 mg/kg) decrease MDMA-appropriate responding to about 20%. The effect of mCPBG is not unlike that seen with DOM. The results suggest a possible modulatory effect by 5-HT$_3$ receptors on the DOM and MDMA stimuli. (It might be noted, for purpose of comparison, that zacopride has essentially no effect on amphetamine-appropriate responding in rats trained to discriminate (+)amphetamine from vehicle). These types of unexpected interactions open up entirely new lines of investigation regarding classical hallucinogens and may hint at a possible role for 5-HT$_3$ antagonists in the treatment of drug abuse involving hallucinogens.

VI. Involvement of Dopamine Receptors

There is no question that LSD can influence dopaminergic function (e.g., CUNNINGHAM and APPEL 1987; WHITE 1986). In fact, the hallucinogenic effects of LSD may involve, at least in part, a dopaminergic mechanism. To date, however, there is little to no evidence that dopamine plays a significant role in the actions of simple tryptamine or phenylalkylamine hallucinogens. It may be the added effect on dopamine (or other neurotransmitter) receptors that accounts for the unusually high potency and unique effects of LSD relative to some of the other classical hallucinogens. A role for dopamine receptors in the actions of these agents requires further investigation. It should be noted, however, that examples of many different classes of dopamine antagonists bind with high affinity at 5-HT$_{2A}$ and 5-HT$_{2C}$ receptors (ROTH et al. 1992). Thus, blockade of the effect of a hallucinogen by one of these agents cannot necessarily be taken as reliable support for involvement of a dopaminergic mechanism. 5-HT$_3$ ligands also appear to indirectly modulate the effect of dopamine (COSTALL et al. 1990; KILPATRICK et al. 1990). This might explain some of the effects noted above when 5-HT$_3$ antagonists were administered in combination with DOM in drug discrimination studies. In contrast, however, 5-HT$_3$ antagonists had essentially no effect on the stimulus properties of (+)amphetamine – an agent thought to produce its stimulus effects primarily via a dopaminergic mechanism (Young and Glennon 1986).

G. Structurally Related Agents and Designer Drugs

The phenylisopropylamine stimulant amphetamine and phenylisopropylamine hallucinogens such as DOM or DOB (Fig. 1) produce effects in animals and in human subjects that are clearly distinguishable from one another (Shulgin and Shulgin 1991). Thus phenylisopropylamines can be divided into at least two distinct classes of psychoactive agents. Structure-activity studies reveal which various substituents influence potency and, occasionally, how they do so. However, as the structure of one of these agents is gradually modified to one of the others, at what point does an amphetamine-like agent become, for example, a hallucinogenic agent? This question is especially pertinent to the investigation of designer drugs, which are agents that share a structural similarity with phenylisopropylamine stimulants and phenylisopropylamine hallucinogens. The α-desmethyl analog of DOB, Nexus, is an example of an agent that seems to retain much of the hallucinogenic character of DOB. Cathinone (or β-ketoamphetamine) and methcathinone (N-methylcathinone) are newly scheduled agents that possess amphetamine-like properties. Are there structures that possess both properties? Is some of the appeal of designer drugs related to a combination of stimulant and hallucinogenic effects? For a review of designer drugs, see Asghar and Desouza (1989).

It has been said that 1-(3,4-methylenedioxyphenyl)-2-aminopropane (MDA; Fig. 6) produces effects in humans that are akin to a combination of cocaine and LSD. It would seem then that MDA is capable of producing both amphetamine-like and hallucinogenic effects. In tests of stimulus generalization with groups of rats trained to discriminate (+)amphetamine

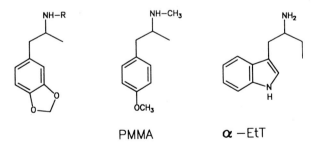

PMMA   α −EtT

Fig. 6. Chemical structures of several designer drugs and related agents: 1-(3,4-methylenedioxyphenyl)-2-aminopropane (MDA) and its N-monomethyl and N-monoethyl derivatives MDMA and MDE (where R = -H, -CH₃, and -C₂H₅, respectively), N-methyl-1-(4-methoxyphenyl)-2-aminopropane (*PMMA*), and α-ethyltryptamine (*α-EtT*)

from vehicle and DOM from vehicle, it has been demonstrated that MDA produces both amphetamine-like and DOM-like effects. The amphetamine-like effect rests primarily with the $S(+)$isomer, whereas the $R(-)$isomer of MDA is the more DOM-like. Rats trained to discriminate MDA from vehicle recognize (+)amphetamine, cocaine, DOM, and LSD. Furthermore, rats can be trained to discriminate 1.25 mg/kg of $S(+)$MDA from 1.25 mg/kg of $R(-)$MDA from saline vehicle using a three-lever operant paradigm (YOUNG and GLENNON 1992), suggesting that the stimulus effects of the isomers of MDA are distinguishable from one another. Furthermore, upon administration of (+)amphetamine, the animals respond in a dose-related fashion on the $S(+)$MDA-appropriate lever, whereas after administration of DOM responding is on the $R(-)$MDA-appropriate lever. Thus, although amphetamine, DOM, and MDA all possess a common phenylisopropylamine skeleton, their actions differ; MDA seems to produce effects similar both to amphetamine and DOM. Therefore it seems that substituents do indeed influence activity.

Attempts have been made to classify phenylisopropylamines as being either amphetamine-like or DOM-like using drug discrimination. Numerous such agents have been already classified (GLENNON 1991, 1993a). Interestingly, certain phenylisopropylamines produce neither amphetamine-like nor DOM-like stimulus effects, and yet they produce MDA-like effects. 1-(3,4-Dimethoxyphenyl)-2-aminopropane (3,4-DMA), and N-ethyl MDA (MDE; "Eve," Fig. 6) are examples of such agents. N-Ethyl MDA is a chain-extended analog of the designer drug N-methyl-1-(3,4-methylene-dioxyphenyl)-2-aminopropane (MDMA) ("Ecstasy," "Adam," Fig. 6). MDMA produces amphetamine-like, but not DOM-like, stimulus effects. Such agents might be capable of producing a unique behavioral effect that is distinct from that of the phenylisopropylamine stimulants and hallucinogens (NICHOLS and OBERLENDER 1989). Initially it was thought that this property might be a consequence of the methylenedioxy group which is present in MDA, MDMA, MDE, and related agents. However, other agents that lack the methylenedioxy group produce MDMA-like effects in animals trained to discriminate MDMA from vehicle. Examples of such agents include para-methoxymethamphetamine (PMMA) (GLENNON and HIGGS 1992) and 1-(3-methoxy-4-methylphenyl)-2-aminopropane (JOHNSON et al. 1991). It would appear then that psychoactive phenylisopropylamines can be divided into at least three categories: (1) stimulants, (2) hallucinogens, and (3) MDMA-like agents. The MDMA-like agents may occasionally retain varying degrees of central stimulant or hallucinogenic activity, but, at least on the basis of drug discrimination studies, it would seem that certain of these agents lack such effects and produce what might be an altogether different psychoactive effect. This poorly understood third category of agents has been extensively investigated by NICHOLS and coworkers (e.g., NICHOLS and OBERLENDER 1989). They have identified several novel agents that lack stimulant and hallucinogen-like effects but that, at the same time, retain MDMA-like effects.

Now that it is realized that a methylenedioxy group is no longer required for MDMA-like activity, there is no reason why investigations must be limited to such derivatives. In fact, there is now evidence that indolylalkylamine derivatives may possess MDMA-like character. For example, α-ethyltryptamine (α-EtT, Fig. 6), a homolog of the hallucinogen α-MeT, is a pyschoactive agent that has defied categorization. Although it was first studied in the early 1960s, and although it has been shown to produce some LSD-like effects in humans (MURPHREE et al. 1961), it has never been classified as a simple hallucinogenic agent. α-EtT has recently appeared on the illicit market as a new designer drug ("ET"), and anecdotal reports suggest that it produces effects similar to MDMA. Studies with MDMA-trained rats have now demonstrated that α-EtT, but not α-MeT, does in fact result in MDMA-stimulus generalization (GLENNON, 1993).

There is a need to continue examination of new structural analogs not only with the intent of formulating and challenging SARs, but also for the purpose of elucidating mechanisms of action and classifying which agents produce which effects.

H. Molecular Modeling of Hallucinogen-Receptor Interactions

Although the exact mechanism of action of classical hallucinogens is unknown at this time, current evidence supports the contention that 5-HT$_{2A}$ and/or 5-HT$_{2C}$ receptors may be involved. This effect may be further modulated by 5-HT$_3$, dopamine, or some other receptor(s). In order to design specific hallucinogen antagonists, to be able to predict the hallucinogenic liability of new therapeutic agents, or simply to unravel the mysteries of neurotransmitter receptors, it would be profitable to have some understanding of how hallucinogen-receptor interactions take place. The ideal situation would be to obtain an X-ray crystal structure of a hallucinogen-bound receptor. Assuming that hallucinogens act as 5-HT$_{2A}$ or 5-HT$_{2C}$ agonists, the target would be an X-ray crystal structure of one of these two receptors. Then, once it is realized which amino acid residues are important for binding, it should be possible to identify what other drugs would utilize these same residues; such agents would likely act either as hallucinogens or as hallucinogen antagonists. Subsequent drug design could be conducted at a molecular level.

The most significant flaw in the above scenario is that, to date, no crystal structure has ever been reported for a membrane-bound G-protein-coupled neurotransmitter receptor. However, several 5-HT receptors, including human 5-HT$_{2A}$ and 5-HT$_{2C}$ receptors (SALTZMAN et al. 1991), have now been cloned and their sequences made available. Also, the three-dimensional structure of the light-sensitive proton pump from bacteria, bacteriorhodopsin, has been determined to near atomic resolution. Although this is not a G-protein receptor, and although there is a low degree of sequence homology between the structures of bacteriorhodopsin and G-protein coupled receptors, it might be possible to deduce structures of neurotransmitter receptors from that of bacteriorhodopsin using molecular modeling techniques. Several investigators are now taking this, or a related, approach to investigate 5-HT$_{2A}$ and 5-HT$_{2C}$ receptors (e.g., Edvardsen et al. 1992; Trumpp-Kallmeyer et al. 1991; Westkaemper and Glennon 1992, 1994; Zhang and Weinstein 1993). Bacteriorhodopsin consists of seven membrane-spanning α-helices connected by intracellular and extracellular loops. By analogy, the neurotransmitter receptors may form a similar arrangement whereby the seven helices form a pore in which drugs can bind. Theoretically, there are a number of different solutions to this problem; that is, a drug can hypothetically interact in more than one orientation with the various receptor features. Proposed binding modes can be tested and validated to some extent by the results of site-directed mutagenesis. That is, if a specific receptor feature is proposed to be important for the binding of a particular agent, a synthetically mutated receptor lacking this feature should display reduced affinity for this agent. Such studies are currently in progress in a number of different laboratories.

Specifically, with regard to a hallucinogen-5-HT$_2$ receptor interaction, it has been proposed that the terminal amine of classical hallucinogens interacts with the asparate moiety in transmembrane helix III (Westakaemper and Glennon 1992, 1994). Other important structural features of the hallucinogens, such as aromatic rings and aryl substituents, interact with other amino acid residues on several of the other helices. One such hypothetical model showing the interaction of $R(-)$DOB with a 5-HT$_{2A}$ receptor is presented in Fig. 7. Since DOB, for example, also interacts with 5-HT$_{2C}$ receptors but not with 5-HT$_{1A}$ receptors, binding features in the 5-HT$_{2A}$ model should also be found in a 5-HT$_{2C}$ receptor model but should be absent in a 5-HT$_{1A}$ receptor model. It should be emphasized that this is only one of several possible models; additional studies are necessary to further support this hypothesis vs other possible modes of binding. Nevertheless, these are the first steps in an entirely novel approach to the investigation of hallucinogen-receptor interactions.

$\longrightarrow$

Fig. 7. Model of 5-HT2 receptors. The *top* and *middle portions* of the figure represent a top (extracellular) and side view, respectively, of a 5-HT2 receptor model. The ribbons denote the helical backbone structures and helix numbering begins from the right most helix (as transmembrane helix 1) and proceeds in a counterclockwise manner. The *bottom portion* of the figure shows a hypothetical mode of interaction of $R(-)$DOB with the 5-HT2 receptor as viewed from the extracellular side. (From Westkaemper and Glennon 1992).

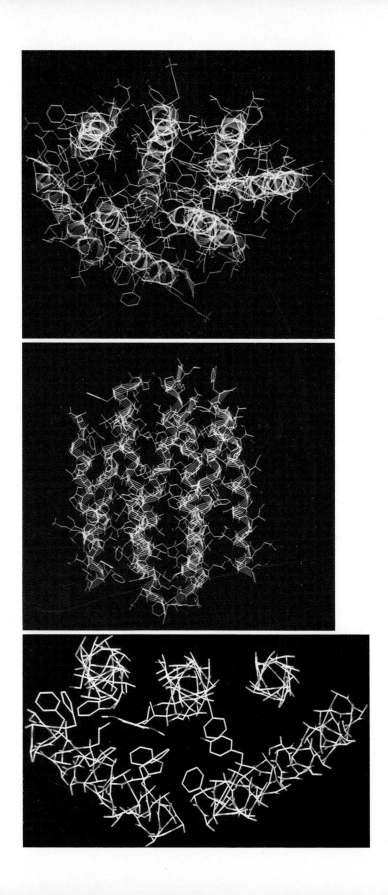

I. Summary

Classical hallucinogens consist of a large number of structurally related indolylalkylamine and phenylalkylamine derivatives that produce preceptural alterations in humans. Certain indolylalkylamines (e.g., DMT, 5-OMe DMT, psilocin) and phenylalkylamines (e.g., mescaline) are naturally occurring, but most are synthetic derivatives. Although the effects produced by these agents may not be identical, many seem to share some behavioral similarity. The results of drug discrimination studies, particularly those with DOM as training drug, support this suggestion. And yet the effects of specific agents can also be rather different (see SHULGIN and SHULGIN 1991). In the absence of comprehensive clinical data, various animal models and behavioral assays have been developed to study these agents. While such studies allow for detailed and comparative investigations of large numbers of agents, in some instances animal studies may be a poor substitute for clinical studies when examining drugs whose primary effect in humans is subjective in nature. Nevertheless, numerous animal assay techniques, though not foolproof, have provided an enormous amount of information that would have been otherwise lacking. Indeed, where both human and animal data are available, SARs, for example, have been remarkably consistent. Animal studies have also implicated a role for 5-HT_{2A} and/or 5-HT_{2C} receptors as mediating the effects that classical hallucinogens have in common. Other neurotransmitters may also play a role in their actions, or at least in the actions of some of the agents; but, convincing evidence has yet to be obtained and additional studies are required.

Structurally, the phenylisopropylamine hallucinogens are derivatives of amphetamine. However, these hallucinogens typically lack amphetamine-like stimulant properties. Conversely, amphetamine is not a hallucinogen. Certain phenylisopropylamine designer drugs and related agents are hallucinogenic, some are amphetamine-like, and yet others are both. Some agents produce a unique nonhallucinogenic nonamphetamine effect. Depending upon the presence and nature of substituent groups, phenylisopropylamine derivatives can produce a combination of these effects. Likewise, some indolylalkylamines are hallucinogenic; some, such as α-MeT, are hallucinogenic with stimulant properties (MURPHREE et al. 1961), and at least one, α-EtT, can produce MDMA-like effects.

Finally, apart from any drug abuse-related issues, classical hallucinogens are proving to be useful neurochemical tools. For example, [³H]DOB, [⁷⁷Br]DOB, [¹²⁵I]DOI, and [³H]LSD have found application in radioligand binding and autoradiographic studies for labeling 5-HT_{2A}, 5-HT_{2C} and other receptors. MDMA and its analogs have prompted investigations of neurotoxicity. Various animal assays are being used with the idea that antagonism of hallucinogen-induced effects may identify novel psychotherapeutic agents. Clearly, the classical hallucinogens and their structural relatives are an interesting, yet complex, group of agents.

Acknowledgements. Work from the author's laboratory was supported in part by PHS grant DA 01642.

References

Asghar K, De Souza E (1989) Pharmacology and toxicology of amphetamine and related designer drugs. US Government Printing Office, Washington DC

Brimblecombe RW, Pinder RM (1975) Hallucinogenic agents. Wright-Scientechnica, Bristol

Burris KD, Sanders-Bush E (1988) Hallucinogens directly activate serotonin 5-HT1C receptors in choroid plexus. Soc Neurosci Abstr 14:553

Costall B, Naylor RJ, Tyers MB (1990) The psychopharmacology of 5-HT3 receptors. Pharmacol Ther 47:181–202

Cunningham KA, Appel JB (1987) Neuropharmacological reassessment of the discriminative stimulus properties of d-lysergic acid diethylamide (LSD). Psychopharmacology (Berl) 91:67–73

Edvardsen O, Sylte I, Dahl SG (1992) Molecular dynamics of serotonin and ritanserin interacting with the 5-HT2 receptor. Mol Brain Res 14:166–178

Efron DH, Holmstedt B, and Kline N (1967) Ethnopharmacologic search for psychoactive drugs. US Government Printing Office, Washington DC

Furst PT (1976) Hallucinogens and culture. Chandler and Sharpe, Novato

Glennon RA (1984) Hallucinogenic phenylisopropylamines: stereochemical aspects. In: Smith DF (ed) Handbook of stereoisomers: drugs in psychopharmacology. CRC Press, Boca Raton, pp 327–368

Glennon RA (1990) Do hallucinogens act as 5-HT2 agonists or antagonists? Neuropsychopharmacology 56:509–517

Glennon RA (1991) Discriminative stimulus properties of hallucinogens and related designer drugs. In: Glennon RA, Jarbe T, Frankenheim J (eds) Drug discrimination: applications to drug abuse research. US Government Printing Office, Washington DC, pp 25–44

Glennon RA (1992) Animal models for assessing hallucinogenic agents. In: Boulton AA, Baker GB, Wu PH (eds) Animal models for the assessment of psychoactive drugs. Humana, Clifton, pp 345–381

Glennon RA (1994) Glassical hallucinogens: an introductory overview. NIDA Res Monogr 146:4–32

Glennon RA (1993b) MDMA-like stimulus effects of α-ethyltryptamine. Pharmacol Biochem Behav 46:459–462

Glennon RA, Higgs R (1992) Investigation of MDMA-related agents in rats trained to discriminate MDMA from saline. Pharmacol Biochem Behav 43:759–763

Glennon RA, Young R, Rosecrans JA (1983) Antagonism of the stimulus effects of the hallucinogen DOM and the purported serotonin agonist quipazine by 5-HT2 antagonists. Eur J Pharmacol 91:189–192

Glennon RA, Titeler M, McKenney JD (1984) Evidence for the involvement of 5-HT2 receptors in the mechanism of action of hallucinogenic agents. Life Sci 35:2505–2511

Glennon RA, Seggel MR, Soine WH, Lyon RA, Davis K, Titeler M (1988) [125I]-1-(2,5-Dimethoxy-4-iodophenyl)-2-aminopropane: an iodinated radioligand that labels the agonist high-affinity state of 5-HT2 receptors. J Med Chem 31:5–7

Glennon RA, Darmani NA, Martin BR (1991) Multiple populations of serotonin receptors may modulate the behavioral effects of serotonergic agents. Life Sci 48:2493–2498

Glennon RA, Teitler M, Sanders-Bush E (1992a) Hallucinogens and serotonergic mechanisms. NIDA Res Monogr 119:131–135

Glennon RA, Raghupathi R, Bartyzel P, Teitler M, Leonhardt S (1992b) Binding of phenylalkylamine derivatives at 5-HT1C and 5-HT2 serotonin receptors: evidence for a lack of selectivity. J Med Chem 35:734–740

Harner MJ (1973) Hallucinogens and shamanism. Oxford University Press, London

Herndon JL, Pierson ME, Glennon RA (1992) Mechanistic investigation of the stimulus properties of 1-(3-trifluoromethylphenyl)piperazine. Pharmacol Biochem Behav 43:739–748

Hollister LE (1968) Chemical psychoses. Thomas, Springfield

Horowski R (1986) Psychiatric side-effects of high-dose lisuride therapy in Parkinsonism. Lancet 30:150

Ismaiel AM, De Los Angeles J, Teitler M, Glennon RA (1993) Antagonism of 1-(2,5-dimethoxy-4-methylphenyl)-2-aminopropane stimulus with a newly identified 5-HT2-versus 5-HT1C-selective antagonist. J Med Chem 36:2519–2525

Jacobs BL, Trulson ME (1978) An animal behavioral model for studying the actions of LSD and related hallucinogens. In: Stillman RC, Willette RE (eds) The psychopharmacology of hallucinogens. Pergamon, New York, pp 301–314

Johnson MP, Frescas SP, Oberlender R, Nichols DE (1991) Synthesis and pharmacological examination of 1-(3-methoxy-4-methylphenyl)-2-aminopropane and 5-methoxy-6-methyl-2-aminoindan: similarities to 3,4-(methylenedioxy)methamphetamine (MDMA). J Med Chem 34:1642–1668

Kier LB, Glennon RA (1978) Progress with several models for the study of SAR of hallucinogenic agents. In: Barnett G, Trsic M, Willette RE (eds) QuaSAR: quantitative structure activity relationships of analgesics, narcotic antagonists, and hallucinogens. US Government Printing Office, Washington DC, pp 159–185

Kilpatrick GJ, Bunce KT, Tyers MB (1990) 5-HT3 receptors. Med Res Rev 10: 441–475

Koek W, Jackson A, Colpaert FC (1992) Behavioral pharmacology of antagonists at 5-HT2/5-TH1C receptors. Neurosci Biobehav Rev 16:95–105

Lyon RA, Davis KH, Titeler M (1987) [^{3}H]DOB (4-bromo-2,5-dimethoxyphenylisopropylamine) labels a guanyl nucleotide-sensitive state of cortical 5-HT2 receptors. Mol Pharmacol 31:194–199

McCleod WR, Sitaram BR (1985) Bufotenine reconsidered. Acta Psychiatr Scand 72:447–452

Meert TF (1991) Application of drug discrimination with drugs of abuse to develop new therapeutic agents. In: Glennon RA, Jarbe T, Frankenheim J (eds) Drug discrimination: applications to drug abuse research. US Government Printing Office, Washington DC, pp 307–324

Meert TF, De Haes P, Janssen PAJ (1989) Risperidone (R 64 766), a potent and complete antagonist in drug discrimination by rats. Psychopharmacology (Berl) 97:206–212

Murphree HB, Dippy RH, Jenney EH, Pfeiffer CC (1961) Effects in normal man of α-methyltryptamine and α-ethyltryptamine. Clin Pharmacol Ther 2:722–726

Naranjo C (1973) The healing journey. Pantheon, New York

Nelson DL, Glennon RA, Wainscott DD (1994) Comparison of the affinities of hallucinogenic phenylalkylamines at the cloned 5-HT$_{2A}$, 5-HT$_{2B}$ and 5-HT$_{2C}$ receptors. Third IUPHAR Satellite Meeting on Serotonin 30 July–3 August 1994. Chicago

Nichols DE, Glennon RA (1984) Medicinal chemistry and structure-activity relationships of hallucinogens. In: Jacobs BL (ed) Hallucinogens: neurochemical, behavioral, and clinical perspectives. Raven, New York, pp 95–142

Nichols DE, Oberlender R (1989) Structure-activity relationships of MDMA-like substances. In: Asghar K, De Souza E (eds) Pharmacology and toxicology of amphetamine and related designer drugs. US Government Printing Office, Washington DC, pp 1–29

Otis LS, Pryor GT, Marquis WJ, Jensen R, Petersen K (1978) Preclinical identification of hallucinogenic compounds. In: Stillman RC, Willette RE (eds) The psychopharmacology of hallucinogens. Pergamon, New York, pp 126–149

Pazos A, Hoyer D, Palacios JM (1984) Binding of serotonergic ligands to the porcine choroid plexus. Characterization of a new type of 5-HT recognition site. Eur J Pharmacol 106:539–546

Pierce PA, Peroutka SJ (1988) Antagonism of 5-hydroxytryptamine 2 receptor-mediated phosphatidylinositol turnover by d-lysergic acid diethylamide. J Pharmacol Exp Ther 247:918–925

Pierce PA, Peroutka SJ (1989) Evidence for distinct 5-hydroxytryptamine$_2$ receptor binding site subtypes in cortical membrane preparations. J Neurochem 52: 656–658

Roth BL, Ciaranello RD, Meltzer HY (1992) Binding of typical and atypical antipsychotic agent sto transiently expressed 5-HT1C receptors. J Pharmacol Exp Ther 260:1361–1365

Saltzman AG, Morse B, Whitman BB, Ivanchenko Y, Jaye M, Felder S (1991) Cloning of the human serotonin 5-HT2 and 5-HT1C receptor subtypes. Biochem Biophys Res Commun 181:1469–1478

Sanders-Bush E (1994) Neurochemical evidence that hallucinogenic drugs are 5-HT1C receptor agonists: what next? NIDA Res Monogr 146:203–213

Shulgin AT, Shulgin A (1991) Pihkal. Transform, Berkeley, p xxi

Stoff DM, Gillin JC, Wyatt RJ (1978) Animal models of drug-induced hallucinations. In: Stillman RC, Willette RE (eds) The psychopharmacology of hallucinogens. Pergamon, New York, pp 259–267

Teitler M, Leonhardt S, Weisberg EL, Hoffman BJ (1990) 4-[^{125}I]Iodo-(2,5-dimethoxyphenyl)isopropylamine and [^{3}H]ketanserin labeling of 5-hydroxytryptamine2 (5-HT2) receptors in mammalian cells transfectd with a rat 5-HT2 cDNA: evidence for multiple states and not multiple 5-HT2 receptor subtypes. Mol Pharmacol 38:594–598

Titeler M, Lyon RA, Glennon RA (1988) Radioligand binding evidence implicates the brain 5-HT2 receptors as a site of action for LSD and phenylisopropylamine hallucinogens. Psychopharmacology (Berl) 94:213–216

Todes CJ (1986) At the receiving end of the lisuride pump. Lancet 5:36–37

Trumpp-Kallmeyer S, Bruinvels A, Hoflack J, Hibert M (1991) Recognition site mapping and receptor modeling: Application to 5-HT receptors. Neurochem Int 19:397–406

Weinshank RL, Adham N, Zgombick J, Bard J, Branchek T, Hartig PR (1992) Molecular analysis of serotonin receptor subtypes. In: Langer SZ, Brunello N, Racagni G, Mendlewicz J (eds) Serotonin receptor subtypes: pharmacological significance and clinical implications. Karger, Basel, pp 1–12

Westkaemper RB, Glennon RA (1991) Approaches to molecular modeling studies and specific application to serotonin ligands and receptors. Pharmacol Biochem Behav 40:1019–1030

Westkaemper RB, Glennon RA (1992) Hypothetical 3-D models of 5-HT receptors. 2nd international symposium on serotonin, Houston, Texas,15–18 Sep 1992, abstract, p 40

Westkaemper RB, Glennon RA (1994) Molecular modeling of the interaction of LSD and other hallucinogens with 5-HT2 receptors. NIDA Res Monogr 146: 263–283

White FJ (1986) Comparative effects of LSD and lisuride: clues to specific hallucinogenic drug action. Pharmacol Biochem Behav 24:365–379

Woolley DW (1962) The biochemical bases of psychoses. Wiley, New York

Young R, Glennon RA (1986) Discriminative stimulus properties of amphetamine and structurally related phenalkylamines. Med Res Rev 6:99–130

Young R, Glennon RA (1992) Discrimination of R(−)MDA from S (+)MDA using a three-lever operant paradigm. 2nd international symposium on serotonin, Houston, Texas, 15–18 Sep 1992, abstract, p 68

Zhang D, Weinstein H (1993) Signal transduction by a 5-HT2 receptor: a mechanistic hypothesis from molecular dynamics simulations of the three-dimensional model of the receptor complexed to ligands. J Med Chem 36:934–938

CHAPTER 11
Alcohol

B. TABAKOFF, K. HELLEVUO, and P.L. HOFFMAN

A. Introduction

Ethanol is molecule of simple chemical structure and complex metabolic, pharmacologic and pathologic actions. The two-carbon backbone of ethanol is efficiently converted by most animals, including human beings, to carbon dioxide and water, and a concomitant of this metabolism is the generation of calories. The calories derived from the metabolism of one gram of ethanol exceed the calories that can be derived from the metabolism of one gram of protein. The 1990 per capita annual consumption of ethanol in the USA by individuals age 14 and older (SEVENTH SPECIAL REPORT TO CONGRESS 1990) can be calculated to account for 12% of the caloric intake of these individuals. This figure is similar to the calories derived by the US population from protein and, thus, ethanol can be considered to be a major foodstuff in both the USA and many other countries in which beverage ethanol is legally consumed.

Ethanol is a psychoactive drug which has been consumed by humans for millennia for its mood- and thought-altering properties. Ethanol's ability to produce intoxication are detailed in the writings of Hammurabi and in Egyptian hieroglyphics (HARPER 1904; GHALIOUNGUI 1979). Although it is clear that ethanol acts directly on the central nervous system (CNS) to produce changes in mood and intoxication, the mechanisms by which ethanol exerts its CNS effects are still being debated (see below). Yet, it can be said with some certainty that ethanol, unlike other psychoactive drugs, does not produce its CNS effects by binding to specific receptors (proteinaceous molecular recognition sites for ethanol) to initiate its actions. Current work on the actions of ethanol within the CNS has, however, demonstrated the selective sensitivity to ethanol of certain neurotransmitter receptor-controlled systems, and we have previously used the term "receptive elements for ethanol" (TABAKOFF and HOFFMAN 1987) to describe these sensitive systems.

Ethanol can, thus, be easily distinguished from other psychoactive drugs by the fact that it is both a food and a psychoactive molecule and by the fact that ethanol's actions in the brain are not initiated by binding to its own specific receptors. Ethanol can also be distinguished from other psychoactive drugs by the wide range of organ pathologies that it can produce. Excessive ethanol intake can produce damage to the liver, cardiovascular system, gonads and brain, and this damage is generated by the metabolism of ethanol in organs, by the metabolic products of ethanol metabolism and by the actions of ethanol per se. We do not consider it possible to provide the appropriate coverage of the full spectrum of ethanol's actions in the space allotted, and the reader is referred to recent reviews on metabolism and metabolic effects of ethanol (Lieber 1991a), on the pathologic consequences of ethanol use (Lieber 1988), and on the psychoactive actions of ethanol (Tabakoff and Hoffman 1992) for more complete information.

B. The Metabolism of Ethanol

More than 90% of ingested ethanol is metabolized in the body through oxidative pathways, and the majority of this metabolism takes place in the liver. Several enzymatic pathways (Fig. 1) are available to convert ethanol to its initial metabolic product, acetaldehyde, but the primary pathway involves the enzyme alcohol dehydrogenase (ADH).

In the human, ADH exists as a number of isozymes, but all of the isozymes utilize NAD as a cofactor (coenzyme) for the conversion of ethanol to acetaldehyde. Alcohol dehydrogenase exists as a homo- or heterodimer of two protein subunits. The proteins that form the various isoenzymes of ADH are transcribed from genes which code for the subunits of three classes of ADH (see below). The subunits do not interact across classes.

Consideration of alcohol metabolism is important within the context of the discussion of alcohol intoxication, the rewarding or reinforcing effects of ethanol, and the aversive effects of ethanol, for the following reasons. An individual's capacity to metabolize ethanol determines the duration of alcohol's actions within the CNS or other organs of the body. Thus, a good correlation has been determined between the initiation, amplitude, and loss of ethanol's effects on psychomotor processes, or other measures of human or other animal performance, and the measures of circulating blood ethanol levels (Carpenter 1962). A caveat to this relationship between ethanol levels in the body and ethanol's effect is that the presence of acute or chronic CNS tolerance to ethanol can change the relationship between the magnitude of ethanol's actions and the blood or brain ethanol levels in an individual (Tabakoff et al. 1986). The average rate of ethanol elimination in humans (70 kg man) has been reported to be approximately 10 g ethanol/h (Kalant 1971). This rate of metabolism can vary up to two- to threefold between individuals because of the genetics of the alcohol metabolizing enzymes (Martin et al. 1985) (and see below), gender (Marshall et al. 1983), the nutritional state of the individual (Lieber 1991b), hormonal (Mardh et al. 1986) and various environmental factors. One can expect, however, that an individual (even a tolerant one) consuming large doses of ethanol (e.g., $\geq 1\,g/kg$) over a short period of time will exhibit measurable changes in mood and behavior and that these effects of ethanol will be evident for one to several hours.

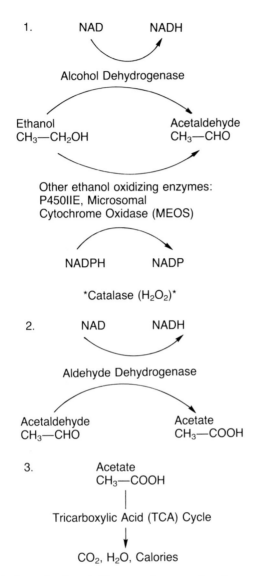

Fig. 1. The metabolism of ethanol. Enzyme classes and cofactor requirements are noted for the initial two steps of the metabolic pathway (see text for details). The actions of catalase are not discussed in the text due to the nominal contributions of this enzyme in vivo where the amounts of H_2O_2 necessary for catalase action are not normally present

The circulating levels of ethanol, and the initial rate of rise of ethanol in blood (and brain) after ingestion of ethanol, are determined by factors related to the absorption of ethanol in the stomach and upper intestine (e.g., the presence of food) and its metabolism (KALANT 1971). A certain amount of ethanol is metabolized by ADH present in the stomach prior to its entry into portal circulation (Moreno and Pares 1991), and there is evidence that the capacity of gastric ethanol metabolizing systems differs between man and women. Lieber and coworkers have reported that women have significantly lower levels of ADH than men in stomach and intestinal mucosa (Frezza et al. 1990). Once ethanol has entered the portal circulation it must traverse the liver prior to becoming available for systemic circulation. The characteristics of the individual's stomach and liver alcohol metabolizing enzymes (see below) make a measurable contribution to the initial (first pass) metabolism of ethanol. Thus, the amount of ethanol initially available to the circulation is controlled in a significant way by the rate of ethanol absorption and the metabolic characteristics of the individual. The initial rate of rise of circulating ethanol levels and the rate of initial delivery of ethanol to the brain has been shown to be an important determinant of an animal's perception of ethanol's actions. A rapid rise in circulating blood ethanol has been associated with the ability to demonstrate the rewarding or reinforcing effects of ethanol (Smith et al. 1976; Lukas and Mendelson 1986).

Metabolic factors also can control the aversive nature of ethanol ingestion in certain individuals. The individual's characteristics with regard to the enzymes that produce and degrade the ethanol metabolic intermediate acetaldehyde can determine a noxious or aversive response to ethanol ingestion. A significant number of individuals of Oriental descent are genetically imbued with a metabolic system that generates high circulating levels of acetaldehyde after ingestion of even small doses of ethanol (Harada et al. 1981). Acetaldehyde can produce vasodilation, flushing, headache and nausea (Feiman 1979), and these physiologic consequences of ethanol ingestion in individuals generating high circulating acetaldehyde levels after ingestion of ethanol have been postulated to limit the ethanol intake of these individuals (Ohmori et al. 1986).

I. Ethanol-Induced Metabolic Anomalies

The metabolic anomalies and pathologic consequences associated with ethanol ingestion in organs that metabolize ethanol, such as liver and testes, have been ascribed to the metabolism of ethanol in these organs. The phenomena described in this section would not apply to brain since it contains little (Bühler et al. 1983) or no (Giri et al. 1989) ADH of the type (see below) which would oxidize ethanol during episodes of non-life-threatening intoxication.

Figure 1 indicates the necessity of the conversion of the co-factor NAD to NADH concomitantly with the oxidation of ethanol to acetaldehyde and acetaldehyde to acetate when the enzymes ADH and aldehyde dehydrogenase participate as catalysts. The production of NADH and the increase in NADH/NAD ratios in organs such as liver and testis has a significant effect on the metabolism of other molecules in these organs. For instance, the oxidation of fatty acids in the liver is greatly diminished in the presence of ethanol and fats are shunted to storage depots in liver (BARAONA and LIEBER 1979). Such metabolic shunting of fats has been used to explain the development of fatty liver in individuals consuming large quantities of ethanol (LIEBER 1988; BARAONA and LIEBER 1979). Glucose metabolism is also significantly affected by ethanol. In the liver of a healthy individual of normal nutritive status, ethanol metabolism and the generation of NADH inhibit glucose oxidation and can produce hyperglycemia (LIEBER 1985). In a malnourished individual in whom blood glucose levels are being maintained by the degradation of amino acids from body proteins, ethanol metabolism and NADH production will inhibit glucose synthesis from amino acids and the result will be hypoglycemia (LIEBER 1985; ARKY 1971). The shift in the NADH/NAD ratio in testis, and in liver, has been used to explain ethanol-induced alterations in steroid hormone metabolism in these organs (CRONHOLM and SJÖVALL 1970). In the testis the production of the androgenic steroid testosterone is inhibited due to the metabolism of ethanol (CICERO 1981) and this phenomenon is considered to be a factor leading to the gynecomastia seen in many male alcoholics (VAN THIEL and LESTER 1976).

The inhibition of the metabolism of steroids and/or other compounds in organs such as liver and testis has also been ascribed to the generation of acetaldehyde (CICERO and BELL 1980) and the competition of acetaldehyde with other substrates for metabolism by aldehyde dehydrogenases. For instance, the aldehyde intermediates of the metabolic degradation of serotonin and epinephrine are substrates for aldehyde dehydrogenase in liver, which oxidizes these intermediates of 5-hydroxyindoleacetic acid (5-HIAA) and vanillylmandelic acid (VMA), respectively. The presence and metabolism of ethanol in the human shifts the metabolism of the "biogenic" aldehydes to their reduced metabolites, 5-hydroxytryptophol and phenyle-thyleneglycol (CICERO and BELL 1980; DAVIS et al. 1967b). The ratio of 5-hydroxytryptophol to 5-HIAA appearing in urine has been proposed for use as a marker of recent ethanol ingestion (VOLTAIRE-CARLSSON et al. 1993).

The generation of acetaldehyde during the metabolism of ethanol in organs such as liver also leads to the complexing or binding of the acetal-dehyde to proteins within these organs (LIN et al. 1988; BEHRENS et al. 1988). This binding of acetaldehyde to proteins has been considered an etiologic factor in liver pathology because acetaldehyde could alter the metabolic activity of the complexed proteins and because the protein-acetaldehyde adducts are antigenic and could generate an autoimmune reac-tion (NIEMELÄ et al. 1987, 1991). Acetaldehyde can also promote the binding to tissue proteins of other aldehydes, which are usually substrates for aldehyde dehydrogenase. For example, protein adducts of 5-hydroxyindo-leacetaldehyde (the serotonin metabolite) have been isolated during in vitro incubations of tissue with acetaldehyde and serotonin (TABAKOFF et al.

1972). Since aldehyde dehydrogenase is expressed in brain (Erwin and Deitrich 1966; Braun et al. 1987), one might suspect that, if acetaldehyde were to reach brain tissue, it would compete with other substrates for metabolism by brain aldehyde dehydrogenase and lead to its own, or other aldehyde, binding to brain proteins or other anomalous reactions.

Using methods which carefully control for the artifactual production of acetaldehyde in blood, Eriksson (1983) has demonstrated that, after an individual consumes ethanol, peripheral blood acetaldehyde concentrations are below the detection limits of $2 \mu M$ in nonalcoholic Caucasians. This demonstration would indicate that minimal levels of acetaldehyde escape from the liver of nonalcoholic individuals during metabolism of ethanol and little acetaldehyde would be available for delivery to organs such as brain. We have already noted that brain has little or no capacity to generate acetaldehyde from ethanol in situ because brain lacks ADHs with high affinity for ethanol (Giri et al. 1989; Tabakoff and von Wartburg 1975). In alcoholics, however, the circulating acetaldehyde levels are significantly higher than in nonalcoholic individuals. Lindros et al. (1980) have reported that concentrations of acetaldehyde as high as $100 \mu M$ may be found in the peripheral blood of alcoholics ingesting ethanol. Under such conditions, one may expect that acetaldehyde generated in the liver would reach the brain in significant concentrations. Whether acetaldehyde, even when present at high concentrations in blood, is able to reach neurons is still being debated. The metabolic barrier for acetaldehyde in ependymal cells and the lumen of the blood vessels is quite active, and little acetaldehyde has been noted in brain tissue or cerebrospinal fluid (CSF) of animals even during experimental manipulations which have generated high circulating levels of acetaldehyde (Sippel 1974; Stowell et al. 1980).

The issue of whether brain tissue is exposed to significant concentrations of acetaldehyde is of consequence because of a proposal that a metabolic anomaly produced by acetaldehyde in brain is a major factor determining escalating ingestion of ethanol by animals and the generation of physical dependence on ethanol. In experiments in which acetaldehyde was added directly to homogenized brain tissue, condensation products of acetaldehyde with dopamine (salsolinol), norepinephrine (hydroxysalsolinol), and sero-tonin (β-carboline derivatives) have been isolated (Cohen and Collins 1970; Melchior and Collins 1982). The condensation of dopamine with its own aldehyde metabolite to form the compound tetrahydropapaveroline (THP) has also been described, and the production of THP by brain homo-genates is increased after addition of acetaldehyde (Davis and Walsh 1970). Myers and coworkers demonstrated that administration of THP to rats or monkeys produced a substantial increase in the acceptance and intake of ethanol (Myers and Melchior 1977a; Myers et al. 1982); they proposed that THP generated in brain during the metabolism of ingested ethanol leads to increased ethanol intake and ethanol physical dependence (Myers and Melchior 1977b). Numerous attempts using sensitive technology have, however, failed to demonstrate the presence of THP in the brain of alcohol consuming animals (BLOOM et al. 1982). This failure, coupled with the difficulty of demonstrating the effects of THP on alcohol intake in other laboratories (AMIT et al. 1982) has diminished enthusiasm for the "THP hypothesis" of alcohol dependence (DAVIS and WALSH 1970; MYERS and MELCHIOR 1977a).

II. Alcohol Dehydrogenase and Microsomal Ethanol Oxidizing Systems

Although the role of acetaldehyde in enhancing ethanol intake is currently quite tenuous, the participation of acetaldehyde in the aversive consequences of ethanol intake, as stated above, is well documented. The genetically determined rate of production and elimination of acetaldehyde has been the subject of numerous studies. As already noted, ADH is the primary enzyme generating acetaldehyde during the course of ethanol metabolism. Human liver ADH activity has been assigned to three enzyme classes (I, II, and III) based on the structural differences and the substrate metabolizing properties of these proteins (WAGNER et al. 1984). The ADH enzyme of each of these classes is composed of two peptide chains. The peptide chains composing the class I enzymes are the products of three different gene loci (ADH_1, ADH_2 and ADH_3) (SMITH 1986). Class II ADH is a product of the gene locus designated as ADH_4 (JÖRNVALL et al. 1987) and the class III ADH is derived from the ADH_5 locus (KAISER et al. 1988). Although class II and class III ADHs are homodimers of the peptides π and χ, respectively, which are products of the genes coding for these proteins, class I ADHs can be either homo- or heterodimers of the peptide products of the ADH_1, ADH_2 or ADH_3 genes (SMITH 1986). The variety of the possible ADH isozymes in humans is enhanced by the presence of polymorphisms (genetic variants) at the ADH_2 and ADH_3 loci. The polymorphic peptides at the ADH_2 locus have been designated β_1, β_2 and β_3 (BOSRON et al. 1983b), while the ADH_3 locus peptides are designated γ_1 and γ_2 (SMITH 1986). There is only one product of the ADH_1 gene, referred to as α.

Both class I and class II ADHs can participate in the oxidation of ethanol under conditions in which liver ethanol levels are reflective of nonanesthetic consumption of ethanol (i.e., concentrations up to 100 mM). Ethanol concentrations of 100 mM ethanol (460 mg%) are still tenfold below the K_m for ethanol for the class III (WAGNER et al. 1984), χ, ADH, and this class of ADH would not contribute significantly to the metabolism of ethanol in normal circumstances.

The class I ADH isozymes (with the exception of the β_3 homodimer) have a low K_m for ethanol (<4 mM) and these enzymes would be easily saturated at liver levels of ethanol which approach are forensic legal level for intoxication of 100 mg% or ~20 mM (BOSRON et al. 1983a). The saturation of these forms of ADH with normally encountered blood ethanol concentrations gives rise to the pseudo-zero-order (linear) ethanol disappearance kinetics until ethanol concentrations approach and fall below 5 mM (WIDMARK 1933). The β_3 homodimer and the class II, π, homodimer, have K_ms for ethanol of approximately 36 mM (BURNELL et al. 1987; CRABB et al. 1983). The π enzyme and the β_3 variant would be particularly active only when ethanol levels approach and exceed these enzymes' K_m values for ethanol (i.e., ethanol $\geq$150 mg%).

The velocity with which the various isoforms of ADH metabolize ethanol also varies significantly. VON WARTBURG et al. (1965) initially described an "atypical" ADH activity in a post-mortem liver sample obtained in Switzerland. The ADH from this "atypical" liver could metabolize ethanol at rates up to five times those of the usually encountered liver ADH activities. Currently, it is known (EHRIG et al. 1988) that the "atypical" ADH activity described by VON WARTBURG et al. (1965) was due to the presence of the β_2 subunits in the examined livers. Due to the greater velocity with which the β_2 isozyme metabolizes ethanol to acetaldehyde, individuals possessing the alleles coding for the β_2 form of class I ADH would be expected to generate greater amounts of acetaldehyde when they consume ethanol (VON WARTBURG and SCHURCH 1968). The ADH_2^2 allele, which encodes the β_2 isozyme, is rarely encountered in Caucasian populations (allelic frequency $\approx$0.05 + 0.10) (SMITH et al. 1971). By contrast, the frequency of the ADH_2^2 allele in the Oriental population has been calculated to be approximately 0.85 (HARADA et al. 1980). The high frequency of an allelic form of an ADH gene which codes for an enzyme capable of quickly generating large amounts of acetaldehyde was initially thought (VON WARTBURG 1976) to be the basis of an acetaldehyde-dependent flushing and other noxious responses seen in a number of Orientals after they imbibe ethanol. Although the presence of the β_2 isozyme contributes to the generation of high acetaldehyde levels in Orientals, the major contributor to the high circulating acetaldehyde levels in certain Orientals who imbibe ethanol is an inactive form of aldehyde dehydrogenase (see below). The ethnic and racial distribution of the isoforms of ADH and aldehyde dehydrogenase have been extensively studied (see references in AGARWAL and GOEDDE 1990), and current opinion links the phenotype which generates high circulating acetaldehyde levels with an aversive response to ethanol (HARADA et al. 1980; THOMASSON et al. 1991).

Acetaldehyde can also be produced from ethanol by a microsomal drug oxidase system designated as the P450IIE1 (LIEBER 1988). LIEBER and coworkers (LIEBER and DECARLI 1972) initially implicated the microsomal drug oxidizing systems in the metabolism of ethanol and referred to these systems as the *microsomal ethanol oxidizing systems* (MEOS). The P450IIE1 system would be expected to metabolize <20% of the ethanol as compared to the quantity metabolized by the ADH system. The chronic ingestion of ethanol, or other drugs, which induce the proliferation of the reticuloendothelial system, leads to a doubling of MEOS activity in experimental animals (LIEBER and DECARLI 1972) and the MEOS system has been proposed to be of greater consequence in ethanol metabolism under such circumstances. The metabolism of ethanol by MEOS is accompanied by the conversion of the high energy compound NADPH to NADP, while the metabolism of ethanol by ADH generates the high energy compound NADH from NAD (Fig. 1). The proliferation of MEOS systems and the utilization of high energy compounds for ethanol metabolism by MEOS has been used to explain weight loss in alcoholics during isocaloric substitution of ethanol for carbohydrates (LIEBER 1988).

III. Aldehyde Dehydrogenase

The acetaldehyde generated from ethanol by either ADH or by the MEOS system is metabolized by aldehyde dehydrogenases (ALDHs). As discussed, the metabolism of acetaldehyde to acetate by ALDH in the liver is, in many cases, so efficient that little or no acetaldehyde escapes into the general circulation (ERIKSSON 1983). However, an ALDH "deficiency" has been noted in Orientals and certain populations of South American Indians (AGARWAL and GOEDDE 1990). Liver ALDH activity has been proposed to be decreased in alcoholics (PALMER and JENKINS 1982, 1985), and erythrocyte ALDH activity has also been found to be low in alcoholics (HELANDER 1993). The phenotype of "low ALDH activity," whether genetically determined (e.g., Orientals) or environmentally derived (e.g., history of high ethanol consumption), results in the appearance of substantial amounts of acetaldehyde in the circulation of an ethanol-drinking individual (HARADA et al. 1981; LINDROS et al. 1980).

The human ALDHs are subcategorized into class I (ALDH1, E1) and class II (ALDH2, E2) enzymes (WEINER and FLYNN 1988). ALDH2 has a low K_m (high affinity) for acetaldehyde and is most probably located in the mitochondria of most ograns, including the liver (TOTTMAR et al. 1973). The importance of ALDH2 in the metabolism of acetaldehyde is demonstrated by the fact that a mutation in the ALDH2 locus, which renders this form of the enzyme inactive (YOSHIDA et al. 1984), results in the accumulation of acetaldehyde in individuals bearing this mutant form of ALDH (HARADA et al. 1981). In individuals with the inactive form of ALDH2 the levels of acetaldehyde rise into the concentration range at which the high K_m ($30\,\mu M$) form of ALDH (ALDH1) can efficiently metabolize acetaldehyde (ERIKSSON et al. 1975).

The mutant form of ALDH2 is noted with high frequency in Orientals ($\approx 44\%$) (AGARWAL and GOEDDE 1990). The presence of this mutant form of ALDH, together with the β_2 variant of ADH, is particularly conducive to generation of high circulating levels of acetaldehyde. The examination of genotype frequencies for ADH and ALDH in alcoholics and nonalcoholics has led to the proposal that genotypes which produce high circulating acetaldehyde levels after consumption of ethanol provide protection against the development of alcoholism (THOMASSON et al. 1990, 1991). Oriental alcoholics have been noted to display significantly lower frequencies of the ADH_2^2 and the mutant $ALDH_2$ than frequencies found in reference Oriental nonalcoholic populations (HARADA et al. 1982; THOMASSON et al. 1991). Caution, however, needs to be exercised in accepting a simple assumption that increased levels of acetaldehyde in the circulation of an alcohol-consuming individual are a priori aversive and directly responsible for avoidance of ethanol. A significant number of alcoholics who continue to drink have high circulating acetaldehyde levels after they consume ethanol (LINDROS et al. 1980; PALMER and JENKINS 1982).

IV. Acetate Metabolism and the Metabolic Effects of Acetate

Aldehyde dehydrogenase converts acetaldehyde to acetate which is further metabolized to CO_2 and H_2O via the tricarboxylic acid (TCA) cycle (Fig. 1). Since ethanol metabolism in the liver generates substantial amounts of NADH, and since high NADH levels inhibit the activity of enzymes of the TCA cycle (ONTKO 1973) in the liver, much of the ethanol-derived acetate leaves the liver to be metabolized by the TCA cycle in other tissues, including brain (LUNDQUIST et al. 1962; HANNAK et al. 1985). The importance of this overflow of acetate during ethanol metabolism has only recently sparked the attention of investigators.

The free acetate derived from ethanol needs to be conjugated with coenzyme A (CoA) to form acetyl-CoA prior to the metabolism of the acetyl-CoA via the TCA pathway. The acetylation reaction requires ATP and results in the formation of one molecule of AMP for each molecule of acetate converted to acetyl-CoA (VINAY et al. 1987). The AMP which is generated in this process can, and does, become a source for generation of free adenosine in tissues metabolizing acetate (ORREGO et al. 1988). Adenosine has been demonstrated to be a neurotransmitter/neuromodulatory substance acting through adenosine A_1 and adenosine A_2 receptors coupled to adenylyl cyclase (see below). The adenosine generated as a result of acetate metabolism in peripheral organs has been proposed as the mediator of alcohol's effects on organ hemodynamics (ORREGO et al. 1988). The adenosine neurotransmitter systems of brain have also been implicated in an animal's sensitivity to the incoordinating actions of ethanol (DAR 1990). However, it needs to be determined whether acetate delivery to, and metabolism in, brain during acute periods of ethanol intoxication is of significant consequence for adenosine neurotransmitter function in the CNS. In studies utilizing neuroblastoma × glioma cell hybrids grown in ethanol-containing culture, extracellular adenosine has been postulated to be a required factor for ethanol-induced heterologous desensitization of adenylyl cyclase (see below). Thus, acetate metabolism and related adenosine generation may also need to be considered in examining the chronic actions of ethanol.

A number of interesting hypotheses have been proposed to link certain actions of ethanol in brain to the generation and actions of the ethanol metabolites acetaldehyde and acetate, but the difficulties in relating the witnessed effects of ethanol intoxication to the concentrations and the time course of the presence of the ethanol metabolites in brain have focused research on the pyschoactive properties of ethanol squarely on the actions of ethanol per se in brain tissue. Therefore, the remainder of this chapter will detail the actions of ethanol on neurochemical processes which are thought to be important for ethanol-induced intoxication and the development of ethanol tolerance and ethanol physical dependence.

C. Actions of Ethanol on the CNS

I. Membrane Hypothesis of Ethanol's Actions

Since ethanol is a molecule that possesses both hydrophilic and hydrophobic properties and does not appear to interact with its own specific receptors in the CNS, it has been proposed that ethanol acts by perturbing the order of neuronal membrane lipids, resulting in a more "fluid" membrane ("membrane lipid hypothesis" of ethanol action) (HUNT 1985). The idea that ethanol produces its intoxicating and/or anesthetic actions by altering membrane lipid fluidity derives from studies by MEYER and colleagues in the 1900s (MEYER 1901). These experiments showed that the anesthetic potency of several classes of chemicals, including alcohols, was correlated with the lipid solubility of the examined chemicals. Demonstrations that ethanol, at nonanesthetic concentrations (e.g., $25–100\,mM$; $115–460\,mg\%$), can actually increase membrane fluidity are more recent and are based on the techniques of electron paramagnetic resonance (EPR), fluorescence polarization and nuclear magnetic resonance (NMR). These investigations have provided more detail regarding ethanol's effects on biological membranes and have given evidence that the effects of ethanol are not uniform throughout the neuronal plasma membrane (e.g., HITZEMANN et al. 1986; WOOD and SCHROEDER 1988; TREISTMAN et al. 1987). A recent analysis using NMR spectroscopy suggested that ethanol may produce an increase in lipid "ordering" at the cellular membrane surface, while increasing "fluidity" (lipid disorder) at the interior of the membrane (HITZEMANN et al. 1986). Such variations in response to ethanol in different areas of neuronal membranes may reflect the presence of domains of particular lipid composition (e.g., cholesterol, phospholipid moieties) within the membrane. WOOD and colleagues have demonstrated that the lipids of the outer (exofacial) leaflet of the membrane bilayer are more sensitive to disordering produced by ethanol than those of the inner (cytofacial) leaflet, perhaps as a result of differences in phospholipid composition of the two leaflets (SCHROEDER et al.

1988). Membrane components other than lipids have also been suggested to contribute to sensitivity to the lipid disordering effects of ethanol: for example, addition of gangliosides to model membranes increased the fluidizing effect of ethanol (HARRIS et al. 1984).

However, while ethanol can affect the fluidity of neuronal and other cell membranes, it has been difficult to demonstrate a convincing causal relationship between the lipid-perturbing effects of ethanol and its nonanesthetic pharmacological effects. The explanation of the anesthetic actions of ethanol derived by invoking the membrane lipid hypothesis has also been challenged. Perhaps the strongest evidence that the effects of ethanol on membrane fluidity may be related to its anesthetic effect was the demonstration that neuronal cell membranes obtained from mice selectively bred to be sensitive to the hypnotic effect of ethanol (long sleep, LS, mice) were more sensitive to ethanol-induced fluidity changes than membranes from mice bred to be resistant to ethanol (short sleep, SS, mice) (GOLDSTEIN et al. 1982). The initial study used EPR to demonstrate the differences in membrane fluidity, but a subsequent study, using fluorescence polarization, did not find the same results (PERLMAN and GOLDSTEIN 1984). Since the fluorescence and EPR probes monitored the order of different parts of the neuronal membrane (GOLDSTEIN et al. 1982; PERLMAN and GOLDSTEIN 1984), and since more recent studies indicate regional differences in membrane sensitivity to ethanol-induced perturbation, it could be concluded from these experiments that the site that influences the hypnotic response of LS and SS mice to ethanol may be localized within a restricted area of the membrane (PERLMAN and GOLDSTEIN 1984). However, the SS and LS mice do not display differences in sensitivity to longer-chain alcohols, either in terms of behavioral responses (HOWERTON et al. 1983) or in the neuronal membrane fluidizing effect of these long chain alcohols (HARRIS et al. 1988). While these results have been interpreted as supporting the hypothesis that differential sensitivity to ethanol-induced membrane lipid perturbation underlies the differential sensitivity to the hypnotic effect of ethanol in the two lines of mice (HARRIS et al. 1988), one may suggest that they indicate the opposite. That is, neuronal membrane lipid perturbation cannot totally account for the hypnotic effects of anesthetics and alcohols. The Meyer-Overton hypothesis (MEYER 1901; OVERTON 1901) states that greater hydrophobicity leads to greater anesthetic potency of a drug. Given this postulate, the behavioral difference between LS and SS mice, if it depends on sensitivity or resistance of neuronal membranes to ethanol-induced changes in membrane fluidity, should remain evident for all compounds that affect membrane fluidity. Instead, the results suggest that there may be an action specific to ethanol that underlies the difference between these lines of mice. The argument against a bulk lipid-perturbing mechanism of action of ethanol is also based on the small extent of the changes in membrane fluidity produced by ethanol concentrations that are not lethal in mammals. The ethanol-induced membrane fluidity changes are comparable to those that occur upon changes in body temperature of 1°C, which is within the normal human diurnal variation for body temperature (HARRIS et al. 1988; TABAKOFF et al. 1988a).

Another hypothesis to explain the pharmacologic effects of ethanol proposes a protein site of action for ethanol. As pointed out by FRANKS and LIEB (1984), the inhibition of activity of the soluble, lipid-free enzyme luciferase by anesthetics, including alcohols, correlates well with the solubility of the anesthetics in a lipid bilayer. These authors proposed that the targets for anesthetics may be certain proteins with amphiphilic "pockets" that contain both hydrophobic and polar components (FRANKS and LIEB 1984). This suggestion is supported to some extent by studies of the effects of long-chain alcohols and general anesthetics on specific proteins (e.g., ion channels, receptor-coupled signal transduction systems) that are discussed in detail below. In some instances, the effects of the various compounds on protein function are not positively correlated with their effects on bulk membrane order (e.g., RABIN et al. 1986; LUTHIN and TABAKOFF 1984).

A critical experiment regarding the role of membrane lipids in the effects of ethanol was one in which proteins that were shown to be differentially sensitive to ethanol in situ were placed in identical lipid environments. The γ-aminobutyric acid (GABA$_A$) receptor in brain membranes of the LS and SS mice is one such protein: the potentiating effects of ethanol on GABA$_A$ receptor function (see below) are greater in assays of membrane receptor systems derived from brains of the LS mice than in membrane receptor systems derived from the SS mice (ALLAN and HARRIS 1986). When mRNA was extracted from brains of the LS and SS mice and expressed in a Xenopus oocyte system, the differential sensitivity of the GABA$_A$ receptors was maintained (WAFFORD et al. 1990). One would expect that in the frog oocyte expression system any difference in lipid composition or native membrane lipid sensitivity to ethanol between LS and SS mice would be eliminated.

Therefore, although the hydrophobicity of ethanol and other alcohols contributes to their pharmacologic and biochemical/molecular effects, the correlation of hydrophobicity with biologic effects per se does not prove a membrane bulk lipid site of action for ethanol. The site(s) of ethanol's action could be the various proteins themselves ("receptive elements," TABAKOFF and HOFFMAN 1987), or could be located at a lipid-protein or water-protein interface. One interesting theory that has been proposed to account for the hypnotic effects of ethanol and other anesthetics is based on studies of the hydrogen bond-breaking activity of alcohols as measured by Fourier transform infrared spectroscopy (YURTTAS et al. 1992). This "dehydration" theory proposes that the "bipolar nature of alcohol allows it to be attracted to the lipophilic domain of a membrane, yet simultaneously be trapped at the membrane surface by hydrogen bonding with polar groups of surface macromolecules, Alcohol would thus be in competition with surface-bound water, and some of the surface-bound water molecules would be displaced" (YURTTAS et al. 1992). The authors presented data showing that butanol could displace bound water from lipid micelles and suggested that the resulting dehydration would not only disrupt the normal support of membrane surface molecules, resulting in disordering of the membrane interior, but would also affect the microenvironments in which proteins sensitive to alcohol are localized.

Adaptive changes in membrane lipids have been proposed to account for some of the chronic effects of ethanol. Initially, it was shown that neuronal and other cell membranes obtained from chronically ethanol-treated (tolerant) animals displayed resistance to the fluidizing effect of ethanol (CHIN and GOLDSTEIN 1977). However, although some authors have attributed this change to a decreased ablility of ethanol to "bind" to or partition into cell membranes (ROTTENBERG 1987), the mechanism for this increased resistance to ethanol is not celar. In some instances a decreased baseline fluidity (increased "rigidity") can be shown in cellular membranes obtained from ethanol-treated animals, but the demonstration of such changes depends on the probe used to assess fluidity (CHIN and GOLDSTEIN 1981; LYON and GOLDSTEIN 1983; WARING et al. 1981) as well as on the origin of the cell membrane. In one study, liver plasma membranes from chronically ethanol-treated animals were reported to be more fluid than those of controls (POLOKOFF et al. 1985). Studies of membrane lipid composition have not yielded consistent results that could account for the altered resistance of the membranes to the acute, fluidizing effect of ethanol (e.g., LYON and GOLDSTEIN 1983; CHIN et al. 1978; ALLING et al. 1982; LITTLETON et al. 1979; SUN and SUN 1977; CREWS et al. 1983). In part, this difficulty may arise because of a lack of consideration of distinct membrane lipid domains. For example, it was found that, while total cholesterol content did not differ between synaptic plasma membranes of control and chronically ethanol-treated mice, there was a change in sterol distribution between the cytofacial and exofacial leaflets of the membranes (WOOD et al. 1990). Since the leaflets of the membrane are differentially sensitive to the fluidizing effect of ethanol (SCHROEDER et al. 1988; WOOD et al. 1989), this and other changes in lipid distribution could contribute to the resistance to the fluidizing effect of ethanol in particular areas of a neuronal membrane in chronically treated animals (WOOD et al. 1989, 1991). There is also some evidence for changes in the fatty acid substituents of certain phospholipids that could affect the response to ethanol of membranes from chronically treated animals (ALOIA et al. 1985), and it was reported that the resistance to the fluidizing effect of ethanol in liver microsomal membranes from chronically ethanol-treated rats could be reproduced in the membranes of control animals by incorporating into the control rat membranes a small amount of phosphatidylinositol or cardiolipin extracted from the membranes of the ethanol-treated rats (HOEK et al. 1988). More recently, it has been reported that all phospholipid classes, when extracted from liver microsomal membranes of ethanol-treated rats and used to create phospholipid vesicles in combination with phospholipids extracted from control animals, conferred resistance to the fluidizing effect of ethanol (ROTTENBERG et al. 1992). This resistance was approximately proportional to the percentage of phospholipids coming from the ethanol-treated animals. In this later study, there was no clear relationship of fatty acid composition of the phospholipids to their ability to produce resistance to ethanol's effect (ROTTENBERG et al. 1992), and the mechanism contributing to resistance to ethanol is not known.

Thus, while there are measurable changes in the biochemical properties of neuronal and other celll membranes after chronic ethanol treatment, the relationship between changes in membrane characteristics and functional (pharmacodynamic) tolerance to the behavioral effects of ethanol remains unclear. It is important to note, for example, that resistance to the neuronal membrane fluidizing effect of ethanol dissipates relatively quickly after ethanol withdrawal (1–4 days, TARASCHI et al. 1986), while tolerance to behavioral effects is evident for a longer time (TABAKOFF and RITZMANN 1977). While the role of neuronal membrane lipids per se in the behavioral effects of ethanol remains uncertain, some membrane-bound proteins that are involved both pre- and postsynaptically in neurotransmission are demonstrably sensitive to ethanol and may, more specifically, mediate some of ethanol's acute and chronic behavioral effects.

II. Voltage-Sensitive Calcium Channels

Intracellular Ca^{2+} plays a crucial role in exocytosis, synaptic transmission and regulation of many cellular signalling processes (BERTOLINO and LLINÁS 1992), and, as a result, the role of Ca^{2+} in the actions of ethanol has been studied extensively. While earlier work focused on examining behavioral interactions of ethanol and administered Ca^{2+} (e.g., HARRIS 1979; ERICKSON et al. 1980; PALMER et al. 1987), a more recent focus has been on the effect of ethanol on Ca^{2+} influx into cells and particularly on influx through voltage-sensitive Ca^{2+} channels (VSCCs). There are at least four different types of VSCC: one low voltage activated "T" (transient) channel and three high voltage activated channels. The latter group includes "L" (long-lasting), "N" (neuronal) and "P" (Purkinje) channels (BERTOLINO and LLINÁS 1992). L-channels are the best characterized and are found in all excitable cells. They are sensitive to dihydropyridine (DHP) agonists and antagonists and to other drugs (DASCAL 1990). In the CNS, L-channels are proposed to be localized on cell bodies and to be involved in the generation of action potentials and in signal transduction events (BERTOLINO and LLINÁS 1992). These channels can be modulated by phosphorylation, both by cyclic AMP-dependent protein kinase (protein kinase A, PKA) and by the calcium/phospholipid-dependent protein kinase, protein kinase C (PKC) (BERTOLINO and LLINÁS 1992; DASCAL 1990). In addition, L-channels can be regulated by guanine nucleotide-binding coupling proteins (G proteins), either directly or via generation of second messengers (DASCAL 1990). The other VSCCs have not been studied in as much detail. T-channels, found in both neurons and skeletal and smooth muscle, are insensitive to DHPs and have been suggested to be involved in maintenance of the resting potential (Bertolino and Llinás 1992; Dascal 1990). N-channels are found in neurons, and their properties are in many ways similar to L-channels. However, they are inhibited by ω-conotoxin, a toxin from sea snails, and the N-channels may be insensitive to DHPs, although this property is controversial. The N-channels, like the L-channels, may be regulated by phosphorylation and/or G proteins (Bertolino and Llinás 1992; Dascal 1990) and have been suggested to play a role in neurotransmitter release (Bertolino and Llinás 1992). P-channels were originally discovered in cerebellar Purkinje neurons and seem to be the most widely distributed Ca^{2+} channels in the CNS (Bertolino and Llinás 1992). They may also be involved in neurotransmitter release (Bertolino and Llinás 1992). The structure of the VSCC is still being elucidated, but there is evidence for a heteromeric complex of α, β, γ and δ subunits that forms all subtypes of functional VSCCs in the brain and other tissues (Catterall 1991).

Most studies of ethanol have focussed on L-channels in the brain or in cells grown in culture, although there have been some analyses of ethanol's actions on other subtypes of VSCCs. Biochemical studies have generally demonstrated that concentrations of ethanol greater than $50\,mM$ inhibit Ca^{2+} entry into synaptosomal preparations in vitro (see Leslie 1986). More recent electrophysiologic patch clamp studies have allowed analysis of the ethanol sensitivity of various types of VSCCs. In the pineal gland, using a whole cell patch clamp technique, ethanol was found to reduce the amplitude of depolarization-induced L-channel currents, but in this study ethanol did not alter DHP agonist-induced increases in intracellular Ca^{2+} (Chik et al. 1992). In a whole cell patch clamp study of PC12 cells, a high concentration ($200\,mM$) of ethanol also decreased L-channel currents and net Ca^{2+} influx (Grant et al. 1993). In nerve terminals isolated from the rat neurohypophysis, ethanol ($10–50\,mM$) inhibited current through both L- and N-channels, with the L-channels being more sensitive to inhibition (Wang et al. 1991). Interestingly, although ethanol also inhibited current through L-channels in cultures of neurally derived transformed cells (N1E-115 and NG108-15 cells) (Twombly et al. 1990), these channels, which are located on the cell soma, were less sensitive to ethanol than those in the neurohypophysis. This difference could be due either to the differential cellular localization of the channels or to the different source of tissue (cells) used for analysis. In Aplysia neurons, L-channels were also relatively insensitive to inhibition by ethanol (Treistman and Wilson 1991). In addition to its effects on L- and N-channels, ethanol ($30\,mM$) was reported to inhibit T-channel current in the transformed neuronal cell cultures (Twombly et al. 1990).

Ethanol has been shown to *increase* intracellular Ca^{2+} levels through release of Ca^{2+} from intracellular stores (e.g., microsomes) (Daniell et al. 1987; Shah and Pant 1988). For the most part, this effect of ethanol occurs at higher concentrations than those necessary to inhibit Ca^{2+} currents through VSCCs, although CARLEN and WU (1988) reported that 20 mM ethanol inhibited voltage-sensitive Ca^{2+} currents in hippocampal cells and that this effect was blocked by intracellular administration of the Ca^{2+} chelator EGTA. These authors interpreted their results as showing that the effects of ethanol on Ca^{2+} influx were mediated by increases in intracellular Ca^{2+} release (CARLEN and WU 1988). It appears that ethanol, at low concentrations, may decrease Ca^{2+} influx into cells both by a direct action at VSCCs and by altering the Ca^{2+} gradient, but that this latter effect is more prominent at higher ethanol concentrations.

There is little evidence for a clear behavioral correlate of the acute inhibition of L-channel function by ethanol, and studies of the interaction of ethanol and L-channel antagonists on locomotor activity have produced conflicting results (WHITE and SMITH 1992; ENGEL et al. 1988). Furthermore, a study in humans failed to demonstrate any effect of the DHP antagonist nifedipine on the intoxicating effect of ethanol (PEREZ-REYES et al. 1992). However, the changes in VSCCs after chronic ethanol administration, and the role of these channels in ethanol tolerance and physical dependence, have received a significant amount of attention. Since the acute effect of ethanol appears to be inhibition of the function of various VSCCs, the possibility existed that chronic administration of ethanol might result in an increased function of these channels as an adaptive response, and such an increase in L-channel function has been found in a number of studies. In rats treated chronically with ethanol by inhalation or by adding ethanol to their drinking water, an increased number of binding sites for DHPs was found in whole brain (DOLIN et al. 1987) and in striatum (LUCCHI et al. 1985). The increased binding in one study was associated with increased biochemical responses to the L-channel agonist Bay-K 8644 (DOLIN et al. 1987). However, in another study there was no significant increase in nitrendipine (DHP) binding in striatal tissue of ethanol-fed rats (WOODWARD et al. 1990), and these authors found a decreased response to Bay K-8644. This same study found a decreased inhibition of Ca^{2+} flux and dopamine release by ethanol (i.e., resistance to ethanol) when the ethanol was added to striatal tissue excised from the ethanol-fed rats.

A number of studies have also examined the chronic effects of ethanol on cells maintained in culture. Chronic ethanol exposure resulted in an increase in DHP binding to adrenal chromaffin cells (HARPER et al. 1989), and exposure of PC12 cells to ethanol for several days resulted in increased DHP binding and increased DHP-sensitive Ca^{2+} flux (SKATTEBOL and RABIN 1987; GREENBERG et al. 1987), (MESSING et al. 1986; GRANT et al. 1993). In some instances, acute ethanol inhibition of Ca^{2+} influx was not altered after chronic ethanol treatment, while in one study sensitivity to ethanol was increased (SKATTEBOL and RABIN 1987; MESSING et al. 1986; GRANT et al. 1993). The overall results suggest that both in brain and in cells in culture, chronic ethanol exposure results in a quantitative "up-regulation" of L-channels. The mechanism of this ethanol-induced L-channel up-regulation has also been investigated. In PC12 cells, the chronic ethanol-induced increase in DHP binding was blocked by concomitant treatment of the cells with a PKC inhibitor (MESSING et al. 1990). In further studies, chronic ethanol treatment of these cells was shown to increase PKC activity, the binding of phorbol ester and immunoreactivity of several isoforms of PKC (MESSING et al. 1991). Chronic ethanol exposure also increased PKC activity in NG-108-15 cells (MESSING et al. 1991). These findings suggested that increased PKC activity may mediate the chronic effects of ethanol on L-channels in certain cell lines, although, acutely, ethanol was not found to alter PKC activity in the PC12 cells (MESSING et al. 1991).

It has been postulated that the increase in L-channels after chronic ethanol treatment could reflect an adaptive response to the initial inhibition of channel activity by ethanol. However, concomitant chronic treatment with DHP antagonists of L-channel function and ethanol prevented the ethanol-induced increase in DHP binding sites both in chromaffin cell cultures and in animals (HARPER et al. 1989; DOLIN and LITTLE 1989). Furthermore, chronic exposure of animals or PC12 cells to DHP-type L-channel antagonists alone does not result in an up-regulation of L-channels (DOLIN and LITTLE 1989; MARKS et al. 1989). In fact, in the PC12 cells, chronic exposure to the L-channel agonist Bay-K 8644 produced an increase in DHP binding, similar to results seen with ethanol (MARKS et al. 1989). Since L-channels are regulated by a number of intracellular second messengers, and ethanol has effects on second messenger systems, it seems likely that effect of chronic ethanol treatment on L-channels is mediated by the second messenger responsive systems (e.g., PKC).

The up-regulation of L-type VSCCs may play a role in generating signs of ethanol withdrawal. For example, it was reported that administrtion of L-channel antagonists reduced ethanol withdrawal seizures in rats (LITTLE et al. 1988a) and that concurrent treatment of animals with L-channel antagonists and ethanol not only blocked the up-regulation of the L-channels, but also blocked ethanol withdrawal seizures (DOLIN and LITTLE 1989; LITTLE et al. 1988b). A calcium channel antagonist also reduced ethanol withdrawal signs in humans (KOPPI et al. 1987). More recent studies have shown that concomitant ethanol and nitrendipine (a DHP antagonist) administation to mice lessened the ethanol withdrawal hyperexcitability noted in hippocampal tissue, as compared to the hyperexcitability seen in hippocampal slices taken from mice that had received ethanol alone (WHITTINGTON and LITTLE 1991). This ethanol-induced hyperexcitability, measured by extracellular recordings of the CA1 area of the hippocampus, could also be reduced by application of a DHP antagonist directly to the hippocampal slices (WHITTINGTON and LITTLE 1991). All of these data are consistent with the hypothesis that the increase in L-channel number/activity after chronic ethanol treatment may contribute to ethanol withdrawal hyperexcitability and seizures. In addition, it was reported (BRENNAN et al. 1990) that DHP binding sites were increased after chronic ethanol exposure to a greater degree in brains of mice selectively bred for susceptibility to ethanol withdrawal seizures (WSP mice) than in those bred for resistance to these seizures (WSR mice). This biochemical difference between selected lines of animals strongly implicates L-channels in the selected trait, i.e., ethanol withdrawal seizure severity. It is important to note, however, that L-channel antagonists do not block all symptoms of ethanol withdrawal (FILE et al. 1991).

L-channels may also be involved in the development of tolerance to ethanol. Rats given injections of the DHP antagonist nifedipine during chronic ethanol administration developed tolerance to the motor-impairing effect of ethanol more slowly than vehicle-treated rats (WU et al. 1987). In rats or mice given the DHP antagonists, nitrendipine or isradipine during chronic ethanol exposure, the development of tolerance ot the ataxic or hypnotic effect of ethanol, respectively, was blocked (DOLIN and LITTLE 1989). The antagonist treatments were similar to those shown to prevent the ethanol-induced up-regulation of VSCCs. The data are compatible with the suggestion that ethanol-induced increased L-channel function in the CNS may contribute both to the development of ethanol tolerance and to the occurrence of ethanol withdrawal hyperexcitability.

III. Neurotransmitter Release

1. Dopamine

In the studies of ethanol's effects on the neuronal terminals in the rat neurohypophysis described above, the ability of ethanol to inhibit VSCC activity was positively correlated with its ability to inhibit vasopressin release from the neurohypophyseal nerve terminals (WANG et al. 1991). If all of ethanol's effects on neurotransmitter/neuromodulator release were a result of its actions on Ca^{2+} influx, then release of all neurotransmitters would be expected to be similarly influenced, but this is not the case. The effect of ethanol on neurotransmitter release appears to reflect a combination of the effects on Ca^{2+} influx and more indirect effects on a neuron through converging neuronal systems. The effect of ethanol on the release of dopamine has been studied in some detail, because of the evidence implicating dopamine systems in the reinforcing effects of ethanol and other psychoactive drugs (WISE 1978; KOOB 1992). Initial investigations relied on the measurement of dopamine precursors (dihydroxyphenylalanine, DOPA) or metabolites (dihydroxyphenylacetic acid, DOPAC; HVA) in brain to determine the effect of ethanol on dopamine metabolism. The levels and rate of accumulation of the dopamine precursors and/or metabolites were used as a measure of neuronal activity. In spite of significant caveats related to the use of such measurements (see HOFFMAN and TABAKOFF 1985), the general conclusion of such studies was that acute ethanol administration to animals produces an increase in the synthesis and release of dopamine in brain (see

HOFFMAN and TABAKOFF 1985). The view that this change reflects an increase in activity of dopaminergic neurons was supported by the results of electrophysiological studies that found increased firing of nigrostriatal and mesolimbic neurons after an acute dose of ethanol to rats (GESSA et al. 1985; MEREU et al. 1984). However, the effect of ethanol on dopamine metabolism is not uniform throughout the brain; the sensitivity to ethanol varies in different brain areas (BACOPOULOS et al. 1978; BUSTOS et al. 1981; FADDA et al. 1980), and there may also be genetically determined differences in the response of the dopaminergic systems to ethanol. For example, the inbred mice of the DBA and Balb strains, which are sensitive to behavioral (locomotor stimulation) effects of ethanol (TABAKOFF and KIIANMAA 1982), showed a biphasic response of striatal dopamine metabolism to ethanol: low doses decreased dopamine release and higher doses increased release (KIIANMAA and TABAKOFF 1983). By contrast, in C57BL mice, which are more resistant to ethanol's activating effects (TABAKOFF and KIIANMAA 1982), there was no change in dopamine release after low doses of ethanol, but an increase in dopamine release, as in the other strains, after higher ethanol doses (KIIANMAA and TABAKOFF 1983). Using a microdialysis technique, which allows more direct measurement of dopamine release, IMPERATO and DI CHIARA (1986) demonstrated that systemically administered doses of ethanol (0.25–0.5 g/kg) stimulated dopamine release in the nucleus accumbens, while higher doses were needed to increase striatal dopamine release. The differential sensitivity of dopamine neurons in the two brain areas may be related to the different regulation of dopamine metabolism in these areas (BUSTOS et al. 1981) and was of particular interest because of the postulated role of mesolimbic dopamine systems in the reinforcing effects of drugs (WISE 1978; KOOB 1992). Electrophysiological studies also indicated that activation of neuronal activity in the ventral tegmental area occurred after lower systemic doses of ethanol than activation of nigrostriatal neurons (GESSA et al. 1985; MEREU et al. 1984). The microdialysis technique has also been used to study dopamine release after direct administration of ethanol into the brain. Ethanol, administered into the nucleus accumbens or corpus striatum, produced dopamine release, but with this mode of ethanol administration there was no apparent difference in sensitivity of the two brain regions (WOZNIAK et al. 1991).

One explanation for the different results obtained after systemic vs intracerebral ethanol administration may be that ethanol, when given systemically, acts through other neurotransmitter systems that impinge on dopaminergic neurons and affect dopamine release. There are several neurotransmitter systems which can be influenced by ethanol that do impinge on dopaminergic neurons. For example, as discussed in further detail below, ethanol is a potent inhibitor of the function of the N-methyl-D-aspartate (NMDA) subtype of glutamate receptor. NMDA itself has been found to stimulate catecholamine release, and, in this case, inhibition of NMDA receptor function by ethanol would result in a decrease in dopamine release

(FINK et al. 1989; WOODWARD and GONZALES 1990). This phenomenon is obviously not compatible with the in vivo findings discussed above. There is also evidence that specific NMDA receptor antagonists given in vivo can, like ethanol, increase dopamine release and metabolism (IMPERATO et al. 1990), and one may postulate the existence of an NMDA receptor-containing interneuronal pathway that could modulate dopamine release and contribute to the observed effects of ethanol on the dopaminergic systems.

Another relevant candidate for regulation of dopaminergic activity systems is serotonin. When the serotonin (5-HT_3) receptor antagonist ICS-205-930 was administered to rats prior to systemic ethanol administration, the effect of ethanol on dopamine release was reduced (CARBONI et al. 1989). This antagonist also reduced the effects of intracerebrally administered ethanol, but a higher dose of the antagonist was needed (WOZNIAK et al. 1990). Ethanol has been found to exert direct effects on 5-HT_3 receptor function and to alter serotonin release and metabolism (see below). Furthermore, agonists acting at the 5-HT_3 receptor can increase dopamine release (BLANDINA et al. 1989). Therefore, the effect of ethanol on dopamine release could be mediated by activation of a serotonergic pathway.

It has also recently been reported that the ability of ethanol to release dopamine in the nucleus accumbens was prevented by intracerebral administration of an antagonist at the δ opiate receptor (ACQUAS et al. 1993). Ethanol has been found to affect the synthesis of opioid peptides and the function of the δ and μ opiate receptors (see below), and these actions of ethanol may also contribute to its effects on dopamine release.

The acute effect of ethanol on dopamine release, particularly in the mesolimbic areas, has been suggested to be important for the reinforcing effect of ethanol, in analogy to the role of dopamine in the reinforcing effects of other drugs (WISE 1978; KOOB 1992). This hypothesis is based primarily on studies of ethanol self-administration or ingestion, which generally find that dopamine receptor agonists reduce ethanol intake, presumably by substuting for ethanol stimulation of the dopaminergic system (SAMSON and HOFFMAN 1995). Results with dopamine receptor antagonists have been more difficult to interpret, since they often have effects similar to dopamine agonists (SAMSON and HOFFMAN 1995). However, these antagonists can act not only at postsynaptic receptors to inhibit dopamine responses, but also at presynaptic receptors to alter doamine release, and the inability to distinguish these effects significantly complicates the interpretation of results. It is also important to note that the method used to determine ethanol intake (e.g., restricted vs unrestricted access to ethanol; free-choice drinking vs operant paradigms of self-administration) can affect the results witnessed with various dopamine receptor agonists and antagonists (SAMSON and HOFFMAN 1995). At present, it can be concluded that the dopaminergic systems may play a role in ethanol reinforcement, but the mechanism of this behavioral effect is less clear for ethanol than for some other drugs (SAMSON and HOFFMAN 1995; TABAKOFF and HOFFMAN 1992).

After chronic ethanol ingestion, tolerance to the dopamine-releasing effect of ethanol occurs, and dopamine metabolism in the striatum is decreased (see Hoffman and Tabakoff 1985). A decrease in the dopamine metabolite HVA was also reported in CSF of alcoholics undergoing withdrawal, in comparison to those not displaying withdrawal symptoms (Major et al. 1977). In addition, there is evidence from animal studies that the function of dopamine (D_1) receptors (mainly in the striatum) was reduced after chronic ethanol exposure (see Hoffman and Tabakoff 1985). The combination of a decrease in dopaminergic activity and decreased receptor function may contribute to the occurrence of some withdrawal symptoms (e.g., tremors), and certain symptoms were reported to be lessened by administration of dopamine agonists to alcoholics undergoing withdrawal (Borg and Weinholdt 1982). Furthermore, tolerance to the effect of ethanol on dopamine release, especially if it occurs in the mesolimbic system, could alter the reinforcing effect of ethanol in animals or humans exposed chronically to ethanol.

2. Serotonin, Norepinephrine and Acetylcholine

Ethanol has also been shown to affect the release and metabolism for a number of other neurotransmitters, and these effects have been previously reviewed (Hoffman and Tabakoff 1985; Tabakoff and Hoffman 1992) and will be discussed only briefly here. Experiments to investigate the effect of ethanol on serotonin metabolism and release have yielded diverse results, which were at least partially dependent on variations in the experimental methods. An overall assessment of the work (Tabakoff and Hoffman 1992) yielded the suggestion that higher doses of ethanol (2–4 g/kg) result in increases in the production of the major metabolite of serotonin, (5-HIAA), in brain with a reduction or lack of change in serotonin levels. These data would indicate an increased release and metabolism (turnover) of serotonin after acute ethanol administration. After chronic ethanol ingestion, a decrease in serotonin metabolism, i.e., decreases in brain 5-HIAA and decreases in brain serotonin turnover, have been reported (Tabakoff et al. 1977; Nutt and Glue 1986; Ballenger et al. 1979). The acute actions of ethanol on serotonin metabolism and the specific effects on the 5-HT$_3$ receptor subtype (see below) may underlie the findings that treatment of animals and humans with serotonin uptake inhibitors (to increase concentrations of serotonin at the synapse) can reduce ethanol ingestion (Naranjo et al. 1984,1987; Murphy et al. 1988; Samson et al. 1990). The explanation for this effect would be that the increase in synaptic serotonin levels mimics the acute effects of ethanol and "substitutes" for them in the individual. The general role of serotonergic systems in ingestive behavior (Blundell 1984) supports the idea that modulation of serotonergic activity may provide a means to alter ethanol intake.

Ethanol also has effects on the noradrenergic systems in brain, and the activity of locus coeruleus neurons, which send noradrenergic projections to rostral brain areas, is very sensitive to modulation by ethanol. In electrophysiologic studies, low concentrations of ethanol $(1-10\,mM)$ had variable effects (some neurons increased firing rates, others decreased) on locus coeruleus neurons, while higher concentrations $(>30\,mM)$ were consistently inhibitory (SHEFNER and TABAKOFF 1985). Biochemical studies have produced positive correlates for the electrophysiologic studies: after low doses of ethanol, or at early time points after higher doses, norepinephrine (NE) release and metabolism were increased (HUNT and MAJCHROWICZ 1974; BACOPOULOS et al. 1978), while higher doses produced a decrease in NE metabolism (HUNT and MAJCHROWICZ 1974). These effects were dependent on the brain area studied, and the low dose effects of ethanol were most prominent in the brainstem (BACOPOULOS et al. 1978). In humans, after ingestion of 1 g/kg of ethanol, an increase in CSF levels of the NE metabolite MOPEG was reported, and this result was interpreted to reflect an increase in brain NE turnover (BORG et al. 1983). The levels of MOPEG in CSF may be influenced by shifts in peripheral NE metabolism during periods of acute ethanol intoxication (DAVIS et al. 1967a), and although the results found with humans ingesting ethanol are qualitiatively similar to those reported in brains of other animals, the confounding effects of peripheral metabolism of NE need to be carefully considered.

Chronic ethanol ingestion has generally been found to result in an increase in NE metabolism in brain, during the time that an animal is intoxicated and for up to 24 h after withdrawal (HOFFMAN and TABAKOFF 1985). The fact that NE metabolism is already increased prior to withdrawal suggests that the ethanol-induced increase in NE metabolism during the withdrawal period is not simply a response to the stress of withdrawal. Ethanol also affects NE receptor function, and these effects are discussed in the section on receptor-effector coupling mechanisms (Sect. V.1). Several reports have appeared on the efficacy of β-adrenergic antagonists in relieving certain ethanol withdrawal signs, e.g. tremor (GESSNER 1979). The effects of the β-adrenergic antagonists may be mediated through peripheral mechanisms (outside of the CNS) and the effects of agents such as propranolol on the metabolism of ethanol (FORMAN et al. 1988; YUKI and THURMAN 1980) should not be ingored. The ability of propranolol to slow ethanol metabolism may contribute to a decreased rate of elimination of ethanol and a more graded period of withdrawal, if propranolol is given while ethanol is still present in a withdrawing individual.

Ethanol, acutely, decreases acetylcholine release, and this is one of the most consistently reported effects of ethanol on neurotransmitter dynamics (see HOFFMAN and TABAKOFF 1985 for review). The sensitivity to ethanol varies among brain regions and was found to be greater in subcortical than in cortical regions (ERICKSON and GRAHAM 1973). The effect of ethanol may be mediated by a change in cholinergic neuron signal conduction or by a presynaptic effect on the amount of acetylcholine available for release. Ethanol does inhibit the uptake of choline (a precursor for acetylcholine synthesis), but this effect of ethanol is dependent on the brain area and strain or species of animal studied (Durkin et al. 1982; Howerton et al. 1982; Hunt et al. 1979). It has also been proposed, as for dopamine metabolism, that the effect of ethanol on acetylcholine release is indirect (Durkin et al. 1982; Howerton et al. 1982; Hunt et al. 1979).

Tolerance develops rapidly to the effect of ethanol on acetylcholine release and can be observed in the rat even during the time that a single dose of ethanol is present (Sinclair and Lo 1978). After longer-term chronic ethanol exposure, release of acetylcholine has been reported to be increased (Clark et al. 1977), perhaps as an adaptive response to the inital inhibitory effect of ethanol. Similary, choline uptake was enhanced in striatum of chronically ethanol-treated rats (Hunt et al. 1979). There is also a change in the properties of the muscarinic cholinergic receptors in brains of ethanol-treated animals, with a significant, though short-lived, increase in the number of receptors occurring in hippocampus and cortex of ethanol-withdrawn mice (Tabakoff et al. 1979; Rabin et al. 1980). This increase was shown to be a result of an increase in the M_1 subtype of muscarinic receptors in the cerebral cortex (Hoffman et al. 1986). The increased cholinergic activity in brains of ethanol-withdrawn animals may also be related to certain signs and symptoms of ethanol withdrawal, since other procedures, such as kindling, that result in CNS hyperexcitability have been shown to be associated with cholinergic supersensitivity in certain brain areas (Burchfiel et al. 1979).

IV. Neuropeptides

1. Opiates

In addition to "classical" biogenic amine neurotransmitters, the brain contains a large number of peptides which are believed to serve either as neurotransmitters or neuromodulators (Brownstein et al. 1989). For the most part, the effects of ethanol on these neuropeptides have not been well characterized, but the growing number of recognized neuropeptides and their involvement in many aspects of behavior is leading to increased interest in the actions of ethanol on neuropeptide dynamics in brain. One family of peptides that has come under scrutiny is the opioid peptides, in part because of the suggestion that endogenous opioids might be involved in the reinforcing effect of ethanol (see Gianoulakis 1989). However, this work has been fraught with difficulty because of the complex dynamics of the synthesis and release of neuropeptides, and the results are often equivocal. For example, in one study, an acute injection of ethanol resulted in increased hypothalamic content of β-endorphin (Schulz et al. 1980), while in another study there was no effect (SEIZINGER et al. 1983). Although one in vitro study did not show an effect of ethanol on β-endorphin release from the rat neurointermediate lobe (GIANOULAKIS and BARCOMB 1987), exposure of dispersed anterior pituitary cells to ethanol increased β-endorphin release (KEITH et al. 1986). Ethanol has also been reported to increase β-endorphin release from perfused rat hypothalamus and from cultured hypothalamic cells (GIANOULAKIS 1990; SARKAR and MINAMI 1990). Several studies have demonstrated that, in vitro, ethanol generally increase adrenocorticotrophic hormone (ACTH) release from pituitary cells (see GIANOULAKIS 1989), and it is expected that there should be similar effects on β-endorphin release, since the two peptides are contained within the same precursor molecule (SIEGEL et al. 1989). However, there are instances when changes in plasma levels of ACTH and β-endorphin have been found to be dissociated (VESCOVI et al. 1992).

Effects of ethanol on the enkephalins have also been evaluated. Some investigators have demonstrated an increase in methionine-enkephalin levels in certain brain areas after acute ethanol injection (SCHULZ et al. 1980; SEIZINGER et al. 1983), while another group found no change in leucine enkephalin levels after a higher dose of ethanol (RYDER et al. 1981). No change in the levels of dynorphin or α-neo-endorphin was observed after acute ethanol injection (SEIZINGER et al. 1983).

There has also been a significant amount of work directed at determining the chronic effect of ethanol on opioid peptides, much of its subject to the same difficulties described above. In addition, varying methods of alcohol administration appear to complicate the results. Overall, the data suggest that chronic ethanol administration in vivo results in a decrease in β-endorphin levels in the pituitary (GIANOULAKIS 1989). However, the effect of ethanol varies depending on the method of ethanol administration and the brain and/or hypothalamic area and species studied (GIANOULAKIS 1989). Some of the variation in the results may be a consequence of differences among laboratories in the methods for isolating and measuring the peptides. In some experiments, incorporation of radioactive amino acids into peptides has been used as a measure of peptide synthesis, and, when ethanol was administered to rats in the drinking water for 3 weeks, there was decreased incorporation of radioactively labeled amino acids into β-endorphin-related peptides in the pituitary (SEIZINGER et al. 1984a). Another means of assessing the effect of ethanol on β-endorphin is to measure mRNA levels for the precursor proopiomelanocortin (POMC). Measurement of mRNA levels is believed to provide an indication of the rate of peptide synthesis. In rats exposed to ethanol by inhalation, one study showed a decrease in the pituitary levels of mRNA for POMC and a decreased release of β-endorphin from the pituitary gland (DAVE et al. 1986). In more recent work, no change in pituitary POMC mRNA was found, but there was a decrease in POMS mRNA in the mediobasal hypothalamus (SCANLON et al. 1992). By contrast, chronic treatment of rats with ethanol in a liquid diet, or by intubation, produced an increase in pituitary peptide processing (Gianoulakis et al. 1981, 1983, 1988; Seizinger et al. 1984b). Similarly, increases in hypothalamic POMC mRNA and incorporation of radioactive phenylalanine into hypothalamic β-endorphin were found in a recent study of rats fed ethanol in a liquid diet for 15 days (Angelogianni and Gianoulakis 1993). In ethanol-sensitive LS mice given repeated ethanol injections, POMC mRNA in the pituitary was increased after 4 days of ethanol treatment and returned to normal after 7 days. In SS mice, POMC mRNA levesl remained elevated at 7 days (Wand and Levine 1991). It is possible that the variation in results is related to differences in stress caused by the ethanol administration procedures and/or ethanol withdrawal, and these are additional factors that significantly complicate interpretation of the data. It has also been reported that chronic ethanol treatment increases N-Acetylation of β-endorphin, a posttranslational modification that reduces neuropeptide activity (Seizinger et al. 1984a).

Studies of the effect of chronic ethanol treatment on enkephalin levels have been somewhat more consistent and indicate a decrease in enkephalin levels in several brain areas (Gianoulakis 1989). The content of dynorphin and α-neo-endorphin was also reduced in certain brain areas of ethanol-treated rats (Seizinger et al. 1983). In one study prodynorphin mRNA levels in the dentate gyrus were reported to be decreased in rats that received repeated intragastric ethanol administrations (Przewlocka et al. 1992). However, when mice were fed ethanol in a liquid diet for 7 days, prodynorphin mRNA levels were increased in a number of brain areas (Gulya et al. 1993). Since prodynorphin mRNA levels have been reported to be decreased by seizure activity and the intubation method for ethanol administration produces severe physical dependence and the opportunity for repeated episodes of withdrawal, it is possible that the decrease in prodynorphin mRNA in the first study is a result of periods of ethanol withdrawal hyperexcitability occurring between the intoxication episodes. The liquid diet method appears to cause dehydration, which in and of itself increases prodynorphin mRNA levels (Gulya et al. 1993).

In humans, some studeis indicated no change in plasma β-endorphin levels in chronic alcoholics (Brambilla et al. 1988; Genazzani et al. 1982), while in others plasma β-endorphin levels were found to be low when alcoholics were admitted to the hospital (Vescovi et al. 1992; Aguirre et al. 1990). In one patient, endorphin levels returned to normal after 5 weeks of abstinence (Vescovi et al. 1992). It has also been reported that sober alcoholics have low CSF levels of β-endorphin and exhibit a trend toward low CSF levels of enkephalin (Genazzani et al. 1982; Borg et al. 1982). Individuals demonstrating a positive family history of alcoholism were reported to have lower plasma levels of β-endorphin than those with a negative family history; in this study, chronic alcoholics who had been abstinent for 6 months also had low β-endorphin levels (Glanoulakis 1989). These data suggested an inherent difference between alcoholics (or family history-positive individuals) and control subjects with respect to β-endorphin levels, in line with the postulate that differences in endogenous opiate levels may be a genetic factor contributing to alcoholism (BLUM et al. 1980). In the family history-postitive individuals, a low dose of ethanol (0.5 g/kg) produced a small increase in plasma β-endorphin levels (GLANOULAKIS 1989). Similarly, in another study of human volunteers, without regard to family history for alcoholism, a high dose of alcohol increased plasma β-endorphin (NABER et al. 1981). It must be reemphasized, however, that β-endorphin is normally coreleased with ACTH in response to stress, and it is difficult to separate the effect of stress from a direct effect of ethanol.

In addition to the effect of ethanol on opioid peptide synthesis, there has been a considerable amount of research regarding ethanol effects on opiate receptors. In general, ethanol in vitro inhibits ligand binding to opiate receptors, and, in membranes from whole rat brain and mouse striatum, the δ receptor (high affinity for enkephalin) was more sensitive to the effects of ethanol that the μ receptor (high affinity for morphine) (HILLER et al. 1981, 1984; TABAKOFF and HOFFMAN 1983). However, in mouse frontal cortical preparations, ethanol had a similar effect on ligand binding to μ and δ receptors and no effect on κ receptors (high affinity for dynorphin) (KHATAMI et al. 1987). While in all of these studies the major effect of high concentrations of ethanol was to inhibit ligand binding to opiate receptors, a lower concentration of ethanol (50 mM) was reported in increase dihydromorphine binding to the μ receptor in mouse striatum (TABAKOFF and HOFFMAN 1983).

In vitro ligand binding studies are normally carried out in very artificial systems, and it is of interest that, when more "physiological" conditions were used (i.e., 37°C, in the presence of Na$^+$ and GTP), the μ receptor in mouse striatum was found to be more sensitive to the inhibitory effect of ethanol than the δ receptor. Under these conditions, ligand binding to the μ receptor was significantly decreased by 25 mM ethanol, while binding to the δ receptor was not affected (HOFFMAN et al. 1984). These results suggested that the μ receptor might be more sensitive to in vivo effects of ethanol, a hypothesis that was supported by the finding of significant alterations in μ receptor properties and function (i.e., decreased affinity for an response to morphine) in mice treated chronically with ethanol (TABAKOFF et al. 1981). Acute and chronic effects of ethanol on the μ opiate receptor would be expected to alter the responses to endogenous opiates and may contribute to certain of the behavioral effects of ethanol, including reinforcement.

An increase in ligand binding to the δ receptor and/or a change in δ receptor properties was also reported in brain membranes of rats treated chronically with ethanol (PFEIFFER et al. 1981; GIANOUAKIS 1983; HYNES et al. 1983). There was, however, no apparent change in the function of this receptor (i.e., inhibition of adenylate cyclase activity) in brains of mice fed ethanol chronically (HOFFMAN and TABAKOFF 1986). Studies with neuronal (NG108-15) cells in culture, by contrast, clearly demonstrated that chronic exposure to ethanol resulted in a significant increase in ligand binding to the δ receptor and a significant enhancement of opiate-induced inhibition of adenylate cyclase activity (Charness et al. 1983, 1986). The increase in opiate receptors in the NG108-15 cells depended on uninterrupted protein synthesis within these cells, presumably required for generation of additional receptor protein (Charness et al. 1986). In other cell lines, chronic ethanol exposure also increased ligand binding to the δ receptor, although the effects varied in magnitude (Charness 1989). In these later studies, the change in function of the δ receptor was attributed not only to increases in receptor protein, but also to changes in the receptor-effector coupling G proteins, which added to the variability among cell lines (Charness 1989). The relationship of the changes in transformed cell lines to those in the brain needs further elucidation particularly since many of the cell lines used contain only δ opiate receptors and interactions among different types of cells and receptors were lacking.

For a variety of reasons, some of which were mentioned above, the effects of ethanol on opioid peptide synthesis and opiate receptors remain controversial, although there is evidence to support contentions that there is some interaction. There has long been an interest in possible commonalities between alcohol and opiate addiction. Recent work suggesting that naloxone is effective in reducing "craving" in alcoholics (Volpicelli et al. 1992) may generate further enthusiasm for elucidating the interaction of ethanol and opiates, particularly with regard to the reinforcing effects of ethanol.

2. Arginine Vasopressin

Another neuropeptide that has received some attetnion with respect to the effects of ethanol is arginine vasopressin (AVP). AVP is the antidiuretic hormone in most species (lysine vasopressin is the form of the hormone in pigs), and release of AVP is stimulated by increases in plasma osmolality or decreases in extracellular fluid volume. The peptide is synthesized primarily in the hypothalamus and released from the neurohyphysis. Upon release, there is an increase in hypothalamic AVP synthesis, reflected in increased mRNA levels and changes in mRNA properties, that maintains peptide levels in the posterior pituitary (Burbach et al. 1984; Zingg et al. 1988). AVP is also synthesized in other parts of brain and appears to have centrally mediated actions on learning and memory (Buijs 1978; Caffé et al. 1987; De Vries et al. 1985; Hoffman 1987). This peptide was also found to modulate ethanol tolerance (Hoffman et al. 1978; Lê et al. 1982; Szabó et al. 1988), and studies with specific vasopressin antagonists suggested that the endogenous neuropeptide played a role in the maintenance of ethanol tolerance (Szabó et al. 1988).

It is generally accepted that, acutely, ethanol inhibits vasopressin release from the pituitary (Beard and Sargent 1979), although there is one study suggesting that low ethanol concentrations may increase vasopressin release

(COLBERN et al. 1985). The few available studies suggest that plasma vaso-pressin is increased in alcoholics, although this increase could be a response to the stress of alcohol withdrawal (BEARD and SARGENT 1979). To more directly evaluate the effect of chronic ethanol exposure on vasopressin synthesis, levels of hypothalamic mRNA for vasopressin were measured in mice that were fed ethanol chronically in a liquid diet. In these animals, AVP mRNA in both the paraventricular and supraoptic nuclei of the hypothalamus was significantly reduced at the time of ethanol withdrawal (ISHIZAWA et al. 1990). The ethanol-fed mice appeared to be dehydrated, based on their increased plasma osmolality, but there was no corresponding increase in plasma vasopressin levels. These data suggested that chronic ethanol ingestion interfered with the vasopressin synthesis-secretion coupling in the mice (ISHIZAWA et al. 1990). A similar pattern was found in rats that were exposed to ethanol by inhalation, with the exception that plasma vasopressin levels were increased in the rats, possibly due to a greater pituitary store in these animals (HOFFMAN and DAVE 1991). A decreased number of immunoreactive vasopressin-containing neurons in the hypotha-lamus of ethanol-treated rats has also been reported, in agreement with the decrease in mRNA levels (KOZLOWSKI et al. 1989). In addition to the decreased mRNA levels in the paraventricular and supraoptic nuclei, a decrease in mRNA levels for vasopressin in several other hypothalamic nuclei, and in an extrahypothalamic brain area has been reported in ethanol-fed mice (HOFFMAN et al. 1990). Thus, chronic ethanol treatment appears to generally reduce the levels of vasopressin mRNA (vasopressin synthesis). With respect to hypothalamic AVP, finding of a deficit in synthesis-secretion coupling may have important implications for alcoholics, since although plasma AVP levels may be increased in these individuals, the results suggest that they would be unable to respond normally to osmotic stimuli for vasopressin release.

3. Neurotensin

The neuropeptide neurotensin has also been studied with regard to the effects of ethanol. Central administration of this peptide potentiated ethanol-induced hypothermia and hypnosis in mice (LUTTINGER et al. 1981). Further-more, mice bred selectively for insensitivity to the hypnotic effect of ethanol (SS mice) were more sensitive to the effect of neurotensin than ethanol-sensitive mice (LS mice (ERWIN and JONES 1989). The greater sensitivity of the SS mice was associated with a higher number of neurotensin receptors in cerebellum, cerebral cortex and striatum and with a lower amount of immunoreactive neurotensin in the hypothalamus (ERWIN and JONES 1989; ERWIN and KORTE 1988). The finding of differences in the neurotensin system between the selected lines of mice suggested that this peptide might be involved in mediating the selected trait, i.e., sensitivity to the hypnotic effect of ethanol (ERWIN and JONES 1989).

The effects of ethanol on both neurotensin levels and neurotensin receptors in brains of SS and LS mice have also been examined. Acute administration of low doses of ethanol decreased neurotensin levels in the hypothalamus and mesolimbic area of both LS and SS mice, suggesting increased release and degradation of the peptide, but had no effect on neurotensin receptor characteristics (ERWIN et al. 1990). Chronic ethanol administration, by the addition of ethanol to the drinking water for 2 weeks, resulted in a decrease in neurotensin receptor number in several brain areas of LS and SS mice and changes in the proportions of high and low affinity receptors (CAMPBELL and ERWIN 1993). In addition, chronic ethanol ingestion resulted in an increase in neurotensin levels in brain areas of both lines of mice (ERWIN et al. 1992). As for other neurotransmitters and neuropeptides, changes in the levels of neurotensin are difficult to interpret, but the sensitivity of this system to perturbation by ethanol suggests that this peptide may play a role in certain of the behavioral effects of ethanol.

V. Postsynaptic Receptors

A single dose of ethanol ingested acutely, or moderate concentrations of ethanol ($\sim 50\,\text{m}M$) added to in vitro assays, have little effect on the properties of neurotransmitter receptors per se, although there are some instances (e.g., opiate receptors) in which ethanol in vitro can affect ligand binding to receptors (see above). Chronic ethanol ingestion has been reported to alter the number or function of some receptors and these effects have been noted in other sections of this chapter. However, evidence has been presented that the inherent characteristics of certain receptors in brain may predispose an individual to alcoholism. The gene for the D2 subtype of dopamine receptor has been found to be polymorphic, with allelic variants in its noncoding region (GRANDY et al. 1989). The restriction endonuclease *Taq I* has been used to identify an A1 and A2 variant (GRANDY et al. 1989). It was originally reported that 69% of alcoholics carried the A1 allele of this receptor, compared with only 20% of controls (BLUM et al. 1990). Therefore, this allele was suggested as a marker for a predisposition to alcoholism and possibly as a causal factor in alcoholism (BLUM et al. 1990). There have been a number of follow-up studies which have generated confusion regarding the association of the A1 allele with alcoholism (GELERNTER et al. 1991; BOLOS et al. 1990; CLONINGER 1991; COMINGS et al. 1991), primarily because of controversy regarding the prevalence of the A1 allele in control subjects and the diagnostic criteria for "alcoholism" used in the various studies. In addition, a high frequency of the A1 allele has been reported to be associated with other disorders, including Tourette's syndrome, attention deficit hyperactivity disorder, autism and posttraumatic stress syndrome (COMINGS et al. 1991). Although a meta-analysis of currently availabe evidence indicates an association between the A1 polymorphism and alcoholism, analysis in alcoholic families has demonstrated *no linkage* between the A1 allele and alcoholism (UHL et al. 1993). The consensus appears to be that the A1 allele of the D2 receptor might modify the clinical expression of alcoholism (and other disorders) but is neither a necessary nor sufficient cause of alcoholism.

1. Receptor-Effector Coupling Systems

The most significant effects of moderate acute doses of ethanol and chronic ethanol administration at the postsynaptic receptor level have, however, been demonstrated on the processes that couple neurotransmitter receptors to their effectors. Two major intracellular signaling enzymes, adenylyl cyclase (which forms cyclic AMP) and phospholipase C (PLC), which catalyzes phosphoinositide metabolism, will be discussed in detail.

a) Adenylyl Cyclase: Acute Effects of Ethanol

Cyclic AMP is a widely distributed second messenger in nearly all cells in the body, including those in the CNS. It is formed from ATP by the action of the enzyme adenylyl cyclase. In the CNS, a number of neurotransmitter receptors are coupled to adenylyl cyclase, either in a stimulatory or an inhibitory manner, through G proteins (FREISSMUTH et al. 1989; BIRNBAUMER 1990). These proteins are heterotrimeric, consisting of α, β and γ subunits. In general, the α subunit determines the function of the G protein, and these are a number of different forms of this subunit. α_s is the subunit associated with the stimulatory G protein (G_s), while α_i is associated with the inhibitory G protein (G_i). There are four theoretical forms of α_s that can be formed by alternative splicing, although, in general, only two forms are detectable, and these two forms have differential distribution in the brain. However, all forms of α_s appear to have similar activities. In contrast, there are at least three forms of α_i, which are structurally related products of different genes, and it is believed that α_{i2} is involved in receptor-mediated inhibition of adenylyl cyclase activity, while the other forms of α_i subserve different functions that may not influence adenylyl cyclase activity (FREISSMUTH et al. 1989). A protein called α_z has also been described that may mediate hormonal inhibition of adenylyl cyclase activity (WONG et al. 1992). There are also other G proteins that are not associated with adenylyl cyclase; G_o is the most abundant G protein in brain, and may affect ion channel function (DOLPHIN 1990), while G_q couples receptors to phospholipase C activity (see below). In addition to the heterogeneity of the G protein α subunits, four forms of β subunits and five forms of γ subunits have been described in mammalian systems (KLEUSS et al. 1993; LEE et al. 1992).

In systems in which receptor activation stimulates adenylyl cyclase activity, agonist interaction with the receptor promotes the formation of an agonist-receptor-G_s, complex, in which GDP is bound to G_s. This complex has high affinity for the agonist, and its formation enhances the rate of displacement of GDP from the G protein by GTP. When GTP binds to G_s, the protein dissociates into the activated α subunit (with GTP bound) and the $\beta\gamma$ subunit, and the α subunit activates adenylyl cyclase. The receptor also dissociates from the complex and reverts to a form with low affinity for agonist. The action of α_s is terminated by hydrolysis of GTP by an intrinsic GTPase and the cycle can then be repeated. This GTPase can be inhibited by cholera toxin, which causes ADP-ribosylation and permanent activation of α_s (FREISSMUTH et al. 1989).

Although this mechanism of adenylyl cyclase activation is generally accepted, the identification of different forms of adenylyl cyclase has complicated the picture. At present, six forms of adenylyl cyclase have been cloned, and there are partial sequences for three more forms (TANG and GILMAN 1992; KRUPINSKI et al. 1992). While all forms of adenylyl cyclase can be stimulated through G_s, there are forms that can also be stimulated by Ca^{2+}/calmodulin in the absence of G_s and forms that can by stimulated by G protein $\beta\gamma$ subunits in the presence of α_s (TANG and GILMAN 1992; TAUSSIG et al. 1993a). The mechanism of inhibition of adenylyl cyclase by G_i also seems to depend on the form of adenylyl cyclase which is expressed in particular cells. While some forms of recombinant adenylyl cyclases can be inhibited by activated α_i which is coexpressed with the adenylyl cyclase (TAUSSIG et al. 1993b), there is also evidence that certain forms of adenylyl cyclase can be inhibited by $\beta\gamma$ subunits (TANG and GILMAN 1991).

In most instances, the acute addition of ethanol to cells or to cell membranes results in the activation of adenylyl cyclase (see HOFFMAN and TABAKOFF 1990 for review). This effect has been shown in both peripheral tissue and in brain and in cultured neuronal cells (HOFFMAN and TABAKOFF 1990). There is some evidence, however, that the cellular milieu can alter the magnitude, and even the direction, of the response of adenylyl cyclase to ethanol. RABE et al. (1990) reported that the response to ethanol of two subclones of the PC12 cell line was diametrically opposed when ethanol's effects were measured with whole cells, while only stimulation was evident when ethanol was added to membranes prepared from these subclones. Most of this work was carried out before the identification of the different forms of adenylyl cyclase and the demonstrations that posttranslational modifications of adenylyl cyclase can alter this enzyme's responsiveness to stimulatory agents (LUSTIG et al. 1993). There is currently no information as to how intracellular regulatory mechanisms may control the sensitivity of various forms of adenylyl cyclase to ethanol. It is relatively clear, however, that the stimulatory effect of ethanol on adenylyl cyclase depends on the presence of G proteins (LUTHIN and TABAKOFF 1984; BODE and MOLINOFF 1988a). In cell-free preparations of mouse striatum, ethanol in vitro enhanced basal and dopamine-stimulated adenylyl cyclase activity in the presence of guanine nucleotides, but had little effect on adenylyl cyclase activity in their absence (LUTHIN and TABAKOFF 1984). Similarly, in cerebral cortical membranes from mice, ethanol increased isoproterenol-and guanine nucleotide-stimulated adenylyl cyclase activity to a greater extent than basal activity (measured in the absence of guanine nucleotides) (SAITO et al. 1985). The hypothesis that G proteins are involved in the stimulatory effect of ethanol was further supported by the finding that low concentrations of ethanol decreased the affinity for agonist of the β-adrenergic receptor in cerebral cortical membranes, but had no effect on antagonist binding (VALVERIUS et at. 1987). As discussed above, high affinity agonist binding reflects the formation of the complex of receptor and G_s (BIRNBAUMER 1990).

Studies of cultured cells have supplemented the findings in brain tissue regarding the role of G_s in the action of ethanol. In membrane preparations from wild-type S49 lymphoma cells and in membranes from the UNC mutant, in which the β-adrenergic receptors are not coupled to G_s and adenylyl cyclase, ethanol increased guanine nucleotide-and fluoride ion (which acts through G_s)-stimulated adenylyl cyclase activity, but little effect was seen in membranes prepared from cyc^- S49 cells, which have no α_s (RABIN and MOLINOFF 1983; BODE and MOLINOFF 1988a). Ethanol also increased agonist-stimulated adenylyl cyclase activity in N1E-115, NG108–15 and rat pineal cells (GORDON et al. 1986; STENSTROM and RICHELSON 1982; CHIK et al. 1987; CHUNG et al. 1989). The rat pineal cells provide another example of an apparent influence of intracellular regulatory mechanisms on the action of ethanol. Ethanol enhanced the cyclic AMP response in pineal cells when vasoactive intestinal peptide (VIP) was used to activate adenylyl cyclase. In the absence of ethanol, the cyclic AMP response of pineal cells to isoproterenol can be greatly enhanced when VIP is added concurrently (synergistic action) (CHIK et al. 1987). Ethanol added to pineal cells together with isoproterenol and VIP reduced the cyclic AMP response generated by these two agonists (CHIK et al. 1987). Thus, the effect of ethanol in the pineal cell seems to depend on the state of activation of adenylyl cyclase and the intracellular mechanisms by which VIP and isoproterenol receptors interact to activate adenylyl cyclase. The inhibition by ethanol of adenylyl cyclase activity in the presence of VIP and isoproterenol has recently been postulated to result from interference by ethanol with a Ca^{2+}-depenent event that mediates the synergistic effect seen with concurrent agonist administration (CHIK and HO 1991). Therefore, the overall effect of ethanol on cyclic AMP levels in any given cell may depend on a number of inter-acting factors.

The mechanism by which ethanol enhances G protein-activated adenylyl cyclase activity in cell-free preparations is only partially understood. Studies in striatal tissue suggested that ethanol affected the rate of activation (dis-sociation) of the subunits of G_s and the interaction of activated α_s with the catalytic unit of adenylyl cyclase (LUTHIN and TABAKOFF 1984). In cerebral cortical membranes (SAITO et al. 1985), ethanol also increased the rate of activation of G_s by guanine nucleotides (similar to the actions of stimulatory agonists).

There is little evidence that ethanol, acutely, alters inhibitory regulation of adenylyl cyclase activity by G_i. In striatal membranes, ethanol did not alter the inhibition of adenylyl cyclase either by acetylcholine or by opiates (HOFFMAN and TABAKOFF 1986; RABIN 1985). In S49 cells, although ethanol was found to inhibit forskolin-stimulated adenylyl cyclase activity, this inhibition was not blocked by treatment of the cells with pertussis toxin, which blocks the activity of G_i (BODE and MOLINOFF 1988a). Therefore, ethanol did not seem to inhibit the forskolin-activated enzyme activity though and interaction with G_i (BODE and MOLINOFF 1988a). However, ethanol was reported to reduce the inhibition of cerebral cortical adenylyl cyclase achieved by activation of adenosine (A_1) receptors, and this effect of ethanol was attributed to an interaction of ethanol with G_i (BAUCHÉ et al. 1987). Due to the apparent complexity of the mechanisms of adenylyl cyclase inhibition (TAUSSIG et al. 1993b), it may be worthwhile to further explore the effects of ethanol on this pathway.

b) Adenylyl Cyclase: Chronic Effects of Ethanol

Chronic exposure of animals or cells in culture to ethanol results in a decreased responsiveness of adenylyl cyclase to stimultion by agonists, similar to that seen during heterologous desensitization produced by chronic agonist exposure. Agonist-induced desensitization has been suggested to result from "uncoupling" of G proteins from adenylyl cyclase (HARDEN 1983), from receptor down-regulation (LEFKOWITZ et al. 1990), and from changes in G protein levels (MILLIGAN and GREEN 1991), but the mechanism of the changes produced by ethanol seems to be even more complex and is still under investigation.

Some of the earliest studies were carried out using brain tissue from mice and rats that were treated chronically with ethanol and showed that, in striatal tissue, dopamine-stimulated adenylyl cyclase actitivity and the ability of GTP to modify agonist binding were reduced (TABAKOFF and HOFFMAN 1979; SAFFEY et al. 1988; LUCCHI et al. 1983). Early studies in rats also showed a decrease in NE-stimulated adenylyl cyclase activity in the cerebral cortex that was paralleled by changes in antagonist binding to β-adrenergic receptors (FRENCH et al. 1975; BANERJEE et al. 1978). However, more detailed investigations in mice revealed that chronic ethanol ingestion resulted in decreased stimulation of adenylyl cyclase activity not only by isoproterenol (SAITO et al. 1987), but also by guanine nucleotides (SAITO et al. 1987), VIP and forskolin (HOFFMAN and TABAKOFF 1990). Ligand binding studies showed a loss of the high affinity β-adrenergic agonist binding site in the cerebral cortex of ethanol-fed mice, with no change in the number of antagonist binding sites (total receptor number) (VALVERIUS et al. 1987). These data suggested a change in the properties of the receptor-G protein (or receptor-G protein-catalytic unit) complex. A similar change in ligand binding was observed in postmortem brain tissue (cerebral cortex) from alcoholics who had measurable blood or urine alcohol levels at the time of death (VALVERIUS et al. 1988). The changes in adenylyl cyclase activity and

β-adrenergic receptor binding properties were also seen in hippocampal tissue of ethanol-fed mice, but not in tissue taken from the cerebellar cortex (VALVERIUS et al. 1989a). There was also no change in the cerebellum of rats treated chronically with ethanol (RABIN et al. 1987). A study of forskolin binding in brains of ethanol-fed mice revealed decreased basal and guanine nucleotide-enhanced binding of this compound (which is believed to interact both with the catalytic unit of adenylyl cyclase and with the α_s-catalytic unit complex) (SEAMON and DALY 1986) in the same brain areas in which decreased adenylyl cyclase stimulation was observed (VALVERIUS et al. 1989b). In contrast to earlier results, which were obtained using inbred C57BL mice and Sprague-Dawley rats fed liquid diets containing ethanol, a recent report indicated decreased cerebellar agonist and guanine nucleotide-stimulated adenylyl cyclase activity after chronic ethanol treatment in the lines of mice selected for sensitivity or resistance to the hypnotic effect of ethanol (LS and SS mice) (WAND and LEVINE 1991; WAND et al. 1993). In this study, mice were given daily injections of a relatively low dose of ethanol. Since similar agonists were used to stimulate adenylyl cyclase activity in all studies of cerebellar tissue, the difference in results is most likely attributable to the different animals used and/or the differences in ethanol administration procedure. Nevertheless, the general finding of decreased cerebellar adenylyl cyclase activity after chronic treatment with ethanol is compatible with studies of other areas of the brain. These same authors also reported decreased stimulation of anterior pituitary adenylyl cyclase activity in chronically ethanol-injected LS, but not SS, mice (WAND and LEVINE 1991).

There have also been a number of studies of the effect of chronic ethanol exposure on adenylyl cyclase activity in cultured cells. As in brain tissue, chronic exposure of N1E-115 cells, NG108-15 cells or S49 lymphoma cells to ethanol in vitro (100 mM or 200 mM for several days) resulted in a reduced reponse of adenylyl cyclase to prostaglandin E_1 and to adenosine receptor agonists (GORDON et al. 1986; RICHELSON et al. 1986; NAGY et al. 1989; BODE and MOLINOFF 1988b). Similarly, exposure of primary cultures of cerebellar granule cells to 120 mM ethanol for 6 days resulted in a reduced ability of isoproterenol and an adenosine analog to increase intracellular cyclic AMP content (RABIN 1990a). Although exposure of PC12 cells to 75 or 150 mM ethanol for 4 days was initially reported not to alter the activation of adenylyl cyclase by guanine nucleotide, fluoride ion, an adenosine receptor agonist or Mn^{2+} (RABIN 1988), in later studies a decreased response to the adenosine analog was found, even after exposure to only 25 mM ethanol for 7 days (RABIN 1990b). In the studies of cerebellar granule cells and PC12 cells, the in vitro effect of an acute exposure to ethanol (i.e., stimulation of adenylyl cyclase activity in the presence of agonist) was found to be increased after the cells had been chronically exposed to ethanol (RABIN 1988, 1990a). This change in sensitivity to an acute ethanol challenge, however, depended on the agonist added in the presence of alcohol (RABIN 1990a).

Chronic ethanol exposure does not down-regulate the adenylyl cyclase response to receptor-mediated stimulatory signals in all cell types. Chronic ethanol exposure did not affect adenylyl cyclase activity of embryonic chick myocytes or mouse neuroblastoma N18TG2 cells (one of the precursor cell lines for NG108-15 cells) (Blumenthal et al. 1991; Charness et al. 1988). In cultured hepatocytes, chronic ethanol exposure was reported to result in an increase in agonist-stimulated cyclic AMP production (Nagy and DeSilva 1992), while a desensitization of adenylyl cyclase similar to that seen in brain cells was reported in regenerating liver of chronically ethanol-fed rats (Diehl et al. 1992).

The desensitization of adenylyl cyclase produced by chronic ethanol exposure in brain and in many cultured neuronal cell lines has led to investigations of the mechanism of this change. Since the phenomenon appears to be similar in many respects to agonist-induced heterologous desensitization, which has been suggested to result from changes in the synthesis or amounts of G proteins (Milligan and Green 1991), many investigations have centered on measuring quantities of G proteins in ethanol-treated cells or tissue from ethanol-treated animals. The first such study was carried out in ethanol-treated NG108-15 cells, and an approximately 30% decrease in the quantity of α_s, mRNA for α_s and activity of extracted and reconstituted α_s was found (Mochly-Rosen et al. 1988). There was no change in the amount of G_i in these cells (Mochly-Rosen et al. 1988). A subsequent study reported identical results in this same cell line, i.e., that a 2 day exposure to $200\,mM$ ethanol resulted in a reduction in adenylyl cyclase stimulation by agonists (but not forskolin) and a reduction (42%) in α_s content (Charness et al. 1988). In N1E-115 cells, which also showed a reduced response to agonists after 2 days of ethanol exposure, there was a threefold increase in the amount of α_i, in addition to a reduction in the amount of α_s after longer ethanol exposure (Charness et al. 1988). In PC12 cells, there was a decrease (approximately 50%) in the two forms of α_s formed by alternative splicing, as measured both by western blotting and cholera toxin-induced [^{32}P]ADP-ribosylation (Rabin 1993). There was on change in α_{i1} or α_{i2} in these cells (Rabin 1993). In the anterior pituitary of LS mice, where chronic ethanol exposure was reported to produce a decrease in agonist-stimulated adenylyl cyclase activity, there was a decrease (35%) in the amount of α_s, measured by immunoblotting, with no change in α_i (Wand and Levine 1991). In SS mice, ethanol treatment did not alter pituitary adenylyl cyclase activity or G protein content (Wand and Levine 1991). In contrast, when cerebellar tissue of chronically ethanol-treated SS and LS mice was examined, as discussed above, agonist-stimulated adenylyl cyclase activity was reduced in both lines of mice (Wand et al. 1993). In addition, α_{i1} and α_{i2} were increased two- to fourfold, as measured by pertussis toxin-catalyzed ADP-ribosylation in cerebellar tissue from both lines of mice and by western blot analysis in tissue from SS mice (Wand et al. 1993). The amount of α_s was reduced 25% in cerebellar tissue of LS mice and was unchanged in tissue of SS mice (WAND and LEVINE 1991). There was on change in α_o or in G protein β subunits in these mice (WAND et al. 1993). In C57BL mice, although agonist-stimulated adenylyl cyclase activity in the cerebral cortex, hippocampus and striatum was significantly reduced after chronic ethanol ingestion (TABAKOFF and HOFFMAN 1979; SAITO et al. 1987; VALVERIUS et al. 1989a), there was no significant change in the content of either form of α_s, of α_{i2} or α_{i3}, or the β subunits of the G proteins in any of these brain areas or in cerebellum, as measured by western blot analysis (WHELAN et al. 1991). There was a statistically significant increase in α_o and α_{i1} in striatum of ethanol-fed mice, but these changes were not believed to be associated with altered adenylyl cyclase activity, since these G proteins do not seem to be coupled to adenylyl cyclase (see above). The variability in the changes in G protein subunit levels after chronic ethanol exposure, both in terms of the specific subunit(s) altered and the magnitude of the reported changes, raises the question whether these quantitative changes are in fact the basis for the altered sensitivity of adenylyl cyclase to stimulation by agonists or simply represent an epiphenomenon of chronic exposure of brain or cultured cells to ethanol. Based on stoichiometric estimates of the amounts of G proteins, receptors and adenylyl cyclase in brain and other cells (RANSNÄS and INSEL 1988; TANG and GILMAN 1991, 1992; STERNWEIS and ROBISHAW 1984), it does not seem likely that the amount of G protein would represent a rate-limiting step in stimulation of adenylyl cyclase activity, and therefore a small decrease in the quantity of α_s, for example, should not have a significant effect on enzyme activity. In the study of LS and SS mouse cerebellar tissue, treatment of the tissue with pertussis toxin to block α_i activity was reported to normalize adenylyl cyclase activity in the tissue of the ethanol-treated mice (WAND et al. 1993). Although this effect was interpreted as reducing the amount of available α_i, it is also possible that qualitative changes in G proteins could have been produced by the chronic administration of ethanol, and pertussis toxin may have eliminated the influence of the qualitatively altered G proteins. One posttranslational change that has been recognized to alter adenylyl cyclase activity is phosphorylation of G protein (α_s and α_i) subunits (PYNE et al. 1992; KATADA et al. 1985; LUSTIG et al. 1993). In addition, in a recent study of inhibition of recombinant adenylyl cyclases, α_i subunits had to be myristoylated to produce inhibition (TAUSSIG et al. 1993b). If such posttranslational protein modifications are altered by ethanol treatment, they could affect responses of adenylyl cyclase whether or not there are also quantitative changes in the G protein subunits.

Another entity which may be modified by chronic exposure to ethanol is the catalytic unit of adenylyl cyclase itself. In some instances, Mn^{2+}-stimulated adenylyl cyclase activity has been demonstrated to be reduced in tissue of ethanol-treated animals (WAND and LEVINE 1991), and Mn^{2+} is believed to act directly on the catalytic unit of adenylyl cyclase (Ross et al. 1978). Furthermore, forskolin stimulation, especially at high concentration, is thought to be due to a direct action on the catalytic unit of adenylyl cyclase (SEAMON and DALY 1986), and forskolin stimulation of adenylyl cyclase has also been found to be decreased in the brain of ethanol-treated animals (WAND et al. 1993). In a study using chick hepatocytes, agonist-induced heterologous desensitization was suggested to be mediated, in part, by PKA-induced phosphorylation of the catalytic unit of adenylyl cyclase (PREMONT et al. 1992). Similar mechanisms cannot be excluded in chronic ethanol-induced desensitization in some tissues and cells. The ethanol-induced changes could well be dependent on the form(s) of adenylyl cyclase expressed in various cells and on available cellular mechanisms for modifying adenylyl cyclase activity in response to chronic perturbation of this system by ethanol.

The decreased stimulation of adenylyl cyclase observed in cells treated chronically with ethanol has been suggested to represent an adaptive response to the initial direct actions of ethanol on adenylyl cyclase activity (GORDON et al. 1986). However, several studies have noted that the chronic effects of ethanol on adenylyl cyclase acivity may be mediated by ethanol's actions on release and uptake of neurotransmitters. In NG108-15 cells and S49 lymphoma cells, the effects of ethanol on adenylyl cyclase activity were attributed to an ethanol-induced extracellular accumulation of adenosine (NAGY et al. 1989). Coincubation of the cells with ethanol and adenosine deaminase blocked both the acute and chronic effects of ethanol on adenylyl cyclase activity (NAGY et al. 1989). It was proposed that ethanol caused the accumulation of extracellular adenosine by inhibiting adenosine transport into the cells (NAGY et al. 1989, 1990). In S49 cells, it was reported that the inhibition of adenosine transport by ethanol depended on activation of PKA; ethanol did not inhibit adenosine transport in S49 cell mutants that lacked PKA or could not produce cyclic AMP (NAGY et al. 1991). These results would imply that PKA activity is also required for chronic ethanol-induced desensitization of adenylyl cyclase in the S49 and NG108-15 cells (NAGY et al. 1989). A recent report has challenged these earlier results and conclusions, at least for NG108-15 cells. In this recent study, chronic exposure of NG108-15 cells to ethanol ($200\,\mathrm{m}M$ for 2 days) was found to reduce the amounts of α_s and α_i and increase α_0, but not to alter mRNA for α_s or α_{i2} (WILLIAMS et al. 1993). This ethanol treatment resulted in *increased* agonist-stimulated cyclic AMP accumulation, and neither the changes in G protein subunits nor the change in enzyme activity were reversed by adenosine deaminase treatment (WILLIAMS et al. 1993). It is of interest that, in this study, cells were grown in medium containing fetal calf serum, while in the earlier experiments defined medium was used. It is possible that factors present in serum influenced the differential results in the two sets of studies, and, if so, results obtained with cell cultures should be extrapolated to the in vivo situation in animals with significant caution.

The importance of PKA in chronic ethanol-induced decreases in agonist responsiveness of adenylyl cyclase has been further investigated. It has been reported that chronic ethanol treatment does not alter either cyclic AMP production or G protein levels in A126-1B2-1 cells, a mutant strain of PC12 cells which lacks PKA (RABIN 1993; RABIN et al. 1992). These results were interpreted to mean that PKA was necessary for ethanol-induced desensitization, a conclusion that has also been reached for certain forms of agonist-induced heterologous desensitization (CLARK et al. 1988). One must be cautious in making generalizations regarding the mechanisms of ethanol's actions, given the variability of ethanol's effects in different cell lines mentioned above.

c) Adenylyl Cyclase as a Marker for Predisposition to Alcoholism

The findings of altered sensitivity of adenylyl cyclase to stimulation by agonists in brains of animals treated chronically with ethanol suggested that this receptor-effector coupling system might serve as a physiologic marker of ethanol intake in humans ("state marker"). In an initial study, platelet adenylyl cyclase activity was measured in a large population of alcoholics and age-matched control subjects (TABAKOFF et al. 1988b). Prostaglandin E_1, fluoride ion- and guanine nucleotide-induced stimulation of platelet membrane adenylyl cyclase were all significantly lower in alcoholic subjects than in controls (TABAKOFF et al. 1988b). The study controlled for the effects of age, race, smoking, and drug use, and the results have since been repeated in smaller populations from other locations (TABAKOFF and HOFFMAN 1989; LEX et al. 1993). An interesting feature of the original investigation (TABAKOFF et al. 1988b) was the finding that adenylyl cyclase activity was low even in alcoholics who had reportedly abstained from alcohol for up to 4 years. These data suggested that the low responsiveness of adenylyl cyclase in the alcoholic individuals might represent an inherent characteristic ("trait marker"), which could be a marker of a genetic predisposition to alcoholism rather than being a result of chronic ethanol intake. It has recently been found that fluoride-stimulated adenylyl cyclase activity is transmitted as a single major gene in families (DEVOR et al. 1991), which is compatible with the idea of a genetic relationship between this enzyme activity and alcoholism. However, as for the A1 allele of the D2 dopamine receptor described above, low platelet adenylyl cyclase activity was found to be associated with, but not linked to, alcoholism (DEVOR et al. 1991). It is important to note that different "subtypes" of alcoholism have been described (CLONINGER 1987; VON KNORRING et al. 1985), but the studies of alcoholic families did not distinguish between subtypes of alcoholics (DEVOR 1991). It may be important in the future to assess adenylyl cyclase activity according to subtypes of alcoholism.

Low adenylyl cyclase activity has also been reported in lymphocytes of alcoholics (DIAMOND et al. 1987), and when these lymphocytes were kept in culture for four to six generations, the lymphocytes that originally came from the alcoholics were found to be more sensitive to chronic effects of ethanol than those of controls (Nagy et al. 1988). These findings are also consistent with an inherent difference in lymphocyte adenylyl cyclase in alcoholics as compared to control subjects. In a more recent report, adenylyl cyclase activity was measured in lymphocyte membranes from control subjects, "active" alcoholics and abstinent alcoholics (these individuals had been in an inpatient alcoholism treatment program and had been abstinent for approximately 3 weeks) (Waltman et al. 1993). In this analysis, the lymphocytes from abstinent alcoholics showed decreased basal, prostaglandin E_1- and forskolin-stimulated adenylyl cyclase activities, threefold higher protein and mRNA levels for α_{i2} and threefold *higher* mRNA levels for α_s than lymphocytes from control subjects (Waltman et al. 1993). Active alcoholics did not display differences in adenylyl cyclase activity compared to controls, but also had a threefold higher lever of α_s mRNA than controls. There was no difference among groups in protein levels of α_s. The authors (Waltman et al. 1993) concluded that the increased level of α_{i2} protein could account for the decreased adenylyl cyclase activity in the cells of the abstinent alcoholics. Low adenylyl cyclase activity has also been noted in platelets of alcoholic (see above), but in platelet membranes derived from abstinent alcoholics, levels of α_s protein do not correlate significantly with adenylyl cyclase activity (Tabakoff et al. 1990), again suggesting a possible change in α_i or adenylyl cyclase itself.

d) Phosphoinositide Metabolism

Another major signal transduction system in brain and other cells is receptor-activated PLC. This enzyme catalyzes the hydrolysis of inositol phospholipids, leading to the production of inositol (1,4,5) trisphosphate, which releases Ca^{2+} from intracellular stores, and diacylglycerol, which activates PKC (Berridge 1993). There are a number of isoforms of PLC, which have been classified into three families (β, γ and δ) (Berridge 1993; Rhee and Choi 1992). Neurotransmitter receptors are coupled to PLC through a class of G proteins called G_q, which are insensitive to pertussis toxin (Rhee and Choi 1992). This class of G proteins specifically interacts with members of the PLC-β family (Berridge 1993; Rhee and Choi 1992), and receptors which affect phosphoinositide metabolism via this pathway include those for bradykinin, angiotensin, histamine, vasopressin (V_1), acetylcholine (muscarinic M_1, M_3 and M_5), NE (α_{1b}) and serotonin (5-HT$_{1c}$) (Berridge 1993; Rhee and Choi 1992). In some cells, receptor-mediated activation of phosphoinositide metabolism is inhibited by pertussis toxin, and it has been suggested that, in these cases, G protein $\beta\gamma$ subunits, which have also been shown to activate the β-isoforms of PLC (Park et al. 1993), may be involved. Since relatively high concentrations of $\beta\gamma$ subunits were required for phospholipase activation, it was proposed that this pathway would be activated through receptors that induce the dissociation of G_0 or G_i, which are present at high concentrations in brain (Tang and Gilman 1991, 1992; Sternweis and Robishaw 1984).

In vitro studies of the effect of ethanol on receptor-stimulated phosphoinositide metabolism in brain slices of adult animals indicated that nonanesthetic concentrations of ethanol had little effect on this system. In preparations from various brain areas, inhibitory effects on NE- or carbachol (acting at muscarinic receptors)-stimulated phosphoinositide metabolism were observed only at ethanol concentrations of $300-500\,mM$ (over 1%) (HOFFMAN et al. 1986; GONZALES et al. 1986). However, there appears to be an age-related effect of ethanol, since it was reported that $50-75\,mM$ ethanol inhibited carbachol-stimulated phosphoinositide metabolism in brain slices of 7 day old rats, but not in older animals (BALDUINI et al. 1991; CANDURA et al. 1991). Phosphoinositide metabolism has also been studied in vivo, after radioactively labeling the endogenous inositol phospholipids. After treatment of mice with hypnotic/anesthetic doses of ethanol ($4-8\,g/$ kg), there was a decrease in the levels of labeled inositol phosphates in cerebral cortex and cerebellum, consistent with an inhibitory effect of high concentrations of ethanol on phosphoinositide metabolism (LIN et al. 1993c).

Neuronal cells in culture appear to be somewhat more sensitive to the effects of ethanol on PLC than brains of adult animals. Bradykinin- and neurotensin-stimulated phosphoinositide metabolism were inhibited over a concentration range of $50-200\,mM$ ethanol, although ethanol did not affect the response to a guanine nucleotide analog, which presumably activates G_q directly (SMITH 1990). Since ethanol did not affect neurotensin binding to its receptor either in N1E-115 cells or mouse brain (SMITH 1990; ERWIN and KORTE 1988), the effect of ethanol was suggested to occur at the level of receptor-G protein coupling (SMITH 1990). The effect of ethanol on the PLC signaling system has also been studied in some nonneuronal cells. At concentrations of 50 and $100\,mM$, ethanol inhibited the activation of PLC by thrombin in rabbit platelets (RAND et al. 1988), while higher concentrations of ethanol ($200-400\,mM$) increased PLC activity in human platelets and rat hepatocytes (RUBIN and HOEK 1988; HOEK et al. 1987).

The chronic effect of ethanol on receptor-coupled phosphoinositide metabolism has also been found to be age-dependent. Young adult mice were fed ethanol in a liquid diet for 1 week, such that they were tolerant to and physically dependent on ethanol (HOFFMAN et al. 1986), and phosphoinositide metabolism was measured in vitro in brain slices taken from these animals. There was no change in the response to NE in any brain area. In cerebral cortex, the potency of carbachol to stimulate phosphoinositide turnover was increased, and this change was apparently a result of an increase in the number of binding sites for this agonist (HOFFMAN et al. 1986). Other studies have used much longer periods of ethanol exposure. For example, young adult mice given ethanol in liquid diet for 2 months or by gavage for 3 weeks were found to have an increased level of incorporation of labeled phosphate into inositol phosphates in cerebral cortex and hippocampus (SUN et al. 1993). This demonstration was taken to indicate a chronic ethanol-induced increase in phosphoinositide lipid turnover.

However, incorporation of radioactively labeled phosphate into inositol phosphates was decreased after ethanol treatment of older mice (Sun et al. 1993).

The chronic effect of ethanol has also been examined in cultured neuronal cells. In N1E-115 cells exposed chronically to ethanol (100 mM for 7 days), the stimulation of phosphoinositide metabolism by bradykinin, but not by neurotensin or a guanine nucleotide analog, was reduced (Smith 1991). Since bradykinin binding to its receptor was not altered, the ethanol-induced changes were ascribed to changes in bradykinin receptor-G protein coupling (Smith 1991). In NG108-15 cells exposed chronically to ethanol (100 mM for 4 days), there was also a decreased response to bradykinin, which in this case was accompanied by a decreased response to guanine nucleotide stimulation of PLC (Simonsson et al. 1991). The effect of the guanine nucleotide on bradykinin binding to its receptor was also reduced, and the authors suggested that the chronic effect of ethanol was due to quantitative or qualitative changes in the G protein which couples receptors to PLC (Simonsson et al. 1991). In a later study it was reported that the effect of ethanol was no longer apparent if cells were kept for 7 days in 100 mM ethanol, and the normalization of this system during the longer exposure to ethanol also appeared to involve changes in G protein function (Rodriguez et al. 1992).

There is some evidence that ethanol may affect other components of the inositol lipid second messenger system. For example, ethanol at concentrations below 100 mM potentiated inositol trisphosphate-mediated release of Ca^{2+} from intracellular stores and enhanced inositol trisphosphate-stimulated currents in Xenopus oocytes (Ilyin and Parker 1992). It was suggested that ethanol might lower the concentration-response threshold for exogenous inositol trisphosphate by enhancing the production of the endogenous inositol compounds in these cells (Ilyin and Parker 1992).

There has also been some investigation of the effect of ethanol on PKC activity. Ethanol has no effect (Machu et al. 1992) or, at nonanesthetic concentrations, a very slight inhibitory effect (Slater et al. 1993) on the activity of purified PKC. However, there is indirect evidence, primarily based on the attenuation of ethanol's effects by protein kinase inhibitors, that ethanol can activate PKC in certain cell types (DePetrillo and Swift 1992; Kharbanda et al. 1993; Kiss 1991; Skwish and Shain 1991). Since the primary effect of ethanol on receptor-mediated phosphoinositide metabolism is inhibition, it seems unlikely that this pathways (i.e., generation of diacylglycerol) would mediate ethanol-induced activation of PKC. However, in the presence of ethanol, phospholipase D has been shown to form phosphatidylethanol by transphosphatidylation in rat brain microsomal membranes (Gustavsson and Alling 1987), and phosphatidylethanol has been found in brain and other organs of rats treated with ethanol (Alling et al. 1983). In vitro, phosphatidylethanol can substitute for phosphatidylserine as a PKC activator (Asaoka et al. 1988), and if this same mechanism applies in vitro, it could provide a pathway for PKC activation by ethanol. In the following sections describing the effect of ethanol on receptor-gated ion channels, further implications of ethanol/PKC interactions will be noted.

VI. Ligand-Gated Ion Channels

Neurotransmitters interacting with ligand-gated ion channels alter the conformation of these channels in a way that allows the passage of ions across the cell membrane, resulting in changes in cellular membrane potential and in cellular responses to the neurotransmitter signal. In recent years, it has become apparent that several of the ligand-gated ion channels are quite sensitive to perturbation by ethanol, and a significant amount of work has been done to determine the molecular basis for this sensitivity.

Each of the ligand-gated ion channels consists of several subunits in vivo. On the basis of amino acid sequence homology, the protein subunits of channels responding to acetylcholine (nicotinic cholinergic receptor), serotonin (5-HT$_3$ receptor), GABA (GABA$_A$ receptor) and glycine have been shown to have significant similarities, particularly in their hydrophobic (putative membrane-spanning) segments (UNWIN 1993). The glutamate receptor-coupled ion channels α-amino-3-hydroxy-5-methyl-4-isoxazolepropionate (AMPA/kainate, kainate, NMDA receptors) appear to belong to a separate family, with sequence homology among their own subunits but little similarity to members of the other family (UNWIN 1993). However, even in the absence of a large amount of sequence homology, the two families share many topographical characteristics (UNWIN 1993).

1. GABA$_A$ Receptors

The major inhibitory neurotransmitter in the CNS is GABA, and activation of postsynaptic GABA$_A$ receptors generally produces an inward chloride ion current, resulting in membrane hyperpolarization. As described above, the GABA$_A$ receptor is a member of a family of ligand-gated ion channels, and, as such, consists of several protein subunits (DELOREY and OLSEN 1992; PRITCHETT et al. 1989). The function of the receptor is modulated not only by GABA, but also by benzodiazepines and barbiturates, which potentiate GABA function and which have binding sites on the GABA$_A$ receptor subunits (SIEGHART 1992a). To date, molecular cloning studies have identified five different GABA$_A$ receptor subunits, designated α, β, γ, δ and π (LÜDDENS and WISDEN 1991; SIEGHART 1992b). Each of these subunits has several isoforms, including six α four β and three γ (LÜDDENS and WISDEN 1991; SIEGHART 1992b). The actual in vivo structure(s) of the GABA$_A$ receptor is not yet known. However, there have been a number of in vitro studies in which various combinations of subunits have been expressed, and the physiologic and pharmacologic characteristics of the resulting receptors have been determined. Although homomeric receptors (e.g., those formed by expression of the cDNA for a single receptor subunit in Xenopus oocytes or in cells) can respond to GABA, it is necessary to have α, β and γ subunits present in order to observe the potentiation of GABA responses by benzodiazepines that is seen with native receptors (PRITCHETT et al. 1989). One aspect of $GABA_A$ receptor subunit heterogeneity that may be relevant to the effects of ethanol is that receptors containing the α-6 subunit, found exclusively in the cerebellum, selectively bind the benzodiazepine inverse agonist Ro 15-4513 (LÜDDENS et al. 1990). Ro 15-4513 has been shown to antagonize certain actions of ethanol (SUZDAK et al. 1986a; HOFFMAN et al. 1987; BECKER 1988; NUTT et al. 1988). Receptors containing the α-6 subunit, however, are usually not affected by classical benzodiazepines such as diazepam (DeLOREY and OLSEN 1992).

Posttranslational modifications can also affect $GABA_A$ receptor function. In general, treatment of cell or brain preparations with activators of PKA results in decreased activity of $GABA_A$ receptor MacDonald et al. 1992; Moss et al. 1992a) and, in vitro, PKA-mediated phosphorylation of the β-subunit has been shown to occur (BROWNING et al. 1993; Moss et al. 1992b). However, in some cells (e.g., cerebellar Purkinje cells) PKA activation can increase GABA responses (SESSLER et al. 1989). $GABA_A$ receptor function may also be modulated by PKC, and activation of this kinase appears to decrease $GABA_A$ receptor function (MacDONALD et al. 1992; KELLENBERGER et al. 1992), possibly through phosphorylation of the β or γ subunits (KELLENBERGER et al. 1992; LEIDENHEIMER et al. 1993). These results have been interpreted to indicate that receptor phosphorylation may be a general mechanism for modulating the activity of ligand-gated ion channels.

The effect of ethanol on $GABA_A$ receptor function was originally investigated because of the similar behavioral effects of ethanol, barbiturates and benzodiazepines (e.g., sedative/hypnotic, anxiolytic). Although initial behavioral studies showed that certain actions of ethanol could be increased by GABA agonists and decreased by antagonists (LILJEQUIST and ENGEL 1982; COTT et al. 1976), it was difficult to observe a consistent effect of ethanol on $GABA_A$ receptor function in electrophysiologic studies. These studies have been reviewed elsewhere and have led to some confusion regarding the ability of ethanol to enhance the action of GABA (SHEFNER 1990). However, once the structure of the $GABA_A$ receptor began to be elucidated and both the heterogeneous distribution and the pharmacologic specificity, of $GABA_A$ receptor subunits were noted, it became plausible to suggest that the effects of ethanol varied with the structure of the receptor and therefore varied according to the cells studied. This hypothesis was supported by a patch clamp study of mouse hippocampal and cortical cells in culture, in which ethanol was found to potentiate the effects of GABA only in some cells; however, all cells in this study responded to GABA, barbiturates and benzodiazepines (AGUAYO 1990). A similar result was found (i.e., GABA responses in some, but not all, cells in each brain area were potentiated by ethanol) in a voltage clamp study in which brain regional comparisons were made in a variety of species (REYNOLDS et al. 1992).

In contrast to the electrophysiologic studies, biochemical analysis of $GABA_A$ receptor function has revealed a more consistent effect of ethanol, perhaps because most preparations used are heterogeneous with respect to cell type and would be expected to contain at least some receptors that are sensitive to ethanol. Ethanol at nonanesthetic concentrations (i.e., 5–100 mM), has been reported to enhance GABA-stimulated chloride flux in synaptoneurosome and microsac preparations from certain areas of rat and mouse brain and in spinal neurons in culture (ALLAN and HARRIS 1986; SUZDAK et al. 1986b; MEHTA and TICKU 1988). In some cases, ethanol also enhanced basal chloride flux through the $GABA_A$ receptor-coupled channel (SUZDAK et al. 1986b; MEHTA and TICKU 1988). The effects of ethanol were seen in cerebral cortical and cerebellar preparations from brain, but not in the hippocampus (ALLAN and HARRIS 1986).

The fact that ethanol could potentiate the biochemical effect of GABA supported the hypothesis of a role for the $GABA_A$ receptor-ion channel complex in some of the behavioral effects of ethanol. Further evidence for this hypothesis came from studies with the benzodiazepine partial inverse agonist Ro 15-4513. This compound, as mentioned above, was originally reported to reduce the ataxic effect of ethanol in rats and was touted as a general "alcohol antagonist" (SUZDAK et al. 1986a). However, later studies demonstrated that only certain of ethanol's behavioral effects are reversed by Ro 15-4513 (HOFFMAN et al. 1987; BECKER 1988; NUTT et al. 1988). Moreover, the effects of ethanol and the inverse benzodiazepine agonist may be simply algebraically additive, since ethanol and Ro 15-4513 can have opposite effects on $GABA_A$ receptor function (NUTT et al. 1988; NUTT and LISTER 1987). In addition, other inverse agonists can also reverse some of ethanol's actions (LISTER and DURCAN 1989; PALMER et al. 1988; MARROSU et al. 1988). Recently it has been found that rats that were selectively bred for sensitivity to the incoordinating effects of ethanol (ANT rats) have a mutation in the α-6 subunit of the $GABA_A$ receptor which renders the receptors containing this subunit sensitive to the effects of benzodiazepines (KORPI et al. 1993). This mutation was suggested to underlie the heightened sensitivity of the ANT rats to benzodiazepine-induced impairment of postural reflexes (KORPI et al. 1993). Whether this mutation is in fact related to benzodiazepine sensitivity and may also be related to the greater sensitivity of the ANT rats to ethanol needs further investigation. No data are currently available to indicate that the characteristics of the α-6 subunit of the GABA receptor in some way alter the sensitivity of the $GABA_A$ receptor complex or the whole animal to the effects of ethanol. Certainly, the presence of a biochemical difference between two selected lines of animals implies a relationship between the $GABA_A$ receptor characteristics and the selected behavior (incoordination after ethanol administration), but there are a number of caveats regarding selective breeding experiments (CRABBE et al. 1990) that will have to be addressed before any conclusions can be reached.

Another mechanism of action of ethanol at the GABA$_A$ receptor has been suggested from studies with other selected lines of animals, the LS and SS mice that were bred for sensitivity and resistance, respectively, to the hypnotic effect of ethanol (CRABBE et al. 1990; PHILLIPS et al. 1989). It was found that ethanol enhanced GABA-stimulated chloride ion flux in a cerebellar microsac preparation from LS, but not SS, mice (ALLAN and HARRIS 1986). A similar result has recently been reported in lines of rats bred selectively for sensitivity (HAS) and resistance (LAS) to the hypnotic effect of ethanol (ALLAN et al. 1991). As already discussed, when mRNA from the brains of SS or LS mice was expressed in Xenopus oocytes, the difference in ethanol potentiation of the GABA response persisted (WAFFORD et al. 1990). To determine what characteristic of the GABA$_A$ receptor could account for the differential response to ethanol, antisense studies were carried out in which mRNAs coding for various subunits of the GABA$_A$ receptor were "knocked out" before the mRNA was expressed in oocytes. These studies implicated the alternatively spliced γ-2L form of the γ subunit in the action of ethanol. When this form of the subunit was not present, the GABA response was not enhanced by ethanol (WAFFORD et al. 1991). The γ-2L subunit contains a consensus sequence for phosphorylation by PKC (Moss et al. 1992b), and later studies showed that a point mutation that eliminated the phosphorylation site also eliminated sensitivity to ethanol (WAFFORD and WHITING 1992). Thus, these studies indicate a role for receptor subunit phosphorylation in the potentiating actions of ethanol at the GABA$_A$ receptor. However, quantitative differences in the γ-2 subunits do not appear to mediate the differential action of ethanol on GABA receptors from LS and SS mice, since brains of both lines of mice express the γ-2L subunit (ZAHNISER et al. 1992). Similarly, mRNA for both the γ-2S and γ-2L subunits was reported to be expressed in brain areas containing GABA$_A$ receptors that were either sensitive or insensitive to ethanol (ZAHNISER et al. 1992).

However, if phosphorylation by PKC does influence the response of the GABA$_A$ receptor to ethanol, the generation of sensitivity to ethanol could involve activation of PKC by ethanol (see above) or a change in sub-unit phosphorylation state produced by PKC or other mechanisms (e.g., phosphatase-related). Although ethanol and phosphorylation of certain subunits of the GABA receptor by PKC appear, for the most part, to have opposite effects on GABA$_A$ receptor function (see above), it should be kept in mind that, to date, the importance of the phosphorylation of the γ-2L subunit on GABA$_A$ receptor function has not been elucidated. Phosphorylation of the β-1 subunit of the receptor complex by PKA-mediated mechanisms has been proposed to be involved in GABA$_A$ receptor desensitization (Moss et al. 1992a). It is possible that interactions arising from phosphorylation on more than one subunit might produce distinct effects on receptor function. In one study of rat dorsal root ganglion cells in culture, ethanol was reported to enhance the response to GABA at the nondesensitized channel, but to be inactive when desensitized steady state current was measured (NAKAHIRO et al. 1991). Thus, phosphorylation of the γ-2L subunit might influence the desensitization state of the receptor that arises from phosphorylation of other subunits and, in this way, influence the sensitivity of the complex to ethanol.

There is also some evidence that phosphorylation of the $GABA_A$ receptor by PKA can directly influence the response to ethanol in distinct areas of brain. While, as discussed above, in many cases activation of PKA seems to decrease the response to GABA (increase receptor desensitization), in rat cerebellar Purkinje neurons an opposite effect appears to occur (SESSLER et al. 1989). In these cells, it was also reported that administration of NE or isoproterenol (which would be expected to increase intracellular cyclic AMP levels and activate PKA) potentiated the effects of GABA on cell firing rates (SESSLER et al. 1989). Furthermore, this treatment allowed for further enhancement of GABA responses by ethanol (LIN et al. 1991, 1993a,b). Ethanol has been shown (see above) to enhance agonist-stimulated cyclic AMP production in a number of brain areas, and the effects of ethanol on adenylyl cyclase activity in Purkinje cells may interact with phosphorylation mechanisms which play a role in the potentiating effects of ethanol on $GABA_A$ receptor function.

There is evidence that the $GABA_A$ receptor may respond with an adaptive down-regulation of its activity to the long-term enhancement of function that is expected to occur during chronic ethanol treatment. Ethanol-induced potentiation of GABA-stimulated chloride ion flux was reduced in mice given ethanol in a liquid diet for 7 days and in rats exposed to ethanol vapor for 14 days (ALLAN and HARRIS 1987; MORROW et al. 1988). These reports differed with respect to changes in response to GABA per se, and these differences may be a function of the duration of ethanol exposure and the levels of ethanol achieved in brain. When Xenopus oocytes expressing mouse brain mRNA were exposed in vitro to $50 \, \mathrm{m}M$ ethanol for 4 days, a reduced electrophysiologic response to the $GABA_A$ agonist muscimol was observed (BUCK and HARRIS 1991). The results suggest that the function of the $GABA_A$ receptor-coupled ion channel becomes resistant to the effects of ethanol after chronic ethanol exposure and may also become resistant to the effects of GABA agonists. These changes in function are not easily explained by studies which measured the ligand binding properties of the $GABA_A$ receptor complex in brain. There is little evidence for a generalized decrease in receptor number after chronic ethanol treatment (LILJEQUIST et al. 1986; LILJEQUIST and TABAKOFF 1985; KAROBATH et al. 1980), although in one study a decrease in [^{3}H] flunitrazepam binding was found in several brain areas of mice that had ingested ethanol for 6 weeks (BARNHILL et al. 1991). In addition, in another study of rats that were given

ethanol in the drinking fluid for 3 months, a change in the binding characteristics of t-butylbicyclophosphorothionate (TBPS), which binds to a site in
the ion channel, was reported (HILLMANN et al. 1990). In addition, increased
binding of Ro 15-4513 has been reported in the cerebellum of animals
treated chronically with ethanol, and, in brain tissue from ethanol-fed mice,
the ability of Ro 15-4513 to inhibit muscimol-stimulated chloride ion flux
was increased (MHATER et al. 1988; BUCK and HARRIS 1990).

The mechanism for the changes in response to GABA or ethanol after
chronic ethanol treatment has more recently been suggested to reflect a
change in the subunit composition of the $GABA_A$ receptor. For example,
the same ethanol inhalation paradigm that produced a decreased ability of
$GABA_A$ agonists to stimulate chloride ion flux in microsac preparations of
brain tissue also produced a significant decrease in the level of mRNA for
the α-1 and α-2 subunits of the receptor, with no change in the mRNA
coding for the α-3 subunit (MONTPIED et al. 1991). In WSP mice, there was
also a decrease in the level of brain mRNA for α-1, an increase in mRNA
for γ-3, but no change in mRNA for α-3 or γ-2S after chronic ethanol
exposure (BUCK et al. 1991). In WSR mice, chronic ethanol exposure
resulted in no change in mRNA for α-1, and increase in mRNA for γ-3, and
a decrease in mRNA for α-6 (BUCK et al. 1991). In rats fed ethanol in a
liquid diet for 2 weeks or given ethanol chronically by injection, the level of
mRNA for α-5 was decreased, but mRNA for the α-6 subunit was increased
(MHATER and TICKU 1992; MORROW et al. 1992). This change in expression
of mRNA for the α-6 subunit in rat brain may correspond to the increased
binding and function of Ro 15-4513 previously observed in rats treated
chronically with ethanol (MHATER et al. 1988). Overall, the most consistent
data indicate that chronic ethanol treatment results in a decrease in mRNA
levels for α-1 and no change in α-3 mRNA levels. The relationship of these
changes in mRNA levels to changes in protein expression and the significance
of the possible changes in protein levels to the observed changes in response
to GABA are not known. However, these results provide a hypothesis to
account for ethanol-induced changes in $GABA_A$ receptor function at the
molecular level and emphasize the specificity of ethanol's actions.

There are a number of behavioral effects of ethanol that may be attributed, at least in part, to its acute and chronic actions at the $GABA_A$
receptor. The difference in the response of the GABA receptor to ethanol
in brains of the lines of mice and rats selected for sensitivity or resistance to
the hypnotic effect of ethanol suggests that the $GABA_A$ receptor may
mediate this effect. In addition, there is evidence that the anxiolytic effects
of ethanol, like those of benzodiazepines, are related to the ability of low
concentrations of ethanol to potentiate GABA responses (SUZDAK et al.
1986a,b).

The decreased $GABA_A$ receptor function observed after chronic ethanol
administration could be related to the appearance of ethanol withdrawal
symptoms, which include CNS hyperexcitability, especially if the decreased

GABA$_A$ receptor function occurs in conjunction with increased glutamate receptor function (see below). It has been shown by a number of researchers that administration of drugs that act at GABA$_A$ receptors can alter ethanol withdrawal seizure severity (see TABAKOFF and ROTHSTEIN 1983), but the effects of particular drugs activating at the GABA$_A$ receptor may be selective for certain types of seizures (FRYE et al. 1991), and the ameliorative actions of GABA agonists may be confined to certain brain areas. For example, administration of muscimol into the inferior colliculus, but not the medial septum or substantia nigra, reduced ethanol withdrawal-induced audiogenic seizures in rats (FRYE et al. 1983).

2. Glutamate Receptors

Glutamate is believed to be the major excitatory neurotransmitter in the CNS, and there are two categories of glutamate receptor, ionotropic and metabotropic (SOMMER and SEEBURG 1992; NAKANISHI 1992; SCHOEPP and CONN 1993). The latter receptor type is coupled by G proteins to second messenger systems, while the former represents a family of ligand-gated ion channels. Most work with ethanol to date has focused on the ionotropic glutamate receptors. Pharmacologically, ionotropic glutamate receptors have been divided into a number of subtypes, including the AMPA, kainate, and NMDA receptors, based on the affinity of these receptors for specific agonists and antagonists (COLLINGRIDGE and LESTER 1989). The AMPA and kainate receptors have different characteristics from the NMDA receptor and are often referred to as "non-NMDA" glutamate receptors. These receptors were the first in the family to be cloned, and at least seven non-NMDA receptor subunits have been identified and organized into structural classes (NAKANISHI 1992; see KUTSUWADA et al. 1992). The subunits called GluR 1 – GluR 4 form receptors that respond to both AMPA and kainate, while receptors containing GluR 5 – GluR 7 subunits have high affinity for kainate and lower affinity for AMPA (NAKANISHI 1992). The receptors made up of these latter subunits, as well as proteins called KA-1 and KA-2 (NAKANISHI 1992), therefore seem to have the characteristics expected of a kainate receptor, as distinct from the AMPA/kainate receptors.

Recently, a number of subunits of the NMDA receptor have also been cloned. The first subunit identified in rat was NR1, with the homolog in mouse of ξ (NAKANISHI 1992; MORIYOSHI et al. 1991; YAMAZAKI et al. 1992). These subunits are widely distributed in brain (NAKANISHI 1992), and, when mRNA encoding NR1 or ξ is expressed in Xenopus oocytes, a homomeric receptor is formed which displays many of the characteristics of the NMDA receptor, but which produces small amplitude currents in the presence of agonist (NAKANISHI 1992; MORIYOSHI et al. 1991; YAMAZAKI et al. 1992). A second family of NMDA receptor subunits, designated NR2 in the rat and ε in the mouse (KUTSUWADA et al. 1992; MONYER et al. 1992), has also been characterized. There are four subtypes of this subunit, NR2 A–D; ε 1–4

(Nakanishi 1992; Kutsuwada et al. 1992; Monyer et al. 1992; Ikeda et al. 1992; Ishii et al. 1993), and the distribution of mRNA for several of these subunits is localized to particular brain areas (Nakanishi 1992; Kutsuwada et al. 1992; Monyer et al. 1992). The expression in Xenopus oocytes of NR2 or ε subunit mRNA alone does not produce functional receptors; however, expression studies have shown that the combination of NR1 with various members of the NR2 family results in receptors with characteristics that more nearly resemble those of native receptors (Kutsuwada et al. 1992; Monyer et al. 1992). The various heteromeric receptors have different properties, e.g., affinity for agonists or antagonists; sensitivity to glycine, (see below), which depend on the member of the NR2 or ε family which is expressed along with NR1 or ξ (Kutsuwada et al. 1992; Monyer et al. 1992). It has also been reported that eight splice variants of NR1 exist, adding to the complexity of receptor structure and function (Sugihara et al. 1992; Hollmann et al. 1993). In addition to these NMDA receptor subunits, a "glutamate binding protein" has been cloned, and has been suggested to be one subunit of an NMDA receptor that consists of several proteins (Kumar et al. 1991). There is no homology between the glutamate binding protein and the other NMDA receptor subunits, and it has been postulated that the complex of which this protein is a part represents a second type of NMDA receptor (Kumar et al. 1991).

The pharmacology of the NMDA receptor has been studied in some detail, in part because the NMDA receptor plays a role in a number of critical CNS processes, including long-term potentiation (a model for learning); neuronal development; generation of epileptiform seizure activity and excitotoxicity (see Collingridge and Lester 1989). These functions are dependent on the fact that the NMDA receptor-coupled ion channel is permeable both to Ca^{2+} and to monovalent cations (see Collingridge and Lester 1989). The rate of activation of the NMDA receptor is slow, relative to other glutamate receptors, allowing for summation of responses, and large amounts of Ca^{2+} can enter the cell though this ligand-gated ion channel. Another characteristic of the NMDA receptor that contributes to its effects is the voltage-dependence of receptor activation, i.e., receptor activation is increased with cellular depolarization. For example, synaptic release of glutamate can activate the ionotropic kainate or AMPA/kainate receptors, leading to cellular depolarization, which then facilitates activation of the NMDA receptor. This voltage dependence is mediated by Mg^{2+}, which binds to a site within the ion channel and blocks the channel. Mg^{2+} is released from the channel upon cellular depolarization (see Collingridge and Lester 1989).

The NMDA receptor complex also contains binding sites for a number of "allosteric" modulators of receptor function. Glycine binds to a strychnine-insensitive site on the receptor and is required for activation by glutamate or NMDA (Kleckner and Dingledine 1988; Kemp and Leeson 1993). Thus, glycine and NMDA have been designated co-agonists at the receptor

(KLECKNER and DINGLEDINE 1988; KEMP and LEESON 1993). Within the ion channel, there is a binding site for the dissociative anesthetic phencyclidine (PCP), and the same site binds dizocilpine (MK-801) (see COLLINGRIDGE and LESTER 1989). Both PCP and MK-801 act as uncompetitive antagonists of NMDA receptor function. The complex also contains binding sites for Zn^{2+} (HOLLMANN et al. 1993; PETERS et al. 1987) and for polyamines (SCOTT et al. 1993), although the physiologic role of these modulators is not fully understood.

Recent work suggests that the function of the NMDA receptor, like that of other members of the ligand-gated ion channel family, can be modulated by phosphorylation. The NMDAR receptor subunits 1 and 2 (and the corresponding subunits in the mouse) all contain a number of consensus sequences for phosphorylation by Ca^{2+}-calmodulin-dependent protein kinase and PKC (NAKANISHI 1992; KUTSUWADA et al. 1992; MONYER et al. 1992; IKEDA et al. 1992). It has been reported that direct (phorbol ester) or indirect (through agonists that increase polyphosphoinositide metabolism and diacylglycerol production) stimulation of PKC can increase electrophysiologic responses to NMDA in trigeminal neurons (CHEN and HUANG 1991) and in Xenopus oocytes expressing whole brain mRNA or mRNAs coding for particular NMDA receptor subunits (KUTSUWADA et al. 1992; URUSHIHARA et al. 1992). The effect of PKC stimulation appears to vary depending on the subunit composition of the receptor, however. While NMDA activation of expressed heteromeric $\zeta/\varepsilon2$ receptors was increased by phorbol ester treatment, activation of the $\zeta/\varepsilon3$ receptors was not (KUTSUWADA et al. 1992). We and others have recently found that phorbol ester treatment of cerebellar granule cells, which express the NR2C subunit of the NMDA receptor (MONYER et al. 1992), decreases the response to NMDA (SNELL et al. 1994a; COURTNEY and NICHOLLS 1992). It has also been reported that phorbol esters decrease the electrophysiologic response to NMDA in rat CA1 hippocampal neurons (MAKRAM and SEGAL 1992).

The fact that the NMDA receptor plays a key role in learning and memory, epileptiform seizure activity and neurotoxicity suggested that it might be involved in some of the acute and chronic effects of ethanol, including cognitive effects, withdrawal seizures and neuronal degeneration. As a result, most investigations of the effect of ethanol on glutamate receptors have focused on the NMDA receptor. The acute inhibitory effect of ethanol on NMDA receptor function has been well documented both in biochemical and in electrophysiologic studies (TABAKOFF et al 1991; ARCAVA et al. 1991; WEIGHT et al. 1991). Ethanol, at concentrations that are readily reached in vivo, was a potent inhibitor of NMDA-stimulated calcium influx and cyclic GMP production in primary cultures of cerebellar granule cells (HOFFMAN et al. 1989) and of NMDA-stimulated neurotransmitter release in brain slice preparations (GÖTHERT and FINK 1989; GONZALES and WOODWARD 1990; WOODWARD and GONZALES 1990). This inhibitory effect of ethanol was also found in whole cell patch clamp studies of dissociated embryonic hippocampal cells in culture (LOVINGER et al. 1989; LIMA-LANDMAN and ALBUQUERQUE 1989). The initial studies have been followed by a number of biochemical and electrophysiologic investigations that have been consistent in demonstrating an inhibitory effect of ethanol on NMDA receptor function in cultured embryonic brain cells, in neurons and brain slices from adult animals, and in Xenopus oocytes expressing brain mRNA (DILDY and LESLIE 1989; LOVINGER et al. 1990; DILDY-MAYFIELD and HARRIS 1992a; WHITE et al. 1990a).

There have also been studies in which ethanol, administered systemically to rats, inhibited the activation of certain medial septal neurons by ionto-phoretically applied NMDA (SIMSON et al. 1991), inhibited the excitation of hippocampal neurons by NMDA (SIMSON et al. 1993), and inhibited activation of locus coeruleus neurons by NMDA (ENGBERG and HAJÓS 1992). These studies demonstrate that ethanol can inhibit NMDA receptor function in vivo and in vitro, a conclusion that can also be reached based on discriminative stimulus studies which show that the NMDA receptor antagonist dizocilpine substitutes for ethanol in animals trained to discriminate ethanol (GRANT et al. 1991; GRANT and COLOMBO 1992).

Although virtually all reports indicate that ethanol inhibits NMDA receptor function, differential sensitivity to ethanol has been shown in receptors comprising various splice variants of NR1 expressed in Xenopus oocytes (KOLTCHINE et al. 1993). It is likely that the subunit composition of the NMDA receptor, as well as, or in addition to, the splice variants, influences sensitivity to ethanol. For example, it was recently found that systemically administered ethanol inhibited NMDA-invoked neuronal activity in the inferior colliculus and the hippocampus, but not in the lateral septum (SIMSON et al. 1993). This brain regional sensitivity to ethanol may arise from differences in the characteristics of the NMDA receptors in various brain areas and may contribute to the specificity of behavioral effects of ethanol.

Experiments to determine the mechanism and/or site of action of ethanol at the NMDA receptor have focused on interactions with other modulators of receptor function. In most instances, ethanol inhibition does not appear to the "competitive" with NMDA (DILDY-MAYFIELD and LESLIE 1991; RABE and TABAKOFF 1990), although in a few cases, e.g., NMDA-stimulated cyclic GMP production in cerebellar granule cells (HOFFMAN et al. 1989), increasing the NMDA concentration was reported to reduce the effect of ethanol. The ethanol effects were simply additive with the PCP inhibition of NMDA receptor function in cerebellar granule cells (HOFFMAN et al. 1989), although in dissociated brain cells from neonatal rats a more complex interaction of ethanol and dizocilpine was observed (DILDY-MAYFIELD and LESLIE 1991). It may be important to note, however, that at high concentrations dizocilpine can also affect the nicotinic acetylcholine receptor (RAMOA et al. 1990), and such interactions could generate a confound. There is little or no interaction of ethanol with Mg^{2+} effects on NMDA receptor function (RABE and TABAKOFF 1990; MORRISETT et al. 1991).

In certain cells, ethanol does appear to interfere with the action of the co-agonist glycine. In cerebellar granule cells and in dissociated embryonic brain cells, a high concentration of glycine reversed the inhibitory effect of ethanol on NMDA receptor function (RABE and TABAKOFF 1990). Furthermore, in adult mouse cerebral cortical and hippocampal tissue, ethanol was found to inhibit nonequilibrium dizocilpine binding, and this inhibition was reversed by glycine (SNELL et al. 1993). Glycine also reversed the ability of ethanol to inhibit NMDA-stimulated dopamine release in striatal slices (WOODWARD and GONZALES 1990). This reversal by glycine, however, was not seen in cerebral cortical slices in which NMDA-stimulated NE release was measured (GONZALES and WOODWARD 1990). Similarly, in electrophysiologic studies of hippocampal neurons, ethanol inhibition of NMDA receptor function was reported not to be reversed by glycine (WEIGHT et al. 1991; PEOPLES and WEIGHT 1992). It is probable that brain regional NMDA receptor subunit composition influences the interaction of ethanol and glycine. In cerebellar granule cells, ethanol increases the EC_{50} for glycine's effect on the NMDA response (SNELL et al. 1994b); thus, ethanol inhibits the response to NMDA of cerebellar granule cells when the glycine concentration is low, but not when glycine concentrations exceed its EC_{50} value in the presence of ethanol. As mentioned above, cerebellar granule cells have been found to express mRNA for the ξ and $\varepsilon3$ (NR1 and NR2C subunits) (NAKANISHI 1992; KUTSUWADA et al. 1992). In a recent preliminary study in which heteromeric receptors were expressed in Xenopus oocytes, ethanol inhibited NMDA responses in the presence of both high and low concentrations of glycine when the ξ and $\varepsilon1$ or $\varepsilon2$ subunits were expressed. However, when ξ and $\varepsilon3$ were expressed, ethanol inhibition of responses was reversed in the presence of $10\,\mu M$ glycine (BULLER et al. 1993).

The effect of ethanol on glycine potentiation of NMDA responses does not appear to result from a direct interaction of ethanol with the glycine binding site, as determined in equilibrium ligand binding studies using mouse brain tissue (SNELL et al. 1993). However, the activation of PKC (by phorbol esters) in cerebellar granule cells was found to interfere with the potentiation by glycine of the NMDA response in a manner analogous to the results found with these cells and ethanol (SNELL et al. 1994a,b). SNELL et al. (1994b) reported that the effect of ethanol, like that of phorbol ester activators of PKC, can be reversed by inhibitors of PKC. These data suggest that, in cerebellar granule cells, PKC may be involved in the inhibitory effect of ethanol on NMDA receptor function, including the ethanol-induced change in the EC_{50} for glycine. It is clear that there is a unique type of NMDA receptor in the cerebellum, which may be a function of the expression of $\varepsilon3$ (NR2C) subunits in the granule cells (STERN et al. 1992). Therefore, it may not be possible to extrapolate the PKC mechanism of action of ethanol to all cell types. However, especially in light of the potential role of phosphorylation in the actions of ethanol at the $GABA_A$ receptor (see above), it will be of interest to investigate ethanol/protein kinase interactions with respect to NMDA receptor function in a number of brain areas.

The inhibition of NMDA receptor function by acute exposure to ethanol may result in ethanol-induced cognitive deficits. This postulate is supported by studies showing that ethanol, at low concentrations, inhibits long-term potentiation (LTP) in the hippocampus (Blitzer et al. 1990). There is a great deal of evidence implicating NMDA receptor function in the induction of LTP, which is a model for the synaptic plasticity believed to be involved in learning and memory (Ben-Ari et al. 1992). In one study, the dose-dependence of the effect of ethanol on LTP was found to correlate well with the inhibition by ethanol of responses to NMDA (Blitzer et al. 1990).

The NMDA receptor may also play a role in the effects of ethanol during CNS development. The NMDA receptor has been particularly implicated in synaptic plasticity in the developing visual cortex (Kleinschmidt et al. 1987). Furthermore, the finding of developmental changes in sensitivity to NMDA in the cerebellum suggests that NMDA receptors play a role in developmental plasticity in this area (Watanabe et al. 1992). Activity of the NMDA receptor has also been shown to regulate the migration of cerebellar granule cells in slice preparations from 10 day old mice (Komuro and Rakic 1993). Interference by ethanol with these functions of the NMDA receptor could disrupt the establishment of neuronal connections during development, perhaps contributing to the symptoms of the fetal alcohol syndrome (FAS), a pattern of congenital malformations and mental retardation seen in children of alcoholic mothers (Clarren and Smith 1978). Although there has been little investigation of the possibility that ethanol inhibition of NMDA receptor function contributes to FAS, it has been shown that the electrophysiologic response to NMDA is selectively reduced, as is NMDA-sensitive glutamate binding, in hippocampal tissue obtained from the adult offspring of rats that were treated with ethanol during gestation (Morrisett et al. 1989; Savage et al. 1991). These deficits could contribute to the behavioral and cognitive abnormalities associated with FAS.

Chronic ethanol treatment of adult animals leads to and increase in NMDA receptors in brain. It was originally reported that glutamate binding was increased in brains of adult rats exposed chronically to ethanol (Michaelis et al. 1978). More recently, an increase in dizocilpine binding was found in brains of mice that had been fed ethanol in a liquid diet and were tolerant to and physically dependent on ethanol (Grant et al. 1990; Gulya et al. 1991). There was also an increase in NMDA-sensitive glutamate binding in hippocampal tissue of these mice, but no increase in glycine binding or binding of the competitive NMDA receptor antagonist, CGS-19755 (Snell et al. 1993). In a post-mortem study, glutamate binding was also reported to be increased in hippocampal synaptic membranes from human alcoholics (Michaelis et al. 1990). The finding that there are changes in ligand binding to one of the co-agonist recognition sites and to the ion channel associated with the NMDA receptor, but not to the other co-agonist binding site (the glycine binding site), suggests the possibility that the various compounds may bind to sites on different NMDA receptor subunits and that there may be changes in subunit composition following chronic ethanol exposure, similar to subunit expression changes reported for the $GABA_A$ receptor (see above).

In addition to the changes in ligand binding to the NMDA receptor after chronic ethanol treatment, there is evidence for increased receptor function produced by chronic ethanol exposure. When primary cultures of cerebellar granule cells were exposed to ethanol ($100 \, mM$ for 2 or more days; $20 \, mM$ for 3 or more days), NMDA-induced increases in intracellular calcium, measured with fura-2, were enhanced (IORIO et al. 1992). This change appeared to reflect an increase in receptor number rather than a change in NMDA receptor characteristics (IORIO et al. 1992). In contrast to these findings, there was no change in NMDA-stimulated neurotransmitter release in brain slice preparations obtained from animals treated chronically with ethanol (BROWN et al. 1991). These data may suggest that chronic ethanol exposure has differential effects on presynaptic and postsynaptic NMDA receptors.

The up-regulation of the postsynaptic NMDA receptor seen in animals and cells treated chronically with ethanol may represent an adaptation to the initial inhibitory effect of ethanol. Although earlier studies investigating the chronic effect of NMDA receptor antagonists on receptor function were somewhat equivocal, evidence has accumulated that chronic antagonist treatment does produce an up-regulation of receptor function (WILLIAMS et al. 1992). Regardless of the etiology of the ethanol-induced changes, the increased NMDA receptor activity after chronic ethanol exposure has been suggested to contribute to the occurrence of ethanol withdrawal seizures. The time course of the appearance and disappearance of the increases in dizocilpine binding (reflective of quantitative increases in NMDA receptors) in hippocampus of chronically ethanol-treated mice paralleled the time course of ethanol withdrawal seizures (GULYA et al. 1991). These seizures could be attenuated by competitive and uncompetitive antagonists at the NMDA receptor, and the potency of various compounds to inhibit ethanol withdrawal seizure activity was positively correlated with their ability to inhibit dizocilpine binding to brain membranes (GRANT et al. 1990, 1992; MORRISETT et al. 1990; LILJEQUIST 1991). NMDA, at a dose that had no effect in control animals, exacerbated ethanol withdrawal seizures (GRANT et al. 1990). In addition, there were more dizocilpine binding sites in hippocampal tissue of WSP mice than in tissue of WSR mice (VALVERIUS et al. 1990). A recent preliminary report also showed a higher level of mRNA for the ξ subunit of the NMDA receptor in the brains of WSP mice (KEIR et al. 1993). After chronic ethanol ingestion, dizocilpine binding was increased in brains of both WSP and WSR mice, but the difference between the lines was maintained (VALVERIUS et al. 1990). It was also found that the efficacy of glycine to enhance dizocilpine binding was greater in the tissue of the WSP mice (VALVERIUS et al. 1990). The finding of this biochemical difference between selected lines of mice strongly implicates the difference in the characteristics of the NMDA receptor in mediating the selected trait (i.e., ethanol withdrawal seizure susceptibility) (Phillips et al. 1989).

While these data implicate the NMDA receptor in ethanol withdrawal seizures/physical dependence on ethanol, it should be understood that other neurochemical systems, including the GABA$_A$ receptor (which is down-regulated by chronic ethanol treatment; see above), and VSCCS (which are up-regulated; see above) also most likely contribute to the development and/or expression of ethanol withdrawal. The specific contribution of each system and the nature of the interactions among these systems need further elucidation.

It has recently been postulated that the increase in NMDA receptors observed after chronic ethanol treatment can also contribute to neuronal damage. It is generally believed that the NMDA receptor plays an important role in glutamate-induced excitotoxic damage in CNS neurons (Choi 1988). Rats which were exposed chronically to ethanol by inhalation and injected intrahippocampally with NMDA were shown to be more sensitive to NMDA, as measured by mortality and the decrease in glutamate decarboxylase activity in hippocampus of the survivors (Davidson et al. 1993). The lethal effect of NMDA was suggested to be a reflection of enhanced NMDA receptor function in the ethanol-withdrawn animals (Davidson et al. 1993). Furthermore, glutamate-induced neurotoxicity, mediated by the NMDA receptor, was significantly increased in cerebellar granule cells that had been exposed chronically to ethanol, withdrawn from ethanol and then exposed to glutamate (Iorio et al. 1993). The excitotoxic effect of NMDA was also enhanced in cerebral cortical cells that had been chronically exposed to ethanol (Chandler et al. 1993a). It is likely that the enhanced sensitivity to neurotoxic damage is a consequence of ethanol withdrawal, since it has been shown that ethanol, while present in the cellular milieu, inhibits NMDA receptor function and blocks NMDA receptor-mediated neurotoxicity (Chandler et al. 1993b; Takadera et al. 1990). Therefore, the presence of ethanol in the brain of an animal would be expected to protect against excitotoxicity, even in the presence of an increased number of NMDA receptors.

The suggestion that brain damage produced by chronic ethanol exposure and withdrawal may be in part due to increased NMDA receptor function underlines the contention (Tabakoff 1989) that ethanol withdrawal hyperexcitability should be treated with drugs that are antagonists at the NMDA receptor. Work with NMDA receptor antagonists indicates specific therapeutic approaches for ameliorating ethanol withdrawal seizures and preventing neuronal damage (Grant et al. 1992). There is substantial evidence to indicate that ethanol withdrawal symptoms are exacerbated following multiple episodes of withdrawal, i.e., withdrawal seizures result from a kindling phenomenon (Becker and Hale 1993). Most alcoholics undergo withdrawal numerous times throughout their life, and it is likely that multiple episodes of up-regulation of NMDA receptor function play a key role not only in the worsening seizures, but also in the well-described neuronal damage in alcoholics (CHARNESS 1993).

The NMDA receptor has also been suggested to modulate ethanol tolerance (KHANNA et al. 1992a,b; SZABÓ et al. 1994), although the function of the receptor itself may not become "tolerant" to the effects of ethanol (IORIO et al. 1992; WHITE et al. 1990b). In several studies, dizocilpine or ketamine (a dissociative anesthetic that binds to the PCP site within the ion channel), when administered together with ethanol, were demonstrated to block the development of ethanol tolerance (KHANNA et al. 1992a,b; SZABÓ et al. 1994). Although tolerance may be simply defined as an acquired resistance to the effects of ethanol, in reality it is a more complex phenomenon that can include metabolic and functional (pharmacodynamic) components (TABAKOFF et al. 1982). In addition, pharmacodynamic tolerance can be conditioned or associative, meaning that tolerance generated under certain paradigms reflects a conditioned response to the cues associated with ethanol administration (POULOS and CAPPEL 1991). It appears that conditioned tolerance is most susceptible to blockade by NMDA receptor antagonists. When animals were given ethanol using a paradigm in which conditioned tolerance was prominent (repeated ethanol injections in a particular environment), this tolerance was blocked by dizocilpine (KHANNA et al. 1992b; SZABÓ et al. 1994). When a paradigm (chronic ingestion of a liquid diet containing ethanol) was used in which nonassociative tolerance was generated, development of tolerance was not blocked even by higher doses of dizocilpine (SZABÓ et al. 1994). Therefore, it seems that the effect of dizocilpine may not be on ethanol tolerance per se, but may reflect the role of the NMDA receptor in the processes of learning (conditioning) and/or memory.

As mentioned earlier, the effects of ethanol on non-NMDA glutamate receptors are less well characterized than those on NMDA receptors. In initial studies, it was reported that biochemical and/or electrophysiologic responses to kainate and quisqualate in hippocampal and cerebellar granule cells were much less sensitive to inhibition by ethanol than NMDA responses (HOFFMAN et al. 1989; LOVINGER et al. 1989). In these studies, concentrations of kainate and NMDA that produced approximately equivalent responses were used to investigate ethanol sensitivity (HOFFMAN et al. 1989; LOVINGER et al. 1989). A recent study, in which rat hippocampal mRNA was expressed in Xenopus oocytes, showed that a relatively low concentration of ethanol could inhibit responses to a low concentration of kainate, but ethanol was less effective against a higher kainate concentration (DILDY-MAYFIELD and HARRIS 1992b). Thus, in contrast to most results obtained regarding NMDA receptor function, ethanol inhibition of kainate-induced responses was proposed to be "competitive" with agonist, i.e., inhibition could be overcome by increasing the concentration of kainate. A similar pattern of ethanol inhibition of kainate responses was found when cerebellar mRNA was expressed in oocytes (DILDY-MAYFIELD and HARRIS 1992a). In primary cultures of cerebellar granule cells, ethanol also inhibited the response (increase in intracellular Ca^{2+}) to a concentration of kainate close to the EC_{50} value for the kainate, but did not inhibit the response to higher concentrations of kainate (SNELL et al. 1994b). These results suggest that ethanol has a different mechanism of action at the kainate receptor than at the NMDA receptor. In the cerebellar granule cells, it was also noted that phorbol ester treatment did not affect the response to kainate, whereas it did affect the NMDA response (SNELL et al. 1994b).

There is little evidence for a change in function of the kainate or AMPA/kainate receptors after chronic ethanol treatment. In mice fed ethanol chronically, there was no significant change in hippocampal or cerebral cortical kainate binding (SNELL et al. 1993), and, in cerebellar granule cells exposed chronically to ethanol in vitro for 3 days, the kainate-induced increase in intracellular Ca^{2+} did not change (IORIO et al. 1993).

3. 5-HT$_3$ Receptors

The 5-HT$_3$ receptor is a ligand-gated ion channel that, when activated, is permeable to monovalent cations, although in some cells it also conducts Ca^{2+} and other divalent cations (PETERS et al. 1992). A subunit of this receptor has been cloned from NCB20 cells, and it appears to have a topology similar to that of other ligand-gated ion channels, including four hydrophobic putative membrane-spanning domains and a long cytoplasmic loop between transmembrane domains three and four (MARICQ et al. 1991). The 5-HT$_3$ receptor also contains amino acid residues in transmembrane segment 2 that are similar to those in the nicotinic acetylcholine receptor which are thought to regulate channel conductance (MARICQ et al. 1991). Although there is substantial evidence for interspecies variability of the characteristics of the 5-HT$_3$ receptor, there is less evidence for receptor variability within the tissues of a given species (PETERS et al. 1992). However, it seems likely that subtypes of this receptor will be identified, as is the case for the other ligand-gated ion channels. One type of 5-HT$_3$ receptor/channel variability that has been described is the presence of pre- and postsynaptic 5-HT$_3$ receptors (PETERS et al. 1992). The presynaptic receptors are believed to affect the release of a number of CNS neurotransmitters (PETERS et al. 1992).

Electrophysiologic studies have shown that low concentrations of ethanol can potentiate the actions of serotonin at the 5-HT$_3$ receptor, both in NCB20 cells and in rat nodose ganglion neurons (LOVINGER and WHITE 1991). Trichloroethanol was also found to potentiate the response to serotonin at 5-HT$_3$ receptors in nodose ganglion neurons, and the results suggested that trichloroethanol primarily facilitated the transition from the closed to open states of the channel (LOVINGER and ZHOU 1993). The possibility that ethanol and related compounds affect the kinetics of serotonin-induced channel opening is compatible with results showing that ethanol at concentrations of less than 100 mM had no effect on equilibrium ligand binding to 5-HT$_3$ receptors in membrane preparations from rat brain or NCB20 cells (HELLEVUO et al. 1991).

Ethanol may also affect the function of presynaptic 5-HT$_3$ receptor. Agonists acting at the 5-HT$_3$ receptor have been shown to increase dopamine release from slices of rat striatum (BLANDINA et al. 1989) and the ability of systemically administered ethanol to release dopamine in the nucleus accumbens and striatum of the rat has already been discussed. The 5-HT$_3$ receptor antagonist ICS-205-930 reduced the effect of ethanol on dopamine release when ethanol was given systemically or directly administered into the brain (CARBONI et al. 1989; WOZNIAK et al. 1990). These data suggest that ethanol may influence dopamine release by its action at the 5-HT$_3$ receptor. Ethanol's action at the 5-HT$_3$ receptor may also contribute to certain perceptual effects of ethanol, since 5-HT$_3$ antagonists were found to block the discriminative stimulus effects of ethanol (GRANT and BARRETT 1991).

D. Conclusion

The rather simple molecular structure of ethanol belies its complex actions in the brain and other organs of the body. Ethanol is a psychoactive compound which is consumed in doses of tens to hundreds of grams by humans seeking its mood and behavior altering properties. In comparison to the doses of other psychoactive compounds necessary to produce measurable effects in humans, ethanol is orders of magnitude less potent. Yet, because of ethanol's ingrained use in many cultures, its legal status and its accessibility, it is the most frequently used and misused drug in the Western world, and the consumption of ethanol in countries such as Japan and China and other Eastern countries continues to increase.

The caloric contribution from ethanol is not inconsequential in individuals who consume ethanol even at moderate levels of 10–30 g/day. The generation of calories from ingested ethanol, by metabolism of ethanol in the liver and other organs, does alter the metabolism of other nutrients. The ethanol-induced shunting of the metabolism of fats, carbohydrates and proteins can contribute to organ pathology in individuals consuming moderate to high quantities of ethanol during their lifetime. Ethanol metabolism also alters the metabolism of vitamins such as pyridoxal phosphate (B$_6$), and high intake of ethanol can contribute to thiamine deficiency. Thus, the nutrient properties of ethanol are associated with a significant cost to metabolic homeostasis.

The high doses of ethanol that are necessary to produce its actions on mood and behavior have given credence to theories that ethanol does not generate its CNS effects by interaction with a specific ethanol receptor. A multitude of neuronal systems are affected by ethanol and it has proven difficult to disentangle primary from secondary and tertiary events occurring in brains of ethanol-treated animals. The proposals that have been put forth indicating that ethanol's perturbation of neuronal membrane lipid structure can explain the full range of ethanol's actions in brain have been less than satisfactory. High concentrations of ethanol have been used in most cases to demonstrate ethanol's actions on membrane lipid order, and the differential effects of ethanol on genetically different proteins residing in identical lipid environments are difficult to explain by simply considering a bulk lipid site of action for ethanol. The work (described above) with GABA receptors derived from mice selectively bred for sensitivity and insensitivity to the sedative/hypnotic effects of ethanol has shifted the focus of alcohol researchers to examination of proteins whose actions may be particularly sensitive to perturbation by ethanol. Current work has identified families of proteins that may be the "receptive elements" on which ethanol acts, and through which ethanol initiates the cascade of events that lead to its anxiolytic, sedative, cognitive and other actions.

Many of the proteins currently identified as responding to nonanesthetic concentrations of 5–50 mM ethanol belong to families of receptor-gated ion channels (e.g., certain GABA and glutamate receptor systems), VSCCS, and to signal transducing systems such as the receptor-G protein-adenylyl cyclase systems. These systems have the common characteristics of being multisubunit complexes, located within the neuronal membrane and integral for signal transmission and transduction within the CNS. Interestingly, the adaptive responses of these systems to the chronic presence of ethanol in the brain may well explain the development of functional tolerance and physical dependence on ethanol, which is noted after periods of ethanol administration to, or voluntary ingestion by, animals. Since many of these receptive elements for ethanol have been cloned, structural information is accumulating to assess common motifs which predispose these systems to perturbation by ethanol.

As one examines the evolving information regarding the structural characteristics and post-translational modification processes which control the activity of the receptive elements for ethanol's actions, the possible importance of phosphorylation, prenylation and other posttranslational modifications in mediating ethanol's actions becomes evident. Although direct effects of ethanol on the activity of enzymes (e.g., PKC, PKA) which mediate posttranslational events seem to occur only at quite high concentrations of ethanol, a more subtle alteration of the activity of the posttranslational modification machinery is possible. For instance, ethanol-induced changes in the translocation of PKC or changes in intracellular cyclic AMP levels, which control the activity of PKA, may interface with the genetically determined proclivity of receptor-gated ion channels and other ethanol receptive elements to be acted on by PKC or PKA. Posttranslational modifications may be directly responsible for ethanol's perturbation of receptive element activity or the posttranslational modification may predispose the receptive element to ethanol-induced perturbation. Further studies of the receptive elements should reveal not only the mechanism of ethanol's actions within the CNS, but also the basis of individual differences in sensitivity to ethanol and individual predisposition to the problems of ethanol dependence.

References

Acquas E, Meloni M, Di Chiara G (1993) Blockade of σ-opioid receptors in the nucleus accumbens prevents ethanol-induced stimulation of dopamine release. Eur J Pharmacol 230:239–241

Agarwal DP, Goedde HW (eds) (1990) Alcohol metabolism, alcohol intolerance, and alcoholism – biochemical and pharmacogenetic approaches. Springer, Berlin Heidelberg New York

Aguayo LG (1990) Ethanol potentiates the $GABA_A$-activated Cl^- current in mouse hippocampal and cortical neurons. Eur J Pharmacol 187:127–130

Aguirre JC, Del Arbol JL, Raya J, Ruiz-Requena ME, Rico Irles J (1990) Plasma β-endorphin levels in chronic alcoholics. Alcohol 7:409–412

Allan AM, Harris RA (1986) Gamma-aminobutyric acid and alcohol actions: neurochemical studies of long sleep and short sleep mice. Life Sci 39:2005–2015

Allan AM, Harris RA (1987) Acute and chronic ethanol treatments alter GABA receptor-operated chloride channels. Pharmacol Biochem Behav 27:665–670

Allan AM, Mayes GG, Draski LJ (1991) Gamma-aminobutyric acid-activated chloride channels in rats selectively bred for differential acute sensitivity to alcohol. Alcohol Clin Exp Res 15:212–218

Alling C, Liljequist S, Engel J (1982) The effect of chronic ethanol administration on lipids and fatty acids in subcellular fractions of rat brain. Med Biol 60:145–154

Alling C, Gustavsson L, Anggard E (1983) An abnormal phospholipid in rat organs after ethanol treatment. FEBS Lett 152:24–28

Aloia RC, Paxton J, Daviau JS, Van Gelb O, Mlekusch W, Truppe W, Meyer JA, Brauer FS (1985) Effect of chronic alcohol consumption on rat brain microsome lipid composition membrane fluidity and Na^+ K^+-ATPase activity. Life Sci 36:1003–1017

Amit A, Smith BR, Brown ZW, Williams RL (1982) An examination of the role of TIQ alkaloids in alcohol intake: reinforcers, satiety agents or artifacts. In: Bloom F, Barchas J, Sandler M, Usdin E (eds) Progress in clinical and biological research, vol 90. Liss, New York, pp 345–364

Angelogianni P, Gianoulakis C (1993) Chronic ethanol increases proopiomelanocortin gene expression in the rat hypothalamus. Neuroendocrinology 57:106–114

Arcava Y, Fróes-Ferrão MM, Pereira EFR, Albuquerque EX (1991) Sensitivity of N-methyl-D-aspartate (NMDA) and nicotinic acetylcholine receptors to ethanol and pyrazole. Ann NY Acad Sci 625:451–472

Arky RA (1971) The effects of alcohol on carbohydrate metabolism. In: Kissin B, Begleiter H (eds) The biology of alcoholism. Plenum, New York, pp 197–227

Asaoka Y, Kikkawa U, Sekiguchi K, Shearman MS, Kosaka Y, Nakano Y, Satoh T, Nishizuka Y (1988) Activation of a brain-specific protein kinase C subspecies in the presence of phosphatidylethanol. FEBS Lett 231:221–224

Bacopoulos NG, Bhatanger RK, van Orden LS III (1978) The effects of subhypnotic doses of ethanol on regional catecholamine turnover. J Pharmacol Exp Ther 204:1–10

Balduini W, Candura SM, Manzo L, Cattabeni F, Costa LG (1991) Time-, concentration-, and age-dependent inhibition of muscarinic receptor-stimulated phosphoinositide metabolism by ethanol in the developing rat brain. Neurochem Res 16:1235–1240

Ballenger JC, Goodwin FK, Major LF, Brown GL (1979) Alcohol and central serotonin metabolism in man. Arch Gen Psychiatry 36:224–227

Banerjee SP, Sharma VK, Khanna JM (1978) Alterations in β-adrenergic receptor binding during ethanol withdrawal. Nature 276:407–409

Baraona E, Lieber CS (1979) Effects of ethanol on lipid metabolism. J Lipid Res 20:289–315

Barnhill JG, Ciraulo DA, Greenblatt DK, Faggart MA, Harmatz JS (1991) Benzodiazepine response and receptor binding after chronic ethanol ingestion in a mouse model. J Pharmacol Exp Ther 258:812–819

Bauché F, Bourdeaux-Jaubert AM, Giudicelli Y, Nordmann R (1987) Ethanol alters the adenosine receptor N_i-mediated adenylate cyclase inhibitory response in rat brain cortex in vitro. FEBS Lett 219:296–300

Beard JD, Sargent WQ (1979) Water and electrolyte metabolism following ethanol intake and during acute withdrawal form ethanol. In: Majchrowicz E, Noble EP (eds) Biochemistry and pharmacology of ethanol, vol 2. Plenum, New York, pp 3–16

Becker HC (1988) Effects of the imidazobenzodiazepine Ro 15-4513 on the stimulant and depressant actions of ethanol on spontaneous locomotor activity. Life Sci 43:643–650

Becker HC, Hale RL (1993) Repeated episodes of ethanol withdrawal potentiate the severity of subsequent withdrawal seizures: an animal model of alcohol withdrawal "kindling". Alcohol Clin Exp Res 17:94–98

Behrens UJ, Hoerner M, Lasker JM, Lieber CS (1988) Formation of acetaldehyde adducts with ethanol-inducible P450IIE1 in vivo. Biochem Biophys Res Commun 154:584–590

Ben-Ari Y, Aniksztejn L, Bregestovski P (1992) Protein kinase C modulation of NMDA currents: an important link for LTP induction. Trends Neurosci 15:333–339

Berridge MJ (1993) Inositol triphosphate and calcium signalling. Nature 361:315–325

Bertolino M, Llinás RR (1992) The central role of voltage-activated and receptor-operated calcium channels in neuronal cells. Annu Rev Pharmacol Toxicol 32:399–421

Birnbaumer L (1990) G proteins in signal transduction. Annu Rev Pharmacol Toxicol 30:675–705

Blandina P, Goldfarb J, Craddock-Royal B, Green JP (1989) Release of endogenous dopamine by stimulation of 5-hydroxytryptamine$_3$ receptors in rat striatum. J Pharmacol Exp Ther 251:803–809

Blitzer RD, Gil O, Landau EM (1990) Long-term potentiation in rat hippocampus is inhibited by low concentrations of ethanol. Brain Res 537:203–208

Bloom F, Barchas J, Sandler M, Usdin E (eds) (1982) Beta-carbolines and tetra-hydroisoquinolines. Liss, New York (Progress in clinical and biological research, vol 90)

Blum K, Noble EP, Sheridan PJ, Montgomery A, Ritchie T, Jagadeeswaran P, Nogami H, Briggs AH, Cohn JB (1990) Allelic association of human dopamine D2 receptor gene in alcoholism. JAMA 263:2055–2060

Blum K, Briggs AH, Elston SFA, Hirst M, Hamilton MG, Verebey K (1980) A common denominator theory of alcohol and opiate dependence: review of similarities and differences. In: Rigter H, Crabbe JC (eds) Alcohol tolerance and dependence. Elsevier Biomedical, New York, pp 371–391

Blumenthal RS, Flinn IW, Proske O, Jackson DG, Tena RG, Mitchell MC, Feldman AM (1991) Effects of chronic ethanol exposure on cardiac receptor-adenylate cyclase coupling: studies in cultured embryonic chick myocytes and ethanol fed rats. Alcohol Clin Exp Res 15:1077–1083

Blundell JE (1984) Serotonin and appetite. Neuropharmacology 23:1537–1551

Bode DC, Molinoff PB (1988a) Effects of ethanol in vitro on the beta adrenergic receptor-coupled adenylate cyclase system. J Pharmacol Exp Ther 246:1040–1047

Bode DC, Molinoff PB (1988b) Effects of chronic exposure to ethanol on the physical and functional properties of the plasma membrane of S49 lymphoma cells. Biochemistry 27:5700–5705

Bolos AM, Dean M, Lucas-Derse S, Ramsburg M, Brown GL, Goldman D (1990) Population and pedigree studies reveal a lack of association between dopamine D2 receptor gene and alcoholism. JAMA 264:3156–3160

Borg S, Weinholdt T (1982) Bromocriptine in the treatment of the alcohol withdrawal syndrome. Acta Psych Scand 65:101–111

Borg S, Kvande H, Rydberg U, Terenius L, Wahlstrom A (1982) Endorphin levels in human cerebrospinal fluid during alcohol intoxication and withdrawal. Psychopharmacology 78:101–103

Borg S, Kvande H, Mossberg D, Valverius P, Sedvall G (1983) Central nervous system noradrenaline metabolism and alcohol consumption in man. Pharmacol Biochem Behav 18 [Suppl 1]:375–378

Bosron WF, Magnes LJ, Li T-K (1983a) Kinetic and electrophoretic properties of native and recombined isoenzymes of human liver alcohol dehydrogenase. Biochemistry 22:1852–1857

Bosron WF, Magnes LJ, Li T-K (1983b) Human liver alcohol dehydrogenase: ADH$_{Indianapolis}$ results from genetic polymorphism at the ADH$_2$ gene locus. Biochem Genet 21:735–744

Brambilla F, Sarattini F, Gianelli A, Bianchi M, Panerai A (1988) Plasma opioids in alcoholics after acute alcohol consumption and withdrawal. Acta Psychiatr Scand 77:63–66

Braun T, Bober E, Schaper J, Agarwal DP, Singh S, Goedde HW (1987) Human mitochondrial aldehyde dehydrogenase: mRNA expression in different tissues using a specific probe isolated from a cDNA expression library. Alcohol [Suppl 1]:161–165

Brennan CH, Crabbe J, Littleton JM (1990) Genetic regulation of dihydropyridine-sensitive calcium channels in brain may determine susceptibility to physical dependence on alcohol. Neuropharmacology 29:429–432

Brown LM, Leslie SW, Gonzales RA (1991) The effects of chronic ethanol exposure on N-methyl-D-aspartate-stimulated overflow of [^{3}H]catecholamines from rat brain. Brain Res 547:289–294

Browning MD, Endo S, Smith GB, Dudek EM, Olsen RW (1993) Phosphorylation of the GABA$_A$ receptor by cAMP-dependent protein kinase and by protein kinase C: analysis of the substrate domain. Neurochem Res 18:95–100

Brownstein MJ (1989) Neuropeptides. In: Siegel GJ, Agranoff BW, Albers RW, Molinoff PB (eds) Basic neurochemistry: molecular, cellular and medical aspects, 4th edn. Raven, New York, pp 287–309

Buck KJ, Harris RA (1990) Benzodiazepine agonist and inverse agonist actions on GABA$_A$ receptor-operated chloride channels. II. Chronic effects of ethanol. J Pharmacol Exp Ther 253:713–719

Buck KJ, Harris RA (1991) Chronic ethanol exposure of Xenopus oocytes expressing mouse brain mRNA reduces GABA$_A$ receptor-activated current and benzodiazepine modulation. Mol Neuropharmacol 1:59–64

Buck KJ, Hahner L, Sikela J, Harris RA (1991) Chronic ethanol treatment alters brain levels of γ-aminobutyric acid$_A$ receptor subunit mRNAs: relationship to genetic differences in ethanol withdrawal seizure severity. J Neurochem 57:1452–1455

Bühler R, Pestalozzi D, Hess M, von Wartburg J-P (1983) Immunohistochemical localization of alcohol dehydrogenase in human kidney, endocrine organs and brain. Pharmacol Biochem Behav 18:55–59

Bujis RM (1978) Intra- and extrahypothalamic vasopressin and oxytocin pathways in the rat. Cell Tissue Res 192:423–435

Buller AL, Morrisett RA, Seeburg PH, Monaghan DT (1993) Interaction of ethanol (EtOH) with recombinant heteromeric NMDA receptors expressed in Xenopus oocytes. Alcohol Clin Exp Res 17:475

Burbach JPH, DeHoop MJ, Schmale H, Richter D, DeKloet ER, Haaf JAT, DeWied D (1984) Differential responses to osmotic stress of vasopressin-neurophysin mRNA in hypothalamic nuclei. Neuroendocrinology 39:582–584

Burchfiel JL, Duchowny MS, Duffy FH (1979) Neuronal supersensitivity to acetylcholine induced by kindling in the rat hippocampus. Science 204:1096–1098

Burnell JC, Carr LJ, Dwulet FJ, Edenberg HJ, Li T-K, Bosron WF (1987) The human beta 3 alcohol dehydrogenase subunit differs from β_1 by a Cys for Arg-369 substitution which decreases NAD(H) binding. Biochem Biophys Res Commun 146:1127–1133

Bustos G, Liberona JL, Gysling K (1981) Regulation of transmitter synthesis and release in mesolimbic dopaminergic nerve terminals. Effect of ethanol. Biochem Pharmacol 30:2157–2164

Caffé AR, Van Leeuwen FW, Luiten PGM (1987) Vasopressin cells in the medial amygdala of the rat project to the lateral septum and ventral hippocampus. J Comp Neurol 261:237–252

Campbell AD, Erwin VG (1993) Chronic ethanol administration downregulates neurotensin receptors in long- and short-sleep mice. Pharmacol Biochem Behav 45:95–106

Candura SM, Balduini W, Costa LG (1991) Interaction of short chain aliphatic alcohols with muscarinic receptor-stimulated phosphoinositide metabolism in cerebral cortex from neonatal and adult rats. Neurotoxicology 12:23–32

Carboni E, Acquas E, Frau R, Di Chiara G (1989) Differential inhibitory effects of a 5-HT$_3$ antagonist on drug-induced stimulation of dopamine release. Eur J Pharmacol 164:515–519

Carlen PL, Wu PH (1988) Calcium and sedative-hypnotic drug actions. Int Rev Neurobiol 29:161–189

Carpenter JA (1962) Effects of alcohol on some psychological processes. Q J Stud Alcohol 23:274–314

Catterall WA (1991) Perspective: functional subunit structure of voltage-gated calcium channels. Science 253:1499–1500

Chandler LJ, Newsom H, Sumners C, Crews FT (1993a) Chronic ethanol exposure potentiates NMDA excitotoxicity in cerebral cortical neurons. J Neurochem 60:1578–1581

Chandler LJ, Sumners C, Crews FT (1993b) Ethanol inhibits NMDA receptor-mediated excitotoxicity in rat primary neuronal cultures. Alcohol Clin Exp Res 17:54–60

Charness ME (1989) Ethanol and opioid receptor signalling. Experientia 45:418–427

Charness ME (1993) Brain lesions in alcoholics. Alcohol Clin Exp Res 17:2–11

Charness ME, Gordon AS, Diamond I (1983) Ethanol modulation of opiate receptors in cultured neural cells. Science 222:1426–1428

Charness ME, Querimit LA, Diamond I (1986) Ethanol induces the expression of functional δ-opioid receptors in the neuroblastoma × glioma hybrid NG108-115 cell line. J Biol Chem 261:3164–3169

Charness ME, Querimit LA, Henteleff M (1988) Ethanol differentially regulates G proteins in neural cells. Biochem Biophys Res Commun 155:138–143

Chen L, Huang L-YM (1991) Sustained potentiation of NMDA receptor-mediated glutamate responses through activation of protein kinase C by a mu opioid. Neuron 7:319–326

Chik CL, Ho AK (1991) Inhibitory effects of ethanol on the calcium-dependent potentiation of vasoactive intestinal peptide-stimulated cAMP and cGMP accumulation in rat pinealocytes. Biochem Pharmacol 42:1601–1608

Chik CL, Ho AK, Klein DC (1987) Ethanol inhibits dual receptor stimulation of pineal cAMP and cGMP by vasoactive intestinal peptide and phenylephrine. Biochem Biophys Res Commun 147:145–151

Chik CL, Liu Q-Y, Girard M, Karpinski E, Ho AK (1992) Inhibitory action of ethanol on L-type Ca^{2+} channels and Ca^{2+}-dependent guanosine 3′,5′-

monophosphate accumulation in rat pinealocytes. Endocrinology 131:1895–1902

Chin JH, Goldstein DB (1977) Drug tolerance in biomembranes: a spin label study of the effects of ethanol. Science 196:684–685

Chin JH, Goldstein DB (1981) Membrane-disordering action of ethanol: variation with membrane cholesterol content and depth of the spin label probe. Mol Pharmacol 19:425–431

Chin JH, Parsons LM, Goldstein DB (1978) Increased cholesterol content of erythrocyte and brain membranes in ethanol-tolerant mice. Biochim Biophys Acta 513:358–363

Choi DW (1988) Calcium-mediated neurotoxicity: relationship to specific channel types and role in ischemic damage. Trends Neurosci 11:465–469

Chung CT, Tamarkin L, Hoffman PL, Tabakoff B (1989) Ethanol enhancement of isoproterenol-stimulated melatonin and cyclic AMP release from cultured pineal glands. J Pharmacol Exp Ther 249:16–22

Cicero TJ (1981) Neuroendocrinological effects of ethanol. Annu Rev Med 32:123–142

Cicero TJ, Bell RD (1980) Effects of ethanol and acetaldehyde on biosynthesis of testosterone in the rodent testes. Biochim Biophys Res Commun 94:814–819

Clark JW, Kalant H, Carmichael FJ (1977) Effect of ethanol tolerance on release of acetylcholine and norepinephrine by rat cerebral cortex slices. Can J Physiol Pharmacol 55:758–768

Clark RB, Kunkel MW, Friedman J, Goka TJ, Johnson JA (1988) Activation of cAMP-dependent protein kinase is required for heterologous desensitization of adenylate cyclase in S49 wild-type lymphoma cells. Proc Natl Acad Sci USA 85:1442–1446

Clarren SK, Smith DW (1978) The fetal alcohol syndrome. N Engl J Med 298:1063–1067

Cloninger CR (1987) Neurogenetic adaptive mechanisms in alcoholism. Science 236:410–416

Cloninger CR (1991) D2 dopamine receptor gene is associated but not linked with alcoholism. JAMA 266:1833–1834

Cohen G, Collins MA (1970) Alkaloids from catecholamines in adrenal tissue: possible role in alcoholism. Science 167:1749–1751

Colbern DL, ten Haaf J, Tabakoff B, van Wimersma Greidanus TB (1985) Ethanol increases plasma vasopressin shortly after intraperitoneal injection in rats. Life Sci 37:1029–1032

Collingridge GL, Lester RAJ (1989) Excitatory amino acid receptors in the vertebrate central nervous system. Pharmacol Rev 40:143–210

Comings DE, Comings BG, Muhleman D, Dietz G, Shahbahrami B, Tast D, Knell E, Kocsis P, Baumgarten R, Kovacs BW, Levy DL, Smith M, Borison RL, Evans DD, Klein DN, MacMurray J, Tosk JM, Sverd J, Gysin R, Flanagan SD (1991) The dopamine D2 receptor locus as a modifying gene in neuropsychiatric disorders. JAMA 266:1793–1800

Cott JA, Carlson A, Engel J, Lindquist M (1976) Suppression of ethanol-induced locomotor stimulation by GABA-like drugs. Naunyn Schmiedebergs Arch Pharmacol 295:203–209

Courtney MJ, Nicholls DG (1992) Interactions between phospholipase C-coupled and N-methyl-D-aspartate receptors in cultured cerebellar granule cells: protein kinase C mediated inhibition of N-methyl-D-aspartate responses. J Neurochem 59:983–992

Crabb DW, Bosron WF, Li T-K (1983) Steady-state kinetic properties of purified rat liver alcohol dehydrogenase: application to predicting alcohol elimination rates in vivo. Arch Biochem Biophys 224:299–309

Crabbe JC, Phillips TJ, Kosobud A, Belknap JK (1990) Estimation of genetic correlation: interpretation of experiments using selectively bred and inbred animals. Alcohol Clin Exp Res 14:141–151

Crews FT, Majchrowicz E, Meeks R (1983) Changes in cortical synaptosomal plasma membrane fluidity and composition in ethanol-dependent rats. Psychopharmacology 81:208–213

Cronholm T, Sjövall J (1970) Effect of ethanol on redox state of steroid sulphates in man. Eur J Biochem 13:124–131

Daniell LC, Brass EP, Harris RA (1987) Effect of ethanol on intracellular ionized calcium concentrations in synaptosomes and hepatocytes. Mol Pharmacol 32:831–837

Dar MS (1990) Central adenosinergic system involvement in ethanol-induced motor incoordination in mice. J Pharmacol Exp Ther 255(3):1202–1209

Dascal N (1990) Commentary: analysis and functional characteristics of dihydropyridine-sensitive and -insensitive calcium channel proteins. Biochem Pharmacol 40:1171–1178

Dave JR, Eiden LE, Karanian JW, Eskay RL (1986) Ethanol exposure decreases pituitary corticotropin-releasing factor binding, adenylate cyclase activity, proopiomelanocortin biosynthesis, and plasma β-endorphin levels in the rat. Endocrinology 118:280–286

Davidson MD, Wilce P, Shanley BC (1993) Increased sensitivity of the hippocampus in ethanol-dependent rats to toxic effect of N-methyl-D-aspartic acid in vivo. Brain Res 606:5–9

Davis VE, Walsh MJ (1970) Alcohol, amines and alkaloids: a possible biochemical base for alcohol addiction. Science 167:1005–1007

Davis VE, Brown H, Huff JA, Cashaw JL (1967a) Ethanol-induced alterations of norepinephrine metabolism in man. J Lab Clin Med 69:787–799

Davis VE, Brown H, Huff JA, Cashaw JL (1967b) The alteration of serotonin metabolism to 5-hydroxytryptophol by ethanol ingestion in man. J Lab Clin Med 69:132–140

De Vries GJ, Buijs RM, Van Leeuwen FW, Caffé AR, Swaab DF (1985) The vasopressinergic innervation of the brain in normal and castrated rats. J Comp Neurol 233:326–254

DeLorey TM, Olsen RW (1992) γ-Aminobutyric acid$_A$ receptor structure and function. J Biol Chem 267:16747–16750

DePetrillo PB, Swift RM (1992) Ethanol exposure results in a transient decrease in human platelet cAMP levels: evidence for a protein kinase C mediated process. Alcohol Clin Exp Res 16:290–294

Devor EJ, Cloninger CR, Hoffman PL, Tabakoff B (1991) A genetic study of platelet adenylate cyclase activity: evidence for a single major locus effect in fluoride-stimulated activity. Am J Hum Genet 49:372–377

Diamond I, Wrubel B, Estrin W, Gordon A (1987) Basal and adenosine receptor-stimulated levels of cAMP are reduced in lymphocytes in alcoholic patients. Proc Natl Acad Sci USA 84:1413–1416

Diehl AM, Yang SQ, Cote P, Wand GS (1992) Chronic ethanol consumption disturbs G-protein expression and inhibits cyclic AMP-dependent signalling in regenerating rat liver. Hepatology 16:1212–1219

Dildy JE, Leslie SW (1989) Ethanol inhibits NMDA-induced increases in free intracellular Ca^{2+} in dissociated brain cells. Brain Res 499:383–387

Dildy-Mayfield JE, Leslie SW (1991) Mechanism of inhibition of N-methyl-D-aspartate-stimulated increases in free intracellular Ca^{2+} concentration by ethanol. J Neurochem 56:1536–1543

Dildy-Mayfield JE, Harris RA (1992a) Comparison of ethanol sensitivity of rat brain kainate, DL-α-amino-3-hydroxy-5-methyl-4-isoxalone proprionic acid and N-methyl-D-aspartate receptors expressed in Xenopus oocytes. J Pharmacol Exp Ther 262:487–494

Dildy-Mayfield JE, Harris RA (1992b) Acute and chronic ethanol exposure alters the function of hippocampal kainate receptors expressed in Xenopus oocytes. J Neurochem 58:1569–1572

Dolin SJ, Little HJ (1989) Are changes in neuronal calcium channels involved in ethanol tolerance? J Pharmacol Exp Ther 250:985–991

Dolin S, Hudspith M, Pagonis C, Little H, Littleton J (1987) Increased dihydropyridine-sensitive Ca^{2+} channels in rat brain may underlie ethanol physical dependence. Neuropharmacology 26:275–279

Dolphin AC (1990) G protein modulation of calcium currents in neurons. Annu Rev Physiol 52:243–255

Durkin TP, Hashem-Zadeh H, Mandel P, Ebel A (1982) A comparative study of the acute effects of ethanol on the cholinergic system in hippocampus and striatum of inbred mouse strains. J Pharmacol Exp Ther 220:203–208

Ehrig T, von Wartburg J-P, Wermuth B (1988) cDNA sequence of the β_2-subunit of human liver alcohol dehydrogenase. FEBS Lett 234:53–55

Engberg G, Hajós M (1992) Ethanol attenuates the response of locus coeruleus neurons to excitatory amino acid agonists in vivo. Naunyn Schmiedebergs Arch Pharmacol 345:222–226

Engel JA, Fahlke C, Hulthe P, Hard E, Johannessen K, Snape B, Svenson L (1988) Biochemical and behavioral evidence for an interaction between ethanol and calcium channel antagonists. J Neural Transm 74:181–193

Erickson CK, Graham DT (1973) Alterations in cortical and reticular acetylcholine release by ethanol in vivo. J Pharmacol Exp Ther 185:583–593

Erickson CK, Tyler TD, Beck LK, Duensing KL (1980) Calcium enhancement of alcohol and drug-induced sleeping time in mice and rats. Pharmacol Biochem Behav 12:651–656

Eriksson CJ, Marselos M, Koivula T (1975) Role of cytosolic rat liver aldehyde dehydrogenase in the oxidation of acetaldehyde during ethanol metabolism in vivo. Biochem J 152:709–712

Eriksson CJP (1983) Human blood acetaldehyde concentration during ethanol oxidation (update 1982). Pharmacol Biochem Behav 18:141–150

Erwin VG, Deitrich RA (1966) Brain aldehyde dehydrogenase: localization, purification and properties. J Biol Chem 241:3533–3539

Erwin VG, Jones BC (1989) Comparison of neurotensin levels, receptors and actions in LS/Ibg and SS/Ibg mice. Peptides 10:435–440

Erwin VG, Korte A (1988) Brain neurotensin receptors in mice bred for differences in sensitivity to ethanol. Alcohol 5:195–201

Erwin VG, Jones BC, Radcliffe R (1990) Low doses of ethanol reduce neurotensin levels in discrete brain regions from LS/Ibg and SS/Ibg mice. Alcohol Clin Exp Res 14:42–47

Erwin VG, Campbell AD, Radcliffe RA (1992) Effects of chronic ethanol administration on neurotensinergic processes: Correlations with tolerance in LS and SS mice. Ann NY Acad Sci 654:441–443

Fadda F, Argiolas A, Melis MR, Serra G, Gessa GL (1980) Differential effect of acute and chronic ethanol on dopamine metabolism in frontal cortex, caudate nucleus and substantia nigra. Life Sci 27:979–986

Feiman MD (1979) Biochemical pharmacology of disulfiram. In: Majchrowicz E, Noble EP (eds) Biochemistry and pharmacology of ethanol. Plenum, New York, pp 325–348

File SE, Zharkovsky A, Gulati K (1991) Effects of baclofen and nitrendipine on ethanol withdrawal responses in the rat. Neuropharmacology 30:183–190

Fink K, Göthert M, Molderings G, Schlicker E (1989) N-methyl-D-aspartate (NMDA) receptor-mediated stimulation of noradrenaline release, but not release of other neurotransmitters, in the rat brain cortex: receptor location, characterization and desensitization. Naunyn Schmiedebergs Arch Pharmacol 339:514–521

Forman DT, Bradford BU, Handler JA, Glassman EB, Thurman RG (1988) Involvement of hormones in the swift increase in alcohol metabolism. Ann Clin Lab Sci 18:318–325

Franks NP, Lieb WR (1984) Do general anesthetics act by competitive binding to specific receptors? Nature 310:599–601

Freissmuth M, Casey PJ, Gilman AG (1989) G proteins control diverse pathways of transmembrane signalling. FASEB J 3:2125–2131

French SW, Palmer DS, Narod ME, Reid PE, Ramey CW (1975) Noradrenergic sensitivity of the cerebral cortex after chronic ethanol ingestion and withdrawal. J Pharmacol Exp Ther 194:319–326

Frezza M, Di Padova C, Pozzato G, Terpin M, Baraona E, Lieber CS (1990) High blood alcohol levels in women – the role of decreased gastric alcohol dehydrogenase activity and first-pass metabolism. N Engl J Med 322:95–99

Frye GD, McCown TJ, Breese GR (1983) Characterization of susceptibility to audiogenic seizures in ethanol-dependent rats after microinjection of γ-aminobutyric acid (GABA) agonists into the inferior colliculus, substantia nigra or medial septum. J Pharmacol Exp Ther 227:663–670

Frye GD, Mathew J, Trzeciakowski JP (1991) Effect of ethanol dependence on GABA$_A$ antagonist-induced seizures and agonist-stimulated chloride uptake. Alcohol 8:453–459

Gelernter J, O'Malley S, Risch N, Kranzler HR, Krystal J, Merikangas K, Kennedy JL, Kidd K (1991) No association between an allele at the D2 dopamine receptor gene (DRD2) and alcoholism. JAMA 266:1801–1807

Genazzani AR, Nappi G, Facchinetti F, Mazzella GL, Parrini D, Sinforiani E, Petraglia F, Savoldi F (1982) Central deficiency of β-endorphin in alcohol addicts. J Clin Endocrinol Metab 55:583–586

Gessa GL, Muntoni F, Collu M, Vargiu L, Mereu G (1985) Low doses of ethanol activate dopaminergic neurones in the ventral tegmental area. Brain Res 348:201–203

Gessner PK (1979) Drug therapy of the alcohol withdrawal syndrome. In: Majchrowicz E, Noble EP (eds) Biochemistry and pharmacology of ethanol, vol 2. Plenum, New York, pp 375–435

Ghalioungui P (1979) Fermented beverages in antiquity. In: Gastineu CF, Darby WJ, Turner TB (eds) Fermented food beverages in nutrition. Academic, New York, pp 4–18

Gianoulakis C (1983) Long-term ethanol alters the binding of ^{3}H-opiates to brain membranes. Life Sci 33:725–733

Gianoulakis C (1989) The effect of ethanol on the biosynthesis and regulation of opioid peptides. Experientia 45:428–435

Gianoulakis C (1990) Characterization of the effects of acute ethanol administration on the release of the β-endorphin peptides by the rat hypothalamus. Eur J Pharmacol 180:21–29

Gianoulakis C, Barcomb A (1987) Effect of acute ethanol in vivo and in vitro on the β-endorphin system in the rat. Life Sci 40:19–28

Gianoulakis C, Woo N, Drouin JN, Seidah NG, Kalant H, Chrétien M (1981) Biosynthesis of β-endorphin by the neurointermediate lobes from rats treated with morphine or alcohol. Life Sci 29:1973–1982

Gianoulakis C, Chan JSD, Kalant H, Chrétien M (1983) Chronic ethanol treatment alters the biosynthesis of β-endorphin by the rat neurointermediate lobe. Can J Physiol Pharmacol 61:967–976

Gianoulakis C, Hutchison WD, Kalant H (1988) Effects of ethanol treatment and withdrawal on biosynthesis and processing of proopiomelanocortin by the rat neurointermediate lobe. Endocrinology 122:817–825

Giri PR, Linnoila M, O'Neill JB, Goldman D (1989) Distribution and possible metabolic role of class III alcohol dehydrogenase in the human brain. Brain Res 481:131–141

Goldstein DB, Chin JH, Lyon RC (1982) Ethanol disordering of spin-labeled mouse brain membranes-correlation with genetically determined ethanol sensitivity of mice. Proc Natl Acad Sci USA 79:4231–4233

Gonzales RA, Woodward JJ (1990) Ethanol inhibits N-methyl-D-aspartate-stimulated [³H]norepinephrine release from rat cortical slices. J Pharmacol Exp Ther 253:1138–1144

Gonzales RA, Theiss C, Crews FT (1986) Effects of ethanol on stimulated inositol phospholipid hydrolysis in rat brain. J Pharmacol Exp Ther 237:92–98

Gordon AS, Collier K, Diamond I (1986) Ethanol regulation of adenosine receptor-stimulated cAMP levels in a clonal neural cell line: an in vitro model of cellular tolerance to ethanol. Proc Natl Acad Sci USA 83:2105–2108

Göthert M, Fink M (1989) Inhibition of N-methyl-D-aspartate (NMDA)- and L-glutamate-induced noradrenaline and acetylcholine release in the rat brain by ethanol. Naunyn Schmiedebergs Arch Pharmacol 340:516–521

Grandy DK, Litt M, Allen L, Bunzow Jr, Marchionni M, Makam H, Reed L, Magenis RE, Civelli O (1989) The human dopamine D2 receptor gene is located on chromosome 11 at q22–q23 and identifies a Taq I RFLP. Am J Hum Genet 45:778–785

Grant AJ, Koski G, Treistman SN (1993) Effect of chronic ethanol on calcium currents and calcium uptake in undifferentiated PC12 cells. Brain Res 600:280–284

Grant KA, Barrett JE (1991) Blockade of the discriminative stimulus effect of ethanol with 5-HT₃ receptor antagonists. Psychopharmacology 104:451–456

Grant KA, Colombo G (1992) Discriminative stimulus effects of ethanol: effect of training dose on the substitution of N-methyl-D-aspartate antagonists. J Pharmacol Exp Ther 264:1241–1247

Grant KA, Valverius P, Hudspith M, Tabakoff B (1990) Ethanol withdrawal seizures and the NMDA receptor complex. Eur J Pharmacol 176:289–296

Grant KA, Knisely JS, Tabakoff B, Barrett JE, Balster RL (1991) Ethanol-like discriminative stimulus effects of non-competitive N-methyl-D-aspartate antagonists. Behav Pharmacol 2:87–95

Grant KA, Snell LD, Rogawski MA, Thurkauf A, Tabakoff B (1992) Comparison of the effects of the uncompetitive N-methyl-D-aspartate antagonist (±)-5-aminocarbonyl-10,11-dihydro-5H-dibenzo[a,d]cyclohepten-5,10-imine (ADCI) with its structural analogs dizocilpine (MK-801) and carbamazepine on ethanol withdrawal seizures. J Pharmacol Exp Ther 260:1017–1022

Greenberg DA, Carpenter CL, Messing RO (1987) Ethanol-induced component of ^{45}Ca^{2+} uptake in PC12 cells is sensitive to Ca^{2+} channel modulating drugs. Brain Res 410:143–146

Gulya K, Grant KA, Valverius P, Hoffman PL, Tabakoff B (1991) Brain regional specificity and time course of changes in the NMDA receptor-ionophore complex during alcohol withdrawal. Brain Res 547:129–134

Gulya K, Orpana AK, Sikela JM, Hoffman PL (1993) Prodynorphin and vasopressin mRNA levels are differentially affected by chronic ethanol ingestion in the mouse. Mol Brain Res 20:1–8

Gustavsson L, Alling C (1987) Formation of phosphatidylethanol in rat brain by phospholipase D. Biochem Biophys Res Commun 142:958–963

Hannak D, Bartelt U, Katterman R (1985) Acetate formation after short-term ethanol administration in man. Biol Chem Hoppe Seyler 366:749–753

Harada S, Misawa S, Agarwal DP, Goedde HW (1980) Liver alcohol and aldehyde dehydrogenase in the Japanese: isozyme variation and its possible role in alcohol intoxication. Am J Hum Genet 32:8–15

Harada S, Agarwal DP, Goedde HW (1981) Aldehyde dehydrogenase deficiency as cause of facial flushing reaction to alcohol in Japanese. Lancet ii:982

Harada S, Agarwal DP, Goedde HW, Tagaki S, Ishikawa B (1982) Possible protective role against alcoholism for aldehyde dehydrogenase isozyme deficiency in Japan. Lancet ii:827

Harden TK (1983) Agonist-induced desensitization of the β-adrenergic receptor-linked adenylate cyclase. Pharmacol Rev 35:5–32

Harper JC, Brennan CH, Littleton JM (1989) Genetic up regulation of calcium channels in a cellular model of ethanol dependence. Neuropharmacology 28:1299–1302

Harper RF (1904) The Code of Hammurabi, King of Babylon. University of Chicago Press, Chicago

Harris RA (1979) Alteration of alcohol effects by calcium and other inorganic cations. Pharmacol Biochem Behav 10:527–534

Harris RA, Groh GI, Baxter DM, Hitzemann RJ (1984) Gangliosides enhance the membrane actions of ethanol and pentobarbital. Mol Pharmacol 25:410–417

Harris RA, Zaccaro LM, McQuilkin S, McClard A (1988) Effects of ethanol and calcium on lipid order of membranes from mice selected for genetic differences in ethanol intoxication. Alcohol 5:251–257

Helander A (1993) Aldehyde dehydrogenase in blood: distribution, characteristics and possible use as marker of alcohol misuse. Alcohol 28:135–145

Hellevuo K, Hoffman PL, Tabakoff B (1991) Ethanol fails to modify [^{3}H]GR65630 binding to 5-HT$_3$ receptors in NCB-20 cells and in rat cerebral membranes. Alcohol Clin Exp Res 15:775–778

Hiller JM, Angel LM, Simon EJ (1981) Multiple opiate receptors: alcohol selectively inhibits binding to the delta receptor. Science 214:468–469

Hiller JM, Angel LM, Simon EJ (1984) Characterization of the selective inhibition of the delta subclass of opioid binding sites by alcohols. Mol Pharmacol 25:249–255

Hillmann M, Wilce PA, Pietrzak ER, Ward LC, Shanley BC (1990) Chronic ethanol administration alters binding of [^{35}S]t-butylbicyclophosphorothionate to the GABA-benzodiazepine receptor complex in rat brain. Neurochem Int 16:187–191

Hitzemann RJ, Schuéler HE, Graham-Brittain C, Kreishman GP (1986) Ethanol-induced changes in neuronal membrane order. An NMR study. Biochim Biophys Acta 859:189–197

Hoek JB, Thomas AP, Rubin R, Rubin E (1987) Ethanol-induced mobilization of calcium by activation of phosphoinositide-specific phospholipase C in intact hepatocytes. J Biol Chem 262:682–691

Hoek JB, Taraschi TF, Rubin E (1988) Functional implications of the interaction of ethanol with biologic membranes: actions of ethanol on hormonal signal transduction systems. Semin Liver Dis 8:36–46

Hoffman PL (1987) Central nervous system effects of neurohypophyseal peptides. In: Smith CW (ed) The peptides, vol 8: chemistry, biology and medicine of neurohypophyseal hormones and their analogs. Academic, New York, pp 239–295

Hoffman PL, Dave JR (1991) Chronic ethanol exposure uncouples vasopressin synthesis and secretion in rats. Neuropharmacology 30:1245–1249

Hoffman PL, Tabakoff B (1985) Ethanol's action on brain biochemistry. In: Tarter RE, van Thiel DH (eds) Alcohol and the brain: chronic effects. Plenum, New York, pp 19–68

Hoffman PL, Tabakoff B (1986) Ethanol does not modify opiate receptor inhibition of striatal adenylate cyclase. J Neurochem 46:812–816

Hoffman PL, Tabakoff B (1990) Ethanol and guanine nucleotide binding proteins: a selective interaction. FASEB J 4:2612–2622

Hoffman PL, Ritzmann RF, Walter R, Tabakoff B (1987) Arginine vasopressin maintains ethanol tolerance. Nature 276:614–616

Hoffman PL, Chung CT, Tabakoff (1984) Effects of ethanol, temperature, and endogenous regulatory factors on the characteristics of striatal opiate receptors. J Neurochem 43:1003–1010

Hoffman PL, Moses F, Luthin G, Tabakoff B (1986) Acute and chronic effects of ethanol on receptor-mediated phosphatidylinositol 4,5-bisphosphate breakdown in mouse brain. Mol Pharmacol 30:13–18

Hoffman PL, Tabakoff B, Szabó G, Suzdak PD, Paul SM (1987) Effect of an imidazobenzodiazepine, Ro15-4513, on the incoordination and hypothermia produced by ethanol and pentobarbital. Life Sci 41:611–619

Hoffman PL, Rabe CS, Moses F, Tabakoff B (1989) N-methyl-D-aspartate receptors and ethanol: inhibition of calcium flux and cyclic GMP production. J Neurochem 52:1937–1940

Hoffman PL, Ishizawa H, Giri PR, Dave JR, Grant KA, Liu L-I, Gulya K, Tabakoff B (1990) The role of arginine vasopressin in alcohol tolerance. Ann Med 22:269–274

Hollmann M, Boulter J, Maron C, Beasley L, Sullivan J, Pecht G, Heinemann S (1993) Zinc potentiates agonist-induced currents at certain splice variants of the NMDA receptor. Neuron 10:943–954

Howerton TC, Marks MJ, Collins AC (1982) Norepinephrine, gamma-aminobutyric acid and choline reuptake kinetics and the effects of ethanol in long-sleep and short-sleep mice. Subst Alcohol Actions Misuse 3:89–99

Howerton TC, O'Connor F, Collins AC (1983) Differential effects of long-chain alcohols in long- and short-sleep mice. Psychopharmacology 79:313–317

Hunt WA (1985) Alcohol and biological membranes. Guilford, New York

Hunt WA, Majchrowicz E (1974) Alterations in the turnover of brain norepinephrine and dopamine in the alcohol-dependent rat. J Neurochem 23:549–552

Hunt WA, Majchrowicz E, Dalton J (1979) Alterations in high-affinity choline uptake in brain after chronic ethanol treatment. J Pharmacol Exp Ther 210:259–263

Hynes MD, Lochner MA, Bemis KG, Hymson DL (1983) Chronic ethanol alters the receptor binding characteristics of the δ-opioid receptor ligand D-Ala^2-D-Leu^5 enkephalin in mouse brain. Life Sci 33:2331–2337

Ikeda K, Nagasawa M, Mori H, Araki K, Sakimura K, Watanabe M, Inoue Y, Mishina M (1992) Cloning and expression of the ε4 subunit of the NMDA receptor channel. FEBS Lett 313:34–38

Ilyin V, Parker I (1992) Effects of alcohols on responses evoked by inositol trisphosphate in Xenopus oocytes. J Physiol (Lond) 448:339–354

Imperato A, Di Chiara G (1986) Preferential stimulation of dopamine release in the nucleus accumbens of freely moving rats by ethanol. J Pharmacol Exp Ther 239:219–228

Imperato A, Scrocco MG, Bacchi S, Angelucci L (1990) NMDA receptors and in vivo dopamine release in the nucleus accumbens and caudatus. Eur J Pharmacol 187:555–556

Iorio KR, Reinlib L, Tabokoff B, Hoffman PL (1992) Chronic exposure of cerebellar granule cells to ethanol results in increased NMDA receptor function. Mol Pharmacol 41:1142–1148

Iorio KR, Tabakoff B, Hoffman PL (1993) Glutamate-induced neurotoxicity is increased in cerebellar granule cells exposed chronically to ethanol. Eur J Pharmacol 248:209–212

Ishii T, Moriyoshi K, Sugihara H, Sakurada K, Kadotani H, Yokoi M, Akazawa C, Shigemoto R, Mizuno N, Masu M, Nakanishi S (1993) Molecular characterization of the family of the N-methyl-D-aspartate receptor subunits. J Biol Chem 268:2836–2843

Ishizawa H, Dave JR, Liu L, Tabakoff B, Hoffman PL (1990) Hypothalamic vasopressin mRNA levels in mice are decreased after chronic ethanol ingestion. Eur J Pharmacol 189:119–127

Jörnvall H, Höög J-O, von Bahr-Lindström H, Vallee BL (1987) Mammalian alcohol dehydrogenase of separate classes: intermediates between different enzymes and intraclass isozymes. Proc Natl Acad Sci USA 84:2580–2584

Kaiser R, Holmquist B, Hempel J, Vallee BL, Jörnvall H (1988) Class III human liver alcohol dehydrogenase: a novel structure type equidistantly related to the class I and II enzymes. Biochem 27:1132–1140

Kalant H (1971) Absorbtion, diffusion, distribution and elimination of ethanol: effects on biological membranes. In: Kissin B, Begleiter H (eds) The biology of alcoholism, vol I. Plenum, New York, pp 1–62

Karobath M, Rogers J, Bloom FE (1980) Benzodiazepine receptors remain unchanged after chronic ethanol administration. Neuropharmacology 19:125–128

Katada T, Gilman AG, Watanabe Y, Bauer S, Jakobs KH (1985) Protein kinase C phosphorylates the inhibitory guanine-nucleotide-binding regulatory component and apparently suppresses its function in hormonal inhibition of adenylate cyclase. Eur J Biochem 151:431–437

Keir WJ, Al-Ghoul W, Morrow AL (1993) NMDA receptor subunit composition in withdrawal seizure prone and withdrawal seizure resistant mice. Alcohol Clin Exp Ther 17:477

Keith LD, Crabbe J, Robertson LM, Kendall JW (1986) Ethanol stimulated endorphin and corticotropin secretion in vitro. Brain Res 367:222–229

Kellenberger S, Malherbe P, Sigel E (1992) Function of the $\alpha1\beta2\gamma2s$ γ-aminobutyric acid type A receptor is modulated by protein kinase C via multiple phosphorylation sites. J Biol Chem 267:25660–25663

Kemp JA, Leeson PD (1993) The glycine site of the NMDA receptor – five years on. Trends Pharmacol Sci 14:20–25

Khanna JM, Kalant H, Shah G, Chau A (1992a) Effect of (+)MK-801 and ketamine on rapid tolerance to ethanol. Brain Res Bull 28:311–314

Khanna JM, Kalant H, Weiner J, Chau A, Shah G (1992b) Ketamine retards chronic but not acute tolerance to ethanol. Pharmacol Biochem Behav 42:347–350

Kharbanda S, Nakamura T, Kufe D (1993) Induction of the c-jun proto-oncongene by a protein kinase C-dependent mechanism during exposure of human epidermal keratinocytes to ethanol. Biochem Pharmacol 45:675–681

Khatami S, Hoffman PL, Shibuya T, Salafsky B (1987) Selective effects of ethanol on opiate receptor subtypes in brain. Neuropharmacology 26:1503–1507

Kiianmaa K, Tabakoff B (1983) Neurochemical correlates of tolerance and strain differences in the neurochemical effects of ethanol. Pharmacol Biochem Behav 18 [Suppl 1]:383–388

Kiss Z (1991) Cooperative effects of ethanol and protein kinase C activators on phospholipase-D-mediated hydrolysis of phosphatidylethanolamine in NIH 3T3 fibroblasts. J Biol Chem 266:10344–10350

Kleckner NW, Dingledine R (1988) Requirement for glycine in activation of NMDA receptors expressed in Xenopus oocytes. Science 241:835–837

Kleinschmidt A, Bear MF, Singer W (1987) Blockade of NMDA receptors disrupts experience-dependent plasticity of kitten striate cortex. Science 238:355–358

Kleuss C, Scherübl H, Hescheler J, Schultz G, Wittig B (1993) Selectivity in signal transduction determined by γ subunits of heterotrimeric G proteins. Science 259:832–834

Koltchine V, Anantharam V, Wilson A, Bayley H, Treistman SN (1993) Homomeric assemblies of NMDAR1 splice variants are sensitive to ethanol. Neurosci Lett 152:13–16

Komuro H, Rakic P (1993) Modulation of neuronal migration by NMDA receptors. Science 260:95–97

Koob GF (1992) Drugs of abuse: anatomy, pharmacology and the function of reward pathways. Trends Pharmacol Sci 13:177–184

Koppi S, Eberhardt G, Haller R, Konig P (1987) Calcium channel blocking agent in the treatment of acute alcohol withdrawal. Caroverine versus meprobamate in a randomized double-blind study. Neuropsychobiology 17:49–52

Korpi ER, Kleingoor C, Kettenmann H, Seeburg PH (1993) Benzodiazepine-induced motor impairment linked to point mutation in cerebellar $GABA_A$ receptor. Nature 361:356–359

Kozlowski GP, Long S, deSchweinitz JH (1989) Opposite effects of alcohol on numbers of immunoreactive vasopressin (VP) and oxytocin (OT) neurons in the paraventricular nucleus (PVN). Alcohol Clin Exp Res 13:317

Krupinski J, Lehman TC, Frankenfield CD, Zwaagstra JC, Watson PA (1992) Molecular diversity in the adenylylcyclase family: evidence for eight forms of the enzyme and cloning of type VI. J Biol Chem 267:24858–24862

Kumar KN, Tilakaratne N, Johnson PS, Allen AE, Michaelis EK (1991) Cloning of cDNA for the glutamate-binding subunit of an NMDA receptor complex. Nature 354:70–73

Kutsuwada T, Kashiwabuchi N, Mori H, Sakimura K, Kushiya E, Araki K, Meguro H, Masaki H, Kumanishi T, Arakawa M, Mishina M (1992) Molecular diversity of the NMDA receptor channel. Nature 358:36–41

Lê AD, Kalant H, Khanna JM (1982) Interaction between des-9-glycinamide-[8Arg]vasopressin and serotonin on ethanol tolerance. Eur J Pharmacol 80: 337–345

Lee RW, Lieberman BS, Yamane HK, Bok D, Fung BK-K (1992) A third form of the G protein β subunit. J Biol Chem 267:24776–24781

Lefkowitz RJ, Hausdorff WP, Caron MG (1990) Role of phosphorylation in desensitization of the β-adrenoceptor. Trends Pharmacol Sci 11:190–194

Leidenheimer NJ, Whiting PJ, Harris RA (1993) Activation of calcium-phospholipid-dependent protein kinase enhances benzodiazepine and barbiturate potentiation of the GABA$_A$ receptor. J Neurochem 60:1972–1975

Leslie SW (1986) Sedative-hypnotic drugs: interaction with calcium channels. Alcohol Drug Res 6:371–377

Lex BW, Ellingboe J, LaRosa K, Teoh SK, Mendelson JH (1993) Comparison of platelet monamine oxidase and adenylate cyclase activities in female alcoholics and in non-alcoholic control women with and without a family history of alcoholism. Harvard Rev Psych 1:229–237

Lieber CS (1985) Alcohol and liver: metabolism of ethanol, metabolic effects and pathogenesis of injury. Acta Med Scand [Suppl] 703:11–55

Lieber CS (1988) Biochemical and molecular basis of alcohol-induced injury to liver and other tissues. N Engl J Med 319:1639–1650

Lieber CS (1991a) Pathways of ethanol metabolism and related pathology. In: Palmer TN (ed) Alcoholism: a molecular perspective. Plenum, New York, pp 1–25

Lieber CS (1991b) Alcohol, liver and nutrition. J Am Coll Nutr 10:602–632

Lieber CS, DeCarli LM (1972) The role of the hepatic microsomal ethanol oxidizing system (MEOS) for ethanol metabolism in vivo. J Pharmacol Exp Ther 181:279–287

Liljequist S (1991) The competitive NMDA receptor antagonist, CGP 39551, inhibits ethanol withdrawal seizures. Eur J Pharmacol 192:197–198

Liljequist S, Engel J (1982) Effects of GABAergic agonists and antagonists on various ethanol-induced behavioral changes. Psychopharmacology 78:71–75

Liljequist S, Tabakoff B (1985) Binding characteristics of [^{3}H]flunitrazepam and CL 218,872 in cerebellum and cortex of C57BL mice made tolerant to and dependent on phenobarbital or ethanol. Alcohol 2:215–220

Liljequist S, Culp S, Tabakoff B (1986) Effect of ethanol on the binding of ^{35}S-t-butylbicyclophosphorothionate to mouse brain membranes. Life Sci 38: 1931–1939

Lima-Landman MT, Albuquerque EX (1989) Ethanol potentiates and blocks NMDA-activated single-channel currents in rat hippocampal pyramidal cells. FEBS Lett 247:61–67

Lin AM-Y, Freund RK, Palmer MR (1991) Ethanol potentiation of GABA-induced electrophysiological responses in cerebellum: requirement for catecholamine modulation. Neurosci Lett 122:154–158

Lin AM-Y, Bickford PC, Palmer MR (1993a) The effects of ethanol on γ-aminobutyric acid-induced depressions of cerebellar Purkinje neurons: influence of beta adrenergic receptor action in young and aged Fischer 344 rats. J Pharmacol Exp Ther 264:951–957

Lin AM-Y, Freund RK, Palmer MR (1993b) Sensitization of γ-aminobutyric acid-induced depressions of cerebellar Purkinje neurons to the potentiative effects of ethanol by beta adrenergic mechanisms in rat brain. J Pharmacol Exp Ther 265:426–432

Lin RC, Smith RS, Lumeng L (1988) Detection of a protein-acetaldehyde adduct in the liver of rats fed alcohol chronically. J Clin Invest 81:615–619

Lin T-A, Navidi M, James W, Lin T-N, Sun GY (1993c) Effects of acute ethanol administrtion on polyphosphoinositide turnover and levels of inositol 1,4,5-trisphosphate in mouse cerebrum and cerebellum. Alcohol Clin Exp Res 17:401–405

Lindros KO, Stowell A, Pikkarainen P, Salaspuro M (1980) Elevated blood acetaldehyde in alcoholics with accelerated ethanol elimination. Pharmacol biochem Behav 13 [Suppl 1]:119–124

Lister RG, Durcan MJ (1989) Antagonism of the intoxicating effects of ethanol by the potent benzodiazpine receptor ligand Ro 19-4603. Brain Res 482:141–144

Little HJ, Dolin SJ, Halsey MJ (1988a) Calcium channel antagonists decrease the ethanol withdrawal syndrome. Life Sci 39:2059–2065

Little HJ, Dolin SJ, Whittingto MA (1988b) Possible role of calcium channels in ethanol tolerance and dependence. Ann NY Acad Sci 560:465–466

Littleton JM, John GR, Grieve SJ (1979) Alterations in phospholipid composition in ethanol tolerance and dependence. Alcohol Clin Exp Res 3:50–56

Lovinger DM, Zhou Q (1993) Trichloroethanol potentiation of 5-hydroxytryptamine$_3$ receptor-mediated ion current in nodose ganglion neurons from the adult rat. J Pharmacol Exp Ther 265:771–776

Lovinger DM, White G, Weight FF (1989) Ethanol inhibits NMDA-activated ion current in hippocampal neurons. Science 243:1721–1724

Lovinger DM, White G, Weight FF (1990) NMDA receptor-mediated synaptic excitation selectively inhibited by ethanol in hippocampal slice from adult rat. J Neurosci 10:1372–1379

Lovinger DM, White G (1991) Ethanol potentiation of 5-hydroxytryptamine$_3$ receptor-mediated on current in neruroblastoma cells and isolated adult mammalian neurons. Mol Pharmacol 40:263–270

Lucchi L, Covelli V, Anthopoulou H, Spano PF, Trabucchi M (1983) Effect of chronic ethanol treatment on adenylate cyclase activity in at striatum. Neurosci Lett 40:187–192

Lucchi L, Govoni S, Battaini F, Pasinetti G, Trabucchi (1985) Ethanol administration in vivo alters calcium ions control in rat striatal striatum. Brain Res 332:376–379

Lüddens H, Wisden W (1991) Function and pharmacology of multiple GABA$_A$ receptor subunits. Trends Pharmacol Sci 12:49–51

Lüddens H, Pritchett DB, Köhler M, Killisch I, Keinänen K, Monyer H, Sprengel R, Seeburg PH (1990) Cerebellar GABA$_A$ receptor selective for a behavioural alcohol antagonist. Nature 346:648–651

Lundquist F, Tygstrup N, Winkler K, Kresten M, Munck-Peterson S (1962) Ethanol metabolism and production of free acetate in the human liver. J Clin Invest 5:955–961

Lukas S, Mendelson J (1986) Instrumental analysis of ethanol-induced intoxication in human males. Pyschopharmacology 89:8–13

Lustig KD, Conklin BR, Herzmark P, Taussig R, Bourne HR (1993) Type II adenylate cyclase integrates conincident singnals from G$_s$, and G$_q$. J Biol Chem 268:13900–13905

Luthin CR, Tabakoff B (1984) Activation of adenylate cyclase by alcohols requires the nucleotide-binding protein. J Pharmacol Exp Ther 228:579–587

Luttinger D, Nemeroff CB, Mason GA, Frye GD, Breese GR, Prange AJ Jr (1981) Enhancement of ethanol-induced sedation and hypothermia by centrally administered neurotensin, β-endorphin, and bombesin. Neuropharmacology 20:305–309

Lyon RC, Goldstein DB (1983) Changes in synaptic membrane order associated with chronic ethanol treatment in mice. Mol Pharmacol 23:86–91

MacDonald RL, Twyman RE, Ryan-Jastrow T, Angelotti TP (1992) Regulation of GABA$_A$ receptor channels by anticonvulsant and convulsant drugs and by phosphorylation. In: Engel J Jr, Wasterlain C, Cavalheiro EA, Heinemann U, Avanzini G (eds) Molecular neurobilogy of epilepsy (Epilepsy Res [Suppl] 9). Elsevier Science, Amsterdam, pp 265–277

Machu TK, Olsen RE, Browning MD (1992) Ethanol has no effect on cAMP-dependent protein kinase-, protein kinase C-, or Ca^{2+}-calmodulin-dependent protein kinase II-stimulated phosphorylation of highly purified substrates in vitro. Alcohol Clin Exp Res 16:290–294

Major LF, Ballenger JC, Goodwin FK, Brown GL (1977) Cerebrosponal fluid homovanillic acid in male alcoholics: effects of disulfiram. Biol Psych 12:635–642

Makram H, Segal M (1992) Activation of protein kinase C suppresses response to NMDA in rat CA1 hippocampal neurones. J Physiol (Lond) 457:491–501

Mardh G, Falchuk KH, Auld DS, Vallee BL (1986) Testosterone allostericlally regulates ethanol oxidation by homo- and heterodimeric γ-subunit-containing isoenzymes of human alcohol dehydrogenase. Proc Natl Acad Sci USA 83:2836–2840

Maricq AV, Peterson AS, Brake AJ, Myers RM, Julius D (1991) Primary structure and functional expression of the 5HT$_3$ receptor, a serotonin-gated ion channel. Science 254:432–437

Marks SS, Watson DL, Carpenter CL, Messing RO, Greenberg DA (1989) Comparative effects of chronic exposure to ethanol and calcium channel antagonists on calcium channel antagonist receptors in cultured neural (PC12) cells. J Neurochem 53:168–172

Marrosu F, Mereu G, Giorgi O, Corda MG (1988) The benzodiazepine recognition site inverse agonists Ro 15-4513 and G 7142 both antagonize the EEG effects of ethanol in the rat. Life Sci 43:2151–2158

Marshall AW, Kingstone D, Boss M, Morgan MY (1983) Ethanol elimination in males and females-relationship to menstrual cycle and body composition. Hepatology 3:701–706

Martin NG, Perl J, Oakeshott JG, Gibson JB, Starmer GA, Wilks AV (1985) A twin study of ethanol metabolism. Behav Genet 15:93–109

Mehta AK, Ticku MK (1988) Ethanol potentiation of GABAergic transmission in cultured spinal cord neurons involves γ-aminobutyric acid$_A$-gated chloride channels. J Pharmacol Exp Ther 246:558–564

Melchior CL, Collins MA (1982) The route and significance of endogenous synthesis of alkaloids in animals. CRC Crit Rev Toxicol 9:313–356

Mereu G, Fadda F, Gessa GL (1984) Ethanol stimulates the firing rate of nigral dopaminergic neurons in unanesthetized rats. Brain Res 292:63–69

Messing RO, Carpenter CL, Diamond I, Greenberg DA (1986) Ethanol regulates calcium channels in clonal neural cells. Proc Natl Acad Sci USA 83:6213–6215

Messing RO, Sneade AB, Savidge B (1990) Protein kinase C participates in up-regulation of dihydropyridine-sensitive calcium channels by ethanol. J Neurochem 55:1383–1389

Messing RO, Petersen PJ, Henrich CJ (1991) Chronic ethanol exposure increases levels of protein kinase C delta and epsilon and protein kinase C-dedicated phosphorylation in cultured neural cells. J Biol Chem 266:23428–23432

Meyer HH (1901) Zur Theorie der Alcoholnarkose: der Einfluss wechselnder Temperatur auf wirkungsstarke und Teilungskoeffizient der Narkotica. Naunyn Schmiedebergs Arch Exp Pharmacol 46:990–993

Mhatre MC, Ticku MJ (1992) Chronic ethanol administration alters γ-aminobutyric Acid$_A$ receptor gene expression. Mol Pharmacol 42:415–422

Mhatre MC, Mehta AK, Ticku MK (1988) Chronic ethanol administration increases the binding of the benzodiazepine inverse agonist and alcohol antagonist [³H]Ro 15-4513 in rat brain. Eur J Pharmacol 153:141–145

Michaelis EK, Michaelis ML, Freed WK (1978) Effects of acute and chronic ethanol intake on synaptosomal glutamate binding activity. Biochem Pharmacol 27: 1685–1691

Michaelis EK, Freed WK, Galton N, Foye J, Michaelis ML, Phillips I, Kleinman JE (1990) Glutamate receptor changes in brain synaptic membranes from human alcoholics. Neurochem Res 15:1055–1063

Milligan G, Green A (1991) Agonist control of G-protein levels. Trends Pharmacol Sci 12:207–209

Mochly-Rosen D, Change F-H, Cheever L, Kim M, Diamond I, Gordon AS (1988) Chronic ethanol causes heterologous desensitization of receptors by reducing α_s messenger RNA. Nature 333:848–850

Montpied P, Morrow AL, Karanian JW, Ginns EI, Martin BM, Paul SM (1991) Prolonged ethanol inhalation decreases γ-aminobutyric acid$_A$ receptor α subunit mRNAs in the rat cerebral cortex. Mol Pharmacol 39:157–163

Monyer H, Sprengel R, Schoepfer R, Herb A, Higuchi M, Lomeli H, Burnashev N, Sakmann B, Seeburg PH (1992) Heteromeric NMDA receptors: molecular and functional distinction of subtypes. Science 256:1217–1221

Moreno A, Pares X (1991) Purification and characterization of a new alcohol dehydrogenase from human stomach. J Biol Chem 266:1128–1133

Moriyoshi K, Masu M, Ishii T, Shigemoteo R, Mizuno N, Nakanishi S (1991) Molecular cloning and characterization of the rat NMDA receptor. Nature 354:31–37

Morrisett RA, Martin D, Wilson WA, Savage DD, Swartzwelder HS (1989) Prenatal exposure to ethanol decreases the sensitivity of the adult rat hippocampus to N-methyl-D-aspartate. Alchol 6:415–420

Morrisett RA, Rezvani AH, Overstreet, Wilson WA, Swartzwelder HS (1990) MK-801 potently inhibits alcohol withdrawal seizures in rats. Eur J Pharmacol 176:103–105

Morrisett RA, Martain D, Oetting TA, Lewis DV, Wilson WA, Swartzwelder HS (1991) Ethanol and magnesium ions inhibit N-methyl-D-aspartate-mediated synaptic potentials in an interactive manner. Neuropharmacology 30:1173–1178

Morrow AL, Suzdak PD, Karanian JW, Paul SW (1988) Chronic ethanol administration alters γ-aminobutyric acid, pentobarbital and ethanol-mediated $^{36}Cl^-$-uptake in cerebral cortical synaptoneurosomes. J Pharmacol Exp Ther 246: 158–164

Morrow AL, Herbert JS, Montpied P (1992) Differential effects of chronic ethanol administration on GABA$_A$ receptor $\alpha1$ and $\alpha6$ subunit mRNA levels in rat cerebellum. Mol Cell Neurosci 3:251–258

Moss SJ, Smart TG, Blackstone CD, Huganir RL (1992a) Functional modulation of GABA$_A$ receptors by cAMP-dependent protein phosphorylation. Science 257:661–665

Moss SJ, Doherty CA, Huganir RL (1992b) Identification of the cAMP-dependent protein kinase and protein kinase C phosphorylation sites within the major intracellular domains of the β_1, γ_{2L} subunits of the γ-aminobutyric acid type A receptor. J Biol Chem 267:14470–14476

Murphy JM, Waller MB, Gatto GJ, McBride WJ, Lumeng L, Li TK (1988) Effects of fluoxetine on the intragastric self-administration of ethanol in the alcohol preferring P line of rats. Alcohol 5:283–286

Myers RD, Melchior CL (1977a) Alcohol drinking: abnormal intake caused by tetrahydropapaveroline in brain. Science 196:554–556

Myers RD, Melchior CL (1977b) Differential action on voluntary alcohol intake of tetrahydroisoquinolines or a beta-carboline infused chronically in the ventricle of the rat. Pharmacol Biochem Behav 7:381–392

Myers RD, McCalbe ML, Ruwe WD (1982) Alcohol drinking induced in the monkey by tetrahydropapaveroline (THP) infused into the cerebral ventricle. Pharmacol biochem Behav 16:995–1000

Naber D, Soble MG, Pickar D (1981) Ethanol increases opioid activity in plasma of normal volunteers. Pharmacopsychiatry 14:160–161

Nagy LE, DeSilva SEF (1992) Ethanol increases receptor-dependent cyclic AMP production in cultured hepatocytes by decreasing G_i-mediated inhibition. Bochem J 286:681–686

Nagy LE, Diamond I, Gordon A (1988) Cultured lymphocytes from alcoholic subjects have altered cAMP signal transduction. Proc Natl Acad Sci USA 85:6973–6976

Nagy LE, Diamond I, Collier K, Lopez L, Ullman B, Gordon AS (1989) Adenosine is required for ethanol-induced heterologous desensitization. Mol Pharmacol 36:744–748

Nagy LE, Diamond I, Casso DJ, Franklin C, Gordon AS (1990) Ethanol increases extracellular adenosine by inhibiting adenosine uptake via the nucleoside transporter. J Biol Chem 265:1946–1951

Nagy LE, Diamond I, Gordon AS (1991) cAMP-dependent protein kinase regulates inhibition of adenosine transport by ethanol. Mol Pharmacol 40:812–817

Nakahiro M, Arakawa O, Narahashi T (1991) Modulation of γ-aminobutyric acid receptor-channel complex by alcohols. J Pharmacol Exp Ther 259:235–240

Nakanishi S (1992) Molecular diversity of glutamate receptors and implications for brain function. Science 258:597–603

Naranjo CA, Sellers EM, Roach CA, Woodley DV, Sanchez-Craig M, Sykora K (1984) Zimelidine-induced variations in alcohol intake by non-depressed heavy drinkers. Clin Pharmacol Ther 35:374–381

Naranjo CA, Seller EM, Sullivan JT, Woodley DV, Kalec K, Sykora K (1987) The serotonin uptake inhibitor citalopram attentuates ethanol intake. Clin Pharmacol Ther 41:374–266–274

Niemelä O, Klajner F, Orrego H, Vidins E, Blendis L, Israel Y (1987) Antibodies against acetaldehyde-modified protein epitopes in human alcoholics. Hepatology 7:1210–1214

Niemelä O, Juvonen T, Parkkila S (1991) Immunohistochemical demonstration of acetaldehyde-modified epitopes in human liver after alcohol consumption. J Clin Invest 87:1367–1374

Nutt DJ, Glue P (1986) Monoamines and alcohol. Br J Addict 81:327–338

Nutt DJ, Lister RG (1987) The effect of the imidazodiazepine Ro 15-4513 on the anticonvulsant effects of diazepam, sodium pentobarbital and ethanol. Brain Res 413:193–196

Nutt DJ, Lister RG, Rusche D, Bonetti EP, Reese RE, Rufener R (1988) Ro 15-4513 does not protect rats against the lethal effects of ethanol. Eur J Pharmacol 151:127–129

Ohmori T, Koyama T, Chen C, Yeh E, Reyes BV Jr, Yamashita I (1986) The role of aldehyde dehydrogenase isozyme variance in alcohol sensitivity, drinking habits formation and the development of alcoholism in Japan, Taiwan and the Phillippines. Prog Neuropsychopharmacol Biol Psychiatry 10:229–235

Ontko JA (1973) Effects of ethanol on the metabolism of free fatty acids in isolated liver cells. J Lipid Res 14:78–86

Orrego H, Carmichael FJ, Israel Y (1988) New insights on the mechanism of the alcohol-induced increase in portal blood flow. Can J Physiol Pharmacol 66:1–9

Overton E (1901) Studien über die Narkose Zugleich ein Beitrag zur allemeinen Pharmakologie. Fisher, Jena

Palmer KR, Jenkins WJ (1982) Impaired acetaldehyde oxidation in alcoholics. Gut 23:729–733

Palmer KR, Jenkins WJ (1985) Aldehyde dehydrogenase in alcoholic subjects. Hepatology 5:260–263

Palmer MR, Morrow EL, Erwin VG (1987) Calcium differentially alters behavioral and electrophysiological responses to ethanol in selectively bred mouse lines. Alcohol Clin Exp Res 11:457–463

Palmer MR, Van Horne CG, Harlan JT, Moore EA (1988) Antagonism of ethanol effects on cerebellar Purkinje neurons by the benzodiazepine inverse agonists Ro 15-4513 and FG 7142: electrophysiological studies. J Pharmacol Exp Ther 247:1018–1024

Park D, Jhon D-Y, Lee C-W, Lee K-H, Rhee SG (1993) Activation of phospholipase C isozymes by G protein $\beta\gamma$ subunits. J Biol Chem 268:4573–4576

Peoples RW, Weight FF (1992) Ethanol inhibition of N-methyl-D-aspartate-activated ion current in rat hippocampal neurons is not competitive with glycine. Brain Res 571:342–344

Perez-Reyes M, White WR, Hicks RE (1992) Interaction between ethanol and calcium channel blockers in humans. Alcohol Clin Exp Res 16:769–775

Perlman BJ, Goldstein DB (1984) Genetic influences on the central nervous system depressant and membrane-disordering actions of ethanol and sodium valproate. Mol Pharmacol 26:547–552

Peters JA, Malone HM, Lambert JJ (1992) Recent advances in the electrophysiological characterization of 5-HT$_3$ receptors. Trends Pharmacol Sci 13:391–397

Peters S, Koh J, Choi DW (1987) Zinc selectively blocks the action of N-methyl-D-aspartate on cortical neurons. Science 236:589–593

Pfeiffer A, Seizinger BR, Herz A (1981) Chronic ethanol imbibition interferes with δ, but not with μ-opiate receptors. Neuropharmacology 20:1229–1232

Phillips TJ, Feller DJ, Crabbe JC (1989) Selected mouse lines, alcohol and behavior. Experientia 45:805–827

Polokoff MA, Simon TH, Harris RA, Simon FR, Iwahashi M (1985) Chronic ethanol increases liver plasma membrane fluidity. Biochemistry 24:3114–3120

Poulos CX, Cappel H (1991) Homeostatic theory of drug tolerance: a general model of physiological adaptation. Psychol Rev 98:390–408

Premont RT, Jacobowitz O, Iyengar R (1992) Lowered responsiveness of the catalyst of adenylyl cyclase to stimulation by the G$_s$ in heterologous desensitization: a role for adenosine 3',5'-monophosphate-dependent phosphorylation. Endocrinology 131:2774–2784

Pritchett DB, Sontheimer H, Shivers BD, Ymer S, Kettenmann H, Schoifield PR, Seeburg PH (1989) Importance of a novel GABA$_A$ receptor subunit for benzodiazepine pharmacology. Nature 338:582–585

Przewlocka B, Lason W, Przewlocki R (1992) Repeated ethanol administration decreases prodynorphin biosynthesis in the rat hippocampus. Neurosci Lett 134:195–198

Pyne NJ, Freissmuth M, Palmer S (1992) Phosphorylation of the spliced variant forms of the recombinant stimulatory guanine-nucleotide-binding regulatory protein (G$_{\alpha s}$) by protein kinase C. Biochem J 285:333–338

Rabe CS, Tabakoff B (1990) Glycine site directed agonists reverse ethanol's actions at the NMDA receptor. Mol Pharmacol 38:753–757

Rabe CS, Rathna Giri P, Hoffman PL, Tabakoff B (1990) Effect of ethanol on cyclic AMP levels in intact PC12 cells. Biochem Pharmacol 40:565–571

Rabin RA (1985) Effect of ethanol on inhibition of striatal adenylate cyclase activity. Biochem Pharmacol 34:4329–4331

Rabin RA (1988) Differential response of adenylate cyclase and ATPase activities after chronic ethanol exposure of PC12 cells. J Neurochem 51:1148–1155

Rabin RA (1990a) Direct effects of chronic ethanol exposure of β-adrenergic and adenosine-sesitive adenylate cyclase activities and cyclic AMP content in primary cerebellar cultures. J Neurochem 55:122–128

Rabin RA (1990b) Chronic ethanol exposure of PC 12 cells alters adenylate cyclase activity and intracelular cyclic AMP content. J Pharmacol Exp Ther 252:1021–1027

Rabin RA (1993) Ethanol-induced desensitization of adenylate cyclase: role of the adenosine receptor and GTP-binding proteins. J Pharmacol Exp Ther 264:977–983

Rabin RA, Molinoff PB (1983) Multiple sites of action of ethanol on adenylate cyclase. J Pharmarol Exp Ther 227:551–556

Rabin RA, Wolfe BB, Dibner MD, Zahniser NR, Melchior C, Molinoff PB (1980) Effects of ethanol administration and withdrawal on neurotransmitter receptor systems in C57 mice. J Pharmacol Exp Ther 213:491–496

Rabin RA, Bode DC, Molinoff PB (1986) Relationship between ethanol-induced alterations in fluorescence anisotropy and adenylate cyclase activity. Biochem Pharmacol 35:2331–2335

Rabin RA, Baker RC, Deitrich RA (1987) Effects of chronic ethanol exposure on adenylate cyclase activities in the rat. Pharmacol Bocheem Behav 26:693–697

Rabin RA, Edleman AM, Wagner JA (1992) Activation of protein kinase A is necessary but not sufficient for ethanol-induced desensitization of cyclic AMP production. J Pharmacol Exp Ther 262:257–262

Ramoa AS, Alkondon M, Aracava Y, Irons J, Lunt GG, Deshpande SS, Wonacott S, Aronstam RS, Albuquerque EX (1990) The anticonvulsant MK-801 interacts with peripheral and central nicotinic acetylcholine receptor ion channels. J Pharmacol Exp Ther 254:71–81

Rand ML, Vickers JD, Kinlough-Rathbone RL, Packham MA, Mustard JF (1988) Thrombin-induced inositol trisphosphate production by rabbit platelets is inhibited by ethanol. Biochem J 251:279–284

Ransnäs LA, Insel PA (1988) Quantitation of the guanine nucleotide binding regulatory protein G_s in S49 cell membranes using antipeptide antibodies ot α_s. J Biol Chem 263:9482–9485

Reynolds JN, Prasad A, MacDonald JF (1992) Ethanol modulation of GABA receptor-activated Cl^- currents in neurons of the chick, rat and mouse central nervous system. Eur J Pharmacol 224:173–181

Rhee SG, Choi KD (1992) Regulation of inositol phospholipid-specific phsopholipase C isozymes. J Biol Chem 267:12393–12396

Richelson E, Stenstrom S, Forrary C, Enloe L, Pfenning M (1986) Effects of chronic exposure to ethanol on the prostaglandin E1 receptor-mediated response and binding in a murine neuroblastoma clone (N1E-115). J Pharmacol Exp Ther 239:687–692

Rodriguez FD, Simonsson P, Gustavsson L, Alling C (1992) Mechanisms of adaptation to the effects of ethanol on activation of phospholipase C in NG 108-15 cells. Neuropharmacology 31:1157–1164

Ross EM, Howlett AC, Ferguson KM, Gilman G (1978) Reconstitution of hormone-sensitive adenylate cyclase activity with resolved components of the enzyme. J Biol Chem 253:6401–6412

Rottenberg H (1987) Partition of ethanol and other amphiphilic compounds modulated by chronic alcoholism. Ann NY Acad Sci 492:112–124

Rottenberg H, Bittman R, Li H-L (1992) Resistance to ethanol disordering of membranes from ethanol-fed rats is conferred by all phospholipid classes. Biochim Biophys Acta 1123:282–290

Rubin R, Hoek JB (1988) Alcohol-induced stimulation of phospholipase C in human platelets requires G-protein activation. Biochem J 254:147–153

Ryder S, Straus E, Lieber CS, Yalow RS (1981) Cholecystokinin and enkephalin levels following ethanol administration in rats. Peptides 2:223–226

Saffey K, Gillman MA, Cantrill RC (1988) Chronic in vivo ethanol administration alters the sensitivity of adenylate cyclase coupling in homogenates of rat brain. Neurosci Lett 84:317–322

Saito T, Lee JM, Tabakoff B (1985) Ethanol's effects on cortical adenylate cyclase activity. J Neurochem 44:1037–1044

Saito T, Lee JM, Hoffman PL, Tabakoff B (1987) Effects of chronic ethanol treatment on the β-adrenergic receptor-coupled adenylate cyclase system of mouse cerebral cortex. J Neurochem 48:1817–1822

Samson HH, Hoffman PL (1995) The involvement of CNS catecholamines in alcohol self-administration tolerance and dependence: preclinical studies. In: Kranzler H (ed) Handbook of experimemtal pharmacology: the pharmacology of alcohol abuse. Springer, Berlin Heideberg New York, pp 121–137

Samson HH, Tolliver CA, Schwaz-Stevens K (1990) Oral ethanol self-administration: A behavioral pharmacological approach to CNS control mechanism. Alcohol 7:187–197

Sarkar DK, Minami S (1990) Effect of acute ethanol on beta-endorphin secretion from rat fetal hypothalamic neurons in primary cultures. Life Sci 47:31–36

Savage DD, Queen SA, Sanchez CF, Paxton LL, Mahoney JC, Goodlett CR, West JR (1991) Prenatal ethanol exposure during the last third of gestation in rat reduces hippocampal NMDA agonist binding site density in 45-day-old offspring. Alcohol 9:37–41

Scanlon MN, Lazar-Wesley E, Grant KA, Kunos G (1992) Proopiomelanocortin messenger RNA is decreased in the mediobasal hypothalamus of rats made dependent on ethanol. Alcohol Clin Exp Res 16:1147–1151

Schoepp DD, Conn PJ (1993) Metabotropic glutamate receptors in brain function and pathology. Trends Pharmacol Sci 14:13–20

Schroeder F, Morrison WJ, Gorka C, Wood WG (1988) Transbilayer effects of ethanol on fluidity of brain membrane leaflets. Biochim Biphys Acta 946:85–94

Schulz R, Wuster M, Duka T, Herz A (1980) Acute and chronic ethanol treatment changes endorphin levels in brain and pituitary. Psychopharmacology 68:221–227

Scott RH, Sutton KG, Dolphin AC (1993) Interactions of polyamines with neuronal ion channels. Trends Neurosci 16:153–160

Seamon KB, Daly JW (1986) Forskolin: its biological and chemical properties. Adv Cyclic Nucl Prot Phos Res 20:1–150

Seizinger BR, Boverman K, Maysinger D, Höllt V, Herz A (1983) Differential effects of acute and chronic ethanol treatment on particular opioid peptide systems in discrete regions of rat brain and pituitary. Pharmacol Biochem Behav 18:361–369

Seizinger BR, Höllt V, Herz A (1984a) Effects of chronic ethanol treatment on the in vitro biosynthesis of pro-opiomelanocortin and its post-translational processing to β-endorphin in the intermediate lobe of the rat pituitary. J Neurochem 43:607–613

Seizinger BR, Bovermann K, Höllt V, Herz A (1984b) Enhanced activity of the endorphinergic system in the anterior and neurointermediate lobe of the rat pituitary gland after chronic treatment with ethanol liquid diet. J Pharmacol Exp Ther 230:455–461

Sessler FM, Mouradian RD, Cheng JT, Yeh HH, Liu W, Waterhouse BD (1989) Noradrenergic potentation of cerebellar Purkinje cells responses to GABA: evidence for mediation through the beta-adrenoceptor-coupled cyclic AMP system. Brain Res 499:27–38

Seventh Special Report to the US Congress on Alcohol and Health (1990) US Department of Health and Human Services. National Institute on Alcohol Abuse and Alcoholism, DHHS publication no (ADM) 90–1656

Shah J, Pant HC (1988) Spontaneous calcium release induced by ethanol in the isolated rat brain microsomes. Brain Res 474:94–99

Shefner SA (1990) Electrophysiological effects of ethanol on brain neurons. In: Watson RR (ed) Biochemistry and physiology of substance abuse, vol II. CRC Press, Boca Raton, pp 25–52

Shefner SA, Tabakoff B (1985) Basal firing rate of rat locus coeruleus neurons affects sensitivity to ethanol. Alchol 2:239–243

Sieghart W (1992a) GABA$_A$ receptors: ligand-gated Cl$^-$ ion channels modulated by multiple drug-binding sites. Trends Pharmacol Sci 13:446–450

Sieghart W (1992b) Molecular basis of pharmacological heterogeneity of $GABA_A$ receptors. Cell Signal 4:231–237

Simonsson P, Rodriguez FD, Loman N, Alling C (1991) G proteins coupled to phospholipase C: molecular targets of long-term ethanol exposure. J Neurochem 56:2018–2026

Simson PE, Criswell HE, Johnson KB, Hicks RE, Breese GR (1991) Ethanol inhibits NMDA-evoked elecltrophysiological activity in vivo. J Pharmacol Exp Ther 257:225–231

Simson PE, Criswell HE, Breese GR (1993) Inhibition of NMDA-evoked electro-physiological activity be ethanol in selected brain regions: evidence for ethanol-sensitive and ethanol-insensitive NMDA-evoked responses. Brain Res 607:9–16

Sinclair JG, Lo GF (1978) Acute tolerance to ethanol on the release of acetylcholine from the cat cerebral cortex. Can J Physiol Pharmacol 56:668–670

Sippel HW (1974) The acetaldehyde content of rat brain during ethanol metabolism. J Neurochem 23:451–452

Skattebol A, Rabin RA (1987) Effects of ethanol on $^{45}Ca^{2+}$ uptake in synaptosomes and PC12 cells. Biochem Pharmacol 36:2227–2229

Skwish S, Shain W (1991) Ethanol and diolein stimulate PKC translocation in astroglial cells. Alcohol Clin Exp Res 15:1040–1044

Slater SJ, Cox KJA, Lombardi JV, Ho C, Kelly MB, Rubin E, Stubbs CD (1993) Inhibition of protein kinase C by alcohols and anaesthetics. Nature 364:82–84

Smith M (1986) Genetics of human alcohol and aldehyde dehydrogenases. Adv Hum Genet 15:249–290

Smith M, Hopkinson DA, Harris H (1971) Developmental changes and polymor-phisms in human alcohol dehydrogenase. Ann Hum Genet 34:251–271

Smith SG, Werner TE, Davis WM (1976) Comparison between intravenous and intragastric alcohol self-administration. Physiol Psychol 4:91–93

Smith TL (1990) The effects of acute exposure to ethanol on neurotensin and guanine nucleotide-stimulation of phospholipase C activity in intact N1E-115 neuroblastoma cells. Life Sci 47:115–119

Smith TL (1991) Selective effects of acute and chronic ethanol exposure on neuro-peptide and guanine nucleotide stimulated phospholipase C activity in intact N1E-115 neuroblastoma. J Pharmacol Exp Ther 258:410–415

Snell LD, Tabakoff B, Hoffman PL (1993) Radioligand binding to the N-methyl-D-aspartate receptor/ionophore complex: alterations by ethanol in vitro and by chronic in vivo ethanol ingestion. Brain Res 602:91–98

Snell LD, Iorio KR, Tabakoff B, Hoffman PL (1994a) Protein kinase C activation attenuates N-methyl-D-aspartate induced increases in intracellular calcium in cerebellar granule cells. J Neurochem 62:1783–1789

Snell LD, Tabakoff B, Hoffman PL (1994b) Involvement of protein kinase C in ethanol-induced inhibition of NMDA receptor function in cerebellar granule cells. Alcohol Clin Exp Res 18:81–85

Sommer B, Seeburg PH (1992) Glutamate receptor channels: novel properties and new clones. Trends Pharmacol Sci 13:291–296

Stenstrom S, Richelson E (1982) Acute effect of ethanol on prostaglandin E_1-mediated cyclic AMP formation by a murine neuroblastoma clone. J Pharmacol Exp Ther 221:334–341

Stern P, Béhé P, Schoepfer R, Colquhoun D (1992) Single-channel conductances of NMDA receptors expressed from cloned cDNAs: comparison with native receptors. Proc R Soc Lond [Biol] 250:271–277

Sternweis PC, Robishaw JD (1984) Isolation of two proteins with high affinity for guanine nucleotides from membranes of bovine brain. J Biol Chem 259:13806–13813

Stowell A, Hillbom M, Salaspuro M, Lindros KO (1980) Low acetaldehyde levels in blood, breath and cerebrospinal fluid of intoxicated humans as assayed by improved methods. In: Thurman RG (ed) Alcohol and aldehyde metabolizing systems-IV. Plenum, New York, pp 635–645

Sugihara H, Moriyoshi K, Ishii T, Masu M, Nakanishi S (1992) Structures and properties of seven isoforms of the NMDA receptor generated by alternative splicing. Biochem Biophys Res Commun 185:826–832

Sun GY, Sun AY (1977) Effect of chronic ethanol administration on phospholipid acyl groups of synaptic plasma membrane fraction isolated from guinea pig brain. Res Commun Chem Pathol Pharmacol 18:753–756

Sun GY, Navidi M, Yoa FG, Wood WG, Sun AY (1993) Effects of chronic ethanol administration on polyphosphoinositide metabolism in the mouse brain: variance with age. Neurochem Int 22:11–17

Suzdak PD, Glowa JR, Crawley JN, Schwartz RD, Skolnick P, Paul SM (1986a) A selective imidazobenzodiazepine antagonist of ethanol in rat. Science 234: 1243–1247

Suzdak PD, Schwartz RD, Skolnick P, Paul SM (1986b) Ethanol stimulates γ-aminobutyric acid receptor-mediated chloride transport in rat brain synaptoneurosomes. Proc Natl Acad Sci USA 84:4071–4075

Szabó G, Tabakoff B, Hoffman PL (1988) Receptors with V_1 characteristics mediate the maintenance of ethanol tolerance by vasopressin. J Pharmacol Exp Ther 247:536–541

Szabó G, Tabakoff B, Hoffman PL (1994) The NMDA receptor antagonist, dizocilpine, differentially affects environment-dependent and environment-independent ethanol tolerance. Pyschopharmacology 113:511–517

Tabakoff B (1989) Treatment of alcoholism. N Engl J Med 231:400

Tabakoff B, Hoffman PL (1979) Development of functional dependence on ethanol in dopaminergic systems. J Pharmacol Exp Ther 208:216–222

Tabakoff B, Hoffman PL (1983) Alcohol interactions with brain opiate receptors. Life Sci 32:197–204

Tabakoff B, Hoffman PL (1987) Biochemical pharmacology of alcohol. In: Meltzer HY (ed) Psychopharmacology – the third generation of progress. Raven, New York, pp 1521–1526

Tabakoff B, Hoffman PL (1989) Genetics and biological markers of risk for alcoholism. In: Kiianmaa K, Tabakoff B, Saito T (eds) Genetic aspects of alcoholism. The Finnish foundation for alcohol studies, Helsinki, pp 127–142

Tabakoff B, Hoffman PL (1992) Alcohol: Neurobiology. In: Lowinson JH, Ruiz P, Millman RB (eds) Substance abuse: a comprehensive textbook, 2nd edn. Williams and Wilkins. Baltimore, pp 152–185

Tabakoff B, Kiianmaa K (1982) Does tolerance develop to the activating, as well as the depressant, effects of ethanol? Pharmacol Biochem Behav 17:1073–1076

Tabakoff B, Ritzmann RF (1977) The effects of 6-hydroxydopamine on tolerance to and dependence on ethanol. J Pharmacol Exp Ther 203:319–332

Tabakoff B, Rothstein JD (1983) Biology of tolerance and dependence. In: Tabakoff B, Sutker PB, Randall CL (eds) Medical and social aspects of alcohol abuse. Plenum, New York, pp 187–220

Tabakoff B, von Wartburg J-P (1975) Separation of aldehyde reductases and alcohol dehydrogenase from brain by affinity chromatography: metabolism of succinic semialdehyde and ethanol. Biochem Biophys Res Commun 63:957–966

Tabakoff B, Ungar F, Alivisatos SG (1972) Aldehyde derivatives of indoleamines and the enhancement of their binding onto brain macromolecules by pentobarbital and acetaldehyde. Nature New Biol 238:126–128

Tabakoff B, Hoffman PL, Moses F (1977) Neurochemical correlates of ethanol withdrawal: alterations in serotonergic function. J Pharm Pharmacol 29:471–476

Tabkoff B, Munoz-Marcus M, Fields JZ (1979) Chronic ethanol feeding produces an increase in muscarinic cholinergic receptors in mouse brain. Life Sci 25: 2173–2180

Tabakoff B, Urwyler S, Hoffman PL (1981) Ethanol alters kinetic characteristics and function of striatal morphine receptors. J Neurochem 37:518–521

Tabakoff B, Melchior CL, Hoffman PL (1982) Commentary on ethanol tolerance. Alcohol Clin Exp Res 6:252–259

Tabakoff B, Cornell N, Hoffman PL (1986) Alcohol tolerance. Ann Emerg Med 15:1005–1012

Tabakoff B, Hoffman PL, McLaughlin A (1988a) Is ethanol a discriminating substance? Semin Liver Dis 8:26–35

Tabakoff B, Hoffman PL, Lee JM, Saito T, Willard B, De Leon-Jones F (1988b) Differences in platelet enzyme activity between alcoholics and nonalcoholics. N Engl J Med 318:134–139

Tabakoff B, Whelan JP, Hoffman PL (1990) Two biological markers of alcoholism. Banbury Rep 33:195–204

Tabakoff B, Rabe CS, Hoffman PL (1991) Selective effects of sedative/hypnotic drugs on excitatory amino acid receptors in brain. Ann NY Acad Sci 625:488–495

Takadera T, Suzuki R, Mohri T (1990) Protection by ethanol of cortical neurons from N-methyl-D-aspartate-induced neurotoxicity is associated with blocking calcium influx. Brain Res 537:109–115

Tang W-J, Gilman AG (1991) Type-specific regulation of adenylyl cyclase by G protein $\beta\gamma$ subunits. Science 254:1500–1503

Tang W-J, Gilman AG (1992) Adenylyl cyclases. Cell 70:869–872

Taraschi TF, Ellingson JS, Wu A, Zimmerman R, Rubin E (1986) Membrane tolerance to ethanol is rapidly lost after withdrawal; a model for studies of membrane adaptation. Proc Natl Acad Sci USA 83:3669–3673

Taussig R, Quarmby LA, Gilman G (1993a) Regulation of purified type I and type II adenylylcyclases by G protein $\beta\gamma$ subunits. J Biol Chem 268:9–12

Taussig R, Iñiguez-Lluhi JA, Gilman AG (1993b) Inhibition of adenylyl cyclase by $G_{i\alpha}$. Science 261:218–221

Thomasson HR, Li T-K, Crabb DW (1990) Correlation between alcohol-induced flushing, genotypes for alcohol and aldehyde dyhydrogenases, and alcohol elimination rates. Hepatology 12:903

Thomasson HR, Edenberg JH, Crabb DW, Mai X-L, Jerome RE, Li T-K, Wang S-P, Lin Y-T, Lu RB, Yin SJ (1991) Alcohol and aldehyde dehydrogenase genotypes and alcoholism in Chinese men. Am J Hum Genet 48:677–681

Tottmar SOC, Pettersson H, Kiessling K-H (1973) The subcellular distribution and properties of aldehyde dehydrogenases in rat liver. Biochem J 135:577–586

Treistman SN, Wilson A (1991) Effects of chronic ethanol on currents carried through calcium channels in Aplysia. Alcohol Clin Exp Res 15:489–493

Treistman SN, Moynihan MM, Wolf De (1987) Influence of alcohol, temperature, and region on the mobility of lipids in neuronal membrane. Biochim Biophys Acta 898:109–120

Twombly DA, Herman MD, Kye CH, Narahasi T (1990) Ethanol effects on two types of voltage-activated calcium channels. J Pharmacol Exp Ther 254:1029–1037

Uhl G, Blum K, Noble E, Smith S (1993) Substance abuse vulnerability and D_2 receptor genes. Trends Neurosci 16:83–88

Unwin N (1993) Neurotransmitter action: opening of ligand-gated ion channels. Cell 27:31–41

Urushihara H, Tohda M, Nomura Y (1992) Selective potentiation of N-methyl-D-aspartate-induced current by protein kinase C in Xenopus oocytes injected with rat brain RNA. J Biol Chem 267:11697–11700

Valverius P, Hoffman PL, Tabakoff B (1987) Effect of ethanol on mouse cerebral cortical β-adrenergic receptors. Mol Pharmacol 32:217–222

Valverius P, Hoffman PL, Tabakoff B (1988) β-adrenergic receptor binding in brain of alcoholics. Exp Neurol 105:280–286

Valverius P, Hoffman PL, Tabakoff B (1989a) Hippocampal and cerebellar β-adrenergic receptors and adenylate cyclase are differentially altered by chronic ethanol ingestion. J Neurochem 52:492–497

Valverius P, Hoffman PL, Tabakoff B (1989b) Brain forskolin binding in mice dependent on and tolerant to ethanol. Brain Res 503:38–43

Valverius P, Crabbe JC, Hoffman PL, Tabakoff B (1990) NMDA receptors in mice bred to be prone or resistant to ethanol withdrawal seizures. Eur J Pharmacol 184:185–189

van Thiel DH, Lester R (1976) Alcoholism: its effect on hypothalamic pituitary gonadal function. Gastroenterology 71:318–327

Vescovi PP, Coiro V, Volpi, Giannini A, Passeri M (1992) Plasma β-endorphin, but not met-enkephalin levels are abnormal in chronic alcoholics. Alcohol Alcohol 27:471–475

Vinay P, Cardoso M, Tejedor A, Prud'homme M, Levelillee M, Vinet B, Courteau M, Gougoux A, Rengel M, Lapierre L, Piette Y (1987) Acetate metabolism during hemodialysis: metabolic considerations. Am J Nephrol 7:337–354

Volpicelli JR, Alterman AI, Hayashida M, O'Brien CP (1992) Naltrexone in the treatment of alchohol dependence. Arch Gen Psychiatry 49:876–880

Voltaire-Carlsson A, Hiltunen A, Beck O, Borg S (1993) Clinical ratings, self-reports and urinary 5-hydroxytryptophol in relation to relapse in alcohol dependent male patients. Alcohol Alcohol 28:252

von Knorring A-L, Bohman M, von Knorring L, Oreland L (1985) Platelet MAO activity as a biological marker in subgroups of alcoholism. Acta Psychiat Scand 72:52

von Wartburg J-P (1976) Biochemische auswirkungen des alkoholkonsums. Bibl Nutr Dieta 24:7–16

von Wartburg J-P, Schurch PM (1968) Atypical human liver alcohol dehydrogenase. Ann NY Acad Sci 151:937–946

von Wartburg J-P, Papenberg J, Aebi H (1965) An atypical human alcohol dehydrogenase. Can J Biochem 43:889–898

Wafford KA, Whiting PJ (1992) Ethanol potentiation of $GABA_A$ receptors requires phosphorylation of the alternatively spliced variant of the gamma 2 subunit. FEBS Lett 313:113–117

Wafford KA, Burnett DM, Dunwiddie TV, Harris RA (1990) Genetic differences in the ethanol sensitivity of $GABA_A$ receptors expressed in Xenopus oocytes. Science 249:291–293

Wafford KA, Burnett DM, Leidenheimer NJ, Burt DR, Wang JB, Kofuji P, Dunwiddie TV, Harris RA, Sikela JM (1991) Ethanol sensitivity of the $GABA_A$ receptor expressed in Xenopus oocytes requires eight amino acids contained in the γ_{2L} subunit of the receptor complex. Neuron 7:27–33

Wagner FW, Pares X, Holmquist B, Vallee BL (1984) Physical and enzymatic properties of a class III isoenzyme of human liver alcohol dehydrogenase: χ-ADH. Biochemistry 23:2193–2199

Waltman C, Levine MA, McCaul ME, Svikis DS, Wand GS (1993) Enhanced expression of the inhibitory protein $G_{i\alpha}$ and decreased activity of adenylyl cyclase in lymphocytes of abstinent alcoholics. Alcohol Clin Exp Res 17:315–320

Wand GS, Levine MA (1991) Hormonal tolerance to ethanol is associated with decreased expression of the GTP-binding protein, $G_{s\alpha}$, and adenylyl cyclase activity in ethanol-treated mice. Alcohol Clin Exp Res 15:705–710

Wand GS, Diehl AM, Levine MA, Wolfgang D, Samy S (1993) Chronic ethanol treatment increases expression of inhibitory G-proteins and reduces adenylylcyclase activity in the central nervous system of two lines of ethanol-sensitive mice. J Biol Chem 268:2595–2601

Wang X, Lemos JR, Dayanithi G, Nordmann JJ, Treistman SN (1991) Ethanol reduces vasopressin release by inhibiting calcium currents in nerve terminals. Brain Res 551:338–341

Waring AJ, Rottenberg H, Ohnishi T, Rubin E (1981) Membranes and phospholipids of liver mitochondria from chronic alcoholic rats are resistant to membrane disordering by alcohol. Proc Natl Acad Sci USA 78:2582–2586

Watanabe M, Inoue Y, Sakimura K, Mishina M (1992) Developmental changes in distribution of NMDA receptor channel subunit mRNAs. Dev Neurosci 3:1138–1140

Weight FF, Lovinger DM, White G, Peoples RW (1991) Alcohol and anesthetic actions on excitatory amino acid-activated ion channels. Ann NY Acad Sci 625:97–107

Weiner H, Flynn TG (1988) Nomenclature of mammalian aldehyde dehydrogenases. In: Weiner H, Flynn TG (eds) Enzymology and molecular biology of carbonyl metabolism. Liss, New York, pp xix–xxi

Whelan JP, Hoffman PL, Tabakoff B (1991) Effects of chronic ethanol exposure on G-proteins in mouse brain membranes. Alcohol Clin Exp Res 15:333

White GD, Lovinger DM, Weight FF (1990a) Ethanol inhibits NMDA-activated current but does not affect GABA-activated current in an isolated adult mammalian neuron. Brain Res 507:332–336

White GD, Lovinger DM, Grant KA (1990b) Ethanol (EtOH) inhibition of NMDA-activated ion current is not altered after chronic exposure of rats or neurons in culture. Alcohol Clin Exp Res 14:352

White JM, Smith AM (1992) Modification of the behavioral effects of ethanol by nifedipine. Alcohol Alcohol 27:137–141

Whittington MA, Little HJ (1991) Nitrendipine, given during drinking, decreases the electrophysiological changes in the isolated hippocampal slice, seen during ethanol withdrawal. Br J Pharmacol 103:1677–1684

Widmark EMP (1933) Der einfluss der nahrungsbestandteile auf den alcoholgehalt des blutes. Biochem Z 267:135–151

Williams K, Dichter MA, Molinoff PB (1992) Up-regulation of N-methyl-D-aspartate receptors on cultured cortical neurons after exposure to antagonists. Mol Pharmacol 42:147–151

Williams RJ, Veale MA, Horne P, Kelly E (1993) Ethanol differentially regulates guanine nucleotide-binding protein α subunit expression in NG108-15 cells independently of extracellular adenosine. Mol Pharmacol 43:158–166

Wise RA (1978) Catecholamine theories of reward: a critical review. Brain Res 152:215–247

Wong YH, Conklin BR, Bourne HR (1992) G_z-mediated hormonal inhibition of cyclic AMP accumulation. Science 255:339–342

Wood WG, Schroeder F (1988) Membrane effects of ethanol: bulk lipid versus lipid domains. Life Sci 43:467–475

Wood WG, Gorka C, Schroeder F (1989) Acute and chronic effects of ethanol on transbilayer membrane domains. J Neurochem 52:1925–1930

Wood WG, Schroeder F, Hogy L, Rao AM, Nemecz G (1990) Asymmetric distribution of a fluorescent sterol in synaptic plasma membranes: effects of chronic ethanol consumption. Biochim Biophys Acta 1025:243–246

Wood WG, Gorka C, Johnson JA, Sun GY, Sun AY, Schroeder F (1991) Chronic ethanol consumption alters transbilayer distribution of phosphatidylcholine in erythrocytes of Sinclair (S-1) miniature swine. Alcohol 8:395–399

Woodward JJ, Gonzales RA (1990) Ethanol inhibition of N-methyl-D-aspartate stimulated endogenous dopamine release from rat striatal slices: reversal by glycine. J Neurochem 54:712–715

Woodward JJ, Machu T, Leslie SW (1990) Chronic ethanol treatment alters ω-conotoxin and Bay K 8644 sensitive calcium channels in rat striatal synaptosomes. Alcohol 7:279–284

Wozniak KM, Pert A, Linnoila M (1990) Antagonism of 5-HT$_3$ receptors attenuates the effects of ethanol on extracellular dopamine. Eur J Pharmacol 187:287–289

Wozniak KM, Pert A, Mele A, Linnoila M (1991) Focal application of alcohols elevates extracellular dopamine in rat brain: a microdialysis study. Brain Res 540:31–40

Wu PH, Pham T, Naranjo CA (1987) Nifedipine delays the acquisition of tolerance to ethanol. Eur J Pharmacol 139:233–236

Yamazaki M, Mori H, Araki K, Mori KJ, Mishina M (1992) Cloning, expression and modulation of a mouse NMDA receptor subunit. FEBS Lett 300:39–45

Yoshida A, Huang I-Y, Ikawa M (1984) Molecular abnormality of an inactive aldehyde dehydrogenase variant commonly found in Orientals. Proc Natl Acad Sci USA 81:258–261

Yuki T, Thurman RG (1980) The swift increase in alcohol metabolism. Time course for the increase in hepatic oxygen uptake and the involvement of glycolysis. Biochem J 186:119–126

Yurttas L, Dale BE, Klemm WR (1992) FTIR evidence for alcohol binding and dehydration in phospholipid and ganglioside micelles. Alcohol Clin Exp Res 16:863–869

Zahniser NR, Buck KJ, Curella P, McQuilken SJ, Wilson-Shaw D, Miller CL, Klein RL, Heidenreich KA, Keir WJ, Sikela JM, Harris RA (1992) GABA$_A$ receptor function and regional analysis of subunit mRNAs in long-sleep and short-sleep mouse brain. Mol Brain Res 14:196–206

Zingg HH, Lefebre D, Almazan G (1988) Regulation of vasopressin gene expression in rat hypothalamic neurons. J Biol Chem 261:12956–12959

III. Advances in the Pharmacotherapy of Addiction

CHAPTER 12

Pharmacotherapy of Addiction: Introduction and Principles

F.R. Levin and H.D. Kleber

A. Introduction

To some individuals, it may appear inappropriate to treat drug addiction with "other drugs." This assumption views drug dependence as somehow different from other diseases, e.g., diabetes or hypertension, which have been shown to respond effectively to prescribed medication. Drug dependence can be a chronic, relapsing illness that requires different treatment interventions at different periods in the patient's course of illness. Sometimes, a pharmacologic intervention may be required; at other times, medication may not be useful.

At present, the etiology of drug dependence is best understood as a combination of biological, psychological, and social determinants and design of treatment needs to consider all three possibilities. Depending on the clinical assessment, treatment may focus on one of these areas or may be simultaneously directed towards several problem areas. Regardless of the intervention(s) employed, several points (Kleber 1989) need to be emphasized: First, psychological problems or comorbid psychiatric illnesses may worsen treatment outcome. Some investigators have found that treatment outcome is worse among individuals who have a high severity of psychological problems. Certain modalities of treatment, e.g., therapeutic communities, may not be appropriate for addicted individuals with underlying severe psychopathology (McLellan 1984). Individuals with previous episodes of trauma or character pathology may require more intensive treatment with experienced clinicians and/or a psychotropic medication. Second, the patient's motivation for treatment may effect treatment retention as well as other measures of treatment outcome (e.g., amount of drug use, length of abstinence). The patient may be motivated to enter treatment because of intense family, employment or legal pressures, social or occupational dysfunction, psychological distress or physical distress secondary to withdrawal. Some of these factors are short-lived. When the "pain" is removed, the patient may recall the more positive aspects of his/her prior drug use (e.g., the drug-related euphoria, the exciting lifestyle) and drop out of treatment. Thus it is often useful to maintain pressure on the drug user by appropriately involving family members, employers, friends or significant others in the patient's treatment. Use of agents such as disulfiram or naltrexone, given by a reliable spouse or significant other, may provide an important "window" to ensure abstinence while other therapy takes effect. Third, polydrug use is common. Treatment programs rarely find individuals who are abusing one psychoactive substance. This includes licit substances, e.g., alcohol, as well as combinations of illicit drugs. For example, methadone programs have been overwhelmed by individuals who are dependent on both opiates and cocaine (CONDELLI et al. 1991). This change in drug use patterns has important treatment implications, including the need to recognize that a medication that effects one abused substance may shift the abuse pattern. In general, medications to treat substance dependence need to be given in a context of appropriate psychosocial intervention. Without such support the patient is likely either to discontinue the medication or to take it while still using their drug of choice.

B. Pharmacologic Interventions

While almost all treatment programs use psychological interventions to implement change, for some patients this is not sufficient. In the past decade, there has been substantial substance abuse research either evaluating established medications or attempting to develop novel ones. Clearly, medication can serve as an important tool in treating drug dependence. The appropriate selection of medication, amount of medication, and length of use requires clinical acumen and practical experience with the medications available. Except for treating withdrawal or overdose, medications are not curative for substance dependence as penicillin is for infection. They may, however, have important uses in creating a window of opportunity during which appropriate psychosocial interventions may decrease the risk of relapse. Additionally, they can serve as long-term maintenance agents for patients who do not appear to be able to function without them but can lead productive new lives with them.

Upon scrutinizing all of the established and novel pharmacologic treatment interventions, there are several ways that the existing pharmacologic treatments for drug dependence can be classified. One approach is to categorize medications by their intrinsic mechanism of action, e.g. agonists, antagonists, anti-withdrawal agents, and anti-craving agents. Regardless of the drug of abuse, this classification allows one to understand the confirmed (or purported) mechanism of action of different therapeutic agents. The limitation of the system is that it does not fully explicate the clinical framework in which certain medications should be used in treatment.

Using a second approach, medications can be categorized by their clinical purpose. Specifically, medications may be used for acute intoxication, withdrawal, or relapse prevention (GORELICK 1993). This system is particularly useful for clinicians who are faced with emergency situations (e.g., opiate overdose or alcohol withdrawal) and prefer that available therapeutic agents be classified by their targeted clinical function. The limitation of this system is that it may incorrectly imply that medications are used for only one specific clinical situation. In practice, one medication often has several clinical indications.

In order to provide an overview of both the established and novel therapeutic agents being developed for the treatment of drug dependence, a synthesis of these classification approaches will be employed. By combining these approaches, an understanding of the physiologic mechanisms as well as the clinical indications for different pharmacologic agents may be gained. Since the pharmacotherapy of opiate, nicotine, and cocaine dependence are the focus of the following chapters, the rationales employed regarding the pharmacologic treatment of these addictive substances will be highlighted.

I. Agonists

1. General Principles and Clinical Indications

Agonists are agents which bind and activate receptors. A partial agonist binds to a receptor but produces a weak physiologic response. At times, a partial agonist may act as a competitive antagonist because it will "block" the receptor sites which would be used more efficiently by a pure agonist. An agonist-antagonist may also be a weak agonist at one type of receptor and an antagonist at another. Depending on its half-life, mode of delivery, safety, and side effects, an agonist may be used as to treat withdrawal and/or for maintenance.

2. Drug Classes

a) Opiates

Opiate agonists have been used extensively in various treatment settings with considerable success. Opiate-dependent patients seeking treatment are most frequently addicted to heroin. Heroin's short half-life and poor oral absorption make it neither a good detoxification agent nor a good maintenance medication. Frequent injections are required to avoid withdrawal, placing the addicted individual at higher risk for HIV infection. The most commonly used agent to treat withdrawal and prevent relapse has been methadone, a synthetic mu receptor agoinst which is well absorbed orally and has a long half-life. However, methadone has several limitations: (1) daily ingestion is necessary to prevent withdrawal symptoms, (2) euphoria or other opiate effects may occur with additional opiate intake despite being maintained on methadone, (3) detoxification from methadone may produce a protracted withdrawal syndrome, (4) methadone can be diverted for illegal purposes, and (5) it not only does not block other drugs such as cocaine but may facilitate such use by decreasing aversive effects. Thus, investigators have been examining the use of other opiate agonists.

Buprenorphine, a partial mu agonist, has the advantage of producing substantially less withdrawal symptoms when used as a detoxification agent and may produce less respiratory depression at high doses. Since withdrawal symptoms may occur when buprenorphine is taken by a methadone maintained individual, these individuals may be less likely to augment their prescribed dose with diverted buprenorphine. However, for nontreated opiate-dependent individuals, buprenorphine may produce a "high" when it is injected. To combat this problem, a formulation of buprenorphine and naloxone, a narcotic antagonist, is being explored. Based on naloxone being 100 times more potent by intravenous use than orally, a small amount will not block oral buprenorphine but would blunt the euphoria if the combination is taken intravenously. Given the historical precedent established with Talwin, i.e., the street use of Talwin was remedied by the shift in production to Talwin Nx (pentazocine-naloxone), it is expected that the "street" use of a buprenorphine-naloxone formulation would be minimal.

Although there are preliminary data to suggest that buprenorphine may work well when administered every other day, the medication most studied as an alternative to daily methadone dosing is l-α-acetylmethadol (LAAM), recently approved by the FDA. Due to its long-acting active metabolites, LAAM may be given three times a week without producing significant withdrawal symptoms. Because weekend take-home doses are not required, the potential risk of diversion is less. One potential "niche" for LAAM is as a transition agent. As stable methadone patients require less structure and behavioral interventions, they may be shifted to LAAM and, if clinically indicated, eventually detoxified (Goldstein 1976).

b) Nicotine

Cigarette smoking is an efficient way to deliver nicotine to the brain where, by binding to nicotinic receptors in the ventral tegmental area, it may exert its reinforcing effect. Unlike opiate dependence, in which agonists have been targeted at the opiate mu receptor, no alternatives to nicotine which bind to the nicotinic receptor have been developed. Instead, the pharmacologic treatments have focused on the development of less harmful alternative delivery systems that can be used for nicotine detoxification. Four types of nicotine replacement systems, each with distinct advantages and disadvantages, are currently available: (1) nicotine polacrilex gum, (2) transdermal nicotine, (3) a nicotine nasal spray, and (4) a nicotine inhaler (Jarvik and Schneider 1992). Regardless of the product, the strategy is to stabilize the patient on a steady-dose of nicotine and then slowly detoxify the patient over a period of weeks to months. Speed of delivery, nicotine level obtained, and duration of effect are some of the ways in which the individual manufacturers of transdermal systems have attempted to distinguish their highly advertised product from the others. Initial enthusiasm for these products waned as smokers and physicians were confronted with the high rates of relapse. However, as stressed earlier, pharmacologic interventions should be done in combination with nonpharmacologic treatment strategies. Often, other problem areas require intervention. For example, high rates of depression have been found among smokers seen in medical settings (GLASSMAN et al. 1988). Conversely, many patients treated within psychiatric settings are nicotine-dependent. Nicotine may ameliorate distressing psychiatric symptoms, it may reduce the side effects associated with psychotropic medications, and its cessation may produce distressing psychiatric symptoms. Thus, analogous to other medications used for drug dependence, nicotine replacement products are less likely to be helpful if implemented without an understanding of the psychological and social factors promoting the individual's tobacco use.

c) Cocaine

Similar to nicotine, cocaine is a highly addictive substance, particularly when it is taken using a rapid, potent delivery system (e.g., crack cocaine, intravenous cocaine). During the early 1980s, as use escalated within the United States, it became clear that traditional interventions were minimally effective and a widespread search for new pharmacotherapies to treat cocaine addiction was undertaken. Cocaine's highly reinforcing effects are believed to be due to its ability to block dopamine reuptake in the nucleus accumbens, but no dopamine agonists have shown clear effectiveness for the treatment of cocaine addiction. Agents investigated include those that act directly on the pre or postsynaptic dopamine receptors (e.g., amantadine or bromocriptine) or that are precursors of dopamine (e.g., tryptophan, L-dopa).

Generally, dopamine agonists have been suggested for use during time-limited periods (i.e., acute withdrawal or early recovery). While a "methadone-like" medication for cocaine would be desirable, this may not be viable given the subjective effects of stimulants. Whereas methadone may diminish opiate "craving," cocaine substitutes (e.g., methylphenidate) may stimulate craving unless the cocaine-dependent individual has underlying psychiatric comorbidity (e.g., attention deficit hyperactivity disorder).

II. Antagonists

1. General Principles and Clinical Indications

Antagonists, both competitive and noncompetitive, are agents which prevent other compounds from exerting their biologic effects. An antagonist can be competitive or noncompetitive. If it is competitive, the antagonist reversibly binds to the agonist receptor site. If it is noncompetitive, the antagonist irreversibly binds to the receptor site or binds at another site which reduces the ability of the agonist-bound receptor to produce a physiologic response.

Most commonly, antagonists have been used for acute intoxication or overdose but they may also be clinically useful for rapid detoxification when combined with other agents which reduce withdrawal symptoms. Antagonists have also been used to aid relapse prevention. Because they prevent reinforcing effects, the individual quickly realizes that it is futile to use them (or else stops taking the antagonist). Generally, it is recommended that the drug-dependent individual ingest the antagonist medication while he/she is being observed by a reliable person. Unfortunately, most patients are not compliant with antagonist medication. Some reasons to explain poor patient acceptance compared to agonist medication include: (1) the lack of euphoric/ pleasant effect, (2) the lack of withdrawal effect, and (3) the persistence of some craving (KLEBER 1989). Since compliance is low, antagonists are generally effective for only highly motivated patients (e.g., health professionals) or those mandated to take them (e.g., parolees).

2. Drug Classes

a) Opiates

The most commonly used opiate antagonists are naloxone and naltrexone. Naloxone is administered intravenously and is generally given in emergency situations in which a rapid reversal of an opiate overdose is required. Due to its short-half life, close observation and repeated observation are required to prevent the patient from lapsing back into an opiate-induced coma. In order to circumvent this problem, some clinicians administer naloxone using a pump, which allows a slow, steady infusion of the medication. Occasionally, naloxone is also used in clinical situations to assess whether a patient is physiologically dependent on opiates, for example, in order for a patient to qualify for methadone maintenance. If a potential patient experiences opiate withdrawal symptoms when administered naloxone, then physical dependence on opiates can be documented.

Unlike naloxone, naltrexone is a long-acting oral medication. Although, naltrexone has been combined with clonidine for rapid detoxification, it is generally used for relapse prevention. In order to prevent a prolonged uncomfortable withdrawal syndrome, naloxone is often given prior to initiating naltrexone treatment to insure that the patient is *not* physically dependent on opiates. Since naltrexone blocks the reinforcing properties of opiates, it may prevent impulsive opiate use.

Of note, VOLPICELLI et al. (1992) found that naltrexone may be clinically useful for individuals dependent on alcohol. Using a placebo-controlled design, they found that, although alcoholics maintained on naltrexone were not significantly different than the placebo group regarding their sampling of alcohol (i.e., whether or not they took a drink during the study), they were significantly less likely to relapse (i.e., drink 5 or more days within one week or ingest five or more drinks per drinking occasion). Thus, naltrexone may diminish the reinforcing properties of alcohol. Similarly, investigators are studying naltrexone's effect on nicotine and cocaine use among nicotine and cocaine and cocaine-dependent individuals, respectively. It may be that naltrexone's antagonist effect on the opiate receptors within the brain reward circuit reduces the reinforcing effects of several classes of abusable subs-tances. An injectable form of naltrexone is being been developed which could be given only once a month, and thus may minimize the difficulties associated with patient compliance.

b) Nicotine

Mecamylamine is the only nicotinic antagonist currently suggested as medication for smoking cessation (JARVIK and SCHNEIDER 1992), but its efficacy remains unclear and there is poor patient compliance. Some studies have found this medication useful whereas other studies actually found that patients attempted to counter the blocking effect of this medication by increasing their tobacco use (JARVIK and SCHNEIDER 1992).

c) Cocaine

Since euphoria from cocaine appears to be related to increased dopamine availability, dopamine antagonists have been suggested for the treatment of cocaine addiction. In animals, dopamine blocking agents have been found to attenuate stimulant effects (WISE 1984). However, given the considerable use of cocaine among schizophrenics treated with anti-psychotics that are dopamine antagonists, the effects of cocaine may not be completely blocked by traditional neuroleptics. Further, the anhedonia and risk of tardive dys-kinesia associated with these medications severely limit their use, particularly among cocaine abusers who may already suffer from depression or may have some underlying perturbation in their dopaminergic motor system (SATEL and SWANN 1993).

An exciting new area of research entails the use of monoclonal antibodies which act as catalysts for the enzymatic degradation of cocaine (LANDRY et al. 1993). Unlike conventional antagonists which block cocaine's effect at the receptor site, these cocaine antagonists would reduce or eliminate the reinforcing effects of cocaine by converting cocaine into two nonactive metabolites. In the future, it may be possible that a cocaine-dependent individual could be "immunized" with an anti-cocaine vaccine using this technology. Much more animal and subsequent human research needs to be done before such a treatment option becomes available.

III. Anti-withdrawal Agents

1. General Principles and Clinical Indications

Anti-withdrawal agents are medications which facilitate detoxification from drugs which induce physical or psychological dependence. Detoxification can be done abruptly, usually without medication, or performed more slowly.

When detoxification is done gradually, the clinician may use the drug that the individual is dependent upon, use a medication that is cross-tolerant to the drug of addiction, choose a medication to provide symptomatic relief, or impact on the physiologic mechanisms by which. withdrawal is expressed (Kleber 1989). By alleviating the painful physical and psychological symptoms of withdrawal, the addicted individual may be more likely to remain in treatment.

In the following discussion, several pharmacologic agents will be presented which have been suggested for opiate, nicotine or cocaine detoxification. However, it should be stressed that most opiate-dependent individuals are physically dependent on nicotine and many are psychologically dependent on cocaine. Generally, it is recommended that, if an individual is addicted to two or more classes of substances which produce significant physical withdrawal symptoms, then the individual should be detoxified from one substance at a time. Since many treatment programs have become "smoke-free", many nicotine-dependent alcoholics or opiate addicts entering treatment are required to refrain from smoking, or at least limit their tobacco use while receiving treatment for their other drug use. It remains unclear if smoke-free inpatient proprams have lower retention rates than programs which allow patients to smoke within their facility. Clearly, various pharmacologic and nonpharmacologic approaches to detoxify patients from nicotine should be considered when developing a therapeutic plan to treat the patient's dependent on other addictive substances.

2. Drug Classes

a) Opiates

Opiate-dependent individuals can experience severe physical and psychological withdrawal symptoms when they abruptly stop their opiate use, and a variety of agents have been used to ameliorate these symptoms. In most treatment settings, detoxification from opiates is performed using opiate agonists, most commonly methadone. Generally, this is done over a period of 1–2 weeks. More recently, there has been interest in developing medications which effect the underlying physiologic mechanisms associated with opioid withdrawal. Clonidine has been promoted as a potential detoxification agent because, by binding to the $\alpha2$ autoreceptors in the locus coeruleus (LC), clonidine may reduce the overactivation of the LC due to opiate withdrawal. Clonidine has been used alone, towards the end of a withdrawal regimen using methadone, or in combination with naltrexone for a rapid detoxification. Clonidine's disadvantages are that it can produce sedation or hypotension, and it does not alleviate some of the distressing symptoms of opioid withdrawal (e.g., muscle aches, anxiety). Similar to clonidine, lofexidine, an α agonist, is being studied as a treatment for opioid withdrawal. Compared to clonidine, lofexidine has the advantage of producing less sedation and hypotension. Regardless of the manner in which detoxification is accomplished, each of these methods requires close medical supervision.

b) Nicotine

As described earlier, there are several nicotine replacement systems currently available which can be used to slowly reduce the intake of nicotine over a period of weeks to months. Similar to opiates, clonidine has been used for smoking cessation. However, its efficacy in providing symptomatic relief, particularly the anxiety and craving symptoms, remains unclear. Anxiety and depressive symptoms are commonly described by nicotine-dependent patients when they cease smoking, and various antidepressants (e.g., doxepin, fluoxetine) and anxiolytics (e.g., buspirone) have been suggested to ameliorate these symptoms. However, these medications often take several weeks to work and may not diminish all of the symptoms associated with nicotine withdrawal. Further, there are few studies which have targeted psychotropic treatment medication to specific subpopulations of individuals who want to stop smoking (i.e., those who have become depressed when they have attempted to quit smoking in the past). A more detailed review of the promising treatments are provided in Dr. Hughes' chapter on the pharmacotherapy for nicotine dependence.

c) Cocaine

Much debate exists regarding whether there is a withdrawal syndrome associated with cocaine abstinence. Since Gawin and Kleber's (1986) proposal that there may be three withdrawal phases associated with cocaine abstinence, there has been much discussion regarding the accuracy of this model. Within in-patient settings, withdrawal symptoms appear to decrease rapidly without the reoccurrence of certain psychological symptoms (e.g., craving) (Weddington et al. 1990; Satel et al. 1991). Thus, medication to treat cocaine withdrawal symptoms may not be necessary. However, individuals being treated in out-patient programs, usually continue to be barraged by a myriad of cocaine-related cues which may elicit intense craving or other troubling psychological symptoms. Dopamine agonists or neurotransmitter precursors might be clinically useful for these "withdrawal" symptoms. However, at present, the pharmacologic cocaine treatment literature remains unclear regarding when and if these agents are efficacious.

IV. Anti-craving Agents

The concept of developing anti-craving agents to treat drug dependence is a relatively new one. Prior to the recent cocaine epidemic, clinicians often conceptualized craving as a response to the physical symptoms associated with withdrawal. Thus, within the opiate or nicotine treatment literature, various agents are suggested for their anti-withdrawal effects rather than their anti-craving effects. Conversely, because there is some debate as to whether physical dependence to cocaine occurs, the emphasis has been on the development of agents to treat the psychologic symptoms associated with cocaine abstinence (e.g., craving, depression). Craving may be the main cause for continued cocaine use or the major precipitant of relapse after a prolonged period of cocaine abstinence. Since reduction of craving might improve treatment success, clinicians have searched for medications to reduce cocaine craving and allow intractable patients to become engaged in other treatment modalities.

Since anti-craving medications have been targeted for cocaine, rather than other classes of drugs, no specific discussion of anti-craving medications for opiates or nicotine is provided. Although over 20 medications have been used for the treatment of cocaine addiction and many have been touted as effective for cocaine craving, there is little clear-cut evidence to support the use of any particular agent. At present, with the possible exception of desipramine, no medication has been shown to produce sustained reductions in craving. Until additional controlled studies are completed which use commonly accepted measures of craving as well as other appropriate outcome measures, it will be difficult to assert that any medication is effective for cocaine craving.

C. Conclusion

In this overview, a classification system was used to categorize the various pharmacologic treatment strategies for drug dependence. By using this scheme, it is hoped that the rationale for using various standard and novel medications is more comprehensible. This approach emphasizes the similarities, rather than the differences, in the pharmacologic treatments for different abusable psychoactive substances. Although the appropriate clinical use of some of the various pharmacologic treatment interventions was outlined, a more detailed discussion is provided in the subsequent chapters.

References

Condelli WS, Fairbank JA, Dennis ML, Rachal JV (1991) Cocaine use by clients in methadone programs: significance, scope, and behavioral interventions. J Subst Abuse Treat 8:203–212

Gawin FH, Kleber HD (1986) Abstinence symptomatology and psychiatric diagnosis in cocaine abusers: Clinical observations. Arch Gen Psychiatry 43:107–113

Glassman AH, Stetner F, Walsh Raizman PS, Fleiss JL, Cooper TB, Covey LS (1988) Heavy smokers, smoking cessation, and clonidine: results of a double-blind, randomized trial, JAMA 259:2863–2866

Goldstein A (1976) Heroin addiction: sequential treatment employing pharmacologic supports. Arch Gen Psychiatry 33:353–358

Gorelick DA (1993) Overview of pharmacologic treatment approaches for alcohol and other drug addiction: intoxication, withdrawal, and relapse prevention. Psychiatr Clin North Am 16:141–156

Jarvik ME, Schneider NG (1992) Nicotine. In: Lowinson JH, Ruiz P, Millman RB, Langrod G (eds) Substance abuse: a comprehensive textbook, 2nd edn. Williams and Wilkins, Baltimore, pp 334–356

Kleber HD (1989) Treatment of drug dependence: what works? Int Rev Psychiatry 1:81–100

Landry DW, Zhao K, Yang GX-Q, Glickman M, Georgiadis TM (1993) Antibody-catalyzed degradation of cocaine. Science 259:1899–1901

McLellan AT (1984) The psychiatrically severe drug abuse patient: methadone maintenance or therapeutic community? Am J Drug Alcohol Abuse 10:77–95

Satel SL, Swann AC (1993) Extrapyramidal symptoms and cocaine abuse (letter to the editor). Am J Psychiatry 150:347

Satel SL, Price LH, Palumbo JM, McDougle CJ, Krystal JH, Gawin FH, Charney DS, Heninger GR, Kleber HD (1991) Clinical phenomenology and neurobiology of cocaine abstinence: a prospective inpatient study. Am J Psychiatry 148: 1712–1716

Volpicelli JR, Alterman AI, Hayashida M, O'Brien CP (1992) Naltrexone in the treatment of alcohol dependence. Arch Gen Psychiatry 49:876–880

Weddington WW, Brown BS, Haertzen CA, Cone EJ, Dax EM, Herning RI, Michaelson BS (1990) Changes in mood, craving, and sleep during short-term abstinence reported by male cocaine addicts. Arch Gen Psychiatry 47:861–868

Wise R (1984) Neural mechanisms of the reinforcing action of cocaine. NIDA Res Monogr 50:15–33

CHAPTER 13

Development of Medications for Addictive Disorders

F.J. Vocci

A. History of the Regulatory System for New Drugs

The development of medications in the United States is regulated by the Food and Drug Administration (FDA) under the authority granted by the Federal Food Drug and Cosmetic Act (FFDCA). The regulatory powers of the FDA have evolved due to both catastrophic events and technological advances which led to amendments in the FFDCA. The original enactment of the FFDCA occurred in 1906 and prohibited adulterated food or drugs from interstate commerce. The next major development that influenced the current system was the sulfanilamide elixir tragedy in 1938. This untested medication proved fatal in 107 cases before it was removed from the market. The FFDCA was amended to authorize the FDA to require premarketing approval of a new drug product. Manufacturers were required to file New Drug Applications (NDAs) containing evidence that a drug was safe for its claimed use.

The next major event that influenced how drugs are developed in the US was the thalidomide tragedy, which occurred in the late 1950s/early 1960s. Thalidomide, a sedative and hypnotic agent, is a human teratogen when taken during the first trimester of pregnancy. As a result, the US Congress passed the Kefauver-Harris amendments to the FFDCA. These amendments added the requirement that drugs be shown to be effective for their claimed use prior to marketing. An NDA now had to contain evidence that a drug was both safe and effective. Moreover, these amendments also required that all clinical testing of drugs be performed under Investigational New Drug (IND) applications submitted to the FDA. The FDA promulgated the IND regulations (21 CFR 312); these have been amended to include new developments that occurred in the 1980s (CODE OF FEDERAL REGULATIONS 1987) and early 1990s, e.g., subpart E which allows for expedited development of drugs to treat life-threatening diseases (FEDERAL REGISTER 1992).

B. Scheme of New Drug Development

New drug development is complex multidisciplinary process consisting of four stages. It begins with 1–3 years of preclinical effort directed towards the development of testing a new drug in a clinical investigation (Fig. 1).

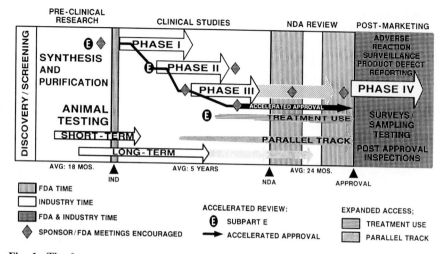

Fig. 1. The four components of the continuum of new drug development and review in the United States: preclinical research, clinical studies, new Drug Application (*NDA*) review, and post-marketing. The average time spent in each component is given in the figure. Actual drug development time occurs during the preclinical and clinical components. NDA review time is a mixture of Food and Drug Administration (*FDA*) review time and industry response time. *IND*, Investigational New Drug application. Subpart *E* refers to the part of the IND regulations (21 CFR 312.80 for *Drugs Intended to Treat Life-Threatening and Severely Debilitating Illnesses*). It is an accelerated development and review process that allows early marketing of medications to treat certain disorders. Briefly, phase I, II and III studies are clinical pharmacology, early efficacy in selected patient populations, and expanded efficacy and safety testing in the intended patient population(s), respectively. Expanded definitions of Phase I, II, and III clinical studies are given in the text

Initially, synthesis and screening of organic compounds by a drug discovery group to determine an effect in a target biological activity yields a prototypic lead compound. This compound and analogues are evaluated for optimization of activity vs toxicity. Following additional synthesis and testing, the drug discovery group seeks approval from its company management to consider the lead candidate for preclinical development. The decision to proceed forward with a compound to the next stage of testing is based on factors such as potency, efficacy, stability, solubility, bioavailability, half-life, metabolism, and excretion as well as business factors related to patent position and competing products in the marketplace. The compound can then be considered a safety assessment candidate. Once a safety assessment candidate is defined, a project team is formed that decides the number, type, and timing of the preclinical tests necessary to support the proposed clinical testing of the safety assessment candidate. The total time spent in this phase is 1–3 years with an average time of 18 months according to an FDA assessment. A more recent estimation from the Office of Technology Assessment (OTA) suggested it takes an average of 3.5 years to get through this phase (OTA 1993).

The second major component of new drug development is the "Clinical Research and Development" phase. During this phase clinical, preclinical chemistry, and manufacturing studies are performed for the purpose of compilation into a NDA. All development proceeds in a step-wise fashion whereby progressively larger patient exposures to drug are supported with an increasing safety detabase that supports the continued administration of the drug to human subjects. The phases of clinical studies in this component of drug development will be described below.

All clinical drug research is conducted under an IND (Fig. 2). (Previously, the technical term for an IND was a "Claimed Investigational Exemption for a New Drug"). An IND is a request from a sponsor to evaluate the effects of an investigational or marketed drug in human subjects. A pharmaceutical sponsor files a "commercial IND." Another type of IND, which will not be discussed here, is an "Investigator IND" in which a qualified individual requests permission to conduct a clinical investigation.

The FDA has codified the regulatory aspects of drug development into regulations (21 CFR 312). Irrespective of the type of sponsor, an IND can be conceived of as an administrative-legal-regulatory framework in which

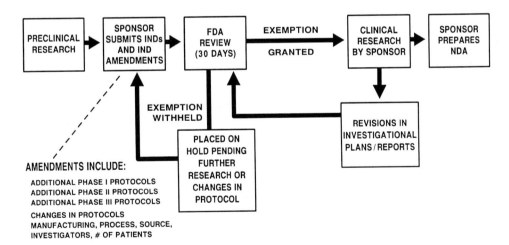

Fig. 2. The interaction of the Food and Drug Administration (*FDA*) and the Investigational New Drug (*IND*) sponsor with respect to the initial filling and subsequent amendments that define the development of a drug product. If the FDA grants an exemption, clinical studies in human subjects may proceed. If an exemption is withheld (denied), the sponsor must modify the protocol or perform further preclinical or pharmaceutical research. Following the initial human study, further work, e.g., additional protocols, is performed and reported to the FDA under amendments. The process continues in an interactive manner until the sponsor prepares and files an *NDA* (New Drug Application) or abandons/closes the IND

the FDA evaluates the safety of an initial clinical trial (Fig. 2). The FDA has 30 days from the time it acknowledges receipt of an application to determine whether the proposed clinical study is relatively safe. If so, it is allowed to proceed, i.e., an exemption is granted. If not, the study is placed on a "clinical hold." This can be removed following the submission and evaluation of requested information in the form of amendments that buttress the safety of the proposed investigation. FDA can place a study on clinical hold for different reasons at different phases of the clinical investigations (see below).

A commercial IND contains information about the synthesis of the drug, its identity, stability, purity, physical characteristics, the formulation of the dosage form to be utilized in the clinical studies (including excipients and preservatives), the composition and excipients of the placebo dosage form (if one is to be used), the pharmacology and toxicology studies performed to support the safety of the drug, and the proposed clinical protocol. A commercial IND also contains information about the administrative management of the study. This includes identifying personnel responsible for assessment of safety and monitoring of the study, the case report forms on which data will be transcribed from source documents, and local site issues (Institutional Review Board approval, the informed consent document and assurance of its proper use, shipment and reconciliation of drug supplies for the study).

Regulatory management of the IND is also a responsibility of the sponsor and is appropriately discharged by monitoring preclinical and clinical studies and documenting their veracity; filing timely reports; reporting serious adverse events in the prescribed time; filing updated reports to investigators of adverse events occurring at other sites; filing new reports (amendments) in the areas of chemistry and manufacturing, pharmacology/toxicology, pharmacokinetics, and clinical studies; filing annual reports of the progress of the IND; and holding meetings with the FDA on issues of major significance at specified times during the development of the drug.

Legal aspects of drug development include responsibility for injury of participants in clinical studies, indemnification of investigators, and the possibility of civil and criminal prosecution for falsification of records, other types of fraud, or withholding information about serious reactions and deaths in patients in clinical studies.

C. Phases of Clinical Investigations

Three separate phases of clinical investigations are defined in the IND regulations. Phase 1 studies comprise the initial investigation of a new drug in a human population. The purpose of the phase 1 studies is to identify the metabolism and pharmacodynamic effects of the new drug in either normal volunteers or a patient population. Most of these clinical pharmacology studies involve a rising-dose design and attempt to determine, within the limits of safety, the maximum tolerated dose in human subjects. Phase 1 studies can be conducted single-blind (when the subject is unaware of dose or whether he/she has received active or placebo drug) or double-blind (neither investigator nor subject are aware of dose, or whether drug or placebo are being administered). The dose-duration of effect relationship and information about variability in response are correlated to pharmacokinetics to determine whether a drug is suitable for further study. Additionally, these studies may investigate a mechanism of action of the drug by challenge studies with pharmacological antagonists, e.g., a naloxone challenge of an opioid agonist. Phase 1 studies are highly controlled, usually conducted in in-patient or residential settings, and involve 20–80 subjects.

Phase 2 studies are the initial controlled clinical studies of the effectiveness of a drug in the patient population for which it is intended. The "early" phase 2 studies are highly controlled investigations; rigid inclusion/exclusion criteria are imposed to select patients with the disorder who are generally otherwise as healthy as possible. Often, dose regimens are compared in early phase 2 investigations. Later phase 2 studies may use broader inclusion/exclusion criteria and admit more subjects to the trials to determine both the drug's effectiveness and its common side effects. Usually, no more than 200 subjects participate in phase 2 studies.

Phase 3 studies are extensions of phase 2 investigations in both controlled and uncontrolled clinical trials. Both open-label studies (in which both subjects and investigators are aware of drug administered) and more controlled trials comparing the drug to placebo, active controls, or a dose-response of several doses of the test drug are performed. In phase 3 studies, exclusion criteria are relaxed relative to phase 2 studies in order to improve the external validity of the study sample of the phase 3 trial to the disease population for which the drug is indicated. Phase 3 trials can involve from several hundred to several thousand subjects.

The IND regulations provide administrative and regulatory requirements for the conduct of clinical trials but are devoid of advice regarding the conduct of clinical investigations in specific disease states. For this purpose, the FDA utilizes guidelines for clinical investigations that are usually disease state specific. The FDA is currently deliberating on the acceptance of a draft *Guidelines for the Development and Evaluation of Drugs for the Treatment of Psychoactive Substance Disorders*. These draft guidelines specify the indications for which medications can be developed as well as the efficacy criteria and statistical models used for analysis of efficacy. The guidelines are in current use by the NIDA Medications Development Division in the development of buprenorphine for the treatment of opiate dependence and other potential medications for the investigational treatment of cocaine dependence.

The duration of the clinical phase of drug development ranges from 2 to 10 years, with an average time of 5 or 6 years according to the FDA or

OTA, respectively. During phase 3, the pharmaceutical sponsor compiles the documentation in chemistry and manufacturing controls regarding the synthesis of the drug substance and the formulation of the drug, the biopharmaceutics/pharmacokinetics of the dosage form, the pharmacology/ toxicology studies, the clinical trials as individual studies and as summaries of safety and efficacy of the drug, and the statistical reports demonstrating efficacy.

FDA has developed regulations on the submission of NDAs (CODE OF FEDERAL REGULATIONS 1985) as well as a general guideline available for advice on formatting, assembling, and submitting NDAs. Specific guidelines are also available regarding the format and content of the chemistry, pharmacology, biopharmaceutics/pharmacokinetics, and clinical and statistical sections of the NDA.

The third major component of drug development is "New Drug Application filing and review" (Fig. 3). Each discipline-specific component of the NDA is evaluated by FDA reviewers who then synthesize a comprehensive overview of the NDA. The NDA review process is iterative with many requests for information sent to the NDA sponsor. Amendments are filed to the FDA and reviewed. If approval is granted, a Summary Basis of Approval is written, labeling is negotiated, preapproval inspections of the manufacturing sites are conducted, and an approval letter is issued. Subsequent NDAs for the same medication are called supplements and can

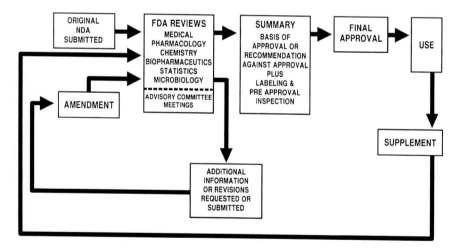

Fig. 3. Following the submission of an New Drug Application *(NDA)*, the Food and Drug Administration *(FDA)* performs discipline specific review of the application. The NDA may be brought before an advisory committee, additional information may be requested of the sponsor, or the application may be approved. If the application is approvable, FDA performs inspections of the manufacturing plant and reviews the product labeling before a final approval letter is issued. Subsequent approvals may be sought in the form of NDA supplements

involve change of manufacturing sites, changes in dosage forms, e.g., addition of a second medication to the product, and additional indications.

NDA approval times range from 2 months to 7 years. The average time to approval is 2 years, according to the FDA and 26.5 months according to the OTA report. The time to approval from the discovery phase to approval is an average of about 8.5 years according to FDA historical review of NDAs submitted between 1980 and 1987, whereas the OTA report states that it can take 12 years from discovery to approval.

D. Special Considerations Related to Medications Development for Addictive Disorders

In addition to the development of NDAs for a drug product to treat an addictive disorder, the abuse liability of the product must be considered for possible scheduling under the US Controlled Substance Act. The evaluation usually involves human subject testing and is described elsewhere (Vocci 1991).

An additional consideration in developing medications for the treatment of addictive disorders is the development of treatment standards in the case where a narcotic drug is to be used for the treatment of opiate dependence. The US Secretary of Health and Human Services (HHS) is required by the US Controlled Substance Act of 1970 to determine the appropriate methods of professional practice for the medical management of narcotic addiction. Moreover, the Narcotic Addict Treatment Act of 1974 specifies that the Secretary of HHS must establish standards regarding the quantities of narcotic drugs that may be provided for unsupervised use by individuals in treatment. This has been operationalized into regulations for the use of narcotic drugs in the treatment of narcotic addiction (FEDERAL REGISTER 1994b).

E. *l*-α-Acetylmethadol: Selected Clinical Studies

The process of medications development has been described in the section above. Here the phases of clinical investigation will be discussed in relation to the development of *l*-α-acetylmethadol (LAAM) for the maintenance treatment of narcotic dependence. Only certain studies will be highlighted for the purposes of illustrating principles of medications development.

LAAM had been evaluated in phase 1 studies by FRASER and ISBELL (1952). These authors performed a number of clinical pharmacology studies in previously morphine-dependent subjects ("post-addicts") and evaluated LAAM's ability to substitute for morphine in currently morphine-dependent subjects. These studies revealed the following properties of LAAM:

1. It is an opiate six to eight times as potent as morphine in suppressing abstinence by both oral and subcutaneous routes.

2. It possesses a slow onset of 1–2 h with a 24–72 h duration of action when orally administered.
3. It has a slower onset and lower peak effect when administered by intravenous and subcutaneous routes.
4. A mild abstinence syndrome ensues following cessation of administration for 14 days.
5. Cumulative toxicity (near-coma, severe emesis and mental confusion) can be produced by bid dosing of LAAM by the subcutaneous route.

The information provided by these authors allowed subsequent phase 2 work to proceed incorporating the concept of every other day or three times a week dosing.

Two early phase 2 studies evaluated the effects of LAAM vs methadone in clinic settings. JAFFE et al. (1970) reported that *d,l*-methadyl acetate (i.e., the enantiomeric pair) substituted for methadone when a 1.2 times dose crossover ratio was used in a double-blind, randomized study in heroin addicts who were in methadone treatment. LAAM had a 72–96 h duration of action as judged by withdrawal onset occuring 3–4 days after the last LAAM dose. LAAM was dosed on a Monday-Wednesday-Friday basis in this study.

The second phase 2 study of LAAM for the treatment of narcotic dependence was reported by JAFFE et al. (1972). In the study *d,l*-methadyl acetate was compared to methadone hydrochloride and subjects were administered the drug in a treatment clinic setting under double-blind, randomized conditions. The *d,l*-methadyl acetate dosage ranged from 36 to 80 mg given on a Monday-Wednesday-Friday schedule. Methadone doses (range = 30–80 mg) were administered daily. All patients received dose increases in the first 2 weeks. Dose increases were rare after 5 weeks. No differences were noted in dropout rate, although more opiate use was noted in the *d,l*-methadyl acetate group than in the methadone group. Clinic attendance was similar in both groups. There was no difference noted in terms of medical safety parameters. These authors correctly deduced that an active metabolite was formed from *d,l*-methadyl acetate.

ZAKS et al. (1972) reported the results of another phase 2 trial in 21 narcotic addicts who were randomly assigned to either *levo*-methadyl acetate (the *l*-isomer or LAAM) or methadone. Subjects received increasing doses of either LAAM or methadone until target doses of 80 mg of LAAM or 100 mg of methadone were reached. Patients were dosed according to clinical experience and results from heroin challenges. Intravenous heroin challenges (50 mg) were administered to four LAAM subjects who had received 30–50 mg on a thrice weekly schedule. Heroin challenges were administered 24 h following LAAM dosing. Three subjects reported a mild euphoria whereas a complete blockade was determined in the fourth subject. Five subjects receiving 80 mg on a three times per week schedule demonstrated a complete cross-tolerance to 50 mg heroin doses administered 24 h after dosing. One of these subjects was also completely blocked when the heroin challenge was administered at 48 h post-LAAM.

A multicenter phase 2 comparison of LAAM vs methadone was conducted in street addicts (LING et al. 1976). In this study 430 narcotic addicts were randomized to either LAAM ($n = 142$) or one of two doses of methadone under double-blind conditions. The initial dose of LAAM or methadone was 30 mg. The dose was increased 10 mg/week to a target dose of 80 mg for LAAM or 50 or 100 mg for methadone. Subjects were maintained for 40 weeks on their respective medications. Measures of efficacy included retention, compliance to scheduled clinic visits, management of withdrawal signs and symptoms, and decreased illicit drug use. The authors pointed out that analysis and interpretation of the efficacy variables is not a simple task.

In retrospect, none of the phase 2 studies systematically explored dose induction regimens. Thus, the inductions were slower than necessary for patient safety and comfort. Early trials in phase 2 should explore dose induction regimens in terms of rates of dose escalation and its positive and negative consequences. It was not until 1979 that JUDSON and GOLDSTEIN compared slow and fast induction regimens. Since induction in the treatment setting can be problematic, as judged by higher dropout rates than any other time in treatment, one of the lessons of LAAM's development would seem to be that more exploration of the starting dose(s) and incremental increases would seem to be warranted.

Another phase 2/phase 3 study was conducted in methadone maintenance patients (LING et al. 1978). This study was designed as a companion study to the previously discussed phase 2 LAAM study (LING et al. 1976) and had similar exclusion criteria but also contained elements of a phase 3 trial. The phase 3 elements are the open-label design and flexible dosing after the second dose of medication. A large cohort of patients ($n = 636$) was admitted to the trial. Methadone maintenance patients were randomly assigned to LAAM or methadone. Those assigned to LAAM ($n = 328$) were transferred on a 1:1 basis. Possible group differences in overall safety were assessed by comparisons of side effects, medical and psychiatric reactions, and dropouts. The largest difference noted was that 62 LAAM patients (and no methadone patients) dropped out due to "medication not holding," a possible indication of relative underdosing which may have been secondary to the 1:1 crossover ratio. Symptoms and side effects were more frequently reported in the LAAM group. The more frequent reporting of signs and symptoms in the LAAM group was reflected in the retention curves. The retention curve for methadone was fitted by a monoexponential function with a rate function of 0.01 per week (i.e., the dropout rate was 1% per week). The retention curve for the LAAM group was more complex and was fitted to a biexponential curve. The fast and slow portions of the curve had rate constants of 0.158 and 0.01 per week, respectively (WRIGHT 1992). In other words, LAAM patients dropped out at 15 times the rate of the methadone group in the beginning of the study but at the same rate as the methadone group in the middle and latter portions of the study. The analysis suggests an incomplete cross-substitution of LAAM for methadone in the early weeks of the study. Again, this points out the problems encountered by a lack of systematic exploration of the dosing crossover ratio prior to embarking on a large scale study.

The large patient cohort in this study allowed a compilation of side effects and evaluation of possible changes in laboratory parameters. Anxiety and irritability were reported as possibly being unique to LAAM but may have been related to the fact that subjects were taking an investigational medication. The authors believed it could have been explained by withdrawal or factors other than LAAM. In the LAAM group, there were no significant changes in the patients' clinical laboratory parameters compared to their baseline values.

A large phase 3 study was conducted by WHYSNER (WHYSNER and LEVINE 1978) at 47 clinics. Five protocols were employed in both street addicts (450 subjects) and methadone maintenance transfer populations (1679 subjects). In the control groups, 369 subjects received methadone and 1760 patients received LAAM. Protocol 1 was an open-label evaluation of a 1.2–1.3 times LAAM crossover ratio in methadone maintenance transfers. Protocol 2 was similar to the first except that 40% of the patients were randomized to methadone. It is interesting to note that protocol 3 was a randomized comparison of three LAAM crossover ratio schedules.

Protocol 4 was a randomized, double-blind evaluation of three LAAM induction schedules in street addicts, and protocol 5 was identical to protocol 4 with the exception that a methadone comparison group was added. These protocols attempted to reconcile many phase 2 issues, especially with respect to dose induction. This was a recognition by WHYSNER and the NIDA that dosing induction issues were unresolved and required further study.

The other phase 3 trial of note was the recent clinical trial sponsored by NIDA, the LAAM Labeling Assessment Study (HERBERT et al. 1994). At the FDA's request, 26 methadone maintenance treatment programs that were not ordinarily involved in research were recruited for this study. The study design differed from standard controlled clinical trials in that it did not involve hypothesis testing. Rather, the objectives of the study were threefold: (1) to determine the adequacy of the product labeling to guide the safe use of LAAM in clinic dispensing; (2) to determine the safety of LAAM in a current addict population; and (3) to develop further experience with LAAM in women. Subjects entering the trial could received LAAM for 64 weeks.

The inclusion/exclusion criteria were highly consistent with a late phase 3 trial. Subjects had to be eligible for or currently receiving methadone maintenance. Exclusion criteria were limited to pregnancy, serious medical problems that would have precluded the subject's participation for the first 12 weeks of the study, and pending incarceration.

A total of 623 subjects were administered LAAM. Retention percentages did not differ by gender, ethnicity, or prior treatment status. No new or unexpected side effects occurred. A high dropout rate was observed in the first 3 weeks of the study; subjects terminating early cited "lack of efficacy" that may have been due to underdosing in the early phase of the study. A slower dropout rate was noted (0.01 or 1% per week) thereafter. The retention rate in this study was similar to the historical clinical trials conducted with LAAM. Retention was positively correlated to weekly dose within both the street addict and methadone maintenance populations.

The clinical experience with LAAM in over 7000 subjects was chronicled in the LAAM NDA which was filed on June 18, 1993. LAAM was approved for the treatment of opiate dependence on July 9, 1993. The approval of the product does not mean that all scientific issues have been adequately addressed. The sponsor agreed to 23 phase 4 commitments, most of which were related to chemistry. Some of the clinical commitments involved the determination of whether hepatic inducing agents such as rifampin or barbiturates and enzyme inhibitors such as erythromycin or ketoconazole affect the metabolism of LAAM and an evaluation of LAAM in patients with severe hepatic or renal disease. Other clinical commitments involve the identification and follow-up of women who become pregnant while taking LAAM.

F. Other Necessary Regulatory Actions Regarding *l-α*-Acetylmethadol

Since LAAM was in schedule 1 of the Controlled Substances Act during its development, the approval of the NDA allowed it to be rescheduled. LAAM rescheduling to CII was affected in August, 1994 (FEDERAL REGISTER 1994a). Since LAAM is an opiate drug development for the treatment of narcotic addiction, treatment standards need to be developed. LAAM treatment standards were published as an interim rule in July, 1994 (FEDERAL REGISTER 1994b).

G. Summary

This chapter reviews the current system for medications development in the US. Specific guidance in developing new drugs is obtained from the draft guideline and FDA staff. The process should be viewed as dynamic as it does not cease with the approval of a NDA for a product. Additionally, there may be parallel considerations involving drug scheduling and the development of treatment standards for certain types of medications.

Selected clinical trials involving LAAM were highlighted to illustrate the differences between clinical trial phases and to point out the issues that were not addressed in the beginning of LAAM's development. Early studies focus on the pharmacodynamics of the medication; phase 2 studies are primarily interested in demonstrating efficacy; and phase 3 trials involve safety and efficacy with the emphasis more on safety in the intended patient population. The obvious lesson from LAAM is that more attention needs to be paid to dose induction schedules when developing medications for addictive disorders. Dropout rates are high in the initial portion of clinical trials in narcotic-dependent populations. There are probably a multitude of reasons that can be ascribed to the early high dropout rate, only a fraction of which are pharmacological or treatment-related. Nonetheless, evaluation and development of optimal induction schedules are critical to the development of as smooth a transition as possible for individuals entering treatment.

References

Code of Federal Regulations (1985) Applications for FDA approval to market a new drug or an antibiotic drug. US Government Printing Office, Washington DC, title 21, part 314-414.445

Code of Federal Regulations (1987) Investigational new drug application. US Government Printing Office, Washington DC, title 21, part 312-312.145

Federal Register (1992) New drug, antibiotic, and biological product regulations; accelerated approval: final rule vol 57. US Government Printing Office, Washington DC, pp 58942–58960

Federal Register (1994a) Schedules of controlled substances: transfer of Levo-acetylmethadol from Schedule I to Schedule II vol 58. US Government Printing Office, Washington DC, pp 43795–43797

Federal Register (1994b) Levo-ALPHA-acetyl-methadol (LAAM) in maintenance; Revision of conditions for use in the treatment of narcotic addiction, vol 58. US Government Printing Office, Washington DC, pp 38704–38711

Fraser HF, Isbell H (1952) Actions and addiction liabilities of alpha-acetylmethadols in man. J Pharmacol Exp Ther 105:458–465

Herbert S, Montgomery A, Fudala P, Vocci F, Gampel J, Mojsiak J, Hill J, Walsh R (1994) LAAM labeling assessment study: retention, dosing, and side effects in a 64 week study. Presented to the College on Problems of Drug Dependence, Palm Beach, Florida

Jaffe JH, Schuster CR, Smith BB, Blachley PH (1970) Comparison of acetylmethadol and methadone in the treatment of long-term heroin users. JAMA 211(11): 1834–1836

Jaffe JH, Senay EC, Schuster CR, Renault PR, Smith B, DiMenza S (1972) Methadyl acetate versus methadone: A double blind study in heroin users. JAMA 224(2):437–442

Judson BA, Goldstein A (1979) Levo-alpha-acetylmethadol (LAAM) in the treatment of heroin addicts: dosage levels for induction and stabilization. Drug Alcohol Depend 4:461–466

Ling W, Charuvastra C, Kaim SC, Klett CJ (1976) Methadyl acetate and methadone as maintenance treatments for heroin addicts: A Veterans Administration Cooperative Study. Arch Gen Psychiatry 33(6):709–720

Ling W, Klett CJ, Gillis RD (1978) A cooperative clinical study of methadyl acetate I. Three-times-a-week regimen. Arch Gen Psychiatry 35:345–353

OTA (1993) Pharmaceutical research and development: costs, risk, and rewards. Office of Technology Assessment, no H-522, Government Printing Office no 05200301315-1, Feb 1993

Vocci FJ (1991) The necessity and utility of abuse liability assessment in human subjects. Br J Addict 86(12):1537–1542

Whysner JA, Levine GL (1978) Phase II clinical trial of LAAM report of current status and analysis of early terminations. National Institute on Drug Abuse Research monograph 19. US Government Printing Office, Washington DC, pp 277–290

Wright CW (1992) Medical officer's review of a multicenter, open label study of the efficacy of LAAM and methadone in patients with chronic opioid addiction. IND 36287

Zaks A, Fink M, Freedman AF (1972) Levomethadyl in maintenance treatment of opiate dependence. JAMA 220(6):811–813

CHAPTER 14a
Long-Term Pharmacotherapy for Opiate (Primarily Heroin) Addiction: Opioid Agonists

M.J. KREEK

A. Introduction

Opiate addiction, primarily addiction to the short-acting synthetic derivative of the natural product morphine, has been recognized as a major health problem in the United States since the mid-1960s (KREEK 1987). The problems of excessive use and addiction to opiate compounds may be traced back to the mid-1800s in the United States and, historically, the medicinal use and misuse of opiates has existed for hundreds and even thousands of years. Yet the recognition of the enormous impact of heroin addiction in the United States, and also of the need to approach this addiction as a medical problem with social and economic ramifications, was not generally embraced until the late 1960s or early 1970s (NYSWANDER 1956; DOLE and NYSWANDER 1965; DOLE et al. 1966a,b; KREEK 1972, 1973a,b, 1975, 1991a, 1992a,b,c; ZWEBEN and PAYTE 1990; GOLDSTEIN 1994). Until this time, fear of legal constraints or sanctions, because of ongoing punitive interpretations of earlier laws, caused most physicians and health care professionals to shun the responsibility of treating of addicts. Then, significant findings based on carefully conducted basic and applied clinical research studies showed the efficacy of a pharmacotherapy for addiction specifically using methadone maintenance treatment (DOLE and NYSWANDER 1965, 1966; DOLE et al. 1966a,b; KREEK 1991a, 1992a,b,c). The research supporting this therapy had been planned with a rationale based on all available knowledge concerning both the pharmacology of specific opioid agents, natural and synthetic, and their effects, as well as clinical observations of their duration of action or kinetics (KREEK 1991a, 1992a,b,c,d,e,f). This research was also based on all available information regarding the successes and failures of previous approaches to managing opiate addiction, including primarily drug-free approaches, incarceration, and a few attempts to treat heroin addiction pharmacologically on a short-term basis. These studies had been conceptualized and conducted with full consideration of the unsuccessful, but legitimate, attempts of some nations to use a variety of approaches of legalization of short-acting opiate drugs, such as heroin, in attempts to manage the problem. With the advent of a successful and appropriate pharmacotherapy, the attitudes of many, but certainly not all, policy makers have shifted gradually towards supporting such therapy, and legal interpretations of earlier laws have been changed to legitimize appropriate pharmacotherapy. Therefore, increasing numbers of physicians and other health care professionals have become involved in treatment of heroin addiction using pharmacotherapies.

Early research efforts were built, in part, upon the early work of the United States Public Health Service at Lexington, Kentucky, a facility which was to become the intramural program of the National Institute on Drug Abuse, and also upon a few research laboratories in universities throughout the nation, which had been studying the problem of drug dependency, either in the laboratory or at the clinical research level (PESCOR 1943). Up until the late 1960s, the interpretation of laws at the time and the reticence of the medical profession to become involved in treatment caused a twofold approach to the management of heroin addicts: either detention or incarceration, or drug-free or abstinence treatment, when any treatment was offered. The opiate addict was viewed either as a criminal (and indeed, most addicts have committed many crimes, primarily larceny, for acquisition of funds to purchase illicit drugs) or as persons with characterological disorders (so-called "addictive personality"). Thus, management by imprisonment or by attempts at behavioral modification were felt to be appropriate. However, the consistent lack of sustained success in such approaches and the perpetuation of addiction in heroin addicts led to the development of a "detoxification" treatment for addiction by the late 1950s and early 1960s, both by the United States Public Health Service group at Lexington and also by a few public facilities in New York City (HIMMELSBACH 1941; PESCOR 1943; NYSWANDER 1956; BERLE and NYSWANDER 1964; VAILLANT 1966; DOLE et al. 1969). By the early 1960s, the pharmacotherapeutic agent used in these detoxification facilities was the synthetic opioid methadone. Detoxification using methadone to prevent primarily the most severe acute signs and symptoms of abstinence was carried out in these public based facilities, where methadone was administered in decreasing doses over a short period of 1–2 weeks, and the detoxified addict was then released, usually within 10–21 days. Following cycles of heroin addiction, opiate withdrawal signs and symptoms usually appear within 4–6 h after the last short-acting opiate dose. The peak in signs and symptoms is at 48–72 h after the last dose, and the acute abstinence period may last for 7–14 days. Long-term or protracted abstinence symptoms have been well characterized by the group at Lexington, the group at Rockefeller, and others. These include physiological abnormalities, such as abnormal pulse, blood pressure, and respiration; abnormal behavior, especially atypical depression; and also abnormalities in neuroendocrine function which may persist for months or even years. The most persistent symptoms include restlessness, irritability, inability to concentrate, and sleep disorders. Each of these abnormalities may persist for months or even years after cessation of heroin addiction in the untreated former addict and may contribute to drug "hunger" or craving and also to relapse to illicit use of opiates.

In methadone detoxification treatment of heroin addiction, the initial methadone dose is usually 20–40 mg/day, given in two to four divided doses.

After 2–4 days of stabilization on that dose, it is usually tapered down over a 7–10 day period, reduced at a daily rate of 10%–20%, about 5 mg each day. In other regimens, the methadone dose is reduced more gradually, usually at the rate of 2.5 mg a day during the last few days of detoxification. Some short-term methadone detoxification programs, which are around 10–14 days in duration, have also been extended for 30 days. Other so-called methadone detoxification programs have extended for 6 months, but these are really de facto methadone maintenance programs, with a limited time period for treatment because of a variety of policy reasons. Irrespective of the intensity with which counseling and rehabilitation efforts are provided with or without psychiatric or behavioral modification treatment, and whether they are provided concomitantly with or following the short-term pharmacological detoxification from heroin, with methadone, the partial agonist, buprenorphine, or with any other pharmacotherapeutic agent, such as an α_2 adrenergic agonist, clonidine, or lofexidine, or with use of prolonged general anesthesia (with all the inherent dangers, including mortality, thereof, for a medical problem with no intrinsic mortality), or with the addition, during or after other pharmacological detoxification, of an opioid antagonist, relapse to illicit opiate use usually occurs within 6 months to 2 years (WIKLER et al. 1953; BERLE and NYSWANDER 1964; VAILLANT 1966; DOLLERY et al. 1976; RESNICK et al. 1977; KLEBER et al. 1985; KOSTEN et al. 1989a; LOIMER et al. 1990, 1991; NIGAM et al. 1993). Short-term detoxification from heroin using methadone as the pharmacotherapeutic agent has been shown, however, to be as or significantly more effective with respect to reducing or eliminating essentially all acute and subacute signs and symptoms of opioid abstinence than all other kinds of pharmacologically supported detoxification methods, because of the long-acting pharmacokinetic profile and site-specific action of methadone.

In the early 1960s, however, it became apparent that "detoxification treatment" was a "revolving door" treatment, since the majority of long-term addicts thus treated soon reappeared, requiring another detoxification procedure. Similar results following short-term detoxification treatment with any pharmacological agent or procedure such as acupuncture, electroacupuncture, general anesthesia, hyperbaric chamber, or behavioral or religious protocols have been found over the last 50 years. Despite the development of many novel and combined nonopioid agonist pharmacotherapeutic approaches to detoxification treatment, including use of α_2 adrenergic agonist drugs such as clonidine or lofexidine to ameliorate some (but not all) of the opioid withdrawal or abstinence symptoms, alone or in combination with an opioid antagonist such as naltrexone to hasten the onset and disappearance of signs and symptoms of opiate withdrawal, relapse back to illicit opiate use is still highly probable. When appropriate long-term follow-up studies have been performed, the relapse rates are the same, with 70%–95% of all long-term heroin addicts relapsing within 6 months to a few years of detoxification by any method and in any of several nations where studies have been conducted. The consistent observation that this approach fails in the management of most "hard-core" heroin addicts also contributed to the recognition of an urgent need for developing pharmacotherapies for long-term management. Indeed, it is exclusively the long-term pharmacotherapies that to this day have been shown to be truly effective in the management of chronic opiate dependency (BALL and Ross 1991).

Although not within the mandate of this chapter, it should be stressed that ever since the beginning of research on pharmacotherapy of addiction, it has been clear that the most effective pharmacotherapeutic programs must also provide counseling, rehabilitation services, and (in the most successful programs) access to both health care and psychiatric care as needed. Most of these programs at this point in time employ methadone in the maintenance treatment of addiction; some (as is to be discussed), soon may also use l-α-acetylmethadol or another currently experimental drug, such as the partial opioid agonist, buprenorphine, or the specific opioid antagonist, naltrexone (which has had limited success in management of unselected heroin addicts) (DOLE and NYSWANDER 1965; DOLE et al. 1966a,b,c, 1968; NYSWANDER and DOLE 1967; KREEK 1991a,b,c, 1992a,b,c,d,e,f; BALL and Ross 1991; McLELLAN et al. 1993). Therefore, although pharmacotherapy is the focus of this chapter, it should be emphasized that pharmacotherapy is most effectively carried out in the setting of immediate access to a wide variety of services which address the medical, psychiatric, behavioral, and psychosocial needs of a patient. Since long-term pharmacotherapies delivered in well-constituted programs provide the most effective treatments for opiate dependency, these will be emphasized herein, with no further discussion of detoxification procedures. Each pharmacotherapeutic approach for opiate addiction, both long-term maintenance approaches and very short-term pharmacotherapaeutic detoxification approaches, have been used primarily for the management of the hard core or long-term heroin addict. A long-term "opiate addict" is defined operationally and officially for Federal guidelines for use of methadone maintenance treatment, as a person with 1 year or more of multiple daily illicit uses of opiates, with the development of tolerance, physical dependence, and drug-seeking behavior, that is, addiction (KREEK 1991a, 1994a,b).

Some of the pharmacotherapeutic approaches should be considered appropriate for use in early intervention, that is, for the management of opiate abusers or of early addicts who do not meet the criteria for maintenance treatment with an opioid agonist or partial agonist. Unfortunately, few studies have been performed to address the relative effectiveness of pharmacotherapies which do not involve the use of opiate agonists or partial agonists (such as methadone, LAAM, or buprenorphine, which are appropriate solely for management of long-term heroin addiction), for possible use in the management of opiate abuse, or for early opiate addiction. Such studies are needed, possibly using an opioid antagonist, such as naltrexone or nalmefene, which simply blocks the effects of any self-administered opiate. Other potentially useful approaches are those which are directed toward enhancing the endogenous opioid system for augmenting the pharmacological management of the signs and symptoms of withdrawal, and especially also for managing the protracted abstinence symptoms, and thus for relapse prevention. Such approaches might have utility in the management of early addiction such as the administration of a natural opioid peptide, such as dynorphin A_{1-17} or its shorter synthetic congener dynorphin A_{1-3}; use of a compound which may block or slow the enzymatic pathways of normal biotransformation and degradation of endogenous opioid peptides; or use of electro-acupuncture or transcutaneous electrical stimulation to enhance the endogenous release of endorphins (CHOU et al. 1993, 1994; TAKEMORI et al. 1993; KREEK et al. 1994; HAN 1993). These are all approaches currently under study for possible future use in early intervention or in relapse prevention for those long-term addicts not able or unwilling to participate in long-term pharmacotherapy with an opioid agonist or partial agonist following acute pharmacological detoxification treatment.

For the purposes of this chapter, long-term opiate or heroin addiction will be defined according to the operational definition used in Federal regulations governing the use of methadone and LAAM and, possibly in the future, buprenorphine: daily, multiple dose use of opiate illicitly obtained, for at least 1 year, with the development of tolerance, physical dependence, and drug-seeking behavior (KREEK 1991a, 1994a). Although this definition is not based on neurobiological data or even necessarily on clinical research, it will be used because it is the definition currently used by the United States Food and Drug Administration in their regulations managing use of methadone and LAAM in the pharmacotherapy of opiate addiction and has been found to be useful for evaluation of neurobiological, treatment, and outcome studies (KREEK 1987, 1991a, 1992a,b,c,d,e,f; RETTIG and YARMOLINSKY 1994). The "point of no return" in the natural history of opiate addiction has not been fully elucidated, either by basic clinical, basic human neurobiological, or applied clinical research studies, nor in any more fundamental laboratory studies. However, applied clinical research and epidemiological studies have provided substantial evidence documenting the natural history of opiate addiction and the proclivity for relapse following the restoration of the drug-free state in long-term opiate addicts. Thus opiate addiction is a chronic relapsing disease. Similarly, much information on the natural history of addiction has been derived through the criminal justice system, including courts, prisons, and probation. All of these data have provided evidence of the chronic relapsing characteristics of opiate addiction and have thus contributed to the operational definition.

In this chapter, the major FDA-approved pharmacotherapeutic approaches for the long-term management of opiate dependency will be discussed, including the use of (1) the long-acting opioid methadone, (2) the recently approved, even longer-acting opiate l-α-acetylmethadol (LAAM), (3) the specific opioid antagonist naltrexone, which has also been approved for the adjunctive treatment of opiate dependency and, very briefly (4) the still experimental, potentially useful partial opioid agonist (mixed agonist-antagonist) buprenorphine (which has been studied to date only in relatively short-term management of opiate dependency). Each of these agents has been used in many different ways, but most effectively when combined, as appropriate, with counseling and rehabilitation efforts and also with primary medical and psychiatric care. Short-term treatment with agents used in the pharmacotherapeutic approaches for short-term detoxification treatment of addiction will not be discussed further; as mentioned above, these approaches include the use of (1) the agonist methadone, (2) the partial agonist buprenorphine, (3) the antagonists naloxone and naltrexone, (4) the α-adrenergic agonist agents clonidine or lofexidine, (5) combined therapy with an α_2-adrenergic agent such as clonidine or lofexidine, coupled with an opioid antagonist, usually naltrexone and clonidine and, (6) an antagonist combined with high doses of a benzodiazepine, barbituate, or with general anesthesia.

The goals of any pharmacotherapy for opioid dependency should include: (1) prevention of signs and symptoms of opiate withdrawal or abstinence; (2) significant reduction or elimination of drug hunger and drug craving, thus preventing a major component contributing ot relapse to illicit use of drugs; and (3) normalization of physiological functions disrupted by chronic use of the short-acting opiate (KREEK 1973a,b, 1987, 1991a, 1992a,b,c). The rationale of the use of any agent for the treatment of opiate dependency should be based on basic laboratory and clinical research scientific information. Ideally, an agent should be targeted to a specific site of action, such as a receptor at which the drug acts, or alternatively at the site or mechanism of some specific neurochemical event which follows receptor binding and activation. Pharmacotherapeutic action should be as specific and selective as possible. Also, an effective therapeutic agent will probably need to be orally effective, both to provide ease in administration and to reduce or eliminate any continuation of behavior involving needle use and needle sharing. This will also eliminate any risk from diseases which are communicated by the parenteral route (CHERUBIN 1967; LOURIA et al. 1967; SAPIRA et al. 1968; KREEK et al. 1972; STIMMEL et al. 1975; KREEK 1978a) and are extremely common in untreated heroin addicts, with prevalences of hepatitis delta up to 30%, of HIV-1 up to 60%, and of hepatitis B and C over 90% in some regions (CHERUBIN et al. 1970; KREEK 1972, 1973a,b, 1978a; KREEK et al. 1972, 1986, 1990a,b; SAPIRA et al. 1968; DESJARLAIS et al. 1984, 1989; NOVICK et al. 1981b; 1985a,b,c, 1986a,b,c, 1995). In addition, pharmacotherapeutic agents should be long acting to allow dose administration to be no more frequent than once a day, thereby increasing compliance and reducing difficulties in therapeutic management in those instances where medication may be administered only in a clinic setting. Of even greater importance, the long-acting agent should provide a steady-state perfusion of receptors or other sites of action and thus permit normalization of disrupted physiology.

B. *dl*-Methadone (and *l*-Methadone) as Used in Long-Term Pharmacotherapy of Opiate Addiction

I. Chemistry, Pharmacokinetics, Pharmacodynamics, and Mechanisms of Action

Methadone is a synthetic pure opioid agonist now known to act exclusively at the μ-type opioid receptors. Methadone (4,4-diphenyl-6-dimethylamino-heptanone-3) was first synthesized by the I.G. Farben Industrie Hoechst-Am-Main in the late 1930s and studied in the United States after World War II for use in the management of pain (EISLER and SCHAUMANN 1939; EDDY 1947; ISBELL et al. 1947a,b,c,d, 1948; KARR 1947; KIRCHHOF and UCHIYAMA 1947; WIKLER and FRANK 1947; WIKLER et al. 1947; WOODS et al. 1947a,b; CHEN 1948). The Technical Industrial Intelligence Committee of the State Department of the United States identified the existence of this compound, details of its synthesis, and early research on it. Because of interest in its possible unique properties, the U.S. State Department brought it to the United States, England, and France for further research in 1946. The chemical structure of methadone was identified as being novel for an analgesic and thus, it was felt to be of interest. Although the Council on Pharmacy and Chemistry of the American Medical Association adopted the generic name "methadone," it became known by different potential trade names while being studied by different compaines. It was referred to as "Dolophine", from "morphine" and the Latin derived word for pain, "dolor," by Eli Lilly; as "Adanon" by Winthrop; as "Miadone" in England; as "L'amidon" in France; as "Amidon" in its clinical testing in Germany; and as "compound 10820" and "Dolantin" as initially synthesized by the I.G. Farben Industrie. The compound underwent extensive early studies at the United States Public Health Service Hospital for research and treatment of opiate addiction in Lexington, Kentucky, where it was usually referred to as "amidone," "methadon," or "compound 10820." In the earliest studies there and else-where, it was found to be an analgesic to which tolerance and physical dependence would develop as with other opioid analgesics such as morphine (BERCEL 1948; CEHN 1948; FINNEGAN et al. 1948; ISBELL and EISENMAN 1948; ISBELL et al. 1948; JENNEY and PFEIFFER 1948; MAY and MOSETTIG 1948; POPKIN 1948; EDDY et al. 1949; POHLAND et al. 1949; SPEETER et al. 1949; FRASER and ISBELL 1951, 1952). The chemical name of methadone is 4,4-diphenyl-6-dimethylamino-3-heptanone. Methadone is usually synthesized as a racemic mixture, and this is the form used in treatment in the United States and most other places. It has been shown that the *l* (more correctly designated *R* or [−]) enantiomer is a pharmacologically active moiety, at least fifty times more potent than the relatively inactive *d* (more correctly *S* or [+]) enantiomer. Racemic methadone is usually synthesized and prepared for clinical use as a hydrochloride salt. Analgesic studies have shown that a dose of 2.5–7.5 mg methadone may relieve pain and is comparable to approximately 10–15 mg morphine sulfate in naive subjects, and that methadone relieves pain with much less sedation at those comparable doses than does morphine. Early studies showed that the duration of action of pain relief was very similar to that of morphine, that is, 4–6 h. However, if multiple doses of methadone were given over a 24-h dosing period, with a cumulative dose of 10–40 mg, respiratory depression would ensue in opiate naive subjects when the higher doses were used. These early clinical research findings, all made prior to the development and availability of sufficiently sensitive and specific chemical analytical techniques for measurement of the pharmacokinetics of methadone, morphine, or any of the opioid analgesics, gave early indication of the possible uniquely long-acting properties of methadone. Of importance, it was also found that euphoria, which occurs following morphine administration, was "rarely present, if at all, with methadone" (Chen 1948). Early work with methadone at the USPHS Hospital in Lexington, Kentucky, the Lilly Research Laboratories, the National Institutes of Health, and a few other places was conducted by Eddy, Isbell, Wikler, Pohland, Chen, May, and others from 1946 to 1949.

The first report of use of methadone for the short-term pharmacologic management of the withdrawal symptoms resulting from abrupt cessation of heroin in opiate addicts was an abstract by Isbell, Eisenman, Wikler, and colleagues at Lexington, Kentucky in 1947 (Isbell et al. 1947b). As summarized in an early review by K.K. Chen from the Lilly Research Laboratories, "although physical dependence develops with methadone, its withdrawal symptoms are so slight that it can be used for treatment of morphine addiction, a need which had been recognized since the opening of the United States Public Health Service resource at Lexington under the initial directorship of Dr. Himmelsbach, resulting in a smooth, milder, abstinence" (Chen 1948; Isbell et al. 1947b). The discovery of the potential for use of methadone in short-term detoxification treatment of opiate addiction was thus a direct reuslt of the early research performed at the USPHS Hospital in Lexington in their primary research mission to determine the abuse liability of new analgesic medications. The properties of any potentially opiate-like chemical agent and the capacity of such an agent to produce tolerance and physical dependence, which Himmelsbach had defined as requisite for the development of addiction to a drug, along with habituation, were studied at the Lexington facility (Himmelsbach 1941). A limited number of opiate-tolerant dependent former morphine or heroin addicts at Lexington were studied using methadone as the test agent; methadone doses as high as 150–200 mg were administered up to four times a day for short durations and no severe side effects noted. However, the fact that in those earliest studies methadone was given four times a day suggests that there was still no appreciation for the long-acting pharmacokinetic properties of methadone. The use of methadone for short-term detoxification treatment of heroin addiction was introduced around 1958–1961 into a very limited number of centers, including the Bellevue Hospital in New York City (then under the joint auspices of Columbia University College of Physicians and Surgeons, Cornell University Medical College, and New York University Medical School), and the Metropolitan Hospital, another city hospital in New York City, and a few city-funded proprietary "detoxification centers" for treatment of addiction, such as the Manhattan General Hospital (later to become the Bernstein Institute of Beth Israel Hospital) along with a limited program using methadone at the USPHS Hospital in Lexington, Kentucky. However, methadone continued to be used in divided multiple daily low doses in the detoxification treatment for opiate addiction.

In 1963, research was planned and initiated in early 1964 to develop a possible long-term pharmacotherapy for opiate addiction at the Rockefeller Institute for Medical Research under the leadership of Professor Vincent P. Dole, who was joined for this initial work by the late Dr. Marie Nyswander, a psychiatrist who had worked in the field of addiction treatment for many years, and also by Dr. Mary Jeanne Kreek, a physician-scientist on a research elective while then still in residency training in internal medicine at the New York Hospital-Cornell Medical Center. The goals and criteria for selection of a potentially appropriate pharmacotherapeutic agent for the management of heroin addiction (and now more broadly applicable for the development of a pharmacotherapy for any addictive disease) were informally and later formally defined (KREEK 1991a, 1992a,b,c). It was obvious that previous attempts, both in short-term detoxification treatment and, in Europe, long-term management of heroin addiction using parenterally (usually intravenously) administered short-acting opiates, such as heroin itself or morphine, had been unsuccessful and in large part because of three fundamental problems: (1) Tolerance develops rapidly following repeated exposures to a short-acting opiate, with rapid onset following intravenous administration but also rapid offset action, in humans; thus the dose of opiate used to manage opiate dependency, even in short-term detoxification and especially in long-term treatment, had to be rapidly escalated simply to prevent the signs and symptoms of narcotic withdrawal. (2) Because of both the rapid onset of action and short-acting pharmacokinetic properties of these opioid agonists following intravenous administration of a dose adequate to prevent abstinence symptoms for 4–6h, there was an immediate onset of euphoria with the related "reinforcing" effects of such action, and somnolence, thus causing a period during which the opiate addict receiving treatment was relatively afunctional for 30min to 2h. If a dose of short-acting opiate sufficient to prevent signs and symptoms of opiate withdrawal for at least 4–6h was employed, the first 4h of the dosing interval would usually be spent in a period of euphoria, followed by somnolence, and then with decreasing narcotic effects and the gradual onset and escalation of the abstinence syndrome, that is the signs and symptoms of opiate withdrawal, thus rendering the person unable to function normally during most or all of the time. (3) The third problem that needed to be countered was that most short-acting narcotics, including heroin and morphine, require parenteral, usually intravenous, administration to achieve their maximum effects. Therefore, a medication was sought which would have a very slow onset of action and a prolonged duration of action, that is, a long-acting pharmacokinetic profile so that the dosing interval could be long with ideally once a day dosing. Such an agent, because of both a slow onset as well as slow offset of action, would provide very little if any reinforcing effects, that is, no euphoria or "high" and other narcotic-like effects, and at the same time would have a long sustained duration of action, thus preventing early appearance of narcotic withdrawal symptoms or the abstinence syndrome for a long dosing interval. Also, hopefully, such an agent would prevent "drug hunger" or craving (DOLE et al. 1966a,b). It was hypothesized that such a longer acting agent with sustained action would lead to a slower development of tolerance, which was later proven with research on methadone as used in maintenance treatment: with providing a steady state of plasma levels of medication, a sustained level of tolerance pertains for very long periods of time. Finally, (4) it was clearly desirable to have an orally effective agent to simplify dose administration, decrease the possibility for unsterile needle use or sharing, which leads to infection with many viruses and bacteria, and to remove the addict from the lifestyle of interacting with other addicts for the use of needles and other injection equipment for the self-administration of heroin or other drugs of abuse.

Based on the available studies, both in animal models and especially in humans, including the limited studies in former addicts conducted at Lexington, Kentucky, the few studies for the potential use of methadone in management of pain, and especially the fairly well-developed clinical experience in the pharmacotherapy of addicts using methadone for short-term detoxification treatment (work which Dr. Nyswander was acquainted with both from her experience at Lexington as well as Bellevue, and which Dr. Kreek also had experience working with as a medical student of Columbia University working at Bellevue), methadone was identified as potentially the only opioid agonist then available which might meet the criteria of oral effectiveness and possibly slow onset and long duration of action. At the time of the early work pertaining to the use of methadone in the long-term pharmacotherapy of heroin addiction, which was initiated in early 1964 at the Rockefeller Institute for Medical Research, there were no sufficiently sensitive and specific chemical analytical techniques for measuring methadone, natural morphine, or its synthetic derivative heroin, in human blood or even in urine. Therefore, careful clinical observations in the in-patient research unit of the Rockefeller Institute Hospital had to be conducted to verify the predicted pharmacokinetic as well as pharmacodynamic properties of methadone when administered orally in a single daily dose to former, long-term heroin addicts in the experimental management of opiate addiction (DOLE and NYSWANDER 1965; DOLE et al. 1966a,b; KREEK 1992a,b,c).

By the early 1970s, three groups working independently had developed sensitive and specific analytical techniques using gas-liquid chromatography for the measurement of methadone in body fluids (INTURRISI and VEREBEY 1972a,b; SULLIVAN and BLAKE 1972; SULLIVAN et al. 1972; KREEK 1973c; DOLE and KREEK 1973; KREEK et al. 1974, 1976a,b). The initial studies of methadone pharmacokinetics suggested that methadone might have a relatively short half-life, due to the limitations in these newly developed analytical methods. Further studies conducted in patients receiving methadone on a daily basis for treatment of heroin addiction and reported by 1972–1973 verified the clinical observations which had been made in 1964 by the Rockefeller research team (DOLE et al. 1966a,b, KREEK 1973c). Those early observations had shown that, whereas methadone provided analgesia for only 4–6h, similar to morphine, when methadone was administered once daily to heroin addicts, it prevented signs and symptoms of opiate withdrawal for 24h (DOLE et al. 1966a,b). Also during such oral administration, "drug hunger" or drug craving were eliminated throughout that dosing interval (DOLE et al. 1966a,b). In addition, after oral administration of methadone to a tolerant and dependent addict, there was no "high," euphoria, or other perceived narcotic effect, even in initial phases of treatment if the dose of methadone was selected to be less than that to which the individual had developed opioid tolerance. Subsequently, it was found in the 1964 research at the Rockefeller Institute Hospital that treatment doses could be slowly escalated up to a full treatment dose level of 60–120 mg/day with no narcotic-like effects occurring because of the gradual development of tolerance, and that these higher doses provided cross-tolerance or "blockade" of the effects of any potentially superimposed short-acting illicit narcotic (see section below) (DOLE et al. 1966a,b).

The pharmacokinetic studies of the early 1970s showed that methadone has a very slow onset of action after oral administration, with peak plasma level reached between 2 and 4h, and with a sustained plasma level over a 24h dosing interval (INTURRISI and VEREBEY 1972a,b; KREEK 1973c; KREEK et al. 1974; ANGGARD et al. 1974). It was also shown that the peak plasma levels are low, usually less than twice the trough or sustained plasma level (KREEK 1973c). Finally, it was shown that the average plasma apparent terminal half-life of methadone was around 24h in study subjects. It is known that methadone accumulates in tissue stores, primarily in hepatic stores which, in the setting of chronic oral dosing, sustains plasma levels in a relatively steady state and thus modifies the apparent terminal half-life as determined in classical pharmacokinetic sutdies (DOLE and KREEK 1973; KREEK 1973c; KREEK et al. 1976b, 1978a). Laboratory studies conducted in an animal model using an isolated perfused rabbit liver preparation, as well as studies conducted in whole rats, showed that methadone is very widely distributed throughout the body after administration by oral or parenteral route and that the liver is the major site of accumulation of methadone (HARTE et al. 1976; KREEK et al. 1978a). Further studies using the perfused rabbit liver showed that methadone is bound primarily in membrane fractions and that it is slowly released in an unchanged form from those membrane fractions back into the circulation (KREEK et al. 1978a). In addition, studies in humans have shown that methadone is over 90% bound to plasma proteins and binds both to albumin and all globulin fractions (KREEK et al. 1976b, 1978a; POND et al. 1985). Thus, there is a hepatic resevoir for subsequent release of unchanged methadone into the circulation, coupled with a buffering of the free or unbound concentrations of methadone in plasma. Methadone is metabolized primarily by successive N-demethylation steps to form a pyrolline and a pyrollidine, which are inactive in metabolites requiring the hepatic p450-related enzymes for oxidative metabolism (SULLIVAN et al. 1972; KREEK et al. 1976a,b).

Using stable isotope-labeled species of the active l-$(R)(-)$enantiomer (pentadeuteromethadone) and a separate labeled species of the relatively inactive d-$(S)(+)$-enantiomer (trideuteromethadone), administered as a single dose intravenously against a background of daily administered unlabeled doses of racemic methadone in patients maintained on methadone, the pharmacokinetics of each species was determined using a separate stable isotope-labeled species (octadeuteromethadone) for quantitation of the dl racemic mixture as used in therapy. In this selective stable isotope labeling, different amounts of deuterium were placed in nonreactive positions in the methadone molecules. Special pharmacokinetic studies have been performed (HACHEY et al. 1977; KREEK et al. 1979; NAKAMURA et al. 1982). Studies conducted in patients receiving methadone on a chronic basis for treatment of addiction have shown that the half-life of the active l-$(R[-])$enantiomer is around 48 h, whereas the half-life of the inactive d-$(S[+])$ enantiomer is around 16 h. Thus, methadone has been shown to have a slow onset of action and a very long-acting pharmacokinetic profile in humans. This is in sharp contrast with the pharmacokinetic profile of heroin and morphine in humans, which have a very fast onset of action and short duration of action, with rapid decline of effectiveness. Heroin, the man-made deacetylated congener of morphine, undergoes successive deacetylation from diacetyl to monoacetyl morphine and on to morphine; the pharmacokinetic half-life of heroin is approximately 1–2 h and the plasma apparent beta terminal half-life of the major metabolite morphine is approximately 4–6 h. Also, the systemic bioavailability of heroin and morphine is very limited, with less than 30% of an orally administered compound reaching the systemic circulation.

Cerebrospinal fluid levels of methadone have also been measured along with concurrent serum plasma methadone levels; the CSF levels ranged from 2% to 73% of the concomitantly measured plasma levels, peak methadone levels in CSF appeared approximately 3–8 h after methadone administration, as contrasted to peak plasma levels which occur 2–4 h after methadone administration (RUBENSTEIN et al. 1978). The highest levels of CSF methadone thus may be observed at a time when plasma levels have already declined toward the steady state levels (RUBENSTEIN et al. 1978). Although orally administered morphine is now currently widely used for the management of chronic pain, morphine and heroin abuse is primarily by the parenteral route, intravenously or subcutaneously ("skinpopping"), or alternatively by the nasal route of administration, either of the salt form and more recently of the free base, which provides higher arterial levels of drug. When used orally, morphine has both reduced bioavailability, but also slower onset of action, somewhat longer apparent terminal half-life during chronic multiple dose administration. In contrast, methadone has an over 90% systemic bioavailability after oral administration. Because of its rapid and wide distribution, as well as extensive tissue binding, even following intravenous administration, as has been used in some of the pharmacokinetic studies, after a transient peak plasma level, the plasma levels then drop to those seen after oral administration, with a prolonged and sustained plasma level. Similar findings have been made in numerous subsequent studies (WOLFF et al. 1991a,b, 1993; BORG et al. 1993). These pharmacokinetic findings support and are essentially identical with the earlier pharmacodynamic observations made in patients undergoing studies with respect to the use of methadone for pharmacotherapy of addiction. Methadone is excreted both in urine and, following biliary secretion, in feces in essentially equal amounts; unlike many other opiate drugs, in the setting of renal disease or failure, it may be secreted essentially completely by the fecal route, preventing accumulation and thus the potential of any toxicity (KREEK et al. 1976b, 1978b, 1980a,b,c, 1983a; BOWEN et al. 1978).

Recently, following the cloning of the first opioid receptor, the δ receptor, in late 1992 independently by Evans and by Kieffer, and using that cloned δ opioid receptor structure, the research groups of Yu, Uhl, Akil and Watson, and others successfully cloned the μ as well as κ opioid receptors in 1993, which led to studies using an opioid receptor transfected cell model (EVANS et al. 1992; KIEFFER et al. 1992; CHEN et al. 1993a,b; WANG et al. 1993; THOMPSON et al. 1993). In these studies, it has been shown that methadone is one of the most highly selective μ opioid receptor agonists, with a selectivity similar to that of fentanyl and possibly even greater selectivity than morphine or heroin. The original investigative team of Dole at the Rockefeller University had hypothesized that methadone was targeted exclusively at opiate receptors and, although such had not been fully defined or documented to exist, the team had hypothesized that methadone would be directed at those opioid receptors which would be very similar to those receptors which bind morphine and heroin (later known to be primarily though not exclusively, the μ opioid receptors). Dole and also Goldstein attempted to delineate the specific opioid receptor in 1968 to 1971, for which he developed an experimental approach of using stereoselective binding of radioactive opioid ligands (DOLE 1970; INGOGLIA and DOLE 1970; GOLDSTEIN et al. 1971). However, it was not until high specific activity radioisotope labeled ligands, naloxone and etorphine, which bind selectively to opioid receptors were available in 1972 that the specific opioid receptors could be identified using similar techniques to those used by Dole and Goldstein, and repeated by three independent groups in 1973, those of Snyder, Simon, and Terenius (PERT and SNYDER 1973; SIMON et al. 1973; TERENIUS 1973). The high selectivity for the μ opioid receptors and the slow onset and long-acting pharmacokinetic profile of methadone probably explain the high level of efficacy of methadone when administered in appropriate doses and used in well constituted treatment programs with well-trained, diverse, and concerned staff members in the long-term pharmacotherapy of opiate addiction. This is in sharp contrast with the very limited efficacy of use of short-acting opioid agonists and also of opioid antagonists in the long-term treatment of opiate addiction in unselected populations of long-term heroin addicts. The pharmacokinetic profile of methadone with its slow onset and prolonged duration of action as defined in vigorous studies supported the earlier observations that methadone could be administered in a single constant daily dose, a dosing pattern in sharp contrast to that which had been used in the early (and still often in current) use of methadone in short-term detoxification treatment.

In the maintenance treatment of heroin addiction, the oral methadone dose which is recommended for an initial dose is similar to that used in short-term detoxification treatments (DOLE et al. 1966a,b; DOLE and NYSWANDER 1966; KREEK 1991a, 1992a,b,c). For most subjects, a dose of 20–40 mg/24 h is appropriate, that is adequate to prevent opiate withdrawal symptoms for most, if not all, of the 24 h dosing interval, yet low enough not to cause any undesired opiate-like effects such as euphoria or somnolence, or any undesirable side effects such as respiratory depression, which might occur only in any modestly narcotic tolerant addicts. The need for accumulation of methadone in tissue stores also explains the differing needs of patients in the first 2 to 3 weeks of treatment with respect to dosing interval, as contrasted to those in long-term treatment. Because of this need to saturate tissue stores over the first several days or up to 2 weeks, in some patients it is essential to give methadone in divided doses, that is, give an additional daily dose of methadone. These additional small doses of methadone (10–20 mg) may be needed at 6–12 h after the usual treatment dose for up to 1 or 2 weeks to prevent signs and symptoms of narcotic withdrawal as well as drug craving. In the optimal treatment paradigm originally developed in 1964, following initial stabilization on 20–40 mg/day of methadone delivered as a single oral dose, with or without an additional dose 6–12 h later for 1–2 weeks, the dose should then be increased gradually by 5–10 mg each week until a full treatment dose providing cross tolerance, thus "blockade" against any illicitly superimposed dosage of short-acting narcotic, is reached (KREEK 1991a, 1992a,b,c; RETTIG and YARMOLINSKY 1994). Original research in the 1960s and repeatedly reproduced in the 1970s, 1980s, and 1990s, have all shown that a full treatment of dose of 60–120 mg/day is optimal, especially during the first year in treatment for most unselected heroin addicts, and that such a dose provides a steady state of plasma levels and receptor perfusion and also a necessary degree of cross tolerance to provide "blockade" against any euphorigenic effects of superimposed short-acting narcotics (DOLE et al. 1966a,b; KREEK 1973a,b, 1991a,b,c, 1992a,b,c; BALL and Ross 1991). For patients who have been in treatment for 1 year or more, who have stopped all illicit opiate use, and who have also responded to treatment with respect to change of lifestyle, that is, rehabilitation, lower methadone doses (30–60 mg/day) may also be effective in some patients.

Recent very long-term follow-up studies of the medical status of patients in continued methadone treatment for over 10 years have shown that there are no adverse medical side effects or toxic effects of very long-term (10 years or more) moderate to high dose methadone treatment (NOVICK et al. 1993). Therefore, there is no medical indication or need for reducing the full treatment of dose of methadone. Studies have shown that doses of methadone should never be varied for behavioral management, that is, for the purposes of contingency contracting, which has been used by many clinics, but also more recently studied and shown to cause more disruption in behaviors and illicit drug use, rather than the hypothesized improvement in behaviors. Now there is consensus and agreement that methadone doses should be kept stable and that only unrelated forms of contingency management (such as medication "take home" privileges, as used by Rockefeller groups from the inception of pharmacotherapeutic treatment research to reinforce abstinence from use of illicit drugs) should be used, as desired, by clinics.

1. Methadone Disposition in Setting of Chronic Diseases

In the setting of severe chronic liver disease such as postviral (hepatitis B, C, or delta) or alcoholic cirrhosis, an anomalous pharmacokinetic situation has been documented to pertain during methadone maintenance treatment: lower, rather than higher, plasma levels of methadone are found in patients with severe liver disease, despite the fact that metabolic biotransformation is retarded due to hepatic impairment and thus there is slow metabolic clearance of the drug (KREEK 1976b; KREEK et al. 1980a,b, 1983; NOVICK et al. 1981a, 1985a). These observed lower plasma levels of methadone may lead to the appearance of early withdrawal symptoms before the end of a 24-h dosing interval. This documented lowering of plasma levels is probably due to a significantly reduced hepatic storage capacity for methadone (KREEK et al. 1978a). Thus in the setting of very severe chronic liver disease, a lowering of methadone dose is certainly not indicated and in some instances, an increased dose may be needed (NOVICK et al. 1981a, 1985a). In the setting of modest liver disease, no abnormalities of methadone disposition are found. Studies of methadone disposition also have been studied in a limited number of patients with chronic renal disease (KREEK et al. 1980c). It has been shown that methadone is normally excreted equally by the urinary and hepatibiliary fecal route of elimination, as contrasted to many opioid agonists which are excreted almost exclusively by urinary excretion (KREEK et al. 1980a,b,c, 1983a). In the setting of anuria, methadone's metabolites may be excreted exclusively by the fecal elimination. Also in these studies, it has been shown that there is no significant accumulation of any of the metabolites of methadone and no side effects due to any metabolites, as may pertain for use of some opioid agonists in the setting of severe renal disease or failure. Thus methadone may be safely used at usual doses for the maintenance treatment of opiate addiction in patients with chronic renal disease and it may become the opioid agonist of choice for the management of pain in the setting of severe chronic renal disease (KREEK et al. 1980c).

2. Methadone Disposition in Pregnancy

Other studies have shown that in the setting of pregnancy, methadone disposition is significantly altered (KREEK 1979b, 1981–2, 1983a; KREEK et al. 1974; POND et al. 1985). In the second half of pregnancy, especially in the third trimester, the plasma levels of methadone have been shown to become significantly reduced following administration of a constant dose of methadone throughout pregnancy. In the early postpartum period, it has been shown that the same doses of methadone result in the normally expected plasma levels, which are much higher than those found in late pregnancy. These studies have suggested that there is an accelerated biotransformation of methadone in late pregnancy (POND et al. 1985). This is most likely due to the effects of the very high levels of progestins which are known to enhance several of the hepatic microsomal P450-related enzyme activities and resulting in accelerated metabolism of several medications during late pregnancy in humans. These findings suggest that doses of methadone should not be reduced during pregnancy, especially not during late pregnancy, as they often are for a well-intentioned concern, but unsubstantiated need, to reduce the exposure of the neonate to this opioid. Many studies have shown that there is no teratogenicity due to methadone in early embryonic development (FINNEGAN et al. 1982). However, it has been shown that around 50%–70% of babies born to pregnant women receiving moderate to high dose methadone treatment will undergo mild, moderate, and severe narcotic withdrawal symptoms in the perinatal period. There has not been a clear cut maternal methadone dose relationship to the appearance or severity of these symptoms in the neonate. Severe symptomatology has been usually observed in babies born to mothers who are using other drugs including alcohol, cocaine, barbiturates, or benzodiazepines during pregnancy. The methadone maintenance patients in general, and including pregnant women, have been shown to be more likely to use other drugs of abuse including alcohol, barbiturates, benzodiazepines, and cocaine in an attempt to self-medicate the real signs and symptoms of opiate withdrawal, which may occur when the doses or plasma levels of methadone are inadequate to provide sustained prevention of opioid abstinence symptoms for a 24 h dosing interval. Therefore, at this time, the consensus of opinion is that pregnant women should continue to be maintained on at least a moderate dose of methadone throughout pregnancy and that the dose of methadone certainly should not be lowered during the third trimester, since this may endanger the pregnant woman by allowing the onset of opioid abstinence and drug craving and in turn possible relapse to illicit opiate or other drug use (FINNEGAN et al. 1982). In related studies, methadone has been shown to appear in breast milk as would be expected, but because of the extensive tissue binding and also plasma protein binding, the amounts which appear in breast milk are extremely small, far less than that which could be anticipated to have any pharmacologic effect in a neonate, if the usual range of 24 h volumes of breast milk were consumed (KREEK 1979b, 81–2, 1983a; KREEK et al. 1974; POND et al. 1985). It is very unlikely that any physiological effects would result from this minute exposure. Also, the amount of methadone transmitted in breast milk is significantly less than the amount of opiate which would be required to manage opioid withdrawal symptoms and therefore would have no effect on ameliorating any signs and symptoms of opiate withdrawal which might appear in a neonate. Several studies have been conducted to show that an opioid agonist should be used in the management of opiate withdrawal when moderate to severe symptoms appear in babies born to mothers maintained on methadone and who are not polydrug abusers. The opioid agonist used should depend upon the experience of the neonatologist, but may be paregoric or tincture of opium or more recently, possibly oral morphine, are probably the opioid agonists of choice, although oral morphine (and methadone itself) have not yet been studied for use in this purpose.

3. Drug and Alcohol Interactions with Methadone

Drug and alcohol interactions with methadone may alter its therapeutic effectiveness (KREEK 1978b, 1981, 1984, 1986a,b, 1988, 1990a; KREEK et al. 1976a,b). Since the effectiveness of methadone, as used in maintenance treatment, primarily due to the establishment of a sustained steady state of opioid at specific opiate receptor sites, any perturbation which alters the steady state perfusion may alter the effectiveness of methadone. Agents which either significantly accelerate the biotransformation of methadone or inhibit its metabolism may result in disruption of the steady state. The first agent which was shown to significantly alter methadone biotransformation and thus its pharmacokinetic profile was the anti-tuberculosis agent rifampin (KREEK et al. 1976a,b; BENDING and SKACEL 1977; BACIEWICZ et al. 1987). The original studies were performed when rifampin was still an experimental drug for the management of resistant strain tuberculosis. Rifampin now is one of the major agents used for the management of tuberculosis, especially in populations at high risk for resistant strain tuberculosis, including active and former drug addicts in treatment, with or without HIV-1, that is, AIDS virus infection. Following initial clinical observations of the appearance of narcotic withdrawal symptoms and drug-seeking behavior in previously well-maintained and rehabilitated methadone maintained patients with tuberculosis receiving rifampin as part of their chemotherapeutic regimen, clinical pharmacologic studies were conducted. These studies showed that during concomitant administration of rifampin, the area under the plasma concentration time curve of methadone is significantly reduced (KREEK et al. 1976a,b). Also, there is an increased urinary excretion of the major pyrrolidine metabolite, suggesting an enhancement of metabolic clearance methadone in this setting (KREEK et al. 1976a,b). These studies have been subsequently corroborated by other research groups. A second related observation was made and studied with respect to the effects of phenytoin, a very commonly used anticonvulsant (Dilantin; FINELLI 1976; TONG et al. 1981). It was seen that patients who were receiving phenytoin seemed to experience early abstinence symptoms during methadone maintenance treatment. Clinical pharmacologic studies were conducted which showed that phenytoin accelerates the biotransformation of methadone with decreased area under the plasma concentration time curve and increased urinary excretion of the major metabolites (TONG et al. 1981). In the experimental subjects receiving either rifampin and methadone or phenytoin and methadone, signs and symptoms of narcotic withdrawal, as well as the appearance of drug craving or "drug hunger" appeared during concomitant chemotherapy. Therefore, when rifampin or phenytoin must be used in a therapeutic regimen of a methadone maintained patient, the doses of methadone must be increased to provide an adequate plasma levels. Also, in some cases, divided doses of methadone may be needed because of the probable shortened pharmacokinetic half-life of methadone in this setting. Rifampin has been shown to significantly increase hepatic microsomal P450-related enzyme activity. Phenytoin may have a similar but more complex effect on drug disposition. It is assumed, though not confirmed in human studies, that barbiturates, well-established enhancers of P450-related enzyme systems, will also accelerate the metabolism of methadone. Another anticonvulsant, valproic acid, has been shown not to have any effect on metabolized methadone (SAXON et al. 1989).

It has been shown by many groups that ethanol has a bimodal effect on biotransformation, metabolism, and pharmacokinetics of medications which normally require the hepatic P450-related microsomal enzymes for their metabolism. Although ethanol is metabolized primarily by alcohol dehydrogenase, when it is present in large amounts, 10%–30% of ethanol may be metabolized by hepatic microsomal P450-related drug metabolizing enzymes. Chronic use of moderate to large amounts of ethanol results in an enhancement in activity of these hepatic enzymes. When ethanol is present in very large amounts (probably in excess of 150 mg per deciliter), it may inhibit the biotransformation of drugs normally requiring these enzymes for biotransformation. However, when ethanol has been cleared from the body, in a person with chronic heavy use of ethanol and thus enhanced microsomal enzyme activity, acceleration of the biotransformation of a second medication requiring these types of enzymes may be seen. A limited number of studies have been performed in humans with respect to the possible metabolic and pharmacokinetic interactions of methadone and ethanol (KREEK 1978b,

1981, 1984, 1988, 1990a). It has been shown that methadone biotransformation may be accelerated in subjects who chronically use very large amounts of ethanol (five or more drinks equivalent a day) but is not changed in subjects whose consumption of ethanol is at a "social" level (four or fewer drinks equivalent per day). In addition, some preliminary studies have shown that when blood ethanol concentration exceeds 150 mg/dl, inhibition of methadone biotransformation may ensue, with the result of a prolongation and increase in the peak levels of methadone.

It has been hypothesized that other agents including benzodiazepines, such as diazepam, and the H_2 receptor blocking agent cimetidine may inhibit the biotransformation of methadone (Preston et al. 1986). Although some clinical observations have suggested that very high doses of benzodiazepines and of cimetidine may inhibit methadone metabolism, conclusive clinical pharmacological studies of these potential interactions using usual therapeutic doses of these agents have not been conducted in humans. Although questions about the possible interaction of various agents used to treat HIV infection and AIDS with methadone have been raised, only one significant drug interaction has been found to date between any of these agents and methadone (Selwyn et al. 1989; Borg et al. 1993, 1994; Borg and Kreek 1995). Methadone may slow the metabolism of AZT in some patients (Selwyn et al. 1989; Borg and Kreek 1995). Also, recent studies have shown that methadone itself may inhibit the metabolism of pharmacotherapeutic agents which metabolize by activity of the cytochrome P450 2D6 (CPY2D6) (Wu et al. 1993). Genetic variability of the activity of cytochrome P450 2D6 has been described, which in turn results in a polymorphic distribution of the capacity for oxidation of medications such as debrisoquine, dextrometrophan, and several other drugs. In vitro studies using human livers have shown that methadone decreases biotransformation by the pathway of O-demethylation by hepatic enzyme activity in vitro. Another agent which inhibits activity of these enzymes is quinidine, which, using single oral doses (50–250 mg), has been shown both in vitro and in vivo to block the activity of the P450 2D6 enzyme activity to convert so-called "extensive metabolizer" into a "poor metabolizer" of specific medications which usually are biotransformed by this enzyme activity.

From the earliest research in 1964 to the present time, numerous studies have shown that administration of an adequate methadone dose is essential for effective treatment, especially during the first year when experimentation with illicit drug use may be attempted. Here, the "blockade" effect, conferred by a high degree of tolerance and cross-tolerance developed during moderate to high dose methadone treatment, is needed (Ball and Ross 1991; Caplehorn and Bell 1991; Caplehorn et al. 1993; Dole 1988; Dole and Nyswander 1965; Dole et al. 1966b). Also, treatment with moderate to high doses of methadone may avoid problems related to modest drug interactions such as those possible during intermittent heavy use of ethanol. The effectiveness of adequate doses, usually 60–120 mg/day, may be best assessed by measuring plasma levels of methadone, as has been recommended since 1973 (KREEK 1973c; BORG et al. 1995b, in press). Now there is ready access to excellent analytical technology, most using gas-liquid chromatography with or without mass spectrometry techniques, which have been shown repeatedly to be superior to radioimmunoassay or enzyme-linked analyses for this purpose (WOLFF et al. 1993; BORG et al. 1995b, in press). The low cost and widespread availability of excellent commercial tests make such monitoring feasible when needed.

II. Use of Methadone in Long-Term Pharmacotherapy of Opiate Addiction: "Methadone Maintenance"

1. Safety, Efficacy, and Effectiveness

The development of methadone maintenance treatment preceded and paralleled much of the more basic pharmacological and neurobiological research discussed in the previous section. However, the hypothesized and later the documented facts upon which the rationale of methadone maintenance treatment was predicated include: (1) that methadone is an orally effective opioid agonist agent in humans; (2) that methadone in humans is a long-acting opioid with a pharmacokinetic profile distinctly different from any other opioid agonist (other than l-α-acetylmethadol) available and approved for use in humans in 1964 when the research began (although in sharp contrast, methadone has a very short-acting pharmacokinetic profile in rodent species and other experimental animals) with a slow onset of action and sustained, steady, and prolonged duration of action in humans; (3) that methadone is a pure opioid agonist; (4) that methadone is a highly selective ligand which is directed at the early hypothesized and later characterized and cloned μ opioid receptors; (5) that methadone, therefore, when used in a single oral daily dose, would provide a steady state of plasma levels; (6) that the action of methadone would be to prevent withdrawal symptoms, prevent drug hunger or drug craving, and also of importance, allow normalization of physiological systems which are profoundly disrupted by the "on-off" effects of short-acting opiates, that is the rapid onset, offset, and decline of the pharmacokinetic profile of heroin and other short-acting opiates; (7) and that this mechanism of action of methadone, because of its long-acting pharmacokinetic profile in the setting of the administration of a single, daily, moderate to high level oral dose in humans, would rely on the steady state perfusion of μ type opioid receptors, both at brain and peripheral sites, and that this action would permit the normalization of physiological functions disrupted during heroin addiction, and the prevention of abstinence symptoms, and the cessation of "drug hunger" or craving, and thus prevent relapse to use of short-acting illicit opiate drugs (KREEK 1991a, 1992a,b,c). In summary, the rationale of methadone was to provide a pharmacotherapy with a highly selectively targeted opioid agonist, acting at the same site of action as heroin or morphine, but in sharp contrast, providing a slow onset of action and steady state perfusion at those specific opioid receptor sites (KREEK 1992c). Thus, from the initial studies, it was hypothesized and later shown that methadone, in effect, provides a specifically targeted chemotherapy. It was also shown by both early and late studies that through the mechanism of development of a high level of tolerance to moderate to high doses of methadone (60–120 mg/day), cross-tolerance is also developed, so that any superimposition of an illicit short-acting opiate such as heroin or morphine in an attempt to achieve a euphoric state, commonly tried by former heroin addicts during the first 3–6 months of methadone treatment, is unsuccessful (DOLE et al. 1966a,b). The failure to achieve a euphoric state on self-administration of doses of short-acting narcotic, primarily heroin, which were previously euphorigenic for that individual, affects to an extent the classical behavioral phenomena of deconditioning, with ultimate extinction of the behavior of self-administration of heroin, since no rewarding or deleterious effect of the superimposed short-acting opiate, is experienced.

In the original research on the possible use of methadone in chronic long-term pharmacotherapy of heroin addiction, conducted at the Rockefeller Institute for Medical Research and Rockefeller Institute Hospital under the leadership of Dr. Vincent Dole in early 1964, divided doses of methadone were used for a very brief period of time, following the dosing regimen used for detoxification treatment (DOLE et al. 1966b,c). However, during the first 2 months of early research, it was hypothesized that a single daily dose of methadone would prevent the onset of sings and symptoms of opiate withdrawal for a 24-h dosing interval and also would prevent drug hunger. It was observed in this clinical research setting that following a single initial oral dose, first administered in doses of 20–40 mg/day and then with gradual increases in doses of 5–10 mg/week, up to full treatment doses of 60–120 mg/day, no narcotic-like effects of any type were perceived by the former heroin addicted patient or observed by clinical observers (DOLE et al. 1966a,b; KREEK 1991a). It was also found that the steady, moderate to high doses of methadone (60–120 mg/day), once reached after slow ascendency of dose and stabilization on that dose, which are essential for optimal treatment of most patients, did not need to be increased further over the study period, in sharp contrast to what pertains with experimental or clinical treatment of opiate addiction with a short-acting opiate such as morphine or heroin, where regular, even daily, dose escalation is required because of rapidly developing tolerance. Further studies and clinical observations conducted over many years have shown that a steady moderate to high dose of methadone may be used effectively for the maintenance treatment of opiate addiction for many years without any need to increase the dose, and that the effectiveness is pharmacological, and not a placebo-type, response since a blind dose reduction leads to signs and symptoms of opiate withdrawal. It has also been observed that when the subjects undergo dose reduction, often for the ultimate goal of medication elimination, the rate of reduction in dose may be quite rapid (10 or even 20 mg/week) until a critical and apparently individual dose threshold is reached, usually between 15 and 35 mg in most patients. A further dose reduction below that level must proceed extremely slowly to prevent the signs and symptoms of opiate withdrawal from appearing, and to prevent drug craving from increasing to uncontrollable limits, which may lead to relapse.

In the early research on methadone maintenance as a potential long-term pharmacotherapy in 1964, following the initial observations of the efficacy of methadone in preventing drug hunger and drug craving, as well as the success in rendering the individual essentially normal from both a behavioral and functional standpoint, further studies were conducted to determine both the potential effectiveness and safety of methadone treatment. Also, a critical question was asked as to what would be the effects and potential adverse reactions if patients should attempt to achieve an euphoria by superimposition of a short-acting narcotic such as heroin against the background of methadone maintenance treatment, an event which was predicted to be likely since methadone itself, in steady oral doses to which the patient has become tolerant, provides no euphoria or other so-called "desired narcotic-like effects" when appropriately used in treatment. Thus, a sequence of studies were conducted to determine the level of tolerance, cross tolerance, and possible "blockade" by methadone of the effects of a superimposed short-acting narcotic (Dole et al. 1966a,b). Two series of double-blinded studies, each 4 weeks in length, using a double-blinded, Latin-squared design were conducted at the Rockefeller Institute Hospital in 1964 in former heroin addicts stabilized for at least 8 weeks in steady dose methadone maintenance treatment (Dole et al. 1966a,b). In these studies of opioid tolerance and cross-tolerance, or "blockade," all subjects were former heroin addicts who had been slowly induced into stable-dose methadone maintenance treatment; each patient was receiving a daily dose of 80–100 mg once a day in a single dose. Four opioid agonists, including heroin, hydromorphone, morphine, and methadone, as well as saline were studied. Each of these four medications was administered intravenously, 1 per day on consecutive weekdays for 4 weeks in random order derived from a Latin square design, a double-blinded manner. The dose of two of the added drugs remained constant, morphine at 30 mg, and methadone at 40 mg, with the doses of heroin ranging from 20 mg up to 80 mg. Both subjective and objective measurements were made by the clinical investigator as well as the patient and nurse observers before, during, and following the double-blinded injection of the opioid agonist or saline. It was determined that when the code was broken, none of the superimposed narcotics (or saline) caused any euphoria or narcotic-like effects; following the injection of morphine, a consistent report of "pins and needles" was made by the study subjects, presumably due to the well documented effect of morphine on histamine release. This was reported as an isolated finding in subjects who had become refractory or cross tolerant to all of the other

Table 1. Heroin versus methadone[a] and l-α-acetyl methadol

	Heroin	Methadone	l-α-acetyl methadol (LAAM)
Route of administration	Intravenous	Oral	Oral
Onset of action	Immediate	30 min	30 min
Duration of action	3–6 h	24–36 h	48–72 h
Euphoria	First 1–2 h	None (if appropriate dose used)	None (if appropriate dose used)
With drawal symptoms	After 3–4 h	After 24 h	After 48 or 72 h

Methadone and LAAM: action is at specific μ type opiate receptor sites; mechanism of action is by providing a steady state perfusion of pure opiate agonist at μ opiate receptor sites over dosing interval.

[a] Effects of high dosages of drug or therapeutic agent in tolerant individuals.
[b] Includes action of major metabolite, morphine.
Adapted and modified from KREEK (1992c).

effects of morphine and heroin. The former heroin addict study subjects maintained on methadone inquired about the absence of a "rush" or "high," that is, the euphoric effects which usually follow this "pins and needles" cue since they experienced no subjective narcotic-like effects. Of importance, no respiratory depression occurred when a short-acting narcotic was superimposed in the long-acting narcotic. At the end of these double-blinded studies, a series of single-blinded studies also carried out primarily to determine the safety of methadone treatment in the setting of potential use of illicit heroin in escalating amounts in any attempt to achieve euphoria. It was shown that only extremely high doses of heroin, far beyond those ordinarily obtained and used by street addicts, could exceed the degree of tolerance and cross tolerance developed by steady dose treatment with 80–100 mg/day of methadone; such doses of heroin produced some narcotic-like or euphorigenic effects in methadone stabilized patients. However, no respiratory depression occurred in this setting, showing the high index of safety of methadone in doses to be used in maintenance treatment in the setting of combined illicit use of heroin. The "blockade studies," which were carried out at the beginning of methadone maintenance treatment research, included the initial subjects entering into experimental methadone maintenance treatment (DOLE et al. 1966a,b).

After the initial studies of the potential use of methadone in the maintenance treatment of addiction were conducted and the cross-tolerance or so-called "blockade" phenomena studies were completed, long-term methadone treatment research progressed into a more "real world" setting, initially utilizing former short-term detoxification centers, and soon after utilizing newly developed specialized clinics affiliated with medical centers involved in the original research (DOLE and NYSWANDER 1965, 1966, 1967, 1976). Computerized tracking of all patients entering treatment from time of the initial studies in 1964 was conducted in New York, providing an enormous data base for analysis of treatment outcome. An independent unit at Columbia University School of Public Health, headed by Dr. Frances Gearing, was identified for assessment of outcome data (DOLE et al. 1968; GEARING 1970; GEARING and SCHWEITZER 1974). Although the initial research on use of the long-acting opioid methadone for pharmacotherapy of opiate addiction, that is, the effects of methadone when given daily to former heroin addicts along with the studies which elucidated the phenomena of cross-tolerance providing "blockade" against illicit heroin and other short-acting opioid studies performed in 1964 were not reported until 1966, the subsequent applied clinical research studies on the actual use of methadone in the maintenance treatment of addiction conducted at facilities away from the Rockefeller Institute Hospital were reported by 1965 (DOLE et al. 1966a,b; DOLE and NYSWANDER 1965). In all the initial treatment research, all patients were stabilized in an inpatient unit for at least 2 months prior to proceeding to outpatient care. The methadone maintenance treatment modality, by early assessment, was found to be extremely effective, further supporting the original observations at the Rockefeller Institute in which methadone seemed to satisfy the goals of any pharmacotherapy for opiate addiction, that is, to prevent withdrawal symptoms, to prevent "drug hunger" or craving, and to block any euphoric effects of superimposed short-acting narcotics. The prospective studies which were initiated in 1964 and were completed and reported in 1972–1973 also documented normalization of physiological functions disrupted by chronic heroin addiction (KREEK 1972, 1973a,b). By 1967, the first methadone maintenance research treatment initiated in an outpatient setting was conducted. In these studies, conducted initially at the then renamed Rockefeller University Hospital Clinic, patients were induced gradually into methadone maintenance treatment in an outpatient clinic, using the same protocol as used earlier for inpatients. It was found that this approach was equally effective for most patients, as long as adequate patient-staff interaction was provided, including three to six hours of observations in clinic by the medical staff each day during the first one to two weeks of induction into treatment, to determine the responses of slowly increasing doses of methadone and to alter the rate of methadone dose increases as indicated.

By July 1, 1966, within 2.5 years after the first research on methadone maintenance treatment, a total of 214 chronic heroin users had been admitted to treatment at the Rockefeller University Hospital and related clinics (KREEK 1973a,b). After 3–5 years of continuous treatment, the overall retention rate of this initial treatment research group was 60%. These first consecutive 214 patients admitted to research treatment were all volunteers, who were all at least 19 years old and had met the early admission criteria imposed by our own research group at the Rockefeller University, of 3 years or more of regular heroin usage, with development of tolerance, dependence, and addiction, failures at other forms of treatment (detoxification or drug-free environments, or both), and usually a criminal record for narcotic-related offenses (KREEK 1972, 1973a,b). A larger cohort of patients, including all admitted from the beginning of treatment research in 1964 up through December 31, 1968, were found to have a 67% retention in treatment for 3 years or more when evaluated at the end of 1971 (GEARING and SCHWEITZER 1974). In related studies, it was found that the 2-year retention rate in treatment in the 1960s and early 1970s was greater than 80%. This high rate of retention was coupled with an abrupt drop in arrest rates and increasing restoration to normal social functioning, including employment, pursuit of education, and/or homemaking. By April, 1972, more than 16 000 patients were enrolled in methadone maintenance treatment programs (MMTP) in the city of New York, and many of these patients had been maintained with methadone for longer than 8 years (DOLE et al. 1968; CHAMBERS et al. 1970; GEARING 1970; BABST et al. 1971; DOLE 1973; KREEK 1973a,b; GEARING and SCHWEITZER 1974; DOLE and NYSWANDER 1976). By 1971, several other experimental methadone maintenance treatment programs had been initiated with a variety of treatment research studies connected in conjunction with those programs, including those of Jaffe in Chicago, Maslansky in Minneapolis, and Goldstein in Palo Alto, (GOLDSTEIN 1971, 1972; JAFFE 1970; JAFFE and SENAY 1971; MASLANSKY 1970). Follow-up studies of this initial research group and subsequent groups have shown that patients receiving 80–100 mg/day of methadone (and in subsequent studies by BALL et al. 60–120 mg/day) have less than a 5%–15% rate of any heroin use after the first 6 months of stabilization in treatment and that 1-year retention rates in effective programs providing counseling and other services as needed as well as using adequate doses of methadone in maintenance treatment rates of 60% to over 80% (BALL and Ross 1991; BALL et al. 1987; DOLE 1988; DOLE and NYSWANDER 1965; DOLE et al. 1966a,b, 1968; KREEK 1973a,b).

a) Medical Safety and Physiological Effects of
Chronic Methadone Treatment

These early studies demonstrated that methadone, when used in maintenance treatment, would be safe with respect to the potential for respiratory overdose if a short-acting narcotic should be superimposed. They also showed that the high level of tolerance to methadone as used in maintenance treatment doses also provided a cross tolerance to any euphoric or other narcotic-like effects from superimposed short-acting narcotics. This suggested that, in addition to the first identified actions of methadone, beneficial for management of opiate addiction, that is, both the prevention of narcotic withdrawal symptoms and the reduction or elimination of drug hunger or drug craving, methadone might also be effective in a mode of classic conditioned behavior modification. When no positive, desired, or reinforcing effect, that is, no euphoric effects were achieved after superimposition of a short-acting opiate such as heroin, the classical phenomena of extinction or deconditioning would occur and subjects ultimately would no longer self-administer an agent that was giving them no reward. Subsequent studies and clinical experience over thirty years have suggested that all three of these mechanisms of action may be occurring. Clearly, methadone does prevent opiate abstinence over a 24-h dosing interval. Also, when adequate doses of methadone are used, drug hunger or drug craving is significantly reduced or eliminated. Although over 50% of all heroin addicts treated with methadone maintenance treatment may try to achieve a euphoric effect from superimposed illicit short-acting narcotics such as heroin, most will soon stop such attempts, as they achieve no desired euphoric results. A limited number (up to 20%) of patients may drop out of treatment voluntarily, usually during the first 6 months, to return to a state in which they can achieve the euphorogenic effects from use of heroin. However, when inadequate doses of methadone are used, as unfortunately has pertained in numerous clinics throughout the United States and the rest of the world, then it is possible, as shown in the original blockade studies, to "override" the cross-tolerance effects of methadone. Numerous studies in each decade up to the present time have shown that when the doses of methadone used are inadequate (that is, lesss than 40–60 mg for patients in the early months of treatment) the percentage of patients continuing to use illicit opiates increases in an inverse dose-dependent manner (BALL and ROSS 1991; BALL et al. 1987).

The side effects observed during the first 6 months of methadone treatment are those which could be anticipated from earlier studies of the effects of morphine or any pure opioid agonist in humans, including initially drowsiness, nausea, vomiting, and more persistently, sleep disorders, primarily insomnia, difficulty in urination, and when the dose of methadone is increased too rapidly, even edema of the lower extremities probably due to inappropriate secretion of antidiuretic hormone; also significant constipation and increased sweating occur (DOBBS 1971; KREEK 1972, 1973a,b, 1978a, 1983b; YAFFE et al. 1973). Neuroendocrine abnormalities of the hypothalamic-pituitary-adrenal axis have been documented during cycles of subacute and chronic opiate addiction, including menstrual irregularities in heroin addicted females and sexual dysfunction in both men and women. However, when studies have been conducted using a repeated measured design, in which patients are followed prospectively from time of entry into methadone maintenance treatment onward, it has been found that within 3 years, full tolerance develops to these effects in over 80% of all patients in treatment (KREEK 1973a,b). Up to 20% of the general population may experience each one of the symptoms related to libido and sexual perofrmance while receiving no medications of any type (KREEK 1983b). Sleep disorders, sexual dysfunction, and related neuroendocrine abnormalities resolve in over 80% of patients within 1 year of treatment. Excessive perspiration is a side effect to which tolerance develops in only around half of all patients after three years of treatment. In patients enrolled in long-term methadone maintenance treatment, minimal side effects and no toxic effects have been found (KREEK 1972, 1973a,b, 1978a, 1983b). Performance effectiveness, attentional and cognitive function, and memory were found to be normal in stabilized methadone maintained patients (APPEL and GORDON 1974, 1976; GORDON 1973, 1976; GORDON and APPEL 1972; GORDON et al. 1967; HENDERSON et al. 1972; BABST et al. 1973). Pain perception was also found to be normal in stabilized methadone maintained patients (HO and DOLE 1979). Also, neurologic assessments have shown no problems attributable to methadone maintenance per se (LENN et al. 1975).

Abnormalities in liver function tests, which are found in about half of patients at time of admission to treatment, have been shown to persist without significant change during treatment, unless alcohol abuse is a causative factor and is reduced during treatment. Most of the liver function abnormalities observed are due to the presence of chronic viral liver disease, primarily hepatitis B and hepatitis C, and/or alcoholic liver disease and sometimes more recently identified as complicated by hepatitis D (CHERUBIN 1967; CHERUBIN et al. 1970, 1972a,b, 1976; KREEK 1972, 1973a,b, 1978a, 1983b; KREEK et al. 1972, 1986, 1990a,b; LOURIA et al. 1967; NOVICK et al. 1981b, 1985a,b,c, 1986a,b, 1988c, 1995; SAPIRA et al. 1968; STIMMEL et al. 1972a,b, 1973, 1975, 1982; WHITE 1973). Also, serum protein and immunoglobulin abnormalities are present in over half of patients at time of entry and also persist, although specific indices of immune function improve with time and with treatment. Opioid induced gastrointestinal dysfunction, primarily slowing of gastrointestinal motility leading to chronic constipation, may also persist for several years, since tolerance develops very slowly to the effects of methadone and all other opioid drugs in the gastrointestinal tract (KREEK 1972, 1973a,b, 1983b). The one death which has been attributed directly to methadone maintenance treatment was that of fecal impaction following chronic methadone ingestion in a patient who refused to accept any of the recommended medications and nutritional modifications to alleviate the chronic constipation which is common during the early months and up to the first 3 years of methadone maintenance treatment (RUBENSTEIN and WOLFF 1976). Therefore, a new approach for managing this problem by oral administration of naloxone, a specific opiate antagonist which has extremely low systemic bioavailability after oral administration (<2%) has been proposed and studied both in animal models and humans with some very interesting, positive results in this early research (KREEK et al. 1983b,c; CULPEPPER-MORGAN et al. 1992, 1995). Follow-up of patients in moderate to high dose treatment for more than 10 years have been conducted (NOVICK et al. 1993). These studies have shown that there are no adverse effects and no medical complications related to long-term methadone maintenance treatment. When compared with a group of active heroin addicts with a similar number of years of addiction, the only abnormalities which were seen with increased frequency in the methadone maintained patients were commonly observed medical problems of relatively sedentary middle-income Americans, including moderate obesity, adult onset diabetes, and the presence of symptomatic hiatus hernia (a common gastrointestinal abnormality, but which is appropriately diagnosed only in those with adequate access to health care, which rarely includes untreated heroin addicts).

Subsequent work in animal models and clinical research in humans have documented that activation of opioid receptors at diverse sites in the body are related to each one of these so-called "side effects" of all opioid agonists, including the long-acting opiate methadone as used in ascending doses during the early weeks of methadone maintenance treatment. Thus, these are actual direct effects of opiates acting at their specific opiate receptors. In the prospective studies of patients entering methadone maintenance treatment, each of these actions and sites of actions were delineated, which later provided important clues concerning the localization of opioid receptors and the sites and roles of action of the specific components of the endogenous opioid system, that is, the opioid peptides which naturally bind to these opiate receptors. In the prospective studies of patients entering methadone maintenance treatment, it was also observed that tolerance develops at very different rates to each of these diverse effects. Tolerance develops very rapidly to the sedative or somnolence-producing and nausea and vomiting-evoking effects, and also analgesic effects of methadone as used in maintenance treatment of addiction. For this reason, it was found early on in treatment research that methadone-maintained patients will experience pain in any setting where a non-opiate dependent person will experience pain, including childbirth, trauma, surgery, and painful diseases (Ho and Dole 1979). It was also shown that such pain in methadone maintained patients could be easily managed by use of the usual pharmacotherapeutic agents, often using short-acting opioid agonists such as morphine or hydromorphone. When opioid agonists are needed, it has been found that the upper range of normal doses are needed, but with shortened dosing intervals required. It was also found, of course, that use of a mixed agonist-antagonist, such as pentazocine (Talwin) is contraindicated, since the antagonist components of such medications will precipitate opiate withdrawal in a methadone-maintained patient. Accidental overdose due to ingestion of a treatment dose of methadone by an opiate-naive or modestly tolerant adult or by a child can easily be managed with continued use of an opioid antagonist such as naloxone for the duration of symptoms, as long as 24 h or more as needed (Hartman and Kreek 1983).

b) Treatment Effectiveness and Outcome

In the early work in this area, the treatment research was performed with the hypothesis that opiate addiction is primarily a "metabolic" disease, that is, a disease in which derangements of normal physiology exist which contribute to, or cause, the drug-seeking behavior and addiction (Dole and Nyswander 1965, 1966, 1967; Dole et al. 1966a,b; Kreek 1972, 1973a,b, 1992a,b,c). It was not known then (or now) whether such a "metabolic basis" might be due in part and in some persons to a genetically determined predisposition, combined with environmental and psychosocial factors leading to exposure and addiction (although it was appreciated from the early research period that environmental factors may play a very important role in all persons with addiction), or alternatively, whether drug use itself causes long-term, persistent, or possibly permanent changes in normal physiology, which then in turn may contribute to the perpetuation of or relapse to addiction (Kreek 1972, 1973a,b, 1992a,b,c). Therefore, for the initial studies, a definition of "hardcore" or "long-term" addiction was defined for research and operational purposes (Dole et al. 1966a,b). This definition has been altered only modestly over the ensuing years, in accord with basic and applied clinical research findings. The initial definition of "hardcore" addiction, which is an essential prerequisite for entry into methadone maintenance research initially included three years or more of daily illicit opiate use, with multiple self administrations per day of short-acting narcotic, primarily heroin, with the development of tolerance, dependence, and drug-seeking and using behavior. Also, for the initial studies, a requirement was added that all study subjects must have attempted drug-free treatment at least one time. Most subjects had attempted drug-free treatment multiple times and also had been incarcerated multiple times and thus kept in a drug-free state (fully or relatively) for several periods. The mean numbers of years of addiction of the first prospecitve research study subject group was 13 years (KREEK 1973a,b). Thus, the initial study subjects were truly very long-term or "hardcore" heroin addicts with tolerance, dependence, drug-seeking behavior, and many failures both at drug-free treatment and relapses following release from prison. Over subsequent years, the official definition, which is enunciated in the United States federal guidelines that govern use of methadone (and now also *l-α* acetylmethadol) for the maintenance treatment of opiate addiction, defines long-term heroin addiction as one year or more of regular use of an opiate (usually heroin), with multiple self-administrations each day and with the development of tolerance, dependence, and drug-seeking behavior. Thus, the time of addiction required to define "hardcore" or long-term addiction has been reduced from 3 years down to 1 year (and from a biological standpoint, possibly should be even much shorter) (KREEK 1994a). This operational definition, which guides treatment, has also allowed for appropriate comparative studies of treatment modalities as well as for the neurobiological correlates of long-term opiate addiction.

In the initial research, the patient-staff interactions and informal as well as formal counseling were conducted primarily by the physicians with help of the research nurses and the staff of the occupational therapy unit at the Rockefeller University Hospital (DOLE and NYSWANDER 1965, 1966, 1967, 1976; DOLE et al. 1966a,b). Subsequent studies conducted by many groups have shown the importance of the clinic staff, although diverse staff patterns may be able to achieve equally fine results, providing that an adequate and stabilized dose of methadone is used, and the attitudes and information base of the staff members are appropriate and that positive contingencies, such as allowing increasing "take-home" privileges (to the extent allowed by the Federal guidelines) are given to those patients with documented evidence of no ongoing illicit drug use, including both absence of illicit opiates (the targeted primary goal of methadone treatment) or of other illicit drugs such as cocaine, while not using negative contingencies such as discharge from clinic (JUDSON et al. 1980; KHURI et al. 1984; MASON et al. 1992; MCLELLAN et al. 1983, 1993; OBRIEN et al. 1984; WOODY et al. 1990). Over the years, both informal observations and well-conducted, rigorous studies have shown that staff insight and knowledge about the addictive disease which they are treating, and its multiple social ramifications as well as physical and mental complications are critical. Thus, staff members, who are either trained or knowledgeable by their long involvement in the area and who are humane and caring (a difficult characteristic to measure, but possibly reflected in the data that have been reported repeatedly), and staff of long duration of service usually provide services which result in a better outcome with respect to both patient retention in treatment and patient rehabilitation (BALL and Ross 1991; DOLE and NYSWANDER 1976; KHURI et al. 1984; KREEK 1991a; MCLELLAN et al. 1993). As discussed before, numerous studies showing that less than 15%–20% of unselected heroin addicts with or without codependency or comorbidity, will continue to use illicit opiates once stabilized in methadone treatment or 6 months or more. Adequate doses of methadone during the first year or more of treatment, doses usually of 60–120 mg/day, displayed a dose-dependent response to treatment (DOLE and NYSWANDER 1965; DOLE et al. 1966a,b; DOLE 1988; KREEK 1991a; BALL and ROSS 1991). Recent studies have shown that there is a dose-dependent increment in favorable patient outcome with the infusion of rehabilitation medical and psychiatric services (MCLELLAN et al. 1993). Also, studies of medical maintenance (discussed below) have documented that patients who have responded to methadone maintenance treatment with cessation of illicit opiate use, along with cessation of other illicit drug use, and have succeeded in rehabilitation efforts, may no longer need regular counseling services, but clearly benefit, as do all persons, by access to ongoing medical care (NOVICK and JOSEPH 1991; NOVICK et al. 1988a,b, 1990, 1994). These findings have suggested that in the future, to assure medically effective as well as socially and cost effective treatment, there will be a need to define both the severity and various types of codependency, comorbidity, and multiple medical problems at time of entry, and to provide on site or ready access to treatment or management for these problems, as well as a need for diverse social services. Reassessment of needs as time in treatment progresses is also essential. Thus, staging of disease at entry will become increasingly important. In addition, reevaluation for response to treatment will be important, with tailoring of ancillary services, including increases when needed, or decreases when no longer needed, as treatment progresses.

c) Impact of Discontinuation of Methadone Maintenance Treatment

From 1973 onward, reports have been made of several studies, conducted to follow up patients who were successfully treated with methadone, with stabilization on pharmacotherapy and with rehabilitation achieved, who then underwent dose reduction, elimination, or detoxification at their own request (CUSHMAN and DOLE 1973; DOLE 1973, 1988; CUSHMAN 1981; DESJARLAIS et al. 1981, 1983; RIORDAN et al. 1976; SENAY et al. 1977a; STIMMEL et al. 1974; STIMMEL and RABIN 1977). The earliest of these studies were conducted specifically to determine the functional status of drug-free former methadone-maintained patients, since in the early research it was not known how many patients would do well in an abstinent state following successful methadone maintenance treatment and rehabilitation. Some of those studies were completed prior to 1973, at which time the Federal regulations governing methadone maintenance treatment were first promulgated and carried out under the authority of the Food and Drug Administration and Drug Enforcement Agency. These earliest regulations demanded that patients leave treatment after 2 years, unless special justification for longer treatment were made in each case. Studies of much larger cohorts of patients who had undergone successful methadone treatment prior to mandated cessation of treatment were then conducted, although these discharges are called "voluntary," in fact they were only voluntary in the sense that the patients were not discharged "involuntarily" because of infraction of rules, unacceptable behaviors, or death. However, the findings from each of these studies are very similar to the earlier studies. Most studies have shown that methadone dose can be reduced relatively rapidly in maintenance patients from usual treatment doses. 60–120 mg/day, down to 30–40 mg/day with very few signs and symptoms of narcotic withdrawal. However, reduction of the dose from that level on down to 0 mg/day dose is accompanied with the onset of classical signs and symptoms of opiate withdrawal of varying degrees of severity with intensity and duration, depending on the rate of withdrawal. It has been shown that the slower the dose is tapered, the milder the symptoms. At very low doses, under 15–20 mg/day, studies have shown that patients may discern a dose reduction (or dose increment if given to prevent or decrease symptoms) of as small as 2 mg/day. However, all patients complete the dose elimination with relative ease and in all cases, abstinent state can be achieved. However, various follow-up studies have shown that only between 10% and 35% of former methadone maintained patients may stay opiate-free for 1 year or more (CUSHMAN and DOLE 1973; DOLE and JOSEPH 1977; STIMMEL et al. 1977, 1978; CUSHMAN 1981; DESJARLAIS et al. 1981, 1983). Also, relapse may occur at any time after achieving the medication-free state, although many of the relapses occur within the first several months. Relapse to alcohol abuse and prescription drug use, especially use of benzodiazopenes dilazopenes, often precede relapse to illicit opiate use. Some studies have shown that an increased duration of treatment with methadone is associated with a greater likelihood of remaining abstinent following cessation of methadone treatment for one year or more (DESJARLAIS et al. 1981, 1983). In the most rigorously conducted follow-up study, in which essentially all subjects were accounted for, only 46% remained free of opiate use for 6 months following discontinuation of methadone treatment, and only 19% remained free of illicit opiate use for 1 year or more (CUSHMAN and DOLE 1973; DOLE and JOSEPH 1977). With many studies replicating these findings over a 25-year period, consensus of most programs is that the decision regarding dose reduction and elimination should be made by physicians, counselors, and patients after interactive discussions and assessment of the individual case, including assessment of the degree of rehabilitation. Communication or full understanding of relapse potential, and the multiple interest risks of such relapse, including possible exposure to blood borne diseases such as HIV-1, hepatitis B, C, and delta infections are disclosed and at the same time, the patient's desires with respect to stopping treatment are also discussed. The reasons for desiring cessation of treatment include the fear of breach of confidentiality; the negative attitudes about methadone maintenance treatment by spouses, family, peers, and the general public; and the inconvenience of the minimum required attendance in treatment which now, according to FDA regulations, is minimally one clinic visit per week in standard programs (however, in many regions and clinics, demands even for rehabilitated patients may be for more frequent visits of two, three, or even six times a week).

d) Medical Maintenance Treatment (Office Based Treatment)

Very long-term methadone maintenance treatment may be desirable and accepted by many or even most heroin addicts entering treatment. It has been recognized that, as patients progress in treatment, their needs for support services, including intense counseling and rehabilitation efforts, may become significantly reduced or eliminated. It is also recognized that some patients may respond fully to methadone treatment with respect to achieving the primary goal of cessation of illicit opiate use, but may continue to require counseling and rehabilitation services. Yet other patients may respond to methadone treatment with respect to cessation of illicit opiate use, but continue to have ongoing problems with other types of chemical dependency (at this time, especially cocaine dependency and alcoholism), medical problems (especially HIV infection, AIDS, or sequellae of viral or alcoholic liver disease), or psychiatric problems (especially depression), and need a special or enhanced treatment for periods of time. Ideally, many different types of treatment programs would evolve to address each of these problem (KREEK 1991a). To date, the availability of physical resources and trained staff have not permitted this to happen, with exception of a few research studies of patients who have achieved both primary and secondary goals of treatment. Studies have been conducted in at least three cities and are possibly ongoing in other cities to determine the feasibility and effectiveness of medical maintenance treatment for long-term methadone-maintained patients who have had a good record of treatment performance to be treated in a general clinic or "office based" practice. In the New York City studies, under the leadership of Novick, patients who have had at least 5 years of methadone maintenance treatment in a standard clinic, with excellent performance, that is, with cessation of all illicit opiate use and no ongoing polydrug or alcohol abuse, all positive indicators for rehabilitation, met the inclusion criteria (NOVICK et al. 1988a,b, 1990, 1994; NOVICK and JOSEPH 1991). In this medical maintenance study, patients are seen as infrequently as once every 4 weeks or not more often than once a week, in a medical office (not a drug abuse treatment program) at which time they provide a urine speciment for drug monitoring, take one daily dose of methadone under observation, and undergo a conventional medical, and also intense drug use history, physical examination, and when appropriate, other laboratory testing. Medication is obtained at a local hospital pharmacy every 1–4 weeks. Patients not ready to start on-going studies of performance are returned to more intensive conventional treatment. Over 95% of long-term patients entering this type of medical maintenance program have been able to successfully remain in that treatment regimen which is conducted in standard, institution based, general medical clinics which are "office based" practices.

Senay in Chicago and the late Richard Lane in Baltimore have conducted similar studies. In Chicago, patients with at least 6 months of good performance have been entered into medical maintenance treatment. Patients who had performed well in conventional methadone maintenance treatment, but for a shorter period of time were included in this study, and thus greater constraints were made during the study. The patients ingest methadone under observation at least twice a month and are seen by a clinic physician at least once a month. The patients are also intermittently called in for random urine checks as well as random medication bottle counts. This program has met with a high degree of success, with over 70% of patients remaining in the medical maintenance program in good standing for at least 1 year. Further development of these types of programs in the future may allow increased utilization of limited resources for patients in need of treatment.

e) Methadone Maintenance Treatment: Current Status

Repeated studies from 1964 through 1994 have shown that in well-constituted appropriate treatment programs, with caring and knowledgeable staff, on-site or ready access to appropriate medical and psychiatric services, a voluntary one-year retention rate of 70%–80% and a 2–5 year retention of over 50%–60% can be achieved and, when adequate methadone doses are also used on a stable dosing pattern, less than 15%–20% of patients will continue to use any illicit opiates after 6 months of stalilization (DOLE et al. 1966a,b; KREEK 1991a, 1992a,b,c; BORG et al. 1995a). Also, multiple studies have shown that arrest rates and incarceration rates related to criminal activity are significantly reduced. Socialization, as evidenced by going to school, running a home, or working, are significantly increased. Other indirect indicators of efficacy in reducing or stopping illicit drug use included a significant reduction in new cases of hepatitis B, as observed in the setting of rapid expansion of effective methadone maintenance programs in New York City in the early 1970s. Although methadone maintenance treatment was widely accepted for a few years in the early 1970s, primarily with a goal of decreasing criminality, subsequent changes in economic times and policy focus forced support for appropriate treatment subsequently diminished. Federal regulations regarding methadone maintenance treatment, which were first promulgated in 1973, initially contained some "guidelines" which were not based on scientific information or clinical related infromation and which were soon found to be counterproductive and hazardous for many patients. One critical guideline was the requirement to discontinue methadone maintenance treatment after 2 years unless specific individual justifications were given. This led many clinics to simply discharge patients at the end of 2 years rather than justify continued treatment. Further research (as detailed above) showed that well-rehabilitated methadone maintained patients thus discharged from treatment would relapse to drug abuse, initially to alcohol and other drug abuse, with over 70% ultimately relapsing to opiate use within 2 years. Thus by 1983, following a major series of consensus meetings sponsored by the National Institute on Drug Abuse, the federal FDA guidelines governing methadone maintenance treatment were changed to recommend treatment for as long as needed (COOPER et al. 1983). The current guidelines may be revised again soon (RETTIG and YARMOLINSKY 1994).

At this time, methadone maintenance programs have been developed in 40 states in United States and in multiple countries of Europe and throughout the world. The outcome results are essentially the same in all appropriately staffed programs using adequate doses of methadone. So-called "poor programs" with inadequate staff and services, but which administer adequate doses of methadone (and also in a recent research model thereof) have relatively modest outcomes, with reduced retention of 45%–55% and high percentages continuing to use other illicit drugs, yet with less than 20% of patients continuing illicit opiate use (McLELLAN et al. 1993). However, clearly even more "poor outcomes" and lower retention rates can be seen if inadequate doses of methadone, with significant increases in illicit opiate use, are used. The "good outcomes" have been reproduced in many states and countries, whenever appropriately staffed clinics, with knowledgeable and caring staff, in which combined utilization of adequate doses of methadone with appropriate counseling along with on-site or ready access to medical and psychiatric services pertains, with repeated documentation of 70% or greater retention, and 1 year cessation of all illicit opiate use in over 80% of patients in treatment (BALL and ROSS 1991; McLELLAN et al. 1993; BORG et al. 1995a; ADELSON et al. 1996, in press). These findings pertain in either the absence or presence of psychiatric comorbidity. Although up to 30% of all patients entering treatment may have a concomitant diagnosis of depression or anxiety and a smaller percentage with other psychiatric diagnoses, the outcome in appropriately constituted clinics is approximately the same for all subjects (MASON et al. 1992; O'BRIEN et al. 1984). From 1964 onward, codependency with alcohol may pertain in 25% to 50% of all heroin addicts entering treatment, and since 1978 codependency with cocaine may range from 20% up to as high as 90% in the New York City area, again success with respect to cessation of illicit opiate use may be achieved (BROWN et al. 1973; JACKSON and RICHMAN 1973; KREEK 1973a,b; PASCARELLI and EATON 1973; PUGLIESE et al. 1975; GORDIS and KREEK 1977; MADDUX and ELLIOT 1975; CUSHMAN et al. 1978; BEVERLY et al. 1980; JACKSON et al. 1982; HARTMAN et al. 1983; McLELLAN et al. 1983; HUNT et al. 1984; KHURI et al. 1984; CUSHMAN 1987; GORDIS 1988; JOHANSON and FISCHMAN 1989; NOVICK et al. 1989a; ROUNSAVILLE et al. 1982). A significant reduction in codependency with cocaine has been observed in clinics in which studies have recently been conducted when the use of cocaine has become very prevalent (Hanbury et al. 1986; Chaisson et al. 1989; Borg et al. 1995a). However, 20%–40% of patients maintained on methadone in otherwise effective programs using both behavioral, psychiatric, and other psychosocial approaches in attempts at the treatment of cocaine addiction may continue to use cocaine and up to 25% may continue to use alcohol to levels of abuse or definition of alcoholism. Ideally new pharmacotherapies can be developed to use for treatment of these refractory patients, which can be given along with methadone maintenance treatment.

An effective methadone maintenance program thus leads to a very high level of retention, as well as a very high level of cessation of illicit opiate use, the primary goals of treatment, along with a highly significant reduction of criminality, reduction in use of other addictive drugs, and an increase in social productivity, important secondary goals. In addition, significant reduction in illicit use of other illicit drugs, as well as excessive use of alcohol, occurs, thus achieving other very important secondary goals. However, methadone does not have a primary action which would be expected to specifically treat alcoholism. Methadone is not targeted at the primary initial sites of action of cocaine, that is, the inhibition of presynaptic reutake of dopamine, serotonin and norepinephrine by binding to specific transporter sites. The enhancement by cocaine of synaptic concentrations of dopamine is now believed to be the primary mechanism of the reinforcing effects of cocaine. Therefore, it is understandable that methadone does not specifically manage cocaine dependency. However, some recent data suggest that the endogenous opiate system also may be perturbed in cocaine dependency (Branch et al. 1992; Unterwald et al. 1992, 1994a,b; Spangler et al. 1993a,b). Rectification of this derangement by steady state methadone treatment may contribute, in part, to some improvement in presence of cocaine dependency in heroin addicts entering methadone maintenance treatment, or treatment with another opioid agent, l-α-acetylmethadol, or with the partial agonist bupreorphine, all three of which have been shown to effect a similar degree of reduction of cocaine abuse.

Currently, there are approximately 720 methadone maintenance clinics, including both public, not-for-profit, and private clinics throughout the United States currently treating over 120 000 patients. In addition, there are several thousand patients in methadone maintenance treatment in other parts of the world. The numbers of patients in methadone maintenance treatment have increased markedly worldwide since 1986, with a recognition of the highly significant positive public health effects of effective methadone maintenance programs in reducing HIV+ infection and AIDS. However, it has been obvious that negative attitudes toward methadone treatment have led to inadequate support for the development of appropriate treatment programs. These attitudes also have led to difficulty in identifying sites to locate new treatment programs ("not in my backyard"). Reduction in numbers of staff and reduction in the training level of staffs, due to reduced amount of funds allocated for staff development and salaries, and often an overemphasis on the regulatory aspects of treatment with an underemphasis on the rehabilitation goals, has ensued over the past 20 years. Also, many programs have continued to prescribe inadequate doses of methadone because of either a fear, ignorance, or reticence to give adequate treatment doses, despite repeated and rigorous scientific and clinical data documenting the need for moderate to high doses. Most programs have inadequate to no access to medical and psychiatric care, and also have limited on-site counseling and access to rehabilitation services as needed. Even with these multiple problems, the few studies performed have shown that 45%–55% of former heroin addicts in programs with inadequate resources which provide minimal services respond to chronic methadone pharmacotherapy alone, a percentage of response far greater than that to any drug-free treatment, with less than 30% responding to the very best drug-free treatment according the few rigorous studies which have been conducted using as a denominator unselected heroin addicts who seek and/or initially enter treatment.

III. Neuroendocrine Immune and Gastroenterological Effects of Methadone (As Contrasted with Heroin Addiction)

1. Neuroendocrine Function

As part of the prospective studies initiated in 1964 by the Rockefeller University research group and carried out prospectively for the next several years, the effects of heroin on physiological functions, as evidenced by status at time of entry into methadone treatment, was evaluated, as well as prospectively reassessed during induction and stabilization in treatment. As part of the prospective studies, the effects of ascending and then stabilized, moderate to high doses (60–120 mg/day) of the long-acting opioid methadone as used in treatment on neuroendocrine function and numerous other aspects of normal physiology, were rigorously assessed. In addition, clinical observations were made by many clinicians and clinical investigators at other institutions involved in early methadone maintenance treatment efforts (BLINICK 1968; CUSHMAN et al. 1970; CUSHMAN 1972, 1973a; DREEK 1972, 1973a,b, 1978a, 1983b; RENAULT et al. 1972; SHENKMAN et al. 1972; MARKS and GOLDRING 1973; SANTEN and BARDIN 1973; WEBSTER et al. 1973; AZIZI et al. 1973, 1974; ESPEJO et al. 1973; PELOSI et al. 1974; CUSHMAN and KREEK 1974a,b; SANTEN 1974; CICERO et al. 1975; HELLMAN et al. 1975; SANTEN et al. 1975). A limited number of earlier studies on the effects of opiate addiction on endocrine function had been conducted at Lexington (EISENMAN et al. 1958, 1961, 1969; MARTIN and JASINSKI 1969; MARTIN et al. 1973). One of the most striking early findings was made at the Beth Israel Medical Center by Blinick and colleagues (BLINICK 1968). By early 1966, the observation was made that in female former heroin addicts with a history of protracted amennorhea or abnormal menses, menses return to normal in those stabilized in methadone treatment for 2 months to 1 year; also, ovulation returned to normal. Several patients became pregnant and in the early clinical practice were maintained on a full treatment dose of 60–100 mg methadone per day. A striking observation was that these babies were of essentially normal birth weight, as contrasted to very low birth weight in infants delivered to mothers actively addicted to heroin. Also, and of great importance, there was minimal evidence of any severe or protracted "withdrawal symptoms" in the infants. Some children required no treatment; others required short-term treatment with a short-acting opioid agonist, such as a tincture of opium or paregoric. Several subsequent studies confirmed these very early findings in studies of well rehabilitated patients. In these early studies, polydrug abuse, including cocaine, benzodiazepine, and barbiturate abuse, as well as alcohol abuse, was rare in the late 1960s. However, problems were subsequently observed in many infants born to methadone-maintained mothers, which were attributed to methadone. Careful review of these case reports indicate that almost all of neonates experiencing severe or protracted withdrawal or atypical symptoms after birth were born to mothers who had been abusing other drugs or alcohol during pregnancy. Subsequent studies have reconfirmed the earlier findings, which observed that patients maintained on moderate or even high doses of methadone give birth to infants of low normal to normal birth weight, with around half but not all of the infants requiring some specific opioid agonist management for withdrawal symptoms in the perinatal period (Finnegan et al. 1982; Kreek 1983a, 1992e). Also, many in vitro and animal studies, as well as human observations, have shown that there are no teratogenic effects of methadone as used in maintenance treatment. Because of the concern about the appearance of withdrawal symptoms, many clinical groups over the years have adopted the policy of lowering the doses of methadone during pregnancy. However, numerous observations have been made of women returning to heroin use or polydrug abuse in the setting of abstinence following dose reduction. Subsequently, rigorous clinical pharmacotherapy studies showed that methadone metabolism accelerated during the second half of pregnancy and especially in the third trimester (Kreek 1979b, 1981–2; Pond et al. 1985; Finnegan et al. 1982; Kreek 1983a, 1992e). Thus, even patients kept on a stable dose will experience lower plasma levels for given dose and lower area under the plasma concentration time curve over a 24-h dosing interval. These findings explain the common complaints of narcotic abstinence symptoms during late pregnancy. Therefore, the general recommendations now are not to lower doses of methadone in the second half of pregnancy and to lower doses, if desired, very gradually during the first trimester. It is also recommended that patients be kept on a dose of methadone which will at least adequately prevent the onset of withdrawal symptoms for a 24-h dosing interval, which is usually 30–60 mg a day.

The clinical observations that menses return to normal in patients maintained on methadone led to studies in animal models and later in humans which found that opiates in general in animal models, and especially the short-acting opiates in humans, such as heroin or morphine, when used acutely or when used even on a chronic basis in cycles of addiction, reduced the pulsatile release of luteinizing hormone from the hypothalamic sites (KREEK 1973a,b; CUSHMAN 1973a; ESPEJO et al. 1973; SANTEN and BARDIN 1973; CUSHMAN and KREEK 1974a,b; PELOSI et al. 1974; SANTEN 1974; CICERO et al. 1975; HELLMAN et al. 1975; MENDELSON and MELLO 1975; MENDELSON et al. 1975a,b; SANTEN et al. 1975; KREEK 1978a; LAFISCA et al. 1981; KREEK and HARTMAN 1982; MENDELSON and MELLO 1982; SMITH et al. 1982; KREEK 1983b). This in turn leads to failure to ovulate and triggers the appearance of abnormal menses in females. In males, the reduction in LH release leads to production of lower than normal levels of plasma testosterone in heroin addicts. Early studies showed that plasma levels of testosterone teturned to normal during steady-dose treatment with a long-acting opioid methadone. However, it was found in several studies that steady-dose methadone treatment for 1 year often was needed to allow normalization of this short-acting effect on LH release. Studies show that hypothalamic-pituitary-gonadal axis remained abnormal during the first 2–3 months of methadone maintenance treatment with observations similar to that made in heroin addicts, but that following stabilization on moderate to high dose methadone for 3 months to 1 year, or in some cases, longer, normalization of LH levels, including LH release, and resultant levels of testosterone in males as well as normal levels of estrogens and as progestins in females with return of normal menstrual cycling and ovulation, occurred.

Effects of heroin on the adrenal function were first observed, but not identified as being part of the entire hypothalamic-pituitary-adrenal axis, by the early studies at Lexington when former addicts under study were given repeated doses of morphine to create cycles of addiction (EISENMAN et al. 1958, 1961, 1969). During these studies, conducted between 1958 and 1969, it was observed by Eisenman and colleagues that urinary levels of 17-ketosteroids and also subsequently, urinary levels of 17-hydroxycorticosteroids were reduced during cycles of morphine administration. Subsequently, in the early treatment research studies of heroin addicts coming into methadone maintenance treatment research, as well as during the induction period of methadone treatment during which time ascending doses were used, prospective special neuroendocrine studies were carried out to assess the status of hypothalamic-pituitary-adrenal axis prior to and during dose induction and stabilization (KREEK 1972, 1973a,b, 1978a). Lowered levels of urinary glucocorticoids were confirmed in heroin addicts entering and patients during early methadone maintenance treatment. However, urine levels of glucocoriticoid soon returned to normal. Subsequently, it was determined that plasma levels of the glucocorticoid, cortisol, also was modestly reduced in heroin addicts, often with disrupted circadian rhythm of levels, but became normal during methadone treatment. In these early studies, an initial sequence of provocative tests of hypothalamic-pituitary-adrenal integrity were conducted (Kreek 1972, 1973a,b). ACTH infusion studies were conducted in methadone-maintained former heroin addicts and an adequate adrenal cortical reserve was demonstrated, as evidenced by normal release of glucocorticoids following ACTH stimulation (Kreek 1973a,b, 1978a).

Also, a provocative neuroendocrine test of hypothalamic-pituitary reserve was performed. In these studies, metyrapone, a compound which selectively blocks 11-β-hydroxylation, the last step of cortisol synthesis by the adrenal cortex, was administered. A blockade of cortisol production results in a cut-off of the normal negative feedback control by cortisol of the hypothalamic-pituitary part of this important stress responsive axis, the major glucocorticoid in humans. Cortisol is now known to act both at hypothalamic and anterior pituitary sites, to prevent the synthesis and release of corticotropin releasing factor from the hypothalamus, as well as processing and release of pro-opiomelanocortin peptides from the anterior pituitary stress responsive axis. The metyrapone test can be used as a test for stress responsivity, as its administration evokes the same neurochemical events in the brain as does any stressor. At the time of the early studies in 1976–1973, it was not possible to measure the neuropeptides involved directly in plasma. Therefore, multiple doses of metyrapone were given over a 24-h interval and 24-h urine collections were made during the day in which metyrapone was administered and 2 consecutive days (Kreek 1972, 1973a,b, 1987, 1992c; Cushman and Kreek 1974a,b). In these early studies, it was found that the hypothalamic-pituitary reserve, as measured by this metyrapone test, is significantly reduced in heroin addicts. It was also found that hypothalamic-pituitary research is reduced in methadone patients receiving ascending doses of methadone, during the first 2–3 months of methadone maintenance treatment. However, after stabilization on moderate to high doses of methadone for 3 months or more, it was shown that the metyrapone responsivity returned to normal, indicating normalization of hypothalamic-pituitary reserve for response to this chemically induced stressor (Kreek 1973a,b). At the time of these early studies, the precursors of cortisol produced and released by the adrenal cortex were measured in the urine. A normal test response was indicated by a significant elevation of urinary excretion of these precursors (the "Porter-Silber chromogens") or cortisol, reflecting increased anterior pituitary release of ACTH in response to blockade of cortisol negative feedback control.

Several years later, after it became possible to measure pituitary peptides in circulating peripheral plasma by radioimmunoassay techniques and after the identification of β-endorphin, a specific endogenous opioid which is released in equimolar amounts, along with ACTH, from propiomelanocortin, a single gene produced and processed for peptide release from the anterior pituitary of humans, the metyrapone studies were repeated. In neuroendocrine studies, conducted in the 1980s and onward, much new information has been learned. Normal metyrapone test results are now defined as a two to four-fold increase in plasma levels of ACTH and/or β-endorphin during the

8-h period following a single dose of metyrapone administration (KREEK et al. 1981; KREEK and HARTMAN 1982; KREEK et al. 1983d, 1984; KENNEDY et al. 1990; KREEK and CULPEPPER-MORGAN 1991; KREEK 1992c). The earlier findings of normalization of metyrapone test results during stabilized methadone maintenance treatment have been reconfirmed. During heroin addiction when opioid is present, decreased activity of the hypothalamic-pituitary-adrenal axis, with reduced or flattened circadian rhythm of levels of the anterior pituitary peptides ACTH and β-endorphin as well as lowered levels of the adrenal glucorticoid, cortisol, are observed. During acute opiate withdrawal in an opiate addict, the opposite findings are observed with prompt activation and inneractivity of the stress response axis. However, during chronic methadone maintenance treatment, normalization of plasma levels of these hormones is observed, along with a normalization of circadian rhythm in their levels (KREEK 1973a,b, 1978a; KREEK et al. 1981, 1983d, 1984). The release of highest levels of ACTH and β-endorphin are observed during the early morning hours, and cortisol levels similarly follow this circadian pattern as they are released from adrenal cortex in response to ACTH. Further use of metyrapone testing has shown very provocative results (KREEK et al. 1984). In medication-free former heroin addicts treated by the abstinence approach and also in medication-free former methadone-maintained patients, a hyper-responsivity to metyrapone testing has been observed (KREEK et al. 1984; KREEK 1992c). This hyperresponsivity to this chemically induced stress may or may not be related to the return of drug craving and relapse to drug use, which occurs in over 80% of drug-free former heroin addicts and former methadone-maintained patients.

In more recent studies, it has been observed that in some, but not all, methadone-maintained patients in whom metyrapone testing is conducted, classical signs and symptoms of opiate withdrawal will occur around thirty minutes after metyrapone is given and persist for not more than 2 h (KENNEDY et al. 1990). The time course of duration of the symptoms is much shorter than the actual pharmacokinetic profile of metyrapone. Metyrapone is not a specific opiate antagonist. Other studies have shown that methadone plasma levels are not altered during metyrapone administration. These findings have suggested that in the setting of a provoked abrupt increase in release of anterior pituitary POMC-derived peptides, either resultant increased levels of a normal product, or atypical processing of POMC peptides, may lead to the production and release of an endogenous opioid antagonist. Such an endogenous antagonist has been hyothesized by some laboratory investigators to exist. Some studies have suggested that fragments of ACTH may be an endogenous opioid antagonist. Alternatively, we have suggested that the normal metyrapone response of dramatic rise in plasma levels of ACTH and β-endorphin may serve as internal cues to the patient of acute opiate withdrawal, since similar significant changes in these hormone levels are consistently observed, along with elevation of levels of the glucocorticoid, cortisol, in the setting of opiate withdrawal. Very recent studies have suggested that activation of the hypothalamic-pituitary axis may be observed objectively prior to the onset of signs and symptoms of opiate withdrawal, suggesting that these hormonal changes are not simply a result of the stress from opiate withdrawal but rather are initial biological events and which may serve as cues and also activation events which lead to the onset of opioid withdrawal symptoms (Culpepper-Morgan et al. 1992; Rosen et al. 1996, in press).

In the early 1970s, we hypothesized that addictive disease may, in part, be due to an atypical responsivity to stress and stressors (Kreek 1972, 1973a,b, 1992c). These findings and many others from different research studies have suggested that the endogenous opioids may play a very important role in modulation of the important stress responsive hypothalamic-pituitary adrenal axis in addictive diseases (Kreek 1973a,b, 1978a, 1992c; Ho et al. 1980; Holmstrand et al. 1981; Kreek et al. 1981, 1983d, 1984, 1994; Borg et al. 1982; Kreek and Hartman 1982; O'Brien et al. 1982–1988; Emrich et al. 1983; Facchinetti et al. 1984; Kosten et al. 1987; Kennedy et al. 1990; Vescovi et al. 1990; Kreek and Culpepper-Morgan 1991; Culpepper-Morgan et al. 1992). In addition, they may play a role in the modulation of the hypothalamic-pituitary-gonadal axis. Other findings suggest that the endogenous opioid system may play a role in mood modulation. Since it had been hypothesized by our Rockefeller University research group that an atypical responsivity to stress may cause or contribute to the development of opiate (and cocaine) addiction, the disruption of the stress responsive axis by naltrexone may explain in part the limited effectiveness of naltrexone in the treatment of opiate dependency. In contrast, appropriate use of a long-acting agonist such as methadone (and presumably also LAAM), which allows normalization of this stress responsive axis, may thereby contribute to the acceptance, efficacy, and effectiveness of such agonist treatment.

2. Immune Function

Many studies have identified significant disruption of specific indices of immune function in heroin addicts (Sapira 1968; Sherwood et al. 1972; Cushman 1973b,c; Geller and Stimmel 1973; Brown et al. 1974; Drusin et al. 1974; Cushman et al. 1974, 1977; Adham et al. 1978; Jacob et al. 1978; Lazzarin et al. 1984; Poli et al. 1985; Donahoe et al. 1986, 1988; Dyke et al. 1986; Falek et al. 1986; Nair et al. 1986; Novick et al. 1986b, 1988c, 1989b, 1990, 1991, 1993; Donahoe and Falek 1988; Kreek 1989, 1991d,e, 1993; Kreek et al. 1989; Ochshorn et al. 1989, 1990; Bodner et al. 1990, 1991, 1992; Bryant and Holaday 1991; Novick and Kreek 1992; Pinto et al. 1993; Eisenstein et al. 1994). In parallel, several laboratory studies have shown effects of opioid agonist drugs such as morphine and, more recently, the endogenous opioids on specific indices of immune function in animal model studies and in in vitro studies. The abnormalities of immune function found in heroin addicts have become increasingly difficult to study because of the very high prevalence of HIV-1 infection in this population. However, when studied in persons who are not HIV-1 positive, findings have included increased numbers of lymphocytes and absolute numbers of T cells, T cell subsets, and B cells, along with elevated levels of immunoglobulins reflecting abnormal B cell function (NOVICK et al. 1989b). Also, abnormal T cell roscetting has been found in active heroin addicts. In addition, significant lowering of natural killer cell activity has been observed in heroin addicts in many studies. Natural killer cell function is the first line of defense against many viral infections and against tumor invasion. Several studies have shown that exogenous opiates and endogenous opioids may also reduce natural killer cell activity in animal models and in vitro. Most of the abnormalities of immune function found in heroin addicts may be due to the multiple infections diseases to which these persons are exposed or infected, and to a long history of the chronic injections of a variety of foreign substances used to extend or "cut" illicit drugs. Nevertheless, some of the effects may be direct or indirect drug effects, that is, specific opioid effects. Studies have been performed to determine if morphine and heroin have any direct effect on natural killer cell activity using human cells in vitro, and no changes have been found until very high doses of opiates, including the active but also inactive enantiomers of morphine and methadone and also the opioid antagonist naloxone, far in excess of those ever achieved pharmacologically, are reached ($>10^{-4}$ molar) (OCHSHORN et al. 1990).

Prospective studies of patients entering methadone maintenance treatment have suggested a reduction in prevalence and degree of abnormalities with specific indices of immune function, but without normalization after 3 years of treatment (KREEK 1972, 1973a,b, 1978a, 1990d, 1991a,b,c,d,e; KREEK et al. 1972, 1989; OCHSHORN et al. 1989). However, in a rigorous study of patients maintained on moderate to high doses of methadone for 11 years or more, in which a direct comparison was made to persons with similar number of years of heroin addiction and similar exposure to hepatitis B virus, all of whom were negative on testing for HIV-1 (AIDS) virus infection, and of healthy volunteers, it was found that the specific indices of immune function studies, including absolute numbers of T cells, T cell subsets, B cells, and quantitative immunoglobulins, were normal in the long-term methadone-maintained patients (NOVICK et al. 1989b). Also, natural killer cell activity was normal in the long-term methadone-maintained patients (NOVICK et al. 1989b). Since these persons in long-term treatment for more than 11 years had been exposed to far greater levels of opioid, with 24-h sustained perfusion of opioid receptors with these high levels of synthetic opioid methadone, as contrasted with heroin addicts, with intermittent peak levels of heroin followed by heroin withdrawal, it was concluded that methadone itself clearly was not suppressing or adversely affecting any aspect of immune function studied (NOVICK et al. 1989b). It was, therefore, hypothesized that any specific heroin effect on immune function in humans might be an indirect effect and possibly mediated through the neuroendocrine system. As discussed above, in the setting of opiate withdrawal, a phenomena which occurs in street heroin addicts usually three or more times each day, there is a marked elevation in hormones of the hypothalamic-pituitary-adrenal axis, including elevated levels of β-endorphin, ACTH, and cortisol. Cortisol has been found by numerable studies to be a profound suppressor of most indices of immune function. It has also been shown that a minimal physiological decrease in cortisol levels will cause an increase in natural killer cell activity (Bodner et al. 1990, 1991, 1992). Thus, it is possible that in the setting of heroin addiction, these elevated levels of glucocorticoids and/or other neuropeptides may contribute to the abnormalities in immune function, including the reductions in natural killer cell activity observed in heroin addicts and that with normalization of neuroendocrine function during methadone maintenance treatment, immune function also normalizes.

A few abnormalities in thyroid function tests have also been observed in heroin addicts and in methadone-maintained patients (Webster et al. 1973; Azizi et al. 1974; Cushman and Kreek 1974a; Bastomsky and Dent 1976; Kreek 1978a; Spagnolli et al. 1987). Initially, it was thought that thyroid function might be deranged; however, further studies revealed that thyroid function is normal in methadone-maintained patients. Less rigorous studies have been carried out in active heroin addicts. However, thyroid binding globulin is significantly increased in both heroin addiction, along with most other specific and nonspecific serum proteins in heroin addicts. Thus elevated levels of thyroid without altered function may pertain. These elevations in plasma proteins persist, with very slow resolution toward normal levels during long-term methadone maintenance treatment.

3. Gastrointestinal Function

The third type of opioid agonist effects, which may be a problem in methadone maintenance treatment, is the well established actions of opiates on altering gastrointestinal motility. Opiate agonists have been used for centuries for the management of diarrhea, and constipation has been a long-term observed and often dose-limiting side effect of use of morphine and other opiate agonists for relief of chronic pain. In prospective studies of persons entering methadone maintenance treatment, constipation is also a major problem affecting most subjects during the first 6 months of treatment (Kreek 1972, 1973a,b; Dobbs 1971; Rubenstein and Wolff 1976). However, tolerance develops to this effect by 3 years or more of steady-dose treatment in 70%–80% of patients (Kreek 1973a,b, 1978a). It remains a side effect of concern for some patients. Recently, a new potentially effective approach for managing this problem has been introduced. Use of the specific opioid antagonist naloxone, which has extremely limited ($<2\%$) systemic bioavailability following oral administration, may be administered in two or three doses a day to patients in a maintenance treatment of addiction on methadone or other opiate agonists for the relief of chronic pain (Kreek 1973c, 1992d; Albeck et al. 1989; Hahn et al. 1983; Kreek et al. 1983b,c; Kreek and

CULPEPPER-MORGAN 1991; CULPEPPER-MORGAN et al. 1992, 1995). Early preliminary studies have shown that orally administered naloxone will reach intestinal opioid receptor sites in the stomach and small and large intestine and will reverse the opiate agonist effects at these sites without reaching central nervous system opiate receptor sites, which would result in the precipitation of narcotic withdrawal symptoms or in a recrudescence of pain.

These three aspects of physiology, neuroendocrine function, immune function, and gastrointestinal function, are all very important functions which have been shown to be profoundly disrupted during cycles of heroin addiction. All three of these functions, which are critical for stress responsivity, reproductive biology, host defense against infections and neoplasm, and gastrointestinal function, have been shown to be normalized during long-term steady-dose methadone maintenance treatment. Similarly, methadone has been shown to have no adverse effect to performance, including both intellectual, cognitive, and motor performance tasks of a variety of types.

IV. Special Issues Related to Methadone Maintenance and All Other Long-Term Treatments of Opiate (Heroin) Addicts

1. Codependency: Alcoholism, Cocaine, and Other Addictions

Codependency with alcohol has been identified as a major problem in heroin addicts entering methadone maintenance treatment since the beginning of the prospective studies in 1964 (KREEK 1972, 1973a,b, 1984, 1987, 1988, 1991c; BIHARI 1973, 1974; BROWN et al. 1973; JACKSON and RICHMAN 1973; PASCARELLI and EATON 1973; ROSEN et al. 1973; SCOTT et al. 1973; MADDUX and ELLIOT 1975; PUGLIESE et al. 1975; CHARUVASTRA et al. 1976; GORDIS and KREEK 1977; CUSHMAN et al. 1978; BEVERLY et al. 1980; MARCOVICI et al. 1980; ROUNSAVILLE et al. 1982; JACKSON et al. 1982; HARTMAN et al. 1983; MCLELLAN et al. 1983; KHURI et al. 1984; JOSEPH 1985; CUSHMAN 1987; NOVICK 1984; GORDIS 1988). Those studies and subsequent ones have shown that 25%–50% of all heroin addicts have an alcohol abuse problem or are alcoholics. Prospective studies have shown that patients who have an alcohol abuse or alcoholism problem do as well in a good methadone maintenance program with respect to cessation of illicit opiate use and adherence to the clinic regimen as do nonalcoholics overall (BEVERLY et al. 1980; KREEK 1984). Some patients, probably due to the efforts of the rehabilitation and counseling staff, reduce overall intake of ethanol; others continue with an ethanol abuse pattern.

More recently, since 1978, cocaine abuse has been a major problem complicating pharmacotherapy for heroin addiction. Presently, up to 90% of heroin addicts in New York City at this time are also cocaine addicts (HUNT et al. 1984; KREEK 1987, 1991c; JOHANSON and FISCHMAN 1989; KOSTEN et al. 1989b; NOVICK et al. 1989a; TABASCO-MINGUILLAN et al. 1990; KOTHUR et al.

1991; CAMBOR et al. 1992; DESJARLAIS et al. 1992; NUNES et al. 1991; Ho et al. 1992). Several studies have shown that the prevalence of cocaine dependence decreases significantly during methadone maintenance treatment with only about 20%–40% continuing to use cocaine during methadone maintenance treatment and up to another 20%–25% initiating cocaine use during treatment (BORG et al. 1995a; CHAISSON et al. 1989; HANBURY et al. 1986). The net result is a significant reduction by over 50% of those continuing to use cocaine during methadone maintenance treatment. Although this is attributed and probably due primarily to the rehabilitation and counseling efforts and proper contingency use of allowing "take-home" doses of methadone for those with no evidence of continuing cocaine use, recent molecular and neurobiological studies by our laboratory group have shown that the endogenous opioid system, especially the μ and κ receptor, may be significantly disrupted during chronic cocaine administration in a "binge" pattern associated with dependency or addiction (BRANCH et al. 1992; SPANGLER et al. 1993a,b; UNTERWALD et al. 1992, 1994a,b). These findings, coupled with several laboratory findings showing that morphine, methadone, both pure μ opioid receptor agonists, and buprenorphine, a partial μ agonist, all decrease self-administration of cocaine in animal models, suggest that each of these opioid agonists or partial agonists may have some pharmacological efficacy in addition, which may augment the effectiveness of the treatment milieu.

2. Psychiatric Comorbidity

Psychiatric comorbidity has been identified since the earliest days of studies of methadone maintenance treatment (LIEBSON and BIGELOW 1972; GRITZ et al. 1975; INWANG et al. 1976; WOODY et al. 1975, 1990; WEISSMAN et al. 1976; PRUSOFF et al. 1977; KREEK and HARTMAN 1982; KAUFMANN et al. 1983, 1984; OBRIEN et al. 1984; CORTY et al. 1988; MASON et al. 1992). Several excellent studies have shown that multiple psychiatric disorder problems confound heroin addiction. The three most common are probably depression, anxiety, and antisocial personality diagnoses. Depression improves in some subjects during pharmacotherapy, worsens in others, and appears a priori in still others during methadone maintenance treatment, with approximately 20%–30% of heroin addicts and 20%–30% of methadone-maintained patients showing some signs and symptoms of clinical depression. The prevalence of antisocial personality type is much more difficult to assess, since the diagnostic criteria for this disorder are met by almost all heroin addicts if applied according to the currently used diagnostic instruments. The subscale for childhood conduct disorders raises questions concerning behaviors at 15 years and earlier. However, many heroin addicts begin their drug abuse history at a much younger age, as early as 8–10 years. Therefore, a more accurate assessment of the true premorbid problem antisocial personality type is ascertained using both drug abuse and personality questions which are each specifically linked with the question of age of onset of each paranoia (MASON et al. 1992).

3. Chronic Liver Disease, HIV Infection, and AIDS: Preexisting Medical Problems in Heroin Addicts and Impact of Methadone Treatment

Chronic liver disease was identified as a major problem in street heroin addicts by the Lexington research group and by the initial research groups studying heroin addicts entering methadone maintenance treatment (LOURIA et al. 1967; CHERUBIN 1967; SAPIRA 1968; SAPIRA et al. 1968; CHERUBIN et al. 1970, 1972a,b, 1976; JERSILD et al. 1970; KREEK et al. 1972; STIMMEL et al. 1972a,b, 1973; KREEK 1973a,b, 1983; WHITE 1973; STIMMEL et al. 1975, 1982; WEBSTER et al. 1977; MILLER et al. 1979; NOVICK et al. 1981b, 1985a,b,c, 1986a,b,c,d; HARTMAN et al. 1983). In those studies, it was found that over 50% of all heroin addicts had abnormalities of liver function. Once hepatitis B was defined and serum markers were available, the prospective studies conducted by the Rockefeller University group found that 80%–90% of all street heroin addicts had been exposed to hepatitis B (KREEK 1972). After the identification of delta hepatitis in the late 1970s and the development of serum markers for infection with this atypical circular RNA viroid, which requires the presence of replicating hepatitis B for its own infectiousness, retrospective studies were conducted by the Rockefeller University group in heroin addicts and former heroin addicts in methadone treatment (NOVICK et al. 1985c, 1986a; KREEK et al. 1986, 1990a,b; KREEK 1991b,c,d; SHI et al. 1994). These studies showed that up to 30% of untreated heroin addicts in patients in methadone maintenance treatment in some locations had some evidence of ongoing or prior delta infection. More recent studies have shown that the increased infectiousness of persons with both hepatitis B and hepatitis delta increased in the setting of AIDS disease (KREEK 1990b,c,d; KREEK et al. 1990a,b). More recently, with the definition of hepatitis C virus and the development of markers for looking for infection with this virus, it has been found that over 90% of heroin addicts and former heroin addicts in methadone maintenance treatment have had prior infection with hepatitis C virus (GIRARDI et al. 1990; NOVICK et al. 1995).

The most important infectious disease afflicting heroin addicts and confounding patients in methadone maintenance treatment is the HIV-1 or AIDS virus. Studies conducted in 1983–1984, which took advantage of prospectively banked sera collected by the Rockefeller University research team from 1969 onward, were able to identify that HIV-1 infection entered the intravenous drug abusing population in New York City around 1978 with a few cases probably infected before that time (DESJARLAIS et al. 1984, 1989; NOVICK et al. 1986a,e; KREEK 1990b,c,d; KREEK et al. 1990a,b). Then the epidemic rose rapidly from 1978 to 1983, at which time it plateaued with around 50%–60% of all untreated heroin addicts infected with HIV-1

(DESJARLAIS et al. 1984, 1989; NOVICK et al. 1986a,e; KREEK 1990b,c,d; KREEK et al. 1990a,b). That same study was the first to document the enormously important and positive public health impact of effective methadone maintenance treatment (DESJARLAIS et al. 1984, 1989). In 1984, a time when over 50% of street untreated heroin addicts were HIV positive, it was found that only 9% of patients who had entered into an effective methadone maintenance treatment prior to the epidemic reaching New York City in 1978 and remaining in effective treatment until the time of study in 1984, were HIV-1 infected. This was because use of illicit drugs and thus sharing of dirty needles had become significantly reduced or had ceased during effective methadone maintenance treatment. The only patients who were found to be HIV-1 infected with those who had continued to use drugs by a parenteral route were primarily intravenous cocaine users (DESJARLAIS et al. 1984; NOVICK et al. 1989a). Further studies in the United States as well as in other countries, including Sweden and Italy, have shown the very positive impact with respect to AIDS risk reduction of effective methadone maintenance treatment programs (NOVICK et al. 1986a,e, 1989a, 1990, 1993; BALL et al. 1988; BLIX 1988; DOLE 1989; CHAISSON et al. 1989; NATHAN and KARAN 1989; BROWN et al. 1990; KREEK et al. 1990a,b,c,d; WEBER et al. 1990; AJULUCHUKWU et al. 1993; BROWN et al. 1993; METZGER et al. 1993; SAWYER et al. 1993; SIDDIQUI et al. 1993a,b). Thus, all of these studies have shown that treatment with methadone, and presuming that any other effective pharmacotherapy which, for the majority of unselected heroin addicts may be limited to μ agonists and possibly partial agonists, results in normalization of physiological functions disrupted by heroin addiction, including neuroendocrine function, gastrointestial function, and immune function, and also effects a decreased exposure to infectious diseases, in particular HIV-1 virus as well as probably also hepatitis B, C, and delta.

C. *l-α*-Acetylmethadol (LAAM)

I. Chemistry, Pharmacokinetics, Pharmacodynamics, and Mechanisms of Action in Humans

Acetylmethadol, as the racemic mixture, or as the opioid active or *l* and also nonopioid active *d* form, was synthesized and first studied for its possible analgesic actions and addiction liabilities in the late 1940s and early 1950s (BOCKMUHL and ERHART 1948; EDDY et al. 1952; FRASER and ISBELL 1951, 1952; KEATS and BEECHER 1952; SUNG and WAY 1954). *l-α*-acetylmethadol (LAAM) is a synthetic, acetylated, single enantiomeric congener of methadone which is even longer-acting in man (48 h as contrasted to 24 h), and like methadone, is orally effective, meeting the two major criteria for a pharmacotherapeutic agent for treatment of an addictive disease (KAIKO and INTURRISI 1973; KREEK 1973c, 1978a; LEVINE et al. 1973). LAAM was approved by the Food and Drug Administration for use in treatment of heroin addiction in the summer of 1993 following a rapidly processed New Drug Application (NDA) developed by the Medications Development Division of National Institute on Drug Abuse in 1992–1993. This very rapid approval followed a hiatus of over 12 years after completion of the majority of clinical studies documenting the actions, pharmacokinetics, and efficacy of LAAM, during which time essentially no clinical research related to the use of LAAM was performed, due to earlier (and primarily administrative) problems related to the preclinical drug testing findings, and 41 years after the first studies in humans of its opioid properties (ABRAMOWIC 1994; ORLAAM package insert 1993).

The pharmacological rationale for the use of LAAM treatment of heroin addiction is similar to that of methadone. LAAM is a pure opioid agonist directed primarily at the μ type opioid receptor. By virtue of its long duration of action, it provides a steady state perfusion of specific opioid receptor sites. Different from methadone, however, is the fact that LAAM depends on the extensive oxidative metabolism by hepatic P450 type enzymes to two active metabolites, norLAAM (N-demethylated LAAM) for its even longer-acting properties, and after a subsequent second N-demethylation, dinorLAAM (BILLINGS et al. 1974; SULLIVAN et al. 1973; SUNG and WAY 1954). Pharmacological studies have suggested that norLAAM has a similar and the dinorLAAM an even slower clearance than does the parent compound, LAAM. However, kinetic studies of pure metabolites in man have not been performed. The half-lives determined in studies which have been accepted by the FDA have shown that the apparent beta terminal half-life in humans for LAAM is 2.6 days, for norLAAM, 2 days, and for dinorLAAM, 4 days (HENDERSON et al. 1976; KAIKO and INTURRISI 1975; ABRAMOWIC 1994; MISRA and MULE 1975; BLAINE and RENAULT 1976; ORLAAM package insert 1993). A unique type of potential problems with LAAM could result from the fact that LAAM depends for its long-acting properties upon successive biotransformations to the other biologically active and long-acting opioid metabolites, norLAAM and dinorLAAM. In the setting of use of a medication which may enhance the hepatic P450 related enzyme systems such as rifampin, phenytoin, phenobarbital, or carbamazepine, or following the chronic use of alcohol, an accelerated biotransformaiton may be seen, as has been documented with respect to methadone (KREEK et al. 1976a,b; KREEK 1990a; TONG et al. 1981). This could lead to rapid production of the active metabolites. However, the net effect on the sustained, overall steady state of perfusion of critical opioid receptors is not known. Similarly, in the presence of a use of an agent which may impair hepatic drug metabolism, such as very large amounts of ethanol, possibly cimetidine, or large doses of benzendiazepines, the biotransformaiton of LAAM could, in theory, be retarded and the effects of this not determined. The effectiveness of LAAM with respect to providing a sustained and steady reduction of drug craving or "hunger" is not known.

II. Use of LAAM in Long-Term Pharmacotherapy
of Opiate Addiction: Safety, Efficacy, and Effectiveness

The rationale underlying the development of LAAM in treatment was essentially identical to that underlying the development of methadone. LAAM is a significantly longer acting congener of methadone. The original rationale was extended to suggest that the even longer acting properties of LAAM might provide a more steady state action which could be beneficial for the small, but real, number of methadone maintained patients who experience abstinence symptoms before the end of the 24-h dosing interval. A second, very different rationale for the development of LAAM was the one offered primarily by policy-makers concerned about the potential for diversion of "take home" doses of methadone. Underlying that concern was the possibility of primary methadone addiction by use of diverted methadone. However, many well conducted studies, for over 30 years, have found that primary methadone addiction is exceedingly rare. This finding is not surprising given the very slow onset of action of methadone and, thus, its very minimal "rewarding" effects in a tolerant individual, as well as its effects of producing simply somnolence or sleep in a modestly tolerant or naive subject. The magnitude of methadone diversion has been a subject of much controversy (outside the scope of this chapter). However, clearly, the diversion of methadone has not proven to be a major public health problem (RETTING and YARMOLINSKY 1994). In contrast and of great importance, the ability of physicians and other health care workers in methadone maintenance programs to allow "take home" doses for methadone maintained former heroin addicts exhibiting increasingly successful treatment, coupled with cessation of any continued illicit drug use, has been a critical factor for appropriate rehabilitation. The legal ability to permit patients to come three, two, or one time a week to a standard methadone maintenance clinic with "take home" doses provided, has allowed patient to return to a normal lifestyle of working, going to school, or running a home, which would be much more difficult if a daily clinic visit were required for all patients. The medical maintenance paradigm (discussed above), in which rehabilitated long-term methadone treated patients may see a physician in an "office based" practice or general medical clinic as infrequently as once every 4 weeks, that is, "medical maintenance," has proven to be effective in experimental studies to date and has underscored the value of take home doses. Although the longer acting properties of LAAM may indeed be shown to provide a significant advantage for patients who do experience early abstinence due to a defined pharmacological or metabolic reason, the rationale that LAAM treatment would preclude the need for any "take home doses" which are now, as the LAAM approval by the FDA was granted, disallowed for LAAM, may be found to be counterproductive for rehabilitation efforts. It is unlikely that LAAM will be accepted by patients who have already undergone rehabilitation in methadone maintenance treatment and who were able to return to a normal lifestyle during pharmacotherapy, unless and until take home privileges with LAAM are allowed. Since LAAM has been approved for use only in the last year, it is premature to evaluate the impact. In the future, it is possible that regulations precluding take home doses of LAAM will be changed.

A third rationale for the development of LAAM was that it would be more cost effective to have to see patients in clinics only three or four times a week, as opposed to six or seven during early induction into methadone maintenance treatment. Again, many clinicians and other therapists involved in the treatment of heroin addiction have recognized the profound early treatment needs of former addicts, who are very often street criminals, disenfranchised from the health care and social service systems. It has long been recognized that daily visits to an established methadone maintenance clinic may be of utmost importance during the early weeks or months of treatment. Again, it is premature to assess the relative effectiveness of LAAM when administered on a three time a week basis as currently proposed. The fourth, and neurobiologic, rationale for LAAM use is identical to that of methadone. Studies have yet to be performed to verify that normalization of physiological functions disrupted by short-acting opiate use occurs as it does with methadone treatment. However, since LAAM depends for its long-acting properties on each of its two major N-demethylated metabolites, norLAAM and dinorLAAM, it is possible that in the setting of other drug use or abuse, or altered physiological states, an uneven biotransformation may occur which could result in nonsteady state perfusion of opioid receptors which, in turn, could prevent normalization of neuroendocrine function and other physiological system disrupted by fluctuating opioid levels. Further studies will be required to define whether or not normalization of physiological function occurs during long-term treatment with LAAM and if so, what the constraints are on this being achieved.

Studies concerning LAAM use for the treatment of pain and to determine its effects and abuse liability in heroin addicts were conducted in the early 1950s by the groups of Beecher at Harvard, Fraser and Isbell at the United States Public Health Service Hospital in Lexington, and later by a few other groups (DAVID et al. 1956; FRASER and ISBELL 1951, 1952; GRUBER and BAPTISTI 1963; KEATS and BEECHER 1952). However, no formal studies for the potential use of LAAM for the treatment of heroin addiction were performed until 1970, when Jaffe, Schuster, Senay, Blatchley, and Renault first studied the possible effectiveness of the racemic dl-acetylmethadol, and subsequently studied the active enantiomer with lesser side effects, LAAM (JAFFE and SENAY 1971; JAFFE et al. 1970, 1972). By the late 1970s, numerous clinical studies, including controlled Veterans Administration based cooperative trials of the efficacy of LAAM, headed by Ling and colleagues, showed that it is essentially as effective as methadone in the long-term pharmacotherapy of opiate addiction when high doses of LAAM are used (LING et al. 1976). Several other well designed and conducted studies were performed in the 1970s (BLAINE et al. 1978, 1981; GREVERT et al. 1977; LEVINE et al. 1973; LING et al. 1975, 1978; MARCOVICI et al. 1981; BLAINE and RENAULT 1976; RESNICK et al. 1976; SCHECTER and KAUDERS 1975a,b; SENAY et al. 1977a,b; ZAKS et al. 1972). However, because of the misleading, probably invalid, findings of hepatic damage in some of the preclinical studies, all further research of LAAM was halted in the late 1970s until the preclinical studies had been repeated, with rigorous replications, showing no hepatotoxicity of any type in appropriate animal models. In the early 1990s under the sponsorship of NIDA, a multiple site study was performed for the sole purpose of determining the most appropriate product labeling for LAAM. The completion of the study led promptly to the submission and approval of NDA for the FDA in the summer of 1993.

Several reported studies have suggested that induction of LAAM may be conducted de novo in heroin addicts. Alternatively, patients currently stabilized in treatment with methadone may be transferred to LAAM treatment. For patients being induced onto LAAM treatment, it is recommended that initial doses of 20–40 mg be given 3 days a week, or every other day; the doses should then be increased gradually by 5 or 10 mg/ week up to a full treatment dose of 60–120 mg (JAFFE et al. 1979; JUDSON and GOLDSTEIN 1979). Because of its long half-life and thus the potential for accumulation of LAAM and its active metabolites, LAAM should never be given on a daily basis or more often than once every 48 h. As with methadone, the goal of induction into treatment is to gradually titrate the initial dose of LAAM upward to both suppress narcotic withdrawal signs and symptoms, and provide a dose which will "block" the euphoric effects of any superimposed short-acting opiate through the mechanism of cross-tolerance, and simultaneously, to avoid any excessive opioid-like effects. During the first week or two of LAAM treatment, signs and symptoms of narcotic withdrawal and also "drug hunger" may occur because of the lack of adequate accumulation of the desired active parent compound and active metabolites. However, it is more prudent to ascend the dose slowly rather than rapidly, which could pose the risk of respiratory arrest or other signs and symptoms of narcotic overdose. When a patient is being transferred from methadone to LAAM, the recommended initial dose of LAAM is 1.2 to 1.3 times the daily dose of methadone, with an initial dose not to exceed 120 mg (ABRAMOWICX 1994; ORLAAM package insert 1993). Ideally, LAAM should be administered on an every other day basis. However, to facilitate patient and clinic schedules, it can be delivered three days a week, on Monday, Wednesday, and Friday, with an increased dose administered on Friday prior to a 72-h dosing interval (LING et al. 1980). It is recommended that the dose prior to a 3-day dosing interval should be approximately 25% but more than 40% greater than the doses before 2-day dosing intervals (ABRAMOWICX 1994; ORLAAM package insert 1993). The safe and effective starting dose of LAAM is 20–40 mg for "hard core" addicts, those who have demonstrated 1 year of regular heroin use with the development of tolerance and physical dependence. This dose may produce some somnolence in patients who do not have a significantly high degree of tolerance) however, such a dose might produce respiratory depression in a nontolerant individual. Therefore, it is imperative that an appropriate history and documentation of addiction by conducted prior to initiating LAAM treatment. At the same time, if too low a dose is used, opiate-dependent patients will undoubtedly have withdrawal symptoms. As with methadone, LAAM maintenance doses are attained by the addition of 5 or 10 mg each week to achieve a full "blockading" dose through the mechanism of opioid tolerance and cross tolerance.

Rigorous pharmacological and cross-tolerance studies, such as those conducted to determine the ability of methadone as used in maintenance treatment to prevent any "drug hunger" or craving for heroin and "block" any opiate effects or other opioid-like signs in the setting of superimposed short-acting opiates, such as heroin, have not been conducted. However, analysis of street anecdotes as well as other types of pharmacokinetic or pharmacodynamic studies would suggest that a dose of 60–120 mg of LAAM three times per week should provide such a level of tolerance and cross-tolerance to provide "blockade" against any superimposed short-acting opiate, just as doses of 60–120 mg of methadone have been shown to be highly effective. It should be emphasized that it has been rigorously shown that administration of lower doses of methadone leads to continued use of heroin. Therefore, it is expected that lower doses of LAAM will also be associated with problems of continued use of heroin. The most important contraindication to the use of methadone or LAAM is the absence of any physiological addiction to heroin or another short-acting opiate.

The side effects of LAAM have been shown to be similar to those on methadone, primarily excessive sweating, increased constipation, and some effects on libido and sexual function (JUDSON et al. 1983; KREEK 1973a,b, 1992a,b,c). It is not known whether tolerance will develop to these effects over time, as this has been shown during 3 years or more of long-term methadone maintenance treatment (KREEK 1973a,b). It has been suggested in some studies that on the day LAAM is received, stimulatory effects are perceived, whereas on the day of no medication, relative lethargy and depressed mood and activity may be observed (CROWLEY et al. 1979; JUDSON and GOLDSTEIN 1982). The long-term medical safety of methadone has been established by both 3 year and 10 year follow-up studies (KREEK 1973a,b; NOVICK et al. 1993). It is assumed that LAAM will be similarly safe; however, long-term follow-up studies have not been performed.

The treatment of any LAAM overdose would require an even longer or repeated administration of a specific opioid antagonist such as naloxone, naltrexone, or nalmefene than does methadone. If naloxone is used, the usual doses of 0.4 mg, or 0.01 mg/kg, should be administered intravenously and repeated in each 3–5 as required. Also, since naloxone has a half-life of only 2 h, readministration of naloxone in similar doses will be necessary every 2 h for up to 72 or even 96 h in the case of a LAAM overdose, as compared with 24–48 h for a methadone overdose. Use of a longer acting opioid antagonist such as naltrexone or nalmefene may prove more useful in such a setting.

The use of LAAM, when combined with appropriate social and psychological rehabilitation services, and access to health care, has been shown in short-term clinical research studies and longer term open clinical trials to be safe and effective for many heroin addicts. LAAM may provide a better treatment for some heroin addicts who have experienced early abstinence symptoms during methadone treatment. LAAM treatment requires only three clinic visits a week, which could decrease clinic loads. However, this advantage is offset by the fact that heroin addicts entering treatment need daily or 6 day/week visits for counseling and rehabilitation services. For some methadone maintained patients, LAAM seems to cause some side effects such as agitation and feelings of nervousness, which, for those patients, make methadone a preferable drug. Some patients receiving LAAM need or wish to return or change to treatment with methadone, and this can be accomplished with relative ease (Ling et al. 1980, 1984). Alternatively, LAAM treatment can be discontinued after slow dose reduction and elimination (Judson et al. 1983; Sorensen et al. 1982). At this time, LAAM, like methadone, may not be used in private practice. Methadone may be used in private practice only for the management of chronic pain, not for the management of chronic addiction. No medical maintenance programs can be developed yet for treatment with LAAM because of the absolute prohibition of any take home doses. In most individual states, the use of LAAM for the treatment of heroin addiction has not yet been approved. Since time of approval by the Food and Drug Administration in the summer of 1993, oversight for LAAM use in treatment of heroin addiction is being provided by the Health and Human Services Department, at this time through the supervision of the Food and Drug Administration using the regulations and guidelines which pertain to methadone as used in treatment of addictions, as well as by the Department of Justice with enforcement by the Drug Enforcement Agency (DEA) following the regulations which govern use of all Schedule II medications.

III. Neuroendocrine Effects of LAAM

Studies have shown that LAAM, like heroin throughout the stages of addiction and methadone during initial weeks of treatment until tolerance or adaptation develops, reduces release of the peptide hormone LH which controls both testosterone and estradiol secretion (Hargreaves et al. 1983; Mendelson et al. 1976, 1984). Studies have not been conducted to determine whether or not the stress responsive hypothalamic-pituitary-adrenal axis is altered during LAAM treatment, as it is during heroin treatment. However, it is anticipated that, as with methadone, sustained alterations or suppression of this axis will be present only during the induction phase, that is, during the first 2–3 months of LAAM treatment, with normalization thereafter. Further studies will have to be determined to assure this. The effects of LAAM on other aspects of neuroendocrine and immune function have not been rigorously assessed.

Acknowledgments. This work was conducted with support from NIH-NIDA P50-DAO5130, NIH-NIDA KO5-DA00049, and the New York State Office of Alcoholism and Substance Abuse Services. Appreciation is given to Ken Leung for manuscript preparation and reference collection and organization, Neil Maniar and Margaret Porter for proofreading and editing, Jennifer Sudul for final preparation of manuscript and Charlotte Kaiser for help in the literature search for the preparation of the bibliography of this chapter.

References

Abramowic M (1994) LAAM: A long-acting methadone for treatment of heroin addiction. Med Lett 36(924):52

Adelson MO, Shiloney E, Mana S, Kreek MJ (1996) Model treatment research unit patterned after most effective treatment facility following early experience in USA. In: Harris LS (ed) Problems of drug dependence 1995. Proceedings of the 57th Annual Scientific Meeting of the College on Problems of Drug Dependence, NIDA Research Monograph Series. Rockville, MD (in press)

Adham NF, Song MK, Eng BF (1978) Hyper-alpha-2-macroglobulinemia in narcotic addicts. Ann Intern Med 88:793–795

Ajuluchukwu DC, Brown LS Jr, Crummey FC, Foster KF Sr, Ismail YI, Siddiqui N (1993) Demographic, medical history and sexual correlates of HIV seropositive methadone maintained women. J Add Dis 12(4):105–120

Albeck H, Woodfield S, Kreek MJ (1989) Quantitative and pharmacokinetic analysis of naloxone in plasma using high performance liquid chromatography with electrochemical detection and solid phase extraction. J Chromatog 488:435–445

Anggard E, Gunne L-M, Holmstrand J, McMahon RE, Sandberg C-G, Sullivan HR (1974) Disposition of methadone in methadone maintenance. Clin Pharm Ther 17(3):258–266

Appel PW, Gordon NB (1974) Attentional function and monitoring of performance of methadone-maintained ex heroin addicts. Am Psych Association, New Orleans

Appel PW, Gordon NB (1976) Digit-symbol performance in methadone-treated ex-heroin addicts. Am J Psych 133:1337–1340

Azizi F, Vagenakis AG, Longcope C, Ingbar SH, Braverman LE (1973) Decreased serum testosterone concentration in male heroin and methadone addicts. Steroids 22:467–472

Azizi F, Vagenakis AG, Portnay GI, Braverman LE, Ingbar SH (1974) Thyroxine transport and metabolism in methadone and heroin addicts. Ann Intern Med 80:194–199

Babst DV, Chambers CD, Warner A (1971) Patients characteristics associated with retention in a methadone maintenance program. Br J Addict 66:195–204

Babst D, Newman S, Gordon NB, Warner A (1973) Driving record of methadone maintenance patients in New York State. NYS Addict Control Commission, Albany, New York

Baciewicz AM, Self TH, Bekemeyer WB (1987) Update on rifampin drug interactions. Arch Intern Med 147:565–568

Ball JC, Corty E, Bond H et al. (1987) The reduction of intravenous heroin use, non-opiate abuse, and crime during methadone maintenance treatment: Further findings. In: Harris LS (ed) Problems of drug dependence. Proceedings of the 49th Annual Scientific Meeting, the Committee on Problems of Drug Dependence Inc. NIDA Monograph DHHS Pub#(ADM)88-1564, Washington, DC. Supt Docs US Govt Print Off 81:224–230

Ball JC, Lange WR, Myers CP, Friedman SR (1988) Reducing the risk of AIDS through methadone maintenance treatment. J Health Soc Behav 29:214–226

Ball JC, Ross A (1991) The effectiveness of methadone maintenance treatment: Patients, programs, services, and outcome. Springer, Berlin Heidelberg New York

Bastomsky CH, Dent RRM (1976) Elevated serum concentrations of thyroxine-binding globulin and ceruloplasmin in methadone-maintained patients. Clin Res 24:655A

Bending MR, Skacel PO (1977) Rifampin and methadone withdrawal. Lancet 1:1211

Bercel NA (1948) Clinical trial of 10820, a new synthetic analgesic. Dis Nerv Sys 9:15

Berle BB, Nyswander M (1964) Ambulatory withdrawal treatment of heroin addicts. NYS J Med 7/15:1846–1848

Beverly CL, Kreek MJ, Wells AO, Curtis JL (1980) Effects of alcohol abuse on progression of liver disease in methadone-maintained patients. In: Harris LS (ed) Problems of drug dependence 1979. Proceedings of the 41st Annual Scientific Meeting of the Committee on Problems of Drug Dependence. NIDA Research Monograph Series, Rockville, MD DHHS Publication #(ADM)27, 399–401

Bihari B (1973) Alcoholism in M.M.T. patients: etiological factors and treatment approaches. In: Proceedings, Fifth National Conference on Methadone Treatment 1:288–295

Bihari B (1974) Alcoholism and methadone maintenance. Am J Drug Alcohol Abuse 1:79

Billings RE, McMahon RE, Blake DA (1974) L-acetylmethadol (LAAM) treatment of opiate dependence: plasma and urine levels of two pharmacologically active metabolites. Life Sci 14(8):1437–1446

Blaine JD, Renault P (eds) (1976) RX 3x a week LAAM: alternative to methadone. DHEW-NIDA Research Monograph, Series #8, Rockville, MD

Blaine JD, Renault P, Levine GL, Whysner JA (1978) Clinical use of LAAM. Ann NY Acad Sci 311:214–231

Blaine JD, Renault PR, Thomas DB, Whysner JA (1981) Clinical status of methadyl acetate (LAAM). Ann NY Acad Sci 362:101–115

Blinick G (1968) Menstrual function and pregnancy in narcotic addicts treated with methadone. Nature 219:180

Blix O (1988) AIDS and IV heroin addicts: the preventive effect of methadone maintenance in Sweden. Proceedings of the 4th International Conference on AIDS, Stockholm

Bockmuhl M, Erhart G (1948) Justus Liebigs Ann Chem 561:52–85

Bodner G, Albeck H, Soda KM, Kreek MJ (1990) Modulation of natural killer cell activity: Possible role of hypothalamic-pituitary-adrenal hormones. In: Van Ree JM, Mulder AH, Wiegant VM, Van Wimersma Greulanus TB (eds) New leads in opioid research. Excerpta Medica, Amsterdam, pp 330–331

Bodner G, Soda KM, Kennedy J, Kreek MJ (1991) Modulation of NK activity: role of neuroendocrine status. In: Harris LS (ed) Problems of drug dependence 1990. Proceedings of the 52nd Annual Scientific Meeting of the Committee on Problems of Drug Dependence. NIDA Research Monograph Series, Rockville, MD, DHHS Publication No. (ADM) 91-1753, 105:412–413

Bodner G, Pinto S, Albeck H, Kreek MJ (1992) Effects of dynorphin peptides on human natural killer cell activity in vitro. In: Harris LS (ed) Problems of drug dependence 1991. Proceedings of the 53rd Annual Scientific Meeting, The Committee on Problems of Drug Dependence, Inc. National Institute on Drug Abuse Monograph, Rockville, MD, DHHS Publication No. (ADM) 92-1888, 119:331

Borg L, Kreek MJ (1995) Clinical problems associated with interactions between methadone pharmacotherapy and medications used in the treatment of HIV-positive and AIDS patients. Current Opinion Psychiatry 8:199–202

Borg S, Kvande H, Rydberg U, Terenius L, Wahlstrom A (1982) Endorphin levels in human cerebrospinal fluid during alcohol intoxication and withdrawal. Psychopharmacology 78:101–103

Borg L, Ho A, Kreek MJ (1993) Availability of reliable serum methadone determination for management of symptomatic patients. In: Harris LS (ed) Problems of drug dependence 1992. Proceedings of the 54th Annual Scientific Meeting of the College on Problems of Drug Dependence. NIDA Research Monograph Series, Rockville, MD, DHHS Publication No. (ADM) 93-3505, 132:221

Borg L, Ho A, Kreek MJ (1994) Tuberculosis testing in an urban VA methadone clinic: Compliance and results. In: Harris LS (ed) Problems of Drug Dependence 1993. Proceedings of the 55th Annual Scientific Meeting of the College on Problems of Drug Dependence. NIDA Research Monograph Series. Rockville, MD NIH Publication #94-3749, 141:180

Borg L, Broe DM, Ho A, Kreek MJ (1995a) Cocaine abuse is decreased with effective methadone maintenance treatment at an urban Department of Veterans Affairs (DVA) Program. In: Harris LS (ed) Problems of drug dependence 1994. Proceedings of the 56th Annual Scientific Meeting of the College on Problems of Drug Dependence. NIDA Research Monograph Series, Rockville, MD, 153:17

Borg L, Ho A, Peters JE, Kreek MJ (1995b) Availability of reliable serum methadone determination for management of symptomatic patients. J Add Dis (in press)

Bowen DV, Smit ALC, Kreek MJ (1978) Fecal excretion of methadone and its metabolites in man: Application of GC-MS. In: Daly NR (ed) Advances in mass spectrometry. Heyden, Philadelphia, 7B:1634–1639

Branch AD, Unterwald EM, Lee SE, Kreek MJ (1992) Quantitation of preproenkephalin mRNA levels in brain regions from male Fischer rats following chronic cocaine treatment using a recently developed solution hybridization procedure. Mol Brain Res 14:231–238

Brown SB, Kozel NJ, Meyers MB, Dupont RL (1973) Use of alcohol by addict and nonaddict populations. Am J Psychiatry 130:599–601

Brown SM, Stimmel B, Taub RN, Kochwa S, Rosenfield RE (1974) Immunologic dysfunction in heroin addicts. Arch Intern Med 134:1001–1006

Brown LS, Kreek MJ, Trepo C, Chu A, Valdes M, Ajuluchukwu D, Phillips R, Primm BJ, Banks S (1990) Human immunodeficiency virus and viral hepatitis seroepidemiology in New York City intravenous drug abusers (IVDAs). In: Harris LS (ed) Problems of drug dependence 1989. Proceedings of the 51st Annual Scientific Meeting of the Committee on Problems of Drug Dependence. NIDA Research Monograph Series, Rockville, MD DHHS Publication No. (ADM) 90-1663, 95:443–444

Brown LS Jr, Hickson MJ, Ajuluchukwu DC, Bailey J (1993) Medical disorders in a cohort of New York City drug abusers: much more than HIV disease. J Add Dis 12(4):11–27

Bryant HU, Holaday JW (1991) Opioids in immunologic processes. In: Herz A (ed) Opioids II. Springer, Berlin Heidelberg New York, pp 361–392 (Handbook of experimental pharmacology, vol 104)

Cambor R, Ho A, Bodner G, Lampert S, Kennedy J, Kreek MJ (1992) Changes in clinical status of newly abstinent hospitalized cocaine users. In: Harris LS (ed) Problems of drug dependence 1991. Proceedings of the 53rd Annual Scientific Meeting, The Committee on Problems of Drug Dependence, Inc., National Institute on Drug Abuse Monograph Rockville, MD, DHHS Publication No. (ADM) 92-1888, 119:440

Caplehorn JR, Bell J (1991) Methadone dosage and retention of patients in maintenance. Med J Aust 154:195–199

Caplehorn JRM, Bell J, Kleinbaum DG, Gebski VJ (1993) Methadone dose and
 heroin use during maintenance treatment. Addiction 88:119–124
Chaisson RE, Bacchetti P, Osmond D, Brodie B, Sande MA, Moss AR (1989)
 Cocaine use and HIV infection in intravenous drug users in San Francisco.
 JAMA 261:561–565
Chambers CD, Babst DV, Warner A (1970) Characteristics predicting long-term
 retention in a methadone maintenance program. In: Proceedings of the Third
 National Conference on Methadone Treatment. US Government Printing Office,
 Washington, DC 140–143
Charuvastra CV, Panell J, Hopper M, Erhmann M, Blakis M, Ling W (1976) The
 medical safety of the combined usage of disulfiram and methadone. Arch Gen
 Psychiatry 33:391–393
Chen KK (1948) Pharmacology of methadone and related compounds. Ann NY
 Acad Sci 51:83–97
Chen Y, Mestek A, Liu J, Hurley JA, Yu L (1993a) Molecular cloning and functional
 expression of a mu opioid receptor from rat brain. Mol Pharm 44:8–12
Chen Y, Mestek A, Liu J, Yu L (1993b) Molecular cloning of a rat kappa opioid
 receptor reveals sequence similarities to the mu and delta opioid receptors.
 Biochem J 295:625–628
Cherubin CE (1967) The medical sequelae of narcotic addiction. Ann Intern Med
 67:23–33
Cherubin CE, Hargrove RL, Prince AM (1970) The serum hepatitis related antigen
 (SH) in illicit drug users. Am J Epidemiol 91:510–517
Cherubin CD, Kane S, Weinberger DR, Wolfe E, McGinn T (1972a) Persistence of
 transaminase abnormalities in former drug addicts. Ann Intern Med 76:385–389
Cherubin CE, Rosenthal WS, Stenger RE, Prince AM, Baden M, Strauss R,
 McGinn TC (1972b) Chronic liver disease in asymptomatic narcotic addicts.
 Ann Intern Med 76:391–395
Cherubin CE, Schaefer RA, Rosenthal WS, McGinn T, Forte F, Purcell R, Walsnsley
 P (1976) The natural history of liver disease in former drug users. Am J Med Sci
 272:244–253
Chou JZ, Kreek MJ, Chait BT (1993) Study of opioid peptides by laser desorption
 mass spectrometry. In: Harris LS (ed) Problems of drug dependence 1992.
 Proceedings of the 54th Annual Scientific Meeting of the College on Problems of
 Drug Dependence. NIDA Research Monograph Series, Rockville, MD DHHS
 Publication #(ADM)93-3505, 132:380
Chou JZ, Kreek MJ, Chait BT (1994) Matrix-assisted laser desorption mass spec-
 trometry of biotransformation products of dynorphin A in vitro. J Am Soc Mass
 Spectrom 5:10–16
Cicero TJ, Bell RD, Wiest WG, Allison JH, Polakoski K, Robins E (1975) Function
 of the male sex organs in heroin and methadone users. N Engl J Med 292:882–
 887
Cooper JR, Altman F, Brown BS, Czechowicz D (eds) (1983) Research on the
 treatment of narcotic addiction: state of the art. US Dept. Health and Human
 Services, Public Health Service, Alcohol, Drug Abuse, and Mental Health
 Administration, NIDA, Rockville, MD. Treatment Research Monograph Series,
 DHHS Publication No. (ADM) 83-1281
Corty E, Ball JC, Myers CP (1988) Psychological symptoms in methadone maintenance
 patients: prevalence and change over treatment. J Conslt Clin Psych 56(5):
 776–777
Crowley TJ, Jones RH, Hydinger-Macdonald MJ, Lingle JR, Wagner JE, Egan DJ
 (1979) Every-other-day acetylmethadol disturbs circadian cycles of human moti-
 lity. Psychopharm (Berlin) 62:151–155
Culpepper-Morgan JA, Inturrisi CE, Portenoy RK, Foley K, Houde RW, Marsh F,
 Kreek MJ (1992) Treatment of opioid induced constipation with oral naloxone:
 a pilot study. Clin Pharm Ther, 23:90–95

Culpepper-Morgan JA, Holt PR, LaRoche D, Kreek MJ (1995) Orally administered opioid antagonists reverse both mu and kappa opioid agonist delay of gastrointestinal transit in the guinea pig. Life Sci 56:1187–1192

Cushman P Jr (1972) Growth hormone in narcotic addiction. J Clin Endocrinol Metab 35:352–358

Cushman P Jr (1973a) Plasma testosterone in narcotic addiction. Am J Med 55: 452–458

Cushman P Jr (1973b) Persistent increased immunoglobulin M in treated narcotic addiction: Association with liver disease and continuing heroin use. J Allergy Clin Immunol 52:122–128

Cushman P Jr (1973c) Significance of hypermacroglobulinemia in methadone maintained and other narcotic addicts. Proceedings, Fifth National Conference on Methadone Treatment 1:515–522

Cushman P Jr (1974) Hyperimmunoglobulinemia in heroin addiction: some epidemiologic observations including some possible effects of route of administration and multiple drug abuse. Am J Epidemiol 99:218–224

Cushman P Jr (1981) Detoxification after methadone mainterance treatment. Ann NY Acad Sci 362:217–230

Cushman P Jr (1987) Alcohol and opioids: possible interactions of clinical importance. Adv Alcohol Subst Abuse 3:33–46

Cushman P, Dole VP (1973) Detoxification of rehabilitated methadone-maintained patients. JAMA 226(7):747–752

Cushman P, Kreek MJ (1974a) Some endocrinologic observations in narcotic addicts. In: Zimmerman E, George R (eds) Narcotic and the hypothalamus. Raven, New York, pp 161–173

Cushman P, Kreek MJ (1974b) Methadone-maintained patients. Effects of methadone on plasma testosterone, FSH, LH and prolactin. NY State J Med 74:1970–1973

Cushman P Jr, Sherman C (1974) Biologic false-positive reactions in serologic tests for syphilis in narcotic addiction. Am J Clin Pathol 61:346–351

Cushman P Jr, Bordier B, Hilton JG (1970) Hypothalamic-pituitary-adrenal axis in methadone-treated heroin addicts. J Clin Endocrinol Metab 30:24–29

Cushman P Jr, Gupta S, Grieco MH (1977) Immunological studies in methadone maintained patients. Int J Addict 12:241–253

Cushman P, Kreek MJ, Gordis E (1978) Ethanol and methadone in man: a possible drug interaction. Drug Acl Dep 3:35–42

David NA, Semler HJ, Burgner PR (1956) Control of chronic pain by DL-alpha acetylmethadol. JAMA 161(7):599–603

DesJarlais DC, Joseph H, Dole VP (1981) Long-term outcomes after termination from methadone maintenance treatment. Ann NY Acad Sci 362:231–238

DesJarlais DC, Joseph H, Dole VP, Schmeidler J (1983) Predicting post-treatment narcotic use among patients terminating from methadone maintenance. Adv Alc Sub Ab 2:57–68

DesJarlais DC, Marmor M, Cohen H, Yancovitz S, Garber J, Friedman S, Kreek MJ, Miescher A, Khuri E, Friedman SM, Rothenberg R, Echenberg D, O'Malley PO, Braff E, Chin J, Burtenol P, Sikes RK (1984) Antibodies to a retrovirus etiologically associated with Acquired Immunodeficiency Syndrome (AIDS) in populations with increased incidences of the syndrome. Morb Mort Wkly Rep 33:377–379

DesJarlais DC, Friedman SR, Novick DM, Sotheran JL, Thomas P, Yancovitz SR, Mildvan D, Weber J, Kreek MJ, Maslansky R, Bartelme S, Spira T, Marmor M (1989) HIV I Infection among intravenous drug users in Manhattan, New York City 1977 to 1987. JAMA 261:1008–1012

Des Jarlais DC, Wenston J, Friedman SR, Sotheran JL, Maslansky R, Marmor M (1992) Crack cocaine use in a cohort of methadone maintenance patients. J Sub Ab Tr 9:319–325

Dobbs WH (1971) Methadone treatment of heroin addicts. JAMA 218:1536–1541

Dole VP (1970) Biochemistry of addiction. Ann Rev Biochem 39:821–840

Dole VP (1973) Detoxification of methadone patients and public policy. JAMA 226:780–781

Dole VP (1988) Implications of methadone maintenance for theories of narcotic addiction (The Albert Lasker Medical Award). JAMA 260:3025–3029

Dole VP (1989) Methadone treatment and the acquired immunodeficiency syndrome epidemic. JAMA 262:1681–1682

Dole VP, Joseph JH (1977) Methadone maintenance: Outcome after termination. NY State J Med 77:1409–1412

Dole VP, Kreek MJ (1973) Methadone plasma level: Sustained by a reservoir of drug in tissue. Proc Natl Acad Sci USA 70(1):10

Dole VP, Nyswander ME (1965) A medical treatment for diacetylmorphine (heroin) addiction. JAMA 193:646–650

Dole VP, Nyswander ME (1966) Rehabilitation of heroin addicts after blockade with methadone. NYS J Med 66:2011–2017

Dole VP, Nyswander ME (1967) Rehabilitation of the street addict. Arch Env Health 14:477–480

Dole VP, Nyswander ME (1976) Methadone maintenance treatment: A ten-year perspective. JAMA 235:2117–2119

Dole VP, Nyswander ME, Kreek MJ (1966a) Narcotic blockade. Arch Intern Med 118:304–309

Dole VP, Nyswander ME, Kreek MJ (1966b) Narcotic blockade: a medical technique for stopping heroin use by addicts. Trans Assoc Am Physicians 79:122–136

Dole VP, Kim WK, Eglitis J (1966c) Detection of narcotic drugs, tranquilizers, amphetamines, and barbiturates in urine. JAMA 198:349–352

Dole VP, Nyswander ME, Warner A (1968) Successful treatment of 750 criminal addicts. JAMA 206:2708–2711

Dole VP, Robinson JW, Orraca J, Towns E, Searcy P, Caine E (1969) Methadone treatment on randomly selected criminal addicts. N Engl J Med 280:1372–1375

Dollery CT, Davies DS, Draffan GH, Dargie HJ, Dean CR, Reid JL, Clare RA, Murray S (1976) Clinical pharmacology and pharmacokinetics of clonidine. Clin Pharm Ther 19(1):11–17

Donahoe RM, Falek A (1988) Neuroimmunomodulation by opiates and other drugs of abuse: relationship to HIV infections and AIDS. Adv Biochem Psychopharmcol 44:145–158

Donahoe RM, Nicholson JKA, Madden JJ, Donahoe F, Shafer DA, Gordan D, Bokos P, Falek A (1986) Coordinate and independent effects of heroin, cocaine and alcohol abuse on T-cell E-Rosette formation and antigenic marker expression. Clin Immunol Immunopathol 41:254–264

Donahoe RM, Bueso-Ramos C, Falek A, McClure H, Nicholson JKA (1988) Comparative effects of morphine on leukocytic antigenic markers of monkeys and humans. J Neurosci Res 19:157–165

Drusin LM, Litwin SD, Armstrong D, Webster BP (1974) Waldenstrom's macroglobulinemia in a patient with a chronic biologic false-positive serologic test for syphilis. Am J Med 56:429–432

Dyke CV, Stesin A, Jones R, Chuntharapai A, Seaman W (1986) Cocaine increases natural killer cell activity. J Clin Invest 77:1387–1390

Eddy NB (1947) Morphine-like analgesic. J Pharm Assoc 8:536

Eddy NB, Touchberry CF, Lieberman JE (1949) Synthetic analgesics: I. Methadone isomers and derivatives 121–137

Eddy NB, May EL, Mosettig E (1952) Chemistry and pharmacology of the methadols and acetylmethadols. Comp Cl Bib 321–326

Eisenman AJ, Fraser HF, Sloan J, Isbell H (1958) Urinary 17-ketosteroid excretion during a cycle of addiction to morphine. JPET 124:305–311

Eisenman AJ, Fraser HF, Brooks JW (1961) Urinary excretion and plasma levels of 17-hydroxycorticosteroids during a cycle of addiction to morphine. JPET 132:226–231

Eisenman AJ, Sloan JW, Martin WR, Jasinski DR, Brooks JW (1969) Catecholamine and 17-hydroxycorticosteroid excretion during a cycle of morphine dependence in man. J Psych Res 7:19–28

Eisenstein TK, Rogers TJ, Bussiere JL, Rojavin M, Szabo I, Meissler JJ, Belkowski SM, Geller EB, Adler MW, Friedman H, Peterson PH, Molitor T, Chao C, Soderberg L, Brown LS, Kreek MJ (1994) Drugs of abuse and immunosuppression. In: Harris LS (ed) Problems of drug dependence 1993. Proceedings of the 55th Annual Scientific Meeting of the College on Problems of Drug Dependence. NIDA Research Monograph Series, Rockville, MD, NIH Publication No. 94-3748, 140:89–92

Eisler O, Schaumann O (1939) Dolantin ein neuartiges Spasmolytikum und Analgetikum. Deut Med Wochenschr 65:967

Emrich HM, Nusselt L, Gramsch C, John S (1983) Heroin addiction: beta-endorphin immunoreactivity in plasma increases during withdrawal. Pharmacopsychiatria 16:93–96

Espejo R, Hogben G, Stimmel B (1973) Sexual performance of men on methadone maintenance. Proceedings, Fifth National Conference on Methadone Treatment 1:490–493

Evans CJ, Keith DE Jr, Morrison H, Magendzo K, Edwards RH (1992) The delta-opioid receptor isolation of a cDNA by expression cloning and pharmacological characterization. Proc Natl Acad Sci 89:12048–12052

Facchinetti F, Grasso A, Petraglia F, Perrini D, Volpe A, Genazzani AR (1984) Impaired circadian rhythmicity of beta-lipotrophin, beta-endorphin and ACTH in heroin addicts. Acts Endocrin 105:149–155

Falek A, Madden JJ, Shafter DA, Donahoe R (1986) Individual differences in opiate-induced alterations at the cytogenetic DNA repair, and immunologic levels: opportunity for genetic assessment. In: Genetic and Biological Markers in Drug Abuse and Alcoholism, National Institute on Drug Abuse Research Monograph No. 66:11–24

Finelli PF (1976) Phenytoin and methadone tolerance. N Engl J Med 294:227

Finnegan JK, Haag HB, Larson PS, Dreyfuss ML (1948) Observations on the comparative pharmacologic actions of 6-dimethylamino-4,4-diphenyl-heptanone-3 (amidone) and morphine. JPET 92:269

Finnegan LP, Chappel JN, Kreek MJ, Stimmel B, Stryker J (1982) Narcotic addiction in pregnancy. In: Niebyl JR (ed) Drug use in Pregnancy. Lea and Febiger, Philadelphia, pp 163–184

Fraser HF, Isbell H (1951) Addiction potentialities of isomers of 6-*di*-methylamino-4-4-diphenyl-3 acetyoxy-heptane (acetylmethadol). JPET 101:12

Fraser HF, Isbell H (1952) Actions and addiction liabilitites of alpha-acetylmethadols in man. JPET 105:210–215

Gearing FR (1970) Evaluation of methadone maintenance treatment programs. Int J Add 5:517–543

Gearing FR, Schweitzer MD (1974) An epidemiologic evaluation of long-term methadone maintenance treatment for heroin addiction. Am J Epidemiol 100:101–112

Geller SA, Stimmel B (1973) Diagnostic confusion from lymphatic lesions in heroin addicts. Ann Intern Med 78:703–705

Girardi E, Zaccarelli M, Tossini G, Puro V, Narciso P, Visco G (1990) Hepatitis C virus infection in intravenous drug users: prevalence and risk factors. Scand J Infect Dis 22:751–752

Goldstein A (1971) Blind comparison of once-daily and twice-daily dosage schedules in a methadone program. Clin Pharmacol Ther 3(1):59–63

Goldstein A (1972) Heroin addiction and the role of methadone in its treatment. Arch Gen Psych 26:291–297

Goldstein A (1994) Addiction: from biology to drug policy. Freeman, New York

Goldstein A, Lowney LT, Pal BK (1971) Stereospecific and nonspecific interactions of the morphine congener levorphanol in subcellular fractions of mouse brain. Proc Nat Acad Sci USA 68:1742–1747

Gordis E (1988) Methadone maintenance patients in alcoholism treatment. Alcohol Alert 1:1–5

Gordis E, Kreek MJ (1977) Alcoholism and narcotic addiction in pregnancy. Current Probl Obstet Gynecol 1:1–48

Gordon NB (1973) The functional status of the methadone maintained person. In: Simmons LRS, Gold MB (eds) Discrimination and the addict. Sage, Beverly Hills, pp 101–121

Gordon NB (1976) Influence of narcotic drugs on highway safety. Accid Anal Prev 8:3–7

Gordon NB, Appel PW (1972) Performance effectiveness in relation to methadone treatment. Proceedings of the 4th National Conference on Methadone Treatment. Ann NY Acad Sci 425–427

Gordon NB, Warner A, Henderson (Ho) A (1967) Psychomotor and intellectual performance under methadone maintenance. Bull Prob Depend 5136–5144

Grevert P, Masover B, Goldstein A (1977) Failure of methadone and levomethadyl acetate (Levo-alpha-acetylmethadol) maintenance to affect memory. Arch Gen Psych 34:849–853

Gritz ER, Shiffman SM, Jarvik ME, Haber J, Dymond AM, Coger R, Charuvastra V, Schlesinger J (1975) Physiological and psychological effects of methadone in man. Arch Gen Psychiatry 32:237–242

Gruber CM, Baptisti A (1963) Acceptability of morphine and noracymethadol. Clin Pharm Ther 4:172–181

Hachey DL, Kreek MJ, Mattson DH (1977) Quantitative analysis of methadone in biological fluids using deuterium-labelled methadone and GLC-chemical-ionization mass spectrometry. J Pharm Sci 66:1579–1582

Hanbury R, Sturiano V, Cohen M, Stimmel B, Aguillaume C (1986) Cocaine use in persons on methadone maintenance. Adv Alcohol Subst Abuse 6:97–106

Hahn EF, Lahita R, Kreek MJ, Duma C, Inturrisi CE (1983) Naloxone radioimmunoassay: an improved antiserum. J Pharm Pharmacol 35:833–836

Han JS (1993) Acupuncture and stimulation-produced analgesia. In: Herz A (ed) Opioids II. Springer, Berlin Heidelberg New York, pp 105–126 (Handbook of Experimental Pharmacology, vol 104)

Hargreeves WA, Tyler J, Weinberg JA, Benowitz N (1983) [−]-alpha acetylmethadol effects on alcohol and diazepam use, sexual function and cardiac function. Dr Alc Dep 12(4):323–332

Harte EH, Gutjahr CL, Kreek MJ (1976) Long-term persistence of dl-methadone in tissues. Clin Res 24:623A

Hartman N, Kreek MJ (1983) Narcotic poisoning. In: Current Therapy 896–898

Hartman N, Kreek MJ, Ross A, Khuri E, Millman RB, Rodriguez R (1983) Alcohol use in youthful methadone maintained former heroin addicts: Liver impairment and treatment outcome. Alc Clin Exp Res 7:316–320

Hellman L, Fukushima DK, Roffwarg H, Fishman J (1975) Changes in estradiol and cortisol production rates in men under the influence of narcotics. J Clin Endocrinol Metabl 41:1014–1019

Henderson A, Nemes G, Gordon NB, Roos L (1972) Sleep and narcotic tolerance (abstract). Psychophysiology 7:346–347

Henderson GL, Wilson K, Lau DHM (1976) Plasma l-α-acetylmethadol (LAAM) after acute and chronic administration. Clin Pharm Ther 21(1):16–25

Himmelsbach CK (1941) The morphine abstinence syndrome, its nature and treatment. Ann Intern Med 15:829–839

Ho A, Dole VP (1979) Pain perception in drug-free and in methadone maintained human ex-addicts. Proc Soc Exp Bio Med 162:392–395

Ho WKK, Wen HL, Ling H (1980) Beta-endorphin-like immunoactivity in the plasma of heroin addicts and normal subjects. Neuropharm 19:117–120

Ho A, Cambor R, Bodner G, Kreek MJ (1992) Intensity of craving is independent of depression in newly abstinent chronic cocaine users. In: Harris LS (ed) Problems of drug dependence 1991. Proceedings of the 53rd Annual Scientific Meeting, The

Committee on Problems of Drug Dependence, Inc. National Institute on Drug Abuse Monograph, Rockville, MD, DHHS Publication No. (ADM) 92-1888, 119:441

Holmstrand J, Gunne LM, Wahlstrom A (1981) CSF-endorphins in heroin addicts during methadone maintenance and during withdrawal. Pharmacopsychiatria 14:126–128

Hunt DE, Lipton DS, Goldsmith D, Strug D (1984) Street pharmacology: Uses of cocaine and heroin in the treatment of addiction. Drug Alcohol Depend 13: 375–387

Ingoglia NA, Dole VP (1970) Localization of *d* and *l*-methadone after intraventricular injection into rat brain. JPET 175:84–87

Inturrisi CE, Verebey K (1972a) A gas-liquid chromatographic method for the quantitative determination of methadone in human plasma and urine. J Chromatogr 65:361–369

Inturrisi CE, Verebey K (1972b) The levels of methadone in the plasma in methadone maintenance. Clin Pharm Ther 13:633

Inwang EE, Primm BJ, Jones FL, Dekirmenjian H, Davis JM, Henderson CT (1976) Metabolic disposition of 2 phenylethylamine and the role of depression in methadone dependent and detoxified patients. Drug and Alcohol Dependence 1(4):295–303

Isbell H, Eisenman AJ (1948) Physical dependence liability of drugs of the methadone series and of 6-methyldihydromorphine. Fed Proc 7:162

Isbell H, Wikler A, Eisenman AJ, Frank K (1947a) Effect of single doses of 10820 (4,4-diphenyl-6-dimethylamino-heptanone-3) on man. Fed Proc 6:341

Isbell H, Eisenman AJ, Wikler A, Daingerfield M, Frank K (1947b) Treatment of the morphine abstinence syndrome with 10820 (4,4-diphenyl-6-dimethylamino-heptanone-3). Fed Proc 6:340

Isbell H, Eisenman AJ, Wikler A, Daingerfield M, Frank K (1947c) Experimental addiction to 10820 (4,4-diphenyl-6-dimethylamino-heptanone-3) in man. Fed Proc 6:264

Isbell H, Wikler A, Eddy NB, Wilson JL, Moran CF (1947d) Tolerance and addiction liability of 6-dimethylamino-4,4-diphenyl-heptanone-3 (methadon). JAMA 135:888

Isbell H, Eisenman AJ, Wikler A, Frank K (1948) The effects of single doses of 6-dimethylamino-4-4-diphenyl-3-heptanone (amidone, methadone, or "10820") on human subjects. JPET 92:83

Jackson GW, Richman A (1973) Alcohol use among narcotic addicts. Alcohol Health Res World 1:25

Jackson G, Korts D, Hanbury R, Sturiano V, Wolpert L, Cohen M, Stimmel B (1982) Alcohol consumption in persons on methadone maintenance therapy, Am J Drug Alcohol Abuse 9:69–76

Jacob H, Charytan C, Rascoff JM, Golden R, Janis R (1978) Amyloidosis secondary to drug abuse and chronic skin suppuration. Arch Inter Med 138:1150–1151

Jaffe JH (1970) Further experience with methadone in the treatment of narcotics users. Int J Addict 5(3):375–389

Jaffe JH, Senay EC (1971) Methadone and L-methadyl acetate: use in management of narcotic addicts. JAMA 216(8):1303–1305

Jaffe JH, Schuster CR, Smith BB, Blachley PH (1970) Comparison of acetylmethadol and methadone in the treatment of long-term heroin users. JAMA 211:1834–1836

Jaffe JH, Senay EC, Schuster CR, Renault PR, Smith B, DiMenza S (1972) Methadyl acetate v. methadone: a double-blind study in heroin users. JAMA 222(4):437–442

Jaffe JH, Schuster CR, Smith BB, Blachley PH (1979) Comparison of acetylmethadol and methadone in the treatment of long-term heroin users. JAMA 1834–1836

Jenney EH, Pfeiffer CC (1948) Comparative analgesic and toxic effects of the optical isomers of methadon and isomethadon. Fed Proc 7:231

Jersild T, Johansen C, Balslov JT, Hojgaard K, Johansen A, Ott C (1970) Hepatitis in young drug users. Scand J Gastroenterol (Suppl) 7:79

550   M.J. KREEK

Johanson CE, Fischman MW (1989) The pharmacology of cocaine related to its abuse. Pharm Rev 41:3–52

Joseph H (1985) Alcoholism and methadone treatment consequences for the patients and program. Am J Drug Alcohol Abuse 11:37–53

Judson BA, Goldstein A (1979) Levo-alpha-acetylmethadol (LAAM) in the treatment of heroin addicts I. Dosage schedule for induction and stabilization. Dr Alc Dep 4(6):461–466

Judson BA, Goldstein A (1982) Symptom complaints of patients maintained on methadone LAAM (methadyl acetate), and naltrexone at different times in their addiction careers. Dr Alc Dep 10(2–3):269–282

Judson BA, Ortiz S, Crouse L, Carney TM, Goldstein A (1980) A follow-up study of heroin addicts five years after first admission to a methadone treatment program. Drug Alcohol Depend 6:295–313

Judson BA, Goldstein A, Inturrisi CE (1983) Methadyl acetate (LAAM) in the treatment of heroin addicts II. Double-blind comparison of gradual and abrupt detoxification. Arch Gen Psych 40(8):834–840

Kaiko RF, Inturrisi CE (1973) A gas-liquid chromatographic method for the quantitative determination of acetylmethadol and its metabolites in human urine. J Chrom 82:315–321

Kaiko RF, Inturrisi CE (1975) Disposition of acetylmethadol in relation to pharmacologic action. Clin Pharm Ther 18(1):96–103

Karr NW (1947) Effects of 6-dimethylamino-4-4-diphenyl-3-heptanone (Dolophine) on intestinal motility. Fed Proc 6:343

Kaufmann CA, Kreek MJ, Raghunath J, Arns P (1983) Methadone, monoamine oxidase and depression: opioid distribution and acute effects on enzyme activity. Biol Psychiatry 18:1007–1021

Kaufmann CA, Kreek MJ, Karoum F, Chuang LW (1984) Depression during methadone withdrawal: no role for beta-phenylethylamine. Drug Alcohol Dependence 13:21–29

Keats AS, Beecher HK (1952) Analgesic activity and toxic effects of acetylmethadol isomers in man. JPET 105:210–215

Kieffer BL, Befort K, Gaveriaux-Ruff C, Hirth CG (1992) The delta-opioid receptor: Isolation of a cDNA by expression cloning and pharmacological characterization. Proc Natl Acad Sci 89:12048–12052

Kennedy JA, Hartman N, Sbriglio R, Khuri E, Kreek MJ (1990) Metyrapone-induced withdrawal symptoms. Br J Addict 85:1133–1140

Khuri ET, Millman RB, Hartman N, Kreek MJ (1984) Clinical issues concerning alcoholic youthful narcotic abusers. In: Advances in alcohol and substance abuse. Haworth, New York 3:69–86

Kirchhof AC, Uchiyama JK (1947) Spasmolytic action of 6-dimethylamino-4-4-diphenyl-3-heptanone (dolophine), a synthetic analgesic. Fed Proc 6:345

Kleber HD, Riordan CE, Rounsaville B, Kosten T, Charney D, Gaspari J, Hogan I, O'Connor C (1985) Clonidine in outpatient detoxification from methadone maintenance. Arch Gen Psych 42:391–394

Kosten TR, Kreek MJ, Swift C, Carney MK, Ferdinands L (1987) Beta-endorphin levels in CSF during methadone maintenance. Life Sci 41:1071–1076

Kosten TR, Krystal JH, Charney DS, Price LH, Morgan CH, Kleber HD (1989a) Letters to the editor: rapid detoxification from opioid dependence. Am J Psych 146(10):1349

Kosten TR, Kleber HK, Morgan C (1989b) Role of opioid antagonists in treating intravenous cocaine abuse. Life Sci 44:887–892

Kothur R, Marsh F, Posner G (1991) Live function tests in nonparenteral cocaine users. Arch Intern Med 151:1126–1128

Kreek MJ (1972) Medical safety, side effects and toxicity of methadone. In: Proceedings of the Fourth National Conference on Methadone Treatment NAPAN-NIMH, pp 171–174

Kreek MJ (1973a) Medical safety and side effects of methadone in tolerant individuals. JAMA 223:665–668

Kreek MJ (1973b) Physiological implications of methadone treatment. In: Procedings, Fifth National Conference on Methadone Treatment. National Association for the Prevention of Addiction to Narcotics 2:85–91

Kreek MJ (1973c) Plasma and urine levels of methadone. NYS J Med 73(23): 2773–2777

Kreek MJ (1975) Methadone maintenance treatment for chronic opiate addiction. In: Richter R (ed) Medical aspects of drug abuse. Harper and Row, New York, pp 167–185

Kreek MJ (1978a) Medical complications in methadone patients. Ann NY Acad Sci 311:110–134

Kreek MJ (1978b) Effects of drugs and alcohol on opiate disposition and action. In: Adler MW, Manara L, Samnin R (eds) Factors affecting the action of narcotics. Raven, New York, pp 717–739

Kreek MJ (1979a) Methadone in treatment: physiological and pharmacological issues. In: Dupont RL, Goldstein A, O'Connell J (eds) Handbook on drug abuse, NIDA-ADAMHA-DEW-ODAP Executive Office of the President, 57–86

Kreek MJ (1979b) Methadone disposition during the perinatal period in humans. Pharm Biochem Behav 11:1–7

Kreek MJ (1981) Metabolic interactions between opiates and alcohol. Ann NY Acad Sci 362:36–49

Kreek MJ (1981–2) Disposition of narcotics in the perinatal period. In: Publication of AMERSA and The Career Teacher Program in Alcohol and Drug Abuse, 3:7–10

Kreek MJ (1983a) Discussion on clinical perinatal and developmental effects of methadone. In: Cooper JR, Altman F, Brown BS, Czechowicz (eds) Research in the Treatment of Narcotic Addiction: State of the Art. NIDA Monograph DHHS Pub #(ADM) 83-1281, pp 444–453

Kreek MJ (1983b) Health consequences associated with use of mathadone. In: Cooper JR, Altman F, Brown BS, and Czechowicz D (eds) Research in the treatment of narcotic addiction: State of the art. NIDA Monograph DHHS Pub#(ADM)83-1281, 456–482

Kreek MJ (1984) Opioid interactions with alcohol. J Add Dis 3:35–46

Kreek MJ (1986a) Exogenous opioids: Drug-disease interactions. In: Foley KM and Inturrisi CE (eds) Advances in pain research and therapy. Raven Press, New York 8:201–210

Kreek MJ (1986b) Drug Interactions with methadone in humans. In: Braude MC and Ginzburg HM (eds) Strategies for research on the interactions of drugs of abuse. NIDA Research Monograph, Rockville, MD 68:193–225

Kreek MJ (1987) Multiple drug abuse patterns and medical consequences. In: Meltzer HY (ed) Psychopharmacology: the third generation of progress. Raven, New York, pp 1597–1604

Kreek MJ (1988) Opiate-ethanol interactions: Implications for the biological basis and treatment of combined addictive diseases. In: Harris LS (ed) Problems of drug dependence 1987. Proceedings of the 49th Annual Scientific Meeting of the Committee on Problems of Drug Dependence. NIDA Research Monograph Series, Rockville, MD DHHS Publication No. (ADM) 88-1564, 81:428–439

Kreek MJ (1989) Immunological approaches to clinical issues in drug abuse. In: Harris LS (ed) Problems of drug dependence 1988. Proceedings of the 40th Annual Scientific Meeting of the Committee on Problems of Drug Dependence. NIDA Research Monograph Series, Rockville, MD DHHS Publication No. (ADM) 89-1605, 90:77–86

Kreek MJ (1990a) Drug interactions in humans related to drug abuse and its treatment. Modern Methods in Pharmacology 6:265–282

Kreek MJ (1990b) Historical and medical aspects of methadone maintenance treatment: Effectiveness in treatment of heroin addiction and implications of such treatment in the setting of the AIDS epidemic. In: Adamsson C, Jansson B, Rydberg U, Westrin C (eds) Evaluation of different programmes for treatment of drug addicts. Mediciniska Forskningsradet, Stockholm, Sweden, pp 61–76

Kreek MJ (1990c) HIV infection and parenteral drug abuse; ethical issues in diagnosis, treatment, research and the maintenance of confidentiality. In: Allebeck P, Jansson B (eds) Proceedings of the Third International Congress on Ethics in Medicine – Nobel Conference Series. Raven, New York, pp 181–187

Kreek MJ (1990d) Immune function in heroin addicts and former heroin addicts in treatment: Pre/post AIDS epidemic. In: Pham PTK, Rice K (eds) Current chemical and pharmacological advances on drugs of abuse which alter immune function and their impact upon HIV infection. NIDA Research Monograph Series, Rockville, MD 96:192–219

Kreek MJ (1991a) Using methadone effectively: Achieving goals by application of laboratory, clinical and evaluation research and by development of innovative programs. In: Pickens R, Leukefeld C, Schuster R (eds) Improving drug abuse treatment. NIDA Research Monograph Series, Rockville, MD 245–266

Kreek MJ (1991b) Methadone maintenance treatment for harm reduction approach to heroin addiction. In: Loimer N, Schmid R, Springer A (eds) Drug addiction and AIDS. Springer, Vienna New York, pp 153–177

Kreek MJ (1991c) Multiple drug abuse patterns: Recent trends and associated medical consequences. Advances in substance abuse: behavioral and biological research. Kingsley, London 4:91–111

Kreek MJ (1991d) Immunological function in active heroin addicts and methadone maintained former addicts: Observations and possible mechanisms. In: Harris LS (ed) Problems of drug dependence 1990. Proceedings of the 52nd Annual Scientific Meeting of the Committee on Problems of Drug Dependence. NIDA Research Monograph Series, Rockville, MD, DHHS Publication No. (ADM) 91-1753, 105:75–81

Kreek MJ (1991e) Heroin, other opiates, and the immune function. In: Drug abuse and drug abuse research. The Third Triennial Report to Congress from the Secretary, Department of Health and Human Services, Rockville, MD DHHS Publication No. (ADM) 91-1704

Kreek MJ (1992a) The addict as a patient. In: Lowinson JH, Ruiz P, Millman RB, Langrod JG (eds) Substance abuse: a comprehensive textbook. Williams and Wilkins, Baltimore, pp 997–1009

Kreek MJ (1992b) Epilogue: medical maintenance treatment for heroin addiction, from a retrospective and prospective viewpoint. In: State methadone maintenance treatment guidelines. Office for Treatment Improvement, Division for State Assistance 255–272

Kreek MJ (1992c) Rationale for maintenance pharmacotherapy of opiate dependence. Res Pub Assoc Res Nerv Ment Dis 70:205–230

Kreek MJ (1992d) Effects of opiates, opioid antagonists, and cocaine on the endogenous opioid system: clinical and laboratory studies. In: Haris LS (ed) Problems of drug dependence 1991. Proceedings of the 53rd Annual Scientific Meeting, The Committee on Problems of Drug Dependence, Inc. National Institute on Drug Abuse Monograph, Rockville, MD, DHHS Publication No. (ADM) 92-1888, 119:44–48

Kreek MJ (1992e) Effects of drugs of abuse and treatment agents in women. In: Harris LS (ed) Problems of drug dependence 1991. Proceedings of the 53rd Annual Scientific Meeting, the Committee on Problems of Drug Dependence Inc. NIDA Monograph, Rockville, MD DHHS Publication #(ADM) 92-1888, 119:106–110

Kreek MJ (1992f) Pharmacological treatment of opioid dependency. In: Buhringer G, Platt J (eds) Drug abuse and treatment: German and American Perspectives. Krieger, Malabar, pp 398–406

Kreek MJ (1993) Immune function in human IVDU's. In: Harris LS (ed) Problems of drug dependence, 1992; Proceedings of the 54th Annual Scientific Meeting of the College on Problems of Drug Dependence. NIDA Research Monograph Series, Rockville, MD, DHHS Publication No. (ADM) 93-3505, 132:72

Kreek MJ (1994a) Pharmacology and medical aspects of methadone treatment. In: Rettig RA, Yarmolinsky A (eds) *Federal regulation of methadone treatment.* National Academy of Sciences, National Academy Press

Kreek MJ (1994b) Pharmacotherapy of opioid dependence: Rationale and update. Regulatory peptides: Proceedings of the 25th International Narcotics Research Conference (INRC) S1:S255–256

Kreek MJ, Hartman N (1982) Chronic use of opioids and antipsychotic drugs: side effects, effects on endogenous opioids, and toxicity. Ann NY Acad Sci 151–172

Kreek MJ, Culpepper-Morgan J (1991) Neuroendocrine (HPA) and gastrointestinal effects of opiate antagonists: possible therapeutic application. In: Harris LS (ed) Problems of drug dependence, 1990: Proceedings of the 52nd Annual Scientific Meeting of the Committee on Problems of Drug Dependence. NIDA Research Monograph Series. Rockville, MD, DHHS Publication No. (ADM) 91-1753, 105:168–174

Kreek MJ, Dodes L, Kane S, Knobler J, Martin R (1972) Long-term methadone maintenance therapy: effects on liver function. Ann Intern Med 77:598–602

Kreek MJ, Schecter A, Gutjahr CL, Bowen D, Field F, Queenan J, Merkatz I (1974) Analyses of methadone and other drugs in maternal and neonatal body fluids: use in evaluation of symptoms in a neonate of mother maintained on methadone. Am J Dr Alc Ab 1:409–419

Kreek MJ, Garfield JW, Gutjahr CL, Giusti LM (1976a) Rifampin-induced methadone withdrawal. N Engl J Med 294:1104–1106

Kreek MJ, Gutjahr CL, Garfield JW, Bowen DV, Field FH (1976b) Drug interactions with methadone. Ann NY Acad Sci 281:350–374

Kreek MJ, Oratz M, Rothschild MA (1978a) Hepatic extraction of long and short-acting narcotics in the isolated perfused rabbit liver. Gastroenterology 75:88–94

Kreek MJ, Gutjahr CL, Bowen DV, Field FH (1978b) Fecal excretion of methadone and its metabolites: A major pathway of elimination in man. In: Schecter A, Alksne H, Kaufman E (eds) Critical Concerns in the Field of Drug Abuse: Proceedings of the Third National Drug Abuse Conference. Dekker, New York, pp 1206–1210

Kreek MJ, Hachey DL, Klein PD (1979) Stereoselective disposition of methadone in man. Life Sci 24:925–932

Kreek MJ, Kalisman M, Irwin M, Jaffery NF, Scheflan M (1980a) Biliary secretion of methadone and methadone metabolites in man. Res Commun Chem Pathol Pharmacol 29:67–78

Kreek MJ, Bencsath FA, Field FH (1980b) Effects of liver disease on urinary excretion of methadone and metabolites in maintenance patients: Quantitation by direct probe chemical ionization mass spectrometry. Biomed Mass Spec 7:385–395

Kreek MJ, Schecter AJ, Gutjahr CL, Hecht M (1980c) Methadone use in patients with chronic renal disease. Drug Alcohol Depend 5:197–205

Kreek MJ, Wardlaw SL, Friedman J, Schneider B, Frantz AG (1981) Effects of chronic exogenous opioid administration on levels of one endogenous opioid (beta-endorphin) in man. In: Simon E, Takagi H (eds) Advances in endogenous and exogenous opioids. Kodansha Ltd. Tokyo, Japan, pp 364–366

Kreek MJ, Bencsath FA, Fanizza A, Field FH (1983a) Effects of liver disease on fecal excretion of methadone and its unconjugated metabolites in maintenance patients: quantitation by direct probe chemical ionization mass spectrometry. Biomed Mass Spectrm 10:544–549

Kreek MJ, Schaefer RA, Hahn EF, Fishman J (1983b) Naloxone, a specific opioid antagonist, reverses chronic idiopathic constipation. Lancet 2/5:261–262

Kreek MJ, Schaefer RA, Hahn EF, Fishman J (1983c) Naloxone in chronic constipation. Letters to the editors. Lancet 758

Kreek MJ, Wardlaw SL, Hartman N, Raghunath J, Friedman J, Schneider B, Frantz AG (1983d) Circadian rhythms and levels of beta-endorphin, ACTH, and cortisol during chronic methadone maintenance treatment in humans. Life Sci Sup I 33:409–411

Kreek MJ, Raghunath J, Plevy S, Hamer D, Schneider B, Hartman N (1984) ACTH, cortisol and beta-endorphin response to metyrapone testing during chronic methadone maintenance treatment in humans. Neuropeptides 5:277–278

Kreek MJ, Khuri E, Fahey L, Miescher A, Arns P, Spagnoli D, Craig J, Millman R, Harte E (1986) Long-term follow-up studies of the medical status of adolescent former heroin addicts in chronic methadone maintenance treatment: Liver disease and immune status. In: Harris LS (ed) Problems of Drug Dependence 1985, Proceedings of the 47th Annual Scientific Meeting of the Committee on Problem of Drug Dependence. NIDA Research Monograph Series, Rockville, MD DHHS Publication #(ADM) 86-1448, 67:307–309

Kreek MJ, Khuri E, Flomenberg N, Albeck H, Ochshorn M (1989) Immune status of unselected methadone maintained former heroin addicts. In: Quirion R, Jhamandas K, Gianoulakis C (eds) International Narcotics Research Conference 1989 (INRC). Liss, New York, pp 445–448

Kreek MJ, DesJarlais DC, Trepo CL, Novick DM, Abdul-Quader A, Raghunath J (1990a) Contrasting prevalence of delta hepatitis markers in parenteral drug abusers with and without AIDS. J Infect Dis 162:538–541

Kreek MJ, Khuri E, Melia D (1990b) Current patterns of hepatitis B and delta markers in former heroin addicts in methadone maintenance treatment: Need for hepatitis B vaccination. Abstracts of International Association for Study of the Liver Meeting, Queensland, Australia

Kreek MJ, Ho A, Borg L (1994) Dynorphin A_{1-13} causes elevation of serum levels of prolactin in human subjects. In: Harris LS (ed) Problems of Drug Dependence 1993, Proceedings of the 55th Annual Scientific Meeting of the College on Problems of Drug Dependence. NIDA Research Monograph Series, Rockville, MD NIH Publication #94-3749, 141:108

Lafisca S, Bolelli G, Franceschetti F, Filicori M, Flamigini C, Marigo M (1981) Hormone levels in methadone-treated drug addicts. Dr Alc Dep 8:229–234

Lazzarin A, Mella L, Trombini M, Ubert-Foppa C, Franzetti F, Mazzoni G, Galli M (1984) Immunological status in heroin addicts: Effects of methadone maintenance treatment. Drug Alcohol Depend 13:117–123

Lenn NJ, Senay EC, Renault PF, Devel RK (1975) Neurological assessment of patients on prolonged methadone maintenance. Drug Alcohol Dependence 1:305–311

Levine R, Zaks A, Fink M, Freedman AM (1973) Levomethadyl acetate: Prolonged duration of opioid effects, including cross tolerance to heroin in man. JAMA 226(3):316–318

Liebson I, Bigelow G (1972) A behavioral-pharmacological treatment of dually addicted patients. Behav Res Ther 10:403

Ling W, Charuvastra VC, Klett CJ (1975) Current status of the evaluation of LAAM as a maintenance drug for heroin addicts. Am J Dr Alc Ab 2(3–4):307–315

Ling W, Charuvastra C, Kaim SC, Klett CJ (1976) Methadyl acetate and methadone as maintenance treatments for heroin addicts: A Veterans Administration cooperative study. Arch Gen Psych 33(6):709–720

Ling W, Klett CJ, Gillis RD (1978) A cooperative clinical study of methadyl acetate I. 3x/week regimen. Arch Gen Psych 35(3):345–353

Ling W, Blakis M, Holmes ED, Klett CJ, Carter WE (1980) Restabilization with methadone after methadyl acetate maintenance. Arch Gen Psych 37(2):194–196

Ling W, Dorus W, Hargreaves WA, Resnick R, Senay E, Tusson VB (1984) Alternative induction and crossover schedules for methadyl acetate. Arch Gen Psych 41(2):193–199

Loimer N, Lenz K, Presslich O, Schmid R (1990) Rapid transition from methadone maintenance to naltrexone. Lancet 335:111

Loimer N, Lenz K, Schmid R, Presslich O (1991) Technique for greatly shortening the transition from methadone to naltrexone maintenance of patients addicted to opiates. Am J Psych 148(7):933–935

Louria DB, Hensle T, Rose J (1967) The major complications of heroin addiction. Ann Intern Med 67:1–22

Maddux JF, Elliot B (1975) Problem drinkers among patients on methadone. Am J Dr Alc Ab 2:245

Marcovici M, McLellan AT, O'Brien CP, Rosenzweig J (1980) Risk for alcoholism and methadone treatment, a longitudinal study. J Nerv Ment Dis 168:556–558

Marcovici M, O'Brien CP, McLellan AT, Kacian J (1981) A clinical, controlled study of L-alpha-acetylmethadol in the treatment of narcotic addiction. AM J Psych 138(2):234–236

Marks CE, Goldring RM (1973) Chronic hypercapnia during methadone maintenance. Am Rev Respir Dis 108:1088–1093

Martin WR, Jasinski DR (1969) Physiological parameters of morphine dependence in man – tolerance, early abstinence, protracted abstinence. J Psychiat Res 7:9–17

Martin WR, Jasinski DR, Haertzen CA, Kay DC, Jones BE, Mansky PA, Carpenter RW (1973) Methadone – a reevaluation. Arch Gen Psychiatry 28:286–295

Maslansky R (1970) Methadone maintenance programs in Minneapolis. Int J Addict 5:391–405

Mason B, Kreek MJ, Kocsis J, Melia D, Sweeney J (1992) Psychiatric comorbidity in methadone maintained patients. In: Harris LS (ed) Problems of drug dependence 1991. Proceedings of the 53rd Annual Scientific Meeting, the Committee on Problems of Drug Dependence, Inc., NIDA Monograph, Rockville, MD DHHS Publication #(ADM) 92-1888, 119:230

May EL, Mosettig E (1948) Some reactions of amidone. J Org Chem 13:459–464

McLellan AT, Luborsky L, Woody GE, O'Brien CP, Druley KA (1983) Predicting response to alcohol and drug abuse treatment. Arch Gen Psych 40:620–625

McLellan AT, Arndt IO, Metzger DS, Woody GE and O'Brien CP (1993) The effects of psychosocial services in substance abuse treatment. JAMA 269(15): 1953–1959

Mendelson JH, Mello NK (1975) Plasma testosterone levels during chronic heroin use and protracted abstinence: a study of Hong Kong addicts. Clin Pharm Ther 17(5):529–533

Mendelson JH, Mello NK (1982) Hormones and psycho-sexual development in young men following chronic heroin use. Neurobehavioral Toxicol Teratol 4: 441–444

Mendelson JH, Mendelson JE, Patch VD (1975a) Plasma testosterone levels in heroin addiction and during methadone maintenance. J Pharmacol Exper Ther 192:211–217

Mendelson JH, Meyer RE, Ellingboe J, Mirim SM, McDougle M (1875b) Effects of heroin and methadone on plasma cortisol and testosterone. J Pharmacol Exp Ther 195:296–302

Mendelson JH, Inturrisi CE, Renault P, Senay EC (1976) Effects of acetylmethadol on plasma testosterone. Clin Pharm Ther 19(3):371–374

Mendelson JH, Ellingboe J, Judson BA, Goldstein A (1984) Plasma testosterone and luteinizing hormone levels during levo-alpha-acetylmethadol maintenance and withdrawal. Clin Pharm Ther 15(4):545–547

Metzger DS, Woody GE, McLellan AT, O'Brien CP, Druley P, Navaline H, DePhilippis D, Stolley P, Abrutyn E (1993) Human immunodeficiency virus seroconversion among intravenous drug users in- and out-of-treatment: an 18-month prospective follow-up. J AIDS 6:1049–1056

Miller DJ, Kleber H, Bloomer JR (1979) Chronic hepatitis associated with drug abuse: significance of hepatitis B virus. Yale J Biol Med 52:135–140

Misra AL, Mule SJ (1975) L-alpha-acetylmethadol (LAAM) pharmacokinetics and metabolism: Current status. Am J Dr Alc Ab 2(3–4):301–305

Nair NMP, Laing TJ, Schwartz SA (1986) Decreased natural and antibody-dependent cellular cytotoxic activities in intravenous drug abuser. Clin Immunopathol 38: 68–78

Nakamura K, Hachey DL, Kreek MJ, Irving CS, Klein PD (1982) Quantitation of methadone enantiomers in humans using stable isotope-labeled[2H3]-[2H5]-, and [2H8] methadone. J Pharm Sci 71(1):39–43

Nathan JA, Karan LD (1989) Substance abuse treatment modalities in the age of HIV spectrum disease. J Psychoactive Drugs 21:423–429

Nigam AK, Ray R, Tripathi BM (1993) Buprenorphine in opiate withdrawal: A comparison with clonidine. J Sub Ab Tr 10:391–394

Novick DM (1984) Major medical problems and detoxification treatment of parenteral drug-abusing alcoholics. Adv Alcohol Subst Abuse 3:87

Novick DM, Joseph H (1991) Medical maintenance: The treatment of chronic opiate dependence in general medical practice. J Sub Ab Tr 8(4):233–239

Novick DM, Kreek MJ (1992) Methadone and immune function. Am J Med 92: 113–115

Novick DM, Yancovitz SR, Weinberg BA (1979) Amyloidosis in parenteral drug abusers. Mt Sinai J Med 46:163–167

Novick DM, Kreek MJ, Fanizza AM, Yancovitz SR, Gelb AM, Stenger RJ (1981a) Methadone disposition in patients with chronic liver disease. Clin Pharmacol Ther 30:353–362

Novick DM, Gelb AM, Stenger RJ, Yancovitz SR, Adelsberg B, Chateau F, Kreek MJ (1981b) Hepatitis B serologic studies in narcotic users with chronic liver diseases. Am J Gastroenter 75:111–115

Novick DM, Kreek MJ, Arns PA, Lau LL, Yancovitz SR, Gelb AM (1985a) Effect of severe alcoholic liver disease on the disposition of methadone in maintenance patients. Alcohol Clin Exp Res 9:349–354

Novick DM, Enlow RW, Gelb AM, Stenger RJ, Fotino M, Winter JW, Yancovitz SR, Schoenberg MD, Kreek MJ (1985b) Hepatic cirrhosis in young adults: association with adolescent onset of alcohol and parenteral heroin abuse. Gut 26:8–13

Novick DM, Farci P, Karayiannis P, Gelb AM, Stenger RJ, Kreek MJ, Thomas HC (1985c) Hepatitis D virus antibody in HBsAg-positive and HBsAg-negative substance abusers with chronic liver disease. J Med Virol 15:351–356

Novick DM, Khan I, Kreek MJ (1986a) Acquired Immunodeficiency Syndrome and infection with hepatitis viruses in individuals abusing drugs by injection. UN Bull Narc 38:15–25

Novick DM, Brown DJC, Lok ASF, Lloyd JC, Thomas HC (1986b) Influence of sexual preference and chronic hepatitis B virus infection on T lymphocyte subsets, natural killer activity, and suppressor cell activity. J Hepatol 3:363–370

Novick DM, Stenger RJ, Gelb AM, Most J, Yancovitz SR, Kreek MJ (1986c) Chronic liver disease in abusers of alcohol and parenteral drugs: a report of 204 consecutive biopsy-proven cases. Alcoh Clin Exp Res 10:500–505

Novick DM, Stenger RJ, Gelb AM, Most J, Yancovitz SR, Kreek MJ (1986d) Chronic liver disease in abusers of alcohol and parenteral drugs: a report of 204 consecutive biopsy-proven cases. Alcoh Clin Exp Res 10:500–505

Novick D, Kreek MJ, Des Jarlais D, Spira TJ, Khuri ET, Raghunath J, Kalyanaraman VS, Gelb AM, Miescher A (1986e) Antibody to LAV, the putative agent of AIDS, in parenteral drug abusers and methadone-maintained patients: Abstract of clinical research findings: Therapeutic, historical, and ethical aspects. In: Harris LS (ed) Problems of drug dependence 1985. Proceedings of the 47th Annual Scientific Meeting of the Committee on Problems of Drug Dependence. NIDA Research Monograph Series, Rockville, MD DHHS Publication #(ADM) 86-1448, 67:318–320

Novick DM, Pascarelli EF, Joseph H, Salsitz EA, Richman BL, Des Jarlais DC, Anderson M, Dole VP, Nyswander ME (1988a) Methadone maintenance patients in general medical practice. JAMA 259:3299–3302

Novick DM, Joseph H, Dole VP (1988b) Methadone maintenance: Response to article "Methadone maintenance patients in general medical practice." JAMA 260:2835–2836

Novick DM, Des Jarlais DC, Kreek MJ, Spira TJ, Friedman SR, Gelb AM, Stenger RJ, Schable CA, Kalyanaraman VS (1988c) The specificity of antibody tests for human immunodeficiency virus in alcohol and parenteral drug abusers with chronic liver disease. Alc Clin Exp Res 12:687–690

Novick DM, Farci P, Croxson TS, Taylor MB, Schneebaum CW, Lai EM, Bach N, Senie RT, Gelb AM, Kreek MJ (1988d) Hepatitis delta virus and human immunodeficiency virus antibodies in parental drug users who are hepatitis B surface antigen positive. J Infect Dis 158:795–803

Novick DM, Trigg HL, DesJarlais DC, Friedman SR, Vlahov D, Kreek MJ (1989a) Cocaine injection and ethnicity in parenteral drug users during the early years of the human immunodeficiency virus (HIV) epidemic in New York City. J Med Virol 29:181–185

Novick DM, Ochshorn M, Ghali V, Croxson TS, Mercer WD, Chiorazzi N, Kreek MJ (1989b) Natural killer cell activity and lymphocyte subsets in parenteral heroin abusers and long-term methadone maintenance patients. J Pharm Exper Ther 250:606–610

Novick DM, Joseph H, Croxson TS, Salsitz EA, Wang G, Richman BL, Poretsky L, Keefe JB, Whimbey E (1990) Absence of antibody to human immunodeficiency virus in long-term, socially rehabilitated methadone maintenance patients. Arch Intern Med 150:97–99

Novick DM, Ochshorn M, Kreek MJ (1991) In vivo and in vitro studies of opiates and cellular immunity in narcotic addicts. In: Friedman H (ed) Drugs of Abuse, Immunity and Immunodeficiency. Plenum, New York, pp 159–170

Novick DM, Richman BL, Friedman JM, Friedman JE, Fried C, Wilson JP, Townley A, Kreek MJ (1993) The medical status of methadone maintenance patients in treatment for 11–18 years. Drug Alcohol Depend 33:235–245

Novick DM, Joseph H, Salsitz EA, Kalin MF, Keefe JB, Miller EL, Richman BL (1994) Outcomes of treatment of socially rehabilitated methadone maintenance patients in physicians' offices (medical maintenance): follow-up at three and a half to nine and a fourth years. J Gen Intern Med 9:127–130

Novick DM, Reagan KJ, Croxson TS, Gelb AM, Stenger RJ, Kreek MJ (1995) Hepatitis C virus serology in parenteral drug users with chronic liver disease. In: Harris LS (ed) Problems of drug dependence 1994. Proceedings of the 56th Annual Scientific Meeting of the College on Problems of Drug Dependence. NIDA Research Monograph Series, Rockville, MD, 153:439

Nunes EV, Quitkin FM, Brady R, Stewart JW (1991) Imipramine treatment of methadone maintenance patients with affective disorder and illicit drug use. Am J Psych 148:667–669

Nyswander M (1956) The drug addict as a patient. Crune and Stratton, New York

Nyswander ME, Dole VP (1967) The present status of methadone blockade treatment. Am J Psych 123:1441–1442

O'Brien CP, Terenius L, Wahlstrom A, McLellan AT, Krivoy W (1982) Endorphin levels in opioid-dependent human subjects: A longitudinal study. Ann NY Acad Sci 398:377–387

OBrien CP, Woody GE, McLellan AT (1984) Psychiatric disorders in opioid-dependent patients. J Clin Psych 45:9–13

O'Brein CP, Terenius LY, Nyberg F, MeLellan AT, Eriksson AT (1988) Endogenous opioids in cerebrospinal fluid of opioid-dependent humans. Bio Psych 24:649–662

Ochshorn M, Kreek MJ, Khuri E, Fahey L, Craig J, Aldana MC, Albeck H (1989) Normal and abnormal natural killer (NK) activity in methadone maintenance treatment patients. In: Harris LS (ed) Problems of Drug Dependence, 1988; Proceedings of the 50th Annual Scientific Meeting of the Committee on Problems of Drug Dependence. NIDA Research Nonograph Series, Rockville, MD, DHHS Publication No. (ADM) 89-1605, 90:369

Ochshorn M, Novick DM, Kreek MJ (1990) In vitro studies of methadone effect on natural killer (NK) cell activity. Israel J Med Sci 26:421–425

Orlaam drug information (1993) Levomethadyl acetate hydrochloride oral solution. (Package insert) 1–10

Pascarelli EF, Eaton C (1973) Disulfiram (Antabuse) in the treatment of methadone maintenance alcoholics. In: Proceedings of the 5th National Conference on Methadone Treatment National Association for the Prevention of Addiction to Narcotics 1:316–322

Pelosi MA, Sama JC, Caterini H, Kaminetzky HA (1974) Galactorrhea-amenorrhea syndrome associated with heroin addiction. Am J Obstet Gynecol 118:966–970

Pert CB, Snyder SH (1973) Opiate receptor: demonstration in nervous tissue. Science 179:1011–1014

Pescor MJ (1943) Follow-up study of treated narcotic drug addicts. Publ Health Rep Suppl 170:1

Pinto S, Yehuda R, Giller E, Kreek MJ (1993) Abnormal natural killer cell activity (NK) in post-traumatic stress disorder (PTSD) subjects. In: Harris LS (ed) Problems of drug dependence 1992. Proceedings of the 54th Annual Scientific Meeting of the College on Problems of Drug Dependence. NIDA Research Monograph Series, Rockville, MD, DHHS Publication No. (ADM) 93-3505, 132:298

Pohland A, Marshall FJ, Carney TP (1949) Optically active compounds related to methadone. 71:460–462

Poli G, Introna M, Zanaboni F, Pen G, Carbonari M, Aiuti F, Lazzarin A, Moroni M, Mantovani A (1985) Natural killer cells in intravenous drug abusers with lymphadenopathy syndrome. Clin Exp Immunol 62:128–135

Pond SM, Kreek MJ, Tong TG, Raghunath J, Benowitz NL (1985) Altered methadone pharmacokinetics in methadone-maintained pregnant women. J Pharmacol Exp Ther 233:1–6

Popkin RJ (1948) Experiences with a new synthetic analgesic, amidone. Its action on ischemic pain of occlusive arterial diseases. Am Heart J 35:793

Preston KL, Griffiths RR, Cone EJ, Darwin WD, Gorodetzky CW (1986) Diazepam and methadone blood levels following concurrent administration of diazepam and methadone. Drug Alcohol Depend 18:195–202

Prusoff B, Thompson WD, Sholomskas D, Riordan C (1977) Psychosocial stressors and depression among former heroin-dependent patients maintained on methadone. J Nerv Ment Dis 165:57–63

Pugliese A, Martinez M, Maselli A, Zalick DH (1975) Treatment of alcoholic methadone-maintenance patients with disulfiram. J Stud Alcohol 36:1584–1588

Renault PF, Schuster CR, Heinrich RL, Van der Kolk B (1972) Altered plasma cortisol response in patients on methadone maintenance. Clin Pharmacol Ther 13:269–273

Resnick RB, Orlin L, Geyer G, Schuyten-Resnick E, Kestenbaum RS, Freedman AM (1976) L-alpha-acetylmethadol (LAAM): prognostic considerations. Am J Psych 133(7):814–819

Resnick RB, Kestenbaum RS, Washton A, Poole D (1977) Naloxone-precipitated withdrawal: a method for rapid induction onto naltrexone. Clin Pharm Ther 21(4):409–413

Rettig RA, Yarmolinsky A (eds) (1994) Federal regulation of methadone treatment. National Academy of Sciences, National Academy Press

Richman A, Jackson G, Trigg H (1973) Follow-up of methadone maintenance patients hospitalized for abuse of alcohol and barbiturates. In: Proceedings, Fifth National Conference on Methadone Treatment. National Association for the Prevention of Addiction to Narcotics, NY 2:1484–1493

Riordan CE, Mezritz M, Slobetz F, Kleber HD (1976) Successful detoxification from methadone maintenance: follow-up study of 38 patients. JAMA 235:2604–2607

Rosen A, Ottenberg DJ, Barr HL (1973) Patterns of previous abuse of alcohol in a group of hospitalized drug addicts. In: Proceedings, Fifth National Conference on Methadone Treatment National Association for the Prevention of Addiction to Narcotics, NY. 1:306–315

Rosen MI, McMahon TJ, Pearsall HR, Hameedi FA, Woods SW, Kosten TR, Kreek MJ (1966 in press) Correlations among measures of naloxone-precipitated opiate withdrawal. In: Harris LS (ed) Problems of drug dependence 1995. Proceedings of the 57th Annual Scientific Meeting of the College on Problems of Drug Dependence, NIDA Research Monograph Series. Rockville, MD

Rounsaville BJ, Weissman MM, Kleber HD (1982) The significance of alcoholism in treated opiate addicts. J Nerv Ment Dis 170:479–488

Rubenstein RB, Wolff WI (1976) Methadone ileus syndrome: report of a fatal case. Dis Colon Rectum 19:357–359

Rubenstein RB, Kreek MJ, Mbawa N, Wolff WI, Korn R, Gutjahr CL (1978) Human spinal fluid methadone levels. Drug Alcohol Depend 3:103–106

Santen RJ (1974) How narcotics addiction affects reproductive function in women. Contemp Obstet Gynecol 3.4:93–96

Santen RJ, Bardin CW (1973) Episodic luteinizing hormone secretion in man, pulse analysis, clinical interpretation, physiologic mechanisms. J Clin Invest 52: 2617–2628

Santen RJ, Sofsky J, Bilic N, Lippert R (1975) Mechanism of action of narcotics in the production of menstrual dysfunction in women. Fertil Steril 26:538–548

Sapira JD (1968) The narcotic addict as a medical patient. Am J Med 45:555–588

Sapira JD, Jasinski DR, Gorodetzky CW (1968) Liver disease in narcotic addicts. II. The role of the needle. Clin Pharm Ther 9:725–739

Sawyer RC, Brown LS Jr, Bailey J, Hickson M, Lee P, McNair D, Rawls J, Skinner A (1993) Drug abuse treatment programs as centers for HIV-related research and treatment. J Add Dis 12(4):121–129

Saxon AJ, Whittaker S, Hawker CS (1989) Valproic acid, unlike other anti-convulsants, has no effect on methadone metabolism: two cases. J Clin Psych 50:228–229

Schecter A, Kauders F (1975a) Methadone and L-alpha-acetylmethadol in a treatment program in Brooklyn. Am J Dr Alc Ab 2(3–4):331–339

Schecter A, Kauders F (1975b) Patient deaths in a narcotic antagonist (naltrexone) and L-alpha-acetylmethadol program. Am J Dr Alc Ab 2(3–4):443–449

Scott NR, Winslow WW, Gorman DG (1973) Epidemiology of alcoholism in a methadone maintenance program. In: Proceedings, Fifth National Conference on Methadone Treatment, National Association for the Prevention of Addiction to Narcotics, NY 1:284–287

Selwyn PA, Feingold AR, Iezza A, Satyadeo M, Colley J, Torres R, Shaw JFM (1989) Primary care for patients with human immunodeficiency virus (HIV) infection in a methadone maintenance treatment program. Ann Intern Med 110:761–963

Senay EC, Dorus W, Goldberg F, Thornton W (1977a) Withdrawal from methadone maintenance. Arch Gen Psych 34:361–367

Senay EC, Dorus W, Renault PF (1977b) Methadyl acetate and methadone, an open comparison. JAMA 237(2):138–142

Shenkman L, Massie B, Mitsuma T et al. (1972) Effects of chronic methadone administration on the hypothalamic-pituitary-thyroid axis. J Cin Endocrinol Metab 35(1):169–170

Sherwood GK, McGinniss MH, Katon RN, DuPont RL, Webster JB (1972) Negative direct Coombs' tests in narcotic addicts receiving maintenance doses of metha-done. Blood 40:902–904

Shi JM, O'Connor PG, Kosten TR, Schottenfeld RS, Culpepper-Morgan J, Kreek MJ (1994) Seroprevalence of viral hepatitis B, C, D in HIV-infected intravenous drug users. In: Harris LS (ed) Problems of drug dependence 1993. Proceedings of the 55th Annual Scientific Meeting of the College on Problems of Drug Dependence. NIDA Research Monograph Series, Rockville, MD, NIH Publication No. 94-3749, 141:62

Siddiqui NS, Brown LS Jr, Makuch RW (1993a) Short-term declines in CD4 levels associated with cocaine use in HIV-1 seropositive, minority injecting drug users. J Natl Med Assoc 85:293–296

Siddiqui NS, Brown LS Jr, Meyer TJ, Gonzalez V (1993b) Decline in HIV-1 seroprevalence and low seroconversion rate among injecting drug users at a methadone maintenance program in New York. J Psychoactive Dr 25(3):245–250

Simon EJ, Hiller JM, Edelman I (1973) Stereospecific binding of the potent narcotic analgesic [3H] etorphine to rat-brain homogenate. Proc Natl acad Sci USA 70:1947–1949

Smith DE, Moser C, Wesson DR, Apter M, Buxton ME, Davison JV, Orgel M, Buffum J (1982) A clinical guide to the diagnosis and treatment of heroin-related sexual dysfunction. J Psychoact Dr 14:91–99

Sorensen JL, Hargreaves WA, Weinberg JA (1982) Withdrawal from heroin in three or six weeks. Comparison of methadyl acetate and methadone. Arch Gen Psych 39(2):167–171

Spagnolli W, DeVenuto G, Mattarei M, Dalri P, Miori R (1987) Prolactin and thyrotropin pituitary response to thyrotropin releasing hormone in young female heroin addicts. Dr Alc Dep 20:247–254

Spangler R, Unterwald EM, Branch A, Ho A, Kreek MJ (1993a) Chronic cocaine administration increases prodynorphin mRNA levels in the caudate putamen of rats. In: Harris LS (ed) Problems of drug dependence 1992 Proceedings of the 54th Annual Scientific Meeting of the College on Problems of Drug Dependence. NIDA Research Monograph Series, Rockville, MD DHHS Publication #(ADM) 93-3505, 132:142

Spangler R, Unterwald EM, Kreek MJ (1993b) "Binge" cocaine administration induces a sustained increase of prodynorphin mRNA in rat caudate-putamen. Mol Brain Res 19:323–327

Speeter ME, Byrd WM, Cheney LC, Binkley SB (1949) Analgesic carbinols and esters related to amidone (Methadon) J Am Chem Soc 71:57–60

Stimmel B, Rabin J (1977) The ability to remain abstinent upon leaving methadone maintenance: A prospective study. Am J Dr Alc Ab 1:379–391

Stimmel B, Vernace S, Tobias H (1972a) Hepatic dysfunction in heroin addicts: The role of alcohol. J AMA 222:811–812

Stimmel B, Vernace S, Tobias H (1972b) Hepatic function in patients on methadone maintenance therapy. Proceedings, Fourth National Conference on Methadone Treatment. National Association for the Prevention of Addiction to Narcotics, New York, pp 419–420

Stimmel B, Vernace S, Heller E, Tobias H (1973) Hepatitis B antigen and antibody in former heroin addicts on methadone maintenance: Correlation with clinical and histological findings. In: Proceedings, Fifth National Conference on Methadone Treatment. National Association for the Prevention of Addiction to Narcotics, New York 1:501–506

Stimmel B, Goldberg J, Rotkopf E, Cohen M (1974) Ability to remain abstinent after methadone detoxification: a six-year study. JAMA 237:1216–1220

Stimmel B, Vernace S, Schaffner F (1975) Hepatitis B surface antigen and antibody: a prospective study in asymptomatic drug abusers. JAMA 234:1135–1138

Stimmel B, Goldberg J, Rotkopf E, Cohen M (1977) Ability to remain abstinent after methadone detoxification: a six-year study. JAMA 237(12):1216–1220

Stimmel B, Goldberg J, Cohen M, Rotkopf E (1978) Detoxification from methadone maintenance: risk factors associated with relapse to narcotic use. Ann NY Acad Sci 311:173–180

Stimmel B, Korts D, Jackson G (1982) The relationship between hepatitis B surface antigen and antibody and continued drug use in narcotic dependency: a randomized controlled study. Drug Alcohol Depend 10:251–256

Sullivan HR, Blake DA (1972) Quantitative determination of methadone concentrations in human blood, plasma, and urine by gas chromatography. Res Comm Chem Path Pharm 3(3):467–478

Sullivan HR, Smits SE, Due SL, Booher RE, McMahon RE (1972) Metabolism of *d*-methadone: Isolation and identification of analgesically active metabolites. Life Sci 11(1):1093–1104

Sullivan HR, Due SL, McMahon RE (1973) Metabolism of alpha-l-methadol. N-acetylation, a new metabolic pathway. Res Comm Chem Path Pharm 6(3): 1072–1078

Sung C-Y, Way EL (1954) The fate of the optical isomers of alpha-acetylmethadol. JPET 110:260–270

Tabasco-Minguillan J, Novick DM, Kreek MJ (1991) Liver function tests in non-parenteral cocaine users Drug Alcoh Depend 26:169–174

Tabasco-Minguillan J, Novick DM, Kreek MJ (1991) Liver function tests in non-parenteral cocaine users. In: Harris LS (ed) Problems of drug dependence 1990 Proceedings of the 52nd Annual Scientific Meeting of the Committee on Problems of Drug Dependence. NIDA Research monograph Series, Rockville, MD, DHHS Publication No. (ADM) 91-1753, 105:372

Takemori AE, Loh HH, Lee NM (1993) Suppression by dynorphin A and [Des-Tyr] dynorphin A peptides of the expression of opiate withdrawal and tolerance in morphine-dependent mice. JPET 266:121–124

Terenius L (1973) Sterospecific interaction between narcotic analgesics and a synaptic plasma membrane fraction of rat cerebral cortex. Acta Pharmacol-Toxicol. 32:317–320

Thompson RC, Mansour A, Akil H, Watson SJ (1993) Cloning and pharmacological characterization of a rat mu opioid receptor. Neuron 11:903–913

Tong TG, Pond SM, Kreek MJ, Jaffery NF, Benowitz NL (1981) Phenytoin-induced methadone withdrawal. Ann Intern Med 94:349–351

Unterwald EM, Horne-King J, Kreek MJ (1992) Chronic cocaine alters brain mu opioid receptors. Brain Res 584:314–318

Unterwald EM, Rubenfeld JM, Kreek MJ (1994a) Repeated cocaine administration upregulates κ and μ , but not δ, opioid receptors. Neuroreport 5:1613–1616

Unterwald EM, Rubenfeld JM, Spangler R, Imai Y, Wang JB, Uhl GR, Kreek MJ (1994b) Mu opioid receptor mRNA levels following chronic naltrexone administration. Regulatory peptides: proceedings of the 25th International Narcotics Research Conference (INRC) 54:307–308

Vaillant GE (1966) A twelve-year follow-up of New York narcotic addicts: I. The relation of treatment to outcome. Am J Psych 22(7):727–737

Vescovi PP, Gerra G, Maninetti L, Pedrazzoni M, Michelini M, Pioli G, Girasole G, Caccavari R, Maestri D, Passeri M (1990) Metyrapone effects on beta-endorphin, ACTH, and cortisol levels after chronic opiate receptor stimulation in man. Neuropep 15:129–132

Wang JB, Imai Y, Eppler CM, Gregor P, Spivak C, Uhl GR (1993) Mu-opiate receptor: cDNA cloning and expression. Proc Nat Acad Sci USA 90:10230–10234

Weber R, Ledergerber B, Opravil M, Siegenthaler W, Luthy R (1990) Progression of HIV infection in misusers of injected drugs who stop injecting or follow a programme of maintenance treatment with methadone. Br J Med 301:1362–1365

Webster IW, Waddy N, Jenkins LV, Lai LYC (1977) Health status of a group of narcotic addicts in a methadone treatment program. Med J Aust 64(2):485–487

Webster JB, Coupal JJ, Cushman P Jr (1973) Increased serum thyroxine levels in euthyroid narcotic addicts. J Clin Endocrinol Metab 37:928–934

Weissman MM, Slobetz F, Prusoff B, Mezritz M, Howard P (1976) Clinical depression among narcotic addicts maintained on methadone in the community. Am J Psychiatry 133:1434–1438

White AG (1973) Medical disorders in drug addicts: 200 consecutive admissions. JAMA 223:1469–1471

Wikler A, Frank K (1947) Tolerance and physical dependence in intact and chronic spinal dogs during addiction to 10820 (4,4-diphenyl-6-dimethylamino-heptanone-3). Fed Proc 6:371

Wikler A, Frank K, Eisenman AJ (1947) Effects of single doses of 10820 (4,4-diphenyl-6-dimethylamino-heptanone-3) on the nervous system of dogs and cats. Fed Proc 6:384

Wikler A, Fraser HF, Isbell H (1953) N-allylnormorphine: Effects of single doses and precipitation of acute "abstinence syndromes" during addiction to morphine, methadone or heroin in man (post-addicts) JPET 109:8–20

Wolff K, Hay A, Raistrick D (1991a) High-dose methadone and the need for drug measurements. Clin Chem 37:1651–1654

Wolff K, Hay A, Raistrick D, Calvert R, Feely M (1991b) Measuring compliance in methadone maintenance patients. Clin Pharmacol Ther 50:199–207

Wolff K, Hay AWM, Raistrick D, Calvert R (1993) Steady-state pharmacokinetics of methadone in opioid addicts. Eur J Clin Pharm 44:189–194

Woods LA, Wyngaarden JB, Seevers MH (1947a) The addiction potentialities of 1,1-diphenyl-1-(dimethylaminoisopropyl)-butane-2 (amidone) in the monkey. Fed Proc 6:387

Woods LA, Wyngaarden JB, Seevers MH (1947b) The addiction potentialities of 1,1-diphenyl-1-(β-dimethylaminopropyl)-butanone-2 hydrochloride (amidone) in the monkey. Proc Soc Exp Bio Med 65:113

Woody GE, O'Brien CP, Rickels K (1975) Depression and anxiety in heroin addicts: a placebo-controlled study of doxepin in combination with methadone. Am J Psychiatry 132:447–450

Woody GE, McLellan TA, OBrien CP (1990) Clinical-behavioral observations of the long-term effects of drug abuse. In: Harris LS (ed) Problems of drug dependence 1990 Proceedings of the 52nd Annual Scientific Meeting of the Committee on Problems of Drug Dependence. NIDA Monograph, Rockville, MD 71–85

Wu D, Otton SV, Sproule BA, Busto U, Inaba T, Kalow W, Sellers EM (1993) Inhibition of human cytochrome P450 2D6(CYP2D6) by methadone. Br J Clin Pharm 35:30–34

Yaffe GJ, Strelinger RW, Parwatiker S (1973) Physical symptom complaints of patients in methadone maintenance. In: Proceedings of the 5th National Conference on Methadone Treatment NAPA, New York 1:507–514

Zaks A, Fink M, Freedman AM (1972) Levomethadyl in maintenance treatment of opiate dependence. JAMA 220(6):811–813

Zweben JE, Payte JT (1990) Methadone maintenance in the treatment of opioid dependence: a current perspective. West J Med 152:588–599

CHAPTER 14b

Long-Term Pharmacotherapy for Opiate (Primarily Heroin) Addiction: Opioid Antagonists and Partial Agonists

M.J. KREEK

A. Naloxone, Naltrexone, and Nalmefene (Mixed Agonists and Antagonists)

I. Chemistry, Pharmacokinetics, Pharmacodynamics, and Mechanisms of Action in Humans

The action of a specific opioid antagonist is to bind to specific opioid receptors and thus prevent the action of exogenous opioids. Use of such an antagonist will prevent the normal physiological action of the endogenous opioids at their receptors. The first pure specific opioid antagonist which was successfully synthesized, characterized, and patented in 1960 by Dr. J. Fishman with M.J. Lewenstein was naloxone, an alkyl derivative of nor-oxymorphone (BLUMBERG et al. 1961; BLUMBERG and DAYTON 1974). Naloxone was thus the first specific opioid antagonist used clinically and remains extremely useful in the management of accidental or medical opioid overdose. Naloxone has very limited systemic bioavailability, with less than 3% reaching the systemic circulation after oral administration; when it is used for management of narcotic overdose (or when the attempt was made to use it in the more chronic management of opiate dependency) by direct receptor blockade of opiate agonist action in the central nervous system, naloxone must be administered parenterally, and ideally, intravenously (KREEK 1973c; KREEK et al. 1983a,b,c, 1984). Although brief sporadic attempts were made to use naloxone in the maintenance treatment of addiction, it was clear that this use would not be suitable because of the need for a parenteral route of administration. However, studies using naloxone showed that a specific opioid antagonist could block all exogenous opioid effects, and moreover, could reverse the effects of both exogenous and endogenous opioids (KOSTERLITZ and WATT 1968; VEREBEY and MULE 1975; VOLAVKA et al. 1979a,b; MENDELSON et al. 1980; NABER et al. 1981; KREEK and HARTMAN 1982; KREEK et al. 1983a,b,c, 1984; COHEN et al. 1983; KREEK and CULPEPPER-MORGAN 1991). In an opioid tolerant and dependent person, administration of an antagonist would be expected and has been shown to precipitate signs and symptoms of opioid (or narcotic) withdrawal. A naloxone test for opioid dependence was developed and used in early research to determine the presence and degree of opioid dependence in heroin addicts entering treatment or treatment research (BLACHLEY 1975; JUDSON et al. 1980). Naltrexone was developed to provide a more orally effective and longer acting specific opioid antagonist than naloxone. In 1967, Dr. I. Blumberg patented the chemical synthesis and substance of the resultant compound naltrexone, a cyclopropyl derivative of oxymorphone (U.S. patent #3,332,950, July 25, 1967; BLUMBERG and DAYTON 1974). A wide variety of animal studies were performed in which it has been shown that both naloxone and naltrexone bind primarily to the μ type opioid receptors, and that these specific opioid antagonists block the effect of both exogenous and endogenous μ receptor ligands with essentially no agonist properties. Therefore, the action both of naloxone and naltrexone is that of a specific opioid antagonist; the mechanism of action is by acting as a competitive antagonist at specific, primarily μ, receptors. Very large doses of naloxone or naltrexone also bind with much weaker affinity to δ, and to an even lesser extent, κ opioid receptors. Thus, very large amounts of naloxone or naltrexone have some action at δ and κ opioid receptor sites as well as at μ opioid receptors. Another specific opioid antagonist, which is currently in clinical research for use in a variety of indications, is nalmefene. Like naltrexone, nalmefene, is a cyclopropyl derivative of noroxymorphone (DIXON et al. 1984, 1986), and like naloxone and naltrexone, it is primarily a μ-type opioid receptor selective antagonist. However, it has been shown in binding studies that nalmefene has greater affinity to the κ type opioid receptors than either naloxone or naltrexone (MICHEL et al. 1985).

Animal studies have shown that chronic treatment with naltrexone (and presumably any specific opioid antagonist) will cause up-regulation of the specific opiate receptor sites, primarily of the μ opioid receptor types, but to a lesser extent, to δ and κ receptors (TEMPEL et al. 1982; ZUKIN et al. 1982). This up-regulation or increase in density of opioid receptors has been shown in animal models, but not reproduced in humans, to cause super-sensitivity to some opioid effects of agonists such as morphine. In human studies, it has been shown that naltrexone and nalmefene, like naloxone, will antagonize all specific effects of exogenous opiates, and when given prior to opiate administration, will prevent opiate effects. When given to an opiate dependent subject, naloxone, naltrexone, and nalmefene will precipitate the acute opioid abstinence syndrome, the signs and symptoms of opiate withdrawal. Since pretreatment with any specific opioid antagonist blocks opiate agonist effects, including both subjective and objective responses, the rationale behind use of naltrexone in the pharmacotherapy of opiate dependency is that naltrexone, by virtue of its blocking opiate agonist effects, can through the phenomenon of conditioned behavior extinguish drug-seeking behavior, since all euphoria and all the reinforcement produced by self-administration of opiates will be blocked by such treatment.

Naloxone is metabolized primarily by nonoxidative (non-energy-dependent) glucuronidation, primarily in the liver, and also possibly by the gastrointestinal mucosa (HAHN et al. 1983; KREEK et al. 1983a,b). Studies of both orally and intravenously administered naloxone have shown that this medication undergoes an extremely large "first pass" metabolism, primarily by the liver, and also possibly by the gastrointestinal mucosa, forming naloxone glucuronide after oral administration, with less than 3% of naloxone reaching the systemic circulaton. Therefore, naloxone must be used by the parenteral route, usually the intravenous route, to get full action at central nervous system sites of specific opioid receptors. Advantage has been taken of this poor systemic bioavailability in recent studies in which naloxone is administered orally to attempt to reverse some of the signs and symptoms of opiate induced hypomotility, by direct action at the specific opioid receptors within the gastrointestinal wall (KREEK et al. 1983a,b; CULPEPPER-MORGAN 1992). Preliminary results of these studies have shown that oral naloxone may be effective in management of opiate induced constipation, both in the setting of chronic pain management, and in the setting of methadone maintenance treatment for opiate addiction. In addition, other studies have shown that naloxone administered orally may be effective in managing idiopathic chronic constipation by displacing the endogenous opioids from receptors in the gastrointestinal wall. A variety of studies have shown that naloxone has a very short half-life. Recent studies using newly developed sensitive techniques of high performance liquid chromatography have shown that the half-life of naloxone is around 2 h, although trace amounts may occupy receptors for much longer periods, since kinetic studies show that there is a very deep pool of naloxone, which is slowly eliminated from the body (HAHN et al. 1983; ALBECK et al. 1989).

Naltrexone is also metabolized primarily by the liver, but requires energy-dependent oxidative metabolism; it undergoes hepatic microsomal enzyme dependent biotransformation to a primary metabolite, 6-β-naltrexole, formed by biotransformation of the keto group at the C-6 position to a hydroxy group maintaining the beta configuration (CONE 1973). Although it may have some opioid antagonist activity, 6-β-naltrexole has been assessed as around 1/12 to 1/80 as active as naltrexone in various species; the potency ratio in humans is unknown. Within 24 h after oral administration, the majority of naltrexone has been converted to 6-β-naltrexol (CONE 1973; VEREBEY and MULE 1975; VEREBEY et al. 1976; INTURRISI 1976). This metabolite, though a much weaker antagonist, may contribute to the long duration of action of naltrexone, since studies suggest that naltrexone itself is rapidly metabolized. Although it has been shown that plasma levels of the β-naltrexol metabolite may be 1.5 to 10 times higher than those of naltrexone, the narcotic antagonism seems to be best correlated with naltrexone levels. The other metabolites include a variety of oxidative metabolites and their glucuronide conjugates.

Most of the studies of naltrexone pharmacokinetics have been performed after subcutaneous administration or, more usually, following oral administration of naltrexone; very few studies (if any) have been reported in which a direct comparison of oral versus intravenous administration of naltrexone in the same individual was performed to determine systemic bioavailability. Thus, precise data with respect to the systemic bioavailability of naltrexone after oral administration is limited. Diverse data have been forthcoming concerning the plasma apparent terminal half-life ($t_{1/2}\ \beta$) and the systemic and renal clearance of naltrexone. One set of studies attempting to determine systemic bioavailability of naltrexone after oral administration of a 100 mg dose suggested that 78%–80% of naltrexone is metabolized in the first pass through the liver (Kogan et al. 1977). Using data from many studies, it is estimated that the systemic bioavailability of naltrexone after oral administration is approximately 5%–30% (Meyer et al. 1984; Gonzalez and Brogden 1988). In one study in which a 50 mg dose of naltrexone was administered both intravenously (in five volunteers) and orally (in six other volunteers), and also subcutaneously in other subjects, the plasma apparent terminal half-life was found to be 2.7 h after intravenous administration and an additional 1.6 h after subcutaneous administration, whereas after oral administration, it was 8.9 h (Wall et al. 1981, 1984). To prolong the half-life of naltrexone, a pro-drug, naltrexone 3-salicylate, has been synthesized and studied in animal models, but not in humans (Hussain and Shefter 1988). Many attempts, recently with some relative success, have been made to provide a sustained release of subcutaneous implant form of naltrexone, most utilizing some type of polymer in which naltrexone is imbedded.

Nalmefene is another specific opioid antagonist which has a very different pharmacokinetic profile than either naloxone or naltrexone. Nalmefene is significantly more potent than both naloxone and naltrexone in precipitating opiate withdrawal in an opiate dependent animal. Nalmefene has been shown to have a duration of action of over 8 h in the reversal of acute opiate induced narcotic anesthesia or analgesia. The apparent plasma terminal half-life of nalmefene after oral administration is around 8–9 h (Dixon et al. 1984, 1986, 1987; Chou et al. 1993). It is metabolized in the liver almost exclusively by the nonoxidative pathway of glucuronidation, and it is excreted primarily in urine as a glucuronide. Nalmefene is more orally effective than naltrexone; it has approximately a 40%–60% systemic bioavailability after oral administration. Nalmefene has been shown to block endogenous opioids, probably both of μ receptor selective and also κ receptor selective types, in addition to blocking or reversing effects of exogenous opiate induced narcotic effects. In high doses, all three of these specific opioid antagonists act at all three types of opioid receptors, although each is primarily active at the μ type receptors.

Several reports have shown that naltrexone is probably a dose-dependent hepatotoxin when administered in very high doses to human subjects. The first observations of the hepatotoxicity of naltrexone were found when doses greater than 100 mg were administered for a variety of experimental indications (Atkinson et al. 1985). However, other studies have shown that the usual treatment protocol for management of opiate addiction, 100 mg for 2 days/week and 150 mg a third day (to achieve an average of 50 mg/day), results in changes in liver function tests reflecting mild hepatocellular damage in patients receiving naltrexone in this regimen for the management of opiate addiction (KOSTEN et al. 1986a,b). It has been shown that these changes are reversible. There has been no case reported yet of clinically significant hepatotoxicity due to naltrexone. Also, there has been no evidence of any hepatotoxicity due to naltrexone when individual doses of 50 mg/day administered once daily or less have been used. It has been suspected that the major metabolite of naltrexone, 6-β-naltrexol, is the hepatotoxin rather than naltrexone itself. However, to date, β-naltrexol apparently has not been administered directly in animal models, either to determine its activity or its potential for hepatotoxicity. There is no evidence to date of any hepatotoxicity due to naloxone or nalmefene.

II. Use of Naltrexone and Other Opioid Antagonists: Safety, Efficacy, and Effectiveness

The hypothesis underlying the development of methadone maintenance treatment was to provide steady state perfusion of opioid receptors with a highly selective opioid agonist providing a slow onset and with a very long-acting pharmacokinetic profile which would prevent any rapid onset of opioid effects, thus eliminating any rewarding effects if the medication is appropriately prescribed, and at the same time, provide sustained action in both preventing opioid (narcotic) withdrawal, prevent "drug hunger" or craving, and also allow normalization of physiological functions disrupted by exogenous opiates (DOLE et al. 1966a,b; KREEK 1992a,b,c). In addition, the use of the opioid agonist provided a "blockade" effect against any superimposed short-acting illicit opiates, such as heroin, through the mechanism of opioid tolerance and cross-tolerance. In contrast, the rationale underlying the development of the use of opioid antagonists for the treatment of opiate addiction was very different. The rationale for use of antagonists does not hypothesize any a priori metabolic, neurobiological, or genetic basis for addictive diseases. Also, the rationale leading to the development of antagonists for treatment does not hypothesize that any of the effects caused by long-term illicit use of short-acting narcotics, such as heroin, are long-term, persistent, or permanent physiological changes which might contribute to drug-seeking behavior. Each of these are central hypotheses to the rationale for development of opioid agonist treatment for opiate addiction. Rather, the rationale underlying the development of antagonist treatment was based on the hypothesis that addiction is exclusively a conditioned behavior, that the reinforcing effects, that is, the so-called pleasurable or euphorigenic effects, of a short-acting opiate such as heroin lead to a behavior of self-administration, which then becomes a learned behavior. This rationale includes the tacit assumption that any changes effected by the drug use are transient and fully reversible. Thus, if addiction were solely a conditioned behavioral phenomenon, then it is logically presumed that use of a highly selective, specific opioid antagonist which would block any of the rewarding or pleasurable effects of a superimposed illicit drug such as heroin, along with blocking effects of endogenous opioids, but without providing any agonist or replacement action would lead ultimately to cessation of opiate use through the classical phenomenon of deconditioning and extinction. These concepts had been proposed in early research efforts primarily by Wikler and Martin of the Lexington group and Resnick and colleagues (Martin et al. 1973; Altman et al. 1976; Goldstein 1976). With later findings, these concepts were directly studied by many groups, with extremely important studies carried out by O'Brien and colleagues, by whom the hypotheses relating to addiction were modified to appreciate that, although indeed conditioned phenomena play a very important role in drug use and especially in relapse to drug use, the management of addiction cannot be successfully achieved through nonpharmacologically based deconditioning and extinction behavioral paradigms alone, which were shown to be ineffective in preventing relapse to illicit opiate use (O'Brien et al. 1975).

Many early research groups had studied a possible utility of cyclazocine, a mixed agonist-antagonist with both μ and κ activity, in the treatment of heroin addiction. However, clinical studies reported varying degrees of side effects, in most cases a dysphoria which was considered unacceptable to the patients (Resnick et al. 1974; Brahen et al. 1977). These side effects were attributed to be due probably to its κ opioid agonist action. Brahen and colleagues later conducted a controlled treatment study comparing 20 mg/day cyclazocine against 20 mg/day naltrexone. In a short term study, it was found that loss of appetite, tiredness, and a variety of gastrointestinal symptoms including increased gaseous distention and abdominal cramps were the most common, affecting 10%–30% of naltrexone treated subjects. Similar but more common and more intense symptoms, along with a feeling of restlessness, dysphoria, or paranoia, were experienced by the cyclazocine treated groups. Very early attempts to use naloxone in treatment of opiate dependency have been carried out by Zaks, Fink, Friedman, and colleagues (Zaks et al. 1971). When naloxone was studied as a potential chronic treatment agent, it was found to be ineffective due to its extremely limited systemic bioavailability after oral administration. Enormous doses of naloxone had to be used to produce a narcotic blockade. Also, naloxone has a very short duration of action which made it relatively ineffective even after parenteral injections. The earliest clinical studies of the use of naltrexone in the treatment of heroin dependency were conducted by Martin and Jasinski at the USPHS resource in Lexington, Kentucky, and by O'Brien and Woody at the University of Pennsylvania (Martin 1973; O'Brien 1975). Also in early naltrexone treatment research, Goldstein proposed a possible "steps" approach, i.e., sequential treatment employing a pharmacological approach for the management of opiate addiction in which progression would be made from cessation of opiate use, to methadone maintenance treatment, on to LAAM (*l-α*-acetylmethadol) treatment, and ultimate withdrawal of LAAM treatment, with the institution of naltrexone treatment to block opiate reward mechanisms, to be ultimately followed by treatment with no pharmacological support (GOLDSTEIN 1976).

In an early study, the group of O'BRIEN and colleagues compared the effects of the longer acting opioid antagonist naltrexone in 54 opiate-addicted subjects with 34 controls who did not receive naltrexone; two-thirds of the subjects were heroin addicts beginning treatment, whereas one-third were former heroin addicts stabilized in methadone maintenance treatment (O'BRIEN 1975). Detoxification of heroin addicts was effected by administration of decreasing doses of methadone over a 5–10 day period. Patients already in methadone maintenance treatment underwent treatment dose reduction and elimination over a 1–2 month period. A placebo controlled naloxone test was performed prior to giving the initial dose of naltrexone to determine if any opiate withdrawal symptoms would ensue; if no objective or subjective signs occurred after placebo administration, then 0.8 mg naloxone was given intravenously. If the naloxone challenge test was clearly negative, with no signs of withdrawal, then administration of placebo naltrexone was started; if the naloxone challenge test was positive, it was repeated 24 h later and before beginning treatment. Naltrexone placebo was given for 3 days in these early studies followed by titration of naltrexone doses beginning with 10 mg and proceeding up to 50 mg/day by the second day, and up to 120 mg/day over the next several days, ultimately with 120 mg administered on Monday and Wednesday and 200 mg on Friday. Later, this was modified to the currently most commonly used protocol for administering naltrexone in treatment of opiate addiction: 100 mg on Monday and Wednesday, and 200 mg on Friday. Take home doses of naltrexone were not given in this early study. In this study, 46% of patients discontinued naltrexone treatment within 2 weeks or less, and only two of the 54 subjects who initially took at least one dose of naltrexone stayed in treatment for 24–28 weeks (O'BRIEN et al. 1975). Hollister and colleagues conducted a study of naltrexone treatment in 108 opiate dependent men who had been recently detoxified from daily opiate use, that is, "street addicts" entering into initial naltrexone treatment; only 10 patients remained in naltrexone treatment for more than 14 weeks, an additional 24 remained in treatment for at least 3 weeks. In this study, after an induction period, patients were stabilized on the most commonly used regimen of 100 mg naltrexone on Monday and Wednesday, and 150 mg on Friday (HOLLISTER et al. 1977).

Double-blind challenge studies using hydromorphone also were conducted by the University of Pennsylvania group (O'BRIEN et al. 1975). Trials were run using 1–4 mg hydromorphone. A "rush" was perceived in 27% and 54% of the subjects receiving the low and high dose of hydromorphone alone against the background of naltrexone treatment 48 h after taking 120–150 mg naltrexone. Even higher doses of 200 mg naltrexone blocked the challenge opioid injections conducted 48 h later.

In self-administration behavioral research in which eight heroin addicts were studied on an in-patient setting during 10 days of treatment, while receiving either naloxone or naltrexone, and again when no antagonist was being administered, it was shown that in the unblocked state with no antagonist administered, the subjects injected all available heroin, whereas they essentially ceased heroin self-administration following antagonist administration (ALTMAN et al. 1976). In another early research experience in which 155 patients were treated with 40–200 mg/day naltrexone for periods up to 8 months, it was shown that 80 mg or more of naltrexone was effective in blockading at least some dose of opiate agonist for 48 h, and that this antagonistic effect lasted upto 72 h after administration of larger doses of 120–200 mg/day (VOLAVKA et al. 1976). The major side effects observed in this study were increases in blood pressure and epigastric pain. Challenge studies to determine the extent of narcotic blockade was conducted in 30 patients by administering 25 mg heroin intravenously in a sequence of 76 tests. These studies showed that within 48 h after the last naltrexone dose of 60–200 mg, the effect of 25 mg heroin was suppressed to an average of 14% of those signs and symptoms experienced without naltrexone. Further analysis of data showed that both the dose of naltrexone and the time following dose were most important with respect to decreasing opiate effects. Other studies have been conducted which have shown that tolerance does not develop to the opioid antagonist effect of naltrexone (KLEBER et al. 1985). Studies were conducted in nine former heroin addicts who had been in naltrexone treatment for a mean of 9.4 months. Placebo controlled morphine challenge studies were performed in which a single dose of 20, 40, or 60 mg morphine was administered intravenously. Two subjects reported a transient "binge" or "rush" at one dose. Every subject identified the "pins and needles" sensation due to morphine administration (which had been described in the methadone cross tolerance studies of 1964, reported in 1966) (DOLE et al. 1966). Actual dysphoria postulated to the expected histamine release was experienced after the administration of 60 mg morphine. No titration studies, similar to those performed in the early studies of methadone maintenance, have been performed to determine what dosage of heroin or morphine would be needed to exceed the degree of opiate receptor blockade provided by 50–100 mg naltrexone, although it is assumed that this dose antagonist would block the effect of very high doses of agonist.

The Special Action Office for Drug Abuse Prevention (SAODAP) made an official decision to explore the use of specific narcotic antagonists for the treatment of opiate addiction, and thus contracted with the National Academy of Sciences to provide a design and guidance for such a study. This was designed by a committee at the National Academy of Sciences under the auspices of the National Research Council and was conducted, evaluated, and reported under the auspices of the National Research Council Committee in the Report of the Committee on Clinical Evaluation of Narcotic Antagonists (National Research Council Committee in the Report of the Committee on Clinical Evaluation of Narcotic Antagonists 1978). Five treatment clinics conducted the carefully designed experimental protocol. Naltrexone was selected as the antagonist to be studied because of its high efficacy when given orally, its relatively long duration of action, its freedom from agonist action, and its lack of any serious side effects in earlier studies. This study was conducted with the following aims: (1) to assess the feasibility and acceptability of narcotic antagonist treatment; (2) to assess in a preliminary way the methods for determining the efficacy of such treatment; and (3) to assess the toxicity of naltrexone under control conditions (National Research Council Committee in the Report of the Committee on Clinical Evaluation of Narcotic Antagonists 1978). In this study, opiate addicts and former addicts in some form of treatment were separated into three groups (1) established opiate users recently detoxified from opiate use, that is, "street addicts;" (2) former heroin addicts undergoing methadone maintenance treatment; and (3) former addicts currently in drug-free programs following incarceration or participation in drug-free programs ("post addicts"). All of the study subjects were 18 years or older, and all had a verified history of past or current opiate dependency, with signs or symptoms of opiate withdrawal, the presence of opiates in urine, or current treatment in a methadone program. Patients who were currently street addicts underwent a detoxification for 21 days using low dose methadone, followed by methadone vehicle for 7-14 days, prior to starting a steady medication. Methadone maintained patients who had been treated for at least six months on a daily dose of 50 mg or less underwent slow dose reduction over a period of 4–8 weeks, followed by 14 days on methadone placebo, before starting naltrexone. These protocols were designed to avoid giving opioid antagonist to someone with on-going or recent opiate agonist use. After the first week of induction, naltrexone was given 6 days/week with 50 mg on Monday–Friday, 100 mg on Saturday, and no drug on Sunday. After 8 weeks of treatment, patients were placed on a three times per week dosage schedule, 100 mg on Monday and Wednesday, and 150 mg on Friday. In this study, 735 candidates were selected: 254 street addicts, 276 former addicts in methadone maintenance treatment, and 205 post addicts who had been maintained in a drug free state for protracted periods of time prior to entering the study. After random assignment to study medication in early 1976, 192 subjects received at least one dose of naltrexone in a study of medication. This included 42 street addicts, 48 former addicts in methadone treatment, and 92 post addicts. Although this study was planned to be conducted for a longer period of time, it had to be terminated early because of the extremely high drop-out rate. Of the 182 subjects who were initiated in naltrexone treatment, 169 dropped out of treatment prior to completing 9 months of pharmacotherapy. Nine patients were still in active treatment but for less than 9 months at the point when data collection was terminated. Of the 13 subjects who completed 9 months of treatment at that time, including seven receiving naltrexone and six receiving placebo, none came from the street addict group; three were subjects who had formerly been maintained on methadone, and ten had come from the former drug-free group. Thus, there was a 7% 9 month retention of the entire study group, with a 0% retention of street addicts inducted into naltrexone placebo treatment and an 11% retention of drug-free former opiate dependent persons who had entered the study. This study reaffirmed the antagonist effect of naltrexone, and also suggested that naltrexone was not an effective treatment for unselected active opiate addicts or former addicts stabilized in methadone maintenance treatment. Most of the side effects seemed to be those related to precipitation of abstinence, even in those study subjects who had been in drug-free treatment for a period of time prior to study. It was not clear whether the gastrointestinal symptoms observed such as nausea and vomiting, loss of appetite, abdominal pains and cramps, "due to the antagonist blocking effects of the endogenous opioid system, experienced by many of the subjects," were due to precipitated abstinence or were direct side effects of naltrexone. Also, in this early study, it was stated that "the few patients with either clinical or laboratory evidence of hepatitis were thought to have acquired these abnormalities prior to drug taking; no direct evidence of liver toxicity of naltrexone was observed over these relatively brief periods," although subsequent studies have shown that higher doses and frequently used therapeutic doses of naltrexone (100–150 mg), similar to those used in this study and in most treatment protocols, may cause alteration in liver function in liver function tests reflecting hepatotoxicity (ATKINSON et al. 1985; KOSTEN et al. 1986a,b).

In a subsequent study, 119 previously heroin addicted patients were entered into naltrexone treatment research. Patients were allowed a maximum of 365 days on naltrexone and a maximum of four admissions each to naltrexone treatment (JUDSON et al. 1981). The goals of this particular study were both to test the hypothesis that retention of patients in naltrexone treatment could be improved by recruiting patients prepared for such treatment during participation in a treatment program using *l*-α-acetylmethadol for detoxification, and also to compare the safety and efficacy of 60 versus 120 mg naltrexone administered three times per week. In this study, no differences were found between the two doses, and there was a very low overall retention in treatment, similar to that found in the earlier reports discussed above. Less than 15% of the 119 patients in the study group were still in the first naltrexone treatment period at 15 weeks, and essentially no subjects remained in treatment after 30 weeks, much shorter that the planned study period. Although 36 patients entered treatment a second time, the mean time in treatment was only 31 days, as compared with 45 days in the first treatment period. Only eight patients entered treatment a third time, and two a fourth time. During treatment, by history, approximately half of the patients used heroin at some time during their naltrexone treatment; however, only 9% of all urines tested were positive for opiates. In this study, many subjects reported that heroin craving decreased significantly by the end of the first week, and no side effects of naltrexone were found. A different study of the use of naltrexone in suburban opiate addicts showed that naltrexone combined with psychotherapy could lead to periods of opiate abstinence, but did not prevent relapse (TENNANT et al. 1984). In a later study, the group at the Veterans Administration Medical Center connected with the University of Pennsylvania evaluated the use of naltrexone in 300 unselected opiate addicts (GREENSTEIN et al. 1984). In the early years of naltrexone treatment research, naltrexone had been given very soon after detoxification from opiate agonist; in these later studies, at least one 5-day drug-free interval was given prior to induction into naltrexone, and ancillary medication was also given more frequently. However, the drop out rates were similar; the mean time in treatment for the entire population was 2 months, with retention in treatment ranging from 1 week to 12 months. Many of the patients who dropped out chose to resume methadone maintenance. It was reported that around 20% of the patients experienced clinically significant anxiety or depression, but most of these had been shown to be symptomatic prior to induction in naltrexone treatment. This study showed that naltrexone coupled with psychotherapy gave better results than naltrexone treatment alone, with the majority of patients requiring ongoing supportive therapy.

Although many clinicians and clinical investigators have reported cases in which naltrexone treatment has been used for addicted health care personnel, there is a paucity of formal reports, probably due to the strict and special confidentiality demands for this group. However, in one report, 60 physicians and other health care professionals with narcotic dependency were treated in private physicians' offices with naltrexone using doses of 300–350 mg/week (LING and WESSON 1984). Patients remained in naltrexone treatment for an average of 8 months. Approximately 50% of the patients were rated as having been improved by naltrexone treatment, whereas the other 50% only slightly or moderately improved. Of interest, the most significant finding was that the patients in the improved group did not drink alcohol and remained in naltrexone treatment for more than 6 months and, for the most part, were older (50–69 years). Only five of the 60 subjects, or 8%, remained in treatment for 2 years. Therefore, despite the very modest success of naltrexone treatment for unselected heroin addicts (0%–15% remain in treatment for 6 months or longer), naltrexone has been shown to be somewhat more effective for special groups, including health care personnel. Also, it has been shown to be effective for short to moderate term (less than 1 year) treatment of other special groups in whom relapse could lead to immediate unwanted circumstances, such as those on parole or probation, whose prison release is contingent on continuing naltrexone treatment.

There have been no studies reported in which use of naltrexone in the management of heroin addicts not meeting the criteria for maintenance treatment with methadone or LAAM (i.e., with less than 1 year of daily, multiple dose use of an illicit opiate or less than 18 or in some regions, 16 years of age) has been attempted. Also, there have been no rigorous studies comparing the use of naltrexone with placebo in the long-term management of drug-free former opiate dependent persons who have undergone voluntary detoxification and drug-free treatment for heroin addiction, or voluntary withdrawal and discontinuation of methadone maintenance or LAAM treatment. Studies of this type in the future may reveal subgroups for whom adjunctive treatment with an opioid antagonist may be helpful.

III. Neuroendocrine Effects of Opioid Antagonists

Some very important findings have been made with respect to the effects of specific opioid antagonists on neuroendocrine function, not only in opiate dependent persons, thus relevant for the treatment of opiate addiction, but also in normal healthy volunteer subjects with no history of opiate abuse. An initial finding by Volavka reported in 1979 has been followed up by many research groups, including studies from the NIH, The Rockefeller University, and Yale University, in which it has been shown that a specific opioid antagonist such as naloxone given intravenously (and in later studies, also naltrexone or nalmefene given either intravenously or orally) results in an increase in levels of ACTH and also of cortisol from the adrenal cortex (VOLAVKA et al. 1979a,b, 1980; NABER 1981; COHEN et al. 1983; KREEK et al. 1984; KOSTEN 1986a,b, 1992b). Naloxone has been shown to evoke a dose dependent increase in ACTH levels. As would be expected, since ACTH and β-endorphin are derived in equimolar amounts from the same single precursor peptide product of a single gene, proopiomelanocortin, later studies have shown that β-endorphin, ACTH, and cortisol are increased following opioid antagonist administration. In one study conducted at NIH, in which extremely high dose naloxone was administered by intravenous infusion (2 mg/kg), impairment in cognitive function was noted, along with dysphoric effects, including tension, anxiety, irritability, and depression, all of which are also symptoms of the opiate withdrawal syndrome (COHEN et al. 1983). In yet other studies, it was found that steady state 24 h infusion of moderate to high doses of naloxone (25–30 mg total) as contrasted to bolus administration did not evoke increases in plasma levels of ACTH, β-endorphin, and cortisol, implying that steady state perfusion of opioid receptors may allow stabilization of the otherwise disrupted normal feedback control, in contrast to the effects of a bolus administration of naloxone, naltrexone, or nalmefene (KREEK et al. 1984). All of these studies suggest that in humans, the release of POMC peptides from the anterior pituitary may be, in part, under tonic inhibition by endogenous opioids. This would be a second physiological regulatory mechanism, in addition to the well established negative feedback control by glucocorticoids, at both the hypothalamic and anterior pituitary sites of the stress response CRF and POMC peptide release in humans. In other studies by The Rockefeller University

Laboratory on the Biology of Addictive Diseases, in collaboration with the Yale University Chemical Dependency Research Unit, it has been shown that even during protracted naltrexone treatment of former opiate addicts with all subjects in treatment for at least 15 weeks, the use of the relatively short-acting antagonist naltrexone continues to disrupt normal function of the hypothalamic-pituitary-adrenal axis (KOSTEN et al. 1986a,b). Elevated levels of β-endorphin and elevated levels of cortisol were observed, with chronic elevations in cortisol levels observed both in the morning and afternoon, thus with disruption of the normal circadian rhythm of this steroid which is released in response to ACTH. In a second study conducted in a subset of naltrexone treated patients who could be studied both during and following cessation of naltrexone treatment in the medication free state, it was documented that this naltrexone effect was reversible; normal levels and normal circadian rhythm of cortisol levels were observed in former naltrexone treated subjects who had elevated levels and disruption of the circadian rhythm during chronic naltrexone treatment (KOSTEN et al. 1986b).

In studies conducted primarily by the Harvard-McLean Chemical Dependency Research group, it has been shown that naltrexone and other opioid antagonists cause a significant increase in LH levels, with restoration to a normal pulsatile pattern of LH release, in former heroin addicts who demonstrate opiate induced abnormalities in pulsatile LH release as well as suppression of testosterone level (MIRIN et al. 1976; MENDELSON et al. 1979, 1980). Also, it has been shown that naltrexone pretreatment prevents opiate induced suppression of both LH and testosterone. These studies have been extended to studies in opioid naive volunteer females and males with abnormal LH release, and it has been shown that administration of an opioid antagonist may restore pulsatile LH release and thus in some cases restore normal ovulation in females and allow normal levels of testosterone to be secreted in males. As part of these studies of the effects of the specific opioid antagonist naltrexone on the hypothalamic-pituitary-gonadal axis, Mendelson and colleagues reported that in adult males without a history of opiate abuse, naltrexone produced dysphoric effects including fatigue, sleepiness, nausea, sweating, and occasional feelings of unreality.

B. Buprenorphine

I. Chemistry, Pharmacokinetics, Pharmacodynamics, and Mechanisms of Action in Humans

Buprenorphine is a synthetic opioid classified as either a mixed agonist-antagonist, or alternatively, as a partial agonist (LEWIS and READHEAD 1970; COWAN et al. 1971, 1977; COWAN 1974; LEWIS 1974). Chemically, buprenorphine is a member of the oripavine series with a C_7 side chain containing a tert-butyl group. Its pharmacological properties are partly those of agonism, primarily at μ type opiate receptors. However, most of the agonist effects plateau at a moderate dose range which can be defined in different species; beyond that level, the agonist effects decline, and antagonist effects occur. Thus, a classical bell-shaped or inverted U-type dose response curve can be demonstrated in classical preclinical tests of analgesia. It has been suggested, based on various in vitro binding studies, as well as laboratory in vivo dose response studies, that there is agonist action at μ and possibly κ opioid receptors of a high affinity type, thus with high intrinsic activity (LEWIS 1985). There may also be low affinity binding (that is, with low intrinsic activity) to δ type opioid receptors. However, in laboratory studies in which rats had been trained to discriminate μ type agonists, it has been shown that buprenorphine has predominantly μ-like activity, with much less κ-like activity and no δ. In rat models, buprenorphine was initially shown to have long-acting antinociceptive action to chemical or pressure stimuli with potency much greater than morphine, but potency similar to morphine when heat stimuli were used. When a tail flick test was used, a classical bell-shaped dose response curve was observed. Similarly, a bell-shaped dose response curve was observed in rodent models for the buprenorphine effects in the inhibition of gastrointestinal motility and respiratory depression. The fact that there is a bell-shaped curve for morphine with respect to respiratory depression and arterial P_aCO_2 documents one of the major assets of buprenorphine, that is, potentially there is increased safety for use in drug naive or in completely tolerant humans. In in vitro studies, buprenorphine has been shown to have a gradual onset and very slow offset of action in assays such as inhibition of the guinea pig ileum twitch. In that assay, it was shown that very high doses of naloxone were unable to antagonize the effect of buprenorphine. Further studies have documented the tightness of the binding of buprenorphine to opioid receptors and thus very slow offset from binding, which confers upon this opioid partial agonist unusual kinetics which are important for its pharmacological effects (LEWIS 1985). A variety of studies have shown that it is difficult to antagonize the effects of buprenorphine with pure opioid antagonist once buprenorphine has bound to opioid receptors, and also, that buprenorphine has a related very long duration of action (HEEL et al. 1979; BULLINGHAM et al. 1983; LEHMANN et al. 1988). Studies in animal models show that following chronic administration and then discontinuation of buprenorphine, abstinence symptoms are less than following chronic administration of and abrupt cessation of morphine (LEWIS 1985). It has been suggested that this may be due to the very slow "off time" of buprenorphine from opioid receptors or, alternatively, that a shift from agonist to antagonist actions may prevent the development of significant physical dependence (JASINSKI et al. 1978).

Buprenorphine was developed initially for treatment of pain (COWAN et al. 1977; HEEL et al. 1979). When administered for pain relief, buprenorphine has an analgesic effect for appoximately 4h following intramuscular intravenous administration. However, some studies have reported a duration of action of as long as 10–14 h. In most studies, buprenorphine administered by a parenteral route appears to be more potent than morphine, but similarly effective. There have been some reports related to the use of buprenorphine in pain in which buprenorphine caused decreased systolic blood pressure and heart rate, and also respiratory depression. In some of those cases, the opiate antagonist naloxone failed to completely reverse the adverse effects of respiratory depression (LEHMANN et al. 1988).

Administration of buprenorphine to humans by intravenous route has shown that the apparent β terminal plasma half-life is approximately 3–5 h (BULLINGHAM 1983; OLLEY and TIONG 1988). The long duration of activity of buprenorphine must be attributed to the slow offset time of buprenorphine from opioid receptors. Buprenorphine, because of "first pass" metabolism, is relatively ineffective after oral administration, as is morphine, although with some effectiveness (as evidenced by the increasingly widespread use of oral morphine, irrespective of its "first pass" metabolism and thus less systemic bioavailability after oral administration). The long-acting pharmacokinetic properties of buprenorphine appear to be due to its persistent effects at the opiate receptor with a very slow "off time" rather than to an actually long pharmacokinetic profile. Buprenorphine has been administered sublingually in some studies of pain relief and in most studies of the treatment of opiate addiction (WEINBERG et al. 1988). This is because buprenorphine is very rapidly metabolized by the liver after oral administration with a "first pass" effect precluding administration by the oral route (WEINBERG et al. 1988). Studies have shown that, although quite variable, around 55% of buprenorphine is systemically available when administered sublingually (BULLINGHAM et al. 1983). Following sublingual administration, the time to peak plasma levels is approximately 20 min. The elimination profile suggests that the apparent plasma terminal half-life is around 3–5 h, but with a much longer final elimination of greater than 24 h. The primary route of metabolism is by N-demethylation. Buprenorphine and its metabolites are excreted predominantly in feces, with a small amount of both buprenorphine and N-demethyl buprenorphine and conjugates excreted in the urine (CONE et al. 1984). Studies are in progress to evaluate the effects of combined small amounts of naloxone (0.2 mg) with buprenorphine (0.2 mg) in a 1:1 ratio, and also in a ratio of 1:6, in an attempt to prevent illicit abuse by drug abusers of sublingual preparation of buprenorphine by the parenteral route. On-going studies will determine the optimal ratio and choice of antagonist (naloxone or naltrexone) to be combined with buprenorphine in the sublingual preparation of buprenorphine for use in the treatment of addictions.

In very early studies of buprenorphine conducted at the intramural program of NIDA, the Addiction Research Center, buprenorphine was found to be recognized by former opioid dependent humans as an opiate and to have some abuse liability (JASINSKI et al. 1978). Also, in those early studies, when buprenorphine was administered daily by a subcutaneous route with increasing doses until a daily maximum dose of 8 mg was reached, which was then administered over an additional 6 weeks, following discontinuation of buprenorphine under double blinded conditions, the signs and symptoms of narcotic withdrawal of significance were not observed until 14 days later when a morphine-like abstinence syndrome was recognized and opiate medications were required for relief. In other early studies, administration of daily doses of buprenorphine, again increasing to 8 mg over 14 days and maintained for 10 days, were associated with reports of opiate-like effects which were characterized as a feeling of contentment, but without a "high" or "rush" (MELLO and MENDELSON 1980; MELLO et al. 1982). In these studies, similarly no significant findings of opiate abstinence were found immediately after discontinuation of buprenorphine treatment.

II. Use of Buprenorphine: Safety, Efficacy, and Effectiveness

The possible utility of the use of the partial agonist buprenorphine for the long-term management of opiate addiction was first suggested by JASINSKI and colleagues in 1978, and shortly thereafter, their idea was advanced by early studies of MELLO and MENDELSON in both human and subhuman primate models (JASINSKI et al. 1978; MELLO and MENDELSON 1980; MELLO et al. 1982; BLAINE 1992; LEWIS and WALTER 1992; WOODS et al. 1992; BIGELOW and PRESTON 1992). The mechanisms of action of buprenorphine in the desired potential beneficial effects for the management of opiate addiction was not fully understood initially and was felt to be that of both an opioid agonist, primarily μ receptor directed, and an antagonist, primarily μ receptor directed, with some κ receptor activity of the antagonist type. The early studies have suggested that buprenorphine's action was one of reducing heroin self-administration, both by heroin addicts in the clinical research setting as well as by subhuman primates (MELLO and MENDELSON 1980, 1992; MELLO et al. 1982, 1985; MENDELSON and MELLO 1992). The mechanism of this action was unclear and initially was interpreted to be due to the opioid antagonist properties of buprenorphine. With further studies, it was increasingly shown that the beneficial effects of buprenorphine in the management of opiate addiction is by action and mechanism of action as a μ opioid receptor agonist. Both the actions of buprenorphine in preventing opiate withdrawal symptoms and decreasing heroin self-administration and in subsequent studies providing blockade of the effects of a short-acting opioid seem to be of prolonged duration of action, up to 24 h and possibly longer. However, unlike methadone and LAAM, this long duration of action is not due to the intrinsic pharmacokinetic properties of buprenorphine in humans, but rather, due to the very slow dissociation or "off-time" of buprenorphine from specific opioid receptors.

The rationale behind the initial and continuing studies, which are still on-going at this time to determine the efficacy and relative effectiveness of buprenorphine in the long-term management of opiate addiction, without and with cocaine dependency, as contrasted to maintenance agonist treatment with methadone or LAAM, or antagonist treatment with naltrexone, included several factors. Early studies suggested that buprenorphine did not induce physical dependence, and although more recent studies have shown that physical dependence does develop during chronic buprenorphine treatment, they still suggest that after cessation of buprenorphine treatment, the signs and symptoms of opiate withdrawal are significantly less severe than following cycles of heroin addiction, or following treatment with methadone (or possibly LAAM). This feature could be important, if abrupt discontinuation of maintenance treatment should be needed, which is rarely the case. In fact, however, slow dose reduction elimination may be beneficial with respect to the psychosocial treatments and adjustments which must occur prior to achieving the medication free abstinent state, especially since relapse to opiate use seems to occur in the majority (70%–90%) of all persons rendered abstinent from opiate abuse following discontinuation of pharmacotherapy, irrespective of the treatment modality or intervention used. Secondly, because of its mixed agonist and antagonist properties, the possibility of a lethal overdose in naive or weakly tolerant drug abusers, or in the setting of accidental poisoning, are postulated to be significantly less. Although respiratory depression has been shown in some studies to result from high dose buprenorphine administration, with a few reported cases of the complication of respiratory depression which are difficult to reverse with classical opioid antagonists such as naloxone, following high dose administration of buprenorphine in an analgesia or anesthesia setting, nevertheless, the magnitude and prevalence of respiratory depression following buprenorphine use in a naive or weakly tolerant drug abusers seems to be significantly less than following methadone or LAAM administration. Thirdly, early studies suggested that buprenorphine also might be beneficial in reducing cocaine dependence. However in most subsequent studies, it was found that the magnitude of reduction or cessation of cocaine use is similar in both moderate (60 mg) dose methadone maintenance treatment and in buprenorphine treatment at a dose of 8 mg sublingually daily. The possible disadvantage of buprenorphine, recognized from the beginning, was its potential abuse liability, which many reports of primary buprenorphine abuse and dependence from countries where buprenorphine is very commonly used in the management in pain. These reports have been primarily in the setting of self-administration of buprenorphine by the intravenous route of administration. When buprenorphine was administered by the intravenous administration in a laboratory research setting, similar reinforcing effects of buprenorphine have been observed, with euphorigenic properties which could lead to abuse liability. However, these effects seem to be much less when buprenorphine is administered by the sublingual route of administration, the only route which has been studied to date for the management of opiate dependency.

The first clinical studies of the abuse potential or liability of the analgesic buprenorphine, and also studies which were directed to the concept that buprenorphine might become a potential agent for treating opiate addiction, were conducted by JASINSKI and colleagues in 1978 (JASINSKI et al. 1978). These studies were conducted at the intramural program of the National Institute on Drug Abuse, the Addiction Research Center at Lexington. Different types of studies were conducted: first, a single dose study to determine the effects of buprenorphine when administered acutely, with the subcutaneous route of administration being used; second, to determine the effects of long-term administration of buprenorphine; and third, to determine whether buprenorphine would block the effects of acutely administered morphine, which was also administered in these studies by the subcutaneous route (as contrasted with the initial cross-tolerance studies of methadone where the short-acting medications were administered intravenously against the background of orally administered methadone) (DOLE et al. 1966a,b; JASINSKI et al. 1978). In the single dose studies in which buprenorphine was administered subcutaneously in doses of 0.2, 0.4, and 0.8 mg and compared with morphine administration doses of 15 and 30 mg, buprenorphine produced typically morphine-like effects. These effects included, in addition to the usual opiate-induced pupillary constriction, significant euphoria or "liking" on the "morphine-benzedrine" group ("MBG") scale used by the Addiction Research Center group for most of their studies of abuse liability. In these studies, it was found that the peak miotic effect of buprenorphine occured 6 h after subcutaneous drug administration, which was 2–3 h later than the peak following either acute morphine or methadone administration. Also, however, like methadone, and unlike morphine, this miosis reflecting continuing opioid agonist action persisted for 72 h after acute drug administration. These studies were conducted in former heroin addicts who were prisoner volunteers at the Addiction Research Center. In this population, it was of interest that euphoria, but little sedation, was observed after administration of any of the opioid agonists, indicating possible residual tolerance to the somnolence producing effects of all opioids when given acutely. In these studies, buprenorphine seemed to be 12–50 times more potent than morphine on a dose, depending on the objective or subjective measurement made. In a second sequence of studies, the initial dose of buprenorphine, 0.5 mg subcutaneously, was doubled after 2 days, and then at 4-day intervals until the 14th day, when a dose of 8 mg/day subcutaneously was achieved. During 30 days of chronic administration of buprenorphine in this earliest study, constipation and difficulty of urination were found. Subjects also reported "liking" for buprenorphine, similar to that observed while receiving morphine on a chronic basis. Buprenorphine also produced decreased diastolic blood pressure, as well as decreased systolic blood pressure. After 10–15 days, pulse rate was also decreased, but there were no changes at any time in body weight, caloric intake, or respiratory rate. No clinically significant laboratory test changes were observed. Of interest in this early study was that naloxone hydrochloride in large doses of 4 mg did not pre-cipitate narcotic withdrawal symptoms in subjects who had received chronic buprenorphine treatment for 45–52 days (JASINSKI et al. 1978). However, abrupt discontinuation of buprenorphine itself did produce delayed and gradual changes in the autonomic indices of opiate withdrawal, and sub-jective signs and symptoms of withdrawal, with peak abstinence symptoms observed on the ninth day following cessation of treatment. In an additional sequence of studies beginning on the eighteenth day of chronic bupre-norphine administration, effects of 15 and 30 mg morphine, administered subcutaneously, were determined with blockade of morphine effect found. In further studies of cross-tolerance, the dose of morphine was increased to 120 mg without effect. This early investigative team suggested that bupre-norphine might have abuse liability, because of its relative ability to produce typical morphine-like euphoric and other reinforcing opiate subjective effects. Also, physical dependence, though with opiate withdrawal symptoms of delayed onset following discontinuation of buprenorphine, possibly less severe than following morphine administration, might predict abuse liability. However, the euphoric producing properties seem to be less, in addition to the withdrawal signs being delayed and of less severity. The findings of attenuation of the effects of superimposed morphine for over 24 h suggested that buprenorphine had a long pharmacodynamic profile of action, even though the pharmacokinetics were later determined to be relatively short. Later studies were to show that this long duration of action also protected opiate dependent persons against the signs and symptoms of opiate with-drawal during a 24-h dosing interval.

Subsequent studies were performed by the Harvard group of MELLO and MENDELSON in a human laboratory setting (MELLO and MENDELSON 1980; MELLO et al. 1982). The subjects studied were ten male volunteers with a history of heroin abuse on a clinical research ward. After 5 days of a drug-free and medication-free period, buprenorphine or placebo was given in ascending doses, from 0.5 mg/day subcutaneously up to 8 mg/day over 14 days, and then maintained on 8 mg/day subcutaneously for 10 days. An operant study of heroin self-administration was then conducted. It was found that buprenorphine maintained subjects took significantly less heroin than subjects maintained on placebo, with those treated with buprenorphine taking only 2%–31% of the available heroin, whereas the placebo main-tained subjects self-administered 90%–100% of the available heroin. Also in these studies, tolerance to the opiate agonist-like effects of bupre-norphine were observed to develop gradually over the treatment period. Later, other studies of buprenorphine effects, including blockade of opioid challenge, were conducted in buprenorphine-treated opioid dependent persons. Buprenorphine was administered by the sublingual route in ascending doses of 2, 4, 8 (REISINGER 1985; SEOW et al. 1986; BICKEL et al. 1988a,b; JOHNSON et al. 1989, 1992; KOSTEN et al. 1989, 1993; KOSTEN and KLEBER 1991; RESNICK et al. 1991, 1992; JOHNSON and FUDALA 1992).

Challenges were carried out using the short-acting opiate hydromorphone in multiple subcutaneous doses, as well as a placebo dose; hydromorphone challenges were conducted after 10–14 days of chronic administration of buprenorphine dose, and 24 h after the last medication was received. It was found that 2 mg buprenorphine did not alter the response to hydromorphone challenge, which thus produced a dose-related change in both physiological and subjective measures. However, with increasing doses of buprenorphine, the hydromorphone effects were attenuated with maximal blockade by the highest dose of buprenorphine studied. Buprenorphine itself had a few dose related effects, including very mild respiratory depression. Overall, the profile of buprenorphine agonist effects were different from those observed by the earlier study of Jasinski, but attributed primarily to the route of administration. All of these studies in humans, coupled with the subhuman primate studies conducted by the group of Mello and Mendelson, suggested that buprenorphine could be effective in the maintenance treatment of addiction.

The first reported use of buprenorphine for the actual treatment of addiction came from Belgium, from the group of Reisinger (Reisinger 1985). An interim treatment program was developed in which buprenorphine was prescribed to 65 heroin addicts with the aim of attempting to develop and evaluate a long-term maintenance treatment using buprenorphine. Some patients were seen only once or twice; 34 patients, however, received buprenorphine for a period of 2–17 months. Of those, five finished treatment with dose reduction, dose elimination, and remained drug-free for 3–12 months. Three left treatment against medical advice and were lost in follow-up, and the remaining 26 were still in regular or irregular treatment at time of the initial report. Buprenorphine was prescribed in each case in progressively increasing doses to avoid the occurrence of side effects such as nausea, vomiting, or respiratory depression. These side effects were not observed when the initial doses were 2–4 mg/day. The advantages of buprenorphine in this study were considered to be that physical dependence was less than that developed on methadone, and thus dose reduction and elimination from buprenorphine was easier. Secondly, though not stated in the published paper, the legal situation in Belgium was that buprenorphine was not under the same rigorous regulation as was methadone as used in maintenance treatment. However, in the report from Reisinger, action of buprenorphine was misinterpreted to be "like naltrexone, a relatively pure antagonist," and thus, "more suitable for subjects of the higher level of psychosocial functioning" (Reisinger 1985). Further studies, including interdisciplinary studies, demonstrate that the major effects of buprenorphine were those of an opioid agonist.

From 1986 onward, several reports have appeared of the studies of potential effectiveness of buprenorphine in the maintenance treatment of opiate addiction. In one study from the University of Western Australia, 32 volunteer heroin addicts were studied who had been prescribed buprenorphine by general practitioners for a mean of 7.4 months, since buprenorphine, unlike methadone, was not under regulatory control in Australia for the management of addiction (SEOW et al. 1986). Before the beginning of the study, all subjects were stabilized on buprenorphine 0.6–1.2 mg intramuscularly each day for 1 week and then were randomly assigned to one of two groups, in which buprenorphine was administered sublingually at doses of 2 mg/day or 4 mg/day. Buprenorphine was given during the first, second, fourth and fifth weeks of the study, with placebo given during the third week of the study. Dextropropoxyphene (200 mg 4x/day) and clonidine (0.15 μg 4x/day) were given for management of any signs or symptoms of opiate withdrawal, and oxazepam (300 mg 4x/day) for anxiety, along with tenazepam (20 mg at night) for symptoms of insomnia. This study is difficult to interpret, with the multiple medications administered. Of the 32 subjects who entered the study, 21 completed the 5 week trial. Opiate-like effects, but also fewer withdrawal symptoms, were experienced in the group receiving buprenorphine at a dose of 4 mg sublingually each day. The opiate-like effects included "nodding off," a "rush," and other euphoric-like signs and symptoms, including feeling relaxed, being talkative, and having constipation, difficulty in urination, and pruritis. Overall, both doses of buprenorphine were well tolerated, but the data analysis suggested that the condition of subjects receiving 4 mg sublingually of buprenorphine per day were better stabilized than those receiving the lower doses. However, the doses of buprenorphine used were insufficient to suppress the effects of superimposed short-acting opiate drugs. During the trial, 50% of the specimens were positive for heroin; in contrast, only 23% of urines of patients being maintained on methadone, 50–80 mg/day, in this same setting were positive for illicit opiates. In another study of the use of buprenorphine in the management of opiate addiction reported from Sweden, the group of Gunne first studied 12 heroin addicts on the eighth day of opiate withdrawal and eight healthy volunteers. Each subject was given a single intramuscular dose of buprenorphine 0.6 mg, and subjective and objective findings were measured (BLOM et al. 1987). Only two opiate-like symptoms were observed, decreased tension and decreased dysphoria. In contrast were buprenorphine's effects on the control group; it caused multiple narcotic-like effects, including sleepiness. These findings from acute buprenorphine administration to heroin addicts suggested possible utility of buprenorphine in treatment of addiction, though further studies were not reported by this group.

From 1988 onward, numerous carefully conducted studies, many of which have been placebo-controlled double-blinded trials, have been reported from various investigators in the United States in which the effectiveness of buprenorphine has been assessed for the possible treatment of heroin addiction. An outpatient 90 day (13 week) "short-term maintenance" study was conducted in 45 heroin addicts (BICKEL et al. 1988a). In the study, 2 mg/day of buprenorphine administered sublingually versus 30 mg/day of methadone for 3 weeks, followed by 3 weeks of dose reduction and 6 weeks of placebo medication were evaluated. Less than 10% of subjects in both groups completed the 13 weeks; 15% remained in a program for 6 weeks and no significant differences were found between the two groups with respect to retention time, other uses of illicit opiates, or symptoms. Challenge studies were performed, in which hydromorphone (6 mg) was administered to some subjects in each of the two treatment groups, buprenorphine sublingually (2 mg) and methadone (30 mg) p.o. (Bickel et al. 1988b). The challenge study was conducted during the second week of stabilization. It was found that methadone attenuated the opiate effects of hydromorphone to a greater extent, both with respect to the physiological and the self report measures, than did buprenorphine at the doses used. Another "detoxification" or short-term maintenance open study, 30 days with abrupt discontinuation, was conducted (Kosten and Kleber 1991). Sixteen opiate dependent patients were assigned to treatment with buprenorphine for 30 days at three different doses: 2 mg, 4 mg, or 8 mg sublingually. Only one of the 16 subjects left treatment before the end of the 30 day period. Illicit opiate use was infrequent, with 22% of the urines containing illicit opiates. At the end of the study, buprenorphine was stopped abruptly, and withdrawal symptoms, observed after stopping buprenorphine, were related to the dose of buprenorphine used. The intramural program of NIDA, the Addiction Research Center in Baltimore, conducted a study which was reported in 1989 and 1990 on use of buprenorphine for the management of opiate and heroin dependency (Johnson et al. 1989; Fudala et al. 1990). In the first report, a rapid dose induction procedure for initiating buprenorphine treatment was described in which ascending doses of buprenorphine from 2 to 4 to 8 mg sublingually were studied. A transition from heroin to buprenorphine in 4 days was achieved, with stabilization on the 8 mg sublingual dose. Overall, the subjects reported decrease in withdrawal symptoms, and improvement in feeling of well-being. These subjects were continued on buprenorphine treatment, 8 mg/day sublingually, until study day 18. They were then divided into two groups: one who received buprenorphine daily, and another who received buprenorphine or placebo on alternate days until day 36. From day 37 until day 52, all subjects received placebo. It was found that the patients who received buprenorphine on alternate days had much greater craving for illicit opiate use, with increased complaints of dysphoria and opiate withdrawal. However, in both groups after termination of buprenorphine, withdrawal symptoms were modest. Mild to moderate withdrawal symptoms appeared initially, peaked at around posttreatment day 3–5 and lasted for approximately 10 days. Daily administration of buprenorphine provided control of the signs and symptoms of opiate withdrawal. In 1991, in a follow-up study of 16 heroin addicts who had been in a trial treatment with buprenorphine for two months in an open study, receiving 0.6 to 3.9 mg/day sublingually of buprenorphine, eight of the 16 subjects were "abstinent" from heroin (Resnick et al. 1991, 1992). The medication in this study was delivered subcutaneously and the authors concurred with on-going studies of others that a sublingual form would be advantageous. A further study from this group described 85 heroin addicts who had been "unwilling to receive methadone maintenance or treatment in a therapeutic community" and who were thus placed on a single-blinded study using low dose sublingual buprenorphine. All subjects were maintained on 8 mg or less of buprenorphine administered sublingually. All doses were administered daily under observation; after 4 to 12 weeks on maintenance treatment, subjects were randomized on a single-blinded basis either to receive dose reduction and elimination, or stable dose buprenorphine for seven additional weeks. Subjects who underwent dose reduction were found to develop abstinence symptoms, with "low energy" being the most common symptom, associated with drug-seeking behavior.

In 1991, the Yale treatment research group of Kosten and Kleber reported a study in methadone maintained patients who had a concomitant problem with heroin addiction and cocaine abuse to determine the possible benefit of buprenorphine treatment (KOSTEN et al. 1989c; KOSTEN and KLEBER 1991). In this initial open study, it was suggested that buprenorphine might be more effective than methadone in managing cocaine abusing former heroin addicts (KOSTEN et al. 1989b,c, 1992a; KOSTEN and KLEBER 1991). In one study, 12 subjects who had been maintained on methadone at a dose of 47 ± 8 mg/day were switched to buprenorphine at a dose of 3.8 ± 0.6 mg/day sublingually for 1 month. While on methadone, 20% of the urines monitored over the 6 months prior to study were positive for cocaine, but during the study itself, when all subjects were on an open trial of buprenorphine, the rate of urine positive for cocaine dropped to 2% in the buprenorphine treated group. In an analysis of three unrelated patient groups, 41 subjects receiving 3.2 ± 1.6 mg/day of buprenorphine sublingually, were compared to 61 patients receiving methadone 43 ± 8 mg daily p.o. and 36 patients receiving naltrexone during the first month of treatment. Urines were positive for cocaine in 24% of the methadone subjects not in any special study group, and 3% in a buprenorphine treatment research study group; overall, illicit drugs in urines did not differ between the two groups, with 33% for the methadone groups and 37% for the buprenorphine groups.

In 1993, the Yale group reported a controlled study in which buprenorphine at two doses, 2 mg and 6 mg sublingually administered daily, was compared with methadone at two doses, 35 mg and 65 mg orally, delivered daily during a 24 week maintenance study period in 125 opiate dependent persons (KOSTEN et al. 1993). It was found that the 6 mg dose of buprenorphine was superior to the 2 mg dose in reducing illicit opiate use; however, the higher dose did not improve treatment retention. Self-reports of illicit opiate use declined significantly in all groups. By the third month, significantly more heroin abuse was reported by subjects receiving 2 mg than by to those receiving 6 mg buprenorphine. Both doses of methadone were superior to the low dose of buprenorphine with respect to reducing illicit pretense. The reduction in illicit opiate use was most satisfactory in the study patients receiving methadone 65 mg/day per month: use of illicit opiates occurred on 1.7 days/month of the 24 week study. In the group receiving 35 mg/day of methadone, illicit opiate use occurred on 2.8 days/ month over the 24 week study. In the high dose buprenorphine study group receiving 6 mg sublingually each day, heroin use occurred 4.0 days/month during the 24 weeks of study, and during the low dose buprenorphine treatment, heroin use occurred 6.6 days/month. These authors had hypo-thesized prior to conducting the study that "buprenorphine would be equi-valent to methadone in efficacy"; this hypothesis "was not supported" (KOSTEN et al. 1993). They also found that treatment retention was sig-nificantly better on either dose of methadone. Abstinence from any illicit opiate use for at least 3 weeks was also more common in subjects receiving either dose of methadone than subjects receiving either dose of buprenor-phine. The conclusions of these authors was "methadone was clearly superior to these two buprenorphine doses" (KOSTEN et al. 1993). However, illicit opiate use was reduced more by the higher than the lower dose of buprenor-phine, suggesting that doses of buprenorphine greater than 6 mg/day sub-lingually should be studied for potential efficacy and effectiveness.

Another study from the intramural program of NIDA at Addiction Research Center in Baltimore assessed the efficacy of buprenorphine again for short-term maintenance or detoxification in a study in which bupre-norphine 8 mg/day sublingually was compared with low dose methadone (20 mg/day orally) or moderate dose methadone (60 mg/day) in a 17 week maintenance phase followed by eight weeks detoxification (JOHNSON and FUDALA 1992). In these studies, volunteers seeking treatment for addiction were enrolled. Retention rates were significantly greater for the 8 mg bupre-norphine (42%) than for low dose, 20 mg/day, methadone (20%), but similar to the retention rates in moderate dose methadone treatment (60 mg/ day), over the 17 week maintenance phase (buprenorphine 8 mg 44% and methadone, 60 mg 42%, respectively). The percentage of urine samples negative for opioids was significantly greater for moderate dose methadone (60 mg/day orally) and buprenorphine (8 mg sublingually) than for the low dose methadone (20 mg/day). During the detoxification phase, no difference was observed among the groups with respect to urine samples negative for opioids, with relapse to illicit opiate use similar in the two groups. The percentage of patients who received ancillary services were the same in each of the groups. Therefore, it was concluded that buprenorphine 8 mg/day sublingually was as effective as methadone 60 mg/day, and both were superior to methadone 20 mg/day (a methadone dose which has been shown since 1964 to be inadequate for most subjects). However, these studies failed to show that buprenorphine was any more effective than methadone in reducing illicit cocaine use.

Illicit cocaine use has been shown by many groups to decrease during effective treatment with methadone maintenance when adequate doses of methadone (60 to 120 mg/day) is combined with appropriate counseling and psychosocial services (HANBURY et al. 1986; CHAISSON et al. 1989; BORG et al. 1995a,b). With similar findings made in appropriately controlled trials of buprenorphine compared with methadone, the question was raised in the human laboratory setting of whether buprenorphine had any effects on the subjective response to cocaine. In a study reported by the group of KOSTEN and colleagues, five inpatient study subjects dependent on both cocaine and heroin were first detoxified from opiates and then given 5 days of double-blinded treatment with active or placebo buprenorphine 2 mg/day sublingually, followed by a cross-over to the opposite treatment regimen; on days 3 and 5 of each treatment, intranasal cocaine was administered at a dose of 2 mg/kg (ROSEN et al. 1993). It was found that buprenorphine treatment "significantly enhanced patients' ratings of cocaine induced pleasurable effects." This buprenorphine enhancement of subjective cocaine effects seemed to be more prominent on day 3 than day 5.

In another study, the group of Mendelson, Mello, and colleagues performed in-patient human laboratory studies on the acute interactions of buprenorphine with intravenous cocaine and morphine, as part of the development of an investigational new drug application for buprenorphine for the maintenance treatment of addiction (TEOH et al. 1993). Although their own preclinical studies had suggested that buprenorphine might be useful in the treatment of dual dependence of cocaine and opiates, with decreased self-administration of cocaine during sustained buprenorphine administration in a monkey model (similar to the findings made by earlier investigators with respect to a reduction of cocaine self-administration in a rodent model by morphine and other short-acting pure opiate agonists), further clinical studies were equivocal with respect to the cocaine self-administration reducing effects of buprenorphine. However, the goal of the later study was to determine the safety of buprenorphine when used concomitantly with cocaine or with morphine. Twenty subjects with diagnoses of concurrent cocaine and opiate dependency were randomly assigned to maintenance treatment, either with a single daily dose of 4 mg or 8 mg of sublingual buprenorphine for 21 days. The physiological effects of a single-blinded challenge dose of cocaine, 30 mg intravenously, and on a different day morphine 10 mg/day intravenously, and also intravenous saline placebo, before and during buprenorphine maintenance, were determined. Subjects had undergone methadone detoxification prior to initiating buprenorphine, followed by a 9 day medication-free period. Three baseline challenges were conducted on separate days during the drug-free period on days 7, 8, and 9. Cardiovascular responses to both cocaine and morphine were similar during the medication-free period and during the subsequent buprenorphine maintenance conditions. Also, respiration and temperature changes were similar before and during buprenorphine treatment. Mild opioid agonist-like effects were noted during buprenorphine induction and maintenance, especially headache, sedation, nasal discharge, abdominal discomfort, and anxiety; these decreased within around 14 days. These findings suggest that daily maintenance on buprenorphine is not associated with any acute or toxic interaction with a single dose of intravenous cocaine or morphine. In other studies, this research group showed that in chronic cocaine polydrug abusers who were also opiate dependent and had undergone buprenorphine research treatment, or who had undergone drug-free treatment, cerebral blood flow abnormalities existed when initially studied by SPECT and that these abnormalities improved during medication-free or buprenorphine treatment (HOLMAN et al. 1993).

The group of Bigelow has attempted to determine the potential abuse liability of buprenorphine in methadone maintained subjects by studying the effects of buprenorphine, hydromorphone, and naloxone in methadone maintained volunteer subjects. In this study six volunteer former heroin addicts maintained continuously on very low dose methadone 30 mg/day underwent pharmacological challenges with buprenorphine 2–3 times each week (STRAIN et al. 1992). The challenges were conducted on a double-blind basis with injection of buprenorphine, doses ranging from 0.5–0.8 mg, hydromorphone at doses from 5–10 mg, and naloxone at doses of 0.1 or 0.2 mg, or saline. The study drug challenge injections were given 20 h after the last dose of methadone. Naloxone and hydromorphone were reported to produce characteristic "antagonist-like and agonist-like effects" on subjective, objective, and physiological indices. As expected, the 0.2 mg dose of naloxone caused far more opiate withdrawal effects than did the 0.1 dose. The 5 mg dose of hydromorphone produced modest euphoric effects. Many more opiate-like effects were identified when 10 mg of hydromorphone was administered to the subjects maintained on 30 mg/day of methadone. However, neither doses of buprenorphine were consistently identified as either opioid agonist or antagonist by any one of the subjective or objective measures; also, buprenorphine seemed to produce few effects in the patients maintained on very low dose methadone. Further challenge or cross-tolerance studies in patients on moderate to high doses of buprenorphine, compared with patients receiving appropriately adequate maintenance doses of methadone of 60–120 mg/day, have not been conducted.

Further studies are in progress to determine the efficacy and effectiveness of buprenorphine in the management of long-term opiate addiction. It is possible that one of the characteristics of this medication, that it apparently produces less tolerance and physical dependence and thus allows more rapid dose reduction-elimination, would permit this medication for treatment of moderate term heroin and other opiate addicts who do not meet the United States federal guidelines criteria for entry into methadone maintenance treatment, that is, persons with less than a 1 year history of daily, multiple dose use of a short-acting opiate with the development of tolerance, physical dependence, and drug-seeking behavior. The original concern at the initiation of methadone maintenance research in 1964 was not to cause, by treatment itself with a long-acting opioid agonist, any alterations in normal physiology, through the mechanisms of neural plasticity and adaptation, which in turn might be persistent or permanent and which had not already been caused by illicit use of a short-acting opiate (or existing before exposure to exogenous opiates). Increasing information over 30 years of rigorous studies have documented that methadone allows normalization of disrupted physiological functions, rather than changes towards abnormalities of any type (KREEK 1992a,b,c). However, there is still understandable reticence to recommend a treatment which may be very long-term or even lifelong treatment for groups of heroin abusers or short-term addicts who may respond to a different short-term form of treatment. For early heroin abusers-addicts not meeting the current guideline criteria for addiction and yet who clearly have long-standing disruption (i.e., more than 6 months of daily multiple use of a short-acting illicit narcotic, with the development of tolerance, physical dependence, and drug-seeking behavior), short to intermediate term treatment with buprenorphine followed by a medication-free period might provide an acceptable alternative. Then any relapse or impending relapse would provide an indication for entry into long-term maintenance treatment with methadone or LAAM. Also, for those heroin addicts meeting the criteria for treatment with methadone or LAAM, but who are themselves reticent to accept treatment for any reason, initial treatment with buprenorphine during a more prolonged decision making period might be found to be appropriate. Such treatment would allow a protected time with potentially no continuing illicit drug use, and thus, no exposure to infectious diseases such as hepatitis and HIV infections. Although no studies have been performed in which a transition from heroin addiction to long-term treatment is made with buprenorphine as the first medication and then with transfer of patients to methadone or LAAM, there is not reason based on currently available information to suggest that this would not be possible; such an approach is both logical and potentially desirable. Also, although initially it was feared that rapid transition from methadone treatment to buprenorphine would not be possible because of the antagonist properties of buprenorphine, it has been repeatedly shown that this transition in many patients is smooth and that the doses of buprenorphine used in treatment studies to date, ranging up to 16 mg sublingually each day and in some cases even more, have not caused precipitation of withdrawal symptoms in methadone maintained patients, as was feared based on the laboratory observations of some antagonist-like properties of buprenorphine and the less well documented inverted bell-shaped or "U-shaped curve" of buprenorphine's pharmacodynamic action in humans.

Buprenorphine was first marketed as an analgesic in Spain in 1988 and is widely used in that country, possibly accounting for the very early reports of primary abuse of buprenorphine in that region. Following initial reports of some buprenorphine abuse, there have been several more recent reports of buprenorphine abuse amongst opiate addicts in several countries. Primary buprenorphine addiction seems to be a problem in several countries, including New Zealand, Australia, Germany, Scotland, and England (e.g. QUIGLEY et al. 1984; Possilpark Group 1993). Further studies by the NIDA intramural group of Cone et al. have reaffirmed the earlier findings of subjective and physiological opiate-like effects of intravenously administered buprenorphine, including euphoric-like effects, thus confirming the abuse liability of buprenorphine (Pickworth et al. 1993). These findings have led to the current active development of a new sublingual form of buprenorphine which will combine buprenorphine with naloxone or naltrexone, specific opiate antagonists. Therefore, any attempt to self-administer the sublingual preparation of buprenorphine by an intravenous route would not yield any euphoric effects since the opioid antagonist would block the opioid agonist effects of buprenorphine and yet, because of the very limited systemic bioavailability of naloxone when administered orally, not adversely affecting the desired action of buprenorphine. A similar formulation had been developed years ago, when it was misperceived that an oral preparation of methadone might have an intravenous abuse liability (Kreek 1973c). However, attempts to develop a combined oral methadone-naloxone preparation were discontinued when it was recognized that illicit use of methadone was not possible given the nonsoluble formulations of the oral preparations used in maintenance treatment and that all illicit use of methadone by street opiate addicts was by the oral route. In early studies of this combined preparation, it was found that the addition of the antagonist precipitated gastrointestinal signs and symptoms of withdrawal at dose ratios added, which at the same time did not usually alter the effectiveness of methadone in the prevention of systemic narcotic withdrawal symptoms and also drug craving (Kreek 1973). However, dose response variability was considerable, so that systemic withdrawal symptoms did occur in some subjects. These are problems which will need to be confronted in the development of a buprenorphine antagonist combination preparation.

III. Neuroendocrine Effects of Partial Agonists (Mixed Agonists/Antagonists)

The earliest reports of the neuroendocrine effects of buprenorphine come from studies performed in conjunction with use of buprenorphine as an analgesic. In one study conducted in subjects undergoing surgery, it was shown that buprenorphine, similar to other opioid agonists, causes an abrupt decrease in plasma cortisol levels when buprenorphine was given at doses of 3 mg intramuscularly or 0.3 mg intravenously (McQuay et al. 1980). This decrease was effected in the setting of significantly increased plasma cortisol levels, similar to previous findings for both morphine and fentanyl in analgesia related studies. It was shown that buprenorphine, presumably acting primarily as a μ opioid agonist, was able to "obtund the stress response during and after surgery" (McQuay et al. 1980). The reduction in cortisol levels persisted for approximately 90 min and plasma cortisol levels rose at a later time point following the administration of buprenorphine.

A subsequent neuroendocrine study was carried out in recently abstinent male heroin addicts who had been induced on buprenorphine or buprenorphine placebo medication for the purpose of studying heroin self-administration in the setting of buprenorphine treatment (MENDELSON et al. 1982). An initial dose of buprenorphine (0.5 mg/day subcutaneously) was gradually increased to a final dose of 8 mg subcutaneously per day. On the 15th day of drug induction, the neuroendocrine studies were performed and compared with similar study results obtained after a 5 day medication-free preinduction period. It was found that buprenorphine causes a significant increase in prolactin levels, coupled with significant suppression of plasma LH levels. During a subsequent 10 day period of buprenorphine maintenance at a dose of 8 mg/day subcutaneously, LH levels remained suppressed and prolactin levels continued to be elevated, with no apparent tolerance or adaptation developing during this short period of buprenorphine administration. These findings in humans were further supported by the fact that buprenorphine at these doses is acting essentially exclusively as a μ opioid agonist in humans. In a further study by Mendelson, Mello and colleagues, the phenomenon of buprenorphine stimulation of plasma prolactin levels was further studied in six healthy male subjects with no history of drug abuse (MENDELSON et al. 1989). Each subject was studied on six occasions over a 27 day period. At each study timepoint, a subject could receive a simultaneous intramuscular injection of either buprenorphine 0.3 mg and saline, or buprenorphine 0.3 mg and naloxone in doses of 0.15, 0.3, 0.45 or 0.6 mg, or alternatively, two simultaneous injections of saline to blind and procedure. This study illustrated that buprenorphine 0.3 mg effected a significant and sustained increase in plasma levels of prolactin, with significant elevations in plasma prolactin levels observed at 30 and 55 min. When naloxone was administered simultaneously with buprenorphine, it was shown that doses of 0.3, 0.45, and 0.6 mg naloxone significantly decreased the buprenorphine induced rise in prolactin levels. However, the lowest dose of naloxone used, 0.15 mg subcutaneously, did not attenuate the buprenorphine induced increase in prolactin levels. These findings that naloxone significantly inhibited prolactin stimulating effects of buprenorphine in a clear dose dependent mode further documented the fact that buprenorphine is acting as a μ opioid agonist in these neuroendocrine effects. In a study conducted by a group in Finland, 12 opiate naive healthy students were studied (SAARIALHO-KERE et al. 1987). Buprenorphine was administered sublingually at a dose of 0.4 mg. Two subjects had been treated for 8 days with amitryptilime 25 or 50 mg/day; these studies were therefore somewhat complicated to interpret. However, buprenorphine seemed to increase plasma levels of prolactin in the amitryptilime-buprenorphine treated group as contrasted to those treated with amitryptilime alone.

   In studies carried out by the combined groups of KOSTEN at Yale and KREEK at the Rockefeller University and their colleagues, the neuroendocrine effects of low dose buprenorphine, as administered in a research treatment mode, with 2 mg of buprenorphine sublingually administered for 1 month to 14 former methadone maintained patients and six former heroin addicts treated initially with buprenorphine, were studied (KOSTEN et al. 1992b). Neuroendocrine indices were measured prior to and after the 1 month of buprenorphine experimental treatment. In this study, it was shown, as had been documented previously, that β-endorphin levels were normal in the stabilized methadone maintained patients prior to study, and remained normal during buprenorphine treatment, with no significant reduction in β-endorphin levels. However, in the group of heroin addicts who had not been previously restored to a steady state by methadone treatment and who were placed directly upon buprenorphine, the plasma levels of β-endorphin remained lowered during the 1 month of buprenorphine treatment. In this study, the effect of low dose buprenorphine on unstabilized heroin addicts appeared to be those of a μ opioid agonist, with normalization not achieved within 1 month. Very early studies from the Rockefeller University showed that it takes 2–3 months of stable dose methadone maintenance treatment to restore levels and circadian rhythms of levels of both of the POMC stress related peptides, ACTH and β-endorphin, as well as the resultant levels of peripheral release of cortisol by the adrenal cortex to normal (KREEK 1972; 1973a,b) Then, during chronic long-term treatment with methadone, the Rockefeller group has shown that all hormones of the stress responsive hypothalamic-pituitary-adrenal axis become and remain normalized, with normal plasma levels and circadian rhythm of levels of ACTH, β-endorphin and also the peripheral responsive steroid cortisol. Thus, it is clear that studies of former heroin addicts in longer term buprenorphine treatment will be needed to see if normalization occurs during use of this pharmacokinetic short-acting, but pharmacodynamic long-acting, partial opioid agonist. Also of great importance will be to conduct similar studies in heroin addicts treated with much higher doses of buprenorphine, which other studies are now suggesting must be used to achieve both efficacy and effectiveness in the management of opiate addiction.

Acknowledgments. This work was conducted with support from NIH-NIDA P50-DAO5130, NIH-NIDA KO5-DA00049, and the New York State Office of Alcoholism and Substance Abuse Services. Appreciation is given to Ken Leung for manuscript preparation and reference collection and organization, Neil Maniar and Margaret Porter for proofreading and editing, Jennifer Sudul for final preparation of manuscript and Charlotte Kaiser for help in the literature search for the preparation of the bibliography of this chapter.

References

Albeck H, Woodfield S, Kreek MJ (1989) Quantitative and pharmacokinetic analysis of naloxone in plasma using high performance liquid chromatography with electrochemical detection and solid phase extraction. J Chromatog 488:435–445

Altman JL, Meyer RE, Mirin SM, McNamee HB (1976) Opiate antagonists and the modification of heroin self-administration behavior in man: an experimental study. Int J Add 11(3):485–499

Atkinson RL, Berke LK, Drake CR, Bibbs ML, Williams FL, Kaiser DL (1985) Effects of long-term therapy with naltrexone on body weight in obesity. Clin Pharm Ther 38:419–422

Bickel WK, Stitzer ML, Bigelow GE, Liebson IA, Jasinski DR, Johnson RE (1988a) A clinical trial of buprenorphine: comparison with methadone in the detoxification of heroin addicts. Clin Pharm Ther 43(1):72–78

Bickel WK, Stitzer ML, Bigelow GE, Liebson IA, Jasinski DR, Johnson RE (1988b) Buprenorphine: dose-related blockade of opioid challenge effects in opioid dependent humans. JPET 247:47–53

Bigelow GE, Preston KL (1992) Assessment of buprenorphine in a drug discrimination procedure in humans. NIDA Research Monograph 121:28–37

Blachley PH (1975) Naloxone for diagnosis in methadone programs. J Am Med Assoc 244:334–335

Blaine JD (1992) Introduction (to buprenorphine: an alternative treatment for opioid dependence). NIDA Research Monograph 121:1–4

Blom Y, Bondesson U, Gunne LM (1987) Effects of buprenorphine in heroin addicts. Dr Alc Depend 20:1–7

Blumberg H, Dayton HB (1974) Naloxone, naltrexone, and related noroxymorphines. In: Braude MC, Harris LS, May EL, Smith JP, Villarreal JE (eds) Narcotic antagonists. Advances in biochemical psychopharmacology, vol 8. Raven, New York, pp 33–43

Blumberg H, Dayton HB, George M, Rapaport DN (1961) N-allylnoroxymorphone: a potent narcotic antagonist. Fed Proc 20:311

Borg L, Broe DM, Ho A, Kreek MJ (1995a) Cocaine abuse is decreased with effective methadone maintenance treatment at an urban Department of Veterans Affairs (DVA) Program. In: Harris LS (ed) Problems of Drug Dependence 1994. Proceedings of the 56th Annual Scientific Meeting of the College on Problems of Drug Dependence. NIDA Research Monograph Series, Rockville, MD 153:17

Borg L, Ho A, Peters JE, Kreek MJ (1995b) Availability of reliable serum methadone determination for management of symptomatic patients. J Add Dis (in press)

Brahen LS, Capone T, Wiechert V, Desiderio D (1977) Naltrexone and cyclazocine: a controlled treatment study. Arch Gen Psych 34:1181–1184

Bullingham RES, McQuay HJ, Moore RA (1983) Clinical pharmacokinetics of narcotic agonist-antagonist drugs. Clin Pharm 8:332–343

Chaisson RE, Bacchetti P, Osmond D, Brodie B, Sande MA, Moss AR (1989) Cocaine use and HIV infection in intravenous drug users in San Francisco. JAMA 261:561–565

Chou JZ, Albeck H, Kreek MJ (1993) Determination of nalmefene in plasma by high-performance liquid chromatography with electrochemical detection and its application in pharmacokinetic studies. J Chrom 613:359–364

Cohen MR, Cohen RM, Pickar D, Weingartner H, Murphy DL (1983) High-dose naloxone infusions in normals: dose-dependent behavioral, hormonal, and physiological responses. Arch Gen Psych 40:613–619

Cone EJ (1973) Human metabolite of naltrexone (N-cyclopropylmethylnoroxymorphone) with a novel C-6 isomorphine configuration. Tetrahedron Lett 28: 2607–2610

Cone EJ, Gorodetzky CW, Yousefnejad D, Buchwald WF, Johnson RE (1984) The metabolism and excretion of buprenorphine in humans. Dr Met Disp 12(5): 577–581

Cowan A (1974) Evaluation in nonhuman primates: Evaluation of the physical dependence capacities of oripavine-thebaine partial agonists in patas monkeys. In: Braude MC, Harris LS, May EL, Smith JP, Villarreal JE (eds) Narcotic antagonists. Advances in biochemical psychopharmacology, vol 8. Raven, New York, pp 427–438

Cowan A, Lewis JW, Macfarlane IR (1971) Analgesic and dependence studies with oripavine partial agonists. Br J Pharmacol 43:461P–462P

Cowan A, Lewis JW, Macfarlane IR (1977) Agonist and antagonist properties of buprenorphine, a new antinociceptive agent. Br J Pharmacol 60:537–545

Culpepper-Morgan JA, Inturrisi CE, Portenoy RK, Foley K, Houde RW, Marsh F, Kreek MJ (1992) Treatment of opioid induced constipation with oral naloxone: a pilot study. Clin Pharm Ther 23:90–95

Des Jarlais DC, Wenston J, Friedman SR, Sotheran JL, Maslansky R, Marmor M (1992) Crack cocaine use in a cohort of methadone maintenance patients. J Sub Ab Tr 9:319–325

Dixon R, Hsiao J, Taaffe W, Hahn E, Tuttle R (1984) Nalmefene: radioimmunoassay for a new opioid antagonist. J Pharm Sci 73(11):1645–1646

Dixon R, Howes J, Gentile J, Hsu H-B, Hsiao J, Garg D, Weidler D, Meyer M, Tuttle R (1986) Nalmefene: intravenous safety and kinetics of a new opioid antagonist 39:49–53

Dixon R, Gentile J, Hsu H-B, Hsiao J, Howes J, Garg D, Weidler D (1987) Nalmefene: safety and kinetics after single and multiple oral doses of a new opioid antagonist. J Clin Pharm 27:233–239

Dole VP, Nyswander ME, Kreek MJ (1966a) Narcotic blockade. Arch Intern Med 118:304–309

Dole VP, Nyswander ME, Kreek MJ (1966b) Narcotic blockade: a medical technique for stopping heroin use by addicts. Trans Assoc Am Physicians 79:122–136

Fudala PJ, Jaffe JH, Dax EM, Johnson RE (1990) Use of buprenorphine in the treatment of opioid addiction. II. Physiologic and behavioral effects of daily and alternate-day administration and abrupt withdrawal. Clin Pharm Ther 47: 525–534

Goldstein A (1976) Heroin addiction: sequential treatment employing pharmacologic supports. Arch Gen Psych 33:353–358

Gonzalez JP, Brogden RN (1988) Naltrexone: a review of its pharmacodynamic and pharmacokinetic properties and therapeutic efficacy in the management of opioid dependence. Drugs 35:192–213

Greenstein RA, Arndt IC, McLellan AT, OBrien CP, Evans B (1984) Naltrexone: a clinical perspective. J Clin Psych 45[9(2)]:25–28

Hanbury R, Sturiano V, Cohen M, Stimmel B, Aguillaume C (1986) Cocaine use in persons on methadone maintenance. Adv Alcohol Subst Abuse 6:97–106

Hahn EF, Lahita R, Kreek MJ, Duma C, Inturrisi CE (1983) Naloxone radio-immunoassay: an improved antiserum. J Pharm Pharmacol 35:833–836

Heel RC, Brogden RN, Speight TM, Avery GS (1979) Buprenorphine: a review of its pharmacological properties and therapeutic efficacy. Drugs 17:81–110

Hollister LE, Schwin RL, Kasper P (1977) Naltrexone treatment of opiate-dependent persons. Dr Alc Dep 2:203–209

Holman BL, Mendelson J, Garada B, Teoh SK, Hallgring E, Johnson KA, Mello NK (1993) Regional cerebral blood flow improves with treatment in chronic cocaine polydrug users. J Nucl Med 34:723–727

Hussain MA, Shefter E (1988) Naltrexone-3-salicylate (a prodrug of naltrexone): synthesis and pharmacokinetics in dogs. Pharm Res 5(2):113–115

Inturrisi CE (1976) Disposition of narcotics and narcotic antagonists. Ann NY Acad Sci 273–287

Jasinski DR, Pevnick JS, Griffith JD (1978) Human pharmacology and abuse potential of the analgesic buprenorphine. Arch Gen Psych 35:501–516

Johnson RE, Fudala PJ (1992) Development of buprenorphine for the treatment of opioid dependence. NIDA Research Monography 121:120–141

Johnson RE, Cone EJ, Henningfield JE, Fudala PJ (1989) Use of buprenorphine in the treatment of opiate addiction. I. Physiologic and behavioral effects during a rapid dose induction. Clin Pharm Ther 46:335–343

Johnson RE, Jaffe JH, Fudala PJ (1992) A controlled trial of buprenorphine treatment for opioid dependence. JAMA 267:2750–2755

Judson BA, Himmelberger DU, Goldstein A (1980) The naloxone test for opiate dependence. Clin Pharm Ther 27(4):492–501

Judson BA, Carney TM, Goldstein A (1981) Naltrexone treatment of heroin addiction: efficacy and safety in a double-blind dosage comparison. Dr Alc Dep 7(4):325–346

Kleber HD, Kosten TR, Gaspari J, Topazian M (1985) Nontolerance to the opioid antagonism of naltrexone. Bio Psych 20:66–72

Kogan MJ, Verebey K, Mule SJ (1977) Estimation of the systemic availability and other pharmacokinetic parameters in altrexone in man after acute and chronic oral administration. Res Comm Chem Path Pharm 18(1):29–34

Kosten TR, Kleber HD (1991) Buprenorphine detoxification from opioid dependence: a pilot study. Life Sci 42:635–641

Kosten TR, Kreek MJ, Raghunath J, Kleber HD (1986a) A preliminary study of beta endorphin during chronic naltrexone maintenance treatment in ex-opiate addicts. Life Sci 39:55–59

Kosten TR, Kreek MJ, Raghunath J, Kleber HD (1986b) Cortisol levels during chronic naltrexone maintenance treatment in ex-opiate addicts. Bio Psych 21:217–220

Kosten TR, Kleber HD, Morgan C (1989a) Role of opioid antagonists in treating intravenous cocaine abuse. Life Sci 44:887–892

Kosten TR, Kleber HD, Morgan C (1989b) Treatment of cocaine abuse with buprenorphine. Bio Psych 26:170–172

Kosten TR, Morgan CJ, Kleber HD (1989c) Buprenorphine treatment of cocaine abuse. NIDA Res Mono 95:461

Kosten TR, Morgan C, Kleber HD (1992a) Phase II clinical trials of buprenorphine: detoxification and induction onto naltrexone. NIDA Research Monograph 121:101–119

Kosten TR, Morgan C, Kreek MJ (1992b) Beta-endorphin levels during heroin, methadone, buprenorphine and naloxone challenges: Preliminary findings. Biolog Psych 32:523–528

Kosten TR, Schottenfeld R, Ziedonis D, Falcioni J (1993) Buprenorphine versus methadone maintenance for opioid dependence. J Nerv Ment Dis 181(6):358–364

Kosterlitz HW, Watt AJ (1968) Kinetic parameters of narcotic agonists and antagonists. In: Committee on Problems of Drug Dependence, Proceedings of the 30th meeting in Indianapolis, IN

Kreek MJ (1972) Medical safety, side effects and toxicity of methadone. In: Proceedings of the Fourth National Conference on Methadone Treatment NAPAN-NIMH, pp 171–174

Kreek MJ (1973a) Medical safety and side effects of methadone in tolerant individuals. JAMA 223:665–668

Kreek MJ (1973b) Physiological implications of methadone treatment. In: Proceedings, Fifth National Conference on Methadone Treatment. National Association for the Prevention of Addiction to Narcotics 2:85–91

Kreek MJ (1973c) Plasma and urine levels of methadone. NYS J Med 73(23):2773–2777

Kreek MJ (1992a) The addict as a patient. In: Lowinson JH, Ruiz P, Millman RB, Langrod JG (eds) Substance abuse: a comprehensive textbook. Williams and Wilkins, Baltimore, pp 997–1009

Kreek MJ (1992b) Epilogue: Medical maintenance treatment for heroin addiction, from a retrospective and prospective viewpoint. In: State methadone maintenance treatment guidelines. Office for Treatment Improvement, Division for State Assistance 255–272

Kreek MJ (1992c) Rationale for maintenance pharmacotherapy of opiate dependence. Res Pub Assoc Res Nerv Ment Dis 70:205–230

Kreek MJ, Culpepper-Morgan J (1991) Neuroendocrine (HPA) and gastrointestinal effects of opiate antagonists: possible therapeutic application. In: Harris LS (ed) Problems of Drug Dependence 1990. Proceedings of the 52nd Annual Scientific Meeting of the Committee on Problems of Drug Dependence. NIDA Research Monograph Series, Rockville, MD, DHHS Pub#(ADM)91-1753, 105:168–174

Kreek MJ, Hartman N (1982) Chronic use of opioids and antipsychotic drugs: side effects, effects on endogenous opioids, and toxicity. Annals NY Acad Sci 151–172

Kreek MJ, Schaefer RA, Hahn EF, Fishman J (1983a) Naloxone, a specific opioid antagonist, reverses chronic idiopathic constipation. Lancet 2/5:261–262

Kreek MJ, Schaefer RA, Hahn EF, Fishman J (1983b) Naloxone in chronic constipation. Letters to the editors. Lancet 758b

Kreek MJ, Wardlaw SL, Hartman N, Raghunath J, Friedman J, Schneider B, Frantz AG (1983c) Circadian rhythms and levels of beta-endorphin, ACTH, and cortisol during chronic methadone maintenance treatment in humans. Life Sci Supp I, 33:409–411

Kreek MJ, Schneider BS, Raghunath J, Plevy S (1984) Prolonged (24-hour) infusion of the opioid antagonist naloxone does not significantly alter plasma levels of cortisol and ACTH in humans. Abstracts of the Seventh International Congress of Endocrinology, International Congress Series 652. Excerpta Medica, Amsterdam, p 845

Lehmann KA, Reichling U, Wirtz R (1988) Influence of naloxone on the postoperative analgesic and respiratory effects of buprenorphine. Eur J Clin Pharm 34:343–352

Lewis JW (1974) Ring C-bridged derivatives of thebaine and oripavine. In: Braude MC, Harris LS, May EL, Smith JP, Villarreal JE (eds) Narcotic antagonists. Advances in biochemical psychopharmacology, vol 8. Raven, New York, pp 123–136

Lewis JW (1985) Buprenorphine. Dr Alc Dep 14:363–372

Lewis JW, Readhead MJ (1970) Novel analgetics and molecular rearrangements in morphine-thebaine group XVIII 3-deoxy-6,14-endo-etheno-6,7,8,14-tetrahydrooripavines. J Med Chem 13:525–527

Lewis JW, Walter D (1992) Buprenorphine: background to its development as a treatment for opiate dependence. NIDA Research Monograph 121:5–11

Ling W, Wesson DR (1984) Naltrexone treatment for addicted health-care professionals: A collaborative private practice experience. J Clin Psych 45[9(2)]:46–48

Martin WR, Jasinski DR, Mansky PA (1973) Naltrexone, an antagonist for the treatment of heroin dependence. Arch Gen Psych 28:784–791

McQuay HJ, Bullingham RES, Paterson GMC, Moore RA (1980) Clinical effects of buprenorphine during and after operation. Br J Anaesth 52:1013–1019

Mello NK, Mendelson JH (1980) Buprenorphine suppresses heroin use by heroin addicts. Science 207:657–659

Mello NK, Mendelson JH (1992) Primate studies of the behavioral pharmacology of buprenorphine. NIDA Research Monograph 121:61–100

Mello NK, Mendelson JM, Kuehnle JC (1982) Buprenorphine effects on human heroin self-administration: an operate analysis. JPET 223:30–39

Mello NK, Lukas SE, Mendelson JH (1985) Buprenorphine effects on cigarette smoking. Psychopharmacology 86:417–425

Mendelson JH, Mello NK (1992) Human laboratory studies of buprenorphine. NIDA Research Monograph 121:38–60

Mendelson JH, Ellingboe J, Keuhnle JC, Mello NK (1979) Effects of naltrexone on mood and neuroendocrine function in normal adult males. Psychoneuroendocrinology 3:231–236

Mendelson JH, Ellingboe J, Kuehnle JC, and Mello NK (1980) Heroin and naltrexone effects on pituitary-gonadal hormones in man: Interaction of steroid feedback effects, tolerance, and supersensitivity. JPET 214(3):503–506

Mendelson JH, Ellingboe J, Mello NK, Kuehnle J (1982) Buprenorphine effects on plasma luteinizing hormone and prolactin in male heroin addicts. JPET 220: 252–255

Mendelson JH, Mello NK, Teoh SK, Lloyd-Jones JG, Clifford JM (1989) Naloxone suppresses buprenorphine stimulation of plasma prolactin. J Clin Psychopharm 9(2):105–109

Meyer MC, Straughn AB, Lo M-W, Schary WL, Whitney CC (1984) Bioequivalence, dose-proportionality, and pharmacokinetics of naltrexone after oral administration. J Clin Psych 45[9(2)]:15–19

Michel ME, Bolger G, Weissman B-A (1985) Binding of a new opiate antagonist, nalmefene, to rat brain membranes. Meth Find Exp Clin Pharm 7(4):175–177

Mirin SM, Mendelson JM, Ellingboe J, Meyer RE (1976) Acute effect of heroin and naltrexone on testosterone and gonadotropin secretion: a pilot study. Psychoneuroend 1:359–369

Naber D, Pickar D, Davis GC, Cohen RM, Jimerson DC, Elchisak MA, Defraites EG, Kalin NH, Risch SC, Buchsbaum MS (1981) Naloxone effects on β-endorphin, cortisol, prolactin, growth hormone, HVA and MHPG in plasma of normal volunteers. Psychopharmacology 74:125–128

National Research Council Committee on Clinical Evaluation of Narcotic Antagonists (1978) Clinical evaluation of naltrexone treatment of opiate-dependent individuals. Arch Gen Psych 35:335–340

O'Brien CP, Greenstein RA, Mintz J, Woody GE (1975) Clinical experience with naltrexone. Am J Dr Alc Ab 2:365–377

Olley JE, Tiong GKL (1988) Letters to the editor: Plasma levels of opioid material in man following sublingual and intravenous administration of buprenorphine: exogenous/endogenous opioid interaction? J Pharm Pharmacol 40:666–667

Pickworth WB, Johnson RE, Holicky BA, Cone EJ (1993) Subjective and physiologic effects of intravenous buprenorphine in humans 53:570–576

The Possilpark Group (1993) Drug injectors in Glasgow: a community at risk? A report from a multidisciplinary group. Health Bulletin 51(6):418–429

Quigley AJ, Bredemeyer DE, Seow SS (1984) A case of buprenorphine abuse. Med J Aust 142:425–426

Reisinger M (1985) Buprenorphine as new treatment for heroin dependence. Dr Alc Depend 16:257–262

Resnick R, Fink M, Freedman AM (1974) High-dose cyclazocine therapy of opiate dependence. Am J Psych 131(5):595–597

Resnick RB, Resnick E, Galanter M (1991) Buprenorphine responders: a diagnostic subgroup of heroin addicts? Prog Neuro-Psychopharmacol Biol Psychiat 15: 531–538

Resnick RB, Galanter M, Pycha C, Cohen A, Grandison P, Flood N (1992) Buprenorphine: an alternative to methadone for heroin dependence treatment. Psychopharm Bull 28(1):109–113

Rosen MI, Pearsall HR, McDougle CJ, Price LH, Woods SW, Kosten TR (1993) Effects of acute buprenorphine on responses to intranasal cocaine: a pilot study. Am J Dr Alc Ab 19(4):451–464

Saarialho-Kere U, Mattila MJ, Paloheimo M, Seppala T (1987) Psychomotor, respiratory, and neuroendocrinological effects of buprenorphine and amitriptyline in healthy volunteers 139–146

Seow SSW, Quigley AJ, Ilett KF, Dusci LJ, Swensen G, Harrison-Stewart A, Rappaport L (1986) Buprenorphine: a new maintenance opiate? Med J Austr 144:407–411

Strain EC, Preston KL, Liebson IA, Bigelow GE (1992) Acute effects of buprenorphine, hydromorphone, and naloxone in methadone-maintained volunteers. JPET 261:985–993

Tempel A, Zukin RS, Gardner EL (1982) Supersensitivity of brain opiate receptor subtypes after chronic naltrexone treatment. Life Sci 31:1401–1404

Tennant FS, Rawson RA, Cohen AJ, Mann A (1984) Clinical experience with
     naltrexone in suburban opioid addicts. J Clin Psych 45[9(2)]:42–45
Teoh SK, Mendelson JH, Mello NK, Kuehnle J, Sintavanarong P, Rhoades EM
     (1993) Acute interactions of buprenorphine with intravenous cocaine and mor-
     phine: An investigational new drug phase I safety evaluation. J Clin Psychopharm
     13(2):87–99
U.S. patent #3,332,950, July 25, 1967
Verebey K, Mule SJ (1975) Naltrexone pharmacology, pharmacokinetics, and
     metabolism: current status. Am J Dr Alc Ab 2(3–4):357–363
Verebey K, Volavka J, Mule SJ, Resnick RB (1976) Naltrexone: disposition,
     metabolism, and effects after acute and chronic dosing. Clin Pharm Ther 20:
     315–328
Volavka J, Resnick RB, Kestenbaum RS, Freedman AM (1976) Short-term effects
     of naltrexone in 155 heroin ex-addicts. Bio Psych 11(6):679–685
Volavka J, Cho D, Mallya A, Bauman J (1979a) Naloxone increases ACTH and
     cortisol levels in man. NE J Med 300(18):1056–1057
Volavka J, Mallya A, Bauman J, Pevnick J, Cho D, Reker D, James B, Dornbush R
     (1979b) Hormonal and other effects of naltrexone in normal men. Adv Exp Med
     Bio 116:291–305
Volavka J, Bauman J, Pevnick J, Reker D, James B, Cho D (1980) Short-term
     hormonal effects of naloxone in man. Psychoneuroend 5:225–234
Wall ME, Brine DR, Perez-Reyes M (1981) Metabolism and disposition of naltre-
     xone in man after oral and intravenous administration. Dr Met Disp 9(4):
     369–375
Wall ME, Perez-Reyes M, Brine DR, Cook CE (1984) Naltrexone disposition in
     man after subcutaneous administration. Dr Met Disp 12(6):677–682
Weinberg DS, Inturrisi CE, Reidenberg B, Moulin DE, Nip TJ, Wallenstein S,
     Houde RW, Foley KM (1988) Sublingual absorption of selected opioid anal-
     gesics. Clin Pharm Ther 44:335–342
Woods JH, France CP, Winger GD (1992) Behavioral pharmacology of buprenor-
     phine: issues relevant to its potential in treating drug abuse. NIDA Research
     Monograph 121:12–27
Zaks A, Jones T, Fink M, Freedman AM (1971) Naloxone treatment of opiate
     dependence. JAMA 215(13):2108–2110
Zukin RS, Sugarman JR, Fitz-Syage ML, Gardner EL, Zukin SR, Gintzler AR
     (1982) Naltrexone-induced opiate receptor supersensitivity. Brain Res 245:
     288–292

CHAPTER 15

Pharmacotherapy of Nicotine Dependence

J.R. Hughes

A. Introduction

Several excellent narrative and meta-analytic reviews of different pharmacotherapies for smoking cessation have recently been published (GLASSMAN
and COVEY 1990; GOLDSTEIN et al. 1991; GOURLAY and McNEIL 1990;
HUGHES 1993a; JARVIK and HENNINGFIELD 1988, 1993; JARVIK and SCHNEIDER
1992; LEE and D'ALONZO 1993; NUNN-THOMPSON and SIMON 1989; PRIGNOT
1989; SACHS 1991; SACHS et al. 1994; SACHS and LEISCHOW 1991a; SCHWARTZ
1987; US DEPARTMENT OF HEALTH AND HUMAN SERVICES 1988). The present
review will differ somewhat from these in that it will focus on promising
medications and areas of future research.

I. A Brief History of Treatment for Smoking Cessation

In the 1930s–1950s, trials of medications for smoking cessation usually
reported negative outcomes or nonreplicable positive results. The 1950s–
1960s began to document the efficacy of formal, group-based educational
programs. In the 1960s–1970s, behavioral therapies were shown to be quite
effective (see Sect. D); however, behavior therapies began to lose favor as
their lengthiness and financial costs made them poorly accepted (HUGHES
1994; SHIFFMAN 1993). In the 1970s, emphasis was on public health approaches (e.g., media campaigns and legal restrictions) and brief treatments
(e.g., physician advice and bibliotherapy) (LICHTENSTEIN and GLASGOW 1992).
Also in the 1970s, basic science and clinical studies on smoking as a nicotine
dependence (see Chap. 10, this volume) prompted the first truly successful
pharmacotherapies – clonidine, nicotine polacrilex (nicotine gum) and
transdermal nicotine (nicotine patches), which were marketed in the 1980s–
1990s. The success of these treatments (both therapeutically and financially)
sparked an interest in pharmacotherapy for smoking cessation.

II. Why Pharmacotherapy?

The long-term success rate for any one given attempt at quitting smoking
without assistance is about 5% (COHEN et al. 1989). However, most smokers
make several attempts to quit before they finally quit; eventually, 45% of

those who ever smoked quit (FIORE et al. 1990). At present, 95% of smokers quit on their own with no formal therapy (FIORE et al. 1990). At most, <5% of smokers quit via over-the-counter (OTC) or prescribed medications (FIORE et al. 1990). Thus, pharmaco therapy has contributed little cessation efforts (FIORE et al. 1990). This scenario will probably change in the future due to the fact that social pressure and public health efforts are not uniformly decreasing smoking across all smoking groups; i.e., at least three selection biases are occurring (COAMBS et al. 1989). First, smokers who are quitting are in all likelihood those less dependent on nicotine (HUGHES 1993a). Thus, remaining smokers will be those with more severe nicotine dependence. This trend suggests that treatments focused on the nicotine dependence aspects of smoking (such as pharmacotherapies) will become more prominent.

Second, smoking initiation is occurring more often and smoking cessation less often among the less educated and the poor (PIERCE et al. 1989). Since the less educated and the poor are less likely to attend psychotherapies, pharmacotherapies may become a more attractive therapy to use with future populations of smokers (PINNEY 1991).

Another group of smokers who are not quitting are those with psychiatric or alcohol/drug abuse problems (HUGHES 1993c). In addition, social pressure has made smoking initiation a deviant behavior and perhaps a marker for alcohol/drug problems or psychopathology. As a result of these trends, the prevalence of psychiatric comorbidity and alcoholism among smokers may rise over time (GLASSMAN 1993; HUGHES 1993b; KANDEL and DAVIES 1986). If so, pharmacological treatment of comorbid psychiatric disorders as a method to help smokers stop smoking may become more prevalent.

III. Methodological Issues

1. Paradigms to Test Efficacy

The paradigms used to examine treatment efficacy in smoking cessation can be divided into three types. First, laboratory studies examine the ability of a medication to reduce cigarette or nicotine intake, craving or withdrawal symptoms. These studies are done both in in-patient and out-patient settings and may or may not use smokers who wish to stop smoking.

Second, pilot clinical studies examine the ability of the treatment to induce short-term cessation (<3 months). These may or may not be placebo-controlled trials.

Third, definitive studies are double-blind, placebo-controlled trials that examine the efficacy of the medication on long-term (≥6 months) cessation, often for long periods of time after the medication has been discontinued. Since research on smoking cessation is easier than that with other drugs of dependence (e.g., smokers are generally more compliant, have less psycho-pathology and have more stable life-styles than alcohol-, cocaine- and opioid-dependent persons), studies of this third category are much more common than with other drugs of dependence. In fact, most researchers in smoking cessation do not consider a medication proven effective until several studies in this last category are consistently positive.

Success rates in smoking cessation treatment are often numerically small, e.g., 20%–30%. In interpreting these rates the readers should consider that: (1) these are usually long-term, continuous, biochemically verified abstinence rates; (2) the abstinence rate among self-quitters is 5%–6% (COHEN et al. 1989; HUGHES et al. 1992); and (3) often, the amount of medication and psychological treatment is small.

Finally, it is important to note that the field of testing medications for smoking cessation has undergone substantial methodological improvements in the last 20 years; thus, whether older, less well-designed negative studies represent true or false negative findings is unclear.

2. Traditions of Testing Medications for Smoking vs for Other Drug Dependencies

Smoking treatment research has several traditions, many of which differ from those of treatment research on other drugs of dependence. The traditions include: (1) exclusive use of out-patient settings, (2) use of subjects with moderate-to-high motivation to quit, (3) exclusion of subjects with psychiatric or non-nicotine drug use problems, (4) interest in detecting small changes in abstinence rates (e.g. 5%), (5) clear operational definitions of abstinence, (6) concomitant measurement of abstinence, craving and withdrawal symptoms, (7) biochemical verification of reports of abstinence, (8) counting dropouts, treatment noncompleters and those lost to follow-up as treatment failures, (9) specification of adjunctive psychological therapy, and (10) reporting of point prevalence rates of abstinence (i.e., percent abstinent at a given point in time) and/or rates of continuous abstinence (i.e., percent abstinent at all follow-ups).

Conversely, treatment studies of non-nicotine drug dependencies utilize methodologies not often seen in smoking cessation trials. These include: (1) use of DSM or WHO criteria for dependence for inclusion or descriptive purposes, (2) use of significant others to verify abstinence, (3) time-to-relapse as an acceptable outcome, and (4) collection of nonefficacy outcomes (e.g., treatment retention and improvement in psychosocial functioning).

Finally, both smoking and traditional durg treatment trials often fail to include two important design features: (1) a priori or post-hoc power analyses, and (2) testing of multiple doses of the medication.

IV. Comparing Medications

For three medications – clonidine, nicotine polacrilex, and transdermal nicotine, several placebo-controlled trials are available. Unfortunately, direct comparisons among these drugs are not available.

Since differences in sample size, follow-up, etc., produce different base rates of smoking cessation (BAILLIE et al., in press), the present review will focus on the odds ratio as the metric of efficacy. The odds ratio is defined here as estimates of the relative probability of cessation for the medicated group vs the control group. Thus, an odds ratio of 2.0 could refer to an outcome of 5% vs 10%, 10% vs 20% or 20% vs 40%. The assumption here is that which of these three outcomes occurs is a function of the subjects selected, time of follow-up, etc., and the effectiveness of the medication produces a multiplicative effect upon these base rates (BAILLIE et al., in press).

B. Medications to Aid Smoking Cessation

Medications to aid smoking cessation can be classified as: (1) replacements, (2) blocking agents, (3) medications that make smoking aversive, (4) medications that abate withdrawal or mimic the putative reinforcing effects of nicotine, and (5) medications to treat associated psychopathology (JARVIK and HENNINGFIELD 1988).

I. Nicotine Replacement

Several recent quantitative and nonquantitative reviews have focused on various nicotine replacement therapies (FAGERSTROM 1994; FIORE et al. 1992; FOULDS 1993; POMERLEAU et al. 1988; ROSE 1991; SILAGY et al. 1994a,b; TANG et al. 1994). Replacement strategies for drug dependence are said to work via: (1) relief of drug withdrawal, (2) reduction in the reinforcing effects of a drug via tolerance, and (3) deconditioning by making intake of a drug independent of environmental events (HUGHES 1993a).

The rationale for nicotine replacement has focused almost exclusively on the first rationale, i.e., withdrawal relief. The specific rationale often cited is as follows. Independent of any pharmacology involved, smoking is a very intense habit. For example, a smoker who smokes 30 cigarettes/day and takes 8 puffs/cigarette and has done so for 20 years, smokes 240 times/day, 87 600 times/year and has smoked a total of 1 752 000 times. When a smoker stops, he/she not only has to overcome this intense habit, but also, at the same time, nicotine withdrawal, which may be causing him/her to be anxious, depressed, hungry, impatient, irritable, and to have problems concentrating and sleeping (HUGHES et al. 1990). The rationale for nicotine replacement is that it decreases these withdrawal effects thereby allowing the smoker to better focus on the behavioral changes necessary to stop smoking.

1. Nicotine Polacrilex (Nicotine Gum)

Some recent reviews have focused on nicotine polacrilex (CEPEDA-BENITO 1993; FAGERSTROM 1988; LAM et al. 1987). Oral nicotine is extensively metabolized on first pass through the liver (BENOWITZ 1988). Nicotine polacrilex overcomes this problem via buccal absorption. The polacrilex preparation is 2 or 4mg of nicotine bound to an ion-exchange resin in a gum. When chewed, the gum releases nicotine which is absorbed buccally. When used properly, nicotine polacrilex takes 15–30 min to reach peak nicotine levels and produces venous nicotine levels which are 1/3–2/3 of between-cigarette nicotine levels (BENOWITZ 1988). However, nicotine via smoking directly enters arterial circulation without mixing with venous blood whereas nicotine via nicotine gum is absorbed via venous circulation. Thus, the resultant arterial nicotine concentrations from polacrilex are probably on the order of 1/10–1/20 that from smoking (HENNINGFIELD et al. 1990, 1993).

Nicotine polacrilex decreases most of the symptoms of nicotine withdrawal (HUGHES et al. 1990; HUGHES and HATSUKAMI 1992). Craving, hunger and weight gain are less effectively controlled (HUGHES and HATSUKAMI 1992), but with adequate dosing appear to be relieved by nicotine polacrilex as well (LEISCHOW et al. 1992; SACHS and LEISCHOW 1991b).

Nine meta-analyses have examined the over 40 controlled studies of nicotine polacrilex (BAILLIE et al., in press; CEPEDA-BENITO 1993; GOURLAY and MCNEIL 1990; HUGHES 1991, 1993a; LAM et al. 1987; SILAGY et al. 1994a,b; TANG et al. 1994; VISWESVARAN and SCHMIDT 1992) (Table 1). These meta-analyses varied in whether they examined only long-term follow-up, only placebo-controlled trials, etc. Of the nine, eight reported either odds ratios or absolute difference in per cent quit in both the setting of a general medical practice, where nicotine polacrilex is given with brief advice, and the setting of a cessation clinic, where nicotine polacrilex is given with intense psychological treatment.

Table 1. Meta-analysis of long-term quit rates with nicotine polacrilex (NP) or transdermal nicotine (TN) vs placebo with and without psychological therapy (PT)

| | NP | | | | TN | | | |
| | No PT | | PT | | No PT | | PT | |
	OR	Δ%	OR	Δ%	OR	Δ%	OR	Δ%
BAILLIE (in press)	2.1	+7%	1.7	+10%				
CEPEDA-BENITO (1993)		+2%		+15%				
FIORE et al. (1994)					2.5	+12%	3.4	+13%
GOURLAY and MCNEIL (1990)	1.5		1.7					
HUGHES (1991)		+4%		+7%				
HUGHES (1994)	1.4		2.1		2.6		2.4	
LAM et al. (1981)		0%		+9%				
SILAGY et al. (1994a)	1.8	+5%	1.4	+8%	2.1	+10%	2.0	+10%
TANG et al. (1994)		+3%		+11%				

OR, odds ratio.
Δ%, increase in percent abstinent.

Meta-analysis of studies in general practice reported odds ratios of 1.4–2.1. Thus, adding polacrilex increases the quit rates 1.4–2.1 times the control group, independent of the two settings. However, the control group rates differ greatly across these two settings. As mentioned earlier, this difference may be attributable to differences in subjects, therapists or intensity of adjunctive counseling, etc. In terms of absolute increases in quit rates, nicotine polacrilex increased quit rates in general practice by 0%–7% and by 7%–15% in cessation clinics. These odds ratio and absolute difference results have been more rigorously documented within two meta-analyses (Fiore et al. 1994; Silagy et al. 1994b).

Side effects with nicotine polacrilex have been minor and are due either to mechanical properties of the gum, i.e., sore jaw, hiccups, belching, or to nicotinic effects, i.e., bad taste, burning throat nausea, etc. (Hughes 1986). Serious adverse effects are very rare.

One problem that has occurred with nicotine polacrilex is that 15%–25% of those who stop smoking while using the gum use it longer than recommended; however, most long-term users are chewing only a few pieces per day (Hajek et al. 1988; Hughes et al. 1991). Initially, the duration of treatment for polacrilex was set at 3 months based on data that most relapse occurs in the first 3 months. More recently, clinicians and investigators appear to have become more comfortable with longer term use, and the notion of "nicotine maintenance," similar to methadone maintenance, has been seriously discussed (Henningfield and Stitzer 1991). There are two major issues. First, is long-term use of nicotine polacrilex harmful? Although nicotine via inhaled cigarette smoke has been associated with increases in several coronary risk factors (e.g., lipid profiles and thrombogenisis), whether the slow onset and low levels associated with long-term use of nicotine polacrilex also produce changes in risk factors has not been well-studied (Hughes 1993d). Second, in a society with calvinistic views toward drug use, is nicotine maintenance as acceptable as lifelong medication for hypertension? Or will it be viewed as similar to methadone maintenance?

In contrast to the minority who use nicotine polacrilex too long, the majority of smokers use too few pieces of gum per day and use the gum for too short a period (Hughes 1986). This may be due to the bad taste of the gum, the complicated instructions for its use (chew the gum only intermittently, do not drink acidic beverages while using the gum, etc.), and its ad libitum use. Recent work using fixed-time rather than ad libitum dosing has shown more promising results (Goldstein et al. 1989; Killen et al. 1990).

Finally, although the efficacy of 2 vs 4 mg of nicotine polacrilex in nonselected smokers is unclear, highly dependent smokers are especially benefitted by the higher doses (Jarvik and Henningfield 1993; Kornitzer et al. 1987; Tonnesen et al. 1988).

2. Transdermal Nicotine Systems (Nicotine Patches)

Several narrative and meta-analytic reviews have focused only on transdermal nicotine (BENOWITZ 1993; FAGERSTROM et al. 1992; FIORE et al. 1992, 1994; FOULDS 1993; GORA 1993; HUGHES and GLASER 1993; PALMER and FAULDS 1992; ROSE 1991). In 1991–1992, four transdermal nicotine patches were marketed. Three were designed to be worn for 24h and the fourth for 16h. Among the 24h patches, one places nicotine in a polymer solution in a pad of viscose (Habitrol). A second uses a similar system but also has a rate-controlling membrane (Nicoderm). The third places nicotine in a gel matrix which is in direct contact with the skin (Prostep). The 16h patch (Nicotrol) places nicotine in a rate-controlling adhesive.

With the recommended initial dose of the marketed transdermal patches, peak nicotine blood levels are reached within 4–9h after the first use, deliver 15–22mg of nicotine/day, and produce blood levels slightly higher than with regular use of nicotine polacrilex (Table 2) (PALMER and FAULDS 1992). Nicotine blood levels with the 24h patches drop 25%–50% over the 24h, whereas blood levels approach zero 8h after removal of the 16h patches. Whether the technical differences in the construction of the patches or the pharmacokinetic differences among the 24h patches produce clinically significant differences in efficacy or side effects is unknown.

Transdermal nicotine has several major advantages over nicotine polacrilex. Transdermal nicotine requires only once per day dosing, less instruction in its use, and is more socially acceptable. All of these make compliance greater with transdermal nicotine than with nicotine polacrilex. Second, transdermal nicotine produces more steady nicotine replacement than that from ad lib use of nicotine polacrilex.

Transdermal nicotine decreases nicotine withdrawal and appears to more readily decrease craving than nicotine polacrilex (FAGERSTROM et al. 1992, 1993; HUGHES and GLASER 1993; PALMER and FAULDS 1992). Whether transdermal nicotine reduces post-cessation weight gain is debatable (HUGHES and GLASER 1993; PALMER and FAULDS 1992).

Daily transdermal nicotine begins on the first day of abstinence after abrupt cessation. Transdermal nicotine systems are to be used for 6–12 weeks on the highest dose of transdermal nicotine, followed by 2–4 weeks

Table 2. Pharmacokinetics of the different transdermal nicotine systems

	Habitrol	Nicoderm	Prostep	Nicotrol
Dose	22 ng/24 h	22 ng/24 h	15 ng/24 h	15 mg/16 h
C_{max} (ng/ml)	17	23	16	13
C_{avg} (ng/ml)	13	17	11	9
T_{max} (h)	6	4	9	8

on the middle dose, and 2–4 weeks on the lowest dose for a total of 10–20 weeks of therapy (FIORE et al. 1992).

One transdermal nicotine system allows for abrupt cessation of high dose therapy (Prostep). Given results with abrupt cessation of nicotine polacrilex (HUGHES 1986), it is likely that abrupt cessation of high dose transdermal nicotine would result in nicotine withdrawal; however, the one study of abrupt cessation did not find significant withdrawal (HURT et al. 1994). In addition, one experimental study reported that abrupt cessation from transdermal nicotine did not increase relapse rates compared to gradual weaning (HILLEMAN et al. 1993). Finally, most, but not all, studies report that rates of relapse to smoking are not higher after abrupt cessation vs gradual cessation of the patch (FIORE et al. 1994; SILAGY et al. 1994b).

Five meta-analyses have examined the over 20 trials of transdermal nicotine (FIORE et al. 1994; FOULDS 1993; Po 1993; SILAGY et al. 1994b; TANG et al. 1994). Of these, three reported their results in odds ratios or differences in quit rates with and without psychological treatment (Table 1). In these meta-analyses, transdermal nicotine consistently increases cessation rates two- to three fold (Table 1). As with nicotine polacrilex, the odds ratios for abstinence from transdermal nicotine were similar in the setting of a clinic with intensive psychological therapy (PT) and a general medical practice with brief advice. The odds ratios (ORs) for transdermal nicotine (TN) appear higher than those for nicotine polacrilex (NP); i.e., for TN without PT, ORs = 2.1 – 2.6; for NP without PT = 1.4 – 2.1; for TN with PT = 2.0 – 3.4; and for NP with PT = 1.7 – 2.1. This result was also reported in two prior meta-analyses (FIORE et al. 1994; SILAGY et al. 1994a,b). The only direct comparison of transdermal nicotine and nicotine polacrilex (published in abstract form) reported a slight but nonsignificant advantage of transdermal nicotine over nicotine polacrilex in short-term abstinence (55% vs 43% at 3 months) (MEIER-LAMMERMANN et al. 1990). This evidence for the superiority of transdermal nicotine over nicotine polacrilex should be interpreted cautiously. It is possible that, when smokers are selected for higher dependence, are well-instructed in gum use, use 4 mg rather than 2 mg gum and take at least ten pieces of nicotine polacrilex per day, their abstinence rates will be similar to if they had used transdermal nicotine. However, in clinical settings, physicians do not select only highly dependent patients, often fail to give instructions, and patients typically use only four to six pieces per day of nicotine gum (CUMMINGS et al. 1988; HUGHES 1986). Thus, in clinical settings it is likely that transdermal nicotine will produce higher abstinence rates than nicotine polacrilex.

Side effects with transdermal nicotine are uncommon and less than 5% of patients discontinue transdermal nicotine due to side effects (FAGERSTROM et al. 1993; FIORE et al. 1992; PALMER and FAULDS 1992). Skin reactions and sleep problems are the major side effects with transdermal nicotine. One type of skin reaction is due to irritation from removing the patch per se. A second type consists of a nonimmunological contact urticaria with minor edema, erythema and pruritus. Over time, patients often become tolerant to this reaction. A third type is a true allergic contanct dermatitis. Over time, this reaction can progress to a more severe skin reaction; tolerance does not occur and patch use must be discontinued.

Sleep problems may occur with 24 h patches and consist of difficulty falling asleep, intense dreams and nightmares. These are often difficult to assess as they can be symptoms of nicotine withdrawal; however, transdermal nicotine appears to interfere with sleep onset whereas withdrawal appears to increase awakenings once one is asleep (HUGHES and GLASER 1993; WETTER et al., in press). Patch-induced sleep problems can be treated by removing the 24 h patch during the evening hours, by switching to a 16 h patch or by simply waiting, as tolerance occurs such that these symptoms usually abate within 1 week (FIORE et al. 1992).

Other, less frequent side effects are symptoms of nicotine excess; i.e., dizziness, headache, nausea, and stomachache (HUGHES 1993d). There have been anecodotal reports of myocardial infarctions in cardiac patients from use of transdermal nicotine. A review of these reports by the FDA questioned causality, and a study of transdermal nicotine in cardiac patients found no evidence of increased cardiac risk (RENNARD et al. 1991). Since many of the MI patients were smoking and using transdermal nicotine concomitantly, cardiac patients may need to be especially warned to not smoke and concurrently wear transdermal nicotine (FIORE et al. 1992). This is because concurrent use of cigarettes and transdermal nicotine can produce higher than normal nicotine levels (FOULDS et al. 1992).

Transdermal nicotine is classified as category D (positive evidence of human fetal risk). The major complication of smoking during pregnancy is low birth weight from hypoxia. It is unclear whether this low birth weight is due to carbon monoxide-induced high levels of carboxyhemoglobin or to nicotine-induced vasoconstriction of the placental vessels (BENOWITZ 1991; HUGHES 1993d). However, given the lower blood nicotine levels and lack of carbon monoxide from transdermal nicotine compared to smoking, the use of transdermal nicotine during pregnancy may be reasonable (BENOWITZ 1991; HUGHES 1993d).

Finally, although not studied, long-term use of transdermal nicotine has not appeared to be a problem as with nicotine polacrilex. This may be because transdermal nicotine is not used ad libitum, which may be critical to engendering long-term use (HUGHES 1986).

Rationales for use of the 16 h patches as opposed to the 24 h patches include: (1) 16 h patches more readily mimic smoking (FAGERSTROM et al. 1991), (2) they do not cause insomnia (FAGERSTROM et al. 1991), (3) they do not provide nicotine during the early morning hours when there is an increased probability of cardiac events (HUGHES 1993d), and (4) they allow the receptor to rest between days, which decreases up-regulation which in turn decreases physical dependence on transdermal nicotine (FAGERSTROM et al. 1991).

One rationale for use of the 24 h patches is that they will decrease early morning craving which is said to be at high levels due to overnight nicotine deprivation. Interestingly, the available data suggest that craving is lowest in the morning (Schneider and Jarvik 1984; Shiffman 1979). In addition, relapses are not more common in the early morning (Shiffman 1982). A second rationale for the 24 h patches is that they are more flexible, in that, by taking them off at night, they become similar to 16 h patches (Hughes and Glaser 1993).

Two studies that have compared 16 and 24 h replacement strategies found similar efficacy in withdrawal reduction and abstinence outcomes (Daughton et al. 1991; Fagerstrom et al. 1991). One of these studies (Fagerstrom et al. 1991) and one other (Leischow et al., in press) found the expected greater sleep disturbances with 24 h delivery. One of these studies also found greater morning craving with 16 h than with 24 h patches (Leischow et al., in press).

With the several advantages of transdermal nicotine, whether nicotine polacrilex has a place in smoking cessation therapy has been questioned. Two recent studies reported that combining transdermal nicotine and nicotine polacrilex reduced withdrawal (Fagerstrom et al. 1993) and increased abstinence (Puska et al. 1994) more than transdermal nicotine alone (Fagerstrom et al. 1993). This result suggests that nicotine polacrilex is able to abate acute episodes of more intense withdrawal not relieved by transdermal nicotine. Whether this result is a simple dose-response issue (see below), a function of using medication ad lib or a function of using a different route of administration is unknown. Finally, although not empirically tested, clinicians have reported using nicotine polacrilex to help taper smokers off transdermal nicotine or to prevent relapse during the post transdermal period.

3. Patient Selection for Nicotine Replacement

Smokers can be divided into high vs low nicotine dependence using the Fagerstrom Tolerance Questionnaire (FTQ) (Fagerstrom and Schneider 1989) or the revised FTQ (Heatherton et al. 1991). The original questionnaire has consistently and robustly predicted response to nicotine polacrilex (Fagerstrom and Schneider 1989). Most studies suggest that high dependent smokers do poorly unless they receive nicotine polacrilex and that low dependent smokers receive a small, and debatable, benefit from nicotine polacrilex. Surprisingly, the FTQ has not consistently predicted response to transdermal nicotine (Fiore et al. 1994; Silagy et al. 1994b). Although some hypotheses have been offered, the reason for this inconsistency is unclear (Hughes 1993a).

4. Other Nicotine Replacement Medications

Nicotine nasal sprays and aerosols are being tested (Rose 1991). The rationale for their use is that they more closely mimic smoking. Such products could serve as a stand alone treatment, as a transition product (between smoking and transdermal nicotine), or as a adjunct to use in addition to transdermal nicotine.

Nasal sprays consist of a solution of nicotine in a nasal inhaler similar to inhalers used with antihistamines. Laboratory studies indicate nasal sprays can produce substantial levels of nicotine and reduce smoking (PERKINS et al. 1992; POMERLEAU et al. 1992; ROSE and BEHM 1987; SUTHERLAND et al. 1992a). Early results using nasal spray alone are encouraging (JARVIS et al. 1987; SUTHERLAND et al. 1992b). Nasal irritation and social acceptability with the spray are problems but do not appear to be major limiting factors. Long-term use of the sprays was observed in 43% of successful quitters (SUTHERLAND et al. 1992b). Their abuse liability by nonsmokers has not been tested (HUGHES 1986).

Nicotine inhalers (also called aerosols, smoke-free cigarettes, and vaporizers) are plugs of nicotine in cigarette-like tubes. When warm air is sucked past the plug, the nicotine is vaporized. Early on inhalers often failed to produce adequate nicotine levels (SEPKOVIC et al. 1987), but this can be overcome with more vigorous puffing (RUSSELL et al. 1987a). These devices have been shown to decrease withdrawal and craving and to mimic the satisfaction from smoking (ROSE and BEHM 1987). By contrast, a smoke-free cigarette was recently marketed in the US, not as a cessation tool but as a cigarette substitute, and was withdrawn in part due to poor acceptability. The single clinical trial of an inhaler in a general practice setting found long-term abstinence rates of 15% in active vs 5% in placebo inhalers (TONNESEN et al. 1993).

Nicotine tablets made to be held in the mouth for buccal absorption have been found to reduce smoking somewhat (JARVIK et al. 1970). Also, a buccal tobacco lozenge has been marketed in the UK (BELCHER et al. 1989). Clinical trials of buccal nicotine via tablets/lozenges have not been reported.

5. Lobeline

Lobeline is a drug that produces cross-tolerance with nicotine and is in several over-the-counter smoking cessation preparations. Many tests of lobeline were done in the 1930s–1950s with varying results. Overall, there is little evidence for the efficacy of lobeline (DAVISON and ROSEN 1972; DEPARTMENT OF HEALTH AND HUMAN SERVICES 1982; KOZLOWSKI 1984).

6. Future Studies on Nicotine Replacement

There are several gaps in the literature on nicotine replacement. First, the only dose-response study of transdermal nicotine found better efficacy at higher doses (TRANSDERMAL NICOTINE STUDY GROUP 1991). Given the benign side effect profile of transdermal nicotine, testing of even higher doses of transdermal nicotine has recently begun (FREDRICKSON et al. 1993; MINNEKER et al. 1989; PAOLETTI et al. 1993), but no conclusions can be reached yet.

Second, present recommendations for the duration of nicotine replacement therapy are theoretical ones derived from the duration of nicotine withdrawal and times of increased relapse. Although several retrospective studies have associated longer use of nicotine polacrilex with better outcomes, the two prospective studies of longer vs shorter durations produced contrary results (Fagerstrom and Melin 1985; Hatsukami et al. 1993). A study of using nicotine replacement as methadone maintenance is used, i.e., for very long periods of time before tapering, is indicated.

Third, the major rationale for nicotine replacement is withdrawal relief; however, no study has demonstrated that this is the therapeutic mechanism and there are reasons to doubt this mechanism (Hughes 1993a). Other mechanisms are possible, e.g., when smokers slip and have a cigarette, those on nicotine patches report less pleasure and more aversion to cigarettes than those on placebo patches (Levin et al. 1994). Consequently, studies into how nicotine replacement increases abstinence are indicated.

Fourth, as stated above, the level of nicotine dependence consistently predicts response to nicotine polacrilex (Fagerstrom and Schneider 1989) but not to transdermal nicotine (Hughes 1993a). Thus, at present, there are no empirically validated data to help select smokers who would benefit from transdermal nicotine. Research into such predictors is sorely needed.

Fifth, nicotine polacrilex and transdermal nicotine are over-the-counter medications in several countries. One small trial reported little efficacy of nicotine polacrilex in a dispensary setting (Schneider et al. 1983). Thus, a randomized trial demonstrating efficacy in the OTC setting is needed.

Finally, there are several reasons to believe nicotine might be therapeutic for Alzheimer's, Parkinson's and Tourette's disorders, and ulcerative colitis (Newhouse and Hughes 1991). Since transdermal nicotine is the first practical nicotine delivery system, pilot studies of nicotine administration in these disorders is indicated.

II. Blocking Agents

1. Mecamylamine

Mecamylamine is a noncompetitive cholinergic blocker which was originally marketed as an antihypertensive (Clarke 1991; Stolerman 1986). Mecamylamine consistently reverses the CNS effects of nicotine in nonhumans and humans (Henningfield et al. 1983b; Pickworth et al. 1988). Interestingly, mecamylamine blockage does not appear to be due to actions at the nicotinic-cholinergic site (Clarke 1992) but rather at an associated ion channel. This may be why mecamylamine does not precipitate withdrawal in humans or nonhumans (Clarke 1991; Stolerman 1986).

In human laboratory studies, mecamylamine blocks smoking satisfaction and increases smoking, presumably because smokers are attempting to override nicotine blockade (Nemeth-Coslett et al. 1986; Pomerleau et al. 1987;

Rose et al. 1989). Mecamylamine also increases nicotine preference (Rose et al. 1989). Two early clinical trials reported as abstracts used fairly large doses of mecamylamine (Tennant et al. 1984; Tennant and Tarver 1985). Although mecamylamine blocked the subjective effects of smoking, it did not increase abstinence rates and produced significant side effects in these trials (e.g., abdominal cramps, dizziness and drowsiness). Although a long-term trial of mecamylamine alone has not been reported, adding mecamylamine to transdermal nicotine has been shown to increase long-term cessation (Rose et al. 1994).

Nicotine blockers still warrant testing for several reasons (Clarke 1991; Stolerman 1986). First, unlike other drug-dependent individuals, smokers are often highly motivated, well-adjusted individuals who might comply with antagonist therapy. Second, blockade treatment might be modified by using lower doses and by using the medication prior to smoking cessation. By this strategy, subjects should develop fewer side effects and, as importantly, would learn that smoking now fails to produce the expected effects. Another strategy would be concomitant use of mecamylamine and nicotine. Although this sounds counterintuitive, a rationale for this type of treatment has been put forward (Rose and Levin 1991). Finally, one could treat initially with the agonist (i.e., nicotine) followed by the antagonist (i.e., mecamylamine) as is often done with opioid dependence. In fact, since mecamylamine does not precipitate withdrawal, the antagonist could directly follow the agonist. Third, more specific CNS nicotinic blockers, e.g., those with high affinity for the CNS nicotinic-cholinergic receptor which have fewer side effects, might be developed (Clarke 1991; Stolerman 1986).

2. Naltrexone

The reinforcing effects of nicotine have been hypothesized to be mediated by endogenous opioids (Karras 1982; Pomerleau and Pomerleau 1984); thus, opioid blockers have been tried as treatments. Naloxone increases smoking (again interpreted as an attempt to overcome nicotine blockade) in two studies (Gorelick et al. 1989; Karras and Kane 1980) but not in another (Nemeth-Coslett and Griffiths 1986). In another study, naloxone failed to reduce nicotine self-administration in humans (Mello et al. 1980), and one case report indicated that naltrexone did not change ongoing smoking (Mello et al. 1980). None of these studies reported whether opioid blockers precipitated nicotine withdrawal. Despite these mixed results, given the few side effects of naltrexone and the recent success of naltrexone for treating alcohol dependence (O'Malley et al. 1992; Volpicelli et al. 1992), further studies of this medication are indicated.

III. Clonidine

Clonidine is an α-2 adrenergic antagonist originally developed as an anti-hypertensive but which was found helpful in abating opioid and alcohol withdrawal (Covey and Glassman 1991). An early study found that clonidine abated craving for cigarettes and nicotine withdrawal symptoms (Glassman et al. 1985). Several short- and long-term clinical trials followed using oral or transdermal clonidine, usually in doses of 0.1–0.4 mg/day for 2–6 weeks with and without behavior therapy. A meta-analytic review of these reported an OR of 4.2 when behavior therapy was included and 1.7 when no behavior therapy was included (Covey and Glassman 1991). Although these ORs appear encouraging, they should be viewed with caution for at least two reasons. First, many of these studies reported only short-term results or were published only in abstract form. Second, more recent short-term and long-term trials have produced contradictory results (Glassman et al. 1993; Gourlay et al. 1994; Niaura et al. 1993; Prochazka et al. 1992; Zikos et al. 1991).

An interesting finding in most studies has been that clonidine is either more effective or only effective in women compared to men (Covey and Glassman 1991; Glassman et al. 1993). Why this is so is unclear. Finally, the rationale for clonidine's ability to relieve withdrawal and craving is that it decreases adrenergic overactivity; however, nicotine withdrawal is not a syndrome of adrenergic overactivity (Hughes et al. 1990).

IV. Agents That Make Smoking Aversive

1. Silver Acetate

Another way a medication can be useful in treating smoking is to make intake aversive, e.g., as disulfiram does in alcohol dependence. Silver acetate combines with chemicals in the smoker's saliva to produce an aversive, metallic taste. Older placebo-controlled, short-term trials showed questionable efficacy for silver nitrate tablets (Department of Health and Human Services 1982). More recent trials have again failed to show benefit of silver acetate over placebo (Hymowitz et al. 1993; Jensen et al. 1991). Unfortunately, neither the older or recent trials reported compliance; thus, whether improving compliance (as is necessary with most aversive therapies for drug dependence) would increase the probability of finding significant results is not clear.

2. Other Aversive Agents

One study reported anecdotally that four alcoholics receiving metronidazole for alcoholism reported "a progressive aversion" to cigarettes that lead to cessation (Bazot et al. 1982). Similar anecdotal observations have been noted in smokers who contract hepatitis; however, in both of the situations, food intake is decreased, thus the reduced desire for cigarettes might be a nonspecific nausea effect.

V. Medications to Abate Withdrawal or Replace the Reinforcing Effects of Nicotine

Several other medications have been tried that can be categorized as either reducing specific symptoms of nicotine withdrawal or mimicking effects of nicotine that are putative reinforcers.

1. Anxiolytics

Anxiety is a prominent symptom of nicotine withdrawal (HUGHES et al. 1990). In addition, smoking decreases some, but not all, measures of anxiety in humans and nonhumans (POMERLEAU and POMERLEAU 1984). Sedative/hypnotics have been shown to decrease tobacco withdrawal (GLASSMAN et al. 1985). However, laboratory work indicates benzodiazepines and barbiturates increase, not decrease ongoing smoking (HENNINGFIELD et al. 1983a).

a) Sedatives/Hypnotics

In a laboratory trial, alprazolam reduced withdrawal (GLASSMAN et al. 1985); however, in a well-done, long-term clinical trial, diazepam did not increase abstinence (WEI and YOUNG 1988). Older trials found that the non-benzodiazepine anxiolytics meprobamate and hydroxyzine were not effective (SCHWARTZ and DUBITSKY 1968).

b) Buspirone

Buspirone is a $5HT_{1A}$ (serotonergic) agonist that acts as an anxiolytic but produces minimal sedation, abuse potential, or physical dependence. Three short-term trials reported that buspirone (15–60 mg/day) reduced withdrawal (ROBINSON et al. 1991) and improved smoking cessation rates (GAWIN et al. 1989; WEST et al. 1991). One trial reported that buspirone improved cessation in high anxiety smokers (CINCIRIPINI et al., in press). Given these encouraging results and the benign side effect profile of buspirone, further studies are indicated, especially in smokers who report anxiety on a daily basis or who report significant anxiety on prior quit attempts.

c) β-Blockers

In some studies, β-blockers act as anxiolytics. Placebo-controlled strials of metroprolol, oxprenolol and propanolol have not found that they decrease craving or increase abstinence rates (DOW and FEE 1984; FAREBROTHER et al. 1980).

2. Antidepressants

Many smokers have a past history of depression (40% in smokers vs 10% in the general population) (BRESLAU et al. 1991; GLASSMAN et al. 1990; HALL et al. 1993). In addition, such smokers have significant withdrawal-induced depression (BRESLAU et al. 1992; COVEY et al. 1990; HALL et al. 1994). Finally, unlike other withdrawal symptoms, depression is clearly linked to relapse (COVEY et al. 1990; HALL et al. 1994; HUGHES 1992). These results have led some to hypothesize that nicotine serves as an antidepressant in certain smokers (GLASS 1990; HUGHES 1988).

The single, long-term clinical trial with an antidepressant (imipramine) reported that the drug had no effect on smoking cessation (JACOBS et al. 1971). A study of the effects of zimelidine and citalopram (serotonin uptake blockers) on drinking in alcoholics found that they had no effect on smoking (SELLERS et al. 1987). Two trials with fluoxetine were completed a while ago, but their results have not been published, suggesting these were negative trials. Recent, short-term placebo-controlled trials with doxepin (EDWARDS et al. 1988), tryptophan (BOWEN et al. 1991) and buproprion (FERRY et al., in press) in unselected groups of smokers have shown promising results.

Many antidepressants have significant side effects and a long delay in efficacy; thus, these treatments may not be acceptable to the general population of smokers. A more focused approach might be antidepressant treatment prior to cessation in the 40% of smokers with a past history of depression.

3. Stimulants

Many smokers endorse improved concentration as a major reason for smoking, and several studies suggest that nicotine improves performance, especially on vigilance tasks (LEVIN 1992). Also, several of the symptoms of nicotine withdrawal are similar to those of Attention Deficit Disorder (ADD) (e.g., difficulty concentrating, impatience, irritability and restlessness) (HUGHES et al. 1990) and those with ADD are much more likely to begin smoking (HARTSOUGH and LAMBERT 1987). Since ADD persists into adulthood (GITTELMAN et al. 1985) and is improved by stimulants (WENDER et al. 1985), there is the possibility that there is a group of adult smokers who have continuing symptoms of ADD and are using nicotine to abate these symptoms.

Laboratory studies of amphetamine have reported both increases and decreases in smoking and satisfaction from cigarettes (Low et al. 1984; SCHUSTER et al. 1979). The single long-term study of a stimulant was done many years ago and found that 9 mg/day of amphetamine did not increase abstinence (MILLER 1941). A recent uncontrolled trial reported that methylphenidate decreases tobacco withdrawal (ROBINSON et al. 1993). Finally, if stimulants were found effective, the issue of transferring dependence from nicotine onto the stimulant would need to be addressed.

4. Anorectics

Hunger and weight gain are two of the most widely cited reasons for difficulty in stopping smoking (KLESGES et al. 1989). Short-term trials of women smokers with postcessation weight problems have indicated that both fenfluramine (SPRING et al. 1991) and phenylpropanolamine (KLESGES et al. 1990) reduced post-cessation weight gain and some withdrawal symptoms, and increased abstinence. The results are noteworthy given the recent data that dieting to combat postcessation weight gain worsens rather than improves abstinence rates (HALL et al. 1992; PIRIE et al. 1992). Given the safe side effect profile of phenylpranolamine, further research on using it in smokers with a history of post-cessation weight gain seems indicated.

5. Other Medications

a) Sodium Bicarbonate

Sodium bicarbonate alkalizes the urine and slows the rate of elimination of nicotine which then might decrease withdrawal (SCHACTER et al. 1977). The only trial of this kind of therapy found a nonsignificant trend for greater abstinence in the bicarbonate than in the placebo group 5 weeks post-cessation (56% vs 37%) (FIXX et al. 1983). Further replications have not appeared, and there are reasons to doubt the utility of alkalization therapy, e.g., alkalization actually changes nicotine excretion very little (GRUNBERG and KOZLOWSKI 1986).

b) Glucose/ACTH

Some have hypothesized that, during withdrawal, smokers may mislabel hunger as cigarette craving; thus supplying glucose might decrease craving and increase abstinence (WEST et al. 1990). In the only study of this, dextrose decreased craving more and reduced nicotine gum use more than sorbitol (a low calorie sweetener) (WEST et al. 1990).

Nicotine via smoking increases cortisol (POMERLEAU and POMERLEAU 1984). Thus, another hypothesis is that when this effect is chronically repeated it blunts ACTH secretion, leading to secondary hypoadrenalism, hypoglycemia, anxiety and other symptoms (BOURNE 1991; TARGOVNIK 1989). Two uncontrolled trials of administering ACTH have found promising results (BOURNE 1985; McELHANEY 1989), but controlled trials have not been reported. In addition, several problems in the logic of this hypothesis have been noted e.g., smokers do not have lower blood glucose levels than nonsmokers (WEST 1985).

c) Anticholinergics

Anticholinergics have been tried on the theory that nicotine actually acts as a blocking agent at cholinergic receptors; thus, nicotine withdrawal represents cholinergic rebound and should be helped by anticholinergics (BACHYNSKY 1986). Although uncontrolled trials have found promising results (BACHYNSKY 1986), a controlled trial has not been reported.

C. Use of Adjunctive Psychological Therapies

I. Psychological Therapies for Smoking Cessation

Many different psychological therapies have been proposed for smoking cessation: acupuncture, behavior therapy, bibliotherapy, cognitive therapy, educational groups, hypnosis, inpatient therapy, supportive therapy, and 12-step disease-model therapies (SCHWARTZ 1987). Among these, only behavioral therapies have a substantial empirical base (SHIFFMAN 1993; BAILLIE et al., in press; SCHWARTZ 1987). More recent behavioral programs focus on relapse prevention via avoidance of high risk situations, engaging in alternate behaviors, etc. (BAILLIE et al., in press; KAMARCK and LICHTENSTEIN 1985).

The major problem with behavior therapies has been their availability and acceptability (HUGHES 1991, 1994). The large majority of primary care physicians are not comfortable with or interested in learning to provide behavioral therapy. In addition, given the lack of reimbursement for behavioral therapy for smoking cessation, most medical practices are reluctant to hire someone to deliver behavior therapy. Most public programs for smoking cessation (e.g., the American Lung Association) are behaviorally oriented. However, in many rural areas, they are not available or, if so, only intermittently (e.g., three to four times/year). As a result, several studies have found that <7% of smokers in general practice will attend a behavioral therapy when referred (HUGHES et al. 1989; JAMROZIK et al. 1984; LICHTENSTEIN and HOLLIS 1992; RUSSELL et al. 1987b). Unlike in other areas of alcohol/drug abuse, methods to increase the acceptance of smoking cessation treatment have rarely been examined (LICHTENSTEIN and HOLLIS 1992).

II. Combining Psychological Therapies and Medications

Most studies combining psychological therapy with medications have used behavioral therapies and nicotine polacrilex (HUGHES 1991). These studies have generally found that adding behavior therapy to nicotine polacrilex increases quit rates, and adding nicotine polacrilex to behavior therapy increases quit rates. Overall, these studies find an additive interaction, i.e., combining the two therapies gives quit rates similar to the sum of quit rates from each therapy (HUGHES 1991). Similar results appear to occur with adding behavior therapy to transdermal nicotine (HUGHES, in press).

Several hypotheses to account for this positive interaction are possible: (1) nicotine polacrilex helps initially with withdrawal symptoms and behavior

Table 3. Typical long-term abstinence rates

Self-quitting	5%
Physician advice	10%
Group behavior therapy	20%
Nicotine replacement and physician advice	20%
Nicotine replacement and group behavior therapy	30%–40%

therapy helps later with relapse prevention, (2) nicotine polacrilex helps the dependent smokers while behavior therapy helps smokers in need of support and skills training, or (3) behavior therapy increases compliance to nicotine polacrilex and nicotine polacrilex increases compliance with behavior therapy (HUGHES 1991, in press). None of these hypotheses have been well-tested. Also interesting is that, despite the fears of some pharmacologists, adding behavior therapy does not appear to "swamp" any nicotine-placebo differences. Adjunctive behavior therapy may, in fact, allow pharmacological effects to occur (HUGHES 1991).

D. Treating Smoking Among Alcohol/Drug Abusers

Over 85% of alcohol/drug abusers smoke (HUGHES 1993b). Despite clinical concerns, the existing correlational data suggest stopping smoking increases, not decreases, success in the treatment of alcohol/drug abuse (SOBELL et al. 1992). Although most patients and therapists recommend treating smoking after treating the other drug dependence, whether this is more or less effective than treating smoking while treating the other drug dependence is unknown (HUGHES 1993b). The only information about pharmacotherapy of smoking in alcohol/drug abusers is a post-hoc finding that smokers with a past history of alcohol/drug problems especially benefited from nicotine replacement (HUGHES 1993b).

E. Conclusions: Typical Quit Rates

Many different cessation rates for pharmacotherapies have been cited in the above sections. Table 3 presents typical quit rates using nicotine replacement compared to or along with other modalities. These figures are "ballpark" estimates but give the reader a feel for the relative contribution made by nicotine replacement. Unfortunately, there is an insufficient number of long-term studies of clonidine or other treatments to cite typical cessation rates for these medications.

Acknowledgement. Preparation of this article was supported, in part, by Research Scientist Development Award K02 DA-00109 from the National Institute on Drug Abuse.

References

Bachynsky N (1986) The use of anticholinergic drugs for smoking cessation: A pilot study. Int J Addict 21:789–805

Baillie A, Mattick RP, Hall W, Webster P (in press) Meta-analytic review of the efficacy of smoking cessation interventions. Drug Alcohol Rev

Bazot M, Fabre JD, Ortolan JM, Schweitzer J (1982) An unanticipated effect of metronidazole. In: Molimard R. (ed) 1 journee de la dependance tabaqique. Biomedicale des Sainte-Peres, Paris, pp 113–122

Belcher M, Jarvis MJ, Sutherland G (1989) Nicotine absorption and dependence in an over the counter aid to stopping smoking. Br Med J 298:570

Benowitz NL (1988) Pharmacologic aspects of cigarette smoking and nicotine addiction. Engl J Med 319:1318–1330

Benowitz NL (1991) Nicotine replacement therapy during pregnancy. JAMA 266:3174–3177

Benowitz NL (1993) Nicotine replacement therapy: what has been accomplished – can we do better? Drugs 2:157–170

Bourne SJ (1985) Treatment of cigarette smoking with short term high dosage corticotrophin therapy: preliminary communication. JR Soc Med 78:649–650

Bourne SJ (1991) Towards a neuro-endocrine explanation of tobacco addiction. J Smoking Relat Dis 2:105–110

Bowen DJ, Spring B, Fox E (1991) Tryptophan and high-carbohydrate diets as adjuncts to smoking cessation therapy. J Behav Med 14:97–110

Breslau N, Kilbey MM, Andreski P (1991) Nicotine dependence, major depression, and anxiety in young adults. Arch Gen Psychiatr 48:1069–1074

Breslau N, Kilbey MM, Andreski P (1992) Nicotine withdrawal symptoms and psychiatric disorders: findings from an epidemiologic study of young adults. Am J Psychiatry 149:464–469

Cepeda-Benito A (1993) Meta-analytical review of the efficacy of nicotine chewing gum in smoking treatment programs. J Consult Clin Psychol 61:822–830

Cinciripini PM, Lapitsky L, Seay S, Wallfisch A, Meyer WJ III, Van Vunakis H (1995) A placebo controlled evaluation of the effects of buspirone on smoking cessation: differences between high and low anxiety smokers. J Behav Med (in press)

Clarke PBS (1991) Nicotinic receptor blockade therapy and smoking cessation. Br J Addict 86:501–505

Clarke PBS (1992) Recent advances in understanding the actions of nicotine in the central nervous system. J Natl Cancer Inst Monogr 2:229–238

Coambs RB, Kozlowski LT, Ferrence RG (1989) The future of tobacco use and smoking research. In: Ney T, Gale A. (eds) Smoking and human behavior. Wiley, New York, pp 337–348

Cohen S, Lichtenstein E, Prochaska JO, Rossi JS, Gritz ER, Carr CR, Orleans CT, Schoenbach VJ, Biener L, Abrams D, DeClemente C, Curry S, Marlatt GA, Cumming KM, Emont SL, Giovino G, Ossip-Klein D (1989) Debunking myths about self-quitting. Am Psychol 44:1355–1365

Covey LS, Glassman AH (1991) A meta-analysis of double-blind placebo-controlled trials of clonidine for smoking cessation. Br J Addict 86:991–998

Covey LS, Glassman AH, Stetner F (1990) Depression and depressive symptoms in smoking cessation. Compr Psychiatry 31:350–354

Cummings SR, Hansen B, Richard RJ, Stein MJ, Coates TJ (1988) Internists and nicotine gum. JAMA 260:1565–1569

Daughton DM, Heatley SA, Prendergast JJ, Causey D, Knowles M, Rolf CN, Cheney RA, Hattelid K, Thompson AB, Rennard SI (1991) Effect of transdermal nicotine delivery as an adjunct to low intervention smoking cessation therapy: a randomized placebo-controlled, double-blind study. Arch Intern Med 151:749–752

Davison GC, Rosen RC (1972) Lobeline and reduction of cigarette smoking. Psychol Rep 31:443–456

Department of Health and Human Services (1982) Smoking deterrent drug products for over-the-counter human use: establishment of a monograph. Federal register, Docket no. 81N-0027, pp 490–500

Dow RJ, Fee WM (1984) Use of beta-blocking agents with group therapy in a smoking withdrawal clinic. J R Soc Med 77:648–651

Edwards NB, Murphy JK, Downs AD, Ackerman BJ, Rosenthal TL (1988) Doxepin as an adjunct to smoking cessation: a double-blind pilot study. J Psychiatr 146:373–376

Fagerstrom K-O (1988) Efficacy of nicotine chewing gum: a review. In: Pomerleau OF, Pomerleau CS, Fagerstrom K-O, Henningfield JE, Hughes JR (eds) Nicotine replacement: a critical evaluation. Liss, New York, pp 109–128

Fagerstrom K-O (1994) Combined use of nicotine replacement products. Health Values 18:15–20

Fagerstrom K-O, Melin B (1985) Nicotine chewing gum in smoking cessation: efficacy, nicotine dependence, therapy duration, and clinical recommendations. In: Grabowski J, Hall SM (eds) Pharmacological adjuncts in smoking cessation, National Institute on Drug Abuse monograph 53, Washington: DHHS publication no (ADM) 85-1553, pp 102–109

Fagerstrom K-O, Schneider NG (1989) Measuring nicotine dependence: a review of the Fagerstrom tolerance questionnaire. J Behav Med 12:159–182

Fagerstrom K-O, Lunell E, Molander L (1991) Continuous and intermittent transdermal delivery of nicotine: blockade of withdrawal symptoms and side-effects. J Smoking Relat Dis 2:173–180

Fagerstrom K-O, Sawe U, Tonnesen P (1992) Therapeutic use of nicotine patches: efficacy and safety. J Smoking Relat Dis 3:247–261

Fagerstrom K-O, Schneider NG, Lunell E (1993) Effectiveness of nicotine patch and nicotine gum as individual versus combined treatments for tobacco withdrawal symptoms. Psychopharmacology 111:271–277

Farebrother MJB, Pearce SJ, Turner P, Appleton DR (1980) Propranolol and giving up smoking. Br J Dis Chest 74:95–96

Ferry LH, Robbins AS, Scariati PD, Peters A (in press) Depression, craving and heavy smoking: a pilot trial using bupropion. Addictive Diseases

Fiore MC, Novotny TE, Pierce JP, Giovino GA, Hatziandreu EJ, Newcomb PA, Surawicz TS, Davis RM (1990) Methods used to quit smoking in the United States. JAMA 263:2760–2765

Fiore MC, Jorenby DE, Baker TB, Kenford SL (1992) Tobacco dependence and the nicotine patch: clinical guidelines for effective use. JAMA 268:2687–2694

Fiore MC, Smith SS, Jorenby DE, Baker TB (1994) The effectiveness of the nicotine patch for smoking cessation. JAMA 271:1940–1947

Fixx AJ, Daughton D, Kass I, Smith JL, Wickiser A, Golden CJ (1983) Urinary alkanization and smoking cessation. J Clin Psychol 39:618–623

Foulds J (1993) Does nicotine replacement therapy work? Addiction 88:1473–1478

Foulds J, Stapleton J, Feyerabend C, Vesey C, Jarvis M, Russell MAH (1992) Effect of transdermal nicotine patches on cigarette smoking: a double-blind crossover study. Psychopharmacology 106:421–427

Fredrickson PA, Lee GM, Wingender LA, Croghan IT, Offord KP, Hurt RD (1993) Safety and tolerance of high dose transdermal nicotine therapy in light (LS) and heavy (HS) smokers. J Addict Dis 12:177

Gawin FH, Compton M, Byck R (1989) Buspirone reduces smoking. Arch Gen Psychiatry 46:288–289

Gittelman R, Mannuzza S, Shenker R, Bonagura N (1985) Hyperactive boys almost grown up, I: psychiatric status. Arch Gen Psychiatry 42:937–947

Glass RM (1990) Blue mood, blackened lungs: depression and smoking. JAMA 264:1583–1584

Glassman AH (1993) Cigarette smoking: implications for psychiatric illness. Am J Psychiatry 150:546–553

Glassman AH, Covey LS (1990) Future trends in the pharmacological treatment of smoking cessation. Drugs 40:1–5

Glassman AH, Jackson WK, Walsh BT, Roose SP (1985) Cigarette craving, smoking withdrawal, and clonidine. Science 226:864–866

Glassman AH, Helzer JE, Covey LS, Cottler LB, Stetner F, Tipp JE, Johnson J (1990) Smoking, smoking cessation, and major depression. JAMA 264: 1546–1549

Glassman AH, Covey LS, Dalack GW, Stetner F, Rivelli SK, Fleiss J, Cooper TB (1993) Smoking cessation, clonidine, and vulnerability to nicotine among dependent smokers. Clin Pharmacol Ther 54:670–679

Goldstein MG, Niaura R, Follick MJ, Abrams DB (1989) Effects of behavioral skills training and schedule of nicotine gum administration on smoking cessation. Am J Psychiatry 146:56–60

Goldstein MG, Niaura R, Abrams DB (1991) Pharmacological and behavioral treatment of nicotine dependence: nicotine as a drug of abuse. In: Stoudemire A, Fogel BS (eds) Medical-psychiatric practice. American Psychiatric Press, Washington DC, pp 541–596

Gora ML (1993) Nicotine transdermal systems. Ann Pharmacother 27:742–750

Gorelick DA, Rose JE, Jarvik ME (1989) Effect of naloxone on cigarette smoking. J Subst Abuse 1:153–159

Gourlay S, Forbes A, Marriner T, Kutin J, McNeil J (1994) A placebo-controlled study of three clonidine doses for smoking cessation. Clin Pharmacol Ther 55:64–69

Gourlay SG, McNeil JJ (1990) Antismoking products. Med J Aust 153:699–707

Grunberg NE, Kozlowski LT (1986) Alkaline therapy as an adjunct to smoking cessation programs. Int J Biosocial Res 8:43–52

Hajek P, Jackson P, Belcher M (1988) Long-term use of nicotine chewing gum. JAMA 260:1593–1596

Hall SM, Tunstall CD, Vila KL, Duffy J (1992) Weight gain prevention and smoking cessation: cautionary findings. Am J Publ Health 82:799–803

Hall SM, Munoz RF, Reus VI, Sees KL (1993) Nicotine, negative affect, and depression. J Consult Clin Psychol 61:761–767

Hall SM, Munoz RF, Reus VI (1994) Cognitive-behavioral intervention increases abstinence rates for depressive history smokers. J Consult Clin Psychol 62:141–146

Hartsough CS, Lambert NM (1987) Pattern and progression of drug use among hyperactives and controls: a prospective short-term longitudinal study. J Child Psychiatry 28:543–553

Hatsukami DK, Huber M, Callies A, Skoog K (1993) Physical dependence on nicotine gum: effect of duration of use. Psychopharmacology 111:449–456

Heatherton TF, Kozlowski LT, Frecker RC, Fagerstrom K-O (1991) The Fagerstrom test for nicotine dependence: a revision of the Fagerstrom Tolerance Questionnaire. Br J Addict 86:1119–1127

Henningfield JE, Stitzer ML (1991) New developments in nicotine-delivery systems. Communications, Ossining

Henningfield JE, Chait LD, Griffiths RR (1983a) Cigarette smoking and subjective responsive in alcoholics: effects of pentobarbital. Clin Pharmacol Ther 33:806–812

Henningfield JE, Miyasato K, Johnson RE, Jasinski DR (1983b) Rapid physiologic effects of nicotine in humans and selective blockade of behavioral effects by mecamylamine. In: Harris LS (ed) Problems of drug dependence, 1982, NIDA research monograph. US Government Printing Office, Washington DC, pp 259–265

Henningfield JE, London ED, Benowitz NL (1990) Arterial-venous differences in plasma concentrations of nicotine after cigarette smoking. JAMA 263:2049–2050

Henningfield JE, Stapleton JM, Benowitz NL, Grayson RF, London ED (1993) Higher levels of nicotine in arterial than in venous blood after cigarette smoking. Drug Alcohol Depend 33:23–29

Hilleman DE, Mohiuddin SM, Delcore MG (1994) Comparison of fixed-dose transdermal nicotine, tapered-dose transdermal nicotine and buspirone in smoking cessation. J Clin Pharmacol 34:222–224

Hughes JR (1986) Problems of nicotine gum. In: Ockene JK (ed) Pharmacologic treatment of tobacco dependence: proceedings of the world congress, 3–5 Nov 1985, Institute for the Study of Smoking Behavior and Policy, Cambridge, pp 141–147

Hughes JR (1988) Clonidine, depression and smoking cessation. JAMA 254: 2901–2902

Hughes JR (1991) Combined psychological and nicotine gum treatment for smoking: a critical review. J Subst Abuse 3:337–350

Hughes JR (1992) Tobacco withdrawal in self-quitters. J Consult Clin Psychol 60:689–697

Hughes JR (1993a) Pharmacotherapy for smoking cessation: Unvalidated assumptions and anomalies, and suggestions for further research. J Consult Clin Psychol 61:751–760

Hughes JR (1993b) Treatment of smoking cessation in smokers with past alcohol/drug problems. J Subst Abuse Tx 10:181–187

Hughes JR (1993c) Possible effects of smoke-free inpatient units on psychiatric diagnosis and treatment. J Clin Psychiatr 54:109–114

Hughes JR (1993d) Risk/benefit of nicotine replacement in smoking cessation. Drug Safety 8:49–56

Hughes JR (1994) Behavioral support programs for smoking cessation. Mod Med 62:22–26

Hughes JR (in press) Combining behavioral therapy and pharmacotherapy for smoking cessation: an update. In: Blaine J, Onken L (eds) Integrating behavior therapies with medication in the treatment of drug dependence, NIDA research monograph. US Government Printing Office, Washington DC (in press)

Hughes JR, Glaser M (1993) Transdermal nicotine for smoking cessation. Health Values 17:24–31

Hughes JR, Hatsukami DK (1992) The nicotine withdrawal syndrome: a brief review and update. Int J Smoking Cessation 1:21–26

Hughes JR, Gust SW, Keenan RM, Fenwick JW, Healy ML (1989) Nicotine vs placebo gum in general practice. JAMA 261:1300–1305

Hughes JR, Higgins ST, Hatsukami DK (1990) Effects of abstinence from tobacco: a critical review. In: Kozlowski LT, Annis H, Cappell HD, Glaser F, Goodstadt M, Israel Y, Kalant H, Sellers EM, Vingilis J (eds) Research advances in alcohol and drug problems, vol 10. Plenum, New York, pp 317–398

Hughes JR, Gust SW, Keenan RM, Skoog K, Fenwick JW, Higgins ST (1991) Long-term use of nicotine versus placebo gum. Arch Intern Med 151:1993–1998

Hughes JR, Gulliver SB, Fenwick JW, Cruser K, Valliere WA, Pepper SL, Shea PJ, Solomon LJ (1992) Smoking cessation among self-quitters. Health Psychol 11:331–334

Hurt RD, Dale LC, Fredrickson PA, Caldwell CC, Lee GA, Offord KP, Lauger GG, Marusic Z, Neese LW, Lundberg TG (1994) Nicotine patch therapy for smoking cessation combined with physician advice and nurse follow-up. JAMA 271:595–600

Hymowitz N, Feuerman M, Hollander M, Frances RJ (1993) Smoking deterrence using silver acetate. Hosp Commun Psychiatr 44:113–117

Jacobs MA, Spiker AZ, Norman MM, Wohlberg GW, Knapp PH (1971) Interaction of personality and treatment conditions associated with success in a smoking control program. Psychosom Med 6:545–556

Jamrozik K, Vessey M, Fowler G, Wald N, Parker G, VanVunakis H (1984) Controlled trial of three different antismoking interventions in general practice. Br Med J 288:1499–1503

Jarvik ME, Henningfield JE (1988) Pharmacological treatment of tobacco dependence. Pharmacol Biochem Behav 30:279–294

Jarvik ME, Henningfield JE (1993) Pharmacological adjuncts for the treatment of tobacco dependence. In: Orleans CT, Slade J (eds) Nicotine addiction: principles and management. Oxford University Press, New York, pp 245–261

Jarvik ME, Schneider NG (1992) Nicotine. In: Lowinson JH, Ruiz P, Millman RB, Langrod JG (eds) Substance abuse: a comprehensive textbook. Williams and Wilkins, Baltimore, pp 334–356

Jarvik ME, Glick SD, Nakamura RK (1970) Inhibition of cigarette smoking by orally administered nicotine. Clin Pharmacol Ther 11:574–576

Jarvis MJ, Hajek P, Russell MAH, West RJ (1987) Nasal nicotine solution as an aid to cigarette withdrawal: a pilot clinical trial. Br J Addict 82:983–988

Jensen EJ, Schmidt E, Pedersen B, Dahl R (1991) Effect on smoking cessation of silver acetate, nicotine and ordinary chewing gum. Psychopharmacology 104:470–474

Kamarck TW, Lichtenstein E (1985) Current trends in clinic-based smoking control. Ann Behav Med 7:19–23

Kandel DB, Davies M (1986) Adult sequelae of adolescent depressive symptoms. Arch Gen Psychiatry 43:255–262

Karras A (1982) Neurotransmitter and neuropeptide correlates of cigarette smoking. In: Essman WR, Valvelli L (eds) Neuropharmacology: clinical applications. New York: Springer, Berlin Heidelberg New York, pp 41–66

Karras A, Kane J (1980) Naloxone reduces cigarette smoking. Life Sci 27:1541–1545

Killen JD, Fortmann SP, Newman B, Varady A (1990) Evaluation of a treatment approach combining nicotine gum with self-guided behavioral treatments for smoking relapse prevention. J Consult Clin Psychol 58:85–92

Klesges RC, Meyers AW, LaVasque ME (1989) Smoking, body weight and their effects on smoking behavior: a comprehensive review of the literature. Psychol Bull 106:204–230

Klesges RC, Klesges LM, Meyers AW, Klem ML, Isbell T (1990) The effects of phenylpropanolamine on dietary intake, physical activity, and body weight after smoking cessation. Clin Pharmacol Ther 47:747–754

Kornitzer M, Kittel F, Dramaix M, Bourdoux P (1987) A double blind study of 2 mg versus 4 mg nicotine gum in an industrial setting. J Psychosom Res 31:171–176

Kozlowski LT (1984) Pharmacological approaches to smoking modification. In: Matarazzo JD, Weiss SM, Herd JA, Miller NE (eds) Behavioral health: a handbook of health enhancement and disease prevention. Wiley, New York, (1987)

Lam WL, Sze PC, Sacks HS, Chalmer TC (1987) Meta-analysis of randomized controlled trials of nicotine chewing gum. Lancet 2:27–29

Lee EW, D'Alonzo GE (1993) Cigarette smoking, nicotine addiction and its pharmacologic treatment. Arch Intern Med 153:34–48

Leischow SJ, Sachs DPL, Hansen MD, Bostrom, AG (1992) Effects of different nicotine replacement doses on weight gain after smoking cessation. Arch Fam Med 1:233–237

Leischow SJ, Valente SN, Hill AL, Otte PS, Aickin M, Kligman EW, Nicotine patch and gum: withdrawal symptoms, side effects, and medication preference. In: Harris LS (ed) Problems of drug dependence 1994, NIDA research monograph. US Government Printing Office, Washington DC (in press)

Levin ED (1992) Nicotinic systems and cognitive function. Psychopharmacology 108:417–431

Levin ED, Westman EC, Stein RM, Carnahan E, Sanchez M, Herman S, Behm FM, Rose JE (1994) Nicotine skin patch treatment increases abstinence, decreases withdrawal symptoms and attenuates rewarding effects of smoking. J Clin Psychopharm 14:41–49

Lichtenstein E, Glasgow RE (1992) Smoking cessation: what have we learned over the past decade? J Consult Clin Psychol 60:518–527

Lichtenstein E, Hollis JF (1992) Patient referral to a smoking cessation program: who follows through? J Fam Pract 34:739–744

Low RB, Jones M, Carter B, Cadoret RJ (1984) The effect of d-amphetamine and ephedrine on smoking attitude and behavior. Addict Behav 9:335–345

McElhaney JL (1989) Repository corticotropin injection as an adjunct to smoking cessation during the initial nicotine withdrawal period: results from a family practice clinic. Clin Ther 11:854–861

Meier-Lammermann E, Meyer M, Boleskei PL (1990) Differences between transdermal nicotine application and nicotine chewing gum in combination with behavioural group therapy in smoking cessation. Eur Respir J 3 [Suppl 10]:1685

Mello NK, Mendelson JH, Sellers ML, Kuehnle JC (1980) Effects of heroin self-administration on cigarette smoking. Psychopharmacology 67:45–52

Miller MM (1941) Benzadrine sulphate in the treatment of nicotinism. Med Rec 153:137–138

Minneker E, Buchkremer G, Block M (1989) The effect of different dosages of a transdermal nicotine substitution system on the success rate of smoking cessation therapy. Methods Find Exp Clin Pharmacol 11:219–222

Nemeth-Coslett R, Griffiths RR (1986) Naloxone does not affect cigarette smoking. Psychopharmacology 89:261–264

Nemeth-Coslett R, Henningfield JE, O'Keefe MK, Griffiths RR (1986) Effects of mecamylamine on human cigarette smoking and subjective ratings. Psychopharmacology 88:420–425

Newhouse P, Hughes JR (1991) The role of nicotine and nicotine mechanisms in neuropsychiatric disease. Br J Addict 86:521–526

Niaura R, Goldstein MG, Brown R, Murphy J, Abrams, D (1993) A double-blind randomized dose-response trial of transdermal clonidine for smoking cessation. Presented at Society of Behavioral Medicine Meeting, San Francisco, March 1993

Nunn-Thompson CL, Simon PA (1989) Pharmacotherapy for smoking cessation. Clin Pharm 8:710–720

O'Malley SS, Jaffe AJ, Chang G, Schottenfield RS, Meyer RE, Rounsaville B (1992) Naltrexone and coping skills therapy for alcohol dependence. Arch Gen Psychiatry 49:881–887

Palmer KJ, Faulds D (1992) Transdermal nicotine: a review of its pharmacodynamic and pharmacokinetic properties, and therapeutic use as an aid to smoking cessation. Drugs 44:498–529

Paoletti P, Fornai E, Maggiorelli F, Corlando A, Puntoni R, Molesti D, Carrozzi L, Viegi G, Martinelli M, Celiano F, Gustavsson G, Sawe U (1993) A double-blind clinical trial on smoking cessation using different doses of 16-hours transdermal nicotine patches after cotinine plasma values stratification. Chest 146:A806

Perkins KA, Grobe JE, Stiller RL, Fonte C, Goettler JE (1992) Nasal spray nicotine replacement suppresses cigarette smoking desire and dehavior. Clin Pharmacol Ther 52:627–634

Pickworth WB, Herning RI, Henningfield JE (1988) Mecamylamine reduces some EEG effects of nicotine chewing gum in humans. Pharmacol Biochem Behav 30:149–153

Pierce JP, Fiore MC, Novotny TE, Hatziandreu EJ, Davis RM (1989) Trends in cigarette smoking in the United States. Educational differences are increasing. JAMA 261:56–60

Pinney JM (1991) Factors affecting availability of treatment. In: Henningfield, JE, Stitzer, ML (eds) New developments in nicotine-delivery systems. Cortlandt Communications, Ossining, pp 79–88

Pirie PL, McBride CM, Hellerstedt W, Jeffery RW, Hatsukami DK, Allen S, Lando H (1992) Smoking cessation in women concerned about weight. Am J Public Health 82:1238–1243

Po ALW (1993) Transdermal nicotine in smoking cessation. Eur J Clin Pharmacol 45:519–528

Pomerleau CS, Pomerleau OF, Majchrzak MJ (1987) Mecamylamine pretreatment increases subsequent nicotine self-administration as indicated by changes in plasma nicotine level. Psychopharmacology 91:391–393

Pomerleau O, Pomerleau C, Fagerstrom K-O, Henningfield JE, Hughes JR (1988) Nicotine replacement: a critical evaluation. Haworth, New York

Pomerleau OF, Pomerleau CF (1984) Neuroregulators and the reinforcement of smoking: towards a biobehavioral explanation. Neurosci Biobehav Rev 8: 503–513

Pomerleau OF, Flessland KA, Pomerleau CS, Hariharan M (1992) Controlled dosing of nicotine via an Intranasal Nicotine Aerosol Delivery Device (INADD). Psychopharmacology 15:519–526

Prignot J (1989) Pharmacological approach to smoking cessation. Eur Respir J 2:550–560

Prochazka AV, Petty TL, Nett L, Silvers GW, Sachs DPL, Rennard SI, Daughton DM, Grimm RH Jr, Heim C (1992) Transdermal clonidine reduced some withdrawal symptoms but did not increase smoking cessation. Arch Intern Med 152:2065–2069

Puska P, Vartiainen E, Korhone H, Urjanheimo EL, Gustavsson G (1994) Combining patch and gum in nicotine replacement therapy: results of a double-blind study in North Karelia. Presented at Society of Behavioral Medicine Meeting, Boston, April 1994

Rennard S, Daughton D, Fortmann S, Killen J, Petty E et al. (1991) Transdermal nicotine enhances smoking cessation in coronary artery disease patients. Chest 100:5S (abstract)

Robinson MD, Smith WA, Cederstrom EA, Sutherland DE (1991) Buspirone effect on tobacco withdrawal symptoms: a pilot study. J Am Board Fam Pract 4:89–94

Robinson MD, Anastasio GD, Little JM, Sigmon JL, Menscer D, Pettice, YJ, Norton HJ (1993) Ritalin for tobacco withdrawal: a pilot study. J Addict Dis 12:178

Rose JE (1991) Transdermal nicotine and nasal nicotine administration as smoking cessation treatments. In: Cocores JA (ed) The clinical management of nicotine dependence. Springer, Berlin Heidelberg New York, pp 196–207

Rose JE, Behm FM (1987) Refined cigarette smoke as a method for reducing nicotine intake. Pharmacol Biochem Behav 28:305–310

Rose JE, Levin ED (1991) Concurrent agonist-antagonist administration for the analysis and treatment of drug dependence. Pharmacol Biochem Behav 41: 219–226

Rose JE, Sampson A, Levin ED, Henningfield JE (1989) Mecamylamine increases nicotine preference and attenuates nicotine discrimination. Pharmacol Biochem Behav 32:933–938

Rose JE, Behm FM, Westman EC, Levin ED, Stein RM, Ripka GV (1994) Mecamylamine combined with nicotine skin patch facilitates smoking cessation beyond nicotine patch treatment alone. Clin Pharmacol 56:86–99

Russell MAH, Jarvis MJ, Sutherland G, Feyerabend C (1987a) Nicotine replacement in smoking cessation. JAMA 257:3262–3265

Russell MAH, Stapleton JA, Jackson PH, Hajek P, Belcher M (1987b) District programme to reduce smoking: effect of clinic supported brief intervention by general practitioners. Br Med J 295:1240–1244

Sachs DPL (1991) Advances is smoking cessation treatment. Curr Pulmonol 12: 139–198

Sachs DPL, Leischow SJ (1991a) Pharmacological approaches to smoking cessation. Clin Chest Med 12:9–10

Sachs DPL, Leischow SJ (1991b) Dose-response effects of nicotine polacrilex. Am Rev Respir Dis 143:A255

Sachs DPL, Bostrom AG, Hansen MD (1994) Relapse hazard functions during and after nicotine patch smoking cessation treatment. Abstract presented at the Society of Behavioral Medicine Meeting, Boston, April 1994

Schacter S, Silverstein B, Kozlowski L, Perlick D, Hermann C, Liebling B (1977) Studies of the interaction of psychological and pharmacological determinants of smoking 106:3–40

Schneider NG, Jarvik ME (1984) Time course of smoking withdrawal symptoms as a function of nicotine replacement. Psychopharmacology 82:143–144

Schneider NG, Jarvik ME, Forsythe AB, Read LL, Elliot ML, Schweiger A (1983) Nicotine gum in smoking cessation: a placebo-controlled double-blind trial. Addict Behav 10:253–261

Schuster CR, Lucchesi BR, Emley GS (1979) The effects of d-amphetamine, meprobamate and lobeline on the cigarette smoking behavior of normal human subjects. In: Krasnegor NA (ed) Cigarette smoking as a dependence process, NIDA research monograph series 23 US Government Printing Office, Washington DC, pp 91–99

Schwartz JL (1987) Review and evaluation of smoking cessation methods: the United States and Canada. US Department of Health and Human Services, Washington DC

Schwartz JL, Dubitsky M (1968) One-year follow-up results of a smoking cessation program. Can J Publ Health 59:161–165

Sellers EM, Naranjo CA, Kadlec K (1987) Do serotonin uptake inhibitors decrease smoking? Observations in a group of heavy drinkers. J Clin Psychopharmacol 7:417–420

Sepkovic DW, Colosimo SG, Axelrad CM, Adams JD, Haley NJ (1987) The delivery and uptake of nicotine from an aerosol rod. Am J Publ Health 76:1343–1344

Shiffman S (1993) Smoking cessation treatment: any progress? J Consult Clin psychol 61:718–722

Shiffman SM (1979) The tobacco withdrawal syndrome. In: Krasnegor NA (ed) Cigarette smoking as a dependence process, NIDA research mongraph 23. US Department of Health and Human Services, Washington DC (HEW publication no (ADM) 79-800)

Shiffman SM (1982) Relapse following smoking cessation: a situational analysis. J Consult Clin Psychol 50:71–86

Silagy C, Mant D, Fowler G, Lodge M (1994a) Meta-analysis on efficacy of nicotine replacement therapies in smoking cessation. Lancet 343:139–142

Silagy C, Mant D, Fowler G, Lodge M (1994b) The effectiveness of nicotine replacement therapies in smoking cessation. Curr Clin Trials 113

Sobell LC, Sobell MB, Toneatto T (1992) Recovery from alcohol problems without treatment. In: Heather N, Miller WR, Greeley J (eds) Self-control and addictive behaviors. MacMallion, Sidney, pp 198–242

Spring B, Wurtman J, Gleason R, Wurtman R, Kessler K (1991) Weight gain and withdrawal symptoms after smoking cessation: a preventive intervention using d-fenfluramine. Health Psychol 10:216–223

Stolerman IP (1986) Could nicotine antagonists be used in smoking cessation. Br J Addict 81:47–53

Sutherland G, Russell MAH, Stapleton J, Feyerabend C, Ferno O (1992a) Nasal nicotine spray: a rapid nicotine delivery system. Psychopharmacology 108:512–518

Sutherland G, Stapleton JA, Russell MAH, Jarvis MJ, Hajek P, Belcher M, Feyerabend C (1992b) Randomised controlled trial of nasal nicotine spray in smoking cessation. Lancet 340:324–329

Tang JL, Law M, Wald N (1994) How effective is nicotine replacement therapy in helping people to stop smoking? Br Med J 308:21–25

Targovnik JH (1989) Nicotine, corticotropin, and smoking withdrawal symptoms: literature review and implications for successful control of nicotine addiction. Clin Ther 11:846–853

Tennant FS, Tarver AL (1985) Withdrawal from nicotine dependence using mecamylamine: comparison of three-week and six-week dosage schedules. In: Harris LS (ed) Problems of drug dependence, 1984, US Government Printing Office, Washington DC, pp 291–297 (NIDA research monograph series 55)

Tennant FS, Tarver AL, Rawson RA (1984) Clinical evaluation of mecamylamine for withdrawal from nicotine dependence. In: Harris LS (ed) Problems of drug dependence. US Government Printing Office, Washington DC (NIDA research monograph 49)

Tonnesen P, Fryd V, Hansen M, Helsted J, Gunnersen A, Forchammer H, Stockner M (1988) Effect of nicotine chewing gum in combination with group counseling in the cessation of smoking. N Engl J Med 318:15–27

Tonnesen P, Norregaard J, Mikkelsen K, Jorgensen S, Nilsson F (1993) A double-blind trial of a nicotine inhaler for smoking cessation. JAMA 269:1268–1271

Transdermal Nicotine Study Group (1991) Transdermal nicotine for smoking cessation: results of two multicenter controlled trials. JAMA 266:3133–3138

US Department of Health and Human Services (1988) Tobacco use compared to other drug dependencies. The health consequences of smoking: nicotine addition: a report of the US Surgeon General, US Government Printing Office, Washington DC, pp 241–375

Viswesvaran C, Schmidt FL (1992) A meta-analytic comparison of the effectiveness of smoking cessation methods. J Appl Psychol 77:554–561

Volpicelli JR, Alterman AI, Hayashida M, O'Brien CP (1992) Naltrexone in the treatment of alcohol dependence. Arch Gen Psychiatry 49:876–880

Wei H, Young D (1988) Effect of clonidine on cigarette cessation and in the alleviation of withdrawal symptoms. Br J Addict 83:1221–1227

Wender PH, Reimherr FW, Wood D et al. (1985) A controlled study of methylphenidate in the treatment of attention deficit disorder, residual type in adults. Am J Psychiatry 142:547–552

West RJ (1985) Corticotrophin injections to treat cigarette withdrawal symptoms. J R Soc Med 78:1065–1066

West RJ, Hajek P, Burrows S (1990) Effect of glucose tablets on craving for cigarettes. Psychopharmacology 101:555–559

West RJ, Hajek P, NcNeill A (1991) Effect of buspirone on cigarette withdrawal symptoms and short-term abstinence rates in a smokers clinic. Psychopharmacology 104:91–96

Wetter DW, Fiore MC, Young TB, Baker TB (in press) Nicotine withdrawal and nicotine replacement influence objective measures of sleep. J Consult Clin Psychol (in press)

Zikos A, Molter G, Romovacek M, Nellis K, Gaudio R (1991) Comparison of nicotine gum, clonidine and no medication in a multidisciplinary smoking cessation program. Chest 100:3S

CHAPTER 16

Pharmacotherapies for Cocaine Dependence

E.A. WALLACE and T.R. KOSTEN

A. Introduction

Cocaine has been used by humankind for centuries. The United States is currently experiencing its second epidemic of cocaine use; the first occurred in the late nineteenth through the early twentieth century and the second started in the 1970s. The most recent surge of cocaine use has been facilitated by several factors. These include an increased availability and drop in the price of cocaine, the development of "crack," or a free base of the drug, and an increase in use via routes that enable rapid brain uptake of cocaine, i.e., through intravenous or smoked routes. Compared to intranasal use, smoking cocaine provides a more rapid brain uptake of the drug followed by a more rapid decrement in blood levels (FISCHMAN 1987). This appears to lead to rapid, repeated cocaine administration and may facilitate quicker development of cocaine dependence (JOHANSON and FISCHMAN 1989).

The individual and societal consequences of endemic cocaine use are profound and treatment modalities are greatly needed. In years past, cocaine addiction was viewed as a psychological problem and treatment modalities were focused in this direction. The standard approach to the treatment of cocaine dependence has included individual and group psychotherapy, behavioral modification paradigms, and self-help group approaches. Unfortunately, these measures have not adequately addressed the needs of all individuals dependent on cocaine (KOSTEN and KLEBER 1992). However, the last 30 years have been marked by investigation of the neurobiologic correlates of psychiatric illnesses and in the attempt to develop pharmacotherapies for these disorders, including cocaine dependence. In order to approach the pharmacotherapy of cocaine dependence, it is important first to understand the pharmacologic effects of cocaine and the neurobiologic sequelae of chronic cocaine use (reviewed by KRUA 1989 and elsewhere in this volume).

B. Pharmacologic Treatment of Cocaine Dependence

I. Introduction

Pharmacotherapy has become increasingly employed as an adjunct to traditional chemical dependence treatment modalities for cocaine dependence.

Generally, pharmacotherapeutic agents are used to address neurobiologic changes and the subsequent clinical manifestations that occur secondary to chronic cocaine abuse. More specifically, pharmacotherapeutic agents are employed in an attempt to block the euphorigenic effects of cocaine, to treat initial cocaine withdrawal symptoms and/or promote abstinence, and to prevent relapse. Finally, these agents may be used to treat premorbid psychiatric disorders (WEISS and MIRIN 1990; BLAINE and LING 1992; KLEBER and GAWIN 1986).

The ideal pharmacologic agent would be orally tolerated, long-acting, reverse the neurochemical abnormalities (and their clinical manifestations) caused by chronic cocaine use, and block the effects of additional cocaine. Although no "magic bullet" currently addresses all of these needs, a variety of medications have been employed with some degree of success (KOSTEN 1993).

The question of physical dependence on cocaine has not been fully resolved. Withdrawal from cocaine and other stimulants is not life threatening and does not require the use of specific pharmacotherapy. However, it is clear that symptoms following cocaine discontinuation occur in some users, increasing their discomfort and propensity for relapse (GAWIN and KLEBER 1986). Cocaine abstinence symptoms include fatigue, sleep disturbance, agitation, diaphoresis, nausea, paranoia, and drug craving. Symptoms are most prominent immediately after stopping chronic cocaine use, decrease gradually with time, and are variable across subjects (GAWIN and KLEBER 1986; SATEL et al. 1991; WEDDINGTON et al. 1990). The primary aim of adjunctive pharmacotherapy is to alleviate these symptoms in order to prevent relapse to cocaine use.

II. Medications Employed in the Treatment of Cocaine Dependence

The two major classes of medications employed for the treatment of cocaine dependence are the dopaminergic agonists and antidepressants. These medications are employed to address the underlying neurochemical abnormalities resulting from chronic cocaine use. Dopaminergic agonists are used during the acute withdrawal period and antidepressants are used primarily following this or during the chronic withdrawal syndrome. For the purposes of this discussion, these medications will be reviewed in the following sections as "acute withdrawal agents" and "chronic withdrawal agents." A variety of other medications have also been employed in a limited number of trials and these will also be reviewed in the following sections (Table 1).

1. Acute Withdrawal Agents

During the period of acute cocaine abstinence, the cocaine user may experience a variety of symptoms including fatigue, sleep disturbance, anhedonia, cocaine craving, and occasionally psychosis. Some acute withdrawal

Table 1. Types of pharmacotherapy for cocaine dependence

Acute withdrawal agents

   Anticraving medications
      Bromocriptine
      Amantadine
      L-Dopa
      Flupenthixol
   Medications for other transient withdrawal symptoms
      Neuroleptics
      Benzodiazepine

Chronic withdrawal agents

   Medications for general prevention of relapse
      Desipramine
      Imipramine
      Ludiomil
      Prozac
      Zoloft
      Trazodone
   Medications for comorbid psychiatric disorders
      Lithium (bipolar disorder, cyclothymia)
      Stimulants (residual ADHD)

symptoms, in particular cocaine craving, are postulated to be related to a dopaminergic deficit, although this postulate has only limited clinical support (MENDELSON et al. 1988; McDOUGLE et al. 1992). Investigators have employed agents that stimulate dopamine (DA) receptors or augment synaptic DA, and numerous reports have documented the efficacy of dopaminergic agents in reducing the effects of cocaine withdrawal (DACKIS and GOLD 1985; DACKIS et al. 1986, 1987; EXTEIN and GOLD 1988; GIANNINI et al. 1987a; GIANNINI and BILLET 1987; KOSTEN et al. 1988). These agents are generally employed on a short-term basis during the first several weeks to months following the initiation of abstinence.

a) Anticraving Medications: Bromocriptine and Amantadine

The medications that have been studied most extensively include bromocriptine mesylate (Parlodel) and amantadine hydrochloride (Symmetrel). Bromocriptine is a DA D2 receptor agonist and has been investigated in a number of studies as a treatment for cocaine withdrawal (DACKIS and GOLD 1985; DACKIS et al. 1987; GIANNINI et al. 1987a; GIANNINI and BILLETT 1987). Bromocriptine has been administered to cocaine addicts in doses varying from 0.625 to 7.5 mg/day and for treatment periods ranging from one day to several weeks (TAYLOR and GOLD 1990). The first evidence for its efficacy in humans came from a single-blind trial by DAKIS and colleagues (1986) evaluating the efficacy of 0.625 mg bromocriptine vs placebo on cocaine craving. Although there were only two subjects in the study, both reported reduction in cocaine craving. A subsequent randomized, double-blind study in 13 hospitalized patients using 1.25 mg bromocriptine vs placebo validated these results. In a double-blind, placebo-controlled study of 24 cocaine addicts, Giannini and colleagues reported that bromocriptine reduced cocaine craving significantly more than placebo (Giannini et al. 1987a). It is important to note, however, that the Brief Psychiatric Rating Scale (BPRS) was used to evaluate efficacy in that study, and the BPRS does not specifically measure cocaine craving (Lacombe et al. 1991). In a pilot study of bromocriptine's efficacy in cocaine-abusing methadone maintenance patients, four of six subjects reported a greater than 50% decrease in cocaine craving using bromocriptine 2.5 mg/day (Kosten et al. 1988). Tennant and Sagherian (1987) noted that patients reported that, when they took cocaine, bromocriptine "almost totally blocked the euphorigenic effects of cocaine" in several addicts. Thus, several small studies have demonstrated bromocriptine's efficacy in reducing cocaine craving and other withdrawal symptoms in recently abstinent addicts. However, the incidence of side effects such as headache, nausea, and hypotension have been a problem, especially at higher doses. For this reason, some have recommended initiating treatment slowly at 0.625 mg/day and titrating carefully upward to the maximal therapeutic effect. It is also important to note that bromocriptine has induced psychosis, even in patients with no such previous history (Turner et al. 1984).

Amantadine is not a DA agonist, but is thought ot release DA and norepinephrine from neuronal storage sites and inhibit their reuptake back into the neuron (Robertson and George 1990; Ransmayr et al. 1992). Therefore, it may increase DA at the synapse after chronic cocaine use. Amantadine appears to have efficacy at reducing cocaine craving while having fewer side effects than bromocriptine (Tennant and Sagherian 1987). Initial uncontrolled trials reported reduction in cocaine craving and use (Handelsman et al. 1988; Morgan et al. 1988). Several additional trials have examined amantadine at 200–300 mg daily and found that it reduces cocaine craving and use for several days to 3 weeks (Morgan et al. 1988; Tennant and Tarver 1985). In an open study of amantadine treatment using doses up to 200 mg b.i.d. in cocaine abusing methadone-maintenance clients, Handelsman and colleagues (1988) noted decreases in urine cocaine metabolite level, use of cocaine self-report, and Beck depression scores. Tennant and Sagherian (1987) performed a double-blind comparison of amantadine 100 mg/day and bromocriptine 2.5 mg/day for the treatment of acute cocaine withdrawal in humans. They found amantadine to be superior to bromocriptine (primarily because of the occurrence of side effects at the relatively high dose of bromocriptine). They suggested that bromocriptine might be better suited to addicts who have already stopped using and who require an agonist, whereas amantadine may be safer and more effective during acute withdrawal. It is important to note, however, that this study did not include a placebo control, and subjects received tyrosine and tryptophan (precursors of norepinephrine, DA, and serotonin) in addition to amantadine and bromocriptine. A placebo-controlled study comparing the efficacy of amantadine 400 mg/day vs placebo found no significant difference between treatment groups on a variety of outcome measures (WEDDINGTON et al. 1991). KOSTEN and colleagues (1992) studied the effects of amantadine 300 mg/day compared to desipramine 150 mg/day and placebo in a 12 week trial in methadone-maintained cocaine users. In that study, treatment retention, cocaine craving, and cocaine-free urine samples did not differ significantly across groups. However, in the Kosten et al. study, the attainment of 2 weeks of abstinence from cocaine was significantly more frequent in the patients treated with amantadine than in those treated with placebo or desipramine. Furthermore, ALTERMAN et al. (1992) found that patients treated with amantadine for 4 weeks had significantly greater rates of cocaine abstinence than those treated with placebo at a 4 week follow-up after stopping the medication. In summary, these dopaminergic medications may have some efficacy both in reducing cocaine withdrawal symptoms and in the initial attainment of cocaine abstinence.

b) Other Dopaminergic Agents: L-Dopa, Pergolide, Mazindol and Flupenthixol

L-Dopa/carbidopa (Sinemet), pergolide mesylate (Permax), mazindol, and flupenthixol decanoate (Fluanxol, Depixol) have demonstrated initial efficacy in open trials during the acute withdrawal period. A third therapy that has been employed to facilitate dopaminergic activity is the administration of L-dopa with carbidopa, a peripheral dopa decarboxylase inhibitor. Normal plasma levels of DA have been reported in the range of 1–2 ng/ml (ZÜRCHER and DA PRADA 1979; DA PRADA 1984). Following the administration of L-dopa/carbidopa, plasma DA levels in the range of 1000 ng/ml have been reported (DA PRADA 1984; DA PRADA et al. 1987); thus, this treatment would be expected to acutely increase DA synthesis and levels in the CNS. Several investigators have reported decreased cocaine craving or facilitated abstinence with these agents (ROSEN et al. 1986; COCORES et al. 1989; WOLFSOHN and ANGRIST 1990). Pergolide is a D2 receptor agonist that has a longer duration of action and greater potency than bromocriptine (GOLDSTEIN et al. 1980; MARTIN et al. 1984). It has been associated with reduced cocaine craving in at least one open clinical trial of cocaine abusers (MALCOLM et al. 1991).

Mazindol is DA reuptake inhibitor that initially was associated with a reduction in cocaine use (BERGER et al. 1989; SEIBYL et al. 1992). However, it later demonstrated no improvement in cocaine craving or use in a placebo-controlled, 1 week crossover study (KOSTEN et al. 1993). In a 12 week placebo-controlled study employing mazindol 1 mg daily in the treatment of cocaine-abusing methadone patients, mazindol has demonstrated some superiority over placebo; however, both groups received weekly counseling sessions and other behavioral intervention (MARGOLIN et al. 1993).

Flupenthixol is a depot xanthine reported to have antidepressant effects at low dosages and neuroleptic activity at higher dosages. Gawin and colleagues (1989) noted reduced cocaine craving and use with flupenthixol in an initial, open-label, out-patient trial. A recent, 6 week, double-blind comparison of placebo, i.m. flupenthixol (10–20 mg 2 weeks), and desipramine in 63 crack cocaine users demonstrated that flupenthixol and desipramine both were associated with decreased depression, cocaine use, and cocaine craving. It is important to note that no formal psychotherapy was employed in this trial, and urine toxicology results were not presented (Gawin et al. 1993). A further risk of this approach is related to tardive dyskinesia with chronic flupenthixol treatment.

c) Medications for Transient Symptom Relief

Other agents used transiently during the acute withdrawal period include antipsychotics for residual psychosis (Gawin 1986a) and benzodiazepines or other sedatives for transient anxiety or insomnia (Kosten 1989; Halikas et al. 1993). However, no controlled clinical trials have been reported with these agents.

2. Chronic Treatment Agents

a) Antidepressants

The second major class of medications used in the treatment of cocaine dependence are medications administered during the period following acute withdrawal. These agents are classified differently by various authors, e.g., "chronic agents" (Kosten 1989) or "abstinence maintenance medications" (Halikas et al. 1993). These medications may be initiated during the period of acute withdrawal and continued with the goal of preventing relapse to cocaine use. The length of treatment may be tailored to the needs of the patient. In general, these agents are prescribed for a number of weeks to months or longer, as clinically indicated.

The major class of medications used in this period are various antidepressants. These agents are thought to down-regulate synaptic catecholamine receptors when used chronically, which is opposite to the presynaptic up-regulation caused by chronic stimulant abuse (Gawin and Ellinwood 1988; Kokkindis et al. 1980). Thus, these agents may facilitate normalization of regulation in monoamine systems and return to normal neuronal function.

Desipramine hydrochloride has been the tricyclic most widely studied for the treatment of cocaine dependence. Early studies demonstrated positive results, but more recent placebo-controlled trials have not been as impressive (Gawin and Kleber 1984; Giannini et al. 1986). There have been several randomized placebo-controlled trials of desipramine as an adjunctive pharmacologic treatment of cocaine dependence (Tennant and Tarver 1985; Giannini et al. 1987b; Giannini and Billett 1988; O'Brien et al. 1988;

ARNDT et al. 1992; GAWIN et al. 1989b; KOSTEN et al. 1992; WEDDINGTON et al. 1991). A meta-analysis of these studies by LEVIN and LEHMAN (1991) revealed that desipramine did not improve retention in treatment, but it did produce greater abstinence from cocaine than placebo and was helpful in promoting abstinence. Gawin and colleagues have suggested that the beneficial effects of desipramine are not limited to those with comorbid modd disorders or to those dependent on multiple substances by excluding these individuals from the treatment sample. They also suggested that desipramine treatment might provide a "window of opportunity" during which the antidepressant reverses the neurochemical changes caused by chronic cocaine use, augmenting the effectiveness of a concomitantly employed behavioral intervention (GAWIN et al. 1989b). Desipramine does not appear to be effective when used for short periods of time (less than 2 weeks) or at low doses (less than 100 mg/day), supporting the theory of neuroreceptor change underlying its beneficial effects.

Other tricyclics, such as imipramine and doxepin, have also been employed for cocaine abuse treatment. ROSECAN (1983) employed imipramine in a 12 week open trial with 25 cocaine-using subjects. Abstinence from cocaine was reported in 80% of the subjects. Since subjects were also treated with tyrosine and tryptophan, the contribution of imipramine's effects was not clear (ROSECAN 1983). Imipramine has also been reported to block cocaine euphoria (ROSECAN and NUNES 1987). NUNES and colleagues (1991) recently completed a randomized, double-blind, placebo-controlled, 12 week trial of imipramine (average dose 150–200 mg/day) in 61 cocaine users. Although there was a high drop-out rate, imipramine treatment was associated with a significant decrease in cocaine craving, "high," and ratings of depression, but urine toxicology did show a significant difference from placebo due to small sample size. The authors noted that imipramine's effects may be greatest in mild (intranasal) abusers and in those with concomitant depression.

Cocaine also blocks the reuptake of serotonin (GAWIN and ELLINWOOD 1988). Several antidepressants thought to work predominantly through serotonergic mechanisms have been employed in the treatment of cocaine dependence. These include fluoxetine (Prozac), sertraline (Zoloft), and trazodone hydrochloride (Desyrel). Fluoxetine and sertraline are serotonin reuptake inhibitors. BATKI and colleagues (1993) reported that fluoxetine at 40 mg/day significantly reduced cocaine abuse in methadone-maintained patients. However, other researchers have not demonstrated significant effects using fluoxetine (COVI et al. 1993; GRABOWSKI et al. 1992). KOSTEN and colleagues (1993) noted a significant reduction in cocaine use in a pilot study employing sertraline. Trazodone is a nontricyclic triazolopyridine derivative. SMALL and PURCELL (1985) reported an immediate improvement and decreased cocaine withdrawal symptoms in a single subject with a 3 year history of cocaine abuse with the use of trazodone 100 mg/day. ROWBOTHAM and colleagues (1984) noted that, when trazodone was administered with cocaine, the cardiovascular effects of cocaine were diminished, but it did not change the subject's ratings of the psychogenic effects of cocaine. No data were reported on trazodone's effect on cocaine use behavior. In summary, serotonergic agents have shown generally poor efficacy for achieving or attaining cocaine abstinence, and further work is needed to identify those patients who may benefit and factors predictive of success with these medications, in order to replicate the success of BATKI et al. (1993).

Bupropion (Wellbutrin) is an atypical antidepressant that has also been investigated as a cocaine abuse treatment. In its initial open trials, bupropion appeared to improve treatment retention, decrease cocaine abuse, and eliminate cue-elicited craving among cocaine abusers (MARGOLIN et al. 1991).

Maprotiline (Ludiomil) has also received attention in the pharmaco-therapy of cocaine dependence. In a 7 week open trial of ludiomil (dose approximately 150 mg/day), eight of 11 cocaine abusers achieved abstinence, verified by urine drug screens, and remained abstinent for 10 weeks. Comparison to 20 subjects who did not receive pharmacologic treatment revealed that maprotiline treated subjects demonstrated improved treatment retention, and diminished anxiety, depression, cocaine craving, and cocaine use (BROTMAN et al. 1988).

Monoamine oxidase inhibitors (MAOIs) have also been suggested as a treatment for cocaine abuse (RESNICK and RESNICK 1985), based on their ability to increase CNS norepinephrine and DA, but a significant danger of hypertensive crisis exists with concurrent use of cocaine and MAOIs. To reduce this risk, studies with selective MAO-B agents, such as selegiline (L-deprenyl), might be considered for clinical investigation.

b) Other Medications: Gepirone, Carbamazepine, Amino Acids

Gepirone is an atypical antianxiety agent with serotonin$_{1A}$ receptor partial antagonism. A 12 week, randomized, double-blind, placebo-controlled multicenter trial utilizing gepirone at an average dose of 16.25 mg/day failed to demonstrate the efficacy of gepirone over placebo in the treatment of cocaine dependence. However, the investigators noted that the dose of gepirone was below that believed to be efficacious in depression and suggested that testing at higher doses may be of benefit (JENKINS et al. 1992).

Carbamazepine (Tegretol) is an anticonvulsant that has been investigated based on it's potential effect on cocaine-induced kindling. Although initial trials showed some promise (HALIKAS et al. 1989, 1990, 1991), a recent double-blind, placebo-controlled trial employing therapeutic doses of carbamazepine demonstrated no significant effect on cocaine craving or cocaine self-administration (MONTOYA et al. 1993).

Cocaine abusers may suffer from nutritional deficiencies and some clinicians have advocated the use of L-tyrosine, the amino acid precursor of DA and norepinephrine, and L-tryptophan, the amino acid precursor of serotonin (SKINNER and THOMPSON 1992; TENNANT and SAGHERIAN 1987; ROSECAN

1983; NUNES and KLEIN 1987). Although some of these agents are employed by a number of physicians for cocaine detoxification (HALIKAS et al. 1993), their efficacy has not been demonstrated in at least one double-blind, placebocontrolled clinical trial (CHADWICK et al. 1990).

c) New Medication Approaches

Additional studies are currently investigating several potential pharmaco-therapies. Agents recently examined include disulfiram (Antabuse) and buprenorphine (Buprenex). Disulfiram inhibits the metabolism of acetaldehyde and is primarily employed in the treatment of alocholism. Some investigators have postulated that disulfiram may alter the reinforcing properties of cocaine. Recent studies have demonstrated trends toward decreased alcohol and cocaine use in cocaine-using subjects, and additional studies with this agent are underway (CARROLL et al. 1993; HAMEEDI et al. 1993; HIGGENS et al. 1993). Buprenorphine is a partial mu opioid receptor agonist that has been investigated for the treatment of individuals who abuse opiates and cocaine. Pilot studies (KOSTEN 1989) have demonstrated decreased cocaine use among methadone-maintained addicts but controlled trials by others (OLIVETTO et al., in press; JOHNSON et al. 1992; FUDALA et al. 1991) failed to demonstrate efficacy against cocaine use in doses of buprenophine up to 8 mg/day. However, Schottenfeld and colleagues noted significantly decreased cocaine abuse among methadone-maintained patients using doses of buprenophine up to 16 mg/day (SCHOTTENFELD et al. 1993). Thus, additional work with this agent at higher doses may prove efficacious in opioid and cocaine abusing patients.

3. Comorbid Psychiatric Disorders

Individuals who suffer from psychiatric disorders are at risk for the abuse of illicit drugs in order to "self-medicate" symptoms. Many of these individuals may derive benefit from specific pharmacotherapies. In regard to cocaine dependence, two groups have been identified as potential responders to specific medications. These include cyclothymia/bipolar disorder and attention deficit hyperactivity disorder-residual type (residual ADHD).

a) Cyclothymia and Bipolar Disorder

Lithium carbonate has been investigated as a treatment for cocaine addiction based on animal studies in which lithium antagonized the effects of cocaine, possibly through a serotonergic mechanism (MANDELL and KNAPP 1977; ROSECAN and NUNES 1987). Early case reports under uncontrolled conditions noted a block of anticipated cocaine-induced euphoria in subjects receiving lithium (CRONSON and FLEMENBAUM 1978). Subsequent open and controlled studies, however, have not demonstrated beneficial effects of lithium in cocaine abusers (KLEBER et al. 1984; GAWIN 1986a; GAWIN et al. 1989b).

However, lithium may be beneficial in reducing cocaine use in subjects with a clear history of cyclothymia or bipolar disorder (Gawin 1986b; Nunes et al. 1990).

b) Attention Deficit Hyperactivity Disorder-Residual Type

Methylphenidate inhibits reuptake of DA and is frequently employed as a Pharmacotherapy for ADHD in childhood (Patrick et al. 1987). Some individuals appear to remain impaired by symptoms into adulthood; these people may receive the diagnosis of residual ADHD. Methylphenidate was initially considered for use as a dopaminergic agent for cocaine addicts during acute withdrawal. Unfortunately, in cocaine addicts without concomitant psychiatric illness such as ADHD, methylphenidate appears ineffective and may increase cocaine craving (Gawin et al. 1985). However, it has reportedly facilitated cocaine abstinence in patients with a childhood history of ADHD (Khantzian 1983; Gawin et al. 1984).

III. Clinical Considerations

1. Indications for Treatment

Pharmacologic agents may be utilized in the treatment of cocaine dependence as adjuncts to a comprehensive treatment plan involving such treatments as individual and group psychotherapy, relapse prevention training, behavioral techniques, and social and familial intervention. Although some cocaine abusers achieve abstinence without formal assistance (Toneatto et al. 1993), and others recover through the use of the traditional therapies cited above, a large number of cocaine users are not able to achieve abstinence through traditional means. Such individuals may be candidates for the use of adjunctive pharmacotherapies.

In general, three categories of patients are recognized as appropriate for pharmacologic intervention. These include those who have developed neuroadaptation to heavy cocaine use, those who display psychiatric vulnerability, and those who have developed substantial medical risks from continued cocaine use (Kosten 1989).

The first category includes those patients who have evolved from occasional intranasal use to more high intensity cocaine use, i.e., patients who have used either intravenously or via smoking large quantities of cocaine. Such high intensity use may lead to neuroadapation or neurobiologic changes in neurotransmitter systems, which might involve both tolerance and sensitization (Gold et al. 1985; Spyraki et al. 1982; Wise 1984). An example of a neurobiologic change that may be helped through pharmacological treatment is the up-regulation of dopaminergic reuptake transporters induced by cocaine (Staley et al. 1992). Treatment with DA agonists could prevent a relative DA deficit and improve cocaine-induced behavioral symptoms.

Many cocaine users who use by the high intensity routes have not been able to achieve abstinence using traditional therapies; thus, additional assistance in the form of pharmacologic intervention would seem warranted.

The second category includes those individuals with psychiatric disorders. Psychiatric disorders are frequently seen in treatment-seeking cocaine abusers. Depression is the most common additional psychiatric disorder concomitant to cocaine abuse, but others are seen as well, including cyclothymia, bipolar disorder, residual ADHD, and schizophrenia (KOSTEN 1989; GAWIN and KLEBER 1986; KHANTZIAN 1983; GAWIN et al. 1984; SEIBYL et al. 1992). These individuals may present for treatment using lower amounts of cocaine and, occasionally, through less intense means (e.g., intranasally instead of i.v. or smoked). It is possible that some of these individuals may be attempting to self-medicate (their psychiatric symptoms) through the use of drugs such as cocaine. It is clear that cocaine use may lead to exacerbation of their comorbid psychiatric illnesses and early intervention is, therefore, warranted.

The third category includes individuals who are at increased medical risk secondary to cocaine use. This would include intravenous users who share needles and who are, therefore, at risk of developing acquired immunodeficiency syndrome (AIDS), and individuals who have sustained major secondary illnesses such as endocarditis, myocardial infarction, stroke, and pneumonia. Additional cocaine use clearly places such individuals at risk of acquiring additional medical complications (KOSTEN 1989).

2. Length of Treatment

Recovery from cocaine abuse is a gradual process that varies among individuals. GAWIN and KLEBER (1986b) conducted a naturalistic study of cocaine withdrawal in out-patients and identified three stages of the process: (1) crash, (2) withdrawal, and (3) extinction. Each stage was associated with a specific time frame and range of symptoms. Others who have studied the phenomenology of cocaine withdrawal experienced by in-patients have not noted the same phenomena (SATEL et al. 1991; WEDDINGTON et al. 1990). In any case, it appears that cocaine users suffer from relatively mild withdrawal symptoms following cessation of cocaine use, and this withdrawal syndrome may vary across individuals and environments.

In regard to pharmacotherapy, it would seem reasonable to employ medications to ameliorate specific symptoms during the recovery process and to treat comorbid psychiatric disorders. A key issue is to differentiate withdrawal phenomena from symptoms of the comorbid disorder (e.g., fatigue during the early cocaine withdrawal period as opposed to that caused by endogenous depression), and to then treat comorbid psychiatric disorders following current treatment standards.

During early cocaine abstinence, individuals may suffer from excessive fatigue but may require no pharmacologic intervention. Other individuals may demonstrate paranoia or agitation and require the transient use of neuroleptics or benzodiazepines. Craving may be present during the later withdrawal and, in appropriate patients, this may warrant the use of dopaminergic agents or other treatments; these are generally employed on a short-term basis (e.g., days to weeks).

The potential for relapse is especially high during the first several months of recovery, and, during this period, adjunctive "chronic treatment agents," such as antidepressants, may be helpful. As antidepressants generally take several days to weeks to become effective, these agents may be started during initial abstinence and continued for up to 6 months, or until clinical circumstances warrant a trial at discontinuation.

The final determinants of when to stop treatment are related to the precipitants that led to treatment initiation. The resolution of legal, medical, or psychosocial problems that led to treatment must occur before a schedule for withdrawal of adjunctive pharmacotherapy. As with other illnesses, if treatment is terminated before the patient achieves stability, a poor outcome may result (KOSTEN 1988).

3. Precautions in Treatment

There are several caveats to the use of adjunctive pharmacotherapy for cocaine dependence. These include side effects or complications from the use of medications, the treatment of the medically ill or pregnant addict, and the use of potentially addictive medications (e.g., benzodiazepines, methylphenidate).

All medications are associated with side effects, and one must continually weigh the benefit of the proposed therapy against the risks imposed by treatment. Before initiating pharmacotherapy, the medical status of the patient should be considered thoroughly to avoid unnecessary complications. A complete medical history, physical examination, laboratory tests, and an electrocardiogram are indicated because of the numerous medical complications caused by cocaine use. Some of the medications employed in the treatment of cocaine dependence are associated with significant side effects. Because bromocriptine side effects, for example, include hypotension, some have advocated the use of bromocriptine in in-patients while using amantadine in an out-patient setting (TENNANT and SAGHERIAN 1987). Tricyclic antidepressants are associated with tachycardia, orthostatic hypotension, cardiac conduction changes, and diminished seizure threshold. Obviously, appropriate cardiac consultation should be obtained before initiating therapy with these agents in addicts with abnormal electrocardiograms, chest pain, or previous stroke. Neuroleptics, such as flupenthixol, carry the risk of tardive dyskinesia and their use should be kept to a minimum in individuals without schizophrenia or related psychotic disorders (TASK FORCE ON LATE NEUROLOGICAL EFFECTS OF ANTIPSYCHOTIC DRUGS 1980). As with other medical treatments, Patients should be informed of the risks and benefits of potential pharmacotherapy.

In regard to pregnancy, pharmacologic intervention may compound neonatal problems that have been associated with cocaine use, especially if the mother continues to use cocaine (CHASNOFF et al. 1985). Because of these issues, adjunctive pharmacologic intervention in pregnant cocaine abusers should generally be avoided (KOSTEN 1989). Pregnant cocaine users who are also addicted to opiates should be considered for methadone maintenance (KOSTEN 1989).

Finally, although benzodiazepines or other sedatives may be helpful for insomnia or agitation during initial abstinence, these medications must be used with caution in individuals with addictive disorders. In such situations, the use of benzodiazepines with a shower rate of increase in plasma levels (e.g., prazepam, oxazepam, clonazepam) should be considered over those that have been associated with abuse liability (e.g., alprazolam, diazepam).

Acknowledgement. Supported by the National Institute on Drug Abuse grants P50-DA04060, R-18-DA06190, R01 DA05626, and K02-DA0112 (TRK).

References

Alterman AI, Droba M, Antelo RE, Cornish JW, Sweeney KK, Parikh GA, O'Brien CP (1992) Amantadine may facilitate detoxification of cocaine addicts. Drug Alcohol Depend 31:19–29

Arndt IO, Dorozynsky L, Woody GE, McLellan AT, O'Brien CP (1992) Desipramine treatment of cocaine dependence in methadone-maintained patients. Arch Gen Psychiatry 49:888–893

Batki SL, Manfredi LB, Jacob P III, Jones RT (1993) Fluoxetine for cocaine dependence in methadone-maintenance: quantitative plasma and urine cocaine/benzoylecgonine concentration. J Clin Psychopharmacol 13:243–250

Berger P, Gawin F, Kosten RT (1989) Treatment of cocaine abuse with mazindol (letter). Lancet 1:283

Blaine JD, Ling W (1992) Psychopharmacologic treatment of cocaine dependence. Psychopharmacol Bull 28:11–14

Brotman AW, Witkie SM, Gelenberg AJ et al. (1988) An open trial of maprotiline for the treatment of cocaine abuse: a pilot study. J Clin Psychopharmacol 8:125–127

Carroll K, Ziedonis D, O'Malley S, McCance-Katz E, Cordon L, Rounsaville B (1993) Pharmacologic interventions for alcohol- and cocaine-abusing individuals. Am J Addict 2:77–79

Chadwick MJ, Gregory DL, Wendling G (1990) A double-blind amino acids, L-tryptophan and L-tyrosine, and placebo study with cocaine-dependent subjects in an inpatient chemical dependncy treatment center. Am J Drug Alcohol Abuse 16:275–286

Chasnoff IJ, Burns WJ, Schnoll SH, Burns KA (1985) Cocaine use in pregnancy. N Engl J Med 313:666–669.

Cocores JA, Gold MS, Pottash ALC (1989) Dopaminergic treatments in cocaine withdrawal. Soc Neurosci Abstr 15:251

Covi L, Hess JM, Kreiter NA, Jaffe JH (1993) Fluoxetine and counseling for PCP abuse. In: Harris LS (ed). Committee on problems of drug dependence, 1992. National Institute on Drug Abuse Research Monograph 132, NIH publication no 93-3505, US Government Printing Office, Washington DC, p 321

Cronson AJ, Flemenbaum A (1978) Antagonism of cocaine highs by lithium. Am J Psychiatry 135:856–857

Dackis CA, Gold MS (1985) Bromocriptine as treatment of cocaine abuse. Lancet 1:1151–1152

Dackis CA, Gold MS, Davies RK, Sweeney DR (1986) Bromocriptine treatment for cocaine abuse: the dopamine depletion hypothesis. Int J Psychiatry Med 15:125–135

Dackis CA, Gold MS, Sweeney DR, Byron JP Jr, Climko R (1987) Single dose bromocriptine reverses cocaine craving. Psychiatry Res 20:261–264

Da Prada M (1984) Peripheral decarboxylase inhibition: a biochemical comparison between benserazide and carbidopa. In: Birkmayerr W, Rinne VK, Worm-Peterson et al. (eds) Parkinson's disease: actual problems and management. Roche, Basel, pp 25–38

Da Prada M, Kettler R, Zürcher R et al. (1987) Inhibition of decarboxylase and levels of DOPA and 3-0-methyldopa: a comparative study of benserazide versus carbidopa in rodents and of madopar standard versus madopar HBS in volunteers. Eur Neurol 27 [Suppl 1]:9–20

Extein IL, Gold MS (1988) The treatment of cocaine addicts: Bromocriptine or Desipramine? Psychiatr Ann 18:535–537

Fischman MW (1987) Cocaine and the amphetamines. In: Meltzer HY (ed) Psychopharmacology: the third generation of progress. Raven, New York, pp 1543–1553

Fudala PJ, Johnson RE, Jaffe JH (1991) Outpatient comparison of buprenorphine and methadone maintenance: II. Effects on cocaine usage, retention time in study and missed clinic visits. In: Harris LS (ed) committee on problems of drug dependence 1990. National Institute on Drug Abuse Research monograpy 105, DHHS publication no (ADM)91-1753. US Government Printing Office, Washington DC, pp 576–588

Gawin F, Kleber HD, Riordan CE (1984) Methylphenidate treatment of cocaine dependence-a preliminary report. J Subst Abuse Treat 1:107–112

Gawin FH (1986a) Neuroleptic reduction of cocaine-induced paranoia but not euphoria? Psychopharmacology (Berl) 90:142–143

Gawin FH (1986b) New uses of antidepressants in cocaine abuse. Psychosomatics 27 [Suppl]:24–29

Gawin FH, Ellinwood EH Jr (1988) Cocaine and other stimulants: Actions, abuse, and treatment. N Engl J Med 318:1173–1182

Gawin FH, Kleber HD (1984) Cocaine abuse treatment: open pilot trial with desipramine and lithium carbonate. Arch Gen Psychiatry 41:903–909

Gawin FH, Kleber HD (1984)Abstinence symptomatology and psychiatric diagnosis in cocaine abusers: clinical observations. Arch Gen Psychiatry 43:107–113

Gawin FH, Riordan C, Kleber HD (1985) Methylphenidate treatment of cocaine abusers without attention deficit disorder: a negative report. Am J Drug Alcohol Abuse 11:193–197

Gawin FH, Allen D, Humblestone B (1989a) Outpatient treatment of "crack" cocaine smoking with flupenthixol decanoate: a preliminary report. Arch Gen Psychiatry 46:322–325

Gawin FH, Kleber HD, Byck R, Rounsaville BJ, Kosten TR, Jatlow PI, Morgan C (1989b) Desipramine facilitation of initial cocaine abstinence. Arch Gen Psychiatry 46:117–121

Gawin FH, Khalsa ME, Brown J, Jatlow P (1993) Flupenthixol treatment of crack users: Initial double-blind results. In: Harris LS (ed) Committee on problems of durg dependence, 1992. National Institute on Drug Abuse Research monograph 132, NIH publication no 93-3505. US Goverment Printing Office, Washington DC, p 319

Giannini AJ, Billett W (1987) Bromocriptine-desipramine protocol in treatment of cocaine addiction. J Clin Pharmacol 27:549–554

Giannini AJ, Malone DA, Giannini MC, Price WA, Loiselle RH (1986) Treatment of depression in chronic cocaine and phencyclidine abuse with desipramine. J Clin Pharmacol 26:211–214

Giannini AJ, Baumgartel P, Dimarzio LR (1987a) Bromocriptine therapy in cocaine withdrawal. J Clin Pharmacol 27:267–270

Giannini AJ, Loiselle RH, Giannini MC (1987b) Space-based abstinence: alleviation of withdrawal symptoms in combative cocaine-phencyclidine abuse. Clin Toxicol 25:493–500

Gold MS, Washton AM, Dackis CA (1985) Cocaine abuse: neurochemistry, phenomenology and treatment. In: Kozel, MS, Adams EH (eds) Cocaine use in America: epidemiologic and clinical perspectives. National Institute on Drug Abuse Research monograph 61, DHHS publication no (ADN)87-1414. US Government Printing Office, Washington DC, pp 130–150

Goldstein M, Lieberman A, Lew JY et al. (1980) Interaction of pergolide with central dopaminergic receptors. Proc Natl Acad Sci USA 77:3725–3728

Grabowski J, Elk R, Rhoades H, Cowan K, Schmitz J (1992) Double-blind placebo-controlled studies of fluoxetine in cocaine dependence. Proceedings, College on the problems of drug dependence, Keystone

Halikas J, Kemp K, Kuhn K, Carlson G, Crea F (1989) Carbamazepine for cocaine addiction? (Letter.) Lancet 1:623–624

Halikas JA, Kuhn KL, Maddux TL (1990) Reduction of cocaine use among methadone maintenance patients using concurrent carbamazepine maintenance. Ann Clin Psychiatry 2:3–6

Halikas JA, Crosby RD, Carlson GA, Crea F, Graves NM, Bowers LD (1991) Cocaine reduction in unmotivated crack users using carbamazepine versus placebo in a short-term, double-blind crossover design. Clin Pharmacol Ther 50:81–95

Halikas JA, Nugent SM, Crosby RD, Carlson GA (1993) 1990–1991 survey of pharmacotherapies used in the treatment of cocaine abuse. J Addict Dis 12: 129–139

Hameedi FA, Rosen MI, McCance-Katz E, Woods SW, Price LH, Kosten TR (in press) Effects of disulfiram on cocaine high. Problems of drug dependence, 1993. National Institute on Drug Abuse Research monograph. US Government Printing Office, Washington DC

Handelsman L, Chordia PL, Escovar IM, Marion IJ, Lowinson JH (1988) Amantadine for treatment of cocaine dependence in methadone-maintained patients (letter). Am J Psychiatry 145:533

Higgins ST, Budney AJ, Bickel WK, Hughes JR, Foerg F (1993) Disulfiram therapy in patients abusing cocaine and alcohol (letter). Am J Psychiatry 150:675–676

Jenkins SW, Warfield NA, Blaine JD, Cornish J, Ling W, Rosen MI, Urschel H III, Wesson D, Ziedonis D (1992) A pilot trial of Gepirone vs. placebo in the treatment of cocaine dependency. Psychopharmacol Bull 28:21–26

Johanson C, Fischman MW (1989) The pharmacology of cocaine related to its abuse. Pharmacol Rev 41:3–52

Johnson RE, Jaffe JH, Fudala PJ (1992) A controlled trial of buprenorphine treatment for opiod dependence. JAMA 267:2750–2755

Khantzian EJ (1983) An extreme case of cocaine dependence and marked improvement with methylphenidate treatment. Am J Psychiatry 140:784–785

Kleber HD, Gawin F (1986) Psychopharmacological trials in cocaine abuse treatment. Am J Drug Alcohol Abuse 12:235–246

Kleber HD, Gawin FH, Cocaine Abuse (1984) A review of current and experimental treatments. In: Grabowski J (ed) Cocaine: pharmacology, effects, and treatment of abuse. National Institute on Drug Abuse Research monograph 50, DHHS publication no (ADM)87-1326. US Government Printing Office, Washington DC, pp 111–129

Kokkinidis L, Zacharko RM, Predy PA (1980) Post-amphetamine depression of self-stimulation responding from the substantia nigra: reversal by tricyclic antidepressants. Pharmacol Biochem Behav 13:379–383

Kosten TA, Kosten TR, Gawin FH, Gordon LT, Hogan I, Kleber HD (1992) An open trial of sertraline for cocaine abuse. Am Addict 1:349–353

Kosten TR (1989) Pharmacotheraputic interventions for cocaine abuse: matching patients to treatment. J Nerv Ment Dis 177:379–389

Kosten TR (1993) Pharmacotherapies for cocaine abuse: neurobiological abnormalities reversed with drug intervention. Psychiatr Times 10:25–26

Kosten TR, Kleber HD (eds) (1992) Clinician's guide to cocaine addiction: theory, research, and treatment. Guilford, New York

Kosten TR, Schumann B, Wright D (1988) Bromocriptine treatment of cocaine abuse in patients maintained on methadone (letter). Am J Psychiatry 145: 381–382

Kosten TR, Morgan CM, Falcioni J, Schottenfeld RS (1992) Pharmacotherapy for cocaine abusing methadone maintained patients using amantadine or desipramine.

Kosten TR, Steinberg M, Diakogiannis IA (1993) Cross-over trial of mazindol for cocaine dependence. Am J Addict 2:161–164

Krug SE (1989) Cocaine abuse: historical, epidimiologic, and clinical perspectives for pediatricians. Adv Pediatr 36:369–406

Lacombe S, Stanislav SW, Marken PA (1991) Pharmacologic treatment of cocaine abuse, DICP. Ann Pharmacother 25:818–823

Levin FR, Lehman AF (1991) Meta-analysis of desipramine as an adjunct in the treatment of cocaine addiction. J Clin Psychopharmacol 11:374–378

Malcolm R, Hutto BR, Phillips JD, Ballenger JC (1991) Pergolide mesylate treatment of cocaine withdrawal. J Clin Psychiatry 52:39–40

Mandell AJ, Knapp S (1977) Neurobiological antagonism of cocaine by lithium. In: Ellinwood EH, Kilby MM (eds) Cocaine and other stimulants. Plenum, New York, pp 187–200

Margolin A, Kosten T[R], Petrakis I, Avants SK, Kosten T[A] (1991) Bupropion reduces cocaine abuse in methadone-maintained patients (letter). Arch Gen Psychiatry 48:87

Margolin A, Kosten TR, Avants SK (in press) A double-blind study of mazindol for the prevention of relapse to cocaine abuse in methadone-maintained patients. Problems of drug dependence 1993. National Institute on Drug Abuse Research Monograph. US Government Printing Office, Washington DC

Martin GE, Williams M, Pettibone DJ et al. (1984) Pharmacologic profile of a novel potent direct-acting dopamine agonist, (+),-4-propye-9-hydroxynaphthoxazine [(+)-PHNO]. J Pharmacol Exp Ther 230:569–576

McDougle CJ, Price LH, Palumbo JM, Kosten TR, Heninger GR, Kleber HD (1992) Dopaminergic responsivity during cocaine abstinence. Psychiatry Res 43:77–85

Mendelson JH, Teoh SK, Lange U, Mello NK, Weiss R, Skupny A, Ellingloc J (1988) Anterior pituitary, adrenal and gonadal hormones during cocaine withdrawal. Am J Psychiatry 145:1094–1098

Montoya ID, Levin FR, Fudala PJ, Gorelick DA (in press) Randomized double-blind clinical trial with carbamazepine versus placebo for treatment of cocaine dependence. Problems of durg dependence, 1993. National Institute on Drug Abuse Research monograph. US Government Printing Office, Washington DC

Morgan C, Kosten T, Gawin F, Kleber H (1988) A pilot trial of amantadine for ambulatory withdrawal for cocaine dependence. In: Harris LS (ed) Problems of drug dependence 1987. National Institute on Drug Abuse Research monograph 81, DHHS publication no (ADM) 88–1564. Supt. of Docs., US Government Printing Office, Washington DC, pp 81–85

Nunes EV, Klein DF (1987) Research issues in cocaine abuse: future directions. In: Spitz HI, Rosecan JS (eds) Cocaine abuse: new directions in treatment and research. Brunner Mazel, New York, pp 273–298

Nunes EV, McGrath PJ, Wager S, Quitkin FM (1990) Lithium treatment for cocaine abusers with bipolar spectrum disorders. Am J Psychiatry 147:655–657

Nunes EV, Quitkin FM, Brady R, Stewart JW (1991) Imipramine treatment of methadone maintenance patients with affective disorder and illicit drug use. Am J Psychiatry 148:667–669

O'Brien CP, Childress AR, Arndt IO, McLellan AT, Woody GE, Maany I (1988) Pharmacological and behavioral treatments of cocaine dependence: controlled studies. J Clin Psychiatry 49 [Suppl]:17–22

Oliveto AH, Kosten TR, Schottenfeld R, Ziedonis D, Falcioni J (in press) Cocaine use in buprenorphine- vs. methadone-maintained cocaine users. Am J Addict

Patrick KS, Mueller RA, Gualtieri CT, Breese GR (1987) Pharmacokinetics and actions of methylphenidate. In: Meltzer HY (ed) Psychopharmacology: the third generation of progress. Raven, New York, pp 1387–1395

Ransmayr G, Kuning G, Gerstenbrand F (1992) Modern therapy of Parkinson's disease. J Neural Transm [Suppl] 38:129–140

Resnick R, Resnick E (1985) Psychological issues in the treatment of cocaine abusers. Presented at Columbia University symposium on cocaine abuse: new treatment approaches, 5 Jan 1985

Robertson DR, George CF (1990) Drug therapy for Parkinson's disease in the elderly. Br Med Bull 46:124–146

Rosecan J (1983) The treatment of cocaine abuse with imipramine, L-tyrosine, and L-tryptophan. Presented at the 7th world congress of psychiatry, Vienna, Austria, 14–19 July 1983

Rosecan JS, Nunes EV (1987) Pharmacologic management of cocaine abuse. In: Spitz HI, Rosecan JS (eds) Cocaine abuse: new directions in treatment and research. Brunner Mazel, New York, pp 255–270

Rosen H, Flemenbaum A, Slater VL (1986) Clinical trial of carbiodopa-L-dopa combination for cocaine abuse (letter). Am J Psychiatry 143:1493

Rowbotham MC, Jones RT, Benowitz NL, Jacob P III (1984) Trazodone-oral cocaine interactions. Arch Cen Psychiatry 41:895–899

Satel SL, Price LH, Palumbo JM, McDougle CJ, Krystal JH, Gawin F, Charney DS, Heninger GR, Kleber HD (1991) Clinical phenomenology and neurobiology of cocaine abstinence: a prospective inpatient study. Am J Psychiatry 148: 1712–1716

Schottenfeld RS, Pakes J, Ziedonis DS, Kosten TR (1993) Buprenorphine: dose related effects on cocaine and opioid use in cocaine abusing opioid dependent humans. Biol Psychiatry 34:66–74

Seibyl JP, Brenner L, Krystal JH, Johnson R, Charney DS (1992) Mazindol and cocaine addiction in schizophrenia (letter). Biol Psychiatry 31:1179–1181

Small GW, Purcell JJ (1985) Trazodone and cocaine abuse (letter). Arch Gen Psychiatry 42:524

Spyraki C, Fibiger HC, Phillips AG (1982) Cocaine-induced place preference conditioning: lack of effects of neuroleptics and 6-hydroxydopamine lesions. Brain Res 253:195–203

Staley J, Tobia R, Ruttenber AJ, Wetli CV, Lee Hearn W, Flynn DD, Mash DC (1992) RIT 55 binding to the dopamine transporter in cocaine overdose deaths. Soc Neurosci Abstr 18:542

Task Force on Late Neurological Effects of Antipsychotic Drugs (1980) Am J Psychiatry 137:1163–1172

Taylor WA, Gold MS (1990) Pharmacologic approaches to the treatment of cocaine dependence. West J Med 152:573–577

Tennant FS Jr, Sagherian AA (1987) Double-blind comparison of amantadine and bromocriptine for ambulatory withdrawal from cocaine dependence. Arch Intern Med 147:109–112

Tennant FS, Tarver AL (1985) Double-blind comparison of desipramine and placebo in withdrawal cocaine dependence. In: Harris LS (ed) Problems of drug dependence, 1984. National Institute on Drug Abuse Research monograph 55, DHHS publication no (ADM) 85–1393. US Goverment Printing Office, Washington DC, pp 159–63

Toneatto T, Rubel E, Sobell LC, Sobell MB (in press) Untreated recovery from cocaine dependence. Problems of drug dependence, 1993. National Institute on Drug Abuse Research monograph. US Government Printing Office, Washington DC

Turner TH, Cookson JC, Wass JAH, Drury PL, Price PA, Besser GM (1984) Psychotic reactions during treatment of pituitary tumours with dopamine agonists. Br Med J 289:1101–1103

Weddington WW, Brown BS, Haertzen CA, Cone EJ, Dax EM, Herning RI, Michaelson BS (1990) Changes in mood, craving, and sleep during short-term abstinence reported by male cocaine addicts: a controlled, residential study. Arch Gen Psychiatry 47:861–868

Weddington WW Jr, Brown BS, Haertzen CA, Hess JM, Mahaffey JR, Kolar AF, Jaffe JH (1991) Comparison of amantadine and desipramine combined with psychotherapy for treatment of cocaine dependence. Am J Drug Alcohol Abuse 17:137–152

Weiss RD, Mirim SM (1990) Psychological and pharmacological treatment strategies in cocaine dependence. Ann Clin Psychiatry 2:239–243

Wise R (1984) Neural mechanisms of the reinforcing action of cocaine. In: Grabowski J (ed) Cocaine: pharmacology effects, and treatment of abuse. National Institute on Drug Abuse Research monograph 50, DHHS publication no (ADM)87-1326. US Government Printing Office, Washington DC, pp 15–33

Wolfsohn R, Angrist B (1990) A pilot trial of levodopa/carbidopa in early cocaine abstinence (letter). J Clin Psychopharmacol 10:440–442

Zürcher G, Da Prada M (1979) Radioenzymatic assay of femtomole concentrations of DOPA in tissues and body fluids. J Neurochem 33:631–639

Subject Index

Springer-Verlag
and the Environment

We at Springer-Verlag firmly believe that an international science publisher has a special obligation to the environment, and our corporate policies consistently reflect this conviction.

We also expect our business partners – paper mills, printers, packaging manufacturers, etc. – to commit themselves to using environmentally friendly materials and production processes.

The paper in this book is made from low- or no-chlorine pulp and is acid free, in conformance with international standards for paper permanency.

Printing: Mercedesdruck, Berlin
Binding: Buchbinderei Lüderitz & Bauer, Berlin